*The Warsaw Colloquium on*
INSTRUMENTAL CONDITIONING
AND BRAIN RESEARCH

*The Warsaw Colloquium on*
# INSTRUMENTAL CONDITIONING AND BRAIN RESEARCH

*Proceedings of the Symposium
to honour the memory of Jerzy Konorski and
60 years of the Nencki Institute,
held in Jabłonna near Warsaw, 1-5 May 1979*

*Edited by*

BOGUSŁAW ŻERNICKI and KAZIMIERZ ZIELIŃSKI

*Department of Neurophysiology,
Nencki Institute of Experimental Biology,
Warsaw, Poland*

1980

MARTINUS NIJHOFF PUBLISHERS
THE HAGUE / BOSTON / LONDON

PWN / POLISH SCIENTIFIC PUBLISHERS
WARSZAWA

*Distributors:*

*for the United States and Canada*

Kluwer Boston, Inc.
190 Old Derby Street
Hingham, MA 02043
USA

*for all other countries*

Kluwer Academic Publishers Group
Distribution Centre
P.O. Box 322
3300 AH Dordrecht
The Netherlands

*for Albania, Bulgaria, Chinese People's Republic, Cuba, Czechoslovakia, German Democratic Republic, Hungary, Korean People's Democratic Republic, Mongolia, Poland, Rumania, the U.S.S.R., Vietnam and Yugoslavia*

Ars Polona
Krakowskie Przedmieście 7
00-068 Warszawa
Poland

ISBN-13: 978-94-009-8227-7    e-ISBN-13: 978-94-009-8225-3
DOI: 10.1007/978-94-009-8225-3

Proceedings also published in Acta Neurobiologiae Experimentalis, 1979, issue 6 and 1980, issues 1 and 4.

Copyright © by PWN — Polish Scientific Publishers — Warszawa 1980.
Softcover reprint of the hardcover 1st edition 1980

All rights reserved, including the right to translate or to reproduce this book or parts thereof in any form.

CONTENTS

Preface
Contributors
AKERT, K. Opening address . . . . . . . . . . . . . . 1
ASRATYAN, E. A. Opening address . . . . . . . . . . . . 3
ZIELIŃSKI, K. Involvement of the partial reinforcement procedure in reward training: opening address . . . . . . . . . . . . . 5

**Perceptual mechanisms**

BARTLETT, J. R. and DOTY, R. W. An exploration of the ability of macaques to detect microstimulation of striate cortex . . . . . . . 17
BRUNK, R., CZIHAK, E., OPPERMANN, B. and SANTIBAÑEZ-H., G. Conditioned audio-visual targeting reflexes in split brain cats . . . . 32
BUDOHOSKA, W. and GRABOWSKA, A. Search for structures involved in integration of letters in pairs . . . . . . . . . . . . 42
BUREŠ, J. and NEVEROV, V. P. Reversed postoptokinetic nystagmus: A model of plasticity in the vestibuloocular system . . . . . . 49
BUREŠOVÁ, O. Spatial memory and instrumental conditioning . . . 63
GRASTYÁN, E. and BUZSÁKI, G. The orienting-exploratory response hypothesis of discriminative conditioning . . . . . . . . . 78
ŁAWICKA, W. Auditory targeting reflexes: their determining role in directional instrumental responding . . . . . . . . . . . . 89
SPRAGUE, J. M., BERKLEY, M. A. and HUGHES, H. C. Visual acuity functions and pattern discrimination in the destriate cat . . . . . . 105
ŻERNICKI, B. The pretrigeminal cat as an instrument for investigation of the ocular fixation reflex . . . . . . . . . . . . . . 145

**Motivational mechanisms**

FONBERG, E. and KOSTARCZYK, E. Motivational role of social reinforcement in dog–man relations . . . . . . . . . . . . . . 150
ŁUKASZEWSKA, I. and MŁODKOWSKA, Z. Exploratory motivation in response-to-change test in rats . . . . . . . . . . . . . 170
ONIANI, T. N. and VARTANOVA, N. G. Factors of extinction of alimentary instrumental conditioned reflex . . . . . . . . . . . . 181
SADOWSKI, B. External stimulus control of self-stimulation behavior . . 207
SIMONOV, P. V. The role of emotions in the formation of instrumental conditioned reflex . . . . . . . . . . . . . . . . 219
SOLOMON, R. L. Recent experiments testing an opponent-process theory of acquired motivation . . . . . . . . . . . . . . . 231
STELLAR, E. Brain mechanisms and hedonic processes . . . . . . 250

## Motor mechanisms

FABRE, M. and BUSER, P. Structures involved in acquisition and performance of visually guided movements in the cat . . . . . . . . 262
GAHÉRY, Y., IOFFE, M., MASSION, J. and POLIT, A. The postural support of movement in cat and dog . . . . . . . . . . . . . 284
STAMM, J. The monkey's prefrontal cortex functions in motor programming 299

## Defensive conditioned reflexes

BIGNAMI, G. Pharmacological evidence on the specialization of CNS mechanisms responsible for motor act inhibition by aversive events 321
BRENNAN, J. F. Prefrontal cortical effects on aversively motivated instrumental conditioning in rats: some ontogenic considerations . . . . 339
BRUSH, F. R. and FRALEY, S. M. Central mediation of hormonal influences on instrumental avoidance conditioning . . . . . . . . 369
OVERMIER, J. B., PAYNE, R. J., BRACKBILL, R. M., LINDER, B. and LAWRY, J. A. On the mechanism of the post-asymptotic CR decrement phenomenon . . . . . . . . . . . . . . . 384
PACUT, A. Mathematical modelling of reaction latency: The structure of the models and its motivation . . . . . . . . . . . . 402
PŁAŹNIK, A. and KOSTOWSKI, W. Locus coeruleus lesions and avoidance behavior in rats . . . . . . . . . . . . . . . . 419
SOŁTYSIK, S. S. and WOLFE, G. Protection from extinction by a conditioned inhibitor . . . . . . . . . . . . . . . . . 428
WYRWICKA, W. Mechanisms of motivation in avoidance behavior . . . 449

## Electrophysiological correlates of conditioning

BATUEV, A. S., PIROGOV, A. A., ORLOV, A. A. and SHEAFER, V. I. Cortical mechanisms of goal-directed motor acts in the Rhesus monkey . . 459
GASSANOV, U. and GALASHINA, A. Specific and unspecific neuronal mechanisms of cellular operant conditioning . . . . . . . . 482
GERSTEIN, G. L., RENFREW, D. H. and PUBOLS, B. H. Jr. Two behavioral paradigms for study of rapid changes in functional grouping of neurons 493
KOTLYAR, B., MAIOROV, V. and SAVCHENKO, E. Neuronal mechanisms of conditioned placing reactions in cats . . . . . . . . . . 505
MERZHANOVA, G. Single unit activity in the visual cortex during conditioning in cats . . . . . . . . . . . . . . . . 525
RADIL-WEISS, T., BOHDANECKÝ, Z. and LÁNSKÝ, P. Pacing of behavioral and electroencephalographic events . . . . . . . . . . 539
SKREBITSKY, V. G. and VOROBYEV, V. S. A study of synaptic plasticity in hippocampal slices . . . . . . . . . . . . . . . 551
STOROZHUK, V. M. Central excitation and inhibition in conditioned reflexes 561
VORONIN, L. L. Microelectrode analysis of the cellular mechanisms of conditioned reflex in rabbits . . . . . . . . . . . . . 571

**Subcortical mechanisms**

ADRIANOV, O. S., MOLODKINA, L. N., MUKHIN, E. I. SHUGALEV, N. P. and YAMSHIKOVA, N. G. Participation of caudate nucleus in different forms of voluntary activity in cats . . . . . . . . . . . 607
DIVAC, I. Functions of the neostriatum: cortex-dependent or autonomous? 619
ILYUTCHENOK, R. Yu. The emotiogenic brain structures in conditioning mechanisms: conditioned evoked potentials and motor responses . . 629
MORRISON, A. R. Relationships between phenomena of paradoxical sleep and their counterparts in wakefulness . . . . . . . . . . 643
SHAPOVALOVA, K. B. and BAZHENOVA, S. I. The role of striatum in the acquisition of instrumental defensive reactions in dogs . . . . 660

**Anatomical considerations**

AKERT, K. and HARTMANN-VON MONAKOW, K. Relationships of precentral, premotor and prefrontal cortex to the mediodorsal and intralaminar nuclei of the monkey thalamus . . . . . . . . . . 680
NITECKA, L., AMERSKI, L., PANEK-MIKUŁA, J. and NARKIEWICZ, O. Thalamoamygdaloid connections studied by the method of retrograde transport . . . . . . . . . . . . . . . . . . 699
Subject Index . . . . . . . . . . . . . . . . . 717
Authors' Index . . . . . . . . . . . . . . . . 720

PREFACE

The book contains the proceedings of *The Warsaw Colloquium on Instrumental Conditioning and Brain Research* held in Jabłonna near Warsaw on May 1-5, 1979. The proceedings are also published in the sixth issue of 1979 and the first issue of 1980 of Acta Neurobiologiae Experimentalis. Three papers published later, in the fourth issue of 1980, are also included.

The aim of this Symposium was to honour the memory of the great Polish neurophysiologist Jerzy Konorski and the sixtieth anniversary of the Nencki Institute, where Professor Konorski has created his school. Forty-five papers, prepared mainly by Konorski's pupils and scientific friends, cover several fields of current brain research. To each of them Professor Konorski contributed significantly. In fact, many presented papers have been strongly influenced by his scientific ideas.

Perhaps the most conspicuous feature of the Symposium was its very international character due to numerous participants from both East and West. As such, the Symposium is a worthy symbol of the achievements of both Jerzy Konorski himself and the Nencki Institute as a whole in the field of cooperation of scientists from East and West.

Scientific sessions were held in the mornings and afternoons. In the evenings the participants mutually read and commented their papers. Moreover, colleagues from the USA and Great Britain corrected the English of the manuscripts. In particular Prof. J. F. Brennan, Prof. G. L. Gerstein and Dr. I. Łukaszewska offered a lot of their time for reading the manuscripts. This effort made possible the quick publication of the symposial proceedings.

During the last day several posters were presented by Polish young research workers. The proceedings of the poster session can be found in the first and the second issues of 1980 of Acta Neurobiologiae Experimentalis.

*Editors*

# CONTRIBUTORS

*ADRIANOV, O. S. Brain Research Institute, Academy of Medical Sciences of the USSR, Per. Obukha 5, Moscow 107120, USSR.
*AKERT, K. Institute for Brain Research, University of Zürich, August Forelstrasse 1, CH-8029 Zürich, Switzerland.
AMERSKI, L. Institute of Medical Biology, School of Medicine, Dębinki 1, 80-211 Gdańsk, Poland.
ASRATYAN, E. A. Institute of Higher Nervous Activity and Neurophysiology, Academy of Sciences of the USSR, Butlerova 5a, Moscow 117485, USSR.
BARTLETT, J. R. Died in 1978.
*BATUEV, A. S. Department of Biophysics, Leningrad State University, 7/9 University Embankment, Leningrad 199164, USSR.
BAZHENOVA, S. I. Pavlov Institute of Physiology, Academy of Sciences of the USSR, Nab. Makarova 6, Leningrad V-164, USSR.
BERKLEY, M. A. Department of Psychology, Florida State University, Tallahassee, Florida, USA.
*BIGNAMI, G. Istituto Superiore di Sanita, Viale Regina Elena 299, 00161 Roma, Italy.
BOHDANECKÝ, Z. Institute of Physiology, Czechoslovak Academy of Sciences, Budéjovická 1083, 14220 Prague 4-KRČ, Czechoslovakia.
*BRENNAN, J. F. Department of Psychology, University of Massachusetts-Boston, Harbor Campus, Boston, Massachusetts 02125, USA.
BRACKBILL, R. M. University of Arkansas, Monticello, Arkansas, USA.
BRUNK, R. Institute of Physiology, Humboldt University, Hessische Str. 3-4, 104 Berlin, GDR.
BRUSH, F. R. Experimental Psychology Laboratory, Syracuse University, Syracuse, New York 13210, USA.
*BUDOHOSKA, W. Nencki Institute of Experimental Biology, Pasteura 3, 02-093 Warsaw, Poland.
BUREŠ, J. Institute of Physiology, Czechoslovak Academy of Sciences, Videňská 1083, 14220 Prague 4-KRČ, Czechoslovakia.
BUREŠOVÁ, O. Institute of Physiology, Czechoslovak Academy of Sciences, Videňská 1083, 14220 Prague 4-KRČ, Czechoslovakia.
*BUSER, P. Laboratoire de Neurophysiologie Comparée, Universite Pierre et Marie Curie, 4 Place Jussieu, 75230 Paris-Cedex 05, France.
BUZSÁKI, G. Institute of Physiology, University Medical School, 7643 Pécs, Hungary.
*CZIHAK, E. Institute of Physiology, Humboldt University, Hessische Str. 3-4, 104 Berlin, GDR.
*DIVAC, I. Institute of Neurophysiology, Juliane Maries Vej 36, DX-2100 Copenhagen, Denmark.

\* Present at the symposium.

DOTY, R. W. Center for Brain Research, University of Rochester, Rochester, New York 14642, USA.
FABRE, M. Laboratoire de Neurophysiologie Comparée, Universite Pierre et Marie Curie, 4 Place Jussieu, 75230 Paris-Cedex 05, France.
*FONBERG, E. Nencki Institute of Experimental Biology, Pasteura 3, 02-093 Warsaw, Poland.
FRALEY, S. M. Experimental Psychology Laboratory, Syracuse University, Syracuse, New York 13210, USA.
GAHÉRY, Y. Départment de Neurophysiologie Générale, INP, CNRS, 31 chemin Joshep Aiguier, 13274 Marseille-Cedex 2, France.
GALASHINA, A. Institute of Higher Nervous Activity and Neurophysiology, Academy of Sciences of the USSR, Butlerova 5a, Moscow 117485, USSR.
*GASSANOV, U. Institute of Higher Nervous Activity and Neurophysiology, Academy of Sciences of the USSR, Butlerova 5a, Moscow 117485, USSR.
*GERSTEIN, G. L. Department of Physiology, School of Medicine, University of Pennsylvania, Philadelphia, Pennsylvania 19104, USA.
*GRABOWSKA, A. Nencki Institute of Experimental Biology, Pasteura 3, 02-093 Warsaw, Poland.
*GRASTÝAN, E. Institute of Physiology, University Medical School, Szigeti 12, Pećs, Hungary.
HARTMANN-VON MONAKOW, K. Institute for Brain Research, University of Zürich, August Forelstrasse 1, CH-8029 Zürich, Switzerland.
HUGHES, H. C. Department of Psychology, Florida State University, Tallahassee, Florida, USA.
*ILYUTCHENOK, R. Yu. Institute of Physiology, Siberian Branch of the Academy of Medical Sciences of the USSR, Zolotodolinskaya 101, Novosibirsk 630090, USSR.
IOFFE, M. Institute of Higher Nervous Activity and Neurophysiology, Academy of Sciences of the USSR, Butlerova 5a, Moscow 117485, USSR.
*KOSTARCZYK, E. Nencki Institute of Experimental Biology, Pasteura 3, 02-093 Warsaw, Poland.
*KOSTOWSKI, W. Psychoneurological Institute, Sobieskiego 1/9, 02-957 Warsaw, Poland.
*KOTLYAR, B. Department of Higher Nervous Activity, Faculty of Biology, Moscow State University, Lenin Hills 18, Moscow 117234, USSR.
LÁNSKÝ, P. Institute of Physiology, Czechoslovak Academy of Sciences, Budéjovická 1083, 14220 Prague 4-KRČ, Czechoslovakia.
*ŁAWICKA, W. Nencki Institute of Experimental Biology, Pasteura 3, 02-093 Warsaw, Poland.
LAWRY, J. A. University of Minnesota, Minneapolis, Minnesota 55455, USA.
LINDER, B. McMaster University, Hamilton, Ontario, Canada.
*ŁUKASZEWSKA, I. Nencki Institute of Experimental Biology, Pasteura 3, 02-093 Warsaw, Poland.
MAIOROV, V. Department of Higher Nervous Activity, Faculty of Biology, Moscow State University, Lenin Hills 18, Moscow 117234, USSR.
*MERZHANOVA, G. Institute of Higher Nervous Activity and Neurophysiology, Academy of Sciences of the USSR, Butlerova 5a, Moscow 117485, USSR.
*MASSION, J. Départment de Neurophysiologie Générale, INP, CNRS, 31 chemin Joshep Aiguier, 13274 Marseille-Cedex 2, France.

*MŁODKOWSKA, Z. Nencki Institute of Experimental Biology, Pasteura 3, 02-093 Warsaw, Poland.
MOLODKINA, L. N. Brain Research Institute, Academy of Medical Sciences of the USSR, Per. Obukha 5, Moscow 107120, USSR.
*MORRISON, A. R. Laboratories of Anatomy, School of Veterinary Medicine, University of Pennsylvania, Philadelphia, Pa. 19104, USA.
MUKHIN, E. I. Brain Research Institute, Academy of Medical Sciences of the USSR, Per. Obukha 5, Moscow 107120, USSR.
*NARKIEWICZ, O. Institute of Medical Biology, School of Medicine, Dębinki 1, 80-211 Gdańsk, Poland.
NEVEROV, V. P. Pavlov Institute of Physiology, Academy of Sciences of the USSR, Makarova 6, Leningrad V-164, USSR.
NITECKA, L. Institute of Medical Biology, School of Medicine, Dębinki 1, 80-211 Gdańsk, Poland.
ONIANI, T. N. Beritashvili Institute of Physiology, Georgian Academy of Sciences, 14 Gotua Str., 380060 Tbilisi, USSR.
OPPERMANN, B. Institute of Physiology, Humboldt University, Hessische Str. 3-4, 104 Berlin, GDR.
ORLOV, A. A. Department of Biophysics, Leningrad State University, 7/9 University Embankment, Leningrad 199164, USSR.
*OVERMIER, J. B. Department of Psychology, University of Minnesota, Eliot Hall, Minneapolis, Minnesota 55455, USA.
*PACUT, A. Institute of Automatics, Technical University, Nowowiejska 15/19, 00 665 Warsaw, Poland.
PANEK-MIKUŁA, J. Institute of Medical Biology, School of Medicine, Dębinki 1, 80-211 Gdańsk, Poland.
PAYNE, R. J. Shippensburg, State College, Shippensburg, Pennsylvania, USA.
PŁAŹNIK, A. Psychoneurological Institute, Sobieskiego 1/9, 02-957 Warsaw, Poland.
PIROGOV, A. A. Department of Biophysics, Leningrad State University, 7/9 University Embankment, Leningrad 199164, USSR.
POLIT, A. Department of Psychology, M.I.T., Cambridge, Massachusetts 02139, USA.
PUBOLS, B. H. Department of Physiology, School of Medicine, University of Pennsylvania, Philadelphia, Pa. 19104, USA.
*RADIL-WEISS, T. Institute of Physiology, Czechoslovak Academy of Sciences, Budějovická 1083, 14220 Prague 4-KRČ, Czechoslovakia.
RENFREW, D. H. Department of Physiology, School of Medicine, University of Pennsylvania, Philadelphia, Pa. 19104, USA.
*SADOWSKI, B. Institute of Genetics and Animal Breeding, 05-551 Mroków, Poland.
*SANTIBAÑEZ-H., G. Institute of Physiology, Humboldt University, Hessische Str. 3-4, 104 Berlin, GDR.
SAVCHENKO, E. Department of Higher Nervous Activity, Faculty of Biology, Moscow State University, Lenin Hills 18, Moscow 117234, USSR.
*SHAPOVALOVA, K. B. Pavlov Institute of Physiology, Academy of Sciences of the USSR, Makarova 6, Leningrad V-164, USSR.
SHEAFER, V. I. Department of Biophysics, Leningrad State University, 7/9 University Embankment, Leningrad 199164, USSR.
SHUGALEV, N. P. Brain Research Institute, Academy of Medical Sciences of the USSR, Per. Obukha 5, Moscow 107120, USSR.

*SIMONOV, P. V. Institute of Higher Nervous Activity and Neurophysiology, Academy of Sciences of the USSR, Butlerova 5a, 117485 Moscow, USSR.
*SKREBITSKY, V. G. Brain Research Institute, Academy of Sciences of the USSR, Per. Obukha 5, Moscow 107120, USSR.
*SOLOMON, R. L. Department of Psychology, University of Pennsylvania, Philadelphia, Pa. 19104, USA.
SOŁTYSIK, S. S. Mental Retardation Research Center, School of Medicine, University of California at Los Angeles, Los Angeles, California 90024, USA.
*SPRAGUE, J. M. Department of Anatomy, School of Medicine, University of Pennsylvania, Philadelphia, Pa. 19104, USA.
*STAMM, J. Department of Psychology, State University of New York, Stony Brook, New York 11794, USA.
*STELLAR, E. Department of Psychology, University of Pennsylvania, Philadelphia, Pa. 19104, USA.
*STOROZHUK, V. M. Bogomoletz Institute of Physiology, 4 Bogomoletz Str., Kiev-24, USSR.
VARTANOVA, N. G. Beritashvili Institute of Physiology, Georgian Academy of Sciences, 14 Gotua Str., 380060 Tbilisi, USSR.
VOROBYEV, V. S. Brain Research Institute, Academy of Sciences of the USSR, Per. Obukha 5, Moscow 107120, USSR.
*VORONIN, L. L. Brain Research Institute, Academy of Medical Sciences of the USSR, Per. Obukha 5, Moscow 107120, USSR.
WOLFE, G. Mental Retardation Research Center, School of Medicine, University of California at Los Angeles, Los Angeles, California 90024, USA.
WYRWICKA, W. Department of Anatomy, UCLA School of Medicine, Los Angeles, California 90024, USA.
YAMSHIKOVA, N. G. Brain Research Institute, Academy of Sciences of the USSR, Per. Obukha 5, Moscow 107120, USSR.
*ŻERNICKI, B. Nencki Institute of Experimental Biology, Pasteura 3, 02-093 Warsaw, Poland.
*ZIELIŃSKI, K. Nencki Institute of Experimental Biology, Pasteura 3, 02-093 Warsaw, Poland.

Opening address delivered at the Warsaw Colloquium on Instrumental
Conditioning and Brain Research
May 1979

## OPENING ADDRESS

Konrad AKERT

Institute for Brain Research, University of Zürich
Zürich, Switzerland

Before opening the first session of this Symposium in honor of Professor Jerzy Konorski, I would like to say a few words about the great master. Professor Konorski was not only a truly outstanding scientist and founder of a school of neurobiologists, but he was also a truly gifted teacher. I recognized him as an unusual man in this respct when we first met on the occasion of his visit to the USA in 1957. For here was a man who reported not only facts and data but who succeeded in integrating his findings into a conceptual structure and he did it in the classical Socratic manner of holding an intensive dialogue with the audience.

The second characteristic of his talent as a teacher was his contagious and vibrating enthusiasm which carried him and his listeners away by arousing an indomitable impetus of going right back to the laboratory and tackle the problem with fresh ideas. But Konorski was also a good friend and a generous person, and this was partly due to his warm heart and partly related to his enormous self discipline which directed his interests primarily to science and research and not to power and glory. I felt his generosity during the first meeting when I criticised him after his lecture on the frontal lobe syndrome for not using more rigid anatomical criteria in the reconstruction of his surgical ablations. He was startled for a moment and I realized that I had been too aggressive towards my senior colleague. But after this incident we discussed the problem in detail in my laboratory and as a consequence began to collaborate and became warm friends ever since.

Finally, I would like to mention another merit: perhaps no neuroscientist has shown greater interest and dedication to the idea of bringing East and West together. He served this important goal in many ways, scientifically and personally. The first monograph (1948) "Conditioned reflexes and neuron organization" was an attempt to mediate between Sherringtonian and Pavlovian Neurophysiology. He became an advocate of collaboration by organizing scientific exchange programs between Warsaw, Bethesda and Philadelphia, which had an impact far beyond those who were directly participating. All these efforts have been immensely fruitful for scientists throughout the world, and we can only hope that Professor Konorski's enthusiasm for international friendship and cooperation will forever replicate among scientists.

Opening address prepared for the Warsaw Colloquium on Instrumental
Conditioning and Brain Research
May 1979

## OPENING ADDRESS

Ezras A. ASRATYAN

Institute of Higher Nervous Activity and Neurophysiology
Academy of Sciences of the USSR
Moscow, USSR

J. M. Konorski, to whom this symposium is dedicated, was for some 40 years my close friend in the highest sense of the word. At the basis of our friendship was our deep devotion to I. P. Pavlov's materialistic teaching and also a certain sharing of our scientific fates. We both, J. M. Konorski who came from Poland, and I — from Armenia, were fascinated by the ideas of I. P. Pavlov; we arrived at Leningrad and had the good fortune to become his students. Each of us in his own way and in accordance with his possibilities, has further developed Pavlov's teachings.

The fundamental and invaluable contribution of J. M. Konorski, from my point of view, was that he included into the Pavlovian theory of conditioned reflexes the extremely important number of brain phenomena, known as operant reactions, which up to now are considered by many foreign investigators as nonreflexive in their origin and their nature.

J. M. Konorski's many years of purposeful and thorough experimental work, initiated together with S. Miller, was later on continued under his leadership by a large group of his coworkers. This provided to the discovery of many valuable facts and relationships in a wide variety of conditioned reflexes which Konorski called conditioned reflexes of the second type. These facts concern the characteristics of their specificity, the rules of their formation, performance and inhibition, the physiological mechanisms, anatomical structures, etc. All this, independent of interpretation, forms an important contribution to the theory of I. P. Pavlov and to the contemporary knowledge of brain activity.

In this connection, I must mention one circumstance. It is known that, together with I. P. Pavlov, I have not shared J. M. Konorski's point of view that conditioned reflexes of the second type contrast sharply with the classical conditioned reflexes. At various conferences and at our various meetings discussions of this question arose frequently, sometimes having a rather sharp character. Nevertheless, our personal relations remained sincere, cordial, and cloudless. I remember this particular warmth, because I feel that such interactions between scientists are particularly valuable for the general development of science, and also because, unfortunately, such interactions are quite infrequently encountered.

J. M. Konorski and I also shared a broad interest in the general achievements of contemporary neurophysiology and their effective use in our particular experimental and theoretical investigations of the higher nervous activity.

One of Konorski's most attractive characteristics was his unbounded love of science, and his total commitment to it. Science was the aim of his life. He was also distinguished by the high principles and sincerity of his communication with colleagues. He always defended his convictions consistently and without any compromise.

The memory of J. M. Konorski is very dear to me also because he had a deep understanding of the need for close contact between scientists from the socialist countries who were studying activity of the brain. An outstanding realization of such mutual contacts between scientists of the socialist countries was the organization of regular international conferences since 1958. These conferences united the efforts of the Nencki Institute of Experimental Biology, directed by J. M. Konorski, the Institute of Higher Nervous Activity and Neurophysiology of the USSR Academy of Sciences, under my direction, and the Institute of Physiology of the Czechoslovak Academy of Sciences, with invitations to scientists from other socialist countries. Without any doubt these conferences can be considered the predecessors of the official international organization "Intermozg" which now serves the same goals on even a larger scale, helping to unite the people of the socialist countries and to promote further progress of brain sciences.

Opening address delivered at the Warsaw Colloquium on Instrumental
Conditioning and Brain Research
May 1979

# INVOLVEMENT OF THE PARTIAL REINFORCEMENT PROCEDURE IN REWARD TRAINING: OPENING ADDRESS

Kazimierz ZIELIŃSKI

Department of Neurophysiology, Nencki Institute of Experimental Biology
Warsaw, Poland

*Abstract.* The analysis of the course of reward training presented in early Konorski and Miller's papers (1933, 1936) indicates that the appearance of active instrumental responses is related with the introduction of the partially reinforced, excitatory classically conditioned procedure. It is postulated that the mechanism underlying the effects of this procedure on the shaping of instrumental responses is similar to that proposed by Konorski (1967) for the effects of inhibitory classically conditioned stimuli on the performance of the instrumental response.

It is a great honor and privilege for me to be given the opportunity to deliver an opening lecture about two events of great importance for this country and for the world of science: the establishment of the Nencki Institute of Experimental Biology in 1918, and the publication of the first of Jerzy Konorski's papers on Type II Conditioned Reflexes in 1928 (9, 10). These two events were intimately related.

For more than a dozen years leaders of the intellectual life in Warsaw, supported by scientists from other countries who were pupils and friends of the great Polish physiologist and biochemist, Marceli Nencki, attempted to create a biological institute in Warsaw. Only after Poland regained independence, however, were the several biological laboratories existing since 1911 along with their Scientific Council and Directorial Board, finally legalized as the Nencki Institute of Experimental Biology. A number of outstanding scientists organized a very efficient establishment, providing the opportunity to conduct research in physiology, biochemistry, protozoology, hydrobiology, ethology and mathematical statistics (16).

The Institute's library, which counted about 30,000 volumes of books and periodicals before its total destruction during the second World War, was the reason for the first contact with the Nencki Institute by Jerzy Konorski and Stefan Miller, students of the Medical Faculty at Warsaw University. Konorski wrote in his autobiography (4): "Instead of medicine, we read and reread Pavlov and discussed every detail of the experimental work of his coworkers. Moreover, we found in the library of the Nencki Institute of Experimental Biology the Russian journals in which the papers on conditioned reflexes were published" (p. 186).

As a result of their studies, these two young gifted men designed and started on 1 February, 1928, in the Laboratory of Psychology of the Free Polish University, experiments on avoidance, punishment, reward, and omission learning, varieties of conditioned reflexes other than those studied by Pavlov and his pupils. Progress in their studies was amazingly rapid. The first two papers concerning the properties of Type II Conditioned Reflexes were delivered to the Warsaw Branch of the French Biological Society by the summer of 1928, and after a few months were published in French in "Comptes Rendus de la Société de Biologiae" (9, 10). Soon after, Konorski and Miller were offered the opportunity to continue their research in the physiological laboratories of the Medical Faculty at Warsaw University, and later in the State Psychiatric Hospital in Pruszków near Warsaw. Their results from this early period of research were summarized in a monograph, "The foundation of the physiological theory of acquired movements", published in Polish in 1933 (16). However, much earlier, in 1928, Konorski and Miller exchanged letters with Pavlov, who invited the two young scientists to his laboratories in Leningrad. Konorski worked in the Institute of Experimental Medicine for two years (1931-1933), and Miller for only several months. The main theme of the research conducted by Konorski during the Leningrad period was concerned with the relations between instrumental and classical conditioned reflexes. The results obtained were published in Russian in the "Transactions of Pavlov's Laboratories" in 1936 (7).

After returning from Leningrad the only place in Poland able to accept the two visitors to Pavlov's laboratories within its staff was the Nencki Institute. In contrast to electrophysiology, there was no tradition of behavioral research in Poland. However, the Nencki Institute had developed a strong school of ethology, conducting research on the adaptive behavior of various species from infusoria to amphibians. The Institute appreciated the importance of neurophysiological research for biological sciences and consistently attempted to incorporate different groups of neurophysiologists. In addition, there was no "Sovietophobia" at the Institute. On the contrary, the Soviet Union's scientific centers were the most numerous among those with which the Institute exchanged publications. This aspect was rather important for Konorski's future, since Konorski and Miller, in both lectures and publications, presented an impartial picture of the development of Soviet science as well as the social changes in that country. And finally, Konorski and Miller returned from the Institute of Experimental Medicine, where the physiological section was built by Ivan Pavlov, but biochemical section was organized by Marceli Nencki during the last decade of his life.

Such were the reasons why Jerzy Konorski was able to work at the Nencki Institute in the Department of Physiology, headed by Professor Kazimierz Białaszewicz. Thereafter, for 40 years, Professor Konorski was a member of the Nencki Institute. He started to organize chambers for experiments on conditioning, while participating in the investigations of Liliana Lubińska on the excitability of peripheral nerves and neuromuscular transmission. When the Physiology Department was destroyed by artillery fire during the first days of the battle for Warsaw, Konorski, like other members of the Institute, tried to salvage apparatus and books from under the wreckage and debris, typing out copies of scientific notes and works still in manuscript, which were then preserved in various places. All of this work was in vain, becoming lost or destroyed by enemy actions. Among the materials lost were chapters of the book prepared by Konorski just before the outbreak of the war. The book was eventually published in altered form in 1948 under the title "Conditioned reflexes and neuron organization" (3). However, the many losses in human life were the heaviest blow to the Institute. Among others who were killed in action at the front, fell in the Warsaw Uprising or were murdered by the Nazis, was Stefan Miller, the closest friend and coworker of Konorski. Tragically, he committed suicide when Nazi troups entered his flat in the Warsaw Ghetto.

Jerzy Konorski and Liliana Lubińska were able to flee Warsaw, and after a short period of work at the Białystok hospital, they received an invitation to Leningrad. From there, they moved to Sukhumi, where Konorski was offered the position of head of the Physiological Department belonging to the Institute of Experimental Medicine. By the beginning of 1945 Konorski began his yourney back home. A small group of the pre-war staff, only six persons, reactivated the Nencki Institute, first at Łódź in 1945 and then in Warsaw since 1954. Konorski organized a Department of Neurophysiology in the Institute, which he headed until the end of his life. Although he did not relish administrative duties, he performed many. Since 1946 he was continously a member of Institute's Board of Directors, and he was the Director of the Institute from 1968 to 1973. No single important event in the Institute was without his participation. He actively influenced the development of other departments of the Institute. Research on human perception as well as in neurochemistry were greatly stimulated by him. Today the Nencki Institute numbers 116 scientific workers, and 50 of them were either his pupils and coworkers or were trained by close collaborators of Konorski. This number illustrates the extent of his influence on the present character of our Institute. The journal "Acta Neurobiologiae Experimentalis", a continuation of the pre-war journal, "Acta Biologiae Experimentalis", acquired an international range under his guidance. Professor Konorski was also a leader in the cooperation of Polish neurophysiologists with scientists of other countries, and the audience of this conference provides the best proof of the strength of ties between the Nencki Institute and many leading neurophysiological centers throughout the world.

Jerzy Konorski was the most distinguished among many prominent scientists and outstanding persons fostered by the Nencki Institute. He united talent with industry, intellectual courage with responsibility. All of us were influenced by his ideas and personality, his arguments in discussions and by his papers and books. All of us know many of his scientific achievements. Thus, I would like to present only one of them, and probably the most important, reported for the first time by Miller and Konorski in 1928 in the three-page paper, "On a particular form of conditioned reflex" (10). This paper contains a description of the experimental procedures leading to the formation of the four varieties of Type II Conditioned Reflexes, now usually refered as instrumental reflexes, and their basic differences from the procedures of classically conditioned reflexes. This discovery was presented and discussed in all monographs written by Jerzy Konorski.

When we compare this paper with corresponding phrases in his last book, "Integrative activity of the brain" (4, p. 389), we find a number of differences in the description of the procedures for instrumental conditioning. These differences reflect changes in the understanding of the processes underlying these conditioning procedures. In a brief form they were summarized by Jerzy Konorski within a Postscript to the 1928 paper, when it was translated and published by Skinner in the Journal of the Experimental Analysis of Behavior (11).

In their early papers Konorski and Miller used not only Pavlovian terminology, but they also applied the type of analysis developed by Pavlov's scientific school. They placed emphasis on the stimulus conditions existing during the formation of instrumental responses. Let us consider the reward training procedure. Konorski and Miller (11) noted that an active movement is learned when a compound of two stimuli, external stimulus (A), presented by the experimenter, and the second stimulus (B), consisting of "all the sensations generated by a particular movement such as lifting the leg — that is, a set of muscular, tactile, and other sensations", is accompanied by the food unconditioned stimulus (R). However, the external stimulus alone, similar to "sensations accompanied this particular movement" without the external stimulus, is not accompanied by food. In other words, training of the instrumental response by the reward method involves the discrimination of the three arrangements of stimulus conditions: (i) the external stimulus plus the proprioceptive stimulus plus food, (ii) the external stimulus alone, and (iii) the proprioceptive stimulus alone (A+B+R versus A versus B). After some number of presentations of these three stimulus conditions to the subject "... the stimulus A will then be capable of provoking by itself the appearance of the stimulus B — in order to complete the conditioned compound" (p. 187). Appearance of the movement of which the proprioception is an element of the conditioned compound is the very basis of type II (instrumental) conditioning.

Pavlov considered the excitatory classically conditioned reflex as an effect of the association between conditioned and unconditioned stimuli. After the association between the two stimuli is formed, according to Pavlov "the conditioned stimulus is a signal, as if it were a substitution of the unconditioned stimulus" (12, p. 117). In the early Konorski and Miller writings, the compound consisting of external and proprioceptive stimuli was considered as a conditioned stimulus for the classically conditioned reflex. Thus, there was a basic uniformity between Pavlov's idea of stimulus substitution and Konorski and Miller's thoughts of the active instrumental response as a source of stimulation which is missed when the passive movement was not elicited. The description of instrumental learning procedures with phrases stressing the complementary role of the proprioceptive stimuli smoothed over the difference between the classical and instrumental conditioning procedures. To my mind, this was one of the reasons responsible for the delay in the acceptance of the specificity of instrumental reflexes by the Pavlovian school. Careful reading of the Pavlov's foreword to the Konorski and Miller paper published in Russian in 1936 seems to confirm this supposition (12).

Later experiments conducted by Górska, Jankowska, Tarnecki and others (see 4, p. 467–479) showed "that proprioception of a trained movement does not play any essential role in type II conditioning" and "passive movement as a rule cannot be instrumentalized" (Konorski in the "Postscript", 11, p. 189).

Konorski's impressions about his relations with Pavlov presented in his Autobiography (5, p. 195) reflected both admiration and some reservations caused by Pavlov's negative attitude toward the specificity of Type II Conditioned Reflexes and calling them "motor conditioned reflexes" or "conditioned reflexes of the motor analyser". However, inspection of the Konorski's bibliography (2) indicates that the titles of the two next papers by Konorski and Miller published before their visit to Leningrad in 1930 included just these very phrases used to indicate their own primary views on the nature of instrumental reflexes.

The contemporary reader of the early papers by Konorski and Miller has an impression that the authors were too Pavlovian, that their early readings exerted a stronger influence on them than their own original findings. Only in the course of experimental work did their understanding of instrumental conditioning become more refined and exact. Paradoxically, the Leningrad period was extremely important for the scientific maturation of Konorski as the founder of instrumental conditioned reflexes. There, he used dogs with long training histories in classically conditioned reflexes, which provided an opportunity to investigate the interrelations between instrumental and classically conditioned reflexes. These experiments confirmed the preliminary data obtained in Warsaw, showing that conditioned stimuli, which elicited classically conditioned alimentary salivation, exerted an inhibitory effect on instrumental responses established by the reward procedure (Fig. 1). On the other hand, conditioned stimuli which inhibited classically conditioned salivation, as a rule, elicited instrumental responses trained by the reward procedure (6, p. 101-128; 7, p. 139-143 and 165-175; 4, p. 369-375). These findings demonstrated the basic differences between classically conditioned and instrumental reflexes. Moreover, they directed the further research of Konorski and his pupils into problems of extinction and inhibition (see, 15). Data on the interrelations between classically conditioned and instrumental reflexes provided arguments in numerous discussions with other interpretations of instrumental conditioning. For instance, they were heavily used in Konorski and Miller's paper published in The Journal of General Psychology in 1937 (8) where they discussed Skinner's paper published by the same journal two years earlier (14).

Our present approach to the distinction between various conditioning procedures is based on the analysis of contingencies rather than on stimulus conditions. From this point of view we may distinguish several different relations among the conditioned stimulus, the conditioned respone and the unconditioned stimulus. In the case of classically conditioned reflexes, the conditioned stimulus signals the definite probability of the unconditioned stimulus presentation which cannot be changed by subject's behavior. In typical Pavlovian experiments the excitatory conditioned stimulus is a signal for food or some other unconditioned stimulus inevitably given at the end of its action, and presentation of the

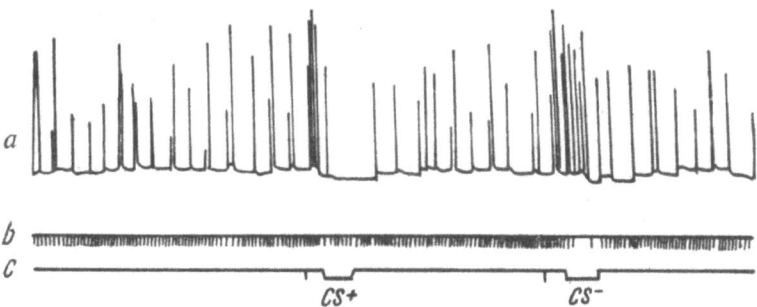

Fig. 1. The effect of classically conditioned stimuli, CS+ (excitatory) and CS− (inhibitory), on instrumental response established in the experimental situation. From above: a, movements of the hindleg; b, salivation; c, CS presentations. The animal repeatedly lifts its right hindleg, each movement being reinforced by food. Then (mark in line c) food is not presented for a few seconds and thereafter CS+ (bell) is given with the 15-s CS-US interval. The dog momentarily stops performing the movement, waits for food and salivates copiously. The food is presented, CS+ is terminated and the instrumental responses are restored. Then again food is not given for a few seconds, and CS− (metronome, 60 cycle/min) is presented. The movements are not stopped, but the salivation is much reduced. From Konorski and Miller (7, p. 141).

inhibitory stimulus is a signal that the unconditioned stimulus will not be given either during or shortly after its action. In the real world outside the conditioning chamber, such strict correlations between conditioned and unconditioned stimuli are rather exceptional. Normally, conditioned stimuli signals the certain probability of an unconditioned stimulus. Performance of a specific behavior is necessary for obtaining an attractive unconditioned stimulus or avoiding an aversive one. When only one kind of overt behavior, let us say, the leg flexion, is observed, four different relations between the conditioned stimulus, conditioned response and unconditioned stimulus may be expected, and all of them were investigated and described by Konorski and Miller in their first experimental paper. In reward training the occurrence of the specific change of behavior (leg flexion) is a necessary condition for presentation of the attractive unconditioned stimulus, whereas in omission training the withholding of this same response during CS action is a condition for obtaining the attractive US after the CS offset. In avoidance training the specific change of behavior is a condition for not presenting the aversive US, and in punishment training the withholding of this very response during CS action is a condition for nonpresentation of the aversive US. The "truly random" pseudoconditioning procedure (13), where CS presentations, US occurrences, and the performance of the CR are not correlated, provides control conditions for both classical and instrumental training procedures.

In the consecutive run of conditioning trials, either classical or instrumental, a subject changes its behavior according to the consequences of this behavior in previous trials. Let us limit our analysis to the alimentary situation. When excitatory classically conditioned reflexes are trained, not only copious salivation, but also the targeting (orienting) reflex to the food bowl become the dominant modes of behavior. The behavior of a well trained dog is immediately changed after the CS onset: all gross bodily movement become inhibited and the animal's posture indicates that the dog is ready to get the food that will be presented toward the end of the CS. During training of inhibitory classically conditioned reflexes, drowsiness in dogs has been frequently described, and when a large proportion of inhibitory trials are presented during an experimental session, full sleep may even be observed. In both varieties of classically conditioned reflexes, a subject learns that its behavior has no consequences on food presentation.

For successful instrumental learning two conditions have to be fulfilled: (i) the change in behavior chosen by the experimenter as an instrumental response must be performed with some minimal initial probability, and (ii) the method employed must provide an opportunity for the subject to recognize that the specific change in its own behavior leads to a change in the probability of the unconditioned stimulus.

It is obvious that only elements of the normal repertoire of an animal's responses may be used as instrumental movements. The higher the behavior chosen by the experimenter as an instrumental response in the hierarchy of responses emitted in a given experimental situation, the easier is the instrumental training. However, responses occurring with a lower probability provide an opportunity for more detailed analysis of the increase in their frequency and for discovering the conditions that lead to stabilization of the instrumental response. For dogs placed on the Pavlovian stand, the leg flexion has a lower probability of occurrence than the pressing a bar located near the food-tray for rats in an operant chamber or a key-pecking response for pigeons. Inspection of the graphs illustrated in the early Konorski and Miller papers indicates that the method of training employed in their experiments inevitably increased the general motor activity of the trained dogs and, among other bodily movements, the leg flexion was also more frequent.

At early stages of reward training conducted according to the method of Konorski and Miller, the dog is confronted with two kinds of trials: (i) passive leg flection elicited during the action of the sporadic external stimulus followed by food presentation, and (ii) action of the same sporadic stimulus without leg flection not followed by the alimentary unconditioned stimulus (6, p. 60–69, 10). Taking into account the discoveries, obtained by Konorski and his collaborators in the 1950's, that proprioceptive feedback is not necessary for instrumental conditioning, we may infer that at the beginning of reward training a dog is confronted with the partially reinforced excitatory, classically conditioned reflex situation. In such a situation the central motor behavior system of a hungry dog is aroused and the trained animal performs various motor responses. Most often movements located high in the hierarchy of responses are performed; for instance, another instrumental movement acquired during the previous experimental history of the trained animal (6, p. 66–69). All unsuccessful behaviors become inhibited, the hierarchy of responses changes in the process of eliminating inappropriate movements and eventually the required behavior occurs.

The importance of the introduction of the partially reinforced excitatory, classically conditioned reflex situation for the course of reward training may be inferred from experiments conducted by Konorski in Leningrad (7, p. 145–187). He showed that during the action of the sporadic stimulus active leg flexions occur only after the introduction of trials in which the sporadic stimulus was not followed by food (Fig. 2). One of these experiments, conducted on dog "Nord", yielded data of special interest for the present analysis. The motor activity of the dog had been drastically increased by caffeine injection. Among various movements, numerous leg flexion responses were performed. Some of them occurred during the action of the sporadic stimulus and according to the rules of reward training they were reinforced by food presentation. However, as seen from Fig. 3, in these circumstances transformation of the excitatory classically conditioned reflex into the instrumental reflex did not occur. The general motor excitement which was independent of the contingencies employed during training did not provide the opportunity for the subject to recognize that performance of the required motor response had any consequences for presentation of the food reinforcement. Only after the introduction of partial reinforcement was a moderate increase of leg flexion responses in the experimental situation observed and on this background instrumental training proceeded successfuly.

When Konorski discussed the effects of classically conditioned stimuli on performance of instrumental responses, his line of thinking was almost identical to that presented above for the analysis of the early stages of reward training. In his last monograph (4) he wrote: "... the inhibitory CS in its early stage of development produces a state which differs from that produced both by the positive CS and by the firmly established inhibitory CS. Whereas the positive CS produces a pure alimentary re-

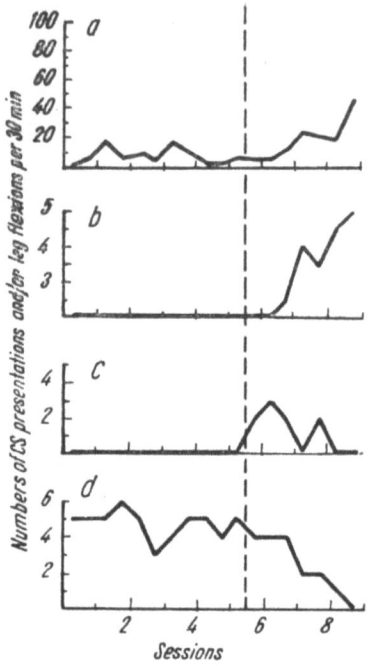

Fig. 2. The course of reward training in dog Fingal. Daily sessions from 14 January 1932 are presented. From above: *a*, active leg flexion in intertrial intervals; *b*, active leg flexion during a noise stimulus reinforced by food; *c*, noise presentations not accompanied by leg flexion or food; *d*, noise accompanied by passive leg flexion and food. Vertical dashed line indicates the introduction of noise presentations not accompanied by passive leg flexion and food presentation. From Konorski and Miller (7, p. 165).

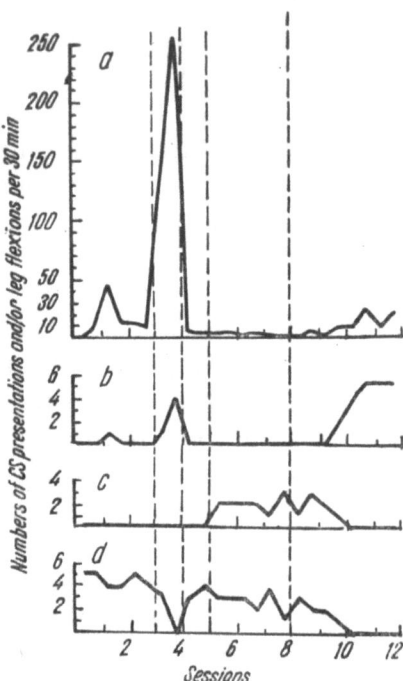

Fig. 3. The course of reward training in dog Nord. Daily sessions from 19 April 1932 are presented. Before the 4th session, 5 cc of 1% caffein solution was injected. Beginning with the 6th session, noise presentations not accompanied by passive leg flection and food presentation were introduced. From the 9th session noise of a higher intensity was used as a conditioned stimulus. Denotations as in Fig. 2. From Konorski and Miller (7, p. 176).

sponse resembling that elicited by the US itself, and the well-established inhibitory CS produces the inhibitory response, the early inhibitory CS given rise to a conflict between the two, resulting in an entirely new state" (p. 375).

This state is assumed to be the increase of the hunger drive, which in turn causes the rise of excitability of the central motor behavior system activating structures responsible for the performance of motor acts most probable in a given situation. An analysis of the early papers by Konorski and Miller suggests that the some mechanism, i.e., the increase of the hunger drive and the excitation of the central motor behavioral system — is active during instrumental reward training at this stage when classically conditioned stimuli are partially reinforced with food presentation.

There are data showing that the increase in the fear drive together with the excitability of the central motor system accompanies the early stages of avoidance training. Experiments by Gibson (1) were especially illustrative. Goats and sheep were trained in a large box enabling observation of the animals' behavior. During classically defensive conditioning, subjects emitted various motor responses, changing their repertoire in each consecutive trial. Similar behavior was observed at early stages of avoidance training, although, the response pattern chosen by the experimenter as an instrumental response soon became dominant. When the situation was changed in such a way that another motor response was now required for successful avoidance of shock, the rise in excitability of the motor system resulted in increased emission of various motor responses came initially, similar to behavior observed during classically defensive conditioning.

All of these data indicate that the behavior of animals is subjected to many changes according to the contingencies employed. Not only during instrumental but also during classical training, changes in the pattern of motor responses occurred. These processes may be overlooked when our attention is directed to only one special class of responses. But once we recognize that our aim is to investigate the integrative activity of the brain and its role in adaptive behavior of humans and animals, we will probably share Konorski's opinion offered in his "Postscript" to the first experimental paper by Miller and Konorski (11): "Finally, the sharp distinction between, not only the procedural side of type I and type II conditioned reflexes, but also between their physiological mechanisms seems to me now largely exaggerated. In fact, further investigation shows with increasing clarity that both types can be explained on the basis of the same general principles of connectionistic processes" (p. 189).

Over the years the scientific interests of Jerzy Konorski became much broader than at the beginning of his scientific career. New lines of research were undertaken and more precise methods were introduced. However, the main problem of the present conference, the central mechanisms of instrumental conditioning, always was the most exciting for him. Professor Konorski presented results of experiments and defended his ideas to many audiences, listened attentively to others opinions and lively discussed new facts and hypotheses. I am sure that the present conference will be both interesting and enjoyable, if we try to emulate the way so typical of Jerzy Konorski, the founder of investigations on instrumential conditioning.

## REFERENCES

1. GIBSON, E. J. 1952. The role of shock in reinforcement. J. Comp. Physiol. Psychol. 45: 18–30.
2. GŁOWACKA, R. 1974. Bibliography of Professor Jerzy Konorski's papers. Acta Neurobiol. Exp. 34: 681–695.
3. KONORSKI, J. 1948. Conditioned reflexes and neuron organization. Univ. Press, Cambridge, 267 p.
4. KONORSKI, J. 1967. Integrative activity of the brain: An interdisciplinary approach. Univ. Chicago Press, Chicago, 531 p.
5. KONORSKI, J. 1974. Jerzy Konorski, In: A history of psychology in autobiography. Vol. 6. Appleton-Century Crofts, New York.
6. KONORSKI, J. and MILLER, S. 1933. Podstawy fizjologicznej teorii ruchów nabytych. Ruchowe odruchy warunkowe. Książnica Atlas TNSW, Warsaw, 168 p.
7. KONORSKI, J. and MILLER, S. 1936. Conditioned reflexes of the motor analyser (in Russian). Tr. Fiziol. Lab. Akad. I. P. Pavlova 6: 119–288.
8. KONORSKI, J. and MILLER, S. 1937. On two types of conditioned reflexes. J. Gen. Psychol. 16: 264–272.
9. MILLER, S. and KONORSKI, J. 1928. Le phénomène de la généralisation motrice. C. R. Séanc. Soc. Biol. 99: 1158.
10. MILLER, S. and KONORSKI, J. 1928. Sur une forme particuliére des réflexes conditionnels. C. R. Seanc. Soc. Biol. 99: 1155-1157.
11. MILLER, S. and KONORSKI, J. 1969. On a particular type of conditioned reflex. J. Exp. Anal. Behav. 12: 187-189 (English transl. of Ref. 10).
12. PAVLOV, I. P. 1936. Foreword to the paper by J. Konorski and S. Miller (in Russian). Tr. Fiziol. Lab. Akad. I. P. Pavlova 6: 115–118.
13. RESCORLA, R. A. 1967. Pavlovian conditioning and its proper control procedures. Psychol. Rev. 74: 71–80.
14. SKINNER, B. F. 1935. Two types of conditioned reflex and a pseudo type. J. Gen. Psychol. 12: 66–77.
15. ZIELIŃSKI, K. 1979. Extinction, inhibition, and differentiation learning. In A. Dickinson and R. A. Boakes (ed.), Associative mechanisms in conditioning. L. Erlbaum Ass., Hillsdale, p. 269–293.
16. ZIELIŃSKI, K. 1979. Sixty Years of the Nencki Institute of Experimental Biology. The Review of the Polish Acad. of Sci. No 4: 47-74.

Kazimierz ZIELIŃSKI, Nencki Institute of Experimental Biology, Pasteura 3, 02-093 Warsaw, Poland.

Lecture prepared for the Warsaw Colloquium on Instrumental
Conditioning and Brain Research
May 1979

# AN EXPLORATION OF THE ABILITY OF MACAQUES TO DETECT MICROSTIMULATION OF STRIATE CORTEX

John R. BARTLETT[1] and Robert W. DOTY

Center for Brain Research, University of Rochester
Rochester, New York, USA

*Abstract.* With its head steadied within a form-fitting mask, a macaque was first taught to signal when it detected the application of 0.2-ms electrical pulses at 50 Hz through electrodes chronically implanted within its striate cortex. Stimuli were then applied via a movable microelectrode and the threshold for the animal's detection determined at intervals of 50–250 µm. With permanently implanted 130– 200-µ diameter electrodes such thresholds range between 50 and 250 µA (and are highly stable), whereas with the microelectrodes sites were encountered, estimated to be primarily within cortical layers V–VI, where the monkey could reliably detect as little as 2–4 µA. The threshold at most sites within striate cortex with the microelectrode, however, was 15–25 µA. Background unit activity recorded with the microelectrode varied greatly in different laminae and survived the microstimulation, but has so far provided no clear basis for predicting threshold. It is tentatively hypothesized that the relatively rare points where the threshold is as much

---

[1] John R. Bartlett died unexpectedly 5 November 1978, full of plans and enthusiasm for the experiments whose beginnings are described herein. We had made the observations jointly for the 13 penetrations in the right striate cortex and produced Figs. 2 and 3. I am thus confident, in making him first author of this paper, that I can represent his views with considerable accuracy; although I suspect we would have had some warm and lengthy arguments over the suitability of exposing our speculations on the cells of Meynert — R. W. Doty.

as 10 times less than that in the surround arise because the giant, solitary cells of Meynert provide the exclusively effective output for the behavioral response. This hypothesis would also explain the singular uniformity of sensation (a "phosphene") evoked in human subjects by such stimuli, and the equivalence of all such stimuli in striate cortex found for the macaque.

## INTRODUCTION

The consequences of applying electric currents to the striate cortex in man are particularly fascinating in two respects. First, is the fact that this crude interjection of massive synchrony into the activity of thousands of otherwise independently active neurons produces a subjective sensation of light, appropriately localized according to the position of the intrusion into the retino-cortical "map" of visual space (7, 15–17). However, given the exquisite organization of the anatomical fabric (e.g., 27, 41) and the intricacy of its physiology (e.g., 4, 25, 26, 34), it is perhaps even more surprising that only a simple, singularly uniform sensation is produced, i.e., a "phosphene" of white, slightly flickering light, wherever the electrode is placed. In other words, in the sensation aroused there is no indication of the color, movement, orientation, binocular disparities and ocular motility which the physiologist finds so clearly represented in activity of the striate cortex nor, seemingly, of the rich interconnectivity of the central visual system.

It must be admitted that most of the relevant anatomical and physiological knowledge of the striate cortex is derived from work on macaques whose subjective experiences can only be inferred. However, the remarkable congruence of anatomy, physiology and psychophysics of the visual system in Old World primates (6, 8, 9, 11–14, 30, 31, 33, 36, 38, 39) gives considerable assurance that the monkey's experience when its striate cortex is stimulated electrically is comparable to that of man. Certainly the monkey finds such stimulation uniform for any locus in striate cortex; for having once learned to signal its detection of such stimulation at one point, it unhesitantly responds to stimulation at any other point so long as that is in area 17: and the monkey remains generally indifferent to stimulation in other cytoarchitectonic areas (18–20, 22) or even the lateral geniculate nucleus or optic tract (37, and Doty, unpublished) unless it is specifically trained to respond to stimulation there. This holds true whether stimuli are applied to the pial surface, as in man, or within the gray or white matter of the striate area, the only difference being that it takes several fold less current to elicit a response with electrodes which penetrate the cortex (5).

For a sample of 200 electrode sites in striate cortex in 14 macaques the threshold for detection of stimulation with 0.2- or 0.5-ms pulses at 50 Hz through chronically implanted 200-μm Pt-Ir electrodes lay between 50 and 250 μA. Assuming that this 5-fold variation in threshold among various locations reflects a true physiological difference rather than a technical flaw (e.g., trauma) or pathology (e.g., variation in tissue encapsulation of the chronically implanted electrodes), it would be of great interest to understand the nature of the differences in the detectability of such interjected excitation, i.e., phosphenes, within the neuronal circuitry of striate cortex. We thus sought to determine whether changes in threshold might be systematically related to position of an electrode within one or another cytoarchitectonic lamina, the white matter, or some other, possibly physiologically definable, location within area 17. As shown by Asanuma and his colleagues (1, 2, 29, 43) stimulation through microelectrodes can provide critical information concerning the capabilities and interconnections of neocortical systems. In the present instance there is the added advantage that a relatively constant psychophysical criterion (detection) can be related not only to anatomical location but to the type of single unit activity in the immediate vicinity of the stimulating microelectrode. By securing the microelectrode in place it should also be possible to follow over the course of several days or weeks any consistent alteration in threshold, thereby gaining some assessment of the possible role played by the connective tissue investiture of the electrode in the threshold obtained.

## METHOD

Only a single, 4.0-kg, male macaque, *Macacca nemestrina*, has been used in these preliminary experiments. It was first trained to make manual contact with a rod. The change in electrical capacitance of the rod activated circuitry to deliver fruit juice if and only if a signal was present such as acoustic clicks or electrical stimulation of striate cortex. Random touching of the rod was discouraged by prolonging the intertrial interval and/or delivering a puff of air to the face. When suprathreshold stimuli were being used consistently, there was essentially no random activity; but when the threshold was being repeatedly probed, as it was with microstimulation, tentative contacting of the rod became somewhat troublesome. In the interest of determining the absolute minimum of current which the animal could detect, one did not wish to punish it too severely for errors, otherwise the monkey would change its criterion to one of a higher level of certainty. On the other hand, great care, obviously, hand to be taken to be certain that the responses arose from genuine detection of the signal by the animal rather than its merely guessing that the stimulus was occurring. In practice, this distinction was rather easily made on the basis of latency and consistency of the response to actual detection. The stimulus lasted for 2 s and the intertrial interval ranged randomly between 5 and 40 s. A criterion of three unequivocal responses, i.e., unhesitant and at appropriate latency, in five presentations was taken as the threshold.

An array of 12 electrodes 3–8 mm apart was implanted within the representation of central vision in striate cortex of the left hemisphere. These were constructed of 92% Pt–8% W enamel-insulated, 127-μm diameter wire cut at a 60° angle with a scalpel to make a sharp point. Depth of insertion beyond the intact dura mater was controlled by gluing a 0.5-mm cuff of Teflon 2.5–3.0 mm above the tip of the wire with cyanoacrylate. A 1-mm hole was drilled through the skull in the anesthetized monkey and the wire inserted until the cuff contacted the dura. It was then cemented in place with methyl methacrylate and brought to a coded receptacle affixed to the skull (4, 5, 18). A piece of platinum foil anchored at the occiput, and a stainless steel screw in midline bone served as reference electrodes for stimulation and recording.

The monkey was trained to respond to stimulation at one cortical locus, using 0.2-ms cathodal pulses at 50 Hz, and subsequently responded to stimulation at the other loci without further training. This gave the animal experience in detecting electrical excitation of striate cortex, and in the following experiments with microstimulation provided an important control on the state of motivation of the animal to make responses to barely supraliminal stimuli.

The animal was then anesthetized with secobarbital, tracheal intubation performed using an "infant" size laryngoscope, and the head was immersed in artist's moulage to make a mold. The mold was then filled with plaster of Paris, making a model of the animal's head, and this model in turn was used to construct a form-fitting mask of fiberglass and resin to restrain the monkey's head during the sessions using microelectrodes. The animal was habituated for a few days to working in this mask. It was then anesthetized again. Under microscopic control a 1-mm diameter hole was carefully drilled through the skull overlying striate cortex. A truncated, plastic hypodermic needle hub was affixed to the skull concentric with this hole. The hub accommodated the microelectrode drive, which was constructed from a glass tuberculin syringe (Fig. 1). This light weight assembly has the advantage, besides its moderate cost, of readily following the slight movements which the monkey makes within the mask; and single units can be held as well with this as with more elaborate arrangements (e.g., 26), i.e., frequently for 10–60 min if desired. The hub opening was sealed with sterile bone wax when not in use, and precautions to maintain sterility were taken whenever it was open. No sign of infection developed.

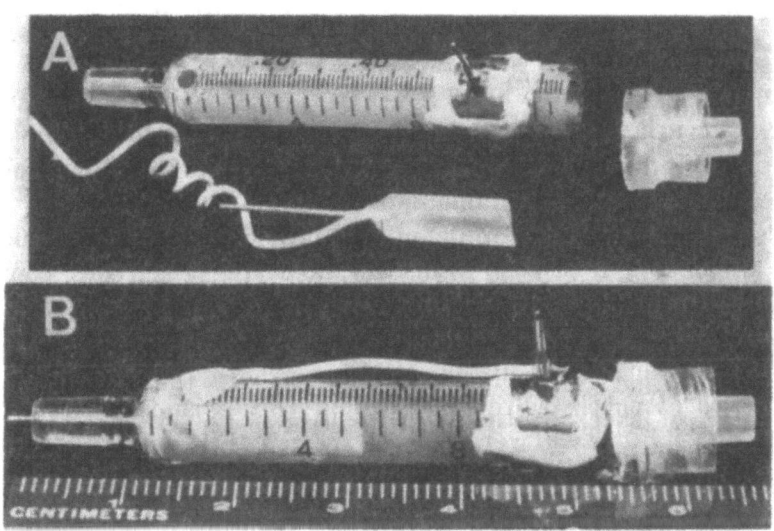

Fig. 1. Microdrive made from glass tuberculin syringe. A: components prior to assembly: above, syringe barrel with hole cut for passage of electrical connection, and mounted gold pin connector; right, Luer-lok tip cut from plastic syringe; below, piston with attached hypodermic needle tubing and highly flexible lead wire. B: assembled microdrive, piston in fully advanced position with hypodermic needle tubing protruding and ready to receive microelectrode which will be inserted into it. Electrical connection passes from the tubing via coiled, flexible wire to mounted connector pin. In use, the microelectrode is withdrawn into the syringe tip and the syringe inserted snugly into plastic needle hub mounted on skull (see text). Microelectrode is advanced by hydraulic pressure on piston, controlled by a micrometer driving a distant, matching syringe via Teflon tubing and a needle fitting the Luer-lok tip. While the short piston length in the Figure allows a travel of about 15 mm, it has some danger of very slow leakage of the hydraulic fluid, 200-centistoke silicone. This can be rectified by using a longer piston (with less travel in this length barrel), or higher viscosity fluid.

With the animal fixed in the mask and performing to stimulation through permanently implanted electrodes, the microelecrode was advanced until electrical contact was signalled by the change in impedance (measured with 1 KHz sinusoidal current, 0.1 µA). The glass-insulated Pt–Ir electrodes (Frederick Haer Co., Brunswick, Maine, 04011) had an initial impedance of roughly 5 MΩ, but this was reduced to about 200 kΩ by stimulation in most instances. The electrode was slowly advanced, particularly at the point where it was estimated it was passing through the dura mater. The threshold at which the monkey could detect stimulation was tested each 50–250 µm, and the spontaneous, background activity was noted or tape recorded at each location. Switching between the recording and stimulation modes with the microelectrode was done remotely via a miniature reed relay in the head stage of the preamplifier. Single unit activity could be selected with a bilevel window discriminator and, via an analog delay line, the entire waveform of the spike could be displayed to provide additional assurance that the same unit was being sampled from one moment to the next.

The constant current stimulator was provided with an "exhauster" circuit which, via a 2N6450 transistor, kept the electrodes connected through 15 kΩ in the interpulse intervals, thereby discharging ("exhausting") the charge which would otherwise quickly accumulate on the microelectrode and carry its operating range to the hydrolytic level (5). A series of checks with and without this exhauster circuit when currents of only 2–4 µA were needed showed that it had no influence on threshold even at these low levels.

Several penetrations which had yielded significant data were marked by passing 35 µA DC, positive current through the microelectrode for 10 s at intervals as it was withdrawn. At the termination of the experiments using the right striate cortex, the monkey was anesthetized and a right occipital lobectomy was performed to obtain the tissue for histological analysis. The experiments then proceeded 5 mos later with the left hemisphere, and finally the monkey was again anesthetized and perfused with 0.9% NaCl followed with 10% formalin. Serial frozen sections were cut horizontally at 50 µm and stained with thionin.

## RESULTS

A total of 18 traverses to various depths through striate cortex with microelectrodes, usually in steps of 250 µm, was made at four loci. The multiple penetrations at each locus over a period of days in these initial explorations has, unfortunately, resulted in considerable uncertainty in most instances as to the exact laminar location at which the observations were made. The most reliable reconstruction of a microelectrode penetration is that in Fig. 2, and here the electrode failed to penetrate deeper than layer III. One of the locations of a series of penetrations was found to be centered on a very well developed external calcarine sulcus in this monkey. This explained some of the otherwise inconsistent results from repeated penetrations at this location. Apparently on some occasions the electrode passed within the sulcus and on others briefly nicked the curvature of the gyrus, but reconstruction of these traverses proved to be impossible.

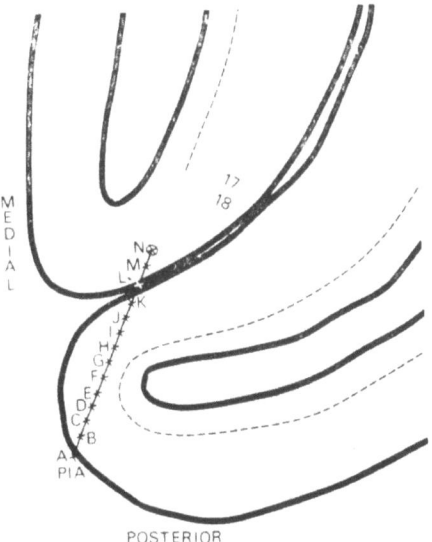

Fig. 2. Reconstruction of microelectrode track passing in 250-µm steps from striate cortex on the medial surface of the occipital pole across the calcarine fissure into area 18. Area 17 designated by heavy dashed line. Letters *A, B, C,* etc. indicate positions at which recordings were made and thresholds determined, as given in Fig. 3.

On several occasions stimuli were applied through the microelectrode as it approached and passed through the dura mater. As judged for location primarily by the level at which single unit activity appears (Fig. 3), but also on those initial occasions at each locus when the surface of the dura had not been obscured by exudate and fibroblasts, the monkey could not detect stimuli of 1.0 mA until the electrode had advanced enough to penetrate the dura. For instance, the penetration illustrated in Figs. 2 and 3 was made 24 h after exposure, and the fresh dura mater was clearly visible. Electrical contact was made, presumably with a thin layer of plasma on the dura mater, at a point 1.7 mm above that labelled "A-pia" in Fig. 2. Stimuli were applied at 1.0 mA there and after each advance of 0.25 mm, but no response was obtained until point "A" (Fig. 2) was reached, 6 min after contact. The threshold there was 250 µA.

Naturally, this high current disrupted the insulation of the microelectrode, and its impedance (Fig. 3) fell to only 0.2 MΩ. Nevertheless, excellent unit activity could still be recorded (Fig. 3).

As also can be seen in Fig. 3, there were very large differences in background activity at different loci along this penetration (Fig. 2). It is clear that the threshold bears no close relation to the magnitude of the background activity. However, at all locations where the threshold was ≦ 10 µA, as occurred somewhere along the course of eight

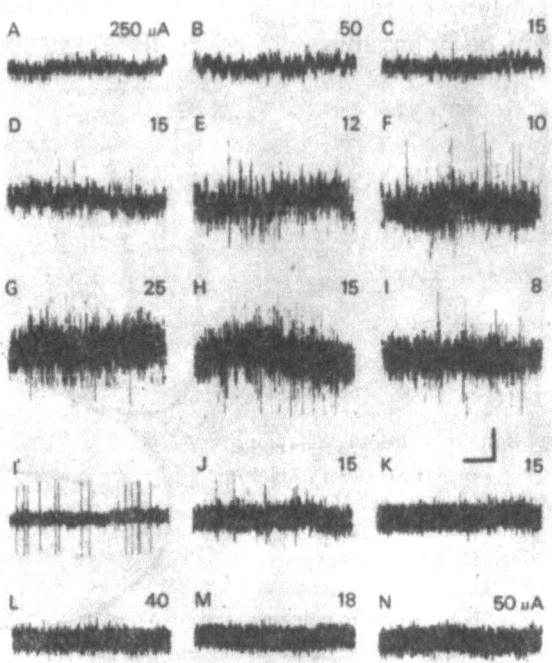

Fig. 3. Electrical activity recorded from microelectrode each 250 μm as it passed along track reconstructed in Fig. 2. Letters A, B, C, etc. correspond between the two Figures. The threshold in μA for the monkey to detect stimulation with 0.2-ms pulses at 50 Hz shown at upper right of each record. Electrode impedance was unchanged at only 0.2 MΩ throughout this run. Clearly, there is little correspondence between the threshold and the background activity at a given point, although the lower thresholds do occur where background is relatively high. Single unit activity was first encountered at "C". Record "I'" from same point as "I", but at lower gain to show single unit whose activity was still unchanged after applying the stimulation about 25 times at 6–15 μA. Note elevation of threshold at point "L", probably near pial surface. The higher threshold at "N" 25 min later, however, may reflect a waning motivation of the animal. Calibration: 20 ms, and 200 μV, except 500 μV for "I'".

of the 18 penetrations made, there was a rich background of single unit discharge. It thus seems likely that the presence of vigorous unit discharge is an indication of the conditions necessary for detection of the weakest stimuli, but that it is not wholly sufficient.

At restricted points along four penetrations the monkey was able with unequivocal reliability to detect stimuli of 2–4 µA. The dimensions of these highly effective points remain uncertain and, of course, probably depend upon the orientation of the penetration with respect to the laminar organization of the cortex. In one instance, however, advancing the electrode 100 µm changed the threshold for detection from 4 to 15 µA. On the other hand, in two instances the threshold remained at 3–4 µA despite an advance of the electrode by 250 µm. Two of these low threshold points were encountered about 1.5 mm below the pia mater, whose position was estimated from the point at which the threshold became $\leq$ 250 µA. In six other instances the position of the low threshold points was compatible with the microelectrode having passed across white matter and into the lower layers of striate cortex subjacent to that on the surface. Four of these eight low threshold points were actually found at intervals $\leq$ 250 µm along a single penetration, which histology showed to have passed through three folds of striate cortex, starting near the foveal representation. All of the data for these eight points are consistent with the location of these low threshold points lying within layers V and/or VI of area 17; and this receives added support from the absence of such points on the three penetrations made at the position illustrated in Fig. 2, which never attained these layers. Two other facts are perhaps relevant: (1) None of these low threshold points were encountered on the first day after anesthetizing the monkey to prepare the site, but were found 3–9 days thereafter; and (2) additional penetrations at the same locus usually did not encounter such a low threshold point even at roughly the same depth within the cortex.

The electrode impedance ranged from 0.2 to 2.1 MΩ on the various occasions when low threshold points were found. Single unit activity was always present at these points, but the level of background activity was consistently lower than that at, e.g., Fig. 3F–H, nor was there a very prominent, well isolated unit in the field of microelectrode recording on any of these occasions. Probable luxotonic units (4, 26, 34) were frequently observed, as could be readily done, of course, simply by turning the room lights on or off (the monkey's mask not being light proof); but the presence or absence of such activity did not appear to be associated with any particular level of threshold.

Thresholds at the majority of points ranged between 15 and 25 µA, and were probably roughly the same for white matter immediately adjacent to striate cortex. On the other hand, a penetration which apparently passed deep into white matter associated with area 18 at a slightly different angle but at the same locus as in Fig. 2, held monotonously a threshold of 100–120 µA as measured each 250 µm over a distance of 5 mm. This high threshold, however, may result more from the monkey's unfamiliarity with stimulation of circumstriate areas (18–20) than with the fact that the penetration passed within white matter.

When the stimuli being applied were in the range of 30 µA or less, single unit activity, as observed within 1–2 s after stimulation, was seemingly unaltered in its background characteristics. This held true even for units with an amplitude on the order of 1 mV or more which, presumably, are rather near to the electrode tip.

On one occasion stimulation was applied for 10 min at 200 µA (with the usual 0.2-ms, 50-Hz pulses). As evidenced by the rapid, small fluctuations in the applied voltage (8 V peak), it could be inferred that hydrolysis (5) occurred throughout this period. The theshold rose from 20 µA to 70 µA, and when tested 30 and 45 min later, was still at 50 µA. Single unit activity was completely eliminated. Subsequently determining the threshold after each advancement of the electrode in 100 µm steps, the following figures were obtained: 50, 40, 40, 30, 20, 15 µA; i.e., within 250 µm the threshold was as low as at the tetanized site prior to the 10 min of tetanization. Unfortunately, return of single unit activity was not monitored.

At two positions along a single penetration the effects of altering stimulus frequency were studied. At the first point about 1 mm into the cortex the threshold at 50 Hz was 10 µA and 25, 30 and 60 µA for 10, 5 and 1 Hz, respectively. At the other point estimated to lie just above white matter (from the abrupt decrease in background and unit activity subsequently observed as the electrode advanced) the threshold was 5 µA at 50 Hz, and 10, 20 and 60 uA for 10, 5 and 1 Hz, respectively.

The effect of changing stimulus polarity was assayed at one point, being 4 µA with cathodal pulses applied via the microelectrode, and 25 µA when anodal pulses were used.

## DISCUSSION

The quick success of these exploratory experiments considerably exceeded what had been anticipated, and while this contributed to certain obvious deficiencies in the data, two major facts nevertheless seem firmly established: (1) the threshold for the monkey to detect microstimulation at one versus another site within striate cortex can vary by more than tenfold, and (2) single unit activity can be consistently recorded via the same electrode as used for stimulation, and it too is grossly variant from one location to another (Fig. 3).

The validity of the experiments depends, of course, on the reliability of the macaque as a psychophysical observer. It is to some degree difficult to convey to the reader the basis for one's confidence that well trained macaques report their detection threshold consistently with a high degree of accuracy. This, however, is readily verified, by repeatedly changing the intensity of the stimulation between threshold and subthreshold levels for a given position of the microelectrode, and by affirming the constancy of the threshold from day to day or moment to moment at any of the implanted electrodes.

The question thus arises as to why there should be such a great difference between thresholds at various points within striate cortex. It had earlier been inferred that it was probably the most excitable elements in the cortex which provided the basis for the macaque's ability to respond to electrical stimuli applied essentially any place in neocortex (18). It would now seem that these "most excitable elements" are rather sparsely distributed, at least in striate cortex. Were this true, it would account for the relative rarity of encountering points at which stimuli of as little as 2–4 µA could be detected.

There is, of course, the problem that the best estimates of the degree to which applied electrical stimuli are elaborated in the neocortical network would not predict such a result (1, 23, 29, 32, 43). The direct excitation of pyramidal (Betz) cells by currents in the range of 2 µA for 0.2 ms is of the order of 50 µm, but the indirect effect, from exciting fibers of passage, dendrites, etc., may range out to 1 mm. Phillips and Porter (32) estimate for the precentral cortex of the macaque that 5 µA will engage a field of roughly five Betz cells and 900 small pyramids. These figures probably give the correct order of magnitude for the population of neurons in the striate cortex of the macaque with a comparable stimulus of 2 µA; for while the effective radius of excitation would be substantially smaller for the 2-µA stimulus, the packing density of neurons in striate cortex is double that of precentral gyrus (35).

There must be some reason for this great discrepancy between the expectation that even a 2-µA stimulus would be diffusely effective, and the fact that it is not. It is thus tempting to suppose that only highly specific and relatively rare neurons in striate cortex exclusively provide the output on which the monkey's response is based. The cells of

Meynert seem highly suitable for such a role. LeGros Clark (27) estimated their entire population in striate cortex as numbering only 1.300, although recent work of Chan-Palay et al. (10) suggests that the figure is nearer to 60,000. The hypothesis, that the cells of Meynert, or similar units are exclusively responsible for the monkey's learned response to stimulation of striate cortex, is attractive not only in explaining the otherwise puzzling rarity of the low threshold points and their seeming clustering in deeper cortical layers, but in providing as well for the surprising uniformity of sensation, the phosphene, produced in man (7, 15–17), and the immediate, inherent equivalence of stimulated striate loci in macaques (18–22).

There are, also, difficulties with this hypothesis; perhaps the major one being that, from the interweaving of the basilar dendrites of the Meynert cells (10), it should not be possible to pass a stimulating microelectrode vertically through the cortex without their coming within the effective radius of stimulation. One would also expect, upon tangential penetration of the cortex, similarly, to bring the apical dendrites within range. Perhaps this is all true, and the relatively gross movements of the microelectrode in 250 μm steps in the present experiments commonly brought it to rest somewhat remote to these critical structures. Since there are upwards of 36,000 synaptic spines on a Meynert cell (10), it would be reasonable to imagine that it might take considerable convergence to excite it indirectly via its afferent inputs, thus requiring a larger current to recruit this population when the electrode lies some distance from the cell, versus the case where the electrode is sufficiently near to excite it directly.

Other problems with the hypothesis, however, do not have even these tentative answers. For instance, Lund et al. (28) are loath to distinguish Meynert cells from the other pyramidal cells in layer V; i.e., perhaps there really is no population of relatively rare, morphologically unique neurons to fit the physiological-psychophysical hypothesis. And, of course, what would there be about the output characteristics and destination of the Meynert cells which would make them such exclusive purveyors of the signal that the ongoing activity of striate neurons had been disrupted by the interjected electrical stimuli? Together with the other pyramidal cells of layer V the cells of Meynert project to the superior temporal cortex, inferior pulvinar and superior colliculus and seem undistinguished in this regard (28, 40, 41).

Finally, the logical thrust of the "Meynert cell" hypothesis leads to the supposition that a single neuron could be responsible for the critical output when the microelectrode is at the point of lowest threshold. The triggering of a behavioral response by even a single quantal neural or receptor event (e.g., 3, 24, 42) is, of course, a possibility, but in the present instance a train of roughly 20–100 pulses was being utilized in the measurements of threshold. The threshold increased greatly (6–12 times) on the two occasions when only a single pulse was used. This suggests that considerable numerical recruitment is required in such circumstance, but whether its effective output might then still all be channelled via one or several Meynert cells or via a different path is problematical, to say the least. However, while it seems rash to speculate that only a single, highly specialized type of cell conveys the information relevant to the behavior, the existence of isolated points with thresholds many times less than the surround forces the consideration of some such degree of specialization; if not of single cells, then of clumped colonies of cells somehow having a preferential, high impact output.

In any event, it is a challenge to understand how nerve impulses interjected by the electrical stimuli can be detected as differing in pattern from among the thousands already occurring each second within a cubic millimeter of striate cortex. With the ability to register the presence or absence of detection, and to manipulate site and parameters of the applied stimuli while recording their effect upon surrounding neural activity, it now seems possible to assay this problem with some chance of success.

Work performed under Contract NIH 70-2279 from the National Institutes of Health. We are grateful to Richard A. Sloane for help in training the monkey, to Scott M. Borden for construction of the mask, and to E. A. DeYoe for much helpful discussion.

## REFERENCES

1. ASANUMA, H. and ROSÉN, I. 1973. Spread of mono- and polysynaptic connections within cat's motor cortex. Exp. Brain Res. 16: 507–520.
2. ASANUMA, H. and SAKATA, H. 1967. Functional organization of a cortical efferent system examined with focal depth stimulation in cats. J. Neurophysiol. 30: 35–54.
3. BARLOW, H. B., LEVICK, W. R. and YOON, M. 1971. Responses to single quanta of light in retinal ganglion cells of the cat. Vision Res. Suppl. 3: 87–101.
4. BARTLETT, J. R. and DOTY, R. W. 1974. Response of units in striate cortex of squirrel monkeys to visual and electrical stimuli. J. Neurophysiol. 37: 621–641.
5. BARTLETT, J. R., DOTY, R. W., LEE, B. B., NEGRÃO, N. and OVERMAN, W. H., JR. 1977. Deleterious effects of prolonged electrical excitation of striate cortex in macaques. Brain, Behav. Evol. 14: 46–66.
6. BOLTZ, R. L., HARWERTH, R. S. and SMITH, E. L. III. 1979. Oriëntation

anisotropy of visual stimuli in rhesus monkey: A behavioral study. Science 205: 511-513.
7. BRINDLEY, G. S. and LEWIN, W. S. 1968. The sensations produced by electrical stimulation of the visual cortex. J. Physiol. (Lond.) 196: 479-493.
8. BROWN, P. K. and WALD, G. 1963. Visual pigments in human and monkey retinas. Nature 200: 37-43.
9. CAVONIUS, C. R. and ROBBINS, D. O. 1973. Relationships between luminance and visual acuity in the rhesus monkey. J. Physiol. (Lond.) 233: 239-246.
10. CHAN-PALAY, V., PALAY, S. L. and BILLINGS-GAGLIARDI, S. M. 1974. Meynert cells in the primate visual cortex. J. Neurocytol. 3: 631-658.
11. COHEN, A. I. 1965. Some electron microscopic observations on interreceptor contacts in the human and macaque retinae. J. Anat. 99: 595-610.
12. COWEY, A. 1979. Cortical maps and visual perception. The Grindley Memorial Lecture. Quart. J. Exp. Psychol. 31: 1-17.
13. DE VALOIS, R. L. and JACOBS, G. H. 1971. Vision. In A. M. Schrier and F. Stollnitz (ed.), Behavior of Nonhuman Primates. 3: Academic Press, New York, p. 107-157.
14. DE VALOIS, R. L., MORGAN, H. C., POLSON, M. C., MEAD, W. R. and HULL, E. M. 1974. Psychophysical studies of monkey vision. I. Macaque luminosity and color vision tests. Vision Res. 14: 53-67.
15. DOBELLE, W. H. and MLADEJOVSKY, M. G. 1974. Phosphenes produced by electrical stimulation of human occipital cortex, and their application to the development of a prosthesis for the blind. J. Physiol. 243: 553-576.
16. DOBELLE, W. H., MLADEJOVSKY, M. G., EVANS, J. R., ROBERTS, T. S. and GIRVIN, J. P. 1974. "Braille" reading by a blind volunteer by visual cortex stimulation. Nature 259: 111-112.
17. DOBELLE, W. H., MLADEJOVSKY, M. G., ROBERTS, T. S. and GIRVIN, J. P. 1974. Artificial vision for the blind: Electrical stimulation of visual cortex offers hope for a functional prosthesis. Science 183: 440-444.
18. DOTY, R. W. 1965. Conditioned reflexes elicited by electrical stimulation of the brain in macaques. J. Neurophysiol. 28: 623-640.
19. DOTY, R. W. 1967. On "butterflies" in the brain (in Russian). In V. S. Rusinov (ed.), Sovremennye problemy elektrofiziologii tsentral'noĭ nervnoĭ sistemy. Izdat. Nauka, Moscow, p. 96-103. On "butterflies" in the brain. In V. S. Rusinov (ed.), Electrophysiology of the central nervous system 1970. Plenum, New York, p. 97-106.
20. DOTY, R. W. 1969. Electrical stimulation of the brain in behavioral context. Ann. Rev. Psychol. 20: 289-320.
21. DOTY, R. W. and NEGRÃO, N. 1973. Forebrain commissures and vision. In R. Jung (ed.), Handbook of Sensory Physiology, Vol. VII/3B. Springer-Verlag, Berlin, p. 543-582.
22. DOTY, R. W., NEGRÃO, N. and YAMAGA, K. 1973. The unilateral engram. Acta Neurobiol. Exp. 33: 711-728.
23. GUSFAFFSON, B. and JANKOWSKA, E. 1976. Direct and indirect activation of nerve cells by electrical pulses applied extracellularly. J. Physiol. (Lond.) 258: 33-61.
24. HENSEL, H. and BOMAN, K. K. A. 1960. Afferent impulses in cutaneous sensory nerves in human subjects. J. Neurophysiol. 23: 564-578.
25. HUBEL, D. H. and WIESEL, T. N. 1977. Functional architecture of macaque monkey visual cortex. Proc. Roy. Soc. Lond. 198: 1-59.

26. KAYAMA, Y., RISO, R. R., BARTLETT, J. R. and DOTY, R. W. 1979. Luxotonic responses of units in macaque striate cortex. J. Neurophysiol. 42: 1495–1517.
27. LE GROS CLARK, W. E. 1941–42. The cells of Meynert in the visual cortex of the monkey. J. Anat. 76: 369–376.
28. LUND, J. S., LUND, R. S., HENDRICKSON, A. E., BUNT, A. H. and FUCHS, A. F. 1975. The origin of efferent pathways from the primary visual cortex, area 17, of the macaque monkey as shown by retrograde transport of horseradish peroxidase. J. Comp. Neurol. 164: 287–304.
29. MARCUS, A., ZARZECKI, P. and ASANUMA, H. 1979. An estimate of effective spread of stimulating current. Exp. Brain Res. 34: 59–72.
30. MARKS, W. B., DOBELLE, W. H. and MAC NICHOLL, E. F., JR. 1964. Visual pigments of single primate cones. Science 143: 1181–1183.
31. PEARLMAN, A. L., BIRCH, J. and MEADOWS, J. C. 1979. Cerebral color blindness: An acquired defect in hue discrimination. Ann. Neurol. 5: 253–261.
32. PHILLIPS, C. B. and PORTER, R. 1977. Corticospinal neurones. Their role in movement. Academic Press, London, 450 p.
33. POLYAK, S. 1957. The vertebrate visual system. University of Chicago Press, Chicago, 1300 p.
34. RISO, R. R., BRUST-CARMONA, H., BARTLETT, J. R. and DOTY, R. W. 1979. Receptive field properties of luxotonic units in macaque striate cortex. Abs. Soc. Neurosci. 5: 805.
35. ROBINS, E. D., SMITH, E. and EYDT, K. M. 1956. The quantitative histochemistry of the cerebral cortex. Architectonic distribution of ten chemical constituents of motor and visual cortices. J. Neurochem. 1: 54–67.
36. SARMIENTO, R. F. 1975. The stereoacuity of macaque monkey. Vision Res. 15: 493–498.
37. SCHUCKMAN, H., KLUGER, A. and FRUMKES, T. E. 1970. Stimulus generalization within the geniculostriate system of the monkey. J. Comp. Physiol. Psychol. 73: 493–500.
38. SCOTT, T. R. and POWELL, D. A. 1963. Measurement of a visual motion after-effect in the rhesus monkey. Science 140: 57–59.
39. SHKOL'NIK-YARROS, E. G. 1971. Neurons and interneuronal connections of the central visual system. B. Haigh (translator), R. W. Doty, (translation ed.) Plenum, New York, 295 p.
40. SPATZ, W. B. 1977. Topographically organized reciprocal connections between areas 17 and MT (visual area of superior temporal sulcus) in the marmoset *Callithrix jacchus*. Exp. Brain Res. 27: 559–572.
41. SPATZ, W. B., TIGGES, J. and TIGGES, MARGARETE. 1970. Subcortical projections, cortical associations, and some intrinsic interlaminar connections of the striate cortex in the squirrel monkey (*Saimiri*). J. Comp. Neurol. 140: 155–174.
42. STEVENS, S. S. 1972. A neural quantum in sensory discrimination. Science 177: 749–762.
43. STONEY, S. D. JR., THOMPSON, W. D. and ASANUMA, H. 1968. Excitation of pyramidal tract cells by intracortical microstimulation: Effective extent of stimulating current. J. Neurophysiol. 31: 659–669.

Robert W. DOTY, Center for Brain Research, University of Rochester, Rochester, New York 14642, USA.

Lecture delivered at the Warsaw Colloquium on Instrumental
Conditioning and Brain Research
May 1979

# CONDITIONED AUDIO-VISUAL TARGETING REFLEXES IN SPLIT BRAIN CATS

Ruth BRUNK, Eva CZIHAK, Barbara OPPERMANN and Guy SANTIBAÑEZ-H.

Institute of Physiology, Humboldt-University, Berlin, GDR

*Abstract.* A conditioned audio-visual targeting reflex was elaborated in 12 freely moving cats. The cats had to localize the loudspeaker emiting by a tone of 1,600 Hz, of 500 ms duration and 80 dB intensity. Each time one of the eight loudspeakers placed in front and behind the cat was activated and the targeting reaction rewarded by food. Normal cats attained the 80%  criterion for the front and rear loudspeakers in 20 and 30 sessions, respectively. The influence of the transection of the corpus callosum on acquisition and retention of the targeting reaction was investigated. The split brain animals with pre-operative experience in the situation did not show any retention, but relearned the targeting of the frontal sources in 20 sessions. No relearning of the posterior sound sources was observed. The animals that had only post-operative training did not reach 80% of correct responses in 50 sessions. Corpus callosum transection influences the integration of the targeting reaction in different ways, depending on the position of the sound source and on pre-operative training.

## INTRODUCTION

In Pavlov's laboratory Bykov and Speransky (1) showed in 1924 that tactile symmetrical differentiation was a matter of considerable difficulty, but after sectioning of the corpus callosum the differentiation was easily established. Sperry (16) described an analogous occurrence in patiens with a surgical transection of the corpus callosum. If a stylus was used to stimulate "a point on a particular joint of a particular finger was held palm up and was screened from the visual field", the patient could find the stimulated point on the same hand, but unlike a normal person he was quite unable to find a corresponding mirror point on the opposite hand, and was also unable to verbally determine the stimulated point if it was on subordinate hand. Gazzaniga et al. (7) showed that performances, in which visual inflow was restricted to one hemisphere and the response involved the hand whose primary cortical representation was on the same hemisphere, were little affected, whereas those performances requiring interaction or direct cooperation of both hemispheres showed marked disruption.

Dobrzecka (3) demonstrated that the transection of the corpus callosum induced a facilitation of symmetrical tactile stimuli differentation in dogs. The facilitation was observed if the right and left side of the animals trunk were stimulated, but not if the distal parts of the fore or hind legs were stimulated. Myers and Sperry (10) trained chiasm-sectioned cats in a visual pattern discrimination task with one eye, and afterwords they bilaterally extirpated the primary visual cortex. They observed that the recall was immediate, or nearly so, for the simpler discrimination. Recall failed however in the case of more complicated discriminations. Dobrzecka et al. (4, 5) showed that the effect of the interaction between the corpus callosum transection and the extirpation of the somatosensory cortex was a consequence of the cortical representation of the receptive field of the stimulated system.

Corballis and Beale (2) reviewed the factual evidence and stated that left-right discrimination tests are specially difficult for animals and humans to solve. They suggested that an interhemispheric fiber system, such as the corpus callosum "symmetrices" memory traces, and thus preserves structural symmetry. Based on the literature, Gazzaniga (6) stated that the interhemispheric exchange of visual, tactile, olfactory, proprioceptive and auditory information is totally disrupted after commissurotomy and that a learned task integrated in both hemispheres is also disrupted.

There is not much information about the role played by the commisural pathway in sound localization. Naumann (11) did not find significant influences after sectioning the corpus callosum. The present study analizes the ability of cats, with a section of the corpus callosum, to target a source of a sound with their eyes.

## METHODS

Subjects were 16 adult male cats divided into three groups: a control *group* (CG, $n = 8$), a group with pre- and post-operative training (PpG, $n = 4$) and a group with post-operative training only (PoG, $n = 4$).

*Training procedure.* Cats on 24 h food deprivation were trained in a cage 100 cm long × 100 cm wide × 80 cm high. Eight loudspeakers oriented toward the center of the cage were placed in the corners. Thus four loudspeakers were in front of the cat (L1, L2, L3 and L4) and four behind it (L5, L6, L7 and L8). The animal was observed through a one-way vision screen. At the beginning of training one or two loudspeakers were used to teach the subject that it would receive food in the feeder every time it looked at the loudspeaker from which a sound was delivered. Usually 16 or 32 trials sufficed for the cat to learn this task. Only then was it submitted to the experimental procedures requiring subjects to visually localize the loudspeaker from which the sound was delivered.

The auditory stimulus was a 1,600 Hz tone of 500 ms duration and 80 dB intensity delivered at random from the eight different loudspeakers. The auditory stimulus was emitted when the subject was looking in the direction of the feeder and especially when its body was oriented perpendicular to the screen. Each loudspeaker was activated twice in a session. The animal received pieces of meat delivered through a rotatory feeder as a reward, and it received up to maximum of 100 g of food during and immediately after the training session.

Each cat was trained during 50 sessions with 16 trials in each session. The intertrial intervals lasted 1–3 min, depending on the time the animal needed to orient its body perpendicular to the screen. A trial was considered correct when the cat was able to localize the activated loudspeaker with its eyes at the first attempt. If the trial had no correct response it was repeated up to a maximum of three times at the same interval as the normal trials. A trial was incorrect when the cat did not look at the activated loudspeaker and reacted in general with a targeting reflex but not always focusing on the activated loudspeaker.

*Surgery and histology.* The transection of the corpus callosum was performed in a stereotaxic instrument. Chloralose anesthesia (70 mg/kg) was injected intraperitoneally. After trepanation, the dura was opened along the interhemispheric fissure to expose the corpus callosum. Two electrode holders were used to make the transection, but instead of an electrode, a needle was placed into each holder. The sharp end of the needle was inserted into the holder and a loose thread was introduced through the eye of the needle. The needles were sterotaxically oriented in such a way that the eyes remained under the corpus callosum level: one at the occipital extreme and the other at the frontal pole. The loose thread was pulled and the corpus callosum sectioned. In order to avoid bending the needles while pulling the thread, a piece of wood was placed between the needles. Its length was exactly that of the required interneedle distance. Antibiotics were subsequently administered. At the end of the experiment the cats were anestethized with Nembutal. The brain was perfused with 10% formaline and subsequently tissue surrounding transection examined. Of the 14 cats operated, only eight received the transection of the corpus callosum without injury to other structures, and these eight subjects are considered in the results. In these cats the anterior commisure was intact. The results were statistically analysed by the t-test.

## RESULTS

*Control group (CG) performance.* In the first five sessions the scores for both sets of loudspeakers were different: 65% of correct responses for the front ones and 57% for the rear set. The animals reached the criterion of 80% correct responses to the speakers placed in front of them (L1–L4) after 20 sessions, while 35 sessions were required to meet the criterion of correct responses to the speakers placed at their rear (L5–L8). At the end of the 50 sessions of training the animals showed 90% positive responses for the front group of loudspeakers while the level of correct responses for the rear speakers remained at 80%. The statistical comparison of the performance seen in Fig. 1 indicated that the differences between responses to each set of speakers were significant ($t = 2,9; P < 0.05$).

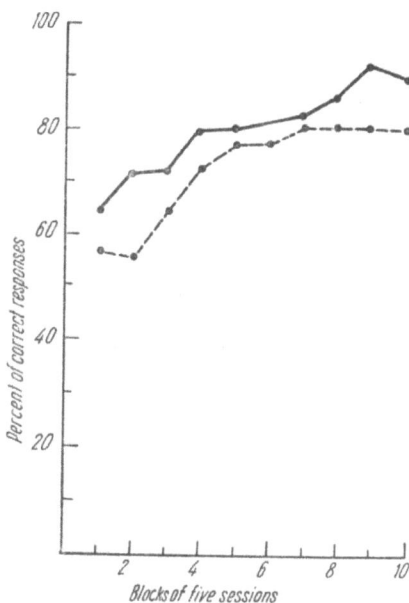

Fig. 1. Course of the learning process in the control group (CG). Solid line, front speakers ($L_1$–$L_4$); dashed line, rear speakers ($L_5$–$L_8$).

*Comparison of learning and post-surgical learning (PpG)*. The transection of the corpus callosum produced a decrease in the retention scores for both sets of speakers the scores for the front set fell to 66%

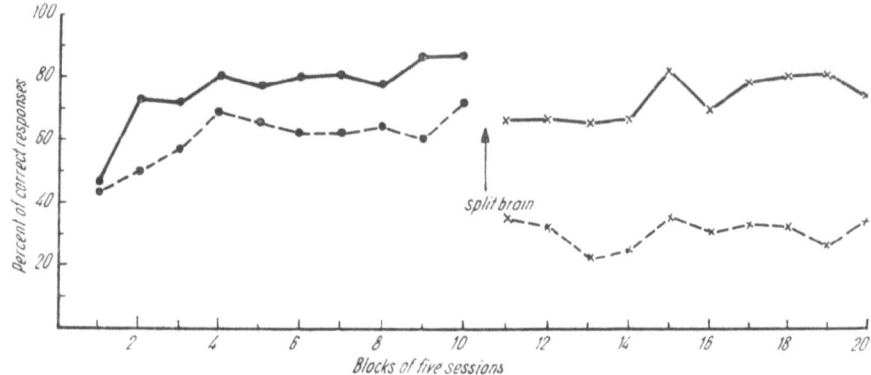

Fig. 2. The course of pre-operative learning and post-operative retention in PpG group. Denotations as in Fig. 1.

while those for the rear set to 36%. Post-operative retraining produced an increase of the scores to 79% for the front speakers until the fourth session, but no further increase was observed until 50th session. Conversely, responses given to the rear speakers did not show any increase

TABLE I

*t* test of comparison of performance in respect to speaker position

| Speaker position | Before-after operation (PpG) | Control-post-operative training (CG-PoG) |
|---|---|---|
| L1 | 0.814 | 2.300* |
| L2 | 0.079 | 5.370** |
| L3 | 0.653 | 5.324** |
| L4 | 2.170 | 2.307* |
| L5 | 2.39* | 3.109* |
| L6 | 3.06* | 3.093* |
| L7 | 3.17* | 2.908* |
| L8 | 5.57** | 2.730* |

during the entire retraining period (Fig. 2). *t*-tests showed no statistical differences between the pre- and post-operative performance for the front speakers, but significant differences emerged for the rear set (Table I).

*Comparison between learning of control group (CG) and operated group (PoG).* During the first ten sessions an increase of correct responses was observed; the curve then tended to stabilize (Fig. 3). Up to the 50th session a slight improvement in the amount of correct responses to the front speakers was seen, whereas a decrease occurred to the rear set. In no case did correct performance reach the levels required by the learning criterion. The comparison between the operated and the normal animals was made for each loudspeaker separately. This analysis indicated that both groups differed significantly for all the speakers (Table I).

*Comparison between control and both surgical groups.* The differences between the CG and both operated groups indicates that the surgical effects were different, depending on the pre-operative experience and on the position of the speakers (Fig. 4). The comparison of the differences CG-PoG-control and without pre-operatory experience groups — between the performance related to the anteriorly (L1–L4) and to the

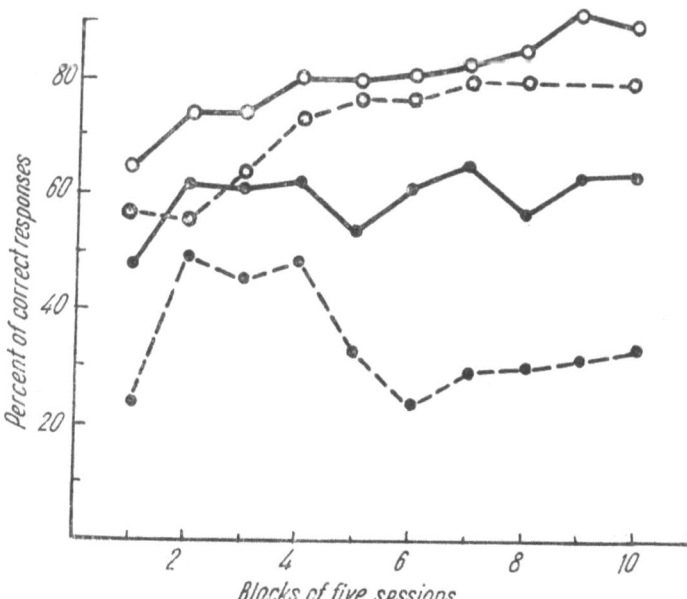

Fig. 3. The course of the learning of control group (CG) and the group which received only post-operative training (PoG). Open circles, CG; filled circles, PoG. Other denotations as in Fig. 1.

posteriorly placed loudspeakers (L5–L8) are statistically significant ($t = 5.00$, $P < 0.01$) and greater than the differences CG-PpG-control and post-operative trained groups — for the same comparison ($t = 3.04$,

$P < 0.01$). However there are not significant differences between CG-PoG and CG-PpG for the loudspeaker placed behind the animals and only a tendency for those placed before them.

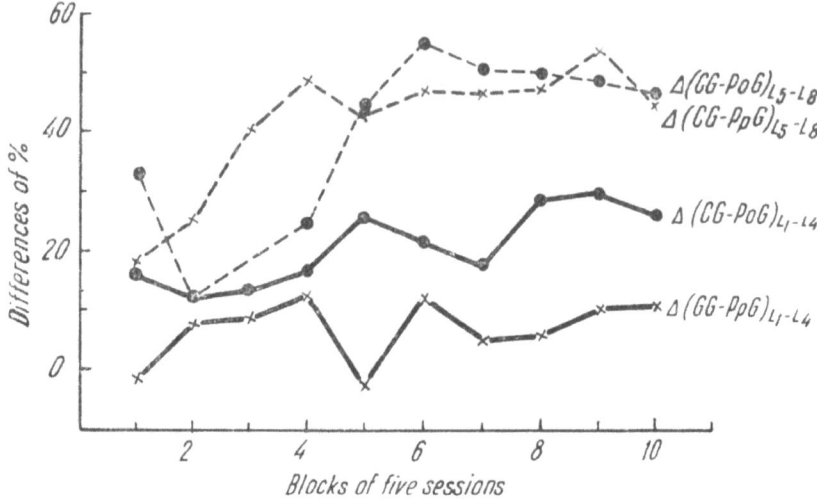

Fig. 4. Differences in percentages of correct responses between the control group (CG) and the group with pre- and post-operative training (PpG) and between the normal group and the group with post-operative training (PoG). Denotations as in Fig. 1.

DISCUSSION

The time taken by the wavefront to reach both ears ($\Delta t$) is, in the opinion of some investigators, the main clue for the localization of sound. If this hypothesis is correct, no differences should exist in the localization of sounds coming from sources placed in front or behind the animals. In general, $\Delta t$ is a very ambiguous clue for localization, since all points lying on a hyperbola of revolution about an axis through both ears will have the same $\Delta t$ (20).

Even if intensity and phase cues contribute to the localization of the sources of a sound the present findings are still puzzling. It is then necessary to try to understand our data by searching for other explanatory hypotheses. It is known that cats do not target the source of a sound directly with their eyes on the basis of direct auditory information impinging upon the neural system, but rather through a complementary feedback coming from the muscles of the pinna activated by the sound (13, 15). The movement of the pinna is controlled by the seventh nerve nucleus. This nucleus receives information from the ipsilateral superior olivary complex, the trapezoid body (9), the inferior colliculus (18), trigeminal nucleus (19) and the telencephalon, more specifically from the sensory motor cortex (14) and perhaps from other structures also. These data suggest that the neurons of the facial nucleus integrate different reflexes activated by a sound, from some very rapid ones, such as contraction of the stapedius muscles, to the movements of the pinna and complex movements of the face which appear with a longer latency. However the activation of coordinated movements of the eyes and head as part of the audio-visual targeting reflex shows that directly or indirectly the auditory information can set motor nuclei other than the facial nucleus into action.

In normal cats, the neural integration processes utilized to target a source of a sound in the frontal plane seem to be different from those involved in the targeting of a source placed in the rear. The pinna movements that serve to localize frontal sources are short and quick, in general performed by the pinna ipsilateral to the sound. To localize sources placed in the rear, the pinna executes long testing tracking and scanning movements and often both auricles are involved. To target a frontal source, small eye or head movements are required, while to target a rear source large head and body movements are needed. The whole posture of the body is involved, including labyrinthic and neck reflexes. These observations suggest that the localization of a rear source is much more difficult. Nevertheless a serious unsolved problem that remains is how the auditory information can activate such a complex pattern of movements.

The animals operated after training did not show retention, but were able to relearn only to target the front set of speakers, whereas animals without pre-operative training did not learn at all. Observations in our laboratory (15) have shown that the unilateral denervation of the auricular muscles produces learning deficits in the localization of tonal sources placed both ipsilateraly and contralateraly to the operated ear. These behavioral deficits are more important on the denervated side, but affect the localization of the set of speakers placed behind the animals more than those placed before them. These findings suggest that symmetrical functions of the hemispheres are required in order to learn an audio-visual instrumental targeting reaction and to relearn to target a source placed behind the animals. The integrative process underlying the targeting reaction directed to a source placed behind the animals is certainly more complicated than that required to localize a source placed in front of them. In the opinion of Myers et al. (10) the interhemispheric connections are more important as the tasks to be solved are more complicated. The differences of our results with those of Naumann (11) could be explained by the differences in training procedures, especially in the position of the speaker. The large deficit of our operated group is mainly related to the localization of the source placed in the rear field, while the localization an the front field is not so deeply disturbed.

REFERENCES

1. BYKOV, K. M. and SPERANSKY, A. D. 1924. The dog with transection of the corpus callosum (in Russian). Tr. Fiziol. Lab. 1: 42–53.
2. CORBALLIS, M. C. and BEALE, I. L. 1970. Bilateral symmetry and behavior. Psychol. Rev. 77: 451–464.
3. DOBRZECKA, C. 1973. The effect of the section of corpus callosum on differentiation of instrumental reflexes to symmetrical tactile stimuli in dogs. Acta Neurobiol. Exp. 33: 543–551.
4. DOBRZECKA, C. 1976. Differentiation of instrumental reflexes to tactile stimuli following dissection of the corpus callosum and a unilateral somatosensory cortex lesion. Activ. Nerv. Super. 18: 123-125.
5. DOBRZECKA, C. KONORSKI, J. STĘPIEŃ, L. and SYCHOWA, B. 1972. The effect of the removal of the somatosensory areas I and II on left-right leg differentiation to tactile stimuli in dogs. Acta Neurobiol. Exp. 32: 14–33.
6. GAZZANIGA, M. S. 1975. Review of the split brain. J. Neurol. 209: 75.
7. GAZZANIGA, M. S., BOGEN, J. E. and SPERRY, R. W. 1965. Observations on visual perception after disconnection of the cerebral hemispheres in man. Brain 88: 221–236.
8. HARRISON, J. M. 1974. The auditory system of the medulla and localization. Fed. Proc. 33: 1901–1903.
9. MASTERTON, R. B. 1974. Adaptation for sound localization in the ear and brainstem for mammals. Fed. Proc. 33: 1904–1910.
10. MYERS, R. E. and SPERRY, R. W. 1958. Interhemispheric connection through the corpus callosum. Arch. Neurol. Psychiatr. 80: 298–303.
11. NAUMANN, G. L. 1958. Sound localization: the role of commissural pathways of the auditory system of the cat. Ph. D. Dissertation, Chicago. Univ. of Chicago.
12. RAVIZZA, R. and DIAMOND, I. T. 1974. Role of auditory cortex in sound localization: a comparative ablation study of hedgehog and bushbaby. Fed. Proc. 33: 1917–1919.
13. SANTIBAÑEZ-H., G. 1976. Targeting reflex. Acta Neurobiol. Exp. 36: 181–203.
14. SANTIBAÑEZ-H., G., ESPINOZA-V., B., ASTORGA-O., L. and STROZZI-V., L. 1974. Macroelectrodic activity of the facial nucleus of the cat. Acta Neurobiol. Exp. 34: 265–276.
15. SANTIBAÑEZ-H. G. and SIEGMUND, H. 1977. Die Targeting Reaktion. 100 Jahre Physiologisches Institut. Wiss. Schrift. Humboldt-Universität Berlin p. 182–193.
16. SPERRY, R. W. 1968. Plasticity of neural transmission. Develop. Biol. Suppl. 2: 306–327.
17. STARR, A. 1974. Neurophysiological mechanisms of sound localization. Fed. Proc. 33: 1911–1914.
18. SYKA, J. and RADIL-WEISS, T. 1971. Electrical stimulation of the tectum in freely moving cats. Brain Res. 28: 567–572.

19. TANAKA, T. Yu. H. and KITAI, S. T. 1971. Trigeminal and spinal inputs to the facial nucleus. Brain Res. 28: 504–508.
20. WHITFIELD, I. C. 1974. A possible neurophysiological basis for the precedence effect. Fed. Proc. 33: 1915–1916.

Ruth BRUNK, Eva CZIHAK, Barbara OPPERMANN and Guy SANTIBAÑEZ-H., Institute of Physiology, Humboldt University, Hessische Str. 3-4, 104 Berlin, GDR.

Lecture delivered at the Warsaw Colloquium on Instrumental
Conditioning and Brain Research
May 1979

# SEARCH FOR STRUCTURES INVOLVED IN INTEGRATION OF LETTERS IN PAIRS

Wanda BUDOHOSKA and Anna GRABOWSKA

Department of Neurophysiology, Nencki Institute of Experimental Biology, Warsaw, Poland

*Abstract.* Two experiments aimed at finding, at what stage of perceptual processing the positive interactions between letters, observed in some previous studies, are established. Single letters and pairs of letters were exposed. The task of the subjects was to recognize letters. In Experiment I bright letters on a black background were used instead of black letters on bright background used in previous experiments. Such change of brightness within the stimuli did not influence the character of interactions between letters. This suggests that the positive interactions are connected with higher level of visual system. In Experiment II patients with damage of various areas of the brain cortex were tested. The facilitation between two letters did not occur in patients with prestriate lesions, although it can be observed in patients with different localization of lesions. This suggests that prestriate cortex is involved in positive interactions between letters.

## INTRODUCTION

Our studies concerning the interactions between elements of compound visual patterns have shown that the nature of these interactions depends on the subject's familiarity with the perceived stimuli. In the case of familiar complex patterns there occurred a facilitation between their components (1-3, 5). In the case of patterns composed of elements that did not combine into a familiar pattern, a negative interaction was observed (2, 5). A task used in some of these studies was to identify letters that were presented very briefly (1, 4). Performance on single letters was compared with performance on the same letters presented in pairs. We found the letters in first position of pairs to be subjected to facilitation from the letters in second position. This finding raised the question as to the level of perceptual analysis at which the interaction takes place.

In Experiment I white letters on the black background were used. The shape of these stimuli was the same as in previous experiments (1, 4) in which black letters on the white background were presented, but overall illumination as well as the distribution of light differed. It seemed interesting to find whether such a change in the experimental conditions could influence the interactions between two letters. This would provide the possibility of inferring whether the observed interactions depend on the stimulus shape or on the light distribution within the stimuli.

In Experiment II we investigated interactions between letters in neurosurgical patients with different localization of brain lesions. The aim of this research was to provide information concerning the structures engaged in the integration of the elements of compound visual patterns.

## METHODS

In Experiment I, 13 subjects with normal vision were run and in Experiments II, 11 patients hospitalized at a neurosurgery clinic: 4 with occipital lesions, 3 with temporal lesions, 2 with parietal lesions and 2 with frontal lesions. The patients did not demonstrate any disorders in visual functions examined by means of standard clinical tests. All subjects were able to perceive, recognize, identify, and appropriately report the graphic characters used in the experiments.

In Experiment I, ten upper case, printed letters (ABCEFLOPTZ), white on a black background, and in Experiment II, seven upper case letters (ABCEFOP), black on a white background, were used. The characters were presented either separately or in pairs (100 combinations of 10 letters in Experiment I and 49 combinations of 7 letters in Experiment II were used). Each of the letters was shown with the same frequency as single stimuli and as the first or second elements within a pair. The letters used in Experiment I were 53' in height and those in Experiment II were 1°10' in height. The space separating two letters was about 9'. The luminance of stimuli was kept constant for a given person and was adjusted by means of Kodak neutral filters to a level which provided 60–80% correct recognition.

A Kodak projector, model Carousel 140 with Laffayet shutter was used. The stimuli were presented foveally. The time of letter exposure was 20 ms in Experiment I and 17 ms in Experiment II. A proper fixation was provided by a light spot projected just above the area in which the test stimuli were to appear. The subjects were to report after each presentation what letter or pair of letters was shown on the screen. Single letters and pairs of letters were presented in separate sessions. The order of sessions was random. Single letters were presented randomly at positions equivalent to the positions of left and right letters in pairs. The right and left side exposures were balanced. The stimuli were presented every 8 s in series of 20 items in Experiment I and of 10 items in Experiment II. The series were separated by 2 min intervals. Each of the Experiments was preceded by 10 min of dark adaptation (Experiment I) or light adaptation (Experiment II).

## RESULTS AND DISCUSSION

### Experiment I

The analysis of data was based on the number of errors made by subjects in the identification of letters exposed in each of the two positions. The principal goal of the experiment was to determine the interaction between the letters in pairs, and so an attempt was made to detect differences in performance on letters presented separately and the same letters presented in pairs. For each subject, the difference between the number of errors made with single letters (sl) and the number of errors made with paired letters (pl), was divided by the sum of the same values to yield the person's identification difference score (IDS) $\frac{(sl-pl)}{(sl+pl)}$. Such a score was computed separately for letters in first and in second positions. The score was positive in case of superior identyfication of paired letters compared with single letters (positive interaction) and it was negative in case of superior identification of single letters (negative interaction). The score equaled zero if there was no difference in number of errors in the two conditions.

The values for each subject are shown in Fig. 1. The scores are positive in 12 subjects for letters in the 1st position and in 8 subjects for letters in the 2nd position; a $t$-test revealed significant facilitation of paired letters only in the case of the first position ($P < 0.01$).

These data are very similar to those obtained in experiments with black latters on white background (1, 4). This shows that facilitation does not depend on the illumination or distribution of light over the stimulus, but rather on the shape of the stimuli. This would suggest that the interactions occur at higher levels of the nervous system. Nevertheless the results of Experiment I give no indication as to which structures of nervous system are involved in this process. Some evidence on these lines seems to emerge from Experiment II which was conducted on patients with brain lesions in different regions of the cortex.

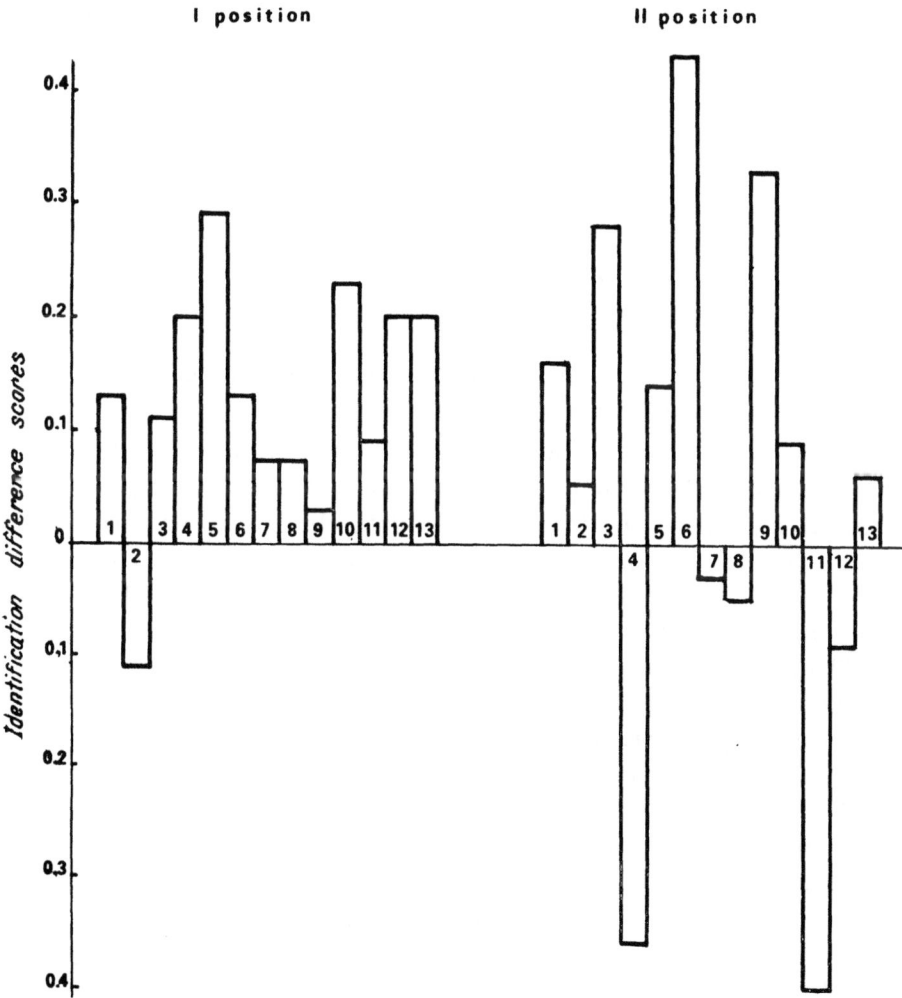

Fig. 1. Identification difference scores for letters exposed in I and II positions in particular subjects (numbers in bars). For detailed explanation see text.

## Experiment II

The data obtained for each of the subjects are given in Table I. The scores were computed in the same way as in Experiment I. Each patient's performance was analyzed individually, since the pooling of data obtained

TABLE I

Identification difference scores for patients with cerebral lesion

| Subject | Hemisphere | Area | Indentification difference scores | |
|---|---|---|---|---|
| | | | I position | II position |
| 1 | left | occipito-parietal | **—0.264** | +0.032 |
| 2 | right | occipito-parietal | **—0.125** | —0.106 |
| 3 | left | occipito-temporal | **—0.241** | 0.000 |
| 4 | right | occipito-temporal | **—0.063** | +0.091 |
| 5 | left | temporal | +0.628 | +0.061 |
| 6 | left | temporal | +0.551 | +0.462 |
| 7 | left | temporal-parietal | +0.511 | **—0.292** |
| 8 | left | parietal | +0.333 | +0.093 |
| 9 | right | parieto-frontal | +0.435 | +0.015 |
| 10 | left | frontal | +0.644 | +0.386 |
| 11 | left | frontal | +0.509 | +0.323 |

from patients with different lesions seemed unwarranted, even for lesions in the same cerebral lobe.

In order to determine whether or not the performance of the patients differed significantly from the performance of normal subjects tested in similar conditions, we tabulated the data obtained from normals run in our laboratory in several experiments of the same kind (e.g., 1, 4), and estimated the hypothetical distribution of their IDS. These scores ranged from —0.066 to +0.658 for letters in first position and from —0.243 to +0.443 for letters in second position. The scores in Table I that exceed these limits (set in bold type) can be taken to differ significantly from the performance of normals with probability below 0.05.

All patients with lesions in occipital cortex identified letters in first position with significantly lower efficiency than normals (the fourth person on the borded of statistical significance). In all these cases their performance on paired letters was inferior to performance on single letters, which means that perception was impaired, not facilitated, as in the case of normals. All remaining patients were better at identifying paired letters than single letters in the first position, the same as in the case of normals. As to identification of letters in the second position, our patients performed about the same as normals except for one case (patient 7), where second position identification was inferior to single letter identification.

The absence of positive interaction between simultaneously presented letters in cases of occipital lesions can be interpreted as reflecting some kind of disorder in the synthesis of visual elements into complex patterns.

These patients seemed to be perceiving the two letters independently of each other, not as component of one pattern. Description of the operated areas and the absence of disorders typical for striate cortex lesions (e. g., decreased acuity of vision or scotoma in the visual field) indicate

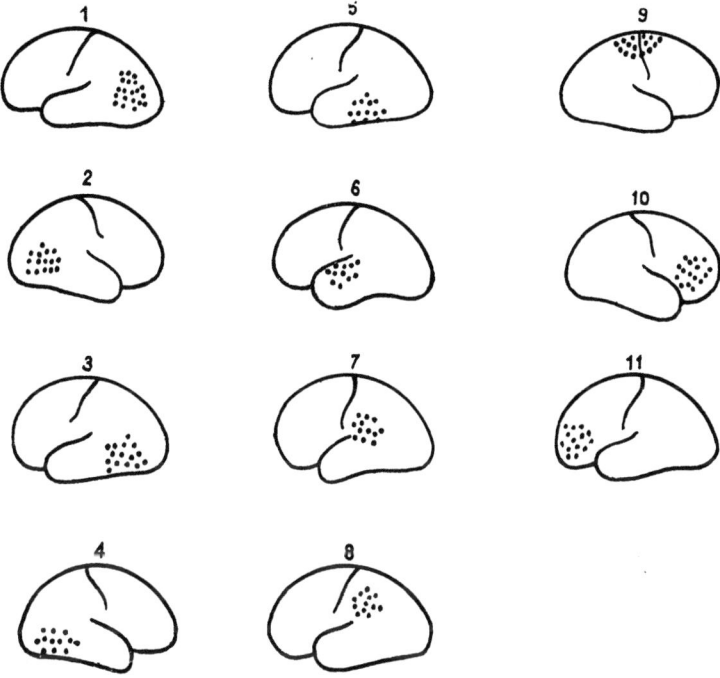

Fig. 2. Damaged area of the brain cortex in 11 patients. The pictures show only the whole area of the surgery (in most cases a tumor was removed), since more precise estimation of the damage region is not possible.

that mainly the prestriate region (18 and 19 area) of the visual cortex was damaged (Fig. 2). This might suggest that the prestriate region serves to integrate the elements of the visual field into complex patterns. Such an interpretation is consistent with clinical data reported by other authors (6) postulating that partial lesions in this region result in impaired ability of integral perception of entire visual patterns.

It is interesting that our patients did not reveal clinical symptoms of visual agnosia: they had no difficulties in identifying objects that were crossed out, or drawn in outline, or overlapping, in naming two objects shown simultaneously, or in describing complicated situational pictures. It may be speculated that the visual impairment recorded in our experiments could be due to the short time of exposure which sufficed for only a near-threshold identification. Normally, when the same patients are free to examine the scene for longer periods, they can take advantage of the redundancy of information and thus perform perfectly. It seems that by throughly examining this kind of interaction between letters we might be able to discover disorders in visual functions which are not detected by ordinary clinical tests.

In summary, our experiments indicte that positive interaction between letters takes place at a level of visual processing at which information about the brightness of stimuli is neglected and the shape of stimuli is analysed. This interaction might occur in the prestriate region of the occipital cortex.

Experiment II was conducted in cooperation with E. Fersten and J. Szumska from Neurosurgery Clinic of Polish Academy of Sciences.

## REFERENCES

1. BUDOHOSKA, W., GRABOWSKA, A. and JABŁONOWSKA, K. 1975. Interaction between two letters. Acta Neurobiol. Exp. 35: 115–123.
2. BUDOHOSKA, W. and CELIŃSKI, M. 1975. Interaction between two simple visual patterns. Acta Neurobiol. Exp. 35: 125–137.
3. BUDOHOSKA, W., GRABOWSKA, A. and JABŁONOWSKA, K. 1979. The effect of interaction between elements of familiar and unfamiliar patterns. Pol. Psychol. Bull. 2 (in press).
4. GRABOWSKA, A. and BUDOHOSKA, W. 1979. Interaction between two spatialy separated letters presented in succession. Acta Neurobiol. Exp. (in press).
5. JABŁONOWSKA, K. and BUDOHOSKA, W. 1976. Hemispheric differences in the visual analysis of the verbal and non-verbal material in children. Acta Neurobiol. Exp. 36: 693–701.
6. LURIA, A. C. 1973. The working brain. An introduction to neuropsychology. Penguin Books Ltd. Harmondsworth, England, 360 p.

Wanda BUDOHOSKA and Anna GRABOWSKA, Nencki Institute of Experimental Biology, Pasteura 3, 02-093 Warsaw, Poland.

Lecture prepared for the Warsaw Colloquium on Instrumental
Conditioning and Brain Research
May 1979

# REVERSED POSTOPTOKINETIC NYSTAGMUS: A MODEL OF PLASTICITY IN THE VESTIBULOOCULAR SYSTEM

J. BUREŠ and V. P. NEVEROV

Institute of Physiology, Czechoslovak Academy of Sciences
Prague, Czechoslovakia

*Abstract.* A simple manifestation of the memorial processes is the so called reversed postoptokinetic nystagmus (RPN), a trace phenomenon elicited in the rabbit by prolonged (60 min) optokinetic (OK) stimulation. Electrophysiological analysis of RPN indicated that the underlying neural trace is weakened, but not suppressed by spreading depression in the cerebral cortex or superior colliculus. An asymmetry of RPN is brought about by unilateral 6-OHDA lesion on substantia nigra. Electroconvulsive shock applied immediately after a period of OK stimulation blocks the subsequent RPN without interfering with OKN. About 70% of neurons in the vestibular complex changed their activity during OKN and RPN. The changes consisted in most cases of an excitation accompanying OKN and inhibition during RPN. The OKN-RPN related reactions were also abundant in flocculus but significant activity changes during RPN were less frequent in this structure. Units in midbrain reticular formation reacted both during OKN and RPN in a similar fashion as the vestibular ones. On the other hand units in the cerebellar deep nuclei and brachium conjunctivum were only weakly influenced by OKN and/or RPN. It is suggested that the neural trace of RPN develops in the vestibular complex and vestibulocerebellum as a part of the process compensating the effect of continued optokinetic stimulation. Flocculus participates in input processing of the optokinetic stimulation whereas reticular formation mediates signal transmission to oculomotor and higher integrating centers. The trace, revealed by sudden cessation of the eliciting stimulus in absence of visual reference signals is probably the neural substrate of the so called motion habituation and visual hallucinations. As other compensatory phenomena in the motor system, RPN has features of instrumental (it improves the organisms control of environment) and classical (it is automatically established and involuntarily emitted) conditioning.

## INTRODUCTION

Present trends in behavioral physiology confirm the correctness of Konorski's (19) prediction that significance of "naturalistic" approaches to the study of behavior will grow. Konorski resented the artificial character of tasks employed in research into classical and instrumental conditioning and warned against the implicit limitations of this line of investigations. "After all, we should not forget that animals' and humans' nervous systems grew in phylogeny and ontogeny under the continuous pressure of the surrounding world which carved them in such a way as to make them best adjusted. In consequence if we substitute artificial factors to act upon the organism for those which really and constantly act upon it, we may lose something very substantial which can contribute largely to understanding the functioning of that organism" (19, p. 214). The rapid development of neuroethology (34), the great attention paid to learning in the feeding system (conditioned food preferences and aversions, 1, 25), the formulation of the principles of belongingness and preparedness (36) and the increasing awareness of evolutionary constraints on learning (37) show that Konorski's views were widely shared.

## PLASTICITY OF THE VESTIBULO-OCULAR REFLEX

The experimental analysis of natural associations controlling fundamental bodily functions has been particularly fruitful in the case of the vestibulo-ocular reflex (see 16, 20, 33). The function of this disynaptic reflex, the anatomy of which was described in studies by Lorente de Nó (21) and Szentagothai (40) is to stabilize the retinal image of stationary targets during head movements. The change of head position detected by labyrinthine receptors generates vestibular signals which produce a compensatory eye movement in the opposite direction, i.e., orienting the gaze to where a stationary target should be. This anticipatory control of gaze is primarily vestibular (it operates also in darkness), but is permanently corrected by vision which evaluates the outcome of each of the thousands of vestibulo-ocular reflexes performed during a single hour of movement. The visual control, accomplished by visual input to vestibular nuclei and the vestibulo-cerebellum, maintains the gain of the vestibulo-ocular reflex close to unity (i.e., the angular displacement of the head is compensated by opposite eye movement of the same amplitude), by a process resembling correction of artillery fire from an observation post. Each miss changes signal transmission through the vestibulo-ocular reflex arc in a way reducing the probability of the particular error, each hit facilitates transmission through the synapses involved.

The plasticity of the vestibulo-ocular reflex shows at the same time that it is not a rigid unconditioned reaction, but that it has many features of instrumental activity, the essence of which is to change the relations of the organism to the external world — in this case to improve the visual perception of the environment during movement. The learning is not limited to maintenance of optimum performance but can even cope with gross experimentally induced distortions of the normal relationship between vestibular and visual signals.

Stratton's (39) ingenious experiments with inverting prismatic spectacles, replicated and extended by Kohler (18) and Gonshor and Melville-Jones (13) are striking examples of the instrumental nature of the vestibulo-ocular plasticity. At the beginning of the experiment head movements trigger a completely maladaptive vestibulo-ocular reflex which shifts the image if the target not closer to but further away from the fovea. This anticompensatory reaction, subjectively experienced as extremely disturbing, gradually diminishes. After several days of inverted vision, the vestibulo-ocular reflex is suppressed and later gradually replaced by a state when the eyes move in the same direction as the head. This development is accompanied by the disappearance of the subjective complaints. Similar training-induced modifications of the vestibulo-ocular reflex, described also in animals (15, 24), require participation of an information storing mechanism which integrates the visual and vestibular inputs and produces counterregulation signals, optimizing the perceptual relationships between the moving organism and stationary environment. The present paper describes a simple manifestation of the underlying process, the reversed postoptokinetic nystagmus (RPN), a trace phenomenon produced by prolonged optokinetic stimulation in the rabbit (17, 26).

## OPTOKINETIC TRACE PHENOMENA

Optokinetic nystagmus (OKN) induced by rotation of the optokinetic drum around the stationary animal is essentially similar to the vestibular nystagmus elicited by rotation of the animal in the opposite direction. Both reactions serve the same function: to stabilize the retinal image of the moving environment during the slow phase of nystagmus. Since no vestibular signals are present during OKN, the stabilization is first achieved by direct activation of the visuo-motor reactions, but more efficient target following is made possible by a predictive mechanism which increases the gain of the OKN response to the whole field rotation to 1.0. According to Cohen et al. (6), activation of this system takes about 5 to 10 s in the monkey. Its manifestation is the so-called positive optokinetic afternystagmus (OKAN), observed when optokinetic stimulation is suddenly terminated by darkness. In spite of the absence of visual stimuli, nystagmus continues to be driven by the inner integrator, the activity of which decays during 20–60 s (6). The positive OKAN is followed after a brief interval by a negative OKAN (22), more appropriately called reversed postoptokinetic nystagmus (RPN) which beats in the opposite direction. In comparison with the positive OKAN, the central excitatory state underlying RPN develops more slowly during optokinetic stimulation and lasts longer in the darkness. In the rabbit 90-min optokinetic stimulation elicits more than 30 min of slowly decaying RPN (26): in humans 15-min OKN produces about 20 s of positive OKAN followed by 60 s of RPN (4).

Whereas the integrator of the positive OKAN directly improves fixation of moving targets, the functional significance of the RPN generating mechanism is less obvious. According to Brandt et al. (4), the RPN integrator accounts for so-called motion habituation which is subjectively experienced as an apparent reduction in velocity during stimulation and as inverted selfmotion after stimulus termination. The transition from positive OKAN to RPN reflects two conflicting tendencies generated by the optokinetic stimulation: the steeply decaying optokinetic tonus and the prolonged countertonus. The first is usually rapidly balanced by the second which becomes the most prominent aftereffect of long-lasting optokinetic stimulation in the rabbit (17). Since RPN is a highly reproducible motor manifestation of a prolonged central state induced by purely visual stimulation, its investigation is directly relevant to the problems of brain plasticity, learning and memory.

## ANATOMICAL SUBSTRATE OF RPN

Attempts to locate the neural centers participating in RPN generation have concentrated so far on various parts of the visual, vestibular, and motor systems. In a typical experiment, the rabbit is subjected to a prolonged (60–90 min) period of optokinetic stimulation followed by 30 to 60 min of darkness. The OKN and RPN reactions are evaluated according to oculographically recorded eye movements. The normal time course of the OKN and RPN reactions is shown in Fig. 1. One to three min after termination of the 90 min optokinetic stimulation, the RPN atains the maximum rate of 60-80 beats/min which gradually decays during 30-40 min to non-measurable rates (less than 5 beats/min).

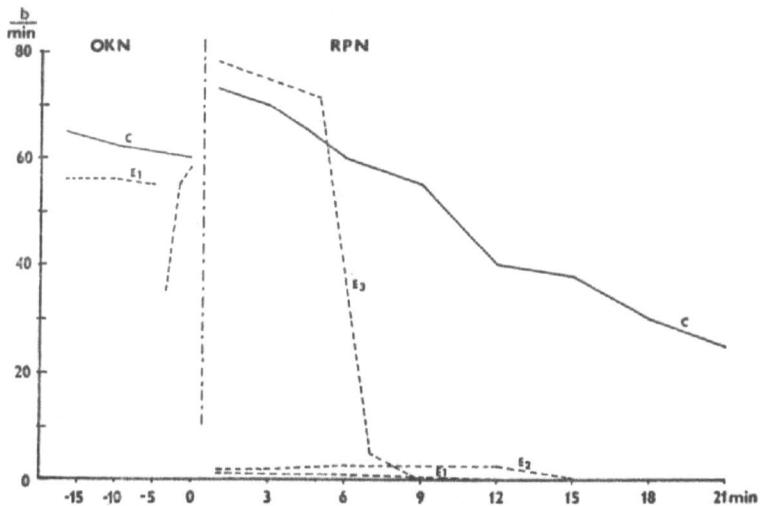

Fig. 1. The effect of electroconvulsive shock (ECS) on OKN and RPN in the rabbit. Ordinate: average rate of OKN or RPN (beats/min). Abscissa: time before (negative values) or after (positive values) termination of 90-min optokinetic stimulation. C: control experiment, E1, E2, or E3: ECS 5 min before, immediately after, or 6 min after termination of the optokinetic stimulation. From Neverov et al. (28).

Surgical decortication does not interfere with OKN and only weakens RPN in the rabbit (30). Ablation of superior colliculi (31) has a similar effect. The possibility that compensatory processes can account for the weak effect of the above interventions was examined by studying the acute consequences of the functional blockade of the visual cortex or the superior colliculus with the spreading depression technique (32). Electrophysiologically monitored single waves of spreading depression elicited in one or both hemispheres during OKN or RPN caused a transient reduction in the frequency of nystagmus which returned to the control level as soon as the activity of the depressed structure was restored.

Unilateral visual neglect can be elicited in rats by damage of the controlateral nigrostriatal system (35). Injection of 10 µg/5 µl of 6-hydroxydopamine into one substantia nigra in the rabbit (27) does not influence the symmetry of monocularly elicited OKN but reduces the RPN evoked by stimulation of the eye contralateral to the lesion. Participation of dopaminergic synapes in the RPN generating system was confirmed by the finding that RPN asymmetry is increased by the systemic application of amphetamine (1 mg/kg) and reversed by the injection of apomorphine (2.5 mg/kg). These results indicate that the cerebral cortex, superior colliculus, and basal ganglia contribute to, but are not indispensable for, RPN generation. Stronger impairment is caused by labyrinthectomy which considerably decreases RPN (29) and permanently abolishes the optokinetic afternystagmus in rabbits (7), monkeys (42) and humans (44). Participation of vestibulo-cerebellum is less imperative since optokinetic afternystagmus can be elicited even after bilateral flocculectomy (41).

## FUNCTIONAL BLOCKADE OF RPN

RPN can only be observed in darkness or when vision is blurred by light-scattering goggles. Switching the light on immediately stops RPN by fixation of the eyes on some stationary target. The RPN interruption does not abolish the corresponding central state, however. After the lights are switched off again, RPN reappears with an intensity decrement corresponding to the duration of the intervening interval (17). Discharging of the RPN generator continues independently of the actual movements of the eyes. The same is true also for charging the RPN generator: RPN develops even after optokinetic stimulation, the frequency of which is so high that it does not elicit OKN. Dissociation between eye movements and the state of the RPN generator indicates that oculomotor centers are at the output of the relevant neural circuits.

Other interventions may discharge the RPN generator completely. This was demonstrated in a series of experiments with electroconvulsive shock (ECS, 50 mA, 1.5 s, transpinnate application) applied after 90 min of optokinetic stimulation (28). No RPN developed after the animal had recovered from the seizure (usually after 2-3 min) and only random eye movements were observed during the subsequent 30 min. Electroconvulsive shock had little effect on OKN which resumed normal frequency during the 2 min after seizure, but abolished subsequent RPN when applied before the termination of the 90-min optokinetic stimulation (Fig. 1). Also in this case the paroxysmal activity obviously discharged the RPN generator by a mechanism akin to retrograde amnesia (for review see 12).

The anterograde effect of ECS on the RPN generator was demonstrated in experiments in which optokinetic stimulation was applied after the seizure. Although OKN was indistinguishable in the control and convulsed animals, RPN development was considerably retarded in the latter group. Similar anterograde interference with memory trace formation was described in a number of learning situations including passive avoidance and conditioned taste aversion (5).

ELECTROPHYSIOLOGICAL CORRELATES OF RPN

The assumption that vestibular nuclei participate in the trace phenomena induced by optokinetic stimulation is supported by electrophysiological evidence. Unit activity of vestibular nuclei is modulated by visual stimuli eliciting OKN and OKAN (8, 23, 43), but experimental conditions usually did not allow for the development of well expressed RPN. In a series of experiments (27), performed in anaesthetized rabbits, the electrophysiological correlates of RPN were investigated under more or less stationary conditions. After initial charging of the RPN integrator by 60-min continuous optokinetic stimulation, the experiment was continued under a maintenance schedule consisting of a regular sequence of 1-min optokinetic stimulation (OKN) followed by 1-min darkness (RPN) and 1-min exposure to the lighted but stationary optokinetic drum (L). Since the RPN decay is partly compensated during the subsequent OKN period, discharging of the RPN generator was very slow. Continuous optokinetic stimulation was applied whenever the RPN response started do decline. Under such conditions the OKN and RPN responses remained stable for hours and could be compared with the ny-

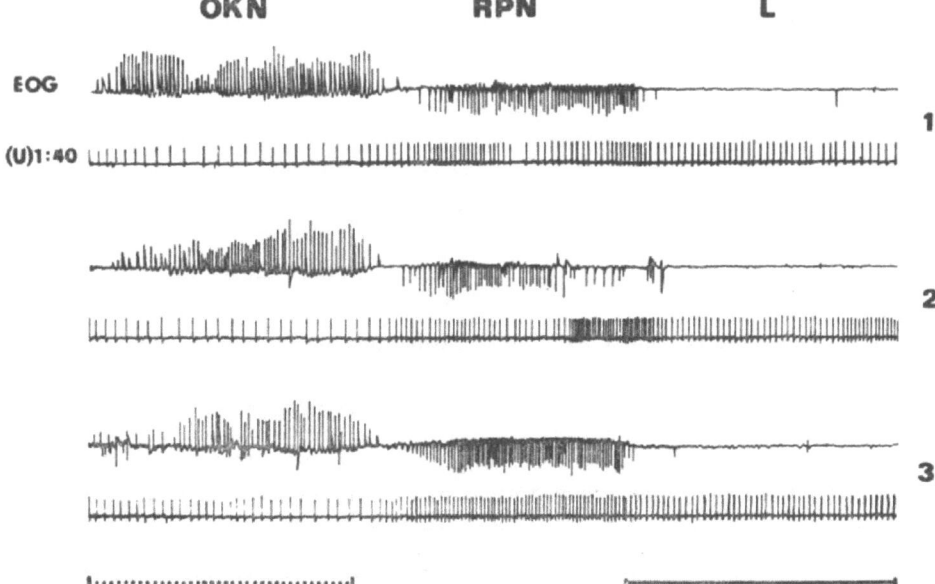

Fig. 2. Polygraph recording of the electrooculogram (EOG) and of vestibular unit activity (U), the frequency of which was reduced forty times (1 : 40), during three successive cycles (1 to 3) of the OKN-RPN-L sequence. The dotted and full lines below the recordings denote the 1-min optokinetic stimulus and 1-min illumination of the stationary optokinetic drum, respectively.

stagmus-free intervals in the illuminated stationary drum. Using this basic arrangement, the activity of 232 neurons in the vestibular nuclei, flocculus and mesencephalic reticular formation was examined by capillary microelectrodes introduced to the brain using a head-mounted mi-

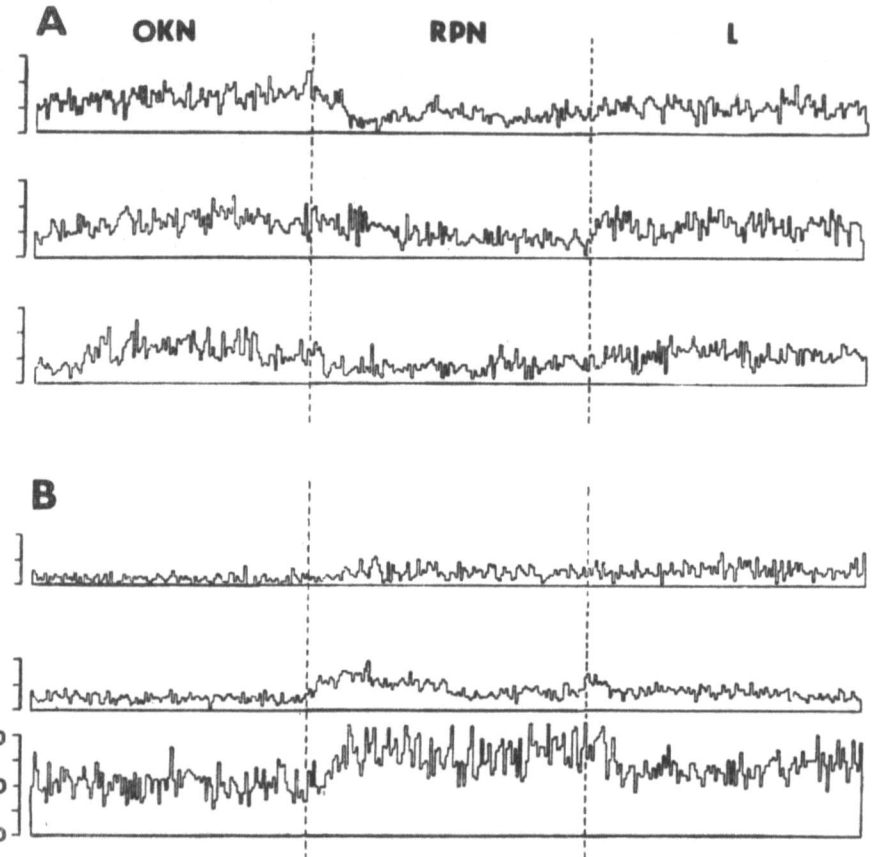

Fig. 3. Examples of computer plotted histograms of vestibular unit reactions to the OKN-RPN-L sequence. A and B, types of reaction. Ordinate: impulses/0.01 min. The bins are 0.01 min. Each histogram is based on 4 to 5 OKN-RPN-L cycles.

crodrive system. Significant changes of activity corresponding to the OKN-RPN-L sequence were found in about 70% of vestibular neurons. A typical experiment is illustrated in Fig. 2 which shows a polygraph record of the eye movements together with the output pulses of the spike counter (1:40). The activity of the neuron decreased during OKN and increased during RPN with respect to the L conditions. The reactions were similar in three successive cycles. Statistical tests found significant ($P < 0.05$) activity changes. Computer plotted poststimulus histograms of units identified with a spike recognition program confirmed the results of the on-line analysis (Fig. 3). The overall results are summarized in Fig. 4 showing the distribution of various types of responses in the population of 91 neurons found within the vestibular complex. About 50% neurons increased their activity during OKN and/or decreased it during RPN (type A). A smaller group of neurons (24%) reacted in an opposite way (type B). The remaining neurons were not significantly

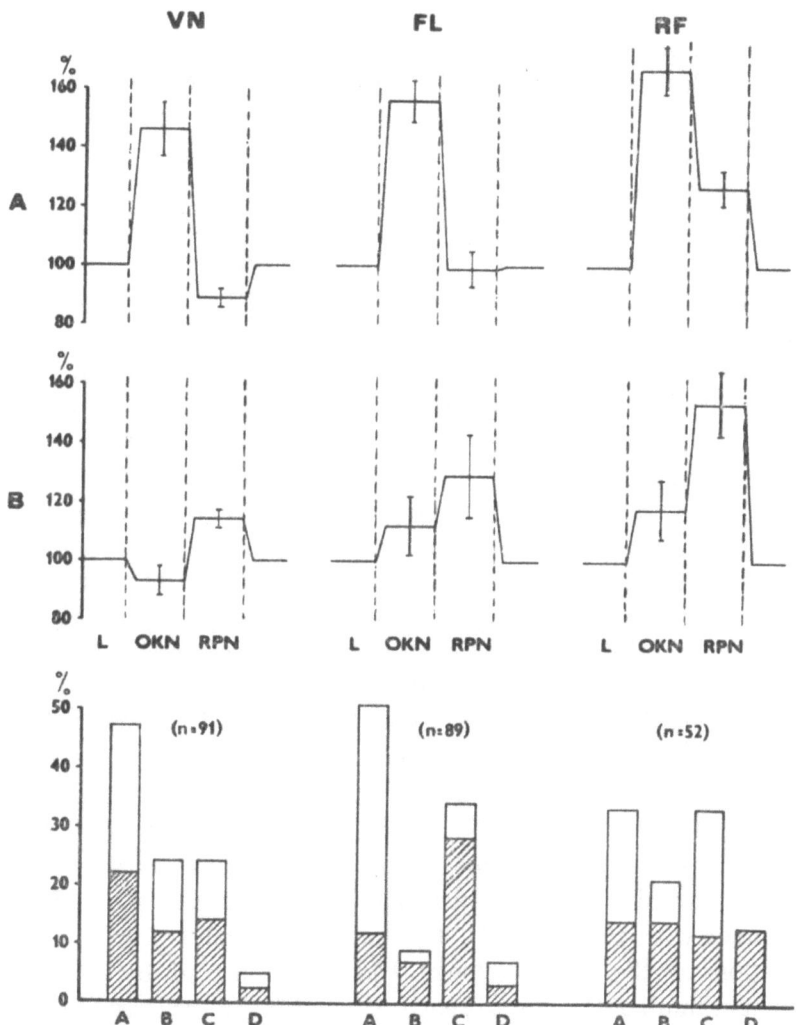

Fig. 4. Unit activity changes of vestibular (VN), floccular (FL) and reticular (RF) neurons during the OKN-RPN-L sequence. A and B, average activity changes in type A and B neurons during OKN and RPN expressed in percentages of the L-firing rate. Vertical bars denote SEM values. C, percentage of neurons displaying reactions of various types (A, B, C, D) to ipsilateral (white part of the columns) and contralateral (black part of the columns) optokinetic stimulation.

affected by the experimental procedure. Ipsilateral optokinetic stimulation (rotation of the drum towards the side of recording) was as effective as the contralateral one.

The same technique was used to examine the reactions of floccular and reticular neurons to the OKN-RPN-L sequence. The distribution of various reactions is shown in Fig. 4. About 50% of floccular units reacted to OKN by a similar activity increase as the vestibular neurons, but their firing rate did not decrease below the light control level during RPN. Type B units showing a significant activity increase during RPN were encountered less frequently in flocculus than in vestibular nuclei. Contralateral optokinetic stimulation was less effective than the ipsilateral one. Inhibitory reactions to OKN and RPN were practically absent.

The so-called nystagmogenic area of the brain stem (3), displayed a similar proportion of type A and B units as the vestibular nucleus, but their activity was above the light control level during both OKN and RPN. There was no difference between the effects of ipsilateral and contralateral optokinetic stimulation in the reticular units. The maximal firing rates in type A and type B units were almost identical.

The results of the electrophysiological experiments are in good agreement with other reports on the reactivity of vestibular neurons to visual stimuli. According to Henn et al. (14) 60% of the vestibular neurons in the monkey are of type I, which is excited by ipsilateral head acceleration (9) or by contralateral optokinetic stimulation, and inhibited by stimuli of opposite direction. Type II units mirror the behavior of type I neurons. From this point of view, type A neurons activated by contralateral and ipsilateral optokinetic stimulation correspond to type I and type II neurons, respectively. Similar considerations apply to the identification of type B neurons with type I and type II neurons. RPN was usually accompanied by less expressed changes of unit activity than OKN but the difference against the light control was significant in both type A and type B units.

Unlike vestibular units, most floccular neurons were activated by optokinetic stimulation and returned to baseline during RPN. This result is consonant with the view (11, 38) that the flocculus participates in the stabilization of the retinal image and is influenced by visual and/or vestibular stimuli (10). The absence of significant responses of floccular neurons to RPN suggests, however, that the flocculus is not directly engaged in the RPN generating mechanism. Also the differential reaction of floccular units to ipsilateral and contralateral optokinetic stimulation indicates that this structure processes visual stimuli in a different way than the vestibular nuclei.

On the other hand, activity of the reticular neurons was enhanced during both OKN and RPN. The distribution of reticular responses resembled the vestibular reactions (symmetry of changes evoked by ipsilateral and contralateral optokinetic stimulation, properties of A and B type neurons). It appears that reticular neurons participate in the transmission of signals generated at the level of vestibular nuclei to oculomotor and higher integrating centres.

CONCLUSION

The results of the lesion experiments and electrophysiological studies point toward vestibular nuclei as the probable site of the RPN integrator. Other structures probably serve as inputs (e.g., the flocculus) or outputs (e.g., the reticular formation) of the system, but are not indispensable for its function. Although RPN is a relatively short-lived phenomenon, it is produced by the same counter-regulation mechanism which accounts for the long-lasting compensation of functional disturbances of the vestibular and visual perception of the body-space relationship. RPN is analogous to Bechterew's (2) nystagmus which appears in a compensated hemilabyrinthectomized rabbit after removal of the remaining labyrinth. Although the vestibular input is symmetric (i.e., nil), after the second labyrinthectomy, nystagmus is produced by imbalance in the activity of vestibular nuclei: the originally deafferented and later compensated side is now more active than the recently lesioned side. Since there seems to be a smooth transition between short-lasting and long-lasting modifications of vestibulo-ocular reactivity the above models are highly relevant for investigation of the mutual relationship of transient and permanent memories. The ECS induced interference with RPN is a promising step in this direction. Finally, analysis of RPN opens an interesting approach to the physiology of perception. In humans, RPN is accompanied by an illusion of inverted self-motion (4). Although the existence of equivalent mental experiences is unverifiable in the rabbit, RPN fully satisfies Konorski's (19) physiological definition of hallucinations manifested by the targeting reflex (slow-phase nystagmus) to the non-operating stimulus (illusion of the moving optokinetic drum). RPN is a model of choice for the investigation of such physiological hallucinations, which are according to Konorski (19) phylogenetically earlier, associative phenomenon than images, and therefore more common in animals than in humans.

REFERENCES

1. BARKER, L. M., BEST, M. R. and DOMJAN, M. (ed.), 1977. Learning mechanisms in food selection. Baylor Univ. Press, Waco, Texas.
2. BECHTEREW, W. 1883. Durchschneidung des nervus acusticus. Pflügers Arch. Ges. Physiol. 30: 312–319.

3. BERGMANN, F., CHAIMOVITZ, M., GUTMAN, J. and ZELIG, S. 1963. Optokinetic nystagmus and its interaction with central nystagmus. J. Physiol. (Lond.) 168: 318-331.
4. BRANDT, Th., DICHGANS, J. and BÜCHELE, W. 1974. Motion habituation: Inverted self-motion perception and optokinetic afternystagmus. Exp. Brain Res. 21: 337-352.
5. BUREŠOVÁ O. and BUREŠ, J. 1979. The anterograde effect of ECS on the acquisition, retrieval and extinction of conditioned taste aversion. Physiol. Behav. 21, 4: 641-645.
6. COHEN, B., MATSUO, V. and RAPHAN, T. 1977. Quantitative analysis of the velocity characteristics of optokinetic nystagmus and optokinetic after-nystagmus. J. Physiol. (Lond.) 270: 321-344.
7. COLLEWIJN, H. 1976. Impairment of optokinetic (after) nystagmus by labyrinthectomy in the rabbit. Exp. Neurol. 52: 146-156.
8. DICHGANS, J. and BRANDT, Th. 1972. Visual-vestibular interaction and motion perception. In J. Dichgans and E. Bizzi (ed.), Cerebral control of eye movements and motion perception. (Bibl. Ophthalmol. 82), Karger, Basel, p. 327-338.
9. DUENSING, F. and SCHAEFER, K. P. 1958. Die Aktivität einzelner Neurone im Bereich der Vestibulariskerne bei Horizontalbeschleunigungen unter besonderer Berücksichtigung des vestibulären Nystagmus. Arch. Psychiatr. Nervenkr. 198: 225-252.
10. DUFOSSÉ, M., ITO, M., JASTREBOFF, P. J. and MIYASHITA, Y. 1978. A neuronal correlate in rabbit's cerebellum to adaptive modification of the vestibulo-ocular reflex. Brain Res. 150: 611-616.
11. GHELARDUCCI, B., ITO, M. and YAGI, N. 1975. Impulse discharges from flocculus Purkinje cells of alert rabbits during visual stimulation combined with horizontal head rotation. Brain Res. 87: 66-72.
12. GIBBS, M. E. and MARK, R. F. 1973. Inhibition of memory formation. Plenum Press, New York.
13. GONSHOR, A. and MELVILL JONES, G. 1973. Changes of human vestibuloocular response induced by vision-reversal during head rotation. J. Physiol. (Lond.) 234: 102-103.
14. HENN, V., YOUNG, L. R. and FINLEY, C. 1974. Vestibular nucleus units in alert monkeys are also influenced by moving visual fields. Brain Res. 71: 144-149.
15. ITO, M. 1972. Neural design of the cerebellar motor control systems. Brain Res. 40: 81-85.
16. JEANNEROD, M. and SCHMID, R. 1978. La plasticité du réflexe vestibulooculaire, In J. Delacour (ed.), Neurobiologie de l'aprentissage. Masson, Paris.
17. KISLYAKOV, V. A. and NEVEROV, V. P. 1966. Reaktsiya glazodvigatelnoi' sistemy na dvizhenie obekt v pole zreniya. Optokineticheskii' nistagm. Izdat. Nauka, Moscow, 51 p.
18. KOHLER, I. 1956. Der Brillenversuch in der Wahrnehmungspsychologie mit Bemerkungen zur Lehre von der Adaptation. Z. Exp. Angew. Psychol. 3: 381-417.
19. KONORSKI, J. 1967. Integrative activity of the brain. An interdisciplinary approach. Univ. Chicago Press, Chicago, 531 p.
20. LLINÁS, R. WALTON, K. 1979. Place of cerebellum in motor learning. In A. M. Brazier (ed.), Brain mechanisms in memory and learning: From the single neuron to man. Raven Press, New York, p. 17-36.

21. LORENTE DE NO, R. 1933. Vestibulo-ocular reflex arc. Arch. Neurol. Psychiatry (Chicago) 30: 245–291.
22. MACKENSEN, G. and WIEGMANN, O. 1959. Untersuchungen zur Physiologie des optokinetischen Nachnystagmus. I. Mitteilung. Die Abhängigkeit des optokinetischen Nachnystagmus von der Winkelgeschwindigkeit des Reizmusters. A. V. Graefes Arch. Ophthal. 160: 497–509.
23. MILES, F. A. 1974. Single unit firing patterns in the vestibular nuclei related to voluntary eye movements and passive body rotation in conscious monkeys. Brain Res. 71: 215–224.
24. MILES, F. A. and FULLER, J. H. 1974. Adaptive plasticity in the vestibulo-ocular responses of the rhesus monkey. Brain Res. 80: 512–516.
25. MILGRAM, N. W., KRAMES, L. and ALLOWAY, T. (ed.) 1977. Food aversion learning. Plenum Press, New York.
26. NEVEROV, V. P. 1963. Prolonged reversive postoptokinetic nystagmus (in Russian). Dokl. Akad. Nauk SSSR 150: 1182–1184.
27. NEVEROV, V. P. and BUREŠ, J. 1979. Asymmetry of optokinetic trace phenomena induced by unilateral 6-OHDA lesion of substantia nigra in the rabbit. Neurosci. Lett. 13: 301–305.
28. NEVEROV, V. P., BUREŠOVÁ, O. and BUREŠ, J. 1977. Effect of ECS on the neural traces underlying the reversive postoptokinetic nystagmus in the rabbit. Physiol. Behav. 18: 7–11.
29. NEVEROV, V. P. and KORYUKIN, V. E. 1979. Changes of optokinetic, postoptokinetic and reversed postoptokinetic nystagmus after bilateral labyrinthectomy (in Russian). Fiziol. Zh. SSSR Im. I. M. Sechenova (in press).
30. NEVEROV, V. P. and KULESHOVA, T. F. 1972. The role of visual cortex in the organization of nystagmic reactions elicited by optokinetic stimulation (in Russian). Zh. Vyssh. Nervn. Deyat. Im. I. P. Pavlova 22: 1055–1060.
31. NEVEROV, V. P., KULESHOVA, T. F. 1976. Nystagmus after unilateral lesions of the superior colliculus in rabbits (in Russian). Fiziol. Zh. SSSR Im. I. M. Sechenova 62: 196–200.
32. NEVEROV, V. P., UEDA, M. and BUREŠ, J. 1976. The effect of cortical and collicular spreading depression on the optokinetic and reversive postoptokinetic nystagmus in rabbits. Brain Res. 115: 318–323.
33. RAPHAN, T. and COHEN, B. 1978. Brainstem mechanisms for rapid and slow eye movements. Ann. Rev. Physiol. 40: 527–552.
34. ROPARTZ, P. 1978. Approche éthologique de la notion d'apprentissage. In J. Delacour (ed.), Neurobiologie de l'apprentissage, Masson, Paris.
35. SECHZER, J. A., ERVIN, G. N. and SMITH, G. P. 1973. Loss of visual placing in rats after hypothalamic injections of 6-hydroxydopamine. Exp. Neurol. 41: 723–737.
36. SELIGMAN, M. E. P. 1970. On the generality of the laws of learning. Psychol. Rev. 77: 406–418.
37. SELIGMAN, M. E. P. and HAGER, J. L. (ed.), 1972. Biological boundaries of learning. Appleton-Century-Crofts, New York, p. 98–102.
38. SIMPSON, J. I. and ALLEY, K. E. 1974. Visual climbing fiber input to rabbit vestibulo cerebellum: a source of direction specific information. Brain Res. 82: 302–308.
39. STRATTON, G. M. 1896. Some preliminary experiments in vision without inversion of the retinal image. Psychol. Rev. 3: 611–617.
40. SZENTAGOTHAI, J. 1950. Recherches expérimentales sur les voies oculogyres. Sem. Hop. (Paris) 26: 2989–2995.

41. TAKEMORI, S., COHEN, B. 1974. Loss of visual suppression of vestibular nystagmus after flocculus lesions. Brain Res. 72: 213–224.
42. UEMURA, T., COHEN, B. 1973. Effects of vestibular nuclei lesions on vestibuloocular reflexes and posture in monkeys. Acta Oto-Laryngol. Suppl. 315: 1–71.
43. WAESPE, W., HENN, V. 1977. Neuronal activity in the vestibular nuclei of the alert monkey during vestibular and optokinetic stimulation. Exp. Brain Res. 27: 523–538.
44. ZEE, D. S., YEE, R. D. ROBINSON, D. A. 1976. Optokinetic responses in labyrinthine-defective human beings. Brain Res. 113: 423–428.

J. BUREŠ, Istitute of Physiology, Czechoslovak Academy of Sciences, Videňská 1083, 142 20 Prague 4, Czechoslovakia.
V. P. NEVEROV, Pavlov Institute of Physiology, Academy of Sciences of the USSR, Leningrad, USSR.

Lecture prepared for the Warsaw Colloquium on Instrumental
Conditioning and Brain Research
May 1979

# SPATIAL MEMORY AND INSTRUMENTAL CONDITIONING

O. BUREŠOVÁ

Institute of Physiology, Czechoslovak Academy of Sciences
Prague, Czechoslovakia

*Abstract.* Short-term spatial memory of rats was studied in a 12-arm radial tubular maze. Correct performance does not depend on maze structure which affects, however, the rat's trajectory through the maze. Rats overtrained in the 12-arm radial maze show considerable transfer of the habit to a maze consisting of 12 parallel alleys entered from a common choice area. When isolated maze channels equipped with one way doors on both ends are randomly scattered over an enclosed area of 2 $m^2$ the rats are capable of visiting them sequentially, even when they encounter this particular configuration for the first time. Rats which were allowed 6 choices in the 12-arm radial maze and were then transferred to linear configuration of the same maze channels do not show significant preference for the yet unvisited channels. This indicates that in absence of spatial cues other sensory properties of the different channels cannot guide the choice behavior of the animal. An essential prerequisite of spatial memory tests is elimination of cues which would allow direct sensory location of the baited alleys. Smelling the food hidden in the food-cups may significantly improve the rats performance unless the maze channels are saturated with food odor. It is concluded that a prerequisite for the striking performance of rats in the tasks of the above type is their thorough familiarization with the list of choices. The spatial location of an item is an important cue for the recognition of the choices already made.

## INTRODUCTION

Professor Konorski was not only one of the great founders of modern physiological psychology, but also an outspoken defender of the view that at least higher mammals experience environmental influences in a similar way as man. He objected to attempts to purge physiological psychology of operationally defined psychological terms and often tried to guess what mental processes and subjective experiences accompany the behavior of his rats, cats or dogs. In this respect his sympathies were with Tolman (12) and Beritov (1) who endowed rats and dogs with beliefs and expectancies, cognitive maps and visual images, as well as with the rudimentary capacity for thought.

This attitude is clearly reflected in Konorski's (5) work on short-term (transient, dynamic) memory, which he defined as a process "enabling the subject to review (imagine) the corresponding stimulus pattern some time after its withdrawal". Transient memory is not a mere aftereffect of sensory stimulation, but is also used to organize sequences of permanent memories required for planned behavior. Konorski attributed transient memory two fundamental functions. The prospective role makes it possible to perform complex behaviors according to a mental plan of action, which lists the sequence of individual acts together with anticipated alternatives much in the same way as a computer program. At the same time the retrospective role of transient memory is indispensable for remembering how far the program execution has already progressed, which of the listed items have already been completed and which have yet to be fulfilled.

At the time of the publication of his fundamental book (5), Konorski (5) complained that the paucity of methods devoted to studies of short-term memory of animals is in clear disproportion to the role of these phenomena in mental processes. The delayed responses (3) and delayed stimulus matching (4) were the most common procedures used in short-term memory investigations. A major contribution to this research is the recent use of the radial maze technique (9–11) for testing the transient spatial memory of rats.

## RADIAL MAZE

The original study by Olton and Samuelson (11) employed an elevated radial maze consisting of a central platform 34 cm in diameter and regularly spaced 7 cm wide and 86 cm long arms. At the far end of each alley was a recessed food cup containing a morsel of food. The animal was placed into the maze and allowed to visit all arms and to eat all pellets. Similar to conventional mazes, the learning minimizes the time and effort needed to obtain the reward. Whereas this is achieved in conventional mazes by learning and remembering the correct start-goal trajectory, successful solution of the radial maze task consists in visiting all alleys without repeating the choices already made. There are n! correct sequences of choices in an n-arm maze. Rats do not prefer any specific sequence but rather find a new one during each trial. Since location of the baited channels by the smell of the hidden food or of the already visited channels by scent marking has been ruled out by carefully controlled experiments (11, 13), the correct solution of the radial maze task is obviously due to continuous comparison of the choice to be made with the short-term memory list of the choices already made. The capacity of this memory file is remarkable: According to Olton et al. (10) the upper limit of the working spatial memory of rats is about twenty five items. However, it is less clear what is remembered. Since rotation of the maze does not substantially deteriorate the performance, extramaze cues seem to be relatively unimportant. Successful solution of the radial maze by blind and anosmic rats (13) supports the assumption that kinesthetic and vestibular signals play a major role in spatial memory (1). The regular plan of the radial maze with constant angular distances between individual alleys can considerably simplify construction of the cognitive map (8, 12) in the rat's brain. The present paper compares performance of rats in the radial maze and in two other mazes composed of the same elements but with different floor plans.

The design of the experiment required a modification of the apparatus. A closed tubular maze (Fig. 1) was used instead of the open elevated maze (6). The channels were equipped with one-way entrance and exit doors so that the animal could pass through the channels only in the centrifugal direction. The choice platform was raised 5 cm above floor and made accessible through a central opening. Whereas in the elevated maze rats return to the choice area by retracing their steps, in the tubular maze they can use a multitude of return trajectories which all converge to the opening in the choice platform. This has the advantage that successive choices are always made from the same starting point. The main reason for using the tubular maze in the present study is, however, that the one-way alleys can be employed as independent blocks suitable for building mazes with various floor-plans.

Thirty male hooded rats aged two months at the beginning of the experiment were used. The animals were maintained on a 24-h food deprivation schedule with food available only in the maze and for a 15-min period after the conclusion of the daily experimental period.

During each session the rat was placed into the apparatus with all channels baited with 100 mg morsels of Larsen's diet. Olfactory recognition of the baited channels was prevented by saturating the maze with food odor released by large pieces of Larsen's diet placed in plastic containers mounted on the perforated ceiling of individual maze channels. The

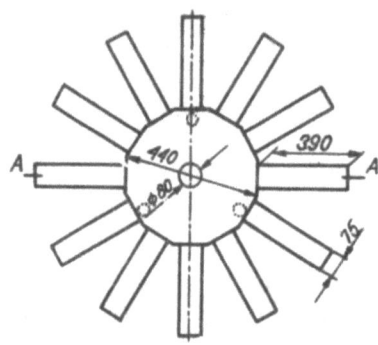

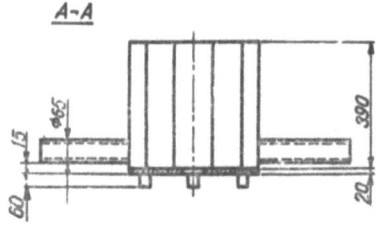

Fig. 1. Schematic drawing of the twelve-arm tubular maze. Above, the floor plan of the apparatus. Below: perpendicular section at the level A-A. The one-way doors, feeders, and additional food containers are not shown.

animals were allowed to eat all pellets or to make 12 choices. The sequence of the visited channels was recorded. One or two sessions were given per day. The performance became asymptotic during the second week of training. All experiments were performed in highly overtrained animals, after 1 to 6 mo of experience with various modifications of the maze.

## MAZE CONFIGURATION AND CHOICE STRATEGY

The correct solution of the task required the animal to visit each channel only once. It is obvious that this was possible if the animal kept a continuous memory record of the choices already made in the current session and avoided their repetition. Memory control of the diminishing list of choices yet to be made is a complementary mechanism contributing to the same effect.

Evaluation of the experiments was based on the comparison of the observed choices with the expected chance performance. The choice behavior of rats can ben viewed as sampling with replacement and modelled as the classic occupancy problem which describes how n identical balls (choices) can be distributed among n boxes (maze alleys). The expected incidence of sessions in which 12, 11, 10, etc. different channels were visited in 12 choices is shown in the curve (Fig. 2), computed for the 12-arm maze according to Feller (2). As follows from this diagram,

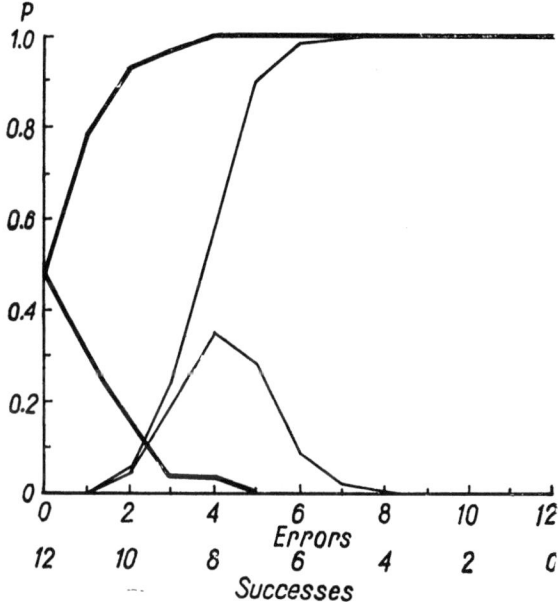

Fig. 2. Choice behavior of a group of 12 rats in the twelve-arm tubular radial maze (heavy line) compared with the mathematically expected random behavior (thin line). Probability of runs with n (0 to 11) errors or (12-n) successes in the first 12 choices and cumulative probability of runs with less than n errors or more than (12-n) successes in the first 12 choices.

probability of a correct solution of the task by random choice is extremely low. It can be expressed by the formula $n!/n^n$ which assumes the value of $5.37 \times 10^{-5}$ for the 12-arm maze. The expected incidence of sessions with one or two errors is also low (0.0035 and 0.0458, respectively), but rapidly increases with increasing number of errors. Sessions with 3 and 4 errors have the probability 0.1994 and 0.3560, respectively, if the choice behavior is random. The expected average number of errors characterizing random choice behavior in the 12-arm maze is 4.18. This theoretical distribution is compared by conventional statistical methods with the distributions obtained in various experiments.

Rats performed in the closed radial maze as reliably as in the elevated maze. The average number of errors per session was 1.25 ± 0.23, the difference against the expected value of 4.18 being highly significant ($P < 0.001$). This result is well comparable with the values reported for the 8-arm and 17-arm elevated mazes (7.5 and 14.5 different arms chosen in the first 8 and 17 choices, respectively).

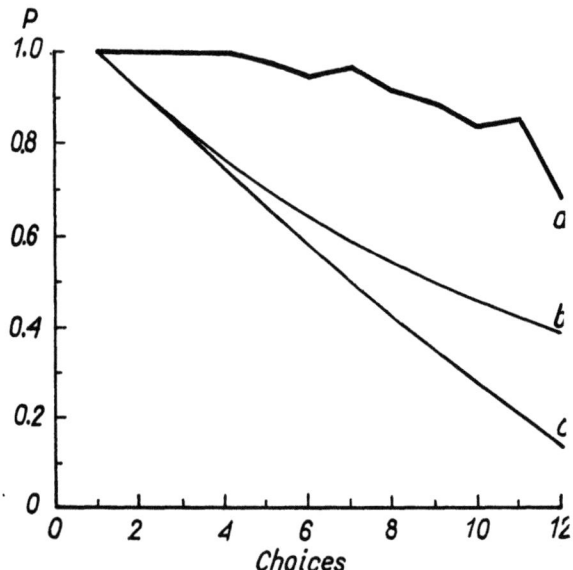

Fig. 3. Choice behavior of a group of 20 rats in the 12-arm tubular radial maze. Experimentally established incidence of successes on choices 1 to 12 (a) is compared with the probability of success on the n-th choice predicted by the random choice model (b) or computed from results of choices 1 to (n-1) (c). Note that the better than chance performance pushes the c-curve far below the b-curve.

Distribution of successes in the first 12 choices is shown in Fig. 3 together with the mathematically expected values. The probability of success on the n-th choice ($P_n$) was computed according to the formula

$$P_n = \frac{12 - n_c}{12}$$

where $n_c$ is the number of different channels visited during the previous (n-1) choices. Practically no errors occurred during the first 5 or 6 choices, and the incidence of errors only gradually increased in the last choices when finding the still unvisited maze arms became more and more difficult.

The radial maze task can be solved correctly even if the individual choices are not remembered. A parsimonious strategy is to visit the adjacent arm on each choice and to move constantly in the clockwise or counterclockwise direction, until the starting point is reached again. Preference for adjacent arms was reported by Olton et al. (10) and by Magni et al. (6) in radial mazes, in which the rat returned from the visited arm to the central platform. Careful analysis of the sequence of the choices in the present apparatus failed to reveal similar response chaining. Figure 4 shows the distribution of the channels visited on

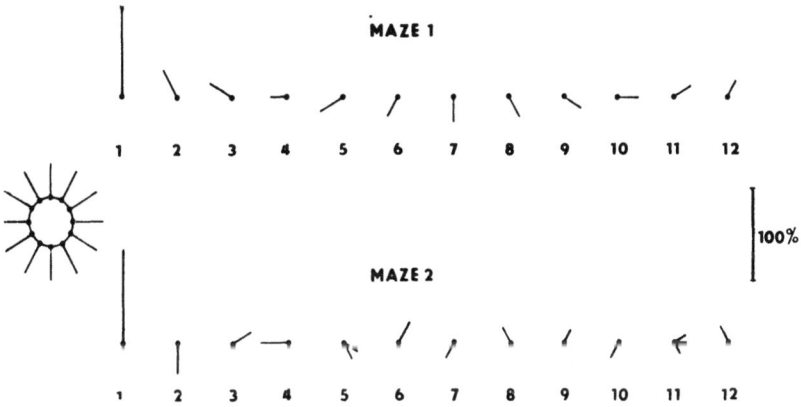

Fig. 4. The channels most frequently selected during choices 1–12 in maze 1 (with blind alleys) and maze 2 (with one-way alleys). The channel selected on the first choice is assigned number 1 and the remaining arms are numbered clockwise. The beams indicate the modal orientation with respect to channel 1 (see the inset maze diagram) and incidence (in %). Note that in maze 1 the modal orientation gradually shifts counterclockwise whereas in maze 2 successive choices tend to be in opposite directions. (From Magni et al. 6).

trials 2 to 12 in errorless sessions. The channel entered on the first choice is assigned No. 1. The beams point in the direction of the channel most frequently selected, the length of the beam denotes the percentage of choices in this direction. The rats showed a clear tendency to enter on their second choice the alley opposite to that selected on the first choice. A channel adjacent to the first choice was usually selected on the third trial, but the choice behavior is soon randomized. In fact, successive choice of two adjacent channels is significantly less frequent than would correspond to random selection from the set of errorless solutions of the task (6). These results indicate that the performance of the rat in the radial maze reflects the animals' capability to remember the choices already made.

## PARALLEL AND RANDOM MAZES

The mazes used by Olton's group were all of the radial type, but the spatial memory involved should not be limited to the abstract floor plan of the regularly spaced arms converging to the center of the apparatus. In fact, the symmetry of the maze and similarity of its elements may rather complicate than facilitate the solution. To test this possibility, animals overtrained in the radial maze were transferred to a maze formed by a parallel array of the same channels (Fig. 5). The maze was

Fig. 5. The twelve-arm parallel maze. Note the triangular choice area to the entrance of which the rats have to return after each choice.

surrounded by a 47 cm high Plexiglass wall with additional partitions that guided the rat after completion of each choice to the entrance of the triangular choice area. Even during the first exposure to this parallel maze, rats showed strong spatial memory; they made significantly more errors, however, than in the radial maze (Fig. 6). During subsequent training, performance slowly improved and in the third session was not significantly different from that in the radial maze. Closer analysis of the sequences of choices did not disclose any consistent system. The rats certainly did not use the obvious strategy of moving from one channel to the next from left to right or vice versa. Adjacent choices were encountered more frequently than in the radial maze, but their incidence was below that predicted by the random choice model.

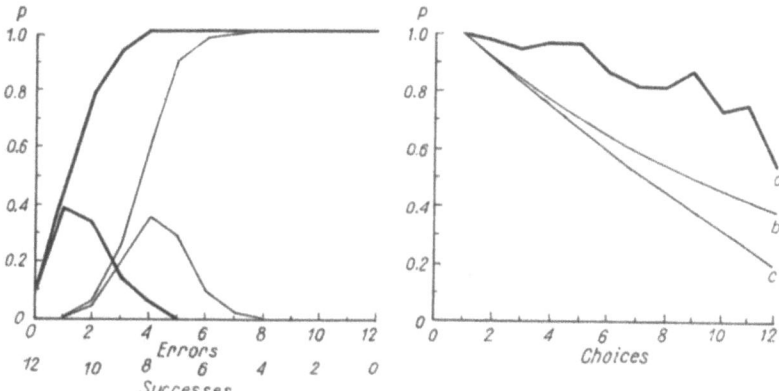

Fig. 6. Choice behavior of a group of 20 rats in the 12-arm parallel maze. Average data from first three maze runs. For description see Figs. 2 and 3.

The errors occurred more frequently in the outer channels 1-4 and 9-12 than in the inner channels 5-8 (Fig. 7), which was probably due to the preference of rats for the channels located close to the ends of the parallel maze array. The distribution of channels visited during the

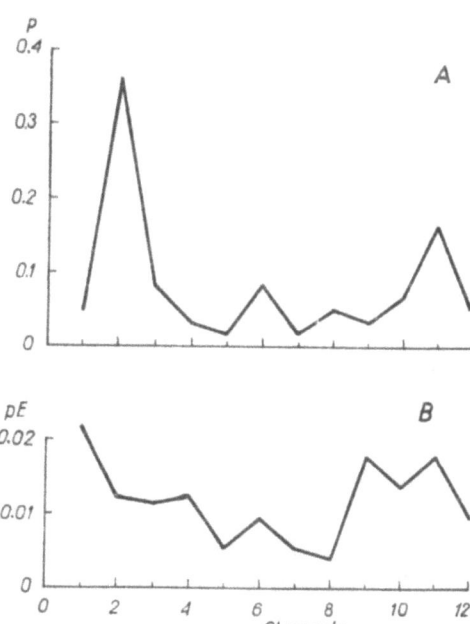

Fig. 7. Probability of selected channels 1-12 ($p$) on the first choice (A) and probability of erroneously visiting ($pE$) the already entered channels 1-12 (B) in the 12-arm parallel maze.

first choice had a similar U-shape with marked peaks at 2 and 11. Errors were more likely to occur around these preferred locations.

The Cartesian floor plan of the parallel maze may facilitate the formation of the cognitive map in the rat's brain in a similar way as the radial configuration. In another experiment the 12 baited channels were randomly scattered over a circular area 1.3 m in diameter as shown in Fig. 8. The overtrained animals were placed into the center of the maze

Fig. 8. The 12-segment random maze. The vertical plexiglas plates mark the channel entrance.

and allowed to visit all channels. During the first exposure the rats already showed strong spatial memory and made only 2.0 ± 0.22 errors in the first 12 choices (Fig. 9). This level of errors was significantly higher ($P < 0.01$) than in the radial maze, but performance improved on subsequent exposures. During the third trial errors were not significantly more frequent than in the radial maze.

Again the rats did not show any obvious tendency to pass through the maze along a definite route. They also did not need more time for making the 12 choices than in the radial or parallel maze. This was partly due to the absence of a fixed starting point to which the rat had to return after each choice. The optimal trajectory required for visiting all channels was thus considerably shorter than in the other two mazes. On the other hand, the maze could be less easily surveyed by the animal that had to rely on the local peculiarities of the channel position (i.e., the distance from and orientation towards the maze wall, the relationship to other channels) rather than on its general location within the maze. The high performance level shown already during the first ex-

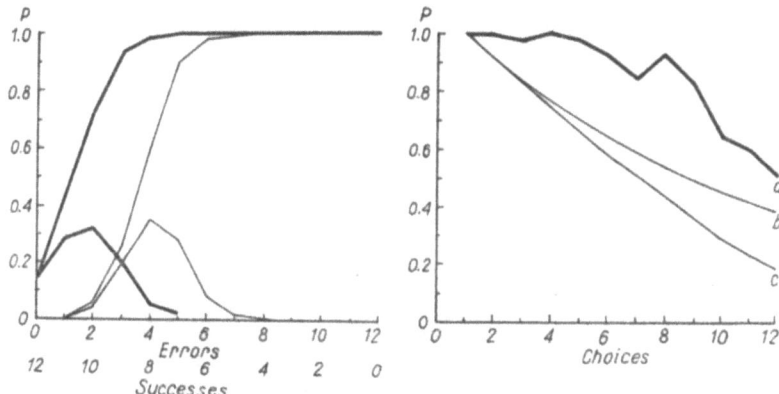

Fig. 9. Choice behavior of a group of 20 rats in the 12-segment random maze. Average data of the first three maze runs. For description see Figs. 2 and 3.

posure to the random maze indicates, that the local signs provide for adequate orientation in a novel environment. The improvement seen with repeated exposures to the random maze is probably due to the increasing familiarity of the rat with the maze floorplan.

## NON-SPATIAL CUES

The dexterity of the animals in solving the task is so impressive that many investigators have looked for additional sensory cues which would allow orientation based on non-spatial information. Maze rotation, exchange of maze arms, elimination of olfaction and/or vision support the spatial memory explanation (11, 13). Another test of this type is the transfer between the radial and parallel mazes.

Rats were given 6 choices in the radial maze and then removed to holding cage. During 5 min, the maze channels were rearranged into the parallel configuration. The animals were then reintroduced into the maze and allowed to make the remaining 6 choices. Since 6 different channels were visited during the first 6 choices, the probability of the seventh choice is 0.5. Random choice behavior should result in 3 errors in choices 7–12. This prediction was confirmed: the rats scored 3.05 ± 0.26 errors in the parallel maze (Fig. 10). This deterioration of performance is not due to forgetting during the 5 min interval required for rebuilding the maze. When the rats were returned after this interval into the radial maze (forgetting control), they scored 1.79 ± 0.18 errors, which is significantly below chance ($P < 0.01$). The effect of forgetting was also significant, however, since animals taken from the maze after the 6th choice and returned there within 15–20 s achieved 0.68 ± 0.19 errors in the 6 post-interval choices.

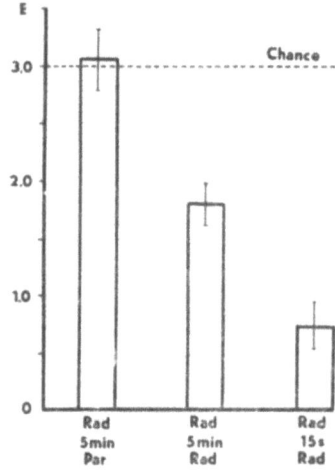

Fig. 10. Transfer from radial (Rad) to parallel (Par) maze and the effect of 5 min or 15 s interruption of the radial maze task. Ordinate: average number of errors (E) in the 6 post-interruption choices (3 errors is the expected chance performance). Vertical bars denote the SEM values.

The failure of the rats to recognise the channels visited in the radial maze, when their configuration was rearranged, indicates that they do not identify the maze arms according to their visual or somesthetic properties. Such a possibility was unlikely, since the maze units were designed and made as identical as possible. The result rules out the more probable hypothesis that the rats use scent marking of the visited channels as a cue.

On the other hand non-spatial factors must not be underestimated. Smell in particular can play an important auxiliary role in solving spatial tasks, when its influence is not effectively eliminated. This was demonstrated in an experiment in which the masking food odor source in the ceiling containers was deliberately omitted. Performance in all 3 types of mazes improved strikingly (Fig. 11). This was most conspicuous in the cases of the less familiar parallel and random mazes, where the number of errors decreased on repeated exposures much faster than when the maze was saturated with food odor. It is obvious that the odor gradient generated in the maze channel by the presence of bait at the end was sufficiently strong to allow olfactory discrimination of the empty and baited maze arms. This factor is more important in the closed tubular mazes used presently, but cannot be entirely neglected even in the open elevated maze. The finding that the radial maze can be mastered by anosmic animals (Zoladek and Roberts 1978) indicates that smell is not necessary for correct performance of the task but this does not prove that smell is not used by intact rats. Both information stored in memory and target generated stimuli contribute to correct solution of natural situations. It is essential that all spatial memory tests assess the role of non-spatial memory cues by appropriate control experiments.

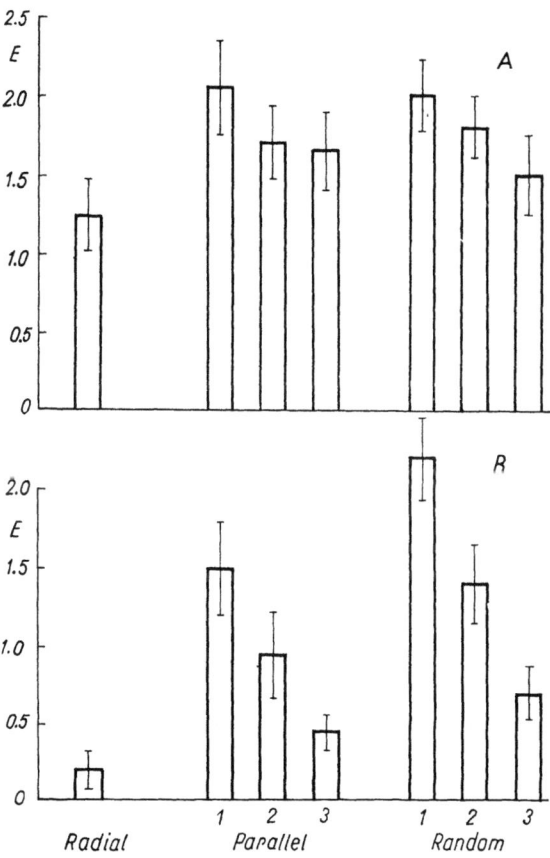

Fig. 11. Average number of errors (± SEM) in the first 12 choices in the 12-arm radial maze (asymptotic performance) and in the first 3 trials in the 12-segment parallel and radial mazes. A, olfactory cues masked by saturation of the maze with food odor. B, the additional food containers not used. The expected random performance is 4.18 errors.

## CONCLUSION

The spatial memory of rats seems to be as efficient as that of dogs and monkeys. The measurement of effective spatial memory is not limited to radial mazes, but can be equally well demonstrated in the case of other geometrical arrangements or for randomly scattered locations. The random maze situation resembles experiments performed by Beritov (1) and by Menzel (7) in dogs and monkeys. The animals were shown how food is being hidden in several places of the room of field. When released, they ran directly to nearest food site, then to the next one and so on. They rarely returned to the place from which the food had already been obtained. The above experiments do not examine only the short-term memory, however. As pointed out by Beritov (1), the image of the food hidden before the eyes of a dog still can influence its behavior after weeks or months. When the dog is brought back after such a long interval into the experimental room, it runs straight to the place, where it remembers the food was hidden during the preceding experiment. The sampling with replacement used in the present experiments is particularly well suited for specific short-term memory examination.

The experiments have also a direct bearing on the historical controversy between the proponents of place learning vs. response learning. The impressive capability of rats to visit a number of different places without repeating the same route strongly supports the contention that rats are guided by a kind of map postulated by Tolman (12). As Konorski (5) pointed out, there is no doubt "that a rat after being acquainted with a given maze, has a clear map of it in his visual gnosis".

## REFERENCES

1. BERITOV, J. S. 1967. Neural mechanisms of higher vertebrate behavior. Translated and edited by W. T. Liberson: Little, Brown and Co., Boston.
2. FELLER, W. 1967. An introduction to probability theory and its applications. Vol. 1. John Wiley and Sons, New York.
3. HUNTER, W. S. 1913. The delayed reaction in animals and children. Behav. Monogr. 2, 1: 1–86.
4. KONORSKI, J. 1959. A new method of physiological investigation of recent memory in animals. Bull. Acad. Pol. Sci. Ser. Sci. Biol. 7: 115–117.
5. KONORSKI, J. 1967. Integrative activity of the brain. An interdisciplinary approach. Univ. Chicago Press, Chicago, 531 p.
6. MAGNI, S., KREKULE, I. and BUREŠ, J. 1979. Radial maze type as determinant of the choice behavior or rats. J. Neurosci. Methods 1 : 343–352.
7. MENZEL, E. W. 1973. Chimpanzee spatial memory organization. Science 182: 943–945.
8. O'KEEFE, J. and NADEL, L. 1978. The hippocampus as a cognitive map. Clarendon Press, Oxford.
9. OLTON, D. S. 1977. Spatial memory. Scientific Am. 236: 82–98.
10. OLTON, D. S., COLLISON, C. and WERZ, M. A. 1977. Spatial memory and radial arm maze performance of rats. Learn. Motiv. 8: 289–314.

11. OLTON, D. S. and SAMUELSON, R. J. 1976. Remembrance of places passed: Spatial memory in rats. J. Exp. Psychol. Anim. Behav. Processes 2: 97–116.
12. TOLMAN, E. C. 1932. Purposive behavior in Animals and man. Appleton-Century, London.
13. ZOLADEK, L. and ROBERTS, W. A. 1978. The sensory basis of spatial memory in the rat. Anim. Learn. Behav. 6 : 77–81.

O. BUREŠOVÁ, Institute of Physiology, Czechoslovak Academy of Sciences, Videňská 1083, 142 20 Prague, Czechoslovakia.

Lecture delivered at the Warsaw Colloquium on Instrumental
Conditioning and Brain Research
May 1979

# THE ORIENTING-EXPLORATORY RESPONSE HYPOTHESIS OF DISCRIMINATIVE CONDITIONING

Endre GRASTYÁN and György BUZSÁKI

Institute of Physiology, University Medical School
Pécs, Hungary

*Abstract.* Behavioral and electrophysiological manifestations occurring under conditions of spatial discontiguity of CS and reinforcement (or response) are described. It is concluded that the conflict or learning difficulties which arise in such situations are the result of a competition between the signal directed conditioned orienting response and the goal directed approach response which require different or opposite directional movements. Considerations are offered to prove that the complex orienting-exploratory response emerging as a result of reinforcement corresponds to the real conditioned response and accordingly the CS to a new independent goal with comparable attractiveness as the primary (consummative) goal. Arguments are brought up against the validity of the stimulus substitution theory of signal directed activities. Findings obtained in fimbria-fornix lesioned animals are presented to show that the orienting-exploratory mode of coping with spatial tasks necessitates the intactness of the limbic system. The conclusion is drawn that an apparently more adaptive mode of behaving in spatial discontiguity, where the CS only releases or triggers the goal response (used occasionally by intact but exclusively by limbic lesioned animals) corresponds to a relatively inferior mode of adaptation.

If fifty years ago you had consulted an expert of conditioning about an experimental design in which CS and US were separated in space you would have been cautioned that no conditioning might occur in such situation. If you insisted to know why, you got an answer not entirely devoid of Aristotelian flavor: because temporal and spatial contiguity are indispensable conditions of association and learning. Times are changing. In the last two decades the paradigm of spatial discontiguity has become increasingly popular because the phenomena which become accentuated by it, signal directed responses, auto-shaping, or sign tracking, seem to offer a new and common approach to the basic problems of classical and instrumental conditioning. Beyond this insight we were also prompted to select this topic for the present occasion because the late Professor Konorski and his coworkers were among the first who paid close attention to and offerred challenging interpretations of such phenomena. Let us introduce the topic by showing our first meeting with that problem.

In the late fifties we reported (6) a correlation between the theta

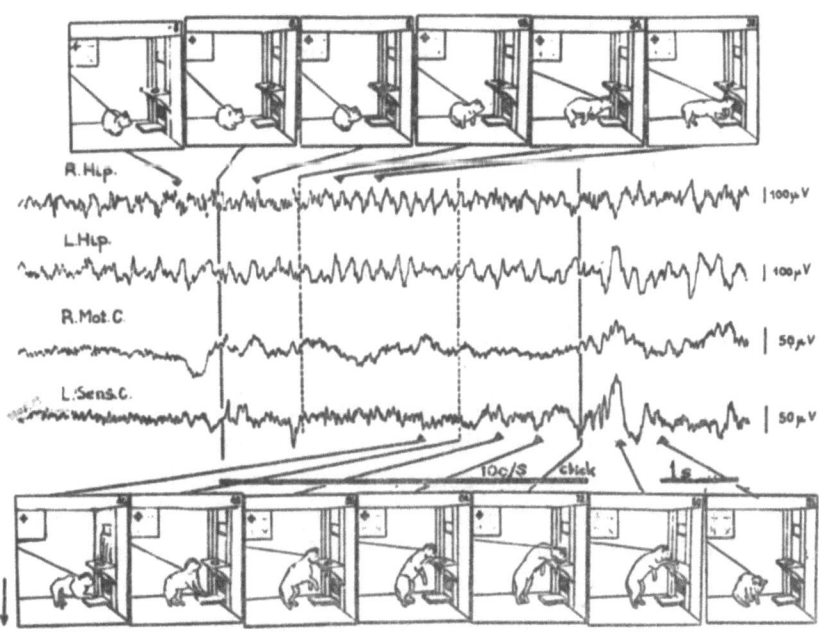

Fig. 1. Simultaneous motion picture and EEG recordings during the performance of a discriminated food reinforced approach response. The acustic CS is delivered through a loudspeaker (black rectangle in the lower right side of each motion picture frame) fixed below the feeding device. The solid horizontal line below the EEG records and the crosses appearing in the left upper corner of each frame mark the presentation of the CS. The numbers in the right upper corner correspond to the number of the motion picture frame copied. R. Hip. and L. Hip., right and left dorsal hippocampus; R. Mot. C., right motor cortex; L. Sens. C., left sensory cortex. From Grastyán et al. (6).

rhythm of the dorsal hippocampus and the orienting response paid to the conditioned signal in discriminative approach conditioning in cats. Actually this finding was an unexpected by-product of the spatial arrangement of the stimuli in the conditioning situation. The CS, an acoustic stimulus was delivered through a loudspeaker placed below the food dispenser, the place of reinforcement. The essence of our findings was that in the initial stage of conditioning an orienting response directed to the loudspeaker was the first and dominating response among all the reactions induced by the CS and strictly coincided with the appearance of hippocampal theta waves. These direct observations were reaffirmed more accurately by motion picture analyses (Fig. 1).

Originally our interest centered around a purely neurophysiological problem, the functional significance and intimate mechanism underlying the hippocampal theta rhythm. In pursuing this interest we attempted later to make a behavioral separation of the orienting response from the rest of conditioned activities by progressively removing the CS from the goal. As an extreme degree of this attempt finally the source of the CS and the feeding device were placed at the two opposite walls of a straight alley, a T or L maze. More precisely the loudspeaker was fixed to the wall behind an elevated platform functioning as the starting place for the approach response.

The findings obtained in that situation (7) considerably surpassed our expectations. Namely, in the first two or three experimental sessions of discriminative approach conditioning in the great majority of cases a progressively increasing orienting and approaching tendency toward the CS appeared. At the same time a progressively decreasing tendency to approach the goal could also be observed. The behavioral observations left no doubt that this was the outcome of competition between the effects of the CS and of the goal. The ensuing conflict reached different degrees in different animals. In about one third of the cases the goal response became entirely suppressed following the first few trials of the 2nd and 3rd experimental sessions and only slowly reappeared in subsequent sessions. Sometimes conflict was so serious that we became suspicious of an intervening sickness of the animals. This suspicion might have gained more weight if we were aware at that time of the findings of Stępień and Stępień (13). The syndrome they described as "magnetoreaction" in a similar situation following the lesions of the precruciate frontal cortex and accordingly regarded as a pathological phenomenon, exactly corresponded to our findings. The consistent emergence of the same phenomenon and the fact that at the climax of the conflict no loss or even decrease of appetite could be observed outside the experimental situation finally convinced us that the competition between the orienting and the goal response was a physiological phenomenon and accordingly what Stępień and Stępień had found was a pathologically exaggerated form of it. In subsequent experimental sessions the signs of conflict gradually vanished and the goal response became restored with progressively decreasing latencies. However, the conditioned orienting response, in most cases in the form of a quick head turn toward the loudspeaker persisted also in the final stage of conditioning after several hundreds of reinforcements (Fig. 2).

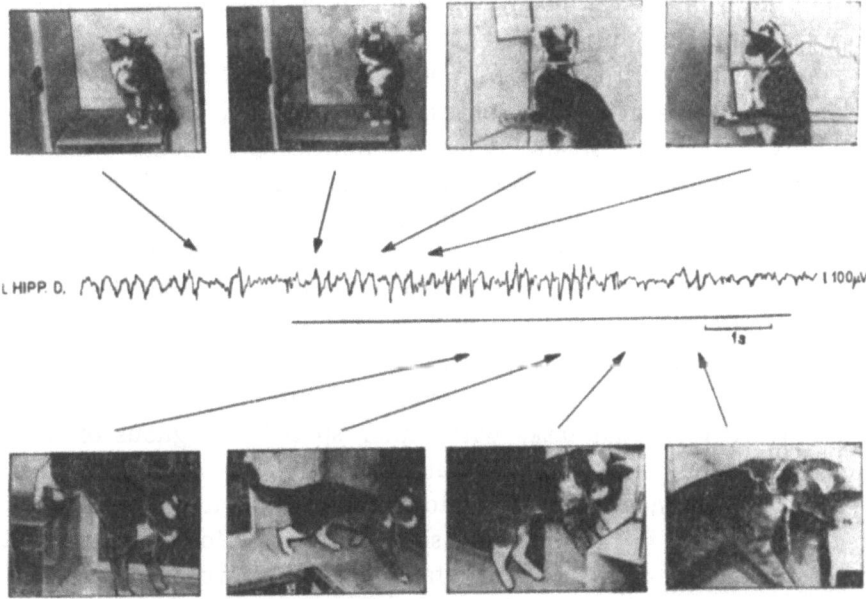

Fig. 2. Behavioral (video-tape recorded) and hippocampal electric manifestations in a L maze induced by a spatially discontiguous acustic CS. Upper row: the conditioned orienting-exploratory response directed to the loudspeaker on the starting platform. Bottom row: the goal response occurring during the second part of presentation of the CS. From Grastyán and Vereczkei (7).

Following this initial study several series of experiments were conducted with the same or modified versions of spatial discontiguity and for reasons of comparison also in the rat. These studies (2–5) confirmed most of the original findings. In addition aversive conditioning was also examined and in this case signal directed responses were not observed. I should like to show some relevant findings from a recent study of ours which make easier to confront our interpretation with others, mainly those based on the auto-shaping phenomenon and conditioning studies operating with spatial variables (3–5). The most conclusive studies of the latter sort were conducted in the Nencki Institute with the leadership of Professor Konorski (9) and by his coworkers (10-12). Before describing our new results let me outline briefly the basic traits of our initial theoretical framework.

The way that we had come to obtain the findings described above made it natural to suppose that the orienting-approaching tendency directed to the spatially discontiguous signal corresponded to the usual conditioned orienting response well known from the classic studies of Pavlov. This is assumed to be an obligatory, or unavoidable, consequence of any kind of discriminative conditioning at least in its initial stage. Under traditional conditions of spatial-contiguity, however, its strength and persistence is hard to assess, because of its spatial and temporal overlap with goal directed responses. Thus, we thought that spatial discontiguity only unveils, and thereby amplifies a response which also exists and has the same significance under conditions of spatial contiguity. The competition between the CS and the goal was partly attributed to their opposing spatial directions. It seemed obvious that under traditional conditions of spatial contiguity the two tendencies coincide in space, and therefore easily merge so that no conflict ensues.

Up to this point our interpretation does not seem to differ significantly from that proposed by Konorski and his coworkers. Namely, in studies involving spatial tasks with either spatial contiguous or discontiguous conditions, they also attribute a positive or negative (parasitic) role to the conditioned orienting or, to use Konorski's term, the "targeting reflex" (10). There is, however, a basic difference as to the nature of the latter between his and our interpretation. Namely, we do not share the traditional view that the orienting response emerging during conditioning is identical with the inborn, unconditioned or consummative form of

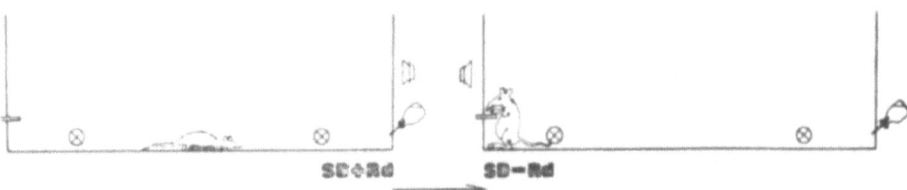

Fig. 3. Schematic of the experimental design. DS, loudspeaker delivering the discriminative signal; Rd, water reward.

orienting which is supposedly induced by any kind of unfamiliar environmental event. The unconditioned orienting reflex is usually restricted to short phasic reactions by organs of the corresponding receptors, like pricking the ears or moving the eyes, and habituates quickly. In contrast, conditioned orientation is a complex, holistic response of a tonic character, which directs the whole organism towards (or away from) a specific object or goal in its environment; it may involve locomotion and, after completion of an approach reaction, even manipulation of the target (Fig. 3). This conceptualization assumes that the conditioned orienting response is a complex learned phenomenon capable to direct the organism in its environment toward new and specific goals (specified by reinforcement).

Accordingly the generalization seems also justified that the orienting response occurring in conditioning is not only a conditioned response, but the only response that is conditioned. Perhaps it was a mistake for us to term this phenomenon "orienting response" because of the traditional connotations it suggests. If it can be identified with the "targeting response" described by Konorski (9) or "sign tracking" used in recent auto-shaping literature (8) then we would not hesitate to accept either of the two terms.

The conflict described earlier also suggests a more feasible interpretation of the above concept of what will be called from now on, the orienting-exploratory response. If it corresponds to the real conditioned response, then the CS is also a new independent goal and the conflict becomes interpretable as the result of competition between a new and an older goal. The objection can, however, be raised that this is hard to reconcile with the requirement of the biological adaptiveness of the learning process. The following reasoning may resolve the paradox. The severity of conflict between CS and consummative goal is proportional to the degree of spatial discontiguity. Spatial discontiguity, however, hardly occurs in the natural settings of the organism, thus we can regard it more or less as a laboratory artefact. Under natural conditions signal and signalled are usually close together or even coincide (e.g., the sight of food). Accordingly the orienting exploratory response directed to the CS leads the organism to the consummative goal, therefore its adaptive nature cannot be questioned.

However, this explanation also fails if we accept the stimulus substitution theory of signal directed conditioned responses. This interpretation which goes back to Pavlov, has became popular recently in the context of auto-shaping. If we assume that the signal becomes a substitute for the reinforcer, then in the case of perfect spatial contiguity the theory works well. However, in the case of even minor spatial discontiguities, which unquestionably also exist in natural circumstances, the unadaptive nature of the signal directed response becomes obvious. Because of such discrepancies we could only accept the substitution hypothesis in a modified form, whereby the signal directed response is viewed as a reinforcement-specific orienting-exploratory response which helps the organism to attain the consummative goal, but the stimulus object does not become identical with it. Fortunately, evidences obtained in recent studies of auto-shaping, mainly in mammals, has considerably devalued the generality of the stimulus substitution theory, thus, even the above mild concession may prove unnecessary (for corresponding literature see the review of Boakes, 1).

We have, however, still another trouble with our interpretation, partly raised by some of our own data which has not been mentioned so far. Namely, we have consistently noticed that a small minority of our animals (about 10% in every experimental group) adapt to spatial discontiguity with an entirely different strategy than that described earlier. No signs of orienting or any kind of signal directed activities were observed during the whole course of conditioning in these cases and consequently no manifestations of conflict could be observed. It seemed that the CS became a potent factor by simply releasing, permitting or triggering the goal directed approach response. The performance of those animals which have shown signal directed responses was significantly slower even in the latest stages of conditioning when the orienting response decreased to its possible minimum level. These facts inevitably raised the critical question of whether there might exist two entirely different learning strategies in solving spatial tasks. An affirmative answer to this question has been given in a recent article by Ławicka and Szczechura (12). The data we are going to show from a recent study (2) seem to confirm their conclusion, but we hope to be able to give a more consistent interpretation within the framework of our own concept.

The study used separate (intact, sham operated, and limbic lesioned) groups of hooded rats that were water deprived in a straight alley to secure water in the presence of an auditory signal (Fig. 4). Before training the fimbria-fornix system was destroyed electrolytically in one third of the animals. The delivery of the auditory signal was a consequence of repeated bar presses at one end of the alley after which the rats had to run for water available at the other end of the alley. Crossing a light beam in the end of the alley turned off the CS and simultaneously activated the water dispenser. In the first stage of the experiment the auditory signal was presented in spatial contiguity with reinforcement, that is at the place of the consummative goal (Fig. 5).

After reaching criterion performance, in the second stage of the experiments, spatial discontiguity was introduced, that is the auditory signal was placed above the bar, i.e., far from the place of reward. In accordance with our previous findings, with the intact animals this arrangement caused abrupt changes characteristic of conflict, that is, increase of latency of the goal response, perseveration of bar pressing and lengthening of duration of single presses. However, quite unexpectedly no signs of conflict appeared in the fimbria-fornix lesioned animals. They continued to respond as if no change had occurred at all.

Identical conditioning procedures were used in the second part of

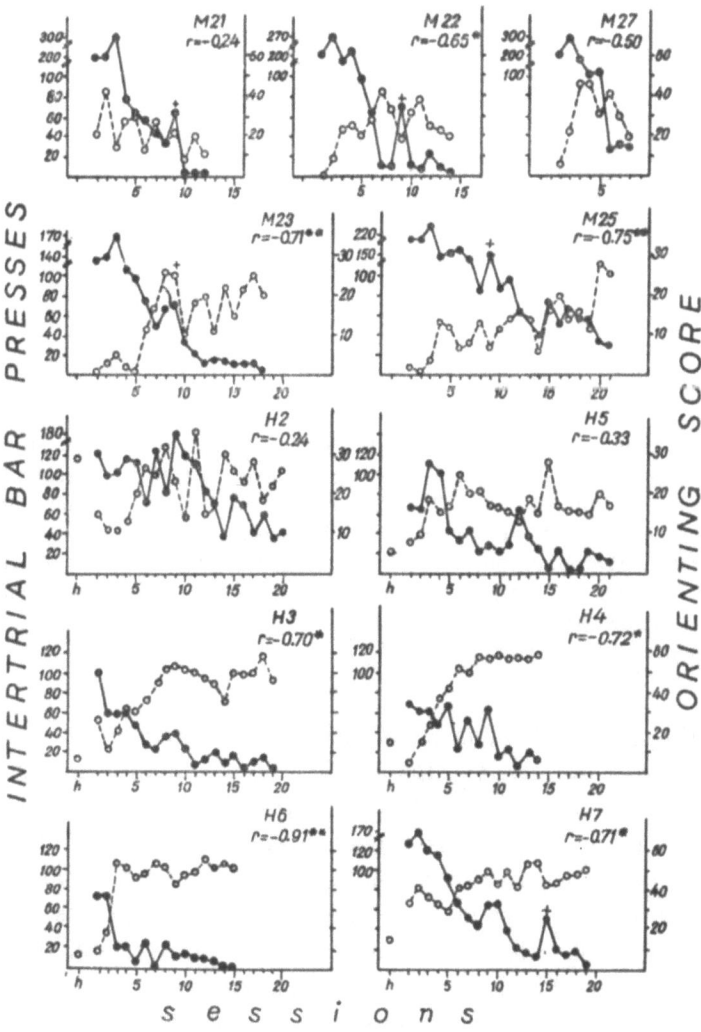

Fig. 4. Relationship between the total number of intertrial bar presses (heavy line) and orienting scores (dashed line) for individual animals; h, habituation session. Asterisks indicate the level of significance. $*P < 0.05$; $**P < 0.01$. Sessions were conducted always in the afternoon except on days marked by $+$.

the study which employed three new groups of rats. In this case the limbic lesions were made surgically and followed initial training under the spatial contiguity condition. After recovery from surgery the introduction of spatial discontiguity caused similar, or even more accentuated, signs of conflict as in the control animals (Fig. 6) in the lesioned animals.

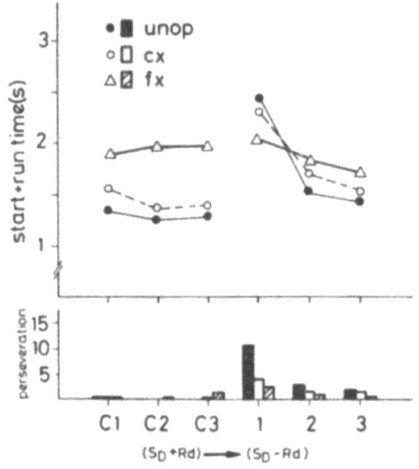

Fig. 5. Total time (start + run times) and lever press perseveration measures on the three criterion sessions (C1, C2, C3) and on the 3 days under $S_D$-Rd spatial discontiguity condition, unop, intact group; cx, sham operated group; fx, fornix lesioned group. Lesions were inflicted before training. From Buzsáki et al. (5).

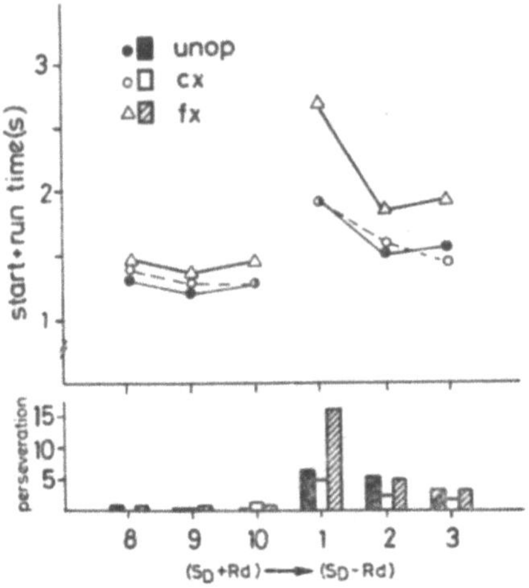

Fig. 6. Total time (start + run times) and lever press preseveration measures on the last three sessions under $S_D +$ Rd spatial contiguity condition and on the 3 test days under $S_D$ — Rd discontiguity condition. Others denotations as in Fig. 5.

The following conclusions are offered by this study:

The findings definitely support the assertion that in coping with spatial tasks two basically different learning modes can occur. In the majority of intact organisms signal-directed responses are obtained, either in spatial contiguity or discontiguity. On the other hand "release", or triggering, of goal responses by the conditioned signal is an exclusive mode of responding after the lesion of the limbic system. Although the adaptive value of this latter mode of responding under conditions of spatial discontiguity seems obvious, we cannot say that it corresponds to a better, superior or even equivalent solution to the former one. This is contradicted by the fact that the releasing mode of action of the CS is a less complex reaction, since it occurs without exception after limbic lesions and is only occasionally used by intact animals. This simple kind of reacting resembles an amplified or stabilized form of sensitization. Its underlying neural mechanisms or its integrative level are unknown. We know, however, that the orienting-exploratory mode of reacting, or the acquisition of this mode, presupposes the intactness of the limbic system, at least in the initial stage of learning. This is clearly demonstrated by the present finding, that the great majority of animals lesioned after training also use this more complex strategy.

If the paradox of unadaptivity of the orienting-exploratory mode under conditions of spatial discontiguity can be resolved by the consideration that the kind of situation containing spatial discontiguity which results in conflict is a completely artificial situation, then there is no obstacle to concluding that the orienting-exploratory mode is the single and highest developed mode of coping with spatial problems of any kind in higher organisms.

Accordingly, we would not regard the two different modes of responding as equivalent learning strategies, between which the organism could chose according to the requirements of the actual spatial task. The relationship between the two modes, however, remains still obscure. We guess that the orienting-exploratory mode of reacting might always be preceded by a short, elusive initial stage when the CS simply releases the goal response. This would become transformed, as discrimination proceeds, into the orienting-exploratory mode. It remains, however, still to be revealed why the minority of intact animals stick to this inferior solution forever.

## REFERENCES

1. BOAKES, R. A. 1977. Performance of learning to associate a stimulus with a positive reinforcer. In H. Davis and H. M. B. Hurwitz (ed.), Operant-Pavlovian interactions. Lawrence Erlbaum Ass., New York.
2. BUZSÁKI, G., GRASTYÁN, E., MOLNÁR, P., TVERITSKAYA, I. N. and HAUBENREISER, J. 1979. Auto-shaping or orienting? Acta Neurobiol. Exp. 39: 179–200.
3. BUZSÁKI, G., GRASTYÁN, E., TVERITSKAYA, I. N. and CZOPF, J. 1979. Hippocampal evoked potentials and EEG changes during classical conditioning in the rat. Electroencephalogr. Clin. Neurophysiol (in press).
4. BUZSÁKI, G., GRASTYÁN, E., WINICZAI, Z. and MÓD, L. 1978. Significance of cue location for intact and fimbria-fornix lesioned rats. Neurosci. Lett. Suppl. 1: 81.
5. BUZSÁKI, G., GRASTYÁN, E., WINICZAI, Z. and MÓD, L. 1979. Maintenance of signal directed behavior in a response dependent paradigm: A systems approach, Acta Neurobiol. Exp. 39: 201–217.
6. GRASTYÁN, E., LISSÁK, K., MADARÁSZ, I. and DONHOFFER, H. 1959. Hippocampal electrical activity during the development of conditioned reflexes. Electroencephalogr. Clin. Neurophysiol. 11: 409–430.

7. GRASTYÁN, E. and VERECZKEI, L. 1974. Effects of spatial separation of the conditioned signal from reinforcement: A demonstration of the conditioned character of the orienting response or the orientational character of conditioning. Behav. Biol. 10: 121–146.
8. HEARST, E. and JENKINS, H. M. 1974. Sign tracking: The stimulus reinforcer relation and directed action. Psychonomic Society, Austin, Texas.
9. KONORSKI, J. 1967. Integrative activity of the brain. An interdisciplinary approach. Univ. Chicago Press, Chicago, 531 p.
10. KONORSKI, J. and ŁAWICKA, W. 1959. Physiological mechanism of delayed reactions. Acta Biol. Exp. 19: 175–197.
11. ŁAWICKA, W. 1969. A proposed mechanism for delayed response impairment in prefrontal animals. Acta Biol. Exp. 29: 401–414.
12. ŁAWICKA, W. and SZCZECHURA, J. 1979. Stimulus-response spatial contiguity vs. S-R spatial discontiguity in auditory spatial tasks. I. Acquisition by normal dogs. Acta Neurobiol. Exp. 39: 1–13.
13. STĘPIEŃ, I. and STĘPIEŃ, L. 1965. The effects of bilateral lesions in precruciate cortex on simple locomotor conditioned response in dogs. Acta Biol. Exp. 25: 387–394.

Endre GRASTYÁN and György BUZSÁKI, Institute of Physiology, University Medical School, 7643 Pécs, Hungary.

Lecture delivered at the Warsaw Colloquium on Instrumental
Conditioning and Brain Research
May 1979

# AUDITORY TARGETING REFLEXES: THEIR DETERMINING ROLE IN DIRECTIONAL INSTRUMENTAL RESPONDING

Wacława ŁAWICKA

Department of Neurophysiology, Nencki Institute of Experimental Biology
Warsaw, Poland

*Abstract.* Two groups of normal dogs have been trained in a task involving one directional alimentary response signalled by the auditory cue. The cue was spatially contiguous with response in one group, and spatially discontiguous in the other. Both groups did not differ in the acquisition rate. However, as it was demonstrated in the test trials, the response of the first group was determined by the location of the cue and not by its auditory quality, while the response of the second group was based on the auditory pattern of the cue and not on its location. On this basis it may be concluded, that depending on stimulus-response spatial contiguity, different properties of an auditory stimulus play a determining role in the performance of directional response. These results support earlier anatomical and electrophysiological data pointing out to the two divisions within the auditory system, one involved in localization and another in pattern recognition of the auditory cue. The difference in the determining role of these two aspects of the auditory cue is also discussed with regard to the system of targeting reflexes, elaborated in Konorski's "Integrative activity of the brain".

## INTRODUCTION

It has been recently shown (4–6, 10–12, 14) that stimulus-response spatial contiguity plays an important role in auditory spatial tasks, resulting in rapid differentiation learning. Since differentiation of directional responses, elaborated under the conditions when the stimuli and responses are spatially separated, requires a higher number of trials to criterion score (5, 6, 9–11), different factors may determine directional responses in each form of differentiation task. The test used here is concerned with the role played by an auditory stimulus, which acquires different determining properties, depending on whether it is spatially contiguous, or discontiguous, with a directional response.

## METHODS

*Subjects.* Eight naive mongrel dogs were used. Dogs lived in individual home cages, where they were fed once daily 15-20 h before testing. Two groups were tested, each containing 4 dogs.

*Apparatus.* All animals were trained in an experimental room, 4 × × 8 m, that contained two food dispensers, $F_l$ and $F_r$, situated on the floor (Fig. 1). Each food dispenser consisted of a box enclosing 10 foodcups, that were mounted on a rotating disc. An opening on the top of the box provided access to one cup. The disc could be activated remotely by the experimenter, who was seated behind the starting platform, which was separated by a wooden, 83 cm high, screen.

The auditory cue was a click series (50-ms pulses at 10/s) produced by a square wave generator. Cues were presented through one of three loudspeakers: with two ($S_l$ and $S_r$) located on the respective foodwells ($F_l$ and $F_r$) and the third ($S_m$) — on the floor opposite the starting platform. The intensity of clicks train (measured by a Bruel Kjaer Pulse

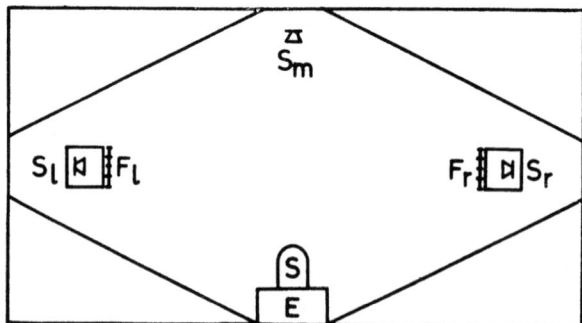

Fig. 1. Schematic diagram of the experimental room. $F_l$, $F_r$, feeders; $S_l$, $S_m$, $S_r$, speakers; S, starting platform; E, place of experimenter. In front of each feeder marked photocell equipment.

Prec. Level Meter) at the starting platform was 48 dB, which was slightly above the background level (42 dB C). Photocells were mounted in front of each foodbox. Each stimulus onset activated an electronic timer which was terminated whenever the animal approached the front of a foodwell.

The time elapsed between the onset of the stimulus and the termination of the timer activation was considered as reaction time. Statistical analysis of the reaction time data was done according to the method described elsewhere (18). The experimental room was illuminated by a centrally situated 500 W overhead electric bulb.

### Testing procedure

*Preliminary training.* On the first day the dog was familiarized with the experimental room for 10 min during which he received one portion of food from each foodwell. Subsequently, the dogs received food only from the right ($F_r$) feeder. They were trained to approach the feeder in response to the sound emitted by the moving bowl, which they soon learned to associate with presentation of food. The animals were also shaped to return to the starting platform several seconds after eating, since standing at the starting place was the necessary condition for food presentation on the next trial. Eighteen food presentation were given in a single session. After 9 successive food presentation there was a 3–5 min intermission (for refilling the feeder), while the animal was kept behind one of the wooden partition in the testing room. Preliminary training was continued until each dog made 18 correct responses in a single session.

*Acquisition.* In this stage of training the auditory cue was introduced and the animals required to approach feeder $F_r$ when it was presented. For the $S_r$ group, the cue was presented through the right speaker, $S_r$ (Fig. 1), being thus spatially contiguous with the response, while for the $S_m$ group the cue was presented from the middle speaker, $S_m$, (Fig. 1), being thus spatially discontiguous with the response.

Each experimental session consisted of 18 trials with 50–70 s intertrial intervals. Also, a few minutes intermission occured in the middle of the session. If in the first session the animal did not approach the feeder to stimulus presentation in 3–5 s after stimulus onset, food was delivered in feeder $F_r$ against the background of the stimulus duration, and the stimulus turned off, as soon as the animal approached the feeder and started eating. If during the stimulus presentation the animal approached the incorrect feeder $F_l$, the stimulus was turned off immediately, no food was presented, the response scored as an error, and the trial repeated with the usual intertrial interval.

After reaching criterion, which was determined as a single session with 18 correct responses performed within the 5 s from the stimulus onset, the animals were submitted to the test trials I procedure. After completing the first type of test trials, 3 regular sessions were presented to each subject, subsequently followed by test trials II.

*Test trials I*. Both groups were submitted to the test trials in which the auditory cue was emitted from speaker $S_l$ (Fig. 1). First, sporadic 4–6 test trials were presented to each subject, distributed among regular trials in 2–3 successive experimental sessions; 22 such sporadic test trials were presented in each group.

In the second stage, the animals were presented with two complete experimental sessions consisting only of test trials. Animals' responses in all test trials were reinforced by food independently of the animal choice, i.e., approaching either $F_l$ or $F_r$ was followed by food presentation in the approached foodwell.

*Test trials II*. During these test trials the auditory cue was presented from a different loudspeaker for each group. For the $S_r$ group the stimulus presentation was transferred to the speaker $S_m$, whereas for the $S_m$ group the stimulus was transferred to the speaker $S_r$. Under these conditions in the first, intermittent stage of test trials procedure, 12 test trials were presented to each subject, distributed among regular trials in three successive experimental sessions. Subsequently, 3 full successive experimental sessions, with exclusive presentation of test trials, were given. As in the test trials I, depending on the animal's choice, the responses to the right, or to the left feeder, were reinforced by food. If during a test trial no response was observed within 5 s of the stimulus duration, the stimulus was turned off. If, however, the animal approached a feeder no later than within 10 s after the stimulus offset, the response was reinforced by food delivery in the approached feeder. The experimental plan, showing successive steps of the experimental procedure, is presented in Table I.

TABLE I

Experimental plan indicating the source of auditory cue during the successive stages of experimenta procedure for the Sr and Sm group. $S_1$, $S_m$, $S_r$: left, midle and right speaker. The bottom row presents numbers of successive sessions

| | Acquisition | Test trials I | | Regular sessions | Test trials II | |
|---|---|---|---|---|---|---|
| | | sporadic | successive | | sporadic | successive |
| Sr group stimulus-response spatially contiguous | $S_r$ | $S_1$ | $S_1$ | $S_r$ | $S_m$ | $S_m$ |
| Sm group stimulus-response spatially discontiguous | $S_m$ | $S_1$ | $S_1$ | $S_m$ | $S_r$ | $S_r$ |
| Session no. | 1–4 | 4–6 | 7–8 | 9–11 | 12–14 | 15–17 |

## RESULTS

*Preliminary stage.* In the preliminary stage of training the animals attained the criterion during the first (5 dogs), or during the second (2 dogs) session. Only one unusually timid animal, that was originally afraid of the sound emitted by the moving bowl, required 4 sessions.

*Acquisition.* On acquisition training no difference in the acquisition rate was observed between the $S_r$ and $S_m$ groups. The animals in both groups started to approach feeder $F_r$ to the presentation of the stimulus alone, i.e., prior to food delivery, within the first 9 trials of the first experimental session. It should be added, that at this stage no responses

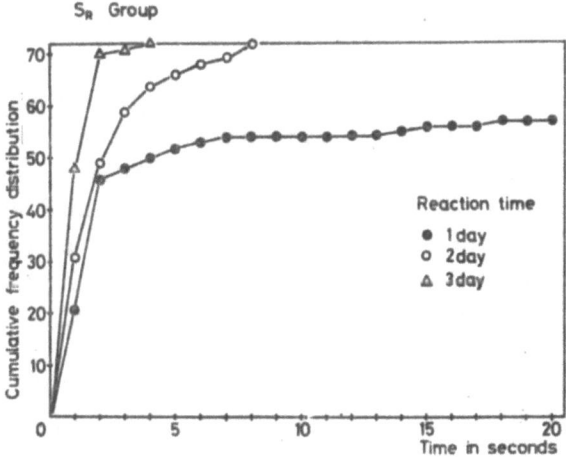

Fig. 2a. Reaction time on three successive acquisition days in the $S_r$ group.

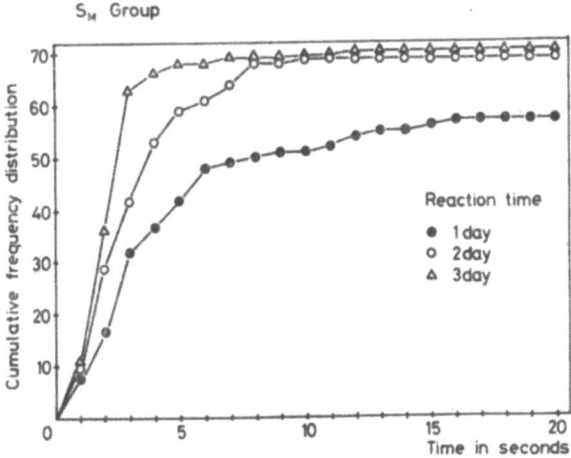

Fig. 2b. Reaction time on three successive acquisition days in the $S_m$ group.

to feeder $F_1$ were observed. Three animals in each group attained the criterion on the third acquisition day; 2 remaining dogs required one more session. However, both groups differed with regard to the reaction time, the number of responses with reaction times within 1–2.9 s being higher for the $S_r$ group. This may be compared in Fig. 2, which illustrates the cumulative frequency distributions of reaction times for the first three successive acquisition days in the $S_r$ (Fig. 2a) and $S_m$ (Fig. 2b) group. The between-group comparisons of the cumulative frequency distributions for each consecutive day showed differences significant at $P < 0.01$ level (Kolmogorov-Smirnov two-tailed test).

*Test trials I.* In these test trials the auditory cue in both groups was emitted from speaker $S_l$, situated on the left feeder. Under these conditions, in the first, 22 dispersed test trials, a striking difference was observed in behavior of these two groups. All animals in the $S_r$ group made 22 total responses to the left, $F_l$, feeder. By contrast, the animals in the $S_m$ group, in spite of the same stimulus re-location, continued to respond to the right feeder, $F_r$, in 22 trials. At the presentation of the auditory cue the animals in both groups demonstrated clearly the targeting response towards the new location; however, only the animals in the $S_r$ group changed their alimentary response by approaching the left feeder, whereas the other, $S_m$, group continued with their regular responses to the right feeder.

In Fig. 3 the cumulative frequency distributions of the reaction times are presented for the $S_r$ and $S_m$ group in 22 dispersed test trials, as oppo-

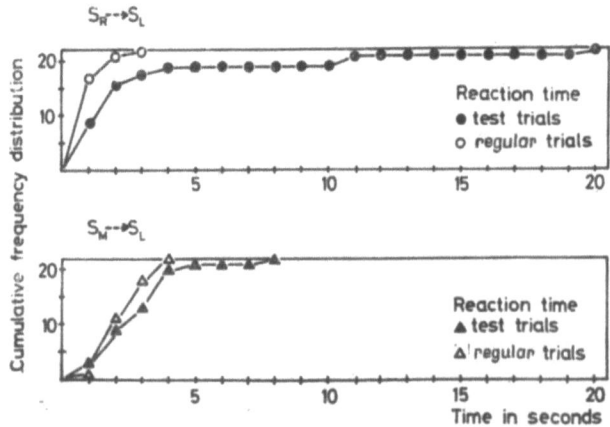

Fig. 3. Reaction time in the $S_r$ and $S_m$ group on 22 dispersed test trials I, on which stimulus presentation was shifted to the speaker $S_l$. For comparison, reaction time for preceding regular trials has been shown.

sed to the preceding regular trials: the frequency distributions for the test and regular trials in each group differs at the $P<0.05$ (Kolmogorov-Smirnov two-tailed test). It may be seen from the curve, illustrating the performance of the $S_r$ group, that the number of responses with fast reaction times (1–1.9 s) decreased during the test trials. This was probably due to the fact that the animals at the end of intertrial intervals were used to face the right feeder. However, since stimulus presentation during the test trials evoked responses towards the left feeder, the turning of the whole body in the opposite direction was required which, resulted in the prolonged reaction time in this group.

Following the intermittent test trial procedure, two complete testing sessions were introduced to each subject, during which the stimulus was presented exclusively from the left loudspeaker. This resulted in 144 responses in each group (Fig. 4); during these two sessions the animals in the $S_r$ group approached the left feeder in 139 trials and the right feeder in 5 trials. Just the opposite relations were observed in the $S_m$

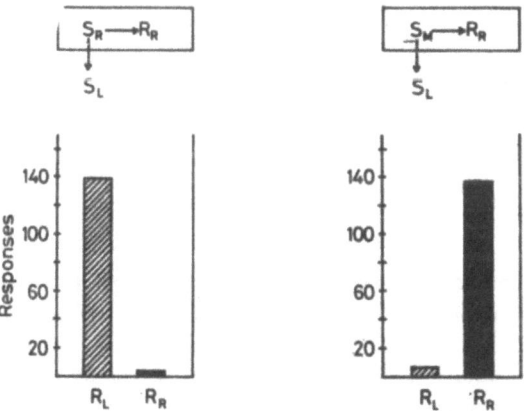

Fig. 4. Response distribution on two successive test trials I sessions, on which the stimulus presentation was shifted to the left speaker in the $S_r$ and $S_m$ group.

group, in which the animals made only 7 responses to the left and 137 responses to the right feeder. It should be added, that those sporadic responses to the right or left feeder occurred irregularly throughout the sessions and did not increase in number, although they were each time reinforced.

The cumulative frequency distributions of reaction times in the $S_r$ and $S_m$ group for these two complete test trials sessions are presented in Fig. 5. In the $S_r$ group, a significant difference in the cumulative distributions of reaction times between the first and second testing sessions has been found ($P<0.001$, Kolmogorov-Smirnov two-tailed test), with the decreased number of responses with the fast reaction time (1-1.9 s) occurring on the first testing session. As indicated earlier for the intermittent test trial procedure, this was associated with the animals' additional turning response, involved in the approach response towards the left feeder. In the second testing session, when the animals started to face the left feeder more consistently, an increased number of responses with short reaction times was observed. The difference in the cumulative frequency distributions of reaction times on the first and second testing day in the $S_m$ group was not significant (Kolmogorov-Smirnov two-tailed test).

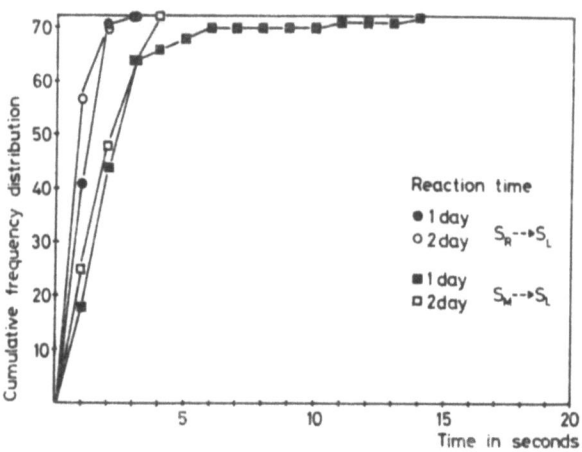

Fig. 5. Reaction time in two successive test trials sessions, in which the stimulus presentation was shifted to the speaker $S_l$, in the $S_r$ and $S_m$ group.

Thus, on the basis of the test trials data it may be concluded, that although both groups had been trained in approaching a single feeder, $F_r$, to the presentation of the same auditory cue, their responses were not determined by the same factor. As revealed by the test trial procedure, the animals' responses in the $S_r$ group were determined predominantly by location of the auditory cue. In contrast, the responses of the $S_m$ group were not determined by location, but presumably by the acoustic feature of the same auditory cue.

Since the difference in factors determining the directional response in both groups was so striking, the second type of test trials was introduced to investigate, whether or not, the responses of the $S_r$ group were also determined by auditory pattern of the cue, although in a comparatively lesser degree.

*Test trials II.* $S_r$ group. In $S_r$ group the stimulus presentation was transferred to the speaker $S_m$, located in the middle of the room. In the first, intermittent stage of these test trials, the behavior of the animals in the $S_r$ group changed drastically, in comparison to that observed during regular trials: in the first test trials the animals were approaching the middle speaker, $S_m$, sniffing it, spending some time in its vicinity, and then coming back to the starting platform. In other test trials — they remained on the starting platform during the stimulus presentation, or were approaching feeder $F_r$ several seconds after the stimulus offset. This performance, as expressed by the reaction time for individual animals during presentation of sporadic test trials, may be seen in Fig. 6a. This

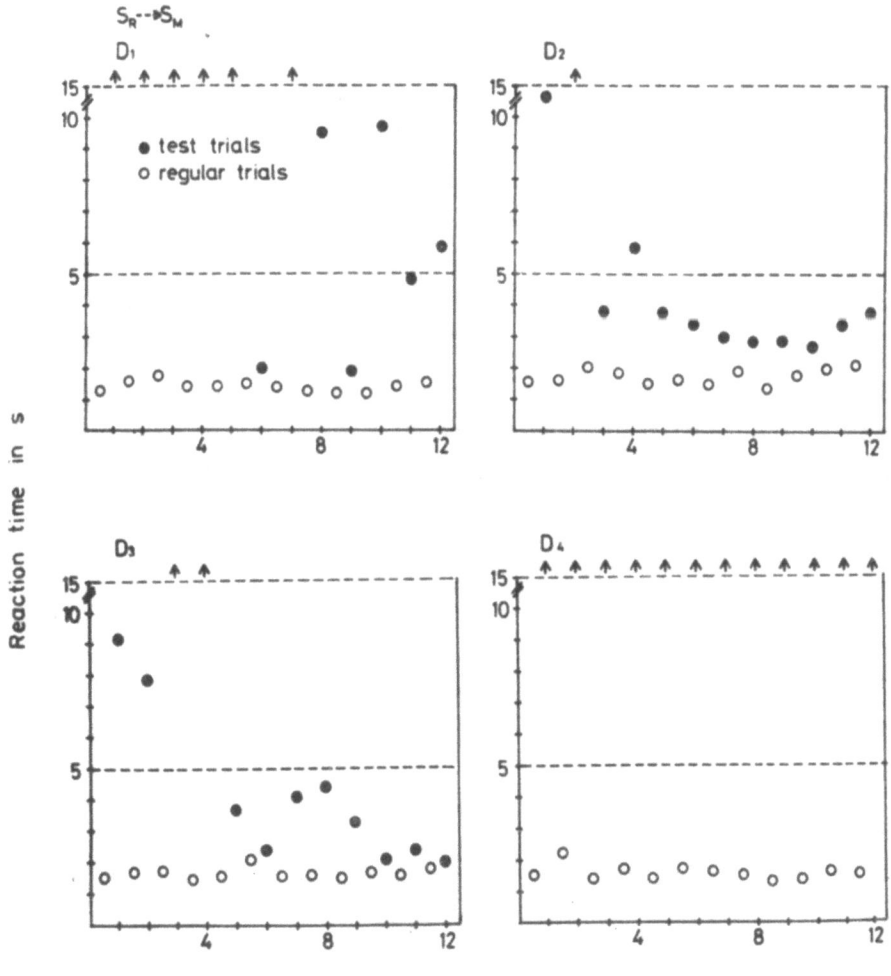

Fig. 6a. Reaction time in the individual animals of the $S_r$ group during 48 sporadic test trial II, on which presentation of the stimulus was shifted to the speaker $S_m$. For comparison, the reaction time of each preceding regular trial has been shown. Arrows indicate no-response trials.

figure illustrates also the comparative reaction time of each preceding regular trial, which is much faster. Within the total 48 sporadic test trials in the whole group, there were 21 no-response trials, 7 trials with responses after the stimulus cessation, and 20 responses during the sti-

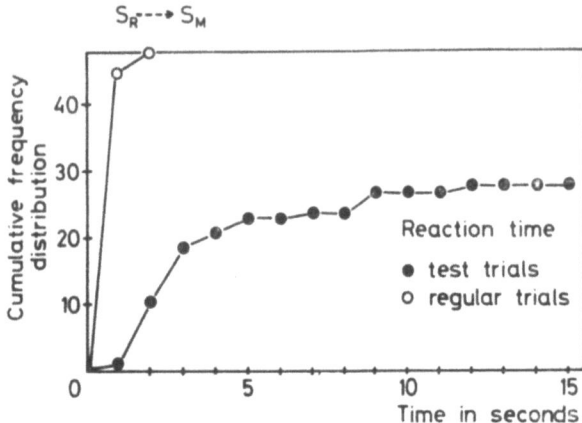

Fig. 6b. The same reaction time data as in Fig. 6a, summarized in a form of cumulative distribution curves for the test- and preceding regular trials.

mulus presentation (comp. Fig. 6a). These data, i.e., the reaction time during sporadic test trials, and the reaction time during each preceding regular trial, are presented by the form of the cumulative frequency

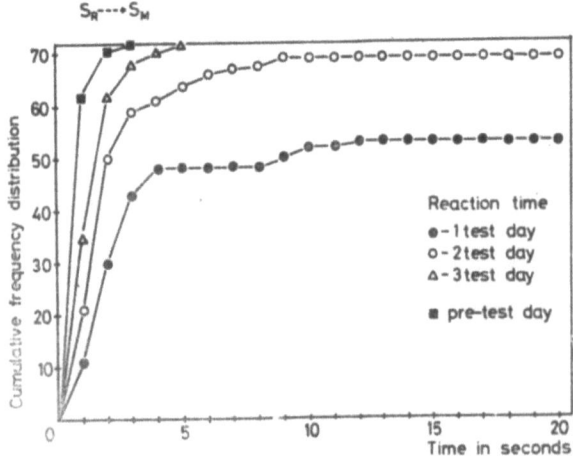

Fig. 7. Reaction time in the $S_r$ group in three test trials II sessions with stimulus presentation shifted to the middle speaker, $S_m$. For comparison, reaction time in one preceding regular session has been shown.

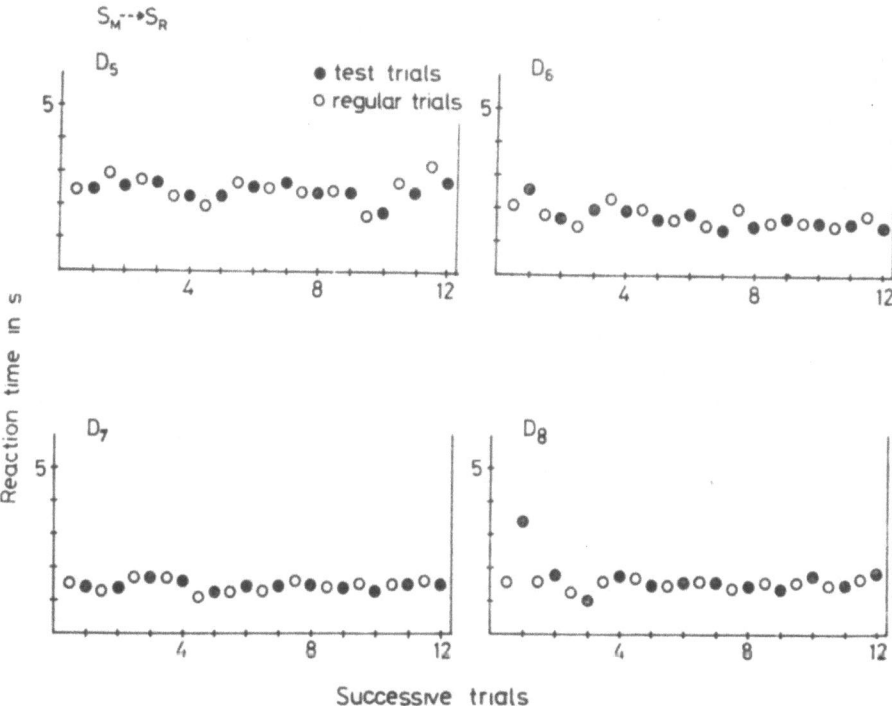

Fig. 8a. Reaction time in the individual animals in the $S_m$ group during 48 sporadic test trials II, in which presentation of the stimulus was shifted to the speaker $S_r$. For comparison, the reaction time of each preceding regular trials has been shown.

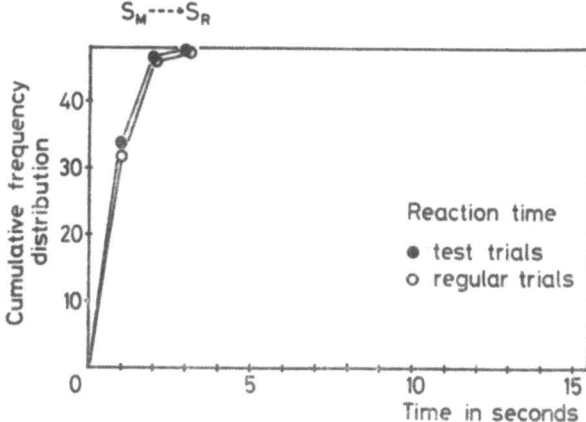

Fig. 8b. The same reaction time data an in Fig. 8a, summarized in a form of cumulative distribution curves for the test- and preceding regular trials.

distribution curves for the whole $S_r$ group in Fig. 6b, with two cumulative distributions differring significantly at the $P < 0.001$ level (Kolmogorov-Smirnov two-tailed test). It may be added, that each individual animal in the $S_r$ group approached feeder $F_r$ in the earlier sessions in more than 100 trials. Following sporadic test trial procedure, three successive experimental sessions were introduced, in which the auditory cue was presented exclusively from the middle speaker, $S_m$. As may be compared in Fig. 7, the number of approach responses with short reaction times was gradually increasing, although even on the third test-day the cumulative frequency distributions of reaction times differred significantly from the pre-test level ($P < 0.001$, Kolmogorov-Smirnov two-tailed test).

$S_m$ group. In the $S_m$ group, test trials II consisted in presentation of the auditory cue from the speaker $S_r$, which proceeded, as in the previous group, in two successive steps. During the intermittent stage of these test trials the dogs manifested no changes in their overt behavior, regularly approaching the $F_r$ feeder, as during the regular trials. The results obtained for each individual subject during the sporadic test trials are presented in Fig. 8a, which illustrates the reaction time of each sporadic test trial, and of an immediately preceding regular trial. These results, summarized in the form of cumulative frequency distribution curves, are compared in Fig. 8b, indicating no difference between the cumulative distributions for the test and regular trials (Kolmogorov-Smirnov two-tailed test). Comparison of performance during the next step, with exclusive presentation of test trials during three successive sessions,

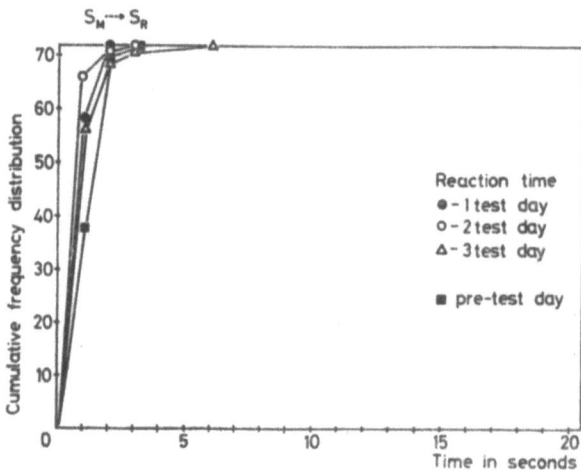

Fig. 9. Reaction time in the $S_m$ group in three test trials II sessions with stimulus presentation shifted to the right speaker, $S_r$. For comparison, reaction time in one preceding regular session has been shown.

revealed also no difference between the sessions (Fig. 9). However, a significant difference has been found between the performance during the test sessions and on the pre-test day ($P < 0.001$, Kolmogorov-Smirnov two-tailed test), with the increased number of fast reaction times (1-1.9 sec) observed during the test sessions (comp. Fig. 9).

Thus, on the basis of test trials II, it may be concluded that if the animals are trained under the conditions of S-R spatial contiguity, as in the case of the $S_r$ group, then inducing the conditions of S-R spatial discontiguity strongly interferes with the animals' performance. The results obtained both in intermittent and repetitive stage of test trials in this group suggest, that when the auditory cue became spatially discontiguous with the response, the animals had to learn the task anew under these changed conditions. The new acquisition seems surprising, since the animals were required to approach the same feeder $F_r$ in response to the same, although differently located auditory cue.

When, however, we changed the conditions of testing in the opposite direction, i.e., from S-R spatial discontiguity to S-R spatial contiguity, as it was in the case of the $S_m$ group, then no interference in animals' performance was observed (comp. Figs. 8a and 8b). Moreover, a comparison of the cumulative distributions of reaction times from the pre-test level with the performance under the conditions when the stimulus and response were spatially contiguous (comp. Fig. 9), revealed more responses with fast reaction times under the latter conditions.

## DISCUSSION

Comparison of aquisition rates of a single directional response, as expressed by performance scores, showed no difference for dogs trained under conditions of S-R spatial contiguity ($S_r$ group) and discontiguity ($S_m$ group). However, measures of reaction time revealed significantly more responses with short reaction time for the spatially contiguous than for the discontiguous group. Also, shifts of the stimulus source to the speaker located in the opposite direction resulted in directional changes of the locomotor response only for the S-R spatially contiguous group.

These results suggest that the animals in the two groups were using different aspects of the same auditory cue for determining their directional alimentary responses. While in the S-R spatially contiguous group the response was determined by the stimulus locus, the response of the S-R spatially discontiguous group was not determined by the same factor, but probably by an acoustic feature of the auditory cue.

This suggestion has been confirmed by the second type of test trials which demonstrated, that performance of the dogs in the S-R spatially contiguous group was disrupted when auditory cue became spatially discontiguous with the alimentary response, thus confirming that stimulus pattern did not acquire a determining role in their performance.

On the other hand, opposite changes in conditions of testing, i.e., inducing the conditions of S-R spatial contiguity in the $S_m$ group during the second type of test trials, produced no disturbing changes in their behavior. The last finding excludes the possibility, that response learning in the $S_m$ group might have been related to the locus of the cue by involving avoidance of approach of the stimulus source (i.e., turning-away movement) established during the course of training.

Thus, it may be concluded, that not only in the auditory differentiation task (10, 11, 13), but even in tests involving one form of directional response, S-R spatial contiguity, or discontiguity, plays an important role, deciding which property of the presented stimulus, its location or auditory feature, will determine the directional alimentary response.

The possibility that the auditory system may, at the lower level, be subdivided into two distinct functional, although not completely isolated subsystems, one ventral involved in localization of sound and reflex function, and another dorsal — involved in the analysis of its acoustic quality, was suggested by Poljak (13) on the basis of his anatomical study. He made this distinction already at the level of cochlear nucleus, dividing it into the ventral and dorsal components, each giving rise to the ventral and dorsal part, respectively. Electrophysiological studies (1-3) dealing with neural analysis of these two auditory subsystems confirmed their differring functional properties, leaving open the question whether they are preserved at the higher levels of auditory system, on which they converge. That these two properties of acoustic signals, sound location and sound quality, may be represented separately at the level of the midbrain in the mesencephalicus lateralis dorsalis, homologous of inferior colliculus in man, has been demonstrated quite recently in the owl by Knudsen and Konishi (7). On this basis it may be assumed, that depending on experimental conditions, the animals in each group may use functionally and anatomically distinct subdivisions of the auditory system for determining the same form of the approach response. The possibility of a functional subdivision within the auditory system does not seem surprising, since a similar dual function has been recognized several years ago within the visual system, one participating in localization, and the other in the pattern recognition of the visual cue (15-17).

These behavioral results may also be considered within the framework of targeting reflexes introduced by Professor Konorski in the "Integrative activity of the brain" (1967). According to his theory, afferent systems are supplied with additional mechanisms, securing the best perception of stimuli falling on receptive surface. One of them is the system of targeting reflexes, whose role consist in adjustment of the lower level of afferent systems and their effectors, often manifested as the head, eyes and ears orientation towards the stimulus source. Targeting reflexes preced each unitary perception and form a necessary condition for the occurrence of the succeeding perceptual process.

Therefore, it may be assumed, that in the $S_r$ group, with the cue and response being spatially contiguous, the actual auditory targeting response, and not the perception of a given pattern, became a sufficient factor to determine the alimentary approach response. That was why in this group, after stimulus displacement during the first type of test trials, the animals were approaching the opposite feeder, with their approach behavior being all the time determined by the targeting reaction. On the other hand, performance in this group was disrupted during the second type of test trials, which required from the animals a directional response unrelated to the targeting response. In that case the dogs had first to extinguish the approach response based on targeting reflex, enabling them to develop the form of instrumental response, with targeting reflex playing no determining role.

In the case of $S_m$ group, trained under the conditions of S-R spatial discontiguity, the targeting reaction played no role in determining the direction of the alimentary approach response. Therefore, shifting the stimulus source during the first type of test trials, resulted in no changes in the direction of instrumental response, although the animals manifested the targeting response to the new stimulus source by head orientation to the sounding speaker.

The result to be commented upon is the higher number of responses with fast reaction time found in regular trials in the S-R spatially contiguous group, as compared to the other group. The auditory cue acquired a food-signalling value in both groups; however, only responses in the S-R spatially contiguous group required no elaboration of determining connections, since directional responses were determined by an unconditioned reflex. In the S-R spatially discontiguous group, on the other hand, the performance required both the cessation of the targeting reflex (8) and the determining of the form of directional response in the process of conditioning, which might account for longer reaction times.

It remains to be seen, if inducing the conditions of S-R spatial contiguity in the $S_m$ group would eventually result in "overtaking" the determining role by the targeting reflex. Increased number of responses with fast reaction time observed in this group during the second type of test trials would rather suggest that possibility.

## REFERENCES

1. EVANS, E. F. 1974. Neural processes for the detection acoustic patterns and for sound localization. In F. O. Schmitt and F. G. Worden (ed.), The neurosciences third study. The MIT Press Cambridge, Massachusetts, p. 131–145.
2. EVANS, E. F. and NELSON, P. G. 1973. The responses of single neurones in the cochlear nucleus of the cat as function of their location and the anaesthetic state. Exp. Brain Res. 17: 402–427.
3. EVANS, E. F. and NELSON, P. G. 1973. On the functional relationship between the dorsal and ventral divisions of the cochlear nucleus of the cat. Exp. Brain Res. 17: 428–442.
4. GRASTYÁN, E. and VERECZKEI, L. 1974. Effects of spatial separation of the conditioned signal from the reinforcement: A demonstration of the conditioned character of the orienting response or the orientational character of conditioning. Behav. Biol. 10: 121–146.
5. HARRISON, J. M. and BRIGGS, R. M. 1977. Orientation and lever responding in auditory discriminations in squirrel monkeys. J. Exp. Analys. Behav. 28: 233–241.
6. HARRISON, J. M., IVERSEN, S. D. and PRATT, S. R. 1977. Control of responding by location of auditory stimuli: Adjacency of sound and response. J. Exp. Anal. Behav. 28: 243–251.
7. KNUDSEN, E. I. and KONISHI, M. 1978. Space and frequency are represented separately in auditory midbrain of the owl. J. Neurophysiol. 41: 870–884.
8. KONORSKI, J. 1967. Integrative activity of the brain. An interdisciplinary approach. Univ. Chicago Press, p. 531.
9. KONORSKI, J. and Ławicka, W. 1959. Physiological mechanism of delayed reactions. I. The analysis and classification of delayed reactions. Acta Biol. Exp. 19: 175–197.
10. ŁAWICKA, W. 1972. Proreal syndrome in dogs. Acta Neurobiol. Exp. 32: 261–276.
11. ŁAWICKA, W. 1975. Auditory quality and location cues in instrumental conditioning. 2nd International Congress of C.I.A.N.S., Abstracts, Praga, Czechoslovakia, p. 19.
12. ŁAWICKA, W. and SZCZECHURA, J. 1979. S-R spatial contiguity vs. S-R spatial discontiguity in auditory spatial tasks. I. Acquisition by normal dogs. Acta Neurobiol. Exp. 39: 1–13.
13. POLJAK, S. 1926. The connections of the acoustic nerve. J. Anat. 60: 465—469.
14. SANTIBÁÑEZ-H., G. 1976. The targeting reflex. Acta Neurobiol. Exp. 36: 181–203.
15. SCHNEIDER, G. E. 1967. Contrasting visuomotor functions of tectum and cortex in the golden hamster. Psychol. Forsch. 31: 52–62.
16. SCHNEIDER, G. E. 1969. Two visual systems. Brain mechanisms for localization and discrimination are dissociated by tectal and cortical lesions. Science 163: 895–902.
17. TREVARTHEN, C. 1968. Two mechanisms of vision in primates. Psychol. Forsch. 31: 299–337.
18. ZIELIŃSKI, K. 1974. Changes in avoidance response latencies after prefrontal lesions in cats: group versus individual data. Acta Neurobiol. Exp. 34: 477–490.

Wacława ŁAWICKA, Nencki Institute of Experimental Biology, Pasteura 3, 02-093 Warsaw, Poland.

Lecture delivered at the Warsaw Colloquium on Instrumental
Conditioning and Brain Research
May 1979

# VISUAL ACUITY FUNCTIONS AND PATTERN DISCRIMINATION IN THE DESTRIATE CAT

J. M. SPRAGUE, M. A. BERKLEY and H. C. HUGHES

Department of Anatomy, University of Pennsylvania, Philadelphia, Pennsylvania, and Department of Psychology, Florida State University, Tallahassee, Florida, USA

*Abstract.* The study provides evidence that the cat is capable of processing visual information of some detail after removal of area 17 and most of area 18. This postoperative discrimination of complex spatial stimuli is not based on the use of such cues as luminance differences or local flux cues. From the deficits which follow extensive lesions of areas 17-18, it appears as though these cortices participate in the detection of fine details. We found a modest increase in the threshold of grating acuity, moderate loss in orientation acuity and extensive deficit in a task requiring topographic alignment of contours (vernier offset). Further evidence for the substantial preservation of spatial vision in the destriate cat is provided by experiments which show that perceptual grouping of rectilinear arrays of figural elements (dots or line segments) into obliquely oriented rows is largely unaffected by the lesion, even when the grouping is initiated by near-threshold proximity cues. Since grouping effects are felt to be involved in the organization of the visual field into the figure-ground dichotomy, these results indicate that the neural mechanisms subserving the initial stage of form perception lie outside of areas 17 and 18. Consistent with the increased acuity thresholds of destriate cats, these animals have deficits in several pattern discriminations that require fine-grained spatial analysis. The results suggest that areas 17-18 serve as a high-spatial frequency analyzer, but are not essential to pattern and form recognition.

## INTRODUCTION

The retinotopic projection and receptive field organization found in the geniculostriate system have strongly suggested that it plays a major role in processing spatial vision information (14, 15, 27, 35, 36, 46). In functional studies of the geniculo-striate system in man, monkey, and lemur, the behavioral findings are consistent with the anatomical and electrophysiological data in that loss of the geniculo-striate system in these primates produces severe and permanent deficits in spatial vision (4, 24), and visuomotor behavior in general (23, 38, 50).

In the cat, although the electrophysiological findings are similar to those observed in monkeys and other primates (1, 14, 35), the functional analysis of the geniculo-striate system has yielded quite different results from those observed in primates. Thus, ablation of area 17, or 17 and 18 in the cat, has relatively little effect on the cat's visuomotor repertory or the ability to discriminate complex patterns and shapes (25, 59, 61). The explanation for this surprising difference between cat and primate is not known, but a comparison of the anatomical organization of the cat and monkey visual systems suggest several plausible hypotheses.

It is now known that in both species the amount of cortex devoted to vision is extensive and consists of multiple areas containing separate representations of the visual field in addition to the classic striate cortex projection (2, 49, 52, 66, 67). Visual input reaches these multiple areas via two major thalamic routes: the dorsal lateral geniculate nucleus ($LGN_d$) and/or several divisions of the pulvinar nuclear mass (PL).

In the monkey, the geniculocortical system ends exclusively or chiefly in area 17 (37, 73), while the tectorecipient zones of the pulvinar (TPL) project chiefly to 18 and 19 (5). Removal of striate cortex (area 17) results in marked, retrograde atrophy of all parts of $LGN_d$, presumably rendering it functionally inactive, with no effect on TPL (23, 38, 50).

In the cat, on the other hand, both $LGN_d$ and TPL project to cortical areas 17, 18, 19 and parts of the lateral suprasylvian cortex (LSA) (20, 27, 29, 34, 43, 46, 52). Thus, removal of only area 17, or 17 and 18, leaves intact parts of both thalamic visual systems, and could account for the substantial degree of spatial vision which survives in the cat after this extensive lesion. Addition of area 19 to this lesion results in several deficits in pattern and form discrimination (58, 74).

The contribution of each of the individual thalamocortical pathways to various aspects of visual behavior is still not well understood, but the new maps derived from anatomical and physiological findings provide additional experimental possibilities for the analysis of the function of these pathways. Thus, properly delineated, the contribution of individual subsystems may be evaluated by selectively ablating them or isolating them by removal of the adjacent pathways. An additional prerequisite in the analysis of individual pathways is the functional isolation of the pathway. This may be achieved by choosing appropriate stimuli which can only be processed (or best processed) by one pathway. The use of psychophysical threshold testing procedures, in which one stimulus dimension is systematically varied, fulfills the requirement of functional isolation. Thus, threshold testing permits specifying the stimulus dimension that the animal is using to make the discrimination, and requires the organism to use the neural system most sensitive to the dimension being varied.

The present study compares the effects of selective ablation of (i) areas 17 and 18 or (ii) visual areas beyond area 17-18 on a variety of visual capacities. In the *first part* of this paper, the effect of these lesions on a series of visual acuity problems was examined. These problem included (i) grating acuity, (ii) topographic alignment (vernier offset) and (iii) orientation acuity. In addition, a form discrimination was used.

The *second part* describes the performance of normal and de-striate cats on a higher-order aspect of spatial vision: the ability to perceive orientation in patterns composed of rectilinear arrays of dots or oriented line segments. The percept of obliquely oriented rows in the patterns used in this study is an example of the Gestalt principle of perceptual grouping according to the spatial proximity of the elements. This process has been regarded as fundamental to the organization of the visual field into the figure-ground dichotomy, one of the first stages in pattern and form perception.

PART I. ACUITY EXPERIMENTS

*Methods*

*Subjects.* Fourteen randomly bred adult cats, 8 males and 6 females were used. Cats were housed in individual cages in the animal colony and during testing were kept at 80% of their ad lib weight. They were maintained on a diet of dry cat chow, vitamin supplements and the small amounts of beef puree received in the testing apparatus.

*Apparatus.* The test apparatus (Fig. 1) consisted of a two-choice discrimination chamber described in detail elsewhere (6). The test chamber is a box (50 cm × 37.5 cm × 25 cm) with a small, round opening at one end. Covering this opening from the outside is small plexiglas cylinder which has two transparent plexiglas response keys at its outer end. Below the keys is a pneumatically-operated reward terminal which dispenses small portions of a pureed-beef baby food (8). The distance between the centres of the response keys is approximately 10 cm, but only 18.8 cm separates their medial edges. Stimuli to be viewed are placed at varying distances behind the nosekeys. The cats are trained to sit or stand in the chamber, extend their heads into the plastic cylinder, inspect the stimuli and make choices by pressing the left or right response keys with their noses. Data collection and stimulus manipulations were performed by a specially adapted PDP-8/1 computer interfaced with several stimulus display devices.

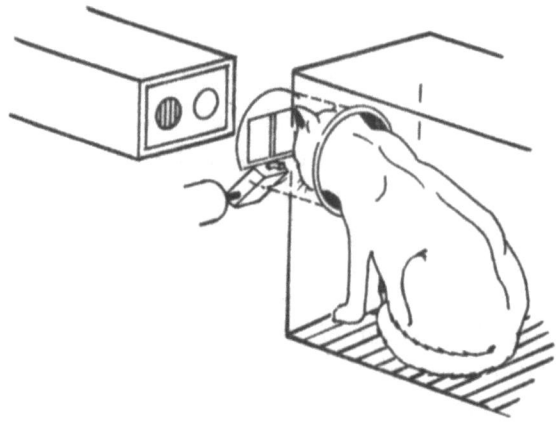

Fig. 1. Schematic representation of two choice simultaneous discrimination testing apparatus employed in testing the visual acuity of cats.

*Training program.* Each animal received 200 trials/day of discrimination testing. A test trial consisted of the presentation of a pair of visual stimuli. If the key pressed was on the same side as the stimulus designated as correct, a small food reward was delivered and a four-second intertrial interval ensued before the next stimulus presentation. Responses to the stimulus designated as incorrect extinguished the stimuli and lengthened the intertrial interval to eight seconds. Consecutive errors increased the intertrial interval to a maximum of 12 s. To encourage the cat to examine the stimuli before responding, a one-second mandatory observation interval was used. This observation contingency required that no responses occur within one-second following stimulus onset. Responses during this interval did not extinguish the stimuli nor activate the feeder, but simply reset the one-second timer and all animals quickly learned not to respond during this interval.

The left-right position of the stimuli followed a random sequence with several minor constraints. The first constraint was that no more than four stimuli be programmed consecutively on the same side. The second constraint concerned the animal's behavior. If 8 consecutive responses were made to the same side, a position habit was considered to have occurred. The program then insured that the correct stimulus appeared on the side opposite to the preferred position response until two consecutve responses were made to that side. At that time, the program returned to its random presentation mode. This procedure helped to prevent development of perseverative position habits in both the initial training and in subsequent testing.

In all cases, the animals were considered to have learned the discrimination problem when they achieved a criterion of either 90% correct in 200 trials (one day) or three days (200 trials/day) of achieving between 85 and 90% correct choices.

*Test stimuli.* The test stimuli used preoperatively fell into two categories: (i) form or shape stimuli, and (ii) threshold detection stimuli.

Form stimuli. The form stimuli were displayed on a small, closed-circuit television monitor and consisted of the following stimulus pairs: vertical versus horizontal bars, circle vs. plus, upright vs. inverted outlined triangles, and polygon outlines vs. circle.

Threshold estimation stimuli. The acuity stimuli consisted of three different types of targets: (i) gratings, (ii) vernier offset lines, and (iii) differently oriented lines.

Gratings. Gratings with square-wave luminance profiles of varying spatial frequencies (one light-dark line-pair per unit of visual angle) were generated on a Tektronix 602 display oscilloscope using a modification of the technique described by Campbell and Green (21). The input to the display oscilloscope was so arranged that high frequency gratings of at least 12 cycles/deg were generated on one-half of the display while a lower frequency grating was generated on the other half. Since, at the luminance employed (44 $cd/m^2$), the high frequency grating was above the cat's estimated grating threshold, it appeared to the cat as a uniform field of equal average luminance as that of the test grating. The spatial frequency of the low frequency half of the grating display could be varied electronically.

The display oscilloscope was placed on a horizontal rail so it could be moved to various distances behind the response keys. Several small tungsten filament lights were attached to the sides and top of the test chamber (arranged so they did not illuminate the target gratings) to provide 75 $cd/m^2$ ambient light to maintain a state of moderate light adaptation.

Vernier offset. To test vernier acuity, two vertical line targets were generated on a display oscilloscope behind the circular apertures. One of the targets consisted of a continuous vertical straight line 5° in length while the other stimulus consisted of two vertical lines whose ends were offset from each other where they met in the middle of the 5° apertures. The width of the lines was 10'. Threshold vernier offset was taken as that horizontal offset between the two halves of the broken line which the cat could not distinguish from the single, unbroken vertical line.

Orientation acuity. In this task, the stimuli consisted of two pairs of lines, also generated on the display oscilloscope, and presented within the 5° apertures of the visual display. One stimulus of the display consisted of a pair of lines that were parallel while the other stimulus consisted of a pair of non-parallel lines. One line of the latter set was parallel to the original pair while the other line was not (see Fig. 8). The orientation of the nonparallel line could be changed electronically and threshold was taken as that angle of the nonparallel line (relative to the parallel line) which the cat could not distinguish from the parallel pair of lines.

Angular acuity was measured in another way using an outlined circle and one of a series of outlined polygons with number of sides increasing from 3–12. These forms, generated by computer, were of equal circumference and were made up of line segments 10' in thickness. These targets were presented via the TV display system and used for form discriminations.

The stimuli for all animals used in this study, with some exceptions as noted in the postoperative training section, were placed at a viewing distance which was greater than the measured near point of accommodation of the cat (18). Thus, the targets minimally were placed at 18 cm but in some cases were moved as far as 54 cm from the cat's eye. In some cases, the stimuli were placed at close (6–12 cm) distances for initial postoperative tests but were returned to preoperative testing distances for threshold measurements.

*Preoperative testing.* Animals that were trained on gratings were first given grating discrimination training with a low frequency grating (0.3 c/d) and a homogeneous field of equal average luminance. When they reached criterion performance, they were presented with higher spatial frequencies. Two methods were used to determine grating acuity thresholds: (1) a modified tracking procedure in which the spatial frequency of the low frequency grating was increased if the cat made a correct choice and was decreased if the cat made in incorrect choice (with this procedure any three-minute period during which a plateau of responding, i.e., an equal number of correct and incorrect responses were made, was taken as an estimate of threshold); and (2) a frequency-of-seeing method derived by plotting the percentage of correct choice on different training days using different spatial frequencies. The 56% correct point was taken as an estimate of threshold. Further details of the training methods and threshold estimating procedures may be found in Bloom and Berkley (18).

In addition to the tests described above, all animals were given a general neurological evaluation and several were tested in a two-choice runway (59).

*Surgery.* Lesions were placed in cats deeply anesthetized with sodium pentabarbital using strict aseptic precautions. Surface landmarks as described by Otsuka and Hassler (48), Sanides and Hoffmann (54), Tusa, et al. (68), Tusa (66) and Palmer, et al. (49), were used to guide the placement of the lesions which were made by gentle subpial aspiration. All animals were permitted a 30-day recovery period before postoperative retesting was begun.

*Postoperative testing.* All postoperative acuity testing was preceded by testing on the light-dark discrimination task. When this test was completed, the animals were evaluated for any changes in threshold; tests were given in the same order as used preoperatively.

In addition to these tests, all animals were tested postoperatively on a variety of visuomotor tasks such as visual placing, visual localization of objects, visual tracking of moving objects, jumping across gaps and extent of visual fields as measured by a perimetry test (56, 63).

*Reconstruction of lesions.* The extent of each cortical lesion was reconstructed in detail by projection-drawings of selected coronal sections through the cortex and the thalamus stained with cresyl violet and Mahon or Woelcke-Heidenhein (44, 54) (see 61 for details). Topography and terminology of the thalamic nuclei were taken largely from Niimi and Kuwahara (45) and of the dorsolateral geniculate from Laties and Sprague (40), Niimi and Sprague (46), Guillery (30), and Hickey and Guillery (32).

## Results

*Extent of lesions.* The extent of 7 lesions is plotted in surface views in Figs. 2 and 3. Striate lesions were made in four cast: in two of these (Zelda, Scarlett) the lesions were large, and involved most of the areas 17 and 18; in the other two (Francis, Streak) the lesion was chiefly in area 17 and involved 18 only in and near the representation of the vertical meridian (Fig. 2). Representative coronal sections through the lesions as well as the resultant retrograde degeneration in $LGN_d$ for Scarlett and Streak are shown in Fig. 4 through 6.

In 3 cats, the lesions were placed outside of areas 17-18 (Fig. 3): in Arlo, the lesion was large, approximating the volume of tissue removed in the 17-18 cats, and included areas 19, 21, 7 and part of 20. In Earl and BJ the lesions were smaller, involving chiefly areas 19 and 21.

*Visuomotor behavior.* As reported previously, none of the cats with area 17–18 lesions showed any lasting deficits in general neurological tests (25, 59, 61). These animals all showed normal responses in the following:

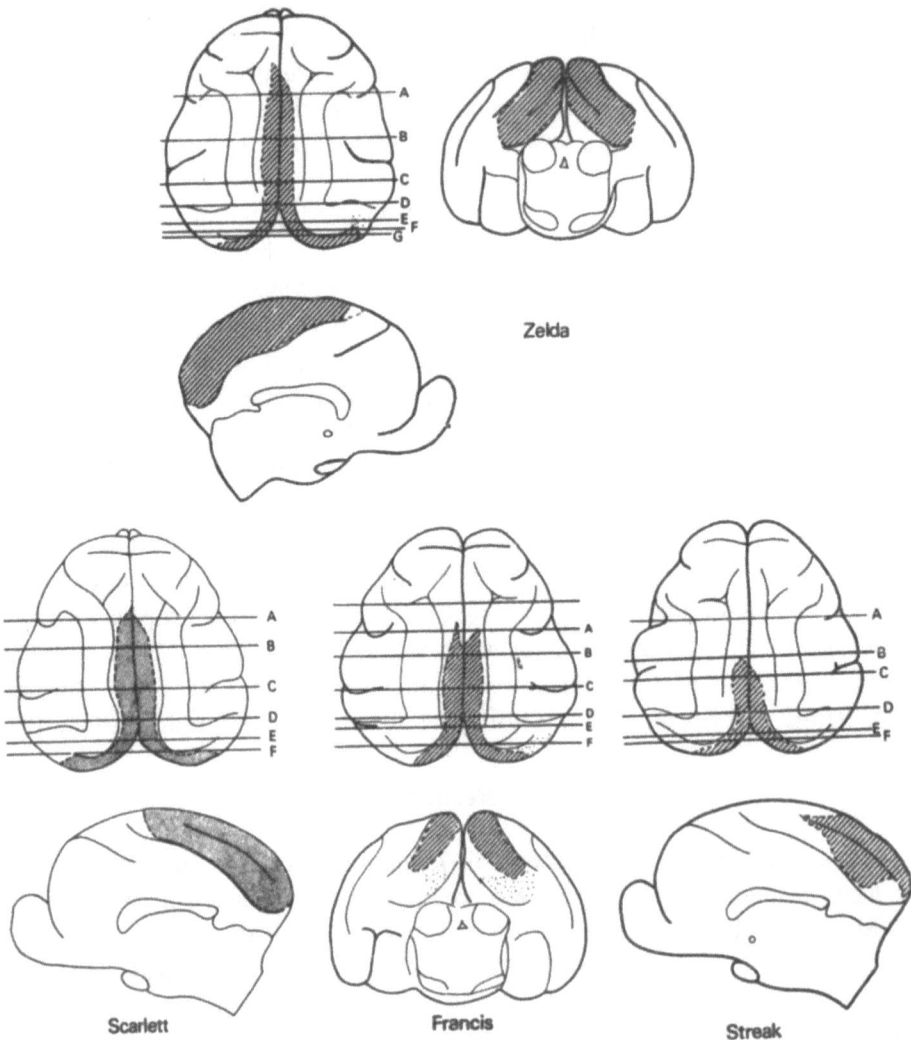

Fig. 2. Surface drawings of the brains of 4 cats showing the extent of the lesions of areas 17–18. Each drawing was modified to show the pattern of gyri and sulci found in individual animals. Dashed lines surround the aspirated areas; free stipple indicates intact cortex which was undercut and denervated. Dorsal and medial views are shown for cats Scarlett and Streak; dorsal and posterior views are shown for cats Zelda and Francis. Lettered lines drawn through the dorsal views of the brains indicate the levels of coronal sections selected for reconstruction of the lesions.

1) blink to threat;
2) visual placing to home cage, edge of table, prongs;
3) accurate localization of stationary food objects (cat chow, spleen) on homogeneous (white) or textured (brown) backgrounds;
4) accurate reaching for and snagging of moving food objects with forelegs;
5) tracking of moving food objects with head and eyes;
6) brisk eye movements of full amplitude in all directions;
7) good depth judgment in jumping up and down and across gaps of up to 3 feet; and
8) full visual fields (measured chiefly in the horizontal meridians), binocular and monocular, using food perimetry.

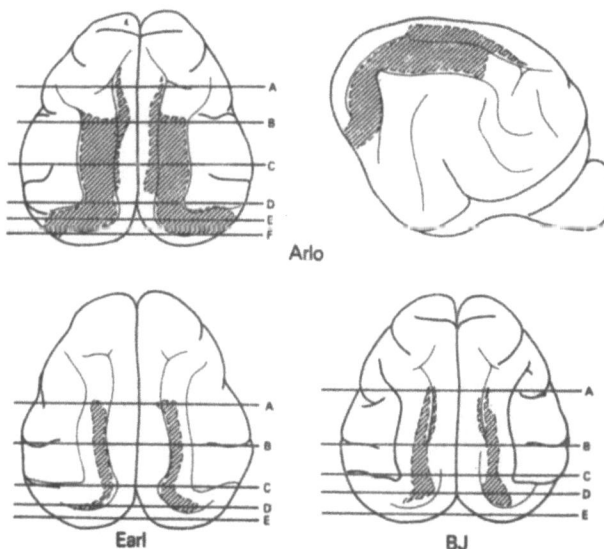

Fig. 3. Surface drawings of the brain of 3 cats showing extent of lateral cortical lesions. Lateral view is also shown for cat Arlo. Details same as in Fig. 2.

The performance on many of these tests strongly suggests that these animals are using the central visual representation present in intact cortex (areas 19, 20, 21 or LSA) and/or superior colliculi and are not operating on peripheral vision mediated by small, residual islands of tissue in area 17 or 18.

*Acuity tests.* Grating acuity. The cats tested on this task had preoperative grating acuity threshold ranging from 3.5 c/deg to 6.0 c/deg (Fig. 7); all were within the range of variability observed in normal cats (18). (Thresholds have been estimated from these curves by taking the intercept of the functions at 56% correct.).

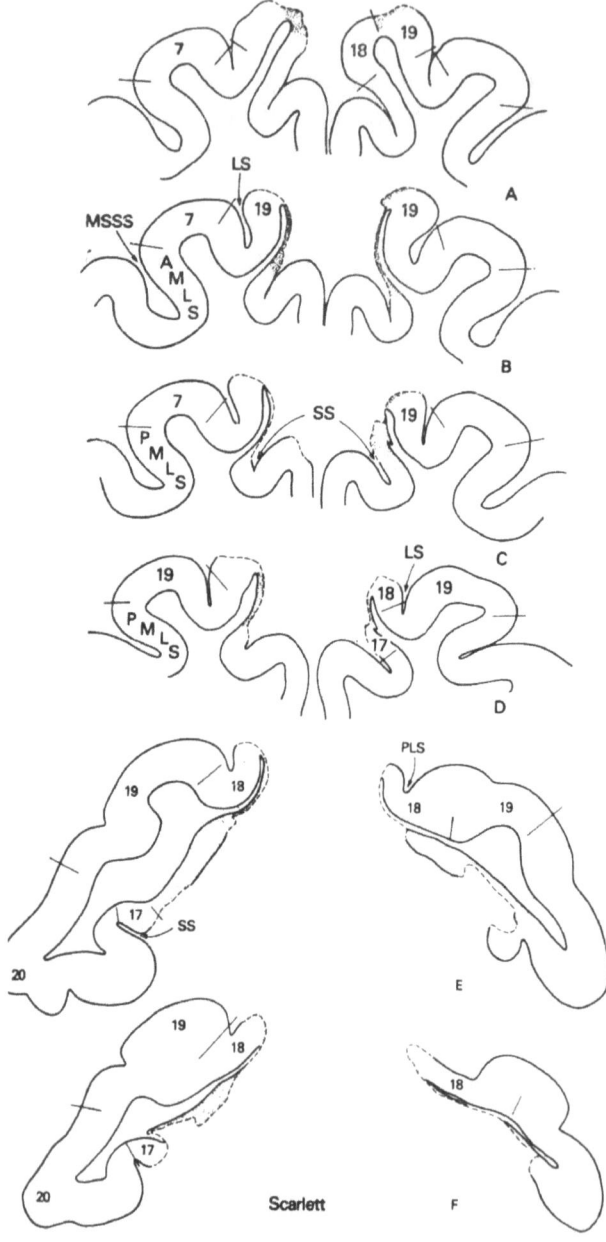

Fig. 4. Selected coronal sections of the cortex of cat Scarlett, taken from the levels indicated by lettered lines in Fig. 2, to show the extent of the lesion. Dashed lines indicate edge of lesion; stipple indicates atrophy and denervation; numerals denote extent of cortical areas delineated anatomically and physiologically (see 61). AMLS, anterior medial lateral suprasylvian area; LS, lateral sulcus; MSSS, medial suprasylvian sulcus; PLS, posterior lateral sulcus; PMLS, posterior medial lateral suprasylvian area (see 49); SS, splenial sulcus.

Nearly complete lesions of areas 17 and 18 in cats Scarlett and Zelda produced a reduction in grating acuity of 33% and 20%, respectively. The results for these two animals are shown graphically in Fig. 7A and 7B. Note the normal postoperative performance levels at low spatial frequencies.

The deficits seen in Zelda and Scarlett were unique to animals with area 17 and 18 ablations. Ablations of comparable size in more lateral cortical regions leaving areas 17-18 intact did not produce a reduction in

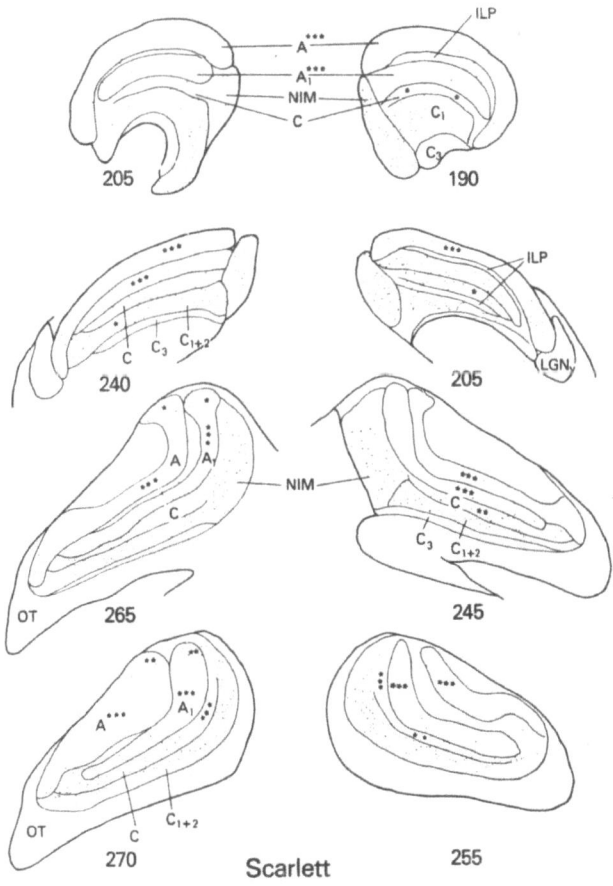

Fig. 5. Sections through the thalamus of cat Scarlett show extent of retrograde atrophy in the dorsal lateral geniculate nucleus ($LGN_d$). The laminae are identified by letters A, A1, C, C1, C2, C3, NIM. Intact cells are shown as dots, degree of atrophy is shown as one asterisk, minimal, two asterisks, moderate, and three asterisks, severe. Numerals are those of individual sections taken in rostro-caudal direction (190-270); since the brain was cut slightly asymmetrically these do not match on the two sides: ILP, posterior interlaminar leaflet $LGN_v$, ventrolateral genicular nucleus, OT, optic tract.

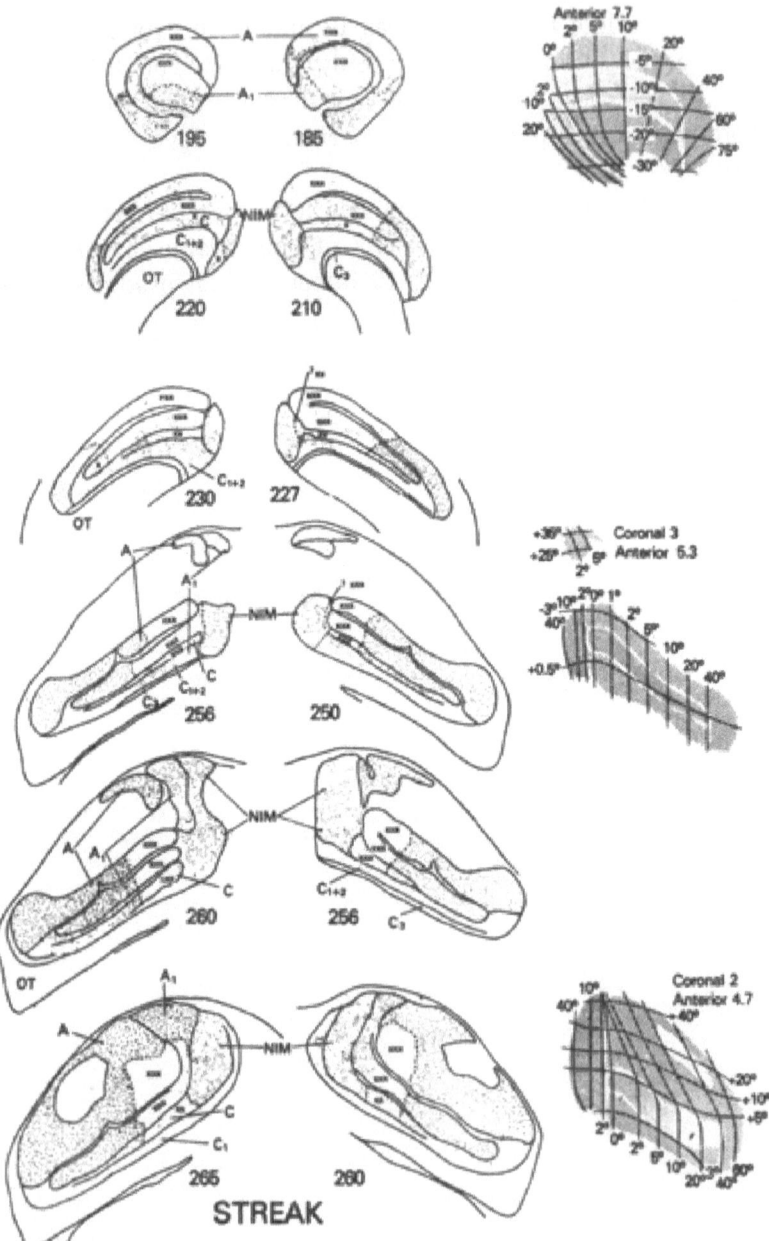

Fig. 6. Coronal sections through $LGN_v$ of cat streak showing extent of retrograde atrophy. Symbols and abbreviations as in Fig. 5; the ? in levels 227 and 250 indicate doubt in identification of atrophied area as part of NIM or adjacent laminae. The three drawings on the right are modified slightly from Sanderson (53) and show the representation of the visual field on $LGN_d$, derived electrophysiologically. Comparison of the middle of these with the adjacent drawing from Streak indicates that $LGN_d$ at a level of area centralis shows laminar atrophy out to 2° eccentricity on the right and 5° on the left. The peripheral parts of these laminae and all of NIM are essentially intact.

grating acuity. An example of this finding is shown in Fig. 7D in which the performance of a cat (Arlo) with a lateral cortical lesion (Fig. 3) is shown. Note that pre- and postoperative performances are essentially identical, i.e., the lesion had no effect on acuity.

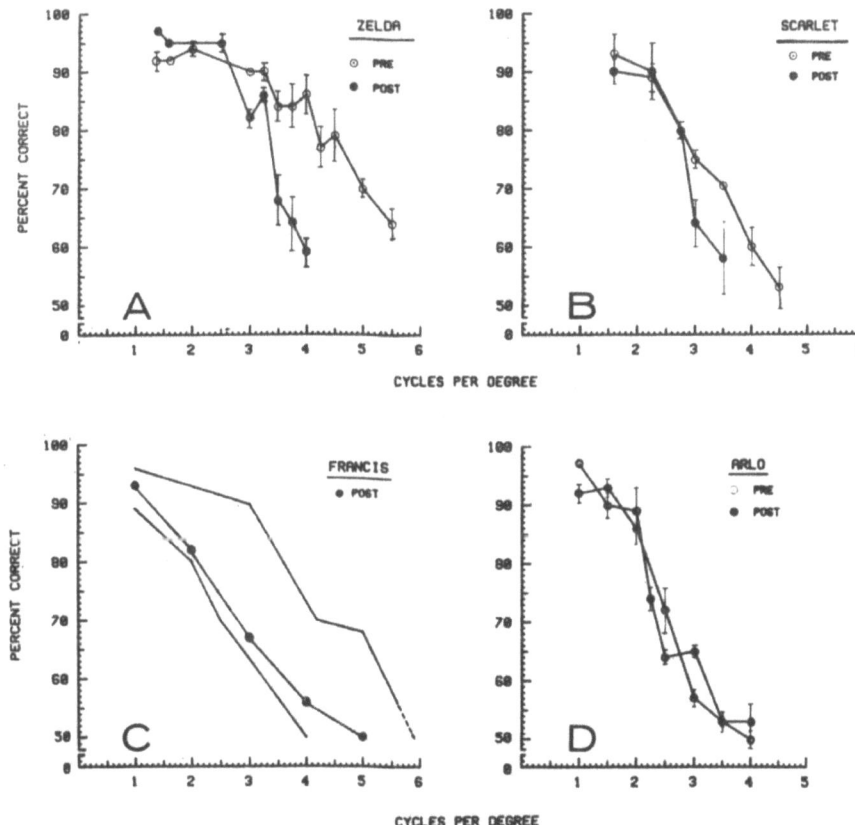

Fig. 7. Pre- and postoperative performance on the grating acuity task. Percentage of correct choices shown on ordinate; grating size in cycles/deg shown on abscissa. Open symbols-preoperative performance; filled symbols-postoperative performance; bars indicate ± 1 standard deviation. Points without vertical bars had SD too small to plot. A, Cat Zelda with large area 17–18 ablation; B, Cat Scarlett with large area 17–18 ablation; C, Cat Francis with smaller area 17–18 ablation tested only postoperatively. Shaded portion of figure depicts range of preoperative values based on 6 cats observed in this study; D, Cat Arlo with ablation of areas 19, 20, 21, LSA (AMLS, PMLS, VLS).

Ablation of area 17–18 also does not preclude the acquisition of a grating discrimination. One animal (Francis) was not trained and tested on grating stimuli until after the area 17–18 ablation was made. The results for this cat are shown in Fig. 7C. Again, note the normal performance levels with low frequency gratings. The threshold acuity for this animal is 4.0 c/deg and is in the low-normal range. For comparison, the shaded area on Fig. 7C depicts the preoperative performance range of 6 cats used in this study.

Line orientation acuity. The ability of normal cats to discriminate

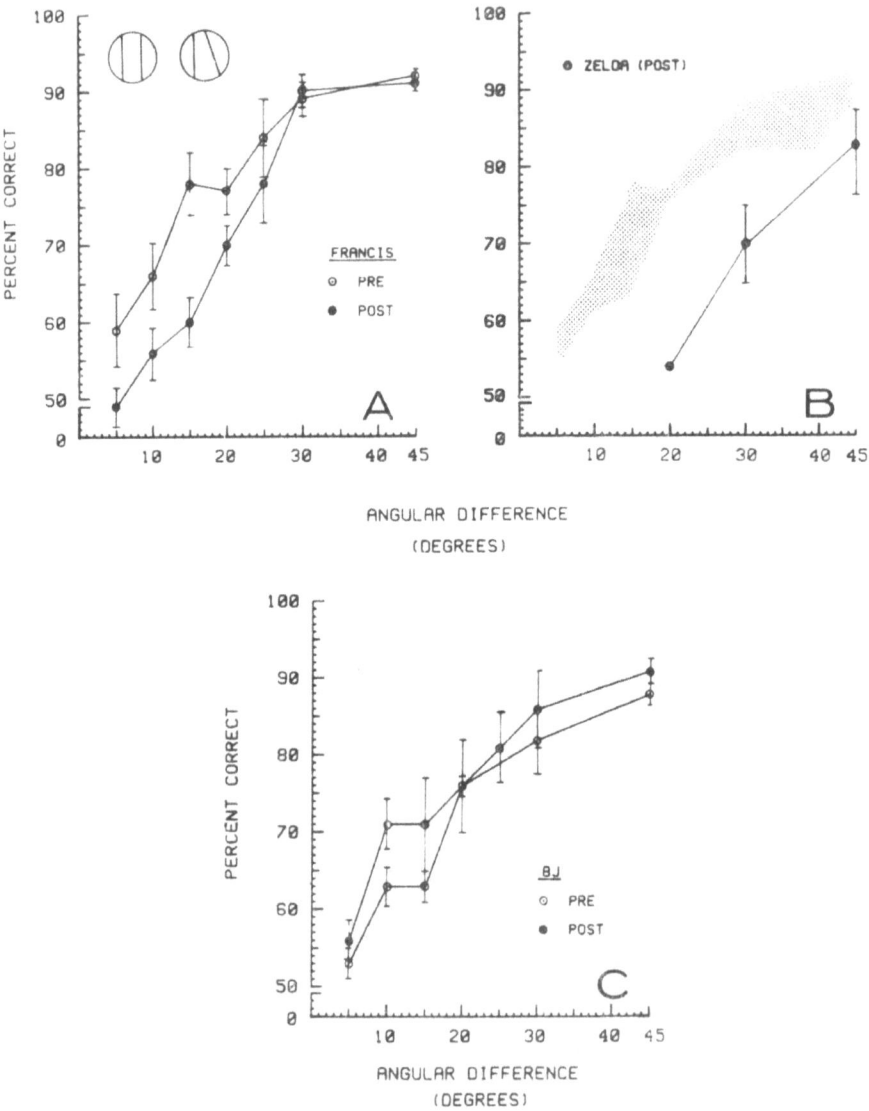

Fig. 8. Pre- and postoperative performance on the line orientation acuity task. Open symbols depict preoperative performance while filled symbols depict postoperative performance. A, Cat Francis with small area 17–18 ablation; B, Cat Zelda with large area 17–18 ablation, only postoperatively tested. Shaded area indicates range of normal (preoperative) performance based on 3 cats observed in this study; C, Cat BJ with area 19–21 ablation. The stimulus configuration used in this test is shown at the upper left of A.

relative line orientations (e.g., deviation from parallelness) is shown in figures 8A-C. The stimulus configuration used in making these measurements is shown at the top left of Fig. 8A. The normal (preoperative) performance on this task is shown by open circles. Note that when the angle between the nonparallel line and the parallel lines is about 30°, the cats perform almost without error. As the angular difference is decreased, performance falls toward chance levels. Threshold angular deviation (angular deviation at 56% correct choices) was found to range from about 3° to 8° in 4 cats. Postoperatively, animals with area 17-18 lesions (Francis and Zelda, Fig. 8A and 8C) showed that they could still discriminate the test stimuli at preoperative levels when large angular deviations were presented but showed a significant increase in threshold. Thus, threshold increased from 3° to 12° for Francis, and was about 20° for Zelda, tested postoperatively only. Since the latter subject had not been tested preoperatively on this task, the range of behavior of normal cats has been shaded in for comparison in Fig. 8B.

Ablation of cortical areas 19–21 (lateral to 17–18) produced no change in performance or threshold performance on this task, as shown by cat BJ in Fig. 8C.

*Polygon test.* As a test to determine if line orientation acuity might be related to the discrimination of geometric shapes, a second test was used in which the cats were required to discriminate small polygons with varying numbers of sides from a circle. While the areas of these line figures were slightly different depending on the number of sides, their perimeters were constant regardless of shape. Figure 9A shows the pre- and postoperative performance on this test series for a cat (Francis) with a small lesion in areas 17–18.

Note that preoperatively, performance declines regularly as the number of sides increases. Polygons of 11 or greater sides could not be discriminated from a circle of equal perimeter. Postoperatively, performance on this task was considerably reduced with all polygons used and the threshold was reached near the 6-sided figure. Compare this deficit with the grating acuity deficit (Fig. 7C) and angular acuity (Fig. 8A) observed in this animal.

A second cat with a smaller ablation of areas 17–18 (Streak) is shown in figure 9B; vertical meridian and central 5° were removed in the representation of area centralis (Figs. 3 and 6). Since this animal was not tested preoperatively on this task, the range for normals ($N = 3$) has been shaded in for comparison. Note the impairment in performance for this animal is similar to that of Francis. Also shown on this figure is the postoperative performance of a cat (Earl) with an ablation lateral to area 17–18, including much of area 19 (Fig. 3). Note that this animal's performance falls clearly within normal range. A third animal with similar lateral lesion (cat BJ) is shown in Fig. 9C. This cat shows a small (nonsignificant) reduction in postoperative performance, but the threshold for discrimination is the same as preoperatively.

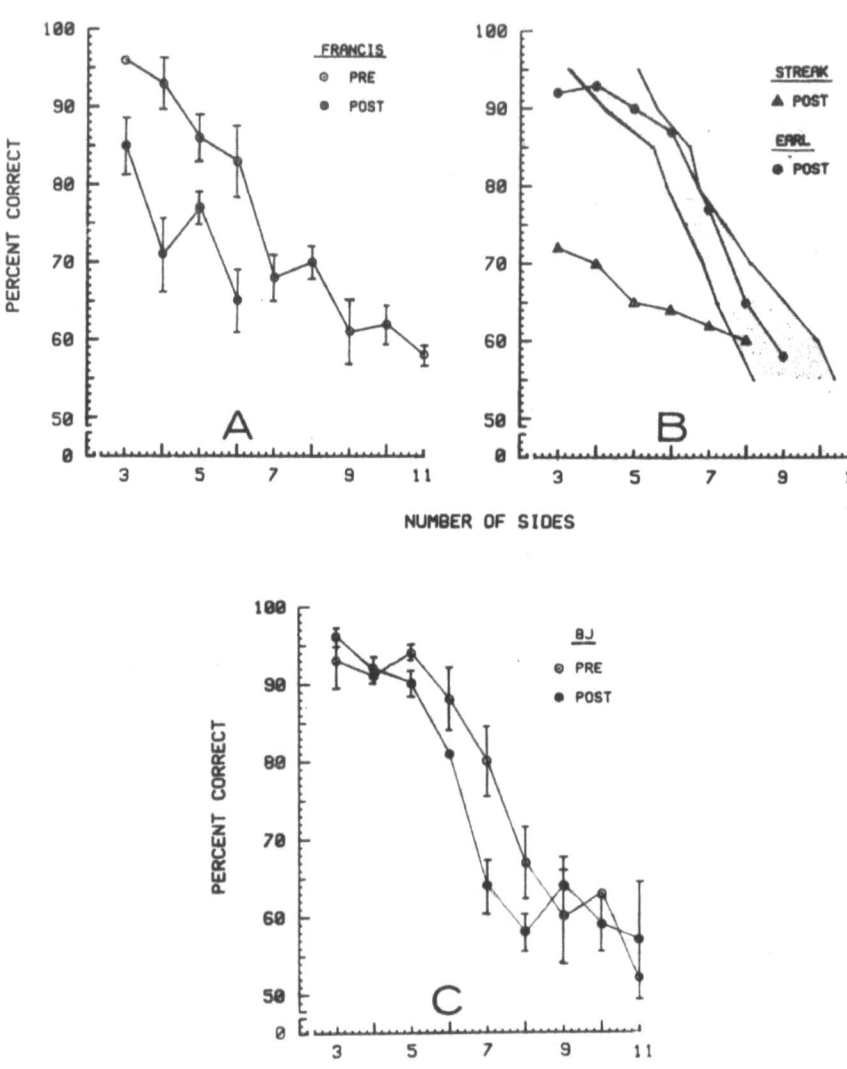

Fig. 9. Pre- and postoperative performance on the polygon discrimination task. Symbols and ordinate as in Figs. 7 and 8. A, Cat Francis with small area 17-18 ablation; B, Cat Streak with small area 17-18 ablation and cat Earl with extrastriate lesion (areas 19, 21, 7). Both cats were trained and tested on this task postoperatively only. Shaded area indicates normal performance levels based on 3 cats. C, Cat BJ with area 19-21 ablation.

Comparison of the results of the polygon tests with those of line orientation acuity suggests that the two are related. It is also likely, however, that some of the postoperative performance deficits seen in Francis and Streak with area 17–18 ablations are related to grating acuity deficits since the thickness of the lines (10') used in the polygon tests are probably near the acuity limit of these animals.

Vernier acuity. Six animals were tested on the vernier offset task. In two normal cats (Streak, Earl), a threshold offset of 5' was obtained. Three cats were tested (Streak, Francis, Zelda) after 17-18 lesions. Streak showed a massive postoperative deficit, going from a preoperative threshold of 6 min of offset (open circles) to essentially no ability to discriminate the targets postoperatively (Fig. 10, filled circles). Zelda and Francis were tested postoperatively only. Francis had a threshold of about

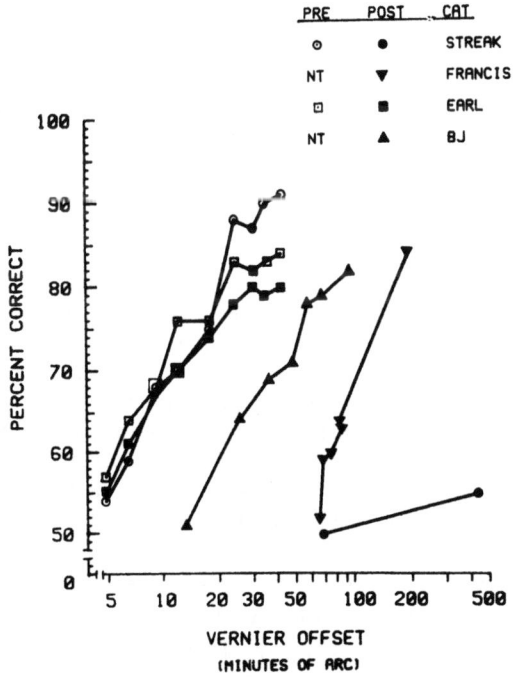

Fig. 10. Pre- and postoperative performance on the vernier offset task. Open symbols indicate postoperative performance. Cats Streak and Francis had small area 17–18 ablations; cats Earl and BJ had ablations of areas 19 and 21. Note that Francis and BJ were not tested on this task preoperatively.

1° (Fig. 10, inverted triangles) and Zelda was never able to learn the discrimination even though she showed only a moderate postoperative impairment on the grating acuity tests (see Fig. 7A).

Two animals with lateral lesions were also tested — Earl and BJ. Earl showed no postoperative deficit (Fig. 10, open and filled square) while BJ (inly tested postoperatively) was somewhat poorer than normal (upright triangles).

Form vision. To compare the visual capacities of the animals used in this study with those used in previous studies, we used a form discrimination task (0+) that has often been used in other studies. In the present study, the polygon tests have already shown that all the animals could discriminate at least some of the figures used in that task. The 0+ test showed that animals with area 17–18 ablations could learn this di-

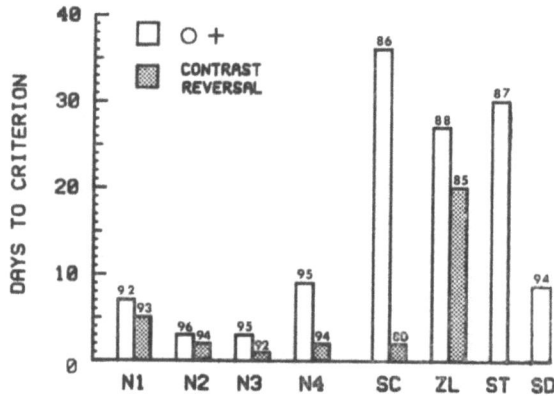

Fig. 11. Histogram showing number of days needed to achieve criterion or asymptotic performance on a form discrimination task for normal and brain-damaged cats trained postoperatively. Days to criterion shown on ordinate. Individual animals are listed on abscissa. N1–N4 are normal cats; SC, Scarlett; Z1, Zelda; ST, Streak and SD, Sid, had ablations of area 17–18. Open bars depict performance with black line stimuli on a white background; shaded bars show performance when contrast is reversed. Numbers at top of bars indicate terminal performance level (percentage of correct choices). The discriminanda were <7° in diameter, and their outlines were 1.5°–1.8° in thickness.

scrimination postoperatively but, for most animals, the rate of acquisition was somewhat slower than normal. A summary of these results is shown in Fig. 11, which shows, in histogram form, the number of days needed to reach criterion by 4 inexperienced normal animals (N1–N4) and four animals with area 17–18 ablations (SC, ZL, ST, SD). Note that one cat with a 17–18 ablation fell within the normal range. As a control for local flux cues, the contrast was reversed after criterion was reached. As can be seen, this maneuver had little effect on one animal (SC) and a great effect on another (Zelda). This is typical even in normal animals (see N1 and N4). From these results we conclude that animals with area 17–18 ablations can learn discriminations based on small forms (7°; line thickness 1.8°) but are likely to be slower than normal in doing so.

PART II. PERCEPTUAL GROUPING AND PATTERN DISCRIMINATION

## Methods

*Subjects.* Four adult cats, 3 females and 1 male, served as subjects.

*Apparatus.* The standard two-choice discrimination box has been described in detail previously (61). It consists of a start box and two stimulus panels placed at opposite ends of an alley 85 cm long. The animals were trained to run down the alley and to choose the one stimulus panel on which was displayed a pattern arbitrarily designated as correct. The visual stimuli were rear-projected onto glass panels mounted on the doors. The animals responded to these stimuli by pressing against the door with their noses. Correct responses would open the door, allowing the subject access to a food reward located midway between the doors on a rear platform. Incorrect responses were punished by delivery of a mild electrical shock through a grid in the runway floor. Upon committing an error, the subject still had to choose the correct alternative, although these "correction responses" were not rewarded. This general procedure will be referred to as "differential reinforcement."

At several points in the course of these experiments, subjects were presented with a series of patterns which contained potentially conflicting orientational cues. During these periods, differential reinforcement was suspended; i.e. responses to either stimulus were rewarded. This procedure is referred to as "equivalence testing."

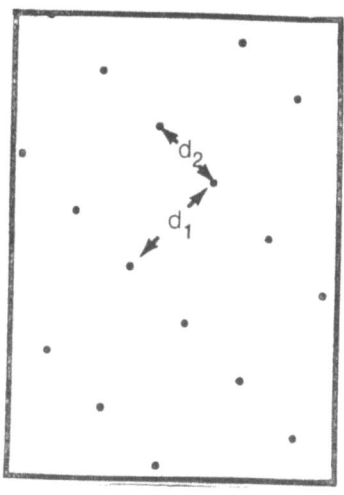

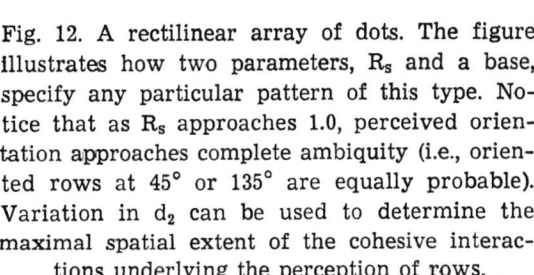

Fig. 12. A rectilinear array of dots. The figure illustrates how two parameters, $R_s$ and a base, specify any particular pattern of this type. Notice that as $R_s$ approaches 1.0, perceived orientation approaches complete ambiquity (i.e., oriented rows at 45° or 135° are equally probable). Variation in $d_2$ can be used to determine the maximal spatial extent of the cohesive interactions underlying the perception of rows.

$$\text{Base} = d_2 \text{ (in mm)}$$

$$R_s = \frac{d_1}{d_2}$$

The animals were given 50–100 trials/day (5 days/week), depending on their appetite and willingness to work. They were fed only in the discrimination box on test days.

*Test stimuli.* Four classes of patterns were used: square-wave gratings, rectilinear arrays of dots, rectilinear arrays of line segments, and randomly distributed arrays of oriented line segments, all oriented at 45° (positive) and 135° (negative) from horizontal. Each was of high contrast.

Square-wave gratings. These consisted of black and white bars 9 mm wide.

Rectilinear dot arrays. Each cat was tested on a series of these patterns, graded in difficulty. Each pair in the series represents a combination of values for the two parameters that specify this class of pattern (Figs. 12 and 13). $R_s$ is regarded as a measure of the magnitude of the proximity cue underlying perceptual grouping. Variation of the base

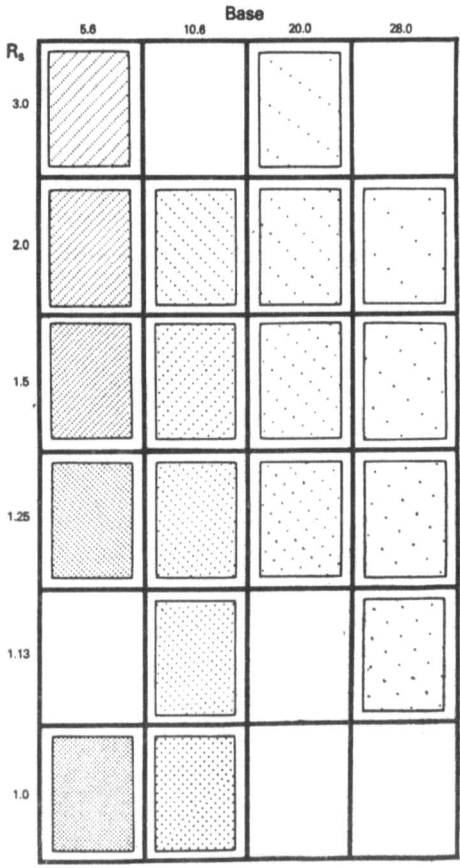

Fig. 13. The complete set of dot patterns.

($d_2$) when $R_s$ is held constant can be used to test the extent of spatial separation over which the cohesive interactions (i.e. grouping) operate. [1]

Rectilinear arrays of oriented line segments. The cats were also tested on a series of rectilinear arrays of line segments in order to determine whether changing the geometry of pattern elements influences the grouping process (Fig. 17).

Since orientation-specific contour-detection is such a prominent aspect of the receptive field organization of cortical neurons (35), it was hypothesized that these patterns might provide conflicting cues with respect to perceived orientation.

If information concerning the orientation-specific aspect of cortical cellular activity were utilized by some hypothetical neural pattern-recognizer, then perceived orientation should be strongly influenced by local geometry of the pattern elements. If on the other hand, information concerning the relative spacing of the elements provided the principle input to the pattern-recognizer, then perceptual grouping would dominate, and perceived orientation would be orthogonal to the orientation of the line segments.

Randomly distributed arrays of oriented line segments. The fourth class of patterns consisted of line segments identical to those above, but whose spatial distribution was random. These patterns were used to measure the ability of cats to detect orientation when the only cue lies in the geometry of pattern elements. Thus, discrimination data using these patterns can be regarded as a control condition meant to establish the capacity of local geometry to support discrimination performance (19).

*Possible sources or orientational cues in each class of pattern.* It will be convenient to adopt a notational scheme for identifying these patterns. The preceding considerations suggest that two levels of orientational cues might be operative. The first involves the relative spacing of the regularly aligned constituent elements, a cue which we shall refer to as a "global structural cue." The second is provided by the geometry of the elements themselves, and will therefore be referred to as a "local structural cue."

The dot arrays in Fig. 13 have global but no local structure while the randomly distributed line segments have local but no global structure. The former are therefore referred to as "structured dot arrays" while the latter are referred to as "unstructured arrays of line segments." Since the rectilinear arrays of line segments have both global and local structural cues, they are referred to as "structured arrays of line segments".

---

[1] While a specific viewing distance was not enforced in these experiments, cats usually scanned the patterns from a position near the end of a partition placed between the doors and extending 10 cm into the runway.

*Procedure.* Each cat was tested on all four types of patterns in the following sequence: (i) oblique gratings; (ii) structured dot arrays; (iii) structured line segment arrays; and (iv) unstructured line segment arrays.

Oblique gratings. All four cats were first trained to discriminate between two gratings of the same spatial frequency but different orientations (45° and 135°, 45° positive). This served to establish percepts of 45° orientation as those that were rewarded. Criterion was 90/100 trials on two successive blocks of 50 trials unless otherwise specified (Fig. 14A).

Structured dot arrays. Subjects were next tested on the series of rectilinear dot arrays. The animals were first tested for their ability to transfer their orientation-discrimination from gratings to either one of two dot-array exemplars. The two patterns for which initial transfer data were obtained are shown in Fig. 14B. Testing on each pair of dot arrays continued until the animals reached criterion or until 300 trials were given. When testing on one pair of stimuli was completed (Fig. 13 $R_s = 3.0$, base = 5.6), testing continued with a new pair in which the value $R_s$ was reduced while the base ($d_2$) remained constant ($R_s = 2.0$, base 5.6). The cats were then required to discriminate progressively more difficult arrays (i.e. $R_s$ approaches 1.0) until discrimination approached chance or util all patterns sharing the same base value had been successfully discriminated. The procedure was then repeated at successively larger bases (10.6, 20.0, 28.0) until the entire series had been presented (Figs. 15 and 16).

Structured arrays of line segments. Cats 3 and 4 were next tested preoperatively on a series of rectilinear arrays of line segments using the "equivalence testing" procedure outlined above (Fig. 17). The data were obtained using a method-of-limits procedure: 3 descending series ($R_s$ reduced in 5 steps) were alternated with 2 ascending series ($R_s$ increased in 5 steps). Within a series, 10 equivalence trials were alternated with 10 differentially reinforced trials using a difficult dot array in order to maintain differential responding.

Randomly distributed arrays of oriented line segments. Cats 3 and 4 were next tested preoperatively for transfer to a randomly distributed array of line segments to determine whether the orientation of these short contours was capable of guiding choice behavior in the absence of global cues. This discrimination was acquired under conditions of differential reinforcement (Fig. 18).

The initial training patterns contained a high density of pattern elements, but it was considered important to assess the degree to which this discrimination depended on the number of the figural elements. Thus, after criterion performance had been reached, performance on low-density patterns was measured using a modification of the method of limits: under conditions of differential reinforcement, 10 test trials on a low-density pattern were alternated with 10 trials on the original high-density unstructured array. Two descending series (numbers gradually reduced) were alternated with two ascending series (numbers gradually increased) to provide an estimate of performance as a function of element-redundancy that was based on 40 trials on each of 7 unstructured line-segment patterns, differing only in the number of constituent elements (Fig. 19).

Retest on structured arrays of line segments. When testing on the unstructured arrays was completed, cats No. 3 and No. 4 were retested for perceived orientation induced by structured arrays of line segments according to the methods previously outlined. This procedure was adopted in order to search for the possible operation of "perceptual set" effects: i.e., since the preceding problem required that attention be paid to the geometry of the pattern elements themselves, it was felt that the results of a retest might differ from the original results, in which case the difference could be attributed to the influence of a perceptual set established by training on the unstructured line segment arrays.

Tests of visuomotor behavior. Each cat was also tested for visuomotor capacity according to the methods outlined in Part I.

Surgery. Surgical procedures were identical to those described in Part I.

Cat 1 died of natural causes after preoperative testing on the set of dot arrays had been completed.

Cat 2 underwent surgery for removal of areas 17–18 after testing on the dot-arrays, and received the entire set of 4 discrimination problems postoperatively.

Cat 3 was tested on the complete set of discrimination problems preoperatively before removal of areas 17–18, but postoperative testing has not yet been completed.

Cat 4 has completed preoperative testing on all tests but has not yet come to surgery.

## Results

*Histology.* Histological data are not yet available for any of these animals, although much experience with this lesion makes it likely that areas 17–18 were completely or extensively removed as planned.

*Visuomotor behavior.* All four cats were within the normal range preoperatively, as was cat 2 postoperatively (see above).

*Preoperative discrimination performance.* Acquisition on obliquely oriented gratings. Figure 14A shows the acquisition functions of cats 1,

2, 3 and 4 for discrimination of oblique gratings. The figure illustrates the uniformly rapid acquisition of this discrimination in all four animals.

Performance on structured dot-arrays. Following a period of overtraining on the gratings, each cat was tested for transfer to one of two

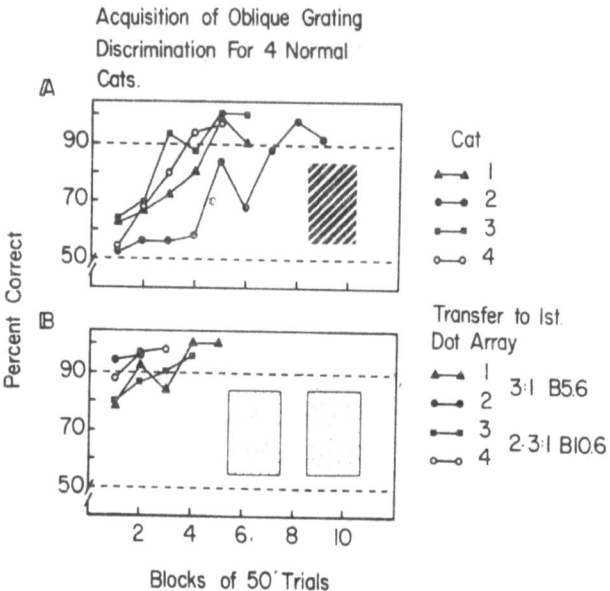

Fig. 14. A, Acquisition curves for oblique gratings discrimination for four normal cats. B. Discrimination performance on first exposure to a dot pattern for the same cats.

dot-array examplars. Transfer data has thus far been obtained on two $R_s$-base combinations — 2.28 : 1, b 10 and 3.0 : 1, b 5. Figure 14B shows the rate at which each cat performed on their initial exposure to a dot-pattern. The curves indicate virtually complete positive transfer in each case.

Upon attainment of criterion performance levels for these two arrays, each cat was tested for their capacity to transfer their discriminations to the remaining dot patterns. The results are summarized in Fig. 15, which illustrates the relationship between discrimination performance and the magnitude of the proximity cue under the various conditions of base spacing. Figure 15 indicates that performance is a direct, monotonic function of $R_s$, a trend that is perhaps clearer in Fig. 16, where the averaged data from all four subjects are plotted. Figure 16 also indicates the influence of a weak inverse relationship between performance and baseline spacing, regardless of $R_s$.

To summarize, the data indicate that perceived orientation (and, by

129

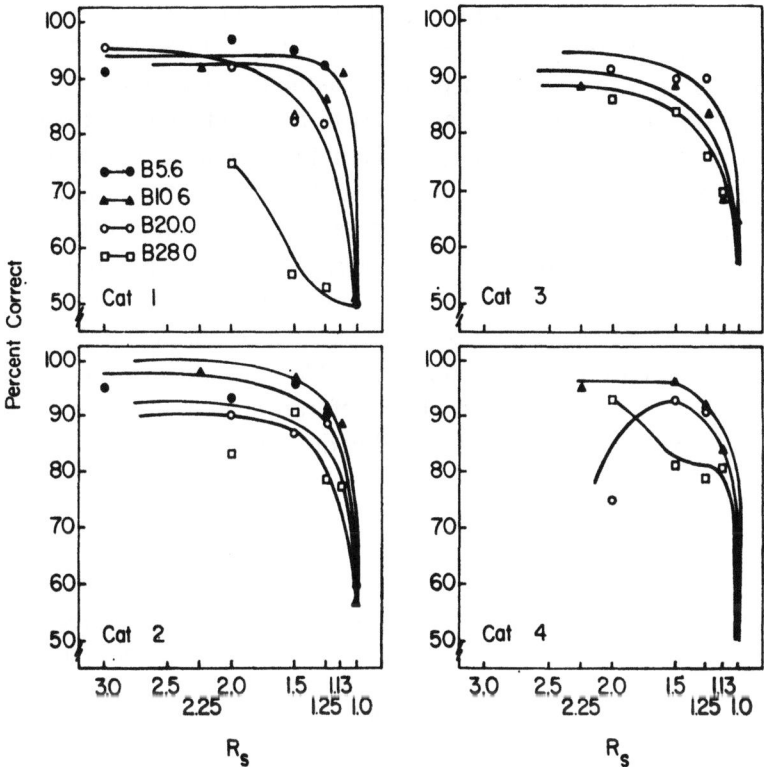

Fig. 15. Discrimination performance as a function of $R_s$ for all four cats. Each data point represents the mean performance level for the $R_s$-base combination indicated.

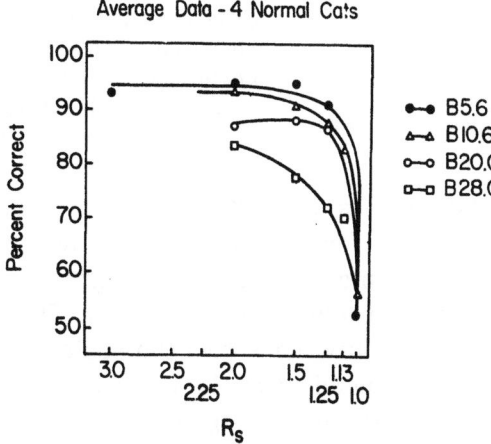

Fig. 16. Discrimination performance as a function of $R_s$-averaged data.

inference, perceptual grouping) is (i) a direct, monotonic function of the relative proximity of pattern elements and (ii) inversely proportional to the overall spacing between pattern elements. It is also clear that the transfer of training from oblique stripes to these dot patterns is excellent; that is, the grouping process is immediate, and apparently, unlearned.

Performance on structured arrays of line segments. Two of the four cats (No. 3 and 4) were tested preoperatively for perceived orientation when viewing a series of rectilinear arrays of line segments. Each pair in this series represented a different level of $R_s$, whose values ranged from 3.5-1.0. These experiments were performed immediately after testing on the dot patterns was completed. The results are summarized in Fig. 17, which is a plot of stimulus preference as a function of $R_s$. The figure indicates that, as was the case for dot patterns, stimulus preference

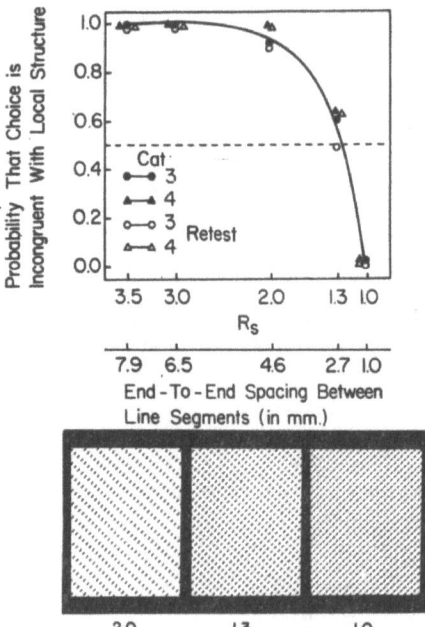

Fig. 17. The stimulus preferences of two normal cats on 5 pairs of structured arrays of line segments. The $R_s$ value differed for each stimulus pair in the set; three examples are shown. The upper scale on the abscissa is for $R_s$; the lower scale shows the end-to-end spacing between adjacent line segments. Recall that all patterns oriented at 45° are rewarded, and thus are preferred by the cats. The probability that the animals selected the pattern which contained line segments in the non-preferred orientation (i.e. 135°) is indicated on the ordinate. The filled symbols indicate the initial results; the open symbols represent the results obtained after training on unstructured arrays of line segments (see text below).

(i.e. perceived orientation) was a direct, monotonic function of $R_s$. Thus, while perceived orientation for both cats was orthogonal to the orientation of the pattern elements for values of $R_s$ greater than 2.0, it was congruent with the local structure when $R_s = 1.0$. The point where the curve crosses the 0.5 probability mark (i.e., the "point of subjective equality") is estimated to correspond to an $R_s$ value of ca. 1.25.

Performance on unstructured line segment arrays. Cats 3 and 4 were next tested for transfer to an array of 120 randomly distributed line segments. It will be recalled that the only discriminative cue in this pattern lies in the orientation of the pattern elements, i.e. in the local structure. The acquisition curves for two normal cats appear in Fig. 18. Cat 4 showed strong evidence of positive transfer from one of the previous tests, while cat 3 did not and required 750 trials to achieve criterion performance levels.

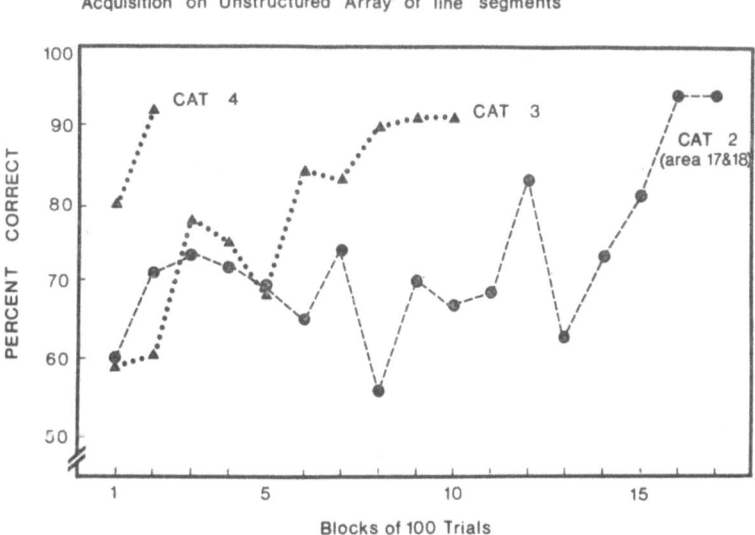

Fig. 18. Acquisition curves for two normal cats (No. 3, 4) and one destriate cat (No. 2) on a discrimination of an array of 120 randomly distributed line segments (see left example, Fig. 19).

Figure 19 illustrates the relationship between discrimination performance and redundancy in unstructured line segment arrays. For the two normal cats performance is independent of the number of segments over a substantial range, but reductions of the number below a critical level lead to performance decrements.

Retest on structured line segment arrays. Once the effectiveness of local structural cues had been established, it was of interest to determine whether such forced attention to local cues would influence perceived orientation in structured arrays of line segments. Equivalence testing on structured arrays was therefore repeated. The results of this retest are indicated by the open symbols in Fig. 17. Comparison between open and filled symbols indicates the lack of consistent "set-effect"; that is, the $R_s$ value associated with ambiguity in perceived orientation shows no consistent evidence of bias towards, or an increase saliency of local structural cues. Thus, in the terminology adopted here, *forced attention* to local cues does not alter subsequently perceived orientation when those same local cues occur in the context of a strong global structure.

*Postoperative discrimination performance (preliminary results).* Complete postoperative data is available only for cat 2. Although the preliminary postoperative results on No. 3 are consistent with these results, only the data on No. 2 will be presented. Testing was resumed 15 days after surgery.

Oblique gratings. Perfect retention of this discrimination was found, i.e. no errors in 100 trials.

Structured dot arrays. Performance of cat 2 on the set of dot-patterns in shown in Fig. 20B. This animal's preoperative data are provided in Fig. 20A to facilitate comparison. Postoperative performance on base

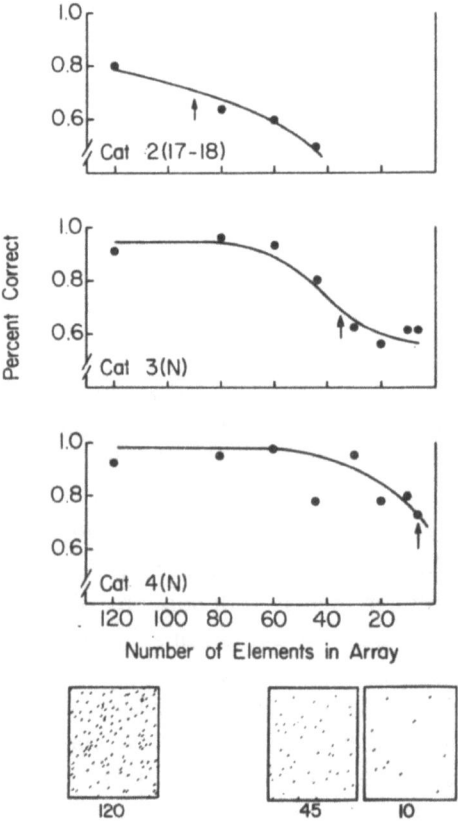

Fig. 19. Effect of redundancy in discriminating differences in local structure. Abscissa, numbers of randomly distributed line segment in the arrays. Ordinate, percent correct responses. The upper curve is the data from a destriate cat (No. 2). The lower two curves were obtained from normal animals. Insets along the bottom are examples of the patterns used.

5.6 and base 10.6 dot patterns was unimpaired compared to preoperative levels; perfect retention occurred at $R_s$ 3.0, 2.0 and 1.5, base 5.6. The animal did have more difficulty in discriminating large-base arrays (20 and 28) than preoperatively.

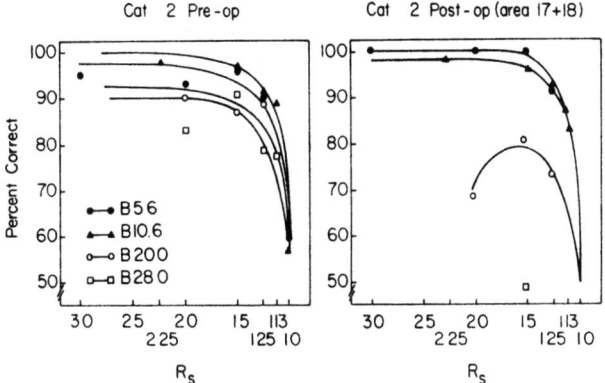

Fig. 20. Discrimination performance on structured arrays of dots for cat 2 before and after removal of areas 17 and 18. Symbols as on Fig. 15.

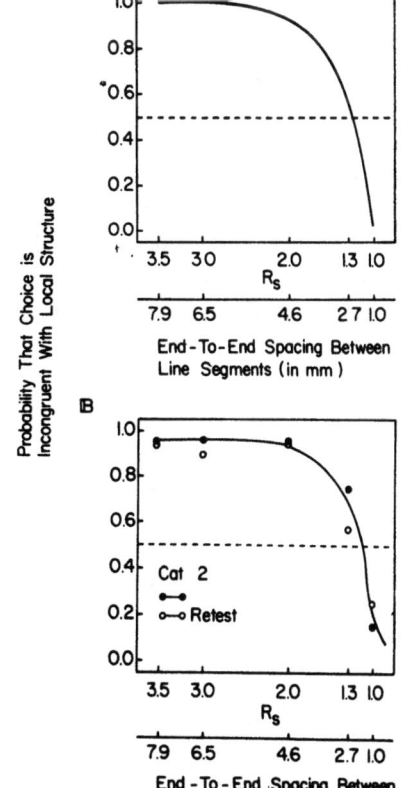

Fig. 21. Stimulus preferences on structured arrays of line segments. Axes are labelled as in Fig. 17. A, This curve is redrawn from Fig. 17, and shows the preferences for two normal cats. B, The bottom graph shows the stimulus preferences of a destriate cat (No. 2). The filled symbols represent the results obtained prior to discrimination training on unstructured arrays of line segments (see Fig. 18) while the open symbols represent the results of a retest performed after training on the unstructured arrays.

Structured line segment arrays. Figure 21 shows that removal of large portions of area 17 and 18 had no detectable effect on perceived orientation in structured line segment arrays.

Unstructured arrays of line segments. Cat No. 2 first learned this discrimination postoperatively. Despite otherwise excellent performance (even on low-$R_s$ dot arrays), there was no positive transfer to this unstructured pattern, and acquisition was slow (see Fig. 18).

This difficulty in discriminating local structural differences was intensified under conditions of reduced redundancy. As Fig. 19 indicates, cat 2 demonstrated a greater dependency on the numerosity in these patterns than did the two normal cats (No. 3 and No. 4).

Summarizing, the evidence indicates that the post-operative deficits appear to be confined to arrays which require sensitivity to local structural differences; detection of global structure is only minimally impaired.[2]

DISCUSSION

*Choice of Lesion Sites.* Many of the animals reported on here received large cortical lesions which included most or all of areas 17 and 18. Rationale for this type of lesion was in part practical: we assumed from the outset that threshold acuity functions might be mediated by the population of simple cells found in areas 17 and 18. We also assumed that the parts of 17–18 in which the area centralis (AC) is represented might be most important. We also studied two cats with smaller lesions limited to the representation of AC to 5–10°. Since the representation of the vertical meridian including AC lies along a common border between 17 and 18 in the lateral (marginal) gyrus, and because the exact position of this border is difficult or impossible to delineate at surgery, a lesion removing both areas was considered essential to fulfill the aims of the experiment.

---

[2] This statement is based on the performance of the destriate animal on low-base dot arrays and on structured arrays of line segments. While the cat clearly had trouble discriminating high-base dot arrays (Fig. 20), control experiments indicate that at least part of the deficit was due to difficulty in *detecting* the small number of figural elements, and not only to the presence of a grouping deficit. Space limitations do not allow us to consider these experiments in detail, so suffice it to say that when the animal was tested on a *successive* discrimination, a deficit in discriminating positive dot-patterns from a uniform field of equal luminance was present but only when the dot pattern had few elements (i.e. high base). We conclude that the deficit seen in the simultaneous discrimination is partly attributable to a difficulty in seeing the dots because of reduced acuity, since it occurred on the same patterns that showed a deficit on the successive discrimination.

Areas 17 and 18 have many similarities in their anatomical connections. The border zone referred to above is the only part of the two areas which receives callosal connnections (12) and similar laminae in both give origin to and receive the callosal fibers (55). In the projections of $LGN_d$, the paired laminae $A-A_1$ project almost exclusively to area 17–18 (27, 28, 34, 43, 52, 61), and the laminar pattern of termination of geniculate afferents to 17 and 18 is essentially the same (43). Moreover, the corticofugal projections to thalamus and midbrain from areas 17 and 18 are similar (39, 69).

Areas 17 and 18 also have many similarities in the physiological properties of their constituent neurons (26, 47, 65) except that area 18 does not appear to receive one functional class of input from the $LGN_d$ (X cells) (33, 64).

Finally, removal of areas 17 and 18 leaves cats with virtually unimpaired visuomotor behavior, including full visual fields as measured in a perimetry test and with excellent brightness and form-discrimination capabilities (25, 59, 61).

*Choice of behavioral tests.* In general, using complex visual stimuli (e.g., shapes) in a discrimination task permits an animal to use any one of a number of dimensions contained in one stimulus (e.g., size, orientation, texture, contour, local flux cues, etc.) to differentiate it from other stimuli. Most form stimuli, even if they are classed as simple, may contain many uncontrolled or unspecified attributes. Thus, what appears as a simple geometric form, e.g., a triangle, contains a variety of dimensions which differentiate it from other relatively simple geometric forms, e.g., circle (72). It is not possible from examination of the figures or behavior of the animal to discover which of the various dimensions available in the stimulus are being used to make the discrimination. The stimuli that were selected in the present study were selected to eliminate or control all except one cue. This was achieved by using a threshold testing procedure in which one dimension of the stimulus complex was varied. If the manipulation of that dimension controlled the animal's behavior, e.g., ability to discriminate between stimuli, it was taken as evidence that it was the cue to which the animal was attending. A second and equally important advantage in using a threshold testing procedure is that it contains within it a control for nonspecific lesion effects (e.g., general sensory impairment). That is, discrimination performance at suprathreshold values should not be affected by the ablation and thus can be used as a performance control.

The choice of the specific tests used was guided by consideration of the large body of data on the electrophysiological properties of neurons in area 17. Specifically, area 17 has been found to be unique in having cells with very small receptive fields which are tightly tuned for contour orientation (31, 35, 70). It also has the largest and the finest-grain map of visual space of all the visual areas that have thus far been studied (66–68). The behavioral tests selected were chosen on the assumption that they might be related to these properties of areas 17 neurons. Specifically, the assumed relationships could be stated in the following way: the ability to discriminate small targets is mediated by area 17 cells with the smallest receptive fields; the limits of contour orientation detection are determined by the cells most tightly tuned for orientation; and contour alignment ability depends upon the fine grain topographic representation of area 17. Similarly, we assumed that the ability to polarize patterns with global structure (rectilinear arrays of dots and line segments) would depend on the fine-grained topographic map present in areas 17–18. Furthermore, it was assumed that the population of simple cells would be best suited for processing local structural information (i.e. orientation of line segments).

With these assumed relationships in mind, the behavioral tests were selected to determine grating acuity, orientation acuity, contour alignment (vernier offset), as well as detection of proximity cues in lattice-like arrays of dots and line segments, and orientation in local structure.

Grating acuity. The preoperative grating acuity estimates (3.5–6.0 c/deg) obtained in this study are comparable to previous estimates of cat acuity obtained with stimuli of comparable luminance and contrast (11, 17, 18).

Postoperatively, all the cats with area 17-18 lesions showed a surprisingly small loss in grating acuity. The final acuity levels achieved by these animals is in the low-normal range for cats and would not be considered deficient if comparisons with preoperative levels hat not been made. Comparison with similar studies in monkey and man with lesions of area 17 reveals a greater degree of preserved grating acuity in the cat when compared with primates (e.g., 71). The reduction of acuity in two cats following area 17–18 ablations is 25–35%, a reduction equivalent to a change in human acuity of from 20/20 (normal) to about 20/30, not an impressive diminution of this visual capacity. The finding that the cats exhibited no significant deficit at suprathreshold grating size both agress with earlier findings indicating retention of the ability to discriminate patterns and forms (25, 61), and indicates that the animals were not suffering from a simple, postoperative performance deficit.

There are several possible explanations of our acuity observations that might be invoked: (1) stimulus target error; (2) an incomplete ablation; and (3) the possibility that preoperative acuity was underestimated.

1. The first possibility, namely that some artifact (e.g., luminance difference) in the target was being used by the animals at the discrimi-

nation cue, not the grating per se, can be ruled out for several reasons. First, luminance matches between the stimuli were continuously checked with a sensitive photometer and by human observers. Second, in all cases the animals showed a reduction in discrimination as grating size was reduced, a finding which would not be expected if luminance cues were present and being used.

2. We do not believe that incomplete ablations of area 17 can account for our results. It seems unlikely that the few remaining cells observed in the histological material could mediate the levels of acuity observed; the healthy cells that were found in the lateral portions of laminae A, $A_1$ and C of $LGN_d$ are too far out in the peripheral projection to predict the observed acuity. That is, the predicted acuity based on the presumed field defect is much lower than that observed if one estimates the decline in acuity to follow the decline in ganglion cells density from area centralis into the periphery (7). Retinal lesions involving a comparable extent of the visual field as that of the area 17–18 ablations on the current study produce a much larger deficit in acuity than observed in the present experiment (9, 16). Finally, the intact cells in portions of $LGN_d$ concerned with central vision were located in portions of the geniculate (NIM, $C_1$ and $C_2$) which project extensively to visual areas lying outside of 17–18.

3. Finally, it could be argued that our preoperative acuity estimates were too low, permitting the animals room to improve postoperatively, and thus reducing the relative magnitude of the "true" postoperative deficits. This seems unlikely since our preoperative acuity estimates were similar to those observed in other studies of normal cats (11, 17, 18) and there is no reason to believe that the cats used in this study deviated from a normal sample.

Vernier offset. One unusual feature of the behavior of cats in the vernier offset task was the difficulty with which cats initially acquired this particular discrimination. It is possible that since the stimulus contains a single relevant cue which is highly localized, it may well take some time for the animals to discover the precise locus necessary to make the discrimination.

A comparison of cat and human vernier offset acuity shows the normal cat to be vastly inferior to man. The three normal cats measured had a mean vernier acuity of about 4 min (240 s while man, under average conditions has, conservatively, a vernier acuity on the order of 10 s, yielding a ratio of 25 times poorer vernier offset acuity in cats than in man (no estimates are currently available for the macaque). Even under conditions of all rod vision, where man might be more comparable to the cat, man's offset threshold is 2.5 times better than the cat (13, 41).

As can be seen from Figure 10, the postoperative deficits is vernier offset after area 17 ablations are considerable. Two of the animals (Streak, Zelda) were completely incapable of performing the task postoperatively and one (Francis) showed 10-20 fold decrease in vernier acuity after area 17 ablation and never was able to achieve a high performance level even at large suprathreshold stimulus values. It could be argued that since the cats never achieved a high performance level even with suprathreshold stimuli they were suffering from a general performance deficit. This is unlikely because the animals were able to perform other suprathreshold tasks at preoperative (normal) levels. There is one consideration, however, that tempers this conclusion and that is that problems other than sensory or perceptual losses may contribute to the deficit since the cue in this task may require accurate fixation. Thus, the poor postoperative performance may be due to the cat's inability to search for and/or to look at (fixate) the essential portion of the stimulus. Although informal observation did not reveal abnormal fixation or stimulus scanning in any of the test animals, these factors cannot be ruled out as a contributor to the acuity deficits observed. Tests specifically designed to examine the role of fixation are currently being performed.

Orientation acuity and polygons (angular acuity). The tests using the orientation and polygon targets show the normal cat to be capable of discriminating relatively small differences in contour orientation on the order of about 5°. This value is not unreasonable in light of the orientation selectivity of neurons in area 17 of this species. Thus, Watkins and Berkley (70) and Rose and Blakemore (51) showed that the most tightly tuned cells for orientation have a half-width at half-height of about 6° (response is reduced 50% when orientation of stimulus is rotated 6° from optimal). While consistent with electrophysiological findings, the observed thresholds again are considerably poorer in the cat than observed in man (19, 42).

Postoperatively, all the animals tested on orientation acuity showed a moderate (3-fold) but consistent loss in the ability to discriminate differences in contour orientation. In this case, it is not possible to rule out the possibility that the small peripheral portions of area 17 that remained intact were mediating the residual capacity because we do not know the relationship between orientation acuity and retinal eccentricity. However, since other residual portions of the system, for example, areas 19, 20, 21 and lateral suprasylvian area are probably also capable of some orientation discrimination (57), which system the animals were using in this particular task is uncertain.

The results using the polygon stimuli are consistent with the reduction in orientation angular acuity observed in using the parallel line stimuli. Thus, animals that showed a large deficit in orientation acuity with the line stimuli also showed a large deficit in the polygon discrimination task (e.g., Francis, Figs. 8A and 9A). This test provides further evidence that parts of the cat visual system outside of area 17-18 can mediate form vision and confirms earlier reports than form discrimination in animals with such lesions (10, 25, 60-62) appear more or less normal except when a dimension (size, angle or thickness of line) of the stimulus approaches the acuity limit.

Form vision. Three of the four animals with 17-18 lesions which were tested in a form discrimination using small stimuli (diameter 7°, line thickness 1.5°) showed prolonged acquisition (Fig. 11). The final level of performance in these animals, although high, was somewhat inferior to that of normal cats and the variability of performance from day to day was greater although one animal with a 17-18 ablation (SD) did perform normally. Subsequent performance on figure-ground (contrast) reversal of the same stimulus pair was variable (e.g., generalization was excellent in cat SC and poor in ZL), but probably within normal range (e.g., excellent in N4, poor in N1) (see 61, p. 452).

Detection of proximity cues and detection of orientation in local structure. It seems clear that cats readily transfer a learned discrimination of oblique gratings to large $R_s$ dot patterns, while subsequent transfer from structured to unstructured arrays of line segments is much less reliable. Ease of transfer from one class of pattern (e.g. oriented square-wave gratings) to another (rectilinear dot arrays) implies that the cues processed by the neural-discriminator exist in both classes of patterns. Similarly, a lack of transfer indicates that the discriminative cue from the original pattern is not present (or is not recognized as being present) in the transfer pattern.

Thus, strong positive transfer can be thought of as an indication of perceived similarity, while a lack of transfer indicates perceived dissimilarity between the original and the transfer-pattern. In this context, the transfer results in these experiments can be interpreted as an indication that global structure is a more important determinant of perceived similarity to square-wave gratings than is local structure.

Additional evidence for the predominance of global structure cues in these experiments can be found in the results using structured arrays of line segments. We have observed that the $R_s$ value associated with orientational ambiguity (Fig. 17, $R_s$ 1.25), does not change after explicit training on unstructured arrays of line segments. This result can be interpreted as evidence that the processing of this local information is suppressed in the presence of a strong global structure.

Thus, it is suggested that the "grain" of the spatial analysis performed by the visual system can be dependent on certain aspects of the input pattern. This apparent change in the visual system's operating mode is a characteristic of a state-determined system in which one of the controlling variables is a step-function (3). In the present example, the controlling step-function would be the presence (or absence) of a global structure in the pattern.

We initially hypothesized that the striate-peristriate cortex (areas 17–18) should process information concerning the spatial relationships in these patterns with more precision and fidelity than any of the other visuo-topic areas, since they contain the most complete representation of the visual field and have the largest cortical magnification factors (68). Accordingly, removal of this cortex would be expected to impair the grouping process, especially when activated by weak proximity cues (low $R_s$ patterns).

However, comparisons between psychometric curves obtained pre- and postoperatively shows that removal of areas 17 and 18 has little if any consequence on the mechanisms subserving the cohesive interactions underlying the grouping effects. This findings contrasts with the substantial deficits in destriate cats when the discrimination requires detection of differences in the local structure of patterns. These deficits are apparent in preoperative acquisition and postoperative retention of discriminations of local structure, and are further evident in the greater redundancy of local structure required by de-striate cats in making these discriminations. The effects of this cortical lesion thus seem to be specifically related to the analysis of local structural information, and thus are consistent with the duality of local and global analysis suggested in the previous paragraphs.

The local structural information is represented in the high-spatial frequency components of the spatial frequency power spectrum of these arrays (cf. 22). It thus seems reasonable to suggest that in these experiments, removal of areas 17 and 18 may be modelled by the imposition of a high-frequency filter on the input channel to a hypothetical neural pattern recognizer. The suggestion that area 17 acts as a high-pass spatial frequency filter is also consistent with the finding that (i) the area is characterized by small receptive fields and (ii) removal of areas 17 and 18 causes the deficits in visual acuity reported here while at the same time producing minimal effects on form vision.

It is a pleasure to acknowledge the technical assistance of D. S. Wermath, Jeanne Levy, Marie Bloom, Mary Jack, Betty Woolsey and T. Chalmers. Research was supported by NEI research grant EY00953 to M. Berkley; EY00577 to J. Sprague. One of us (HCH) was supported by a National Research Service Award (EY05130), and is now a Pennsylvania Plan Scholar.

# REFERENCES

1. ALBUS, K. 1976. The spatial properties of the functional subunit of the orientation domain in area 17 of the cat. Brain Res. 107: 214-215.
2. ALLMAN, J. M. and KAAS, J. H. 1975. The dorsomedial cortical visual area: A third tier area in the occipital lobe of the owl monkey (*Aotus trivirgatus*). Brain Res. 100: 473-487.
3. ASHBY, W. R. 1960. Design for a brain. Chapman and Hall Ltd., London.
4. ATENCIO, F. W., DIAMOND, I. T., and WARD, J. P. 1976. Behavioral study of the visual cortex of *Galago senegalensis*. J. Comp. Physiol. Psychol. 89: 1109-1135.
5. BENEVENTO, L. A. and REZAK, M. 1976. The cortical projections of the inferior pulvinar and adjacent lateral pulvinar in the rhesus monkey (*Macaca mulatta*): An autoradiographic study. Brain Res. 108: 1-24.
6. BERKLEY, M. A. 1970. Visual discriminations in the cat. In W. Stebbins (ed.), Animal psychophysics: The design and conduct of sensory experiments. Appleton-Century-Crofts, New York.
7. BERKLEY, M. A. 1970. Cat visual psychophysics: Neural correlates and comparisons with man. In J. Sprague and A. N. Epstein (ed.), Progress in psychobiology and physiology psychology. Vol. 6. Academic Press, New York.
8. BERKLEY, M. A., CRAWFORD, F. T. and OLIFF, G. 1971. A universal food-paste dispenser for use with cats and other animals. Behav. Res. Methods In strum., 3: 259-260.
9. BERKLEY, M. A., SHERMAN, S. M., WARMATH, D. S. and TUNKL, J. 1978. Visual capacities of adult cats which were reared with a lesion in the retina of one eye and the other occluded. Soc. Neuroscience Abstr. 4: 467.
10. BERKLEY, M. A., SPRAGUE, J. and WARMATH, D. S. 1976. The role of the geniculo-cortical system in form vision: Area 17 and 18 and contour acuities. Soc. Neuroscience Abstr. 2: 1067.
11. BERKLEY, M. A. and WATKINS, D. W. 1971. Visual acuity of the cat estimated from evoked cerebral potentials. Nat. New Biol. 234: 91-92.
12. BERLUCCHI, G. 1972. Anatomical and physiological aspects of visual functions of corpus callosum. Brain Res. 37: 371-392.
13. BERRY, R. N., RIGGS, L. A. and DUNCAN, C. P. 1950. The relation of vernier and depth discrimination to field brightness. J. Exp. Psychol. 40: 349-354.
14. BISHOP, P. O. 1970. Beginning of form vision and depth discrimination in cortex. In The neurosciences, second study program. G. O. Quarton, T. Melnechuk and G. Adelman, (ed.), Rockefeller University Press, New York.
15. BISHOP, P. O. and HENRY, G. H. 1971. Spatial vision. Ann. Rev. Psychol. 22: 119-160.
16. BLAKE, R. and BELLHORN, R. W. 1978. Visual acuity in cats with central retinal lesions. Vision Res. 18: 15-18.
17. BLAKE, R., COOL, S. J. and CRAWFORD, M. L. J. 1974. Visual resolution in the cat. Vision Res. 14: 1211-1217.
18. BLOOM, M. and BERKLEY, M. A. 1977. Visual acuity and the near point of accommodation in cats. Vision Res. 17: 723-730.
19. BOUMA, H. and ANDRIESSEN, J. J. 1968. Perceived orientation of isolated line segments. Vision Res. 8: 493-507.

20. BURROWS, G. R. and HAYHOW, W. R. 1971. The organization of the thalamo-cortical visual pathways in the cat. Brain Behav. Evol. 4: 220-272.
21. CAMPBELL, F. W. and GREEN, D. G. 1965. Optical and retinal factors affecting visual resolution. J. Physiol. 181: 576-593.
22. CORNSWEET, T. N. 1970. Visual perception. Academic Press, Inc., New York.
23. DENNY-BROWN, D. and CHAMBERS, R. A. 1976. Physiological aspects of visual perception. I. Functional aspects of visual cortex. Arch. Neurol. 33: 219-227.
24. DIAMOND, I. T. 1976. Organization of the visual cortex: Comparative anatomical and behavioral studies. Fed. Proc. 35: 60-67.
25. DOTY, R. W. 1971. Survival of pattern vision after removal of striate cortex in the adult cat. J. Comp. Neurol. 143: 341-370.
26. DREHER, B. and COTTEE, L. J. 1975. Visual receptive-field properties of cells in area 18 of cat's cerebral cortex before and after acute lesions in area 17. J. Neurophysiol. 38: 735-750.
27. GAREY, L. J. and POWELL, T. P. S. 1967. The projection of the lateral geniculate nucleus upon the cortex in the cat. Proc. Roy. Soc. B 169: 107-126.
28. GILBERT, C. and KELLY, J. P. 1975. The projections of cells in different layers of the cat's visual cortex. J. Comp. Neurol. 163: 81-106.
29. GLICKSTEIN, M., KING, R. A., MILLER, J. and BERKLEY, M. 1967. Cortical projections from the dorsal lateral geniculate nucleus of cats. J. Comp. Neurol. 130: 55-76.
30. GUILLERY, R. W. 1970. The laminar distribution of retinal fibers in the dorsal lateral geniculate nucleus of the cat: A new interpretation. J. Comp. Neurol. 138: 339-368.
31. HENRY, G. H. DREHER, B. and BISHOP, P. O. 1974. Orientation specificity of cells in cat striate cortex. J. Neurophysiol. 37: 1394-1409.
32. HICKEY, T. and GUILLERY, R. 1974. An autoradiographic study of retinogeniculate pathways in the cat and fox. J. Comp. Neurol. 156: 239-243.
33. HOFFMANN, K. P. and STONE, J. 1971. Conduction velocity of afferents to cat visual cortex: A correlation with cortical receptive field properties. Brain Res. 32: 460-466.
34. HOLLANDER, A. and VANEGAS, T. 1977. The projection from the lateral geniculate nucleus onto the visual cortex in the cat. A quantitative study with horseradish peroxidase. J. Comp. Neurol. 173: 519-536.
35. HUBEL, D. H. and WIESEL, T. N. 1962. Receptive fields, binocular interaction and functional architecture in the cat's visual cortex, J. Physiol. 160; 106-154.
36. HUBEL, D. H. and WIESEL, T. N. 1968. Receptive fields and functional architecture of monkey striate cortex. J. Physiol. 195: 215-243.
37. HUBEL, D. H. and WIESEL, T. N. 1972. Laminar and columnar distribution to geniculo-cortical fibers in the macaque monkey. J. Comp. Neurol. 146: 421-450.
38. HUMPHREY, N. R. 1974. Vision in a monkey without striate cortex. A case study. Perception 3: 241-255.
39. KAWAMURA, S., SPRAGUE, J. M. and NIIMI, K. 1974. Corticofugal projections from the visual cortices to the thalamus, pretectum and superior coolliculus in the cat. J. Comp. Neurol. 158: 339-362.
40. LATIES, A. and SPRAGUE, J. M. 1966. The projection of optic fibers to the visual centers in the cat. J. Comp. Neurol. 127: 35-70.

41. LeGRAND, Y. 1967. Form and space vision. Indiana University Press, Bloomington p. 112.
42. LENNIE, P. 1971. Distortions of perceived orientation. Nat. New Biol. 233: 155–156.
43. LeVAY, S. and GILBERT, C. D. 1976. Laminar patterns of geniculo-cortical projection in the cat. Brain Res. 113: 1–19.
44. LUNA, R. G. 1968. Manual of histologic staining methods of the armed forces institute of pathology. McGraw-Hill, New York.
45. NIIMI, K. and KUWAHARA, E. 1973. The dorsal thalamus of the cat and comparison with monkey and man. J. Hirnforschung. 14: 303–325.
46. NIIMI, K. and SPRAGUE, J. M. 1970. Thalamo-cortical organization of the visual system of the cat. J. Comp. Neurol. 138: 219–249.
47. ORBAN, G. A. and CALLENS, M. 1977. Receptive field types of area 18 neurones in the cat. Exp. Brain Res. 30: 107–123.
48. OTSUKA, R. and HASSLER, R. 1962. Uber Aufbau und Gliederung der corticalen sehsphare bei der Katze. Arch. Psychiatr. u. Ztschr. ges. Neurol. 203; 212–234.
49. PALMER, L. A., ROSENQUIST, A. C. and TUSA, R. J. 1978. The retinotopic organization of lateral suprasylvian visual areas in the cat. J. Comp. Neurol. 177: 237–256.
50. PASIK, T. and PASIK, P. 1971. The visual world of monkeys deprived of striate cortex: Effective stimulus parameters and the importance of the accessory optic system. Vision Res. Suppl. 3: 419–435.
51. ROSE, D. and BLAKEMORE, C. 1974. An analysis of orientation selectivity in the cat's visual cortex. Exp. Brain Res. 20: 1–17.
52. ROSENQUIST, A. C., EDWARDS, S. B. and PALMER, L. A. 1974. An autoradiographic study of the projections of the dorsal lateral geniculate nucleus and the posterior nucleus in the cat. Brain Res. 80: 71–93.
53. SANDERSON, K. J. 1971. The projection of the visual field to the lateral geniculate and the medial interlaminar nuclei in the cat. J. Comp. Neurol. 143: 101–118.
54. SANIDES, F. and HOFFMANN, J. 1969. Cyto- and myeloarchitecture of the visual cortex of the cat and of the surrounding integration cortices. J. Hirnforschung 11: 79–104.
55. SHATZ, C. J. 1977. Anatomy of interhemispheric connections in the visual system of Boston Siamese and ordinary cats. J. Comp. Neurol. 173: 497–518.
56. SHERMAN, S. M. 1973. Visual field defects in monocularly and binocularly deprived cats. Brain Res. 49: 25–45.
57. SPEAR, P. D. and BAUMANN, T. P. 1975. Receptive-field characteristics of single neurons in lateral suprasylvian visual area of the cat. J. Neurophysiol. 38: 1403–1420.
58. SPEAR, P. D. and BRAUN, J. J. 1969. Pattern discrimination following removal of visual neocortex in the cat. Exp. Neurol. 25: 331–348.
59. SPRAGUE, J. M. 1966. Visual, acoustic and somesthetic deficits in the cat after cortical and midbrain lesions. *In* D. P. Purpura and M. Yahr (ed.), The thalamus. Columbia University Press.
60. SPRAGUE, J. M., BERKLEY, M. A., TUNKL, J. and BERLUCCHI, G. 1976. Neural mechanisms mediating pattern and form discriminations in the cat. Soc. Neuroscience Abstr. 2: 1093.

61. SPRAGUE, J. M., LEVY, J., DiBERARDINO, A. and BERLUCCHI, G. 1977. Visual cortical areas mediating form discrimination in the cat. J. Comp. Neurol., 172: 441–488.
62. SPRAGUE, J. M., LEVY, J., DiBERARDINO, A. and CONOMY, J. 1973. Effect of striate and extra striate visual cortical lesions on learning and retention of form discriminations. Abstr. Soc. Neurosci., 3rd Annual Meeting, San Diego.
63. SPRAGUE, J. M. and MEIKLE, Jr., T. H. 1965. The role of the superior colliculus in visually guided behavior. Exp. Neurol. 11: 115-146.
64. STONE, J. and DREHER, B. 1973. Projection of x- and y-cells of the cat's lateral geniculate nucleus to areas 17 and 18 of visual cortex. J. Neurophysiol. 36: 551–567.
65. TRETTER, F., CYNADER, M. and SINGER, W. 1975. Cat parastriate cortex: A primary or secondary visual area? J. Neurophysiol. 38: 1099–1113.
66. TUSA, R. 1978. Retinotopic organization of visual cortex in cat. Ph. D. Thesis, University of Pennsylvania.
67. TUSA, R., PALMER, L. A. and ROSENQUIST, A. C. 1975. The retinotopic organization of the visual cortex in the cat. Soc. Neurosci. Abstr., 1: 52.
68. TUSA, R. J., PALMER, L. A. and ROSENQUIST, A. C. 1978. The retinotopic organization of area 17 (striate cortex) in the cat. J. Comp. Neurol., 177: 213–236.
69. UPDYKE, B. V. 1975. The patterns of projection of cortical areas 17, 18, and 19 onto the laminae of the dorsal lateral geniculate nucleus in the cat. J. Comp. Neurol. 163: 377–396.
70. WATKINS, D. W. and BERKLEY, M. A. 1974. The orientation selectivity of single neurons in cat striate cortex. Exp. Brain Res. 19: 433–446.
71. WEISKRANTZ, L. and COWEY, A. 1963. Striate cortex lesions and visual acuity of the rhesus monkey. J. Comp. Physiol. Psychol. 56: 225–231.
72. WINANS, S. S. 1971. Visual cues used by normal and visual-decorticate cats to discriminate figures of equal luminous flux. J. Comp. Physiol. Psychol. 74: 167–178.
73. WONG-RILEY, M. T. T. 1976. Projections from the dorsal lateral geniculate nucleus to prestriate cortex in the squirrel monkey as demonstrated by retrograde transport of horseradish peroxidase. Brain Res. 109: 595–600.
74. WOOD, C. C., SPEAR, P. D. and BRAUN, J. J. 1974. Effects of sequential lesions of suprasylvian gyri and visual cortex on pattern discrimination in the cat. Brain Res. 66: 443–466.

J. M. SPRAGUE, Department of Anatomy, School of Medicine, University of Pennsylvania, Philadelphia, Pa. 19104, USA.
M. A. BERKLEY and H. C. HUGHES, Department of Psychology, Florida State University, Tallahassee, Florida, USA.

Lecture delivered at the Warsaw Colloquium on Instrumental
Conditioning and Brain Research
May 1979

# THE PRETRIGEMINAL CAT AS AN INSTRUMENT FOR INVESTIGATION OF THE OCULAR FIXATION REFLEX

Bogusław ŻERNICKI

Department of Neurophysiology, Nencki Institute of Experimental Biology
Warsaw, Poland

*Abstract.* The fixation reflex is an important representative of orienting (targeting) reflexes. In the cat with the brainstem transected at the pretrigeminal level the fixation reflex is easy to investigate for the following reasons: (a) all sensory inputs to the isolated cerebrum of the pretrigeminal cat are eliminated, except for visual afferents, (b) the isolated cerebrum is continuously awake, (c) the vertical fixation reflex only is present. The behavioral and neural characteristics of the fixation reflex in the pretrigeminal cat are reviewed.

An important class of behavior is represented by reflexes which adjust a given analyzer to the better perception of a stimulus (13, p. 17). These reflexes, denoted as orienting or targeting (13), operate almost continuously. Good examples are long-term watching of the TV or a lecturer.

In spite of the importance of targeting reflexes, their properties and central mechanisms are poorly known. The obvious reason is that these reflexes are present only in the awake animal, in which it is difficult to record them precisely.

A conspicuous representative of targeting reflexes is ocular fixation reflex. This reflex brings the image of the object on the area centralis and maintains the image in this position for a while. The fixation reflex consists of three phases: (a) a saccadic movement towards the object, (b) maintenance of the fixation, (c) a return saccadic movement. When an object moves slowly and at constant speed in the visual field, the maintenance phase consists of a following (pursuit) movement, which matches that of the stimulus. Such a reflex is called the following reflex (6).

The cat with a brainstem transection at the pretrigeminal level (1, 22) is a valuable instrument for investigating the fixation reflex. The isolated cerebrum of the pretrigeminal cat is awake and controls the vertical fixation reflex, which does not seem to differ significantly from that in the intact cat (25). The three main advantages of the pretrigeminal cat over an intact animal as an instrument for investigating the fixation reflex are as follows:

1. Visual input only to the cerebrum is preserved. Olfactory input is eliminated by tracheotomy which usually precedes the pretrigeminal transection. Thus the effects of a visual stimulus are not obscured by those due to other stimuli. Moreover, the optokinetic reflex (optokinetic nystagmus) is absent (12). The lack of pain is particularly important. The cat can be restrained in the stereotaxic apparatus without local anesthesia, and the position of the eye can be monitored using simple techniques. In our experiments we use a tensometric method (8). The lower margin of the eye is attached with a thread to a hair spring connected to a tensometric transducer. The vertical position of the eye is displayed continuously on an oscilloscope. The sensitivity of the method is 20'.

2. During an experiment the level of wakefulness of the isolated cerebrum usually remains stable. In the acute stage the cerebrum is either alert or drowsy and shifts between these two stages occur infrequently, on average, 1/h (20).

3. Horizontal eye movements are absent in the pretrigeminal cat (the center controlling them is caudal to the transection). The lack of the horizontal fixation reflex is convenient for two reasons: (a) only the vertical position of the eye need to be monitored, (b) a stimulus presented at the lateral or medial part of the visual field is not brought on to the area centralis (see below).

Three disadvantages of the pretrigeminal cat in comparison with an intact cat should be also mentioned:

1. Nursing care of the chronic cats is tedious and the experiments are performed preferably on acute preparations.

2. Some acute preparations (about 30%) are semicomatose or comatose and then the fixation reflex is abortive or absent, respectively (22). In such cases, however, the administration of a small dose of amphetamine (0.5-1 mg i.v.) is usually helpful. The amphetamine also greatly increases the resistance to habituation of the fixation reflex (5).

3. Establishment of a conditioned fixation reflex is less easy than in the intact animal, in which natural reinforcement is available (juice or milk is usually given directly into the mouth). However, by using electrical stimulation of the lateral hypothalamus as a reinforcement, conditioned eye movements can be established in the pretrigeminal cat (10, 11, 18, 26).

In our Laboratory the pretrigeminal cat has been used for several years in investigation of the vertical fixation reflex. The principal findings are as follows:

1. The following reflex can be evoked by a stimulus presented in the lateral or medial hemifield (Fig. 1). During the following phase

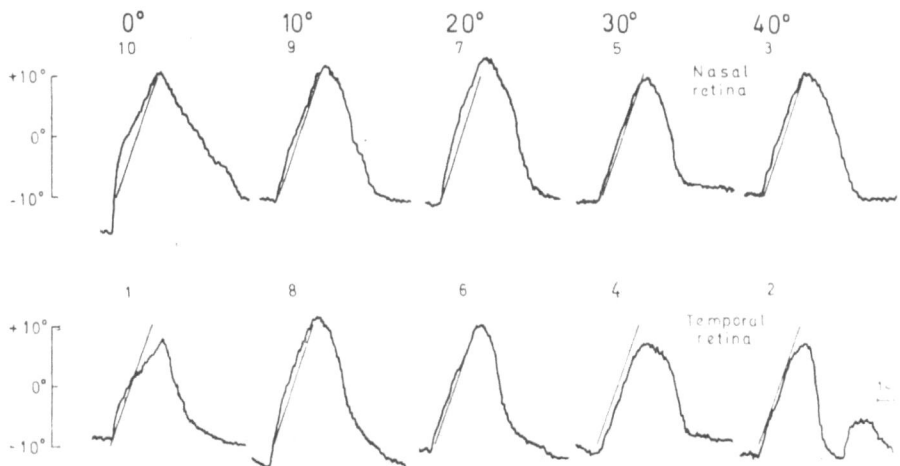

Fig. 1. Following reflexes to the stick (1 × 4°) moving upwards at different distances from the vertical meridian during a representative experimental session. Four reflexes at each distance averaged are indicated by the heavy line and the movement of the stick with a fine line. The small oscillations represent principally noise from the recording system. The small numbers indicate the order of recording. Ordinate: inclination of the eye. Note that the resting position of the eye was continuously low, occasionally even lower than the stimulus onset position. From Michalski et al. (15).

the image of the stimulus is held on the horizontal meridian, i.e. on the visual streak of highest ganglion cell density (9). Thus, the following reflex can be controlled by the peripheral retina.

2. The fixation reflex becomes abortive and habituates easily following ablation of the visual cortex, and it is absent following removal of the superior colliculi (3). On the other hand, ablation of the frontal eye fields (4) enhances the duration of the fixation reflex (5, 21). In particular, its resistance to habituation is increased. In the isolated

midbrain with visual input preserved (the brainstem is transected at the pretrigeminal and "preoptic" levels) the fixation reflex is absent (24). Although in the isolated midbrain the EEG activity is depressed (23, 24), the responses of collicular units to moving visual stimuli are almost normal (2).

3. In cats visually deprived from birth the fixation reflex is only slightly impaired (14). Thus the development of the fixation reflex is predominantly on a genetic basis.

4. During the pursuit (following) stage of the following reflex about 20% of the units in the representation of the area centralis in cortical areas 17 and 18 change their activity, presumably as the effect of a "corollary discharge" from the frontal eye fields (16). In this study the stimulus moved vertically with horizontal eccentricity of 40°, i.e. beyond the receptive fields of the units tested.

5. The intensity of the fixation reflex (as measured by the duration of fixation) is positively correlated with several other ocular and cerebral responses to the visual stimulus: (a) the intensity of pupillary dilatation and cortical EEG desynchronization (25), (b) the level of increase in regularity and frequency of the hippocampal theta activity (17), (c) the level of increase in cerebral blood flow in the occipital and frontal cortex (19), (d) presumably the duration of visual accommodation (7).

## REFERENCES

1. BATINI, C., MORUZZI, G., PALESTINI, M., ROSSI, G. F. and ZANCHETTI, A. 1959. Effects of complete pontine transections on the sleep-wakefulness rhythm: the midpontine pretrigeminal preparation. Arch. Ital. Biol. 97: 1–12.
2. DEC, K., TARNECKI, R. and ŻERNICKI, B. 1978. Single unit responses to moving spots in the superior colliculus of the cat's isolated midbrain. Acta Neurobiol. Exp. 38: 103–112.
3. DREHER, B., MARCHIAFAVA, P. L. and ŻERNICKI, B. 1965. Studies on the visual fixation reflex. II. The neural mechanism of the fixation reflex in normal and pretrigeminal cats. Acta Biol. Exp. 25: 207–217.
4. DREHER, B., SANTIBÁÑEZ-H., G. and ŻERNICKI, B. 1970. Oculomotor cortex localization in the unanesthetized cat. Acta Neurobiol. Exp. 30: 69–77.
5. DREHER, B. and ŻERNICKI, B. 1969. Studies on the visual fixation reflex. III. The effects of frontal lesions in the cat. Acta Biol. Exp. 29: 153–173.
6. DREHER, S. and ŻERNICKI, B. 1969. Visual fixation reflex: behavioral properties and neural mechanism. Acta Biol. Exp. 29: 359–383.
7. ELUL, R. and MARCHIAVAVA, P. L. 1964. Accommodation of the eye as related to behavior in the cat. Arch. Ital. Biol. 102: 616–644.
8. FOLGA, J., MICHALSKI, A., TURLEJSKI, K. and ŻERNICKI, B. 1973. Eye-movement recording with a tensometric method in the pretrigeminal cat. Acta Neurobiol. Exp. 33: 655–658.

9. HUGHES, A. 1975. A quantitative analysis of the cat retinal ganglion cell topography. J. Comp. Neurol. 163: 107–128.
10. IKEGAMI, S., NISHIOKA, S. and KAWAMURA, H. 1979. Operant conditioning of vertical eye movements without visual feedback in the midpontine pretrigeminal cat. Brain Res. 169: 421–431.
11. IKEGAMI, S., NISHIOKA, S. and KAWAMURA, H. 1977. Operant discriminative conditioning of vertical eye movements in the midpontine pretrigeminal cat. Brain Res. 124: 99–108.
12. KING, F. A. and MARCHIAFAVA, P. L. 1963. Ocular movements in the midpontine pretrigeminal preparation. Arch. Ital. Biol. 101: 149–160.
13. KONORSKI, J. 1967. Integrative activity of the brain. An interdisciplinary approach. Univ. Chicago Press, Chicago, 531 p.
14. KOSSUT, M., MICHALSKI, A. and ŻERNICKI, B. 1978. The ocular following reflex in cats deprived of pattern vision from birth. Brain Res. 141: 77–87.
15. MICHALSKI, A., KOSSUT, M. and ŻERNICKI, B. 1977. The ocular following reflex elicited from the retinal periphery in the cat. Vision Res. 17: 731–736.
16. MICHALSKI, A. and MOROZ, B. 1977. The effects of pursuit eye movements on single unit activity in cat visual cortex. Acta Neurobiol. Exp. 37: 261–274.
17. RADIL-WEISS, T., ŻERNICKI, B. and MICHALSKI, A. 1976. Hippocampal theta activity in the acute pretrigeminal cat. Acta Neurobiol. Exp. 36: 517–534.
18. SHLAER, R. and MYERS, M. L. 1972. Operant conditioning of the pretrigeminal cat. Brain Res. 38: 222–225.
19. SKOLASIŃSKA, K., KRÓLICKI, L. and ŻERNICKI, B. 1979. Regional cerebral blood flow and visual attention in the awake isolated cerebrum of the pretrigeminal cat. Acta Neurobiol. Exp. 39: 335–343.
20. ŚLÓSARSKA, M. and ŻERNICKI, B. 1971. Wakefulness and sleep in the isolated cerebrum of the pretrigeminal cat. Arch. Ital. Biol. 109: 287–304.
21. ŻERNICKI, B. 1972. Orienting response hypernormality in frontal cats. Acta Neurobiol. Exp. 32: 431–438.
22. ŻERNICKI, B. 1974. Isolated cerebrum of the pretrigeminal cat. Arch. Ital. Biol. 112: 350–371.
23. ŻERNICKI, B., DEC, K., SARNA, M., JASTREBOFF, P. and VERDEREVSKAYA, N. 1979. Single-unit activity in the cat's isolated midbrain. Acta Neurobiol. Exp. 39: 345–352.
24. ŻERNICKI, B., DOTY, R. W. and SANTIBÁÑEZ-H., G. 1970. Isolated midbrain in cats. Electroenceph. Clin. Neurophysiol. 28: 221–235.
25. ŻERNICKI, B. and DREHER, B. 1965. Studies on the visual fixation. I. General properties of the orientation fixation reflex in pretrigeminal and intact cats. Acta Biol. Exp. 25: 187–205.
26. ŻERNICKI, B., MICHALSKI, A., RADIL-WEISS, T. and KĄCZKOWSKA, E. 1978. Instrumental ocular conditioning in acute pretrigeminal cat. Acta Neurobiol. Exp. 38: 71–77.

Bogusław ŻERNICKI, Nencki Institute of Experimental Biology, Pasteura 3, 02-093 Warsaw, Poland.

Lecture delivered at the Warsaw Colloquium on Instrumental
Conditioning and Brain Research
May 1979

# MOTIVATIONAL ROLE OF SOCIAL REINFORCEMENT IN DOG-MAN RELATIONS

E. FONBERG and E. KOSTARCZYK

Department of Neurophysiology, Nencki Institute of Experimental Biology
Warsaw, Poland

*Abstract.* Lesions of dorsomedial amygdala (DMA) or lateral hypothalamus (LH) produced prominent impairment of the alimentary behavior in dogs. Besides, various behavioral responses, including social, were depressed. To evaluate whether these changes were either specific for alimentary disturbances or dependent on changes in the general reward system, another type of reward, i.e. social contact with experimenter as a reinforcement for several motor responses was introduced. Damage of amygdalo-hypothalamic system, which produced a syndrome of depression, also impaired socially reinforced responses. This impairment lasted much longer then decrease of food intake. Therefore it was assumed that DMA and LH damage produced impairment not limited to alimentary mechanisms, but involving various aspects of positive motivational system and in particular including the social behavior during man-dog relations. Taking into account various theories of reinforcement in instrumental learning the authors are inclined towards a hedonistic theory of reward.

## THE ROLE OF VARIOUS REINFORCEMENTS IN INSTRUMENTAL TRAINING

It was a great scientific step when Konorski and Miller discovered (39, 45) that instrumental performance consists in a separate class of conditioned reactions. Although several properties are common for both classical and instrumental conditioning the main difference is that instrumental performance is guided by values of motivation in respect to rewarding and punishing properties of reinforcement. Konorski and Miller formulated their pioneer discovery in terms of reflexology, motivation was not "a la mode" at that time. However, it is obvious that the basic mechanisms of instrumental conditioning are dependent on the value of motivation. Konorski studied and established several properties of such conditioning. He named instrumental responses conditioned reflexes of the second type to distinguish them from the first type — classical, then he divided the conditioned responses II type into 4 classes. The main feature of these responses is that their performance depends on whether the US is punishing or rewarding. Independently, along the same line proceeded the work of Skinner (67). Since that, instrumental procedure was widely used in the studies of animal behavior, learning, motivation brain mechanisms etc.

Most authors used electric shock as a negative US and food as the positive US. Konorski, as a negative US applied very often also air-puff into the ear. The same procedure was used with success by Fonberg in experiments on avoidance conditioning. Acid injections were also administered as another negative reinforcement (18–20). Other authors also used various negative reinforcements like aversive taste (48), fatigue (32), cold or high temperature (12, 85), concentration of carbon dioxide (42, 92) several kind of painful stimuli and other nociceptive agents. Much less variety of stimuli were used as positive reinforcement, althougt there is evidence that various kinds of long-term deprivation produce the state in which indifferent stimuli, which the subject was deprived of, start to play the role of reward. For example some experimenters successfully utilized sensory reinforcement like changes in light onset (9), changes in auditory stimulation (5), or introduction of tactile stimulation (86). Also the oportunity to explore the environment (46), or to manipulate with objects (30, 37) were used as reward. In his present presentation Stellar reported on the reinforcing values of changes in temperature (73). Such rewards are however very labile and apt to produce the satiation effect. The rewards of a high biological value are the best as positive reinforcers. The most widely used strong positive reward is the alimentary reinforcement (food for hungry animals). Among reinforcements associated with alimentary motivation several authors successfully established the instrumental reactions reinforced with tasty substances which had no nutritive value (78). Some authors used sexual reinforcement (6, 36, 62, 63). Aggression seems also to have reinforcing properties. Lagerspetz (41) found that attacking other mice may be used as reinforcer in the elaboration of instrumental responses (crossing the field connected with electrical current, and overcoming other obstacles). Also utilization of psychotropic drugs was fully sufficient in the elaboration of instrumental conditioned reactions (15).

In our investigation on the effect of amygdalar and hypothalamic lesions on animal behavior it was important to find a strong positive but natural reward other than the alimentary one. We found (22, 23, 25, 58, 59) that dorsomedial lesions of amygdala in dogs similarly to lesions of the lateral hypothalamus (Fig. 1), produced a syndrome of aphagia and impairment of instrumental alimentary responses. We

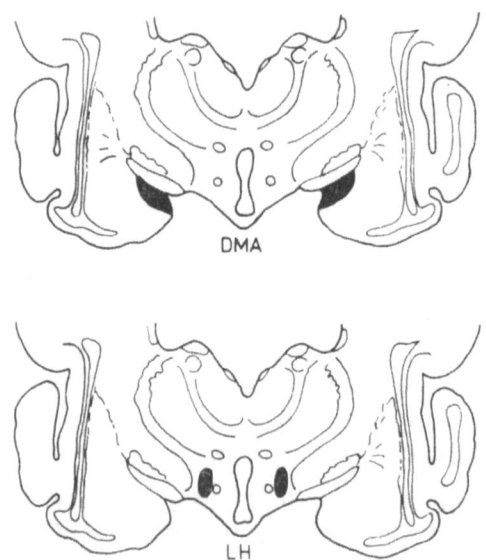

Fig. 1. Damage of either LH or DMA produced similar symptoms: aphagia or hypophagia, vomiting, finickiness, impairment of salivary responses and decrease of instrumental alimentary performance.

assumed that dorsomedial amygdala, parallelly with the lateral hypothalamus consists in alimentary center and that DMA and LH are doubling each other functions (25). But at the same time the dogs exibited several other changes. They were apathetic, atonic, sad, not interested in the surroundings, and unsocial towards humans (Table I). It might be assumed therefore that both DMA and LH lesions impair the general positive reward system hypothetized by Stein (69, 70), Cytawa and Trojniar (16, 17) and accepted by others. On the other hand, as the alimentary reward is both biologically and hedonistically (16, 17, 91) very important, the other changes may be secondary. Lack of rewarding

TABLE I

The effects of medial amygdalar damage on dogs. From Fonberg (25a).

| General apathy | Changes in general motility and posture | Impairment of instrumental performance | Changes of social-emotional behavior | Changes in emotional expression | Changes in alimentary behavior |
|---|---|---|---|---|---|
| A. decrease or absence of orienting reaction to external stimuli | A. often sitting down | A. decrease or loss of retention of instrumental reactions | A. lack of positive emotional responses to known persons | A. sad, "dull" shineless eyes | A. aphagia or hypophagia |
| B. lack of interest in novel environmental stimuli (no approach, sniffing and other exploratory reactions) | B. standing on one place | B. impairment in learning of instrumental reactions | B. no reaction to calling by name | B. no moving of tail and no facial expression of joy | B. aversion for food or finickiness |
| C. lack of interest in well known surrounding (no searching food no directing toward feeding place etc.) | C. standing with legs bent tail under, ears hanging and head down | C. fluctuatings of the instrumental performance | C. no reaction to other commands like "come on" "sit down" "give paw" | C. put outside does not jump, roll on grass or play with objects | C. vomiting |
| | D. walk and other movements slow no jumping no running | | D. negativism (oppose leading and other manipulations, withdrawn) | D. no vocalization of joy, rare vocalization of fear and aggression | |
| | E. cataleptic like responses (holding the passively induced positions) | | E. no aproach for petting | E. characteristic whining | |

values of food intake may produce extinction of all instrumental acts linked with alimentary situation and generalized to other situations. Aversion to food may radiate to the persons which present food and produce the changes in social attitude towards these persons observed after DMA or LH damage.

Our recent experiments were performed especially to elucidate this point. Social changes in our amygdalar and hypothalamic dogs were very striking, therefore in further research we concentrated on social behavior. Several authors described changes in different kinds of social behavior after amygdala and hypothalamic lesions in various species (35, 38, 64, 57, 80, 89). Fuller et al. described socio-affective changes after amygdala lesions in dogs (27). However these authors have not observed the instrumental performance in a social situation and it was not even certain whether it is possible to establish the instrumental reactions reinforced socially. Our recent experiments have brought positive answer to this question.

## INSTRUMENTAL TRAINING SOCIALLY REINFORCED

In our recent experiments (Fonberg, Kostarczyk and Prechtl in preparation) both CS and US were of a social character. In order to dissociate this behavior from alimentary behavior the dogs were never fed by the experimenter and they never obtained food in the experimental compartment. The CSi for the responses were verbal commands in connection with hand gestures. As a reward for correct responses the dog received petting (stroking the back and head of the dog) and vocal acceptance ("good dog") from the experimenter. A more detailed description of the method can be found in another paper (Fonberg, Kostarczyk and Prechtl in preparation). Two series of responses were performed. In the first series a close contact of the dog with the experimenter was important and performance was dependent on the strenght of dog — experimenter bond. Dogs had to react to commands: "sit", "paw", "lay". To these commands sit, paw and lay responses were elaborated. The first response of the dog consisted in sitting down on the floor facing the experimenter. The second response was to place the forepaw in the experimenter outstretched hand. The third responses consisted in lying down on the floor close to experimenter.

In the second series which started when the dogs reached the criterion of performance during the I series of training, sprint and jump responses were trained by the same experimenter but with the help of a technician. Sprint response involved running the 5 m distance (from laboratory technician who held the dog at the other end of the room) to the experimenter on his command "run". For jump responses a wooden barrier of adjustable height was erected on the runway, halfway between the dog and the experimenter which the dog had to jump over to reach the experimenter to the order "jump". During training the barrier was gradually raised in order to reach the top of motor abilities of individual dogs. Each experimental session consisted of 5 consecutive trials of each task. The criterion of learning in both series was a correct response in 90% of the trials completed during 10 consecutive sessions. The number of CS repetitions of commands needed to obtain proper responses and the latencies of responses were recorded by the experi-

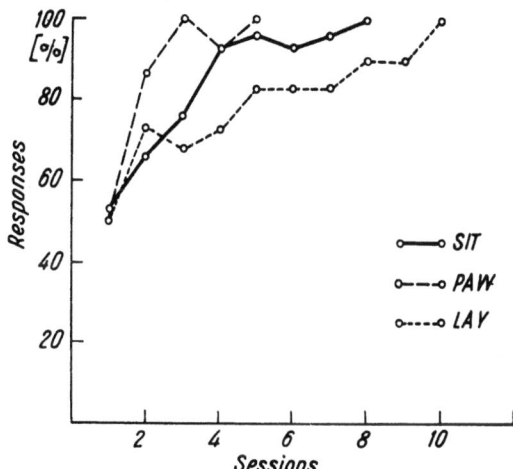

Fig. 2. Course of training to 100% level of performance in 6 dogs.

menter. As shown in Fig. 2 during the first series the dogs very quickly learned all the trained responses. Within 10 sessions all the dogs responded at a 100% level of performance. Some dogs performed proper responses from the very first session, others responded also at a high level in the early period of training. However, latencies of particular reactions varied in individual dogs. In three dogs with a stable and quick performance mean latencies varied from 0.5–2.8 s but in the other dogs they were more fluctuating, ranging from 1.0–11.5 s.

The second series of tasks was introduced when the dogs showed competence in the tasks of the first series. The "sprint" and "jump" approach tasks were used as a measured of both emotional attachment and general motor arousal. The dogs were already familiar with the room and the experimenter (but not the technician) therefore all the dogs at once began to perform these tasks at a level of 100%. In case of jump responses performance dropped occasionally below 100% but it was associated with the barrier height being raised. Most dogs learned without difficulties to jump over the barrier raised gradually (heightened from 50 to 80 cm). In all cases they performed at a 96-100% level with a continuous decrease of response durations. Before operation, all dogs performed all responses in both series at a 100% level and response latencies were very short (0.5 s). Moreover, they often anticipated the command given by the experimenter and tried to perform the responses before a command had been given.

In the course of training the dogs seemed to become gradually emotionally attached to the experimenter and to appreciate more and more the petting reward. During the training of lay responses the number of trials in which the dogs terminated petting were registered. During trials where the dogs remained lying for the whole petting period, we recorded the amount of time spent lying during petting and the amount of time spent lying after petting had been discontinued. In

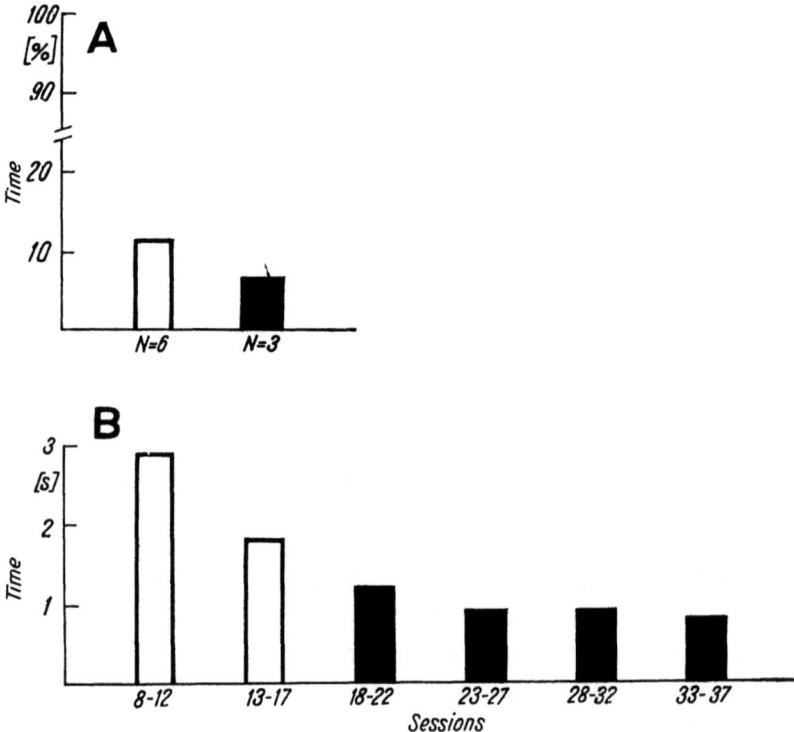

Fig. 3. Shortening of duration of lying response in post-petting period in the course of training. A, Mean percent of the duration of lying for the whole group, for the criterional and overtraining periods (in respect to the total duration of lying). Blocks represent mean from 10 criterional (white bars) and 20 overtraining sessions (black bars). B, Mean duration of lying without petting in a representative dog. Blocks represent mean from five sessions.

a special test the duration of petting was purposely changed in a random way ranging from 5 to 30 s in order to study the reinforcing properties of petting as such. The data showed that in the course of training the frequency of moving away during petting reward decreased. In the first 5 criterion sessions 31.5% of trials were terminated by not retaining the proper position whereas during the last 5 overtraining sessions this percentage dropped to 10.7%. The time of lying after the petting was discontinued also decreased (Fig. 3).

These results show that social reward was sufficient to motivate performance in the training of various instrumental tasks. Moreover, the responses once established were stable, independently of the duration of training, and responses latencies remained low and stable throughout overtraining, without special manipulations in the experimental situation. There arose the question what was the basis for social reward and why its reinforcing value was so high? Various factors may contribute to the reinforcing properties of social reward.

1. The most important is the emotional bond and attachment which was created between the dogs and experimenters in the course of the training. Solomon would probably name it social addiction (68). However we think that the mechanisms of addiction are not the same as in normal instrumental conditioning. We think that social bond has a great biological importance for individual and species survival, whereas addiction — unvalid from biological point of reward and normal homeostasis posses pathological features, although it also can be partly explained by hedonistic theory.

2. Other important factors which may play a rewarding role are pleasant tactile stimuli in the form of petting and handling. These factors seem to be the very potent physiological reinforcing stimuli. Petting may also produce activation of the sensory and motor structures of the brain and therefore facilitate instrumental performance.

3. Additional factor which may also play a rewarding role in these experiments is the sensory enrichment of the environment, especially important in dogs kept all the time in closed boxes with limited amount of stimuli.

All these rewards together produced a "pleasure state", judgeding by the expression of the dogs. In this respect our results agree with the hedonistic learning theory (17, 53, 90) and speak for the existence of a more general positive motivational system including several specific motivations besides alimentary or sexual (31, 51). This supposition is supported by data obtained from our further experiments in which lesion of the amygdala and hypothalamus impaired the social responses of the dogs.

## EFFECT OF LESIONS OF AMYGDALO-HYPOTHALAMIC SYSTEM ON SOCIAL BEHAVIOR

When the social responses had been elaborated and stabilized, lesions of the amygdala and/or hypothalamus were performed. In the present paper 7 dogs from a group of 13 are analyzed which besides social changes, showed post-operatively also severe disturbances of the alimentary behavior. After the operation aphagia was observed in three dogs. It lasted 4-16 days and was accompanied by profound hypophagia (70-84% in percent to preoperative level of food intake). In four dogs hypophagia was less pronounced (22-63%). In five dogs we observed vomiting in the early post-operative period. A marked decrease of general activity was observed in all dogs. They were apathetic and somnolent, looked depressed. Three aphagic dogs showed symptoms of negativism. In all dogs a bizzare stereotyped motor behavior was observed. Similar changes were described by Fonberg after amygdalar and hypothalamic lesions (22, 24, 58, 59). On the 5th day after operation — we could test most dogs only in "sit" and "paw" responses because of their atonic and inert general behavior. None of dogs performed "paw" response, and only one dog performed "sit" response in two trials. Generally they were not interested in our commands, they did not react in a normal way to our attempts to make contact with them (for example: calling by name). They were also quite indifferent towards petting. On the 11th day after operation — in spite of the improvement of the general state (only one dog remained aphagic) four dogs still were not giving any response in series I. The remaining two dogs started to work at a low level of performance and only 1 dog (with the smallest alimentary disturbances) worked at a level of 100% in both responses. In the second series — the first postoperative test session was performed on the 11th day after operation. Because earlier dogs apathy and atony may disturb their motor abilities. All dogs except one aphagic dog (S-3), who did not perform any reactions, performed both sprint and jump responses on 80-100% level of performance (Table II). During later post-operative period — the CRs of both series were recovered sooner or later without special retraining procedures. Only in one dog, Saturn (with hypothalamic lesions) the "lay" responses, in spite of retraining were never recovered. In spite of the appearance of a particular reaction during the first post-operative sessions, the dogs failed to perform it regularly later on. Their responses were fluctuating as a rule. The dogs were deconcentrated, they rarely looked of the experimenter and they did not express their pleasure and joy from the performance, as was usual before the operation. In the second series

TABLE II

Comparison of the post-operative recovery of food intake and social responses

| Dog | Food intake on 10th day after operation (in percent of preoperative level) | Number of proper responses on 11th day | | | |
|---|---|---|---|---|---|
| | | I series | | II series | |
| | | Sit | Paw | Sprint | Jump |
| S-3 | 50  | --- | --- | --- | --- |
| S-5 | 74  | --- | --- | 4 | 4 |
| S-4 | 15  | --- | --- | 5 | 4 |
| S-8 | 72  | 3 | 1 | 4 | 5 |
| S-2 | 136 | --- | --- | 5 | 5 |
| S-9 | 81  | 1 | 1 | 5 | 4 |
| S-1 | 78  | 5 | 5 | 5 | 5 |

(jump and sprint) performance was more regular, only in the jump response the performance dropped when the barrier height was raised. The performance during the early post-operative period is presented in Fig. 4.

We believe that the reason of such behavior was not a decrease of the dogs' motor abilities, because they could jump very well to the same height as before the operation when they were provoked by us in various manners. For instance: we called the dogs shaking the leash, which made a special noise, or we provoked them by jumping or stepping over the barrier, so that they would follow. Nevertheless, though the responses as such were soon recovered, they showed deterioration concerning latency, response duration and the height of barrier jumping. The behavior of dogs was very characteristic: the mean latencies of the first responses were during experimental sessions relatively not prolonged in comparison with the preoperative level, but later, in most trials the dogs walked slowly between technician and experimenter, turning around returning to the starting point, sniffing the floor and weakly responding to repetitions of command. A similar behavior was observed during "jump" and "sprint" responses as well as "sit" and "paw" responses.

Response to petting was also changed. Some dogs accepted the petting, but they lost their behavioral expression of pleasure state evoked by petting, which was always observed before the operation. They did not look at the experimenter, did not touch him, wave tail etc. As a rule the long-lasting indifference towards petting was observed. Dogs during the petting reward were distracted by environmental stimuli and often they terminated petting by moving away from the experimenter (Fig. 5A). It was not caused by hyperactivity as the time of lying without petting in "lay" responses was also prolonged (Fig. 5B).

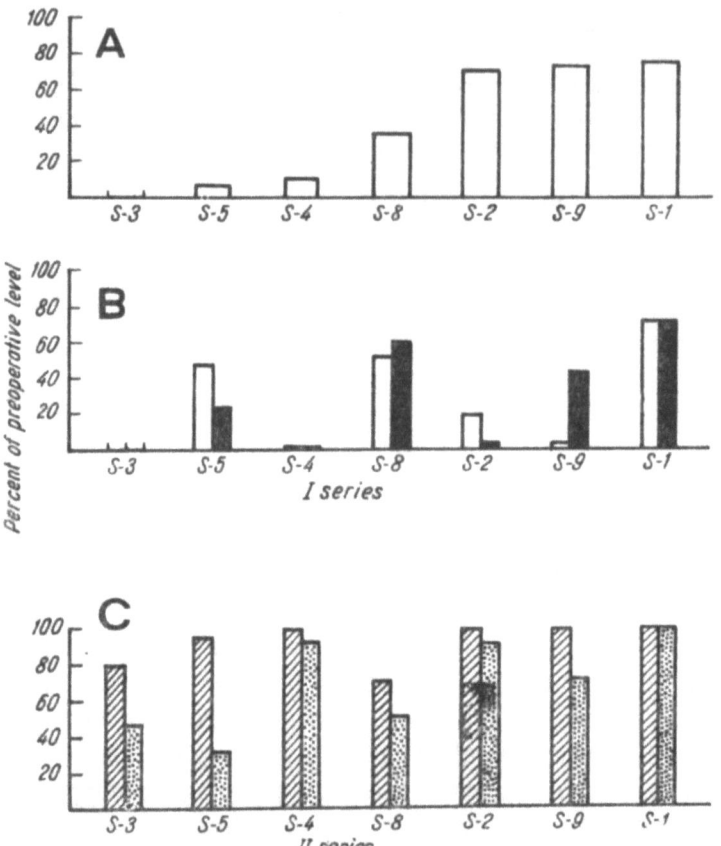

Fig. 4. Severity of alimentary disturbances and socio-emotional behavior in the early post-operative period (5 experimental sessions). A, Percentage of food intake in 10 day's period after operation in comparison to the preoperative level of each dog (100%). In the aphagic dogs amount of food intake was taken into account from days when they started to eat before the 10th day after the operation. B, Percentage of sit (black bars) and paw (white bars) correct responses from the first 5 sessions after operation in respect to criterional preoperative level (100%). C, Percentage of sprint (dashed bars) and jump (dotted bars) correct responses from the first 5 sessions after operation in respect to criterional preoperative level (100%).

## SOCIAL REWARD IN THE VIEW OF HEDONISTIC THEORIES

These results showed that amygdalar or/and hypothalamic lesions not only destroy the alimentary mechanisms as we assumed previously, but also impair other kind of motivation. Apathy, indifference, failure to experience pleasure from social rewards and disruption of social bonds with the experimenters points to deep changes of the "pleasure" system. Alimentary changes are probably — as we believe now, secondary and derive from general discruption of postulated by Stein (69, 70) the "reward circuit".

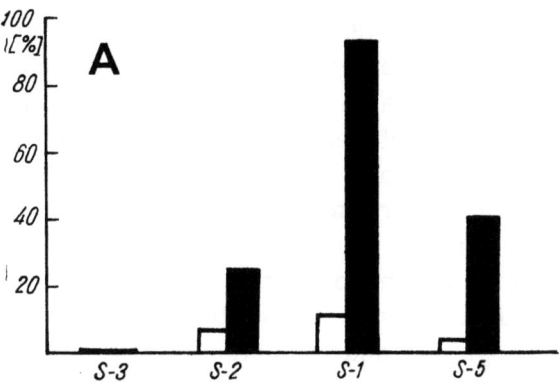

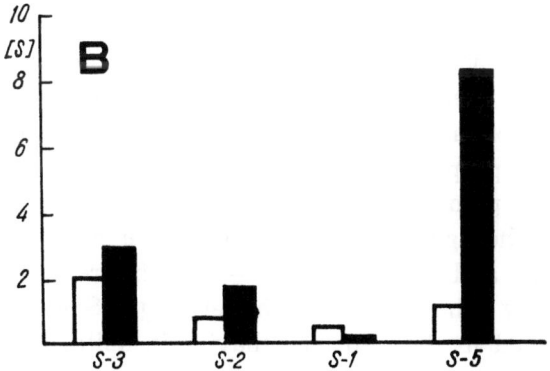

Fig. 5. Behavior related to the petting reward in the later period after operation (after recovery of those responses). A, Percentage of trials in which dogs terminated petting by not retaining the proper position. B, Mean time of lying without petting in "lay" responses. Bars represent 6 experimental sessions. White bars, before operation; black bars, after the operation.

If social reinforcement consists in a real physiological reward, it is possible that it is mediating by the same noradrenergic pathways as it was shown for food reward (1, 52, 76). This may take place especially in case of hypophagic — and aphagic dogs. There is a disagreement between authors as to the nature of reward pathways in the brain. Many transmitters were proposed by various authors as involved in reward mechanisms. They are: norepinephrine (50, 54, 55) dopamine (13, 28) serotonine (29) as well as the nicotine receptor of cholinergic system (51). Stiglick and White (75) hypothetized that reward depends more on the integrations of catecholaminergic systems within the medial forebrain bundle (MFB) than on specific dopaminergic or noradrenergic systems. In spite of this variety — all these transmitters play an important role in the hypothalamo-amygdalar system: dopamine (7, 47, 56, 77) serotonine (2, 3, 26, 79, 83) acetylcholine (8, 40) and norepinephrine (7, 70, 71). Independently from histochemical results, the role of hypothalamus and amygdala in the reward mechanisms was well demonstrated on the basis of self-stimulation experiments (21, 22, 31, 51, 84). If the hedonistic basis of "reward system" is, as assumed by Wyrwicka (91), of a sensory nature, various lesions may differently affect the specific sensory inputs i.e., tactile, taste or smell, which have projections into the hypothalamo-amygdalar system (43, 44, 49, 60, 81, 91). Amygdala as well as hypothalamus are considered to be involved in the recognition and evaluation of stimuli according to their hedonistic and biological values (4, 21, 22, 33, 34, 65, 66, 74, 87), and these various sensory-evaluating inputs to general "reward system" (which, as it is postulated by several authors, may be localized in the MFB) could be differently destroyed by lesions. Therefore, in addition to a general depression of reward mechanisms, some specific particular rewards may lose their hedonistic values more than the others. Such supposition may explain the fact that in spite of the recovery of food intake (even to hyperphagia in one dog), the social responses were still deteriorated. This point of view seems to be supported by our further studies with another group of dogs (which are not yet fully analyzed both behaviorally and anatomically) — which are either not changed in their alimentary behavior or even hyperphagic. Some of them however are deeply changed in their social behavior in spite of hyperphagia. On the other hand, in particular dogs eating may be connected in different ways with pleasure state. It is interesting that the general behavior of lesioned dogs is in many ways similar to some psychotic syndromes observed in psychiatric patients. In addition to disturbances in feeding and in socio-emotional behavior, a decrease of general activity, apathy, atony, negativism, bizzare stereotypies and cataleptic postures are also observed. These psychotic and depressive symptoms were noticed also in our previous investigations (22, 24), and actually attracted our attention because the changes in socio-emotional behavior, demonstrated in the present experiments, are commonly considered as a main feature of psychopathological syndrome in humans. Recently, several studies demonstrated the involvement of noradrenergic reward pathways (61, 70, 72) as well as dopaminergic pathways (10, 14, 82, 88), in schizophrenic — like symptoms. The role of serotonine in human depression was demonstrated earlier and its relation with hypothalamo-amygdalar mechanisms in mediating behavior was documented by Allikmets (2, 3). The behavior of our lesioned dogs seems to be easily explained by a general decrease of pleasure state. Like psychotic patients, they perceive stimuli (for example in "jump" and "sprint" they react to commands), but for them these signals do not have a proper hedonic value. This impairment of hedonic evaluation leads to a lack of interest, a decrease of goal-directed activity and lack of emotional attitude to environmental stimuli, which is reflected in a general deterioration of performance. In this respect our results agree with the hedonistic theory, postulating that a state of pleasure is a sufficient reinforcement for motivated behavior (11, 12, 53, 90, 93), and especially with the more dynamic understanding of hedonistic processes as was proposed by Cytawa and Trojniar (17), who stress the role of the current state of the organism and of previous sensory experience in these processes. On the basis of the hedonesthesia theory we can understand that our lesioned dogs performed better in the second series (jump and sprint), because these responses were less motivated by social reward in the sense of emotional contact and the amount of petting reward given to a dog.

The dogs were not interested in reaching their experimenter (long duration of sprint and jumps) and if they performed these tasks it was rather due to the stimulating effect of the experimental situation per se. Walking and even jumping over a low obstacte within the space of 15 × 5 m belongs rather to the innate behavioral patterns of dogs and the role of the experimenter as reinforcer is less important. On the opposite, the first series of tasks contained more socio-emotional aspects, a longer duration of petting reward, a closer contact of the experimenter with the dog. Therefore, if hedonic values of these stimuli decreased the level of performance in consequence dropped more than in the second series. In conclusion we assume that:

1. Social reward seems to belong to the general reward system including other kinds of rewards, e.g., the alimentary reward,

2. Damage of amygdalo-hypothalamic system results in disturbances in behaviors motivated by both alimentary and social rewards. Our results suggest that amygdala (as well as hypothalamus) are important in mediating hedonistic value of social and alimentary rewards.

We are greatly indebted to Mrs. J. Krakowska and Mrs. A. Kurzaj for technical assistance. This investigation was supported by Foreign Research Agreement 05-001-0 annex 283 A of the US Department of Health, Education and Welfare under PL 480, and partly by Polish Academy of Sciences — Project 10.4.1.01.4.

## REFERENCES

1. AHLSKOG, J. E., RANDALL, P. K. and HOEBEL, B. C. 1975. Hypothalamic hyperphagia dissociation from hyperphagia following destruction of noradrenergic neurons. Science 190 : 399–401.
2. ALLIKMETS, L. H., VAHING, V. A. and LAPIN, I. P. 1968. Effects of direct injection of mediators and chemicals influencing their metabolism into amygdala, septum and hypothalamus in cats. Zh. Vyssh. Nervn. Deyat. 18: 1044–1049.
3. ALLIKMETS, L. H., VAHING, V. A. and LAPIN, I. P. 1969. Dissimilar influences of imipramine, benactyzine and promazine on effects of microinjections of noradrenaline, acetylcholine and serotonine into the amygdala in the cat. Psychopharmacology (Berl) 15: 392–403.
4. ANAND, B. K. 1961. Nervous regulation of food intake. Physiol. Rev. 41: 677–708.
5. BARNES, G. W. and KISH, G. B. 1961. Reinforcing properties of the onset of auditory stimulation. J. Exp. Psychol. 62 : 164–170.
6. BECK, J. 1971. Instrumental conditional reflexes with sexual reinforcement in rats. Acta Neurobiol. Exp. 31: 251–262.
7. BEN-ARI, Y., ZIGMOND, R. E. and MOORE, K. E. 1975. Regional disturbation of tyrosil hydroxylase norepinephrine and dopamine within the amygdaloid complex of the rat. Brain Res. 87: 96–101.
8. BEN-ARI, Y., ZIGMOND, R. E., SHUTE, C. C. and LEWIS, P. R. 1977. Regional distribution of choline acetyltransferaze and acetylcholinesteraze within the amygdaloid complex and stria terminalis system of the rat. Brain Res. 120: 435–445.
9. BERLYNE, D. E. and KOENIG, I. D. V. 1965. Some possible parameters of photic reinforcement. J. Comp. Physiol. Psychol. 60: 276–280.
10. BIRD, E. D., BARNES, J., IVERSEN, L. L., SPOKES, E. G., MACKAY, A. V. P. and SHEPHERD, M. 1977. Increased brain dopamine and reduced glutamic acid decarboxylase and choline acetyl transferase activity in schizophrenia and related psychoses. Lancet 3: 1157-1159.
11. CABANAC, M. 1971. Physiological role of pleasure. Science 173: 1103–1107.
12. CABANAC, M. 1979. Sensory pleasure. The quarterly review of biology. 54: 1–29.
13. CROW, T. J. 1976. Specific monoamine systems as reward pathways: evidence for the hypothesis that activation of the ventral mesencephalic dopaminergic neurons and noradrenergic neurons of the locus coeruleus complex will support self stimulation responding. In A. Wauquier and E. T. Rolls (ed.), Brain stimulation reward. North-Holland Publ. Co., Amsterdam, p. 211–237.
14. CROW, T. J. JOHNSTON, E. C., LONDGEN, A. and OWEN, F. 1978. In P. J. Roberts, G. N. Woodruff and L. L. Iversen (ed.), Dopamine and schizophrenia. Advance Bioch. Psychopharm. Raven Press, New York, 19: 301–310.
15. CYTAWA, J., FRYDRYCHOWSKI, A., LUSZAWSKA and D. TROJNIAR, W. 1974. Instrumental reflexes reinforced with intragastric infusion of morfine. Acta Physiol. Pol. 25: 285–288.
16. CYTAWA, J. and TROJNIAR, W. 1976. The state of pleasure and its role in instrumental conditioning. Act. Nerv. Super. 18: 92–95.

17. CYTAWA, J. and TROJNIAR, W. 1978. Hedonesthesia the nervous process determining motivated ingestive behavior. Acta Neurobiol. Exp. 38: 139–151.
18. FONBERG, E. 1958. Transfer of instrumental avoidance reactions in dogs. Bull. Pol. Acad. Sci. Cl. VI. 6: 353–356.
19. FONBERG, E. 1960. Some features of defensive conditioned reflexes of type II. In E. A. Asratyan (ed.), Tsentralnye i pefericheskie mekhanizmy dvigatelnoi deyatelnosti zhivotnykh. Izad. Akad. Nauk SSSR, Moscow, p. 124–134.
20. FONBERG, E. 1962. Transfer of the conditioned avoidance reactions to the unconditioned noxious stimuli. Acta Biol. Exp. 22: 251–258.
21. FONBERG, E. 1967. The motivational role of the hypothalamus in animal behavior. Acta Biol. Exp. 27: 303–318.
22. FONBERG, E. 1969. The role of the hypothalamus and amygdala in food intake, alimentary motivation and emotional reactions. Acta Biol. Exp. 29: 335–358.
23. FONBERG, E. 1969. Effects of small dorsomedial amygdala lesions on food intake and acquisition of instrumental alimentary reactions in dogs. Physiol. Rev. 4: 739–743.
24. FONBERG, E. 1972. Control of emotional behavior through the hypothalamus and amygdaloid complex. In R. Porter and J. Knight (ed.), Physiology emotion and psychosomatic illness. Elsevier, North-Holland, Amsterdam, p. 131–161.
25. FONBERG, E. 1974. Amygdala functions within the alimentary system. Acta Neurobiol. Exp. 34: 435–466.
25a. FONBERG, E. 1979. The motivational changes following hypothalamic and amygdala lesions and attempt at their treatment. In T. N. Oniani (ed.), Neurophysiology of emotion and wakefulness-sleep cycle. Metsniereba Publ., Tbilisi, 3: 23–48.
26. FULLER, R. W. 1977. Pharmacology of serotonin neurons in the central nervous system. Fed. Proc. 36: 2133–2165.
27. FULLER, J. L., ROSVOLD, H. E. and PRIBRAM, H. K. 1957. The effect on affective and cognitive behavior in the dog of lesions of the piriform-amygdala-hippocampal complex. J. Comp. Physiol. Psychol. 50: 89–96.
28. FOURIEZOS, G., HANSSON, P. and R. A. WISE. 1978. Neuroleptic induced attenuation of brain stimulation reward in rats. J. Comp. Physiol. Psychol. 92: 661–671.
29. GROMOVA, E. A. 1977. Role of the brain monoamine system in learning on various emotional reinforcements. Act. Nerv. Super. 19: 127–128.
30. HARLOW, H. F. and MCCLEARN, G. E. 1954. Object discrimination learned by monkeys of the basis of manipulation motives. J. Comp. Physiol. Psychol. 47: 73–76.
31. HOEBEL, B. G. and TEITELBAUM, P. 1962. Hypothalamic control of feeding and self-stimulation. Science 135: 375–377.
32. HULL, C. L. 1943. Principles of behavior. Appleton-Century, New York, p. 422.
33. ILYUTCHENOK, R. Yu. 1977. Emotions and memory: role of the amygdaloid complex. Act. Nerv. Super. 19: 120–121.
34. ILYUTCHENOK, R. Yu. 1979. The emotiogenic brain structures in conditioning

mechanisms: conditioned evoked potentials and motor responses. Acta Neurobiol. Exp. 39: 503–515.
35. JONASON, K. R., ENLOE, L. J., CONTRUCCI, J. and MEYER, P. M. 1973. Effects of simultaneous and successive septal and amygdaloid lessions on social behavior of the rat. J. Comp. Physiol. Psychol. 83: 54–61.
36. KOGAN, J. 1955. Differential reward values of incomplete and complete sexual behavior. J. Comp. Physiol. Psychol. 48: 59–65.
37. KISH, G. B. and BARNES, G. W. 1961. Reinforcing effects of manipulation in mice, J. Comp. Physiol. Psychol. 54: 713–715.
38. KLING, A., DICKS, D. and GUROWITZ, E. M. 1969. Effects of amygdalectomy on social behavior in the cage ranged vervet (C. aethiop). In C. R. Carpenter and S. Karger (ed.), Proc. 2nd Sec. Int. Congr. Primat., New York, 1: 232–241.
39. KONORSKI, J. and MILLER, S. 1933. Podstawy fizjologicznej teorii ruchów nabytych. Ruchowe odruchy warunkowe. Książnica-Atlas TNSW, Warsaw, 168 p.
40. KOSMAL, A. and NITECKA, L. 1977. Cytoarchitecture and acetylcholinesterase activity of the amygdaloid nuclei in the dogs. Acta Neurobiol. Exp. 37: 363–374.
41. LAGERSPETZ, K. M. J. 1969. Aggression and aggressiveness in laboratory mice. In S. Garattini and E. B. Sigg (ed.), Aggressive behaviour. Excerpta Medica Foundation, Amsterdam, p. 77–85.
42. LEUKEL, F. and QUINTON, E. 1964. Carbon dioxide effects on acquisition and extinction of avoidance behavior. J. Comp. Physiol. Psychol. 57: 267–270.
43. MACHNE, X. and SEGUNDO, J. P. 1956. Unitary responses to efferent volleys in amygdaloid complex. J. Neurophysiology 19: 232–239.
44. MARSHALL, J. F., TURNER, H. B. and TEITELBAUM, P. 1971. Sensory neglect produced by lateral hypothalamic damage. Science 174: 523–525.
45. MILLER, S. and KONORSKI, J. 1928. Sur une forme particuliére des réflexes conditionnels. C. R. Seances Soc. Biol. Fil. 99: 1155–1157.
46. MONTGOMERY, K. C. 1954. The role of exploratory drive in learning. J. Comp. Physiol. Psychol. 47: 60–69.
47. MORA, F. and MEYERS, R. D. 1977. Brain self-stimulation direct evidence for the involvement of dopamine in the prefrontal cortex. Science 197: 1387–1389.
48. NACHMAN, M. 1963. Learned aversion to the taste of litium chloride and generalization to other salts. J. Comp. Physiol. Psychol. 56: 343–349.
49. NORGREN, R. 1976. Taste pathways to hypothalamus and amygdala. J. Comp. Neurol. 166: 17–30.
50. OLDS, J. 1975. Reward and drive neurons. In A. Wauquier and E. T. Rolls (ed.), Brain-stimulation reward. North-Holland Publ. Comp., Amsterdam, p. 1–26.
51. OLDS, J. 1977. Drives and reinforcements: Behavioral studies of hypothalamic functions. Raven Press, New York, Part VII, 140 p.
52. OLTMANS, G. A. and HARVEY, J. A. 1976. Lateral hypothalamic syndrom in rats, a comparison of the behavioral and neurochemical effects of lesions placed in the lateral hypothalamus and nigrostriatal bundle. J. Comp. Physiol. Psychol. 90: 1051–1062.
53. PFAFFMANN, C. 1960. The pleasure of sensation. Physiol. Rev. 67: 253–268

54. POSCHEL, B. P. H. and NINTEMAN, F. 1963. Norepinephrine: A possible excitatory neurohormone of reward system. Life Science 2: 728–788.
55. RITTER, S. and STEIN, L. 1974. Self-stimulation in the mesencephalic trajectory of the ventral noradrenergic bundle. Brain Res. 81: 145–157.
56. RODGERS, R. J. 1977. Attenuation of morphine analgesia in rats by intra-amygdaloid injection of dopamine. Brain Res. 130: 156–162.
57. ROSVOLD, H. E., MIRSKY, A. F., and PRIBRAM, K. H. 1954. Influence of amygdalectomy on social behavior in monkeys. J. Comp. Physiol. Psychol. 47: 173–178.
58. ROŻKOWSKA, E. and FONBERG, E. 1970. The effects of lateral hypothalamic lesions on food intake and instrumental alimentary reflex in dogs. Acta Neurobiol. Exp. 30: 59–68.
59. ROŻKOWSKA, R. and FONBERG, E. 1972. Impairment of salivary reflexes after lateral hypothalamic lesions in dogs. Acta Neurobiol. Exp. 32: 711–720.
60. SAWA, M. and DELGADO, J. M. R. 1963. Amygdala unitary activity in the unrestrained cat. Electroencephalogr. Clin. Neurophysiol. 15: 637–650.
61. SCHALLERT, T., WHISHAW, I. O., De RYCK, M. and TEITELBAUM, P. 1978. The postures of catecholamine-depletion catalepsy: Their possible adaptive value in termoregulation. Physiol. Behav. 21: 817–820.
62. SCHWARTZ, M. 1956. Instrumental and consumatory measures of sexual capacity in the male rat. J. Comp. Physiol. Psychol. 49: 328–333.
63. SHEFFIELD, F. D., WULFF, J. J. and BACKER, R. 1951. Reward value of copulation without sex drive reduction. J. Comp. Physiol. Psychol. 44: 3–8.
64. SHIPLEY, J. E. and KOLB, B. 1977. Neural correlates of species — typical behavior in the Syrian Golden Hamster. J. Comp. Physiol. Psychol. 91: 1056–1073.
65. SIMONOV, P. V. 1972. The role of hypothalamus and amygdala in the regulation of emotions (in Russian). In A. V. Valdman (ed.), Eksperimentalnaya neirofiziologiya emotsii. Nauka, Leningrad, 93–107.
66. SIMONOV, P. V. 1979. The role of emotions in formation of instrumental conditioned reflexes. Acta Neurobiol. Exp. 39: 621–632.
67. SKINNER, B. F. 1938. The behavior of organisms: An experimental approach. Appleton-Century, New York.
68. SOLOMON, R. 1980. Recent experiments testing an opponent — process theory of aquired motivation. Acta Neurobiol. Exp. 40 : 271–289.
69. STEIN, L. 1964. Reciprocal action of reward and punishment mechanims. In R. G. Heath (ed.), The role of pleasure in behavior. Harper and Rov., New York, p. 113–139.
70. STEIN, L. 1969. Chemistry of purposive behavior. In I. Tapp (ed.), Reinforcement and behavior. Academic Press, New York, p. 328–355.
71. STEIN, L. and WISE, C. D. 1969. Release of norepinephrine from hypothalamus and amygdala by rewarding medial forebrain stimulation and amphetamine. J. Comp. Physiol. Psychol. 67: 189–198.
72. STEIN, L. and WISE, C. D. 1971. Possible etiology of schizophrenia. Progressive damage to the noradrenergic reward system by hydroxydopamine. Science 171: 1032–1036.
73. STELLAR, E. 1980. Brain mechanisms and hedonic processes. Acta Neurobiol. Exp. 40 : 313–324.

74. STEVENSON, J. A. F. 1969. Neural control of food and water intake, hypothalamus. Charles C. Thomas, Springfield, 524 p.
75. STIGLICK, A. and WHITE, N. 1977. Effects of lesions of various medial forebrain bundle components on lateral hypothalamic self stimulation. Brain Res. 133: 45–63.
76. STRICKER, E. M. and ZIGMOND, M. J. 1976. Brain catecholamines and the LH syndrome. In D. Novin, W. Wyrwicka and G. A. Bray (ed.), Hunger: Basic mechanisms and clinical implications. Raven Press, New York, p. 19–32.
77. ŚMIAŁOWSKI, A. 1976. The effect of introhippocampal administration of dopamine or apomorphine in EEG of limbic structures in the rabbit brain. Pol. J. Pharmacol. Pharm. 28: 579–585.
78. TAYLOR, C. J. 1969. Effects of palatable nonnutritive bulk as a reinforcement. J. Comp. Physiol. Psychol. 69: 286–290.
79. TELEGDY, C. and VERMES, I. 1977. Effect of environmental changes on activity of the serotonergic system in limbic brain structures and its relation to the function of the pituitary — adrenal system in rats. Acta Nerv. Sup. 19: 306–307.
80. THOMPSON, G. I., BERGLAND, R. M. and TOWFIGHI, J. T. 1977. Social and nonsocial behaviors of adult rhesus monkeys after amygdalectomy in infancy or adulthood. J. Comp. Physiol. Psychol. 91: 533–548.
81. TURNER, B. H. 1973. Sensimotor syndrome produced by lesions of the amygdala and lateral hypothalamus. J. Comp. Physiol. Psychol. 82: 37–37.
82. UNGERSTED, U., LJUNGBERG, T. and SCHULTZ, W. 1978. Dopamine receptor mechanisms behavioral and electrophysiological studies. Adv. Biochem. Psychopharmacol. 19: 311–322.
83. WANG, R. Y. and AGHAJANIAN, G. K. 1977. Inhibition of neurons in the amygdala by dorsal raphe stimulation mediation through a direct serotonergic pathways. Brain Res. 120: 85–102.
84. WAUQUIER, A., MELIS, W., DESMEDT, L. K. C. and SADOWSKI, B. 1976. Self-stimulation in dogs: behavioural effects of anterior basal forebrain, amygdala and lateral hypothalamus implantation. In A. Wauquier and E. T. Rolls (ed.), Brain-stimulation reward. North-Holland Publ. Co., Amsterdam p. 427–430.
85. WEISS, B. 1957. Thermal behavior of the subnourished and pantothenic-acid deprived rat. J. Comp. Physiol. Psychol. 50: 481–485.
86. WENZEL, B. M. 1959. Tactile stimulation as reinforcement for cats and its relation to early feeding experience. Psych. Rep. 5: 297–300.
87. WHITE, N. and WEINGARTEN, H. 1976. Effects of amygdaloid lesions on exploration by rats. Physiol. Behav. 17: 73–79.
88. WISE, C. N. and STEIN, L. 1973. Dopamine-B-hydroxylase deficits in the brain of schizophrenic patients. Science 181: 344–347.
89. WOODS, J. W. 1956. Taming of the wild norway rat by rhinencephalic lesions. Nature 4538: 869.
90. WYRWICKA, W. 1972. The mechanisms of conditioned behavior. Charles C. Thomas, Springfield, Illinois, p. 1–179.
91. WYRWICKA, W. 1976. The problem of motivation in feeding behavior. In

D. Novin, W. Wyrwicka and A. Bray (ed.), Hunger: basic mechanisms and clinical implications. Raven Press, New York, p. 203–213.
92. VAN SOMMERS, P. 1963. Carbon dioxide escape and avoidance behavior in the brown rat. J. Comp. Physiol. Psychol. 56: 584–589.
93. YOUNG, P. T. 1959. The role of effective processes in learning and motivation. Psychol. Rev. 66: 104–125.

E. FONBERG and E. KOSTARCZYK, Nencki Institute of Experimental Biology, Pasteura 3, 02-093 Warsaw, Poland.

Lecture delivered at the Warsaw Colloquium on Instrumental
Conditioning and Brain Research
May 1979

# EXPLORATORY MOTIVATION IN RESPONSE-TO-CHANGE TEST IN RATS

Irena ŁUKASZEWSKA and Zdzisława MŁODKOWSKA

Department of Neurophysiology, Nencki Institute
of Experimental Biology
Warsaw, Poland

Abstract. The tendency to select T-maze arm that has been changed in brightness between two consecutive trials was investigated under different experimental conditions. In trial 1 rat was confined in maze stem for 10 min and allowed to observe black-and-white arms through transparent partitions. In trial 2 the color of one arm was changed, so that both arms were alike (either white or black) and the rat was allowed a free choice. Eight tests separated by 48 h intervals were performed on 4 groups of 20 rats. In the first group rats were removed from the maze immediately after the arm choice. This group showed preponderance of responses to change only in the first test. In successive test repetitions the choices of changed arm dropped to chance level. Rats spent less time on maze arms observation and their choice-latency increased. Two procedural modifications, i.e. providing an exploratory reward for response to change and an opportunity to explore the black and white maze arms independently from test situation, were introduced separately in the second and third group. Each modification resulted in the increase of choice frequency of the changed arm and in the decrease of response latency relative to first group. Application of both modifications in the fourth group led to its superior performance during test repetitions. The effects appeared to result from providing outcome of responses to change and/or attenuation of frustration caused by confinement in trial 1.

## INTRODUCTION

Exploratory behavior appears primarily to give the animal information about its environment. Berlyne (1) distinguished inspective and inquisitive exploratory responses. According to his distinction, inspective exploration concerns stimuli which are available in the animal's visual field, while inquisitive exploration consists in seeking such stimuli. There is much evidence that inspective responses decline as the animal gets information about the particular stimulus. Since inquisitive responses occur with respect to the expectancy of the stimuli not seen by the animal, their persistence should depend on the response outcome. Some support for this supposition comes from the experiment of Sutherland (11), who found that the level of spontaneous alternation could be increased by stimuli that become visible only after the choice. He concluded that spontaneous alternation may occur with respect to the rat's expectancy of the stimuli it would receive beyond the choice point.

It seems likely that a similar mechanism operates in the response-to-change phenomenon. Response-to-change refers to a tendency of the rat to approach the place that has been changed between two successive experiences (3). The testing procedure consists of two trials in an enclosed T-maze. In trial 1 the rat is exposed to arms differing in brightness, but prevented from entering them by transparent partitions. In trial 2, with partitions removed, he is free to enter either arm, both alike in brightness. If the rat's response reflects inspective exploration, choices of either arm would be equally probable. However, it has been found in a number of studies (3–9, 11), that rats tend to select the arm which had been changed between trials. Thus, response-to-change might be viewed as inquisitive response and attributed to the rat's expectancy of some stimuli in the changed arm.

Our unpublished pilot experiment has shown that after a few repetitions of the response-to-change test the proportions of choices of changed arm decreased rapidly to chance level, even when successive tests were separated by 2–3 day intervals. In our procedure the rat was removed from the maze immediately after entering with four feet the particular arm, therefore his supposed expectancy of stimuli could not be satisfied, which might account for a sharp cessation of responses to change. The other possible interpretation of this result refers to conditions of trial 1. Visual stimuli exposed in maze arms presumably evoked inspective responses, but they could not be performed since the rat was prevented from entering the arms. Such situation repeated in successive tests might lead to a decrease in attention to presented stimuli and as a consequence affect the response-to-change frequency.

The present experiment was designed to answer two questions regarding the influence of the mentioned conditions on response-to-change phenomenon: (i) would availability of additional stimuli at the end of maze arm maintain the initial level of responses, and (ii) would exploration of the same maze arm set as used in trial 1, but independently of test situation, have similar effect.

Four groups were used. The first group served as a control and rats were removed from the maze just after the choice, the second and third groups were given a single factor (additional stimuli or exploration), in the fourth group both factors were applied.

## METHODS

*Subjects.* Ss were 80 male Wistar rats housed in groups of 5 with food and water available ad lib. At the time of experiment they were approximately 100 days old. Each rat received intensive handling prior to test. On the day preceding the test directional preferences of rats were examined in a T-maze, painted gray.

*Apparatus.* An enclosed T-maze was employed, which was 30 cm deep and 11 cm wide. The stem was 25 cm long, and the arms were 40 cm long. The end walls of the arms had a one way door leading to the open platform 25 × 25 cm. The arms were detachable. Two sets of black and white arms were used, thus it was possible to present the same color on both sides of the maze. A mirror hung above the maze enabled the observation of rat behavior without disturbing it. The apparatus was located symmetrically in respect to the window and the walls of the room.

*Testing procedure.* Each test consisted of two trials. During the first — exposure trial — the rat was allowed to observe maze arms, one — white, the other — black, but he was prevented from entering the arms by wiremesh partitions. After 10 min exposure, the rat was removed from the maze and reintroduced 1 min later for the second trial, labelled choice trial. Between trials, while the rat was out of the maze, one arm was altered, so that both arms were either white or black. The partitions were removed, so in the second trial the rat was free to enter either arm. An entry was recorded, as well as two other behavioral measures: (i) observing time, which was recorded whenever the rat had his head turned in the direction of one or the other arm, and (ii) choice latency i.e., time elapsed between putting the rat in the stem and his entering the arm with four feet. Both measures were recorded with a stop watch.

*Experimental design.* Four groups were used, each consisting of 20 animals. Group 0 served as a control and rats were removed from the maze immediately after the choice. In Group R, when the rats responded to stimulus change, they could enter the open platform and look around for 30 s. Rats which selected the unchanged arm were removed from the maze. No correction was allowed. In Group E the response-to-change was not followed by access to the open platform, but 2 h after the test the rats were allowed to explore for 3 min the same maze as the one used in the test with black and white arms. The position of white and black colors could be the same as in the test preceding exploration, or different. In Group RE both factors, i.e., access to open platform after the choice of changed arm and maze exploration following the test, were applied. In each group rats were randomly assigned to four conditions resulting from the spatial arrangement of colors in the Exposure trial (white-right, black-left and vice versa) and the change of colors making both arms white or black. Eight successive tests, separated by 2–3 day intervals were performed in each group. In test 1, for each rat the color was changed in the arm which was opposite to his directional preferences, examined on the preceding day. In further tests color was changed randomly.

## RESULTS

*Choice reaction.* Table I shows the frequency of choice of the changed arm in 4 groups during all 8 test applications. In Group 0 only in the first application of the test, the number of rats responding to stimulus change different significantly from chance level (binomial test, $P < 0.001$). In further repetitions of the test, the number of responses to change decreased, becoming close to chance level and furthermore, some rats (number in parenthesis) even refused to move from starting place. On the contrary, Group RE showed in all test applications (except the first one) a significant preponderance of responses to change. In Group R and Group E also a high number of responses to change was observed, although it missed the level of significance in some repetitions. Analysis of variance (Groups x Tests) with row data transformed to arc. sin $\sqrt{P}$, where $P$ denoted percentage of choice of the changed arm, revealed reliable effect of Groups ($F = 4.10$, $df = 3/76$, $P < 0.01$), while effects of test repetitions and interaction (Groups x Tests) were insignificant. Duncan Test indicated a difference between Group 0 and other groups on $P < 0.05$ level. Groups R, E and RE did not differ significantly from each other.

Although in Group 0 response-to-change frequency in particular test

TABLE I

Number of rats selecting the changed arm in 8 tests under 4 different conditions (in each group $n = 20$)

| Test<br>Group and condition | I | II | III | IV | V | VI | VII | VIII | Σ | % | $X^2$ | $P$ |
|---|---|---|---|---|---|---|---|---|---|---|---|---|
| 0 nothing | 18 | 14 | 11(2) | 11(1) | 11(1) | 11(2) | 13 | 11(2) | 100(8) | 62.5 | 10.74 | $P < 0.05$ |
| R reward | 17 | **18** | **16** | 14 | **15** | **16** | 13 | 14 | 123 | 76.9 | 25.12 | $P < 0.001$ |
| E maze exploration | 17 | **19** | **15** | 11 | 13 | 13 | **17** | **16** | 121 | 75.6 | 25.91 | $P < 0.001$ |
| RE both | 14 | **16** | **16** | **17** | **15** | **15** | **17** | **15** | 125 | 78.2 | 28.13 | $P < 0.001$ |

Bold number represent are frequencies which differ significantly from chance level (binomial test). In parenthesis — numbers of rats which refused to make any choice.

was close to chance level, the percentage of choice of the changed arm in all 8 test repetitions was considerably higher than the percentage of choice of the unchanged arm. The significance of this proportion was evaluated by $\chi^2$ test (Table I, last column). The result of $\chi^2$ test indicates that Group 0 responded on chance level throughout the Experiment. For the other three groups, $\chi^2$ test showed a highly significant prevailence of proportions of responses to changed arm. This is in accordance with the result of analysis of variance.

*Observing times.* Figure 1 presents cumulative frequency distributions of observing times in four groups. Data are collected from 8 tests.

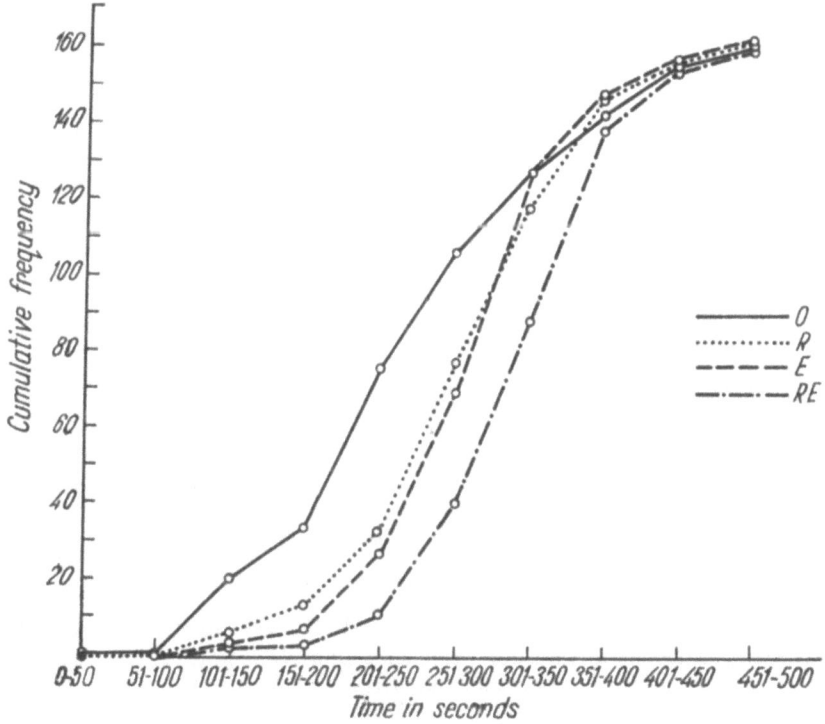

Fig. 1. Cumulative frequency distributions of observing times in different groups. Data of 20 rats from 8 test were analysed, thus total frequency in each group was 160.

In Group 0 the number of short observing times was higher than in other groups, while the opposite holds for observing times in Group RE. For example, it can be seen from Fig. 1 that in Group 0 around half of observing times was below 250 s, whereas in Group RE only 10 so short observing times were noted. The Smirnov two-tailed test (2) indicated the significant difference between the distribution of observing times in Group 0 and those in the remaining groups ($P < 0.01$). The distribution in Group RE differs from that in Group R and Group E on $P < 0.001$ level. Groups R and E did not differ from each other in this respect and occupied the intermediate position.

In successive test repetitions observing time diminished in all groups except the Group RE. Figure 2 shows mean observing times in the first and the last test in particular groups. The Wilcoxon test (two-tailed)

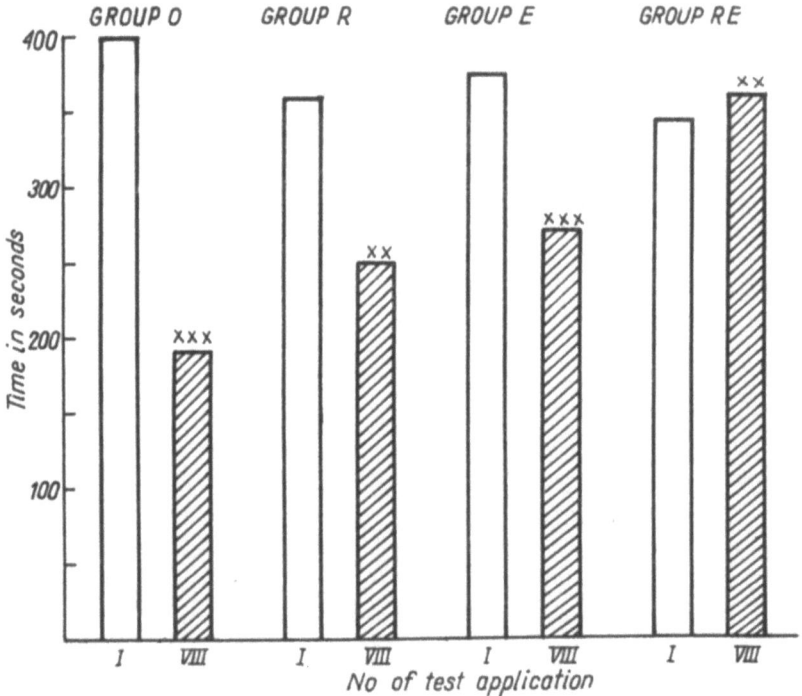

Fig. 2. Mean observing time in I-st and VIII-th test applications in different groups. Significance of differences between I and VIII test: x $P < 0.05$; xx, $P < 0.02$; xxx, $P < 0.01$ (Wilcoxon test, two-tailed).

indicated a significant decrease of observing time in groups 0, R and E ($P_s < 0.01$), while in Group RE observing time slightly but significantly increased ($P < 0.02$).

*Latency of choice.* Figure 3 presents cumulative frequency distribution of choice latencies in four groups. In Group 0 the number of short latencies was lower than in the other groups. The number of latencies below 5 s was two times smaller in Group 0 than in each of the other groups. As indicated by Smirnov two-tailed test, the difference between

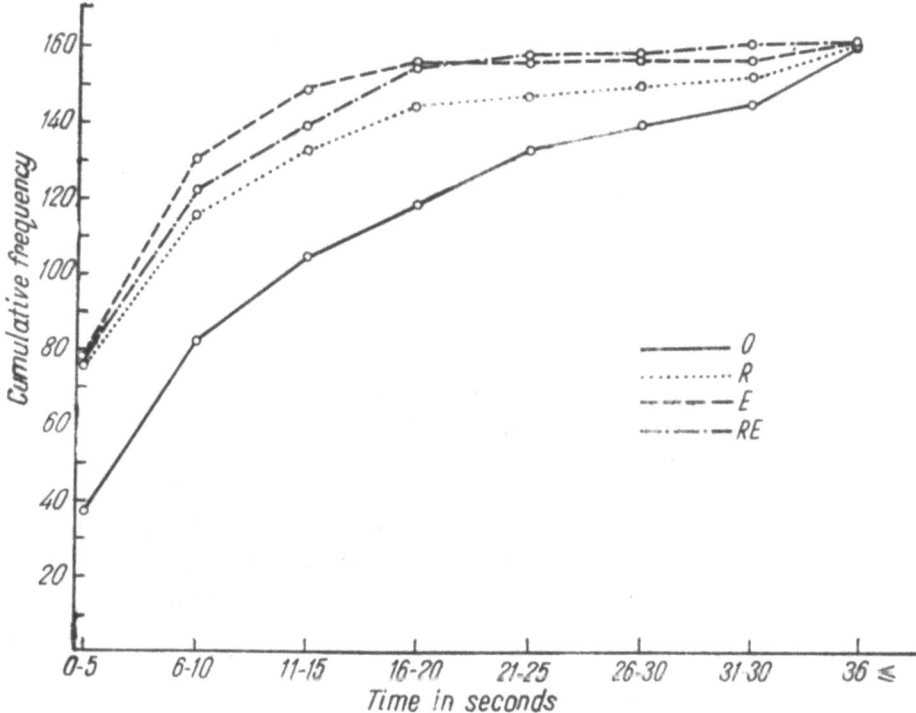

Fig. 3. Cumulative frequency distributions of choice latencies in different groups. Denotations as in Fig. 1.

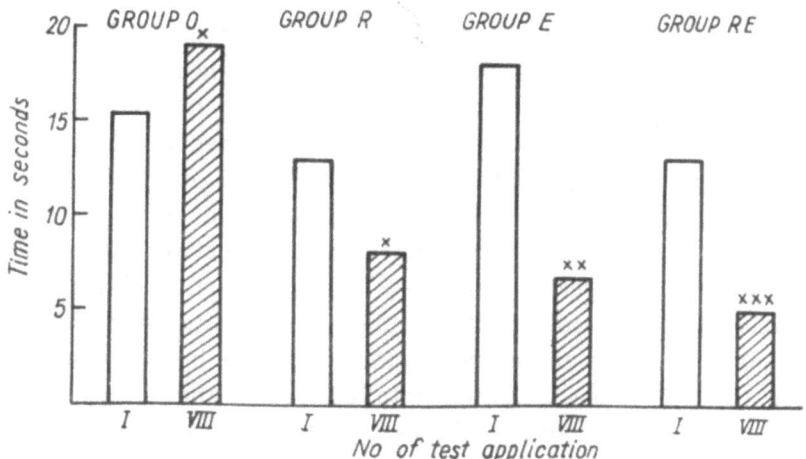

Fig. 4. Mean choice latency in I-st and VIII-th test applications in different groups. Denotations as in Fig. 2.

the distribution of choice latencies in Group 0 and in other groups was significant on $P < 0.001$ level. Groups E, R and RE did not differ from one another.

In the course of test repetition, choice latency in Group E, R and RE decreased. In contrast, choice latency of Group 0 was elongated. The differences between mean values of choice latency of the first and the last test in particular groups are shown in Fig. 4.

## DISCUSSION

Although response-to-change phenomenon was investigated in a number of studies, in none of them it was tested repeatedly on the same subjects. Our data in Group 0 showed that repetitive exposure to change in stimulus pattern abolished the rat's tendency to approach the place of change. It also reduced observing time which reflected decrease of attention to presented stimuli and increased response latency, which pointed to a diminished motivation for the choice of maze arm. These findings indicate that procedure used in the majority of studies on response-to-change, involving confinement of rat between maze arms and no outcome of his response, creates unfavorable conditions for persistent responding to stimulus change.

Alterations in procedure influenced positively the rat's performance. Group R rewarded for selection of changed arm by access to the open platform, as well as Group E, which was allowed to explore the maze after each test, differed significantly from Group 0 in respect of all three measures recorded, i.e. observing time, choice latency and frequency of choice of the changed arm. The effects of these operations were equivalent, since Group R and Group E did not differ markedly from each other in any measure. The effects of these operations were additive. Group RE which was given both of them showed the best performance. This concerned mainly observing times which were significantly longer in Group RE than in Group R or Group E. Besides, only in Group RE observing time increased in successive tests. Although the differences in frequency of response-to-change between groups R, E and RE were insignificant, still, only Group RE showed reliable preponderance of choices of changed arm in all repetitions of the test.

The above findings might be explained on the basis of Berlyne distinction of exploratory behavior into inspective and inquisitive types (1). In Exposure Trial, visual stimuli in maze arms presumably elicited exploratory drive of the inspective type, directed equally toward both arms. Because such drive could not be satisfied by entering the arms, rats became frustrated, which resulted in decreased observation of maze arms in further tests. This may account for a decrease in the number of responses to stimulus change in Group 0. Moreover, seeing free entrance to maze arms in Choice Trial, subjects of this group may perform inspective responses i.e. choose either arm, instead of reacting to change. Exploration of white and black set of arms practised after each test in Group E probably decreased the frustrative effect of Exposure Trial, as well as a tendency to inspective responses in Choice Trial, hence more favorable conditions for response-to-change.

We have realized that the frustrative effect of confinement in Exposure Trial might be also diminished by shortening the time of exposure. Our previous study (7) showed that in single test application, the highest frequency of responses-to-change occurred with exposure time between 5 and 15 min. It is possible that in successive tests the time necessary to perceive the difference between colors of maze arms might decrease. However, we had no indications how much the exposure duration could be lowered, thus extensive preliminary studies would be needed to establish the proper procedure of gradual shortening the Exposure trial.

The suggestion of inquisitive character of response to changed arm in Choice Trial is supported by data obtained in Group R. Improvement of performance due to rewarding the selection of the changed arm by an access to open platform, which brings the animal into visual contact with the environmental stimuli, indicates that rats seek stimuli which are not available at the choice point. The beneficial effect of exposure to environmental visual stimuli following selection of changed arm cannot be attributed to the "reward" as such. We have found (in preparation) that when alimentary reward was applied, rats stopped attending to colors in maze arms and did not respond to stimulus change. Instead, they tended to select the arm, in which they had found food on preceding test trial.

Results obtained on Group E and Group R point to two different exploratory responses occurring in two trials of response-to-change test, namely, inspective responses in Exposure trial and inquisitive ones in Choice trial. Consistent with this interpretation is the superior performance of Group RE that was given both the additional exploration and the reward. Attenuation of frustration caused by unsatisfied inspective exploratory drive and reinforcement of inquisitive responses could maintain the performance in response-to-change test on a considerably high level.

## REFERENCES

1. BERLYNE, D. E. 1960. Conflict, arousal and curiosity. McGraw-Hill Book Co., New York.
2. CONOVER, W. I. 1971. Practical nonparametric statistics. John Wiley and Sons, New York, 462 p.
3. DEMBER, W. N. 1956. Response by the rat to the environmental change. J. Comp. Physiol. Psychol. 49: 93–95.
4. DEMBER, W. N. and MILLBROOK, B. A. 1956. Free-choice by the rat of the greater of two brightness changes. Psychol. Rep. 2: 465–467.
5. FOWLER, H. 1958. Response to environmental change: A positive replication. Psychol. Rep. 4: 506.
6. LEVINE, S., STAATS, S. R. and FROMMER, G. 1958. Studies on "Response by the rat to environmental change". Psychol. Rep. 4: 139–144.
7. ŁUKASZEWSKA, I. 1978. The effects of exposure time and retention interval on response to environmental change in rats. Acta Neurobiol. Exp. 38: 323–331.
8. MARKOWSKA, A. and ŁUKASZEWSKA, I. 1974. Short-term memory of spatio-visual events preserved after frontomedial or frontopolar lesions in rats. Acta Neurobiol. Exp. 34: 715–721.
9. O'CONNEL, R. 1971. The response to stimulus change as a brightness scaling technique. Psychon. Sci. 22: 275–277.
10. SUTHERLAND, N. S. 1975. Spontaneous alternation and stimulus avoidance. J. Comp. Physiol. Psychol. 50: 358–362.
11. WOODS, P. and JENNINGS, S. 1959. Response to environmental change: A further confirmation. Psychol. Rep. 5: 560.

Irena ŁUKASZEWSKA and Zdzisława MŁODKOWSKA, Nencki Institute of Experimental Biology, Pasteura 3, 02-093 Warsaw, Poland.

Lecture prepared for the Warsaw Colloquium on Instrumental
Conditioning and Brain Research
May 1979

# FACTORS OF EXTINCTION OF ALIMENTARY INSTRUMENTAL CONDITIONED REFLEX

T. N. ONIANI and N. G. VARTANOVA

I. S. Beritashvili Institute of Physiology
Georgian Academy of Sciences
Tbilisi, USSR

*Abstract.* Observations were made on the animal's behavior and dynamics of electrical activity in the neo- and archipaleocortex during acquisition of sound discrimination under different experimental conditions and subsequent extinction. On the basis of analysis of the data obtained the following conclusions were drawn: (1) In pre-satiated cats even hundreds of applications of conditioned stimuli without food reinforcement do not lead to extinction of the conditioned reflex. Sound discrimination is not disturbed either. (2) Conditioned reflex and discrimination are not disturbed at repeated (over 300) application of conditioned sounds without food reinforcement provided the animal is hungry and is not allowed to approach the feeders. (3) Extinction of the conditioned reflex is achieved only when the animal is allowed, in response to conditioned stimuli, to approach the feeders where there is no reinforcing portion of food. (4) Stimulation of some mesodiencephalic structures results in the restoration of the extinguished alimentary instrumental reflexes and discrimination. (5) Functional inactivation of the hippocampus by way of induction of epileptiform discharges does not prevent the acquisition and extinction of conditioned feeding behavior. 6. Septal lesions do not prevent the acquisition, but extinction is tangibly delayed (7). It is concluded that the factor of extinction is the recognition that feeders contain no food rather than non-reinforcement, i.e. no food intake. On the basis of analysis of the dynamics of electrical activity in the neo- and archipaleocortex, as well as the mesodiencepalic electrical stimulation effects, some aspects of neurophysiological mechanisms of extinction are discussed.

INTRODUCTION

Scientific interest and ideas on the mechanisms of both the acquisition and extinction of the conditioned reflex arose concurrently and developed in parallel. The general laws governing the essence of the processes which must underlie the acquisition and extinction of conditioned reflexes were formulated by I. P. Pavlov (41). Since then it has been almost unanimously taken for granted that reinforcement of the conditioned stimulus by the unconditioned one is a basic factor in the acquisition of conditioned reflexes of different types, while withdrawal of reinforcement leads to extinction of the acquired habit. The question of the intimate mechanisms of extinction usually gave rise to controversy. Though I. P. Pavlov's viewpoint on the significance of the development of the so-called internal inhibition in the extinction of the conditioned reflex remains the most authoritative, it has not always been shared by leading investigators in the field of the brain mechanisms of learning and memory (24, 45). And even among those regarding the development of internal inhibition resulting from the abolition of reinforcement as the basic process underlying the extinction of the conditioned reflex, difference arose over the nature and localisation of the process in question (5, 28). These are precisely the fundamental questions which, so far, have not got a strict scientific foundation. Further analysis of the mechanisms of extinction apparently calls for a more thorough study of the factors leading to it. Non-reinforcement of the conditioned stimulus with the unconditioned one is a common factor, but it involves a number of components and elucidation of the predominant significance of the latter should undoubtedly help also in clarifying the mechanisms of extinction. From this point of view we have studied the factors of extinction of the alimentary instrumental conditioned reflex.

METHODS

Experiments were carried out on 20 cats in a special experimental chamber (Fig. 1). The rear part of the chamber with the area of 0.3 m² served as a starting section. In the front section (1 m²), at the ends of the side-walls, feeders were placed outside from which the animal could retrieve a piece of meat by raising the suspension door. Alimentary reflexes were elaborated to both feeders. The conditioned signal to one feeder was a tone and clicks to the other. Sources of sound stimuli were located above the feeders in the upper corners of the front chamber. Five seconds after the onset of either signal the door of the starting section was opened and the cat was allowed to perform feeding behavior. After receiving a reinforcing portion of meat the cats were trained to return to the starting section. In a special run of sessions a conditioned alimentary reflex was elaborated with the door of the starting section opened.

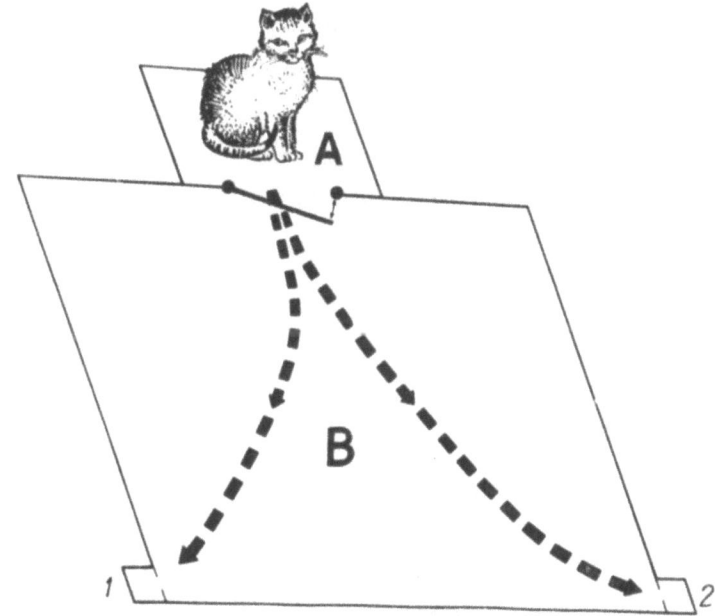

Fig. 1. Floor plan of chamber for elaboration and extinction of instrumental feeding behavior to two feeders. A, starting section; B, front chamber; 1-2, feeders.

When a stable sound discrimination was achieved, the extinction of the conditioned alimentary reflex started. With a view to revealing the role of motivation, different variants of experiments were conducted both in presatiated and in hungry animals. In the course of both acquisition and extinction of the alimentary conditioned reflex observations were made of the dynamics of background electrical activity of the structures of the neo- and archipaleocortex in response to conditioned stimuli. The effect of subcortical electrical stimulation on non-extinguished and extinguished feeding behavior was also studied. To this end, electrodes were chronically implanted by means of a stereotaxic apparatus. The background electrical activity was recorded on the San'ei 13-channel inkwriting electroencephalograph. Frequency-amplitude analysis of delta, theta, alpha, beta₁ and beta₂ rhythms in the neocortex and hippocampus was carried out on a 2-channel analyzer-integrator of the same manufacturer. Rhythms integrated in 5 s epochs were recorded on the electroencephalograph with a special pen. Rectangular pulses from a high frequency output generator were used to stimulate the subcortical structures. Upon completion of the experiments the cats were sacrificed, brains fixed in formalin and the localisation of recording and stimulanting deep electrodes was determined in the serial frontal sections. When required, the data were statistically processed and validity checked by Student's t-test.

## RESULTS

*Effect of non-reinforcement in the absence of food motivation*

In this series of experiments the cats, in which stable conditioned alimentary reflexes to two feeders had been elaborated, were fed ad libitum with meat and then the procedure of extinction was begun. In

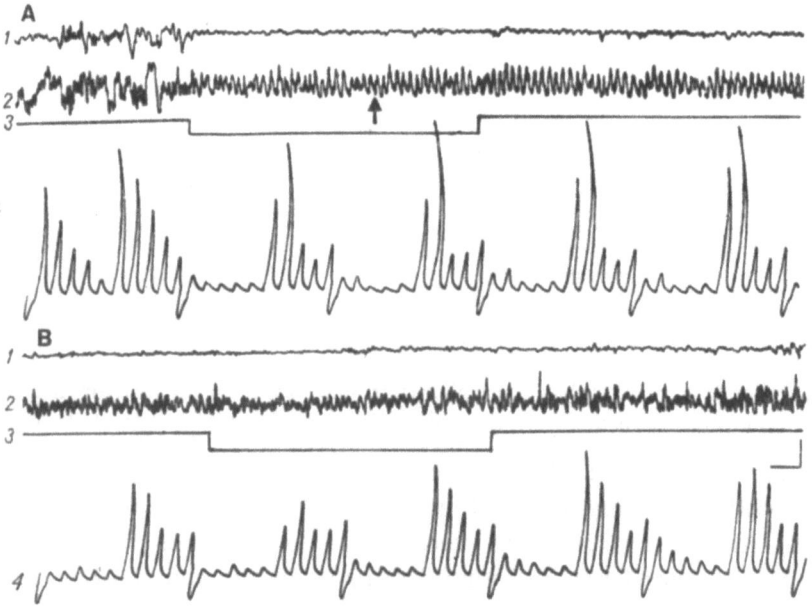

Fig. 2. Dynamics of electrical activity in the neocortex and hippocampus. *A*, in response to a non-extinguished directing conditioned signal and *B*, to an extinguished triggering signal (opening of the door). Leads: 1-sensorimotor cortex, 2-dorsal hippocampus, 3-signal line of directing conditioned stimuli (*A*) and opening of the door (*B*). Arrow on *A* indicates the moment of the door opening. 4-intergrated values of delta, theta, alpha, beta₁ and beta₂ rhythms of neocortex (the first five deflections) and hippocampus (the following five deflections) during 5 s epoch. Calibration: 200μV, 1 s.

response to conditioned stimuli the satiated cats either did not leave the starting section, or did so only 1–2 times and ceased to perform feeding behavior. In spite of this, conditioned stimuli were given at the usual interval (2–3 min) 30 times each per day. If in hungry cats a conditioned stimulus in parallel with desynchronisation of the electroneocorticogram evokes an increase of the hippocampal theta rhythm (Fig. 2A), in the satiated animals desynchronisation of both the neocortical and hippocampal electrical activity is observed without any considerable increase of the theta rhythm (Fig. 2B). Repeated delivery of conditioned stimuli resulted in a progressive reduction of EEG changes up to their total disappearance. The following day, in the pre-satiated animals, the EEG changes to conditioned stimuli recovered and at repeated trials the picture of habituation recurred. Though such experiments on satiated cats were run for 10 days and the conditioned stimuli with non-reinforcement were applied more than 300 times each, we failed to obtain an extinction of feeding behavior. If afterwards conditioned stimuli were applied in a hungry state, the cats were not only active in feeding performance, but 100% sound discrimination was also retained. It follows that non-reinforcement of a conditioned signal with unconditioned stimulation, i.e., the lack of food intake in the absence of relevant motivation cannot play the role of a factor eliciting the extinction of the conditioned alimentary reflex.

*Effect of non-reinforcement in the presence of food motivation and prevention of the instrumental response*

It is known that motivation plays a role in the acquisition of conditioned reflexes (22, 35, 44, 46). It has been demonstrated (1), that in pre-satiated dogs conditioned feeding behavior cannot be elaborated. Hence, it might be assumed that motivation may play a decisive role in the extinction of the conditioned reflex as well, thus explaining the absence of the latter in experiments with pre-satiated animals.

With a view to examining the significance of motivation for the extinction of conditioned feeding behavior, experiments were carried out on hungry animals; these were placed in the starting section and not allowed, while to leave it he conditioned stimuli were administered (30 times a day) without reinforcement. Even before the onset of the conditioned signal, the hungry cats were restless and tried to reach the feeders. In response to the conditioned stimuli their motor activity and emotional tension became sharply enhanced. Activation of the hippocampal theta rhythm during desynchronised electrical activity in the neocortex was also observed.

However, as the trials were repeated, the effect induced by the conditioned stimuli gradually attenuated and eventually the animals sat down quietly in the starting section. Against this background, the opening of the starting section without conditioned sound stimuli caused activation of the instrumental alimentary conditioned reflex and the animal ran up to the feeders. Naturally, this time they were not reinforced. This fact points to the absence of an acute extinction of the conditioned response. On the following experimental day the entire procedure was started anew, though an accelerated habituation to the conditioned stimuli was observed.

In these conditions the extinction of the instrumental response was not observed even after a great number (over 300) of applications of each conditioned sound stimulus without reinforcement. If during the conditioned stimulus the animal was given the opportunity, it correctly approached the relevant feeder, i.e. conditioned sound discrimination was fully retained. It follows that a single non-reinforcement of the conditioned stimulus with the unconditioned one is not sufficient for the extinction of conditioned feeding behavior, even in the presence of a high level of relevant motivation.

*Effect of non-reinforcement of a conditioned signal during a high level of motivation in the absence of a triggering stimulus*

As seen above, in the first two series of experiments, when, between the trials, the cat was kept in the starting section, instrumental feeding behavior towards the feeders was elaborated to a complex of conditioned stimuli. To one feeder the complex consisted of a tone + opening the door, while to the other, clicks + opening the door. In these conditions the tone and clicks clearly serve as directing conditioned stimuli, while the opening of the door acquires the property of a triggering stimulus. During extinction only the first components of these complexes were applied and non-reinforced, while to the second component (opening of the door) which was common for both complexes, actually no extinction occurred and, therefore it (i.e., the second component) could, though indirectly, activate the conditioned reflex mechanisms. On the basis of the above-described experiments one can judge about the absence of extinction only in regard to the directing conditioned stimuli in terms of their discrimination. Naturally, it might be suggested that though in these experiments extinction developed to the directing conditioned stimuli, it was abolished under the influence of the non-extinguished triggering conditioned stimulus, thereby restoring the ability to discriminate the directing conditioned stimuli. With a view to checking this assumption a special series of experiments was carried out, in which the triggering conditioned stimulus was lacking. The conditioned reflexes to the two feeders were elaborated for the same directing stimuli, with the door of the starting section open. The cats were trained after reinforcement of either conditioned stimulus to return to the starting section and wait for the onset of the following conditioned stimulus.

After elaboration of a stable (100%) discrimination of the directing conditioned stimuli, the procedure of extinction was started, which consisted in the following: a hungry cat was placed in the starting section, the door was closed and conditioned sound stimuli were delivered at usual intervals 30 times a day. It is natural that each time, at the beginning of the experiments, the hungry cats kept in the starting section were restless, but afterwards gradually subsided, as if an acute extinction of the conditioned reaction had occurred. However, at the end of the experimental session it sufficed to open the door and, to the conditioned stimuli, the cats correctly chose the previously non-reinforced feeder. Even after many days of experimenting and a great number of non-reinforcing (over 300 trials) of conditioned stimuli with food, no signs of extinction of the alimentary reflexes or disturbance of sound discrimination were observed. This invalidated the assumption regarding the disinhibitory influence of the triggering conditioned stimulus (opening of the door) on the extinction of directing conditioned stimuli, which was supposed to have occurred in the first two series of our experiments. Thus, the third series of experiments strengthened our conviction that non-reinforcement of conditioned stimuli, i.e. absence of food intake, does not lead to extinction of the conditioned reaction, irrespective of whether the conditioned stimuli are applied against the background of high or low level of food activation.

*Dynamics of extinction of conditioned reaction under the conditions of relevant feeding behavior performance*

In this series of experiments the cats were allowed — in response to conditioned stimuli — to accomplish a full cycle of learned feeding behavior, not accompanied by a reward. Most of the cats were found to develop acute extinction of conditioned feeding behavior over a one-day session. In the first trials the cats, let out of the starting section, visit not only the signalled feeder but the other, too, and by repeatedly opening the suspension door persistently demand the reinforcing portion of meat. Whereas under reinforcement they returned to the starting section on their own, now they had to be forced to return there. In the subsequent non-reinforcement trials emotional tension in the cats gradually diminished and after 20–30 trials they ceased to perform feeding behavior. On the following day the conditioned reflex activity was recovered, but acute extinction occurred more readily than the day before. After a few days (60–100 non-reinforcements on the average) extinction of conditioned feeding behavior became chronic, the cats did not respond to conditioned stimuli, though when placed in the chamber at the beginning of the experimental session they could spontaneously visit the feeders and check the presence of meat there. Chronically extinguished reflex could be restored after a few days, if during this time the procedure of non-reinforcement was discontinued. However, re-extinction occurred quickly after 1–2 non-reinforcements.

In the process of acquisition and extinction of the instrumental alimentary conditioned reflexes the electrical activity of the neo- and archipaleocortex undergoes regular dynamics. As soon as sound stimuli become conditioned, their application is attended by an increase of the hippocampal theta rhythm paralleled by desynchronised electrical activity in the neocortex (Fig. 2A). The hippocampal theta rhythm may be particularly enhanced in response to a triggering conditioned stimulus (opening of the door) and during the approach to the feeders. During eating and when the animal is returning to the starting section, the

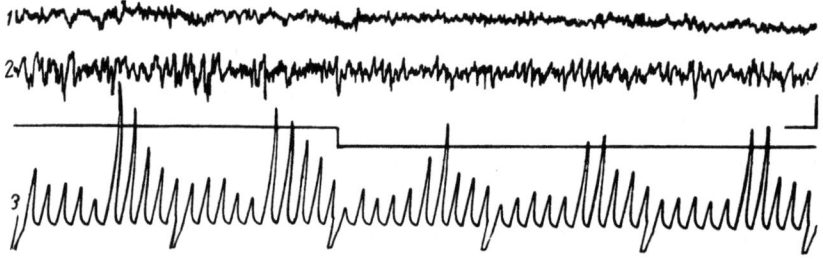

Fig. 3. Changes of electrical activity of neocortex and hippocampus in response to conditioned signals in satiated animal after conditioned reaction. Leads: 1-auditory cortex, 2-dorsal hippocampus, 3-signal line (deflections downwards indicate the onset of conditioned signal, arrow indicates the opening of the door), 4-integrated values of delta, theta, alpha, $beta_1$ and $beta_2$ rhythms of neocortex (the first five deflections) and hippocampus (the following five deflections) during 5 s epoch. Calibration: 200 µV, 1 s.

hippocampal theta rhythm is lacking. After acute extinction of feeding behavior conditioned stimuli evoke only desynchronisation of electrical activity both in the neocortex and hippocampus without an increase of the theta rhythm of the latter (Fig. 3). On the following day, with the recovery of the conditioned feeding behavior, the enhancement of theta rhythm in response to conditioned stimuli is also restored. Desynchronisation of electrical activity in the neo- and archipaleocortex, without the hippocampal theta rhythm increase to conditioned stimuli, may occur in the first trials of non-reinforcement and after deep chronic extinction of the conditioned feeding behavior. We failed to note a development of any sings of synchronization of the elestrocorticogram during the action of sound stimuli to which the conditioned response had been extinguished. On the contrary, as indicated above, conditioned stimuli to which the reaction had been chronically extinguished, may cause an activation of the electrocorticogram, which is observable during both, the slow wave and paradoxical phases. To restore the extinguished conditioned feeding behavior only one or several brief reinforcements were sufficient. Concomitantly, the discrimination of conditioned stimuli was also restored. Characteristically enough, after chronic extinction of conditioned feeding behaviour, a single reinforcement of any directing stimulus causes immediate restoration of the reflex to both feeders.

*Activation of non-extinguished and extinguished feeding behavior by electrical stimulation of subcortical structures*

It is known that electrical stimulation of some subcortical structures results in the activation of the earlier elaborated conditioned reflexes (3, 14, 15, 20, 21, 30, 38, 51). In our experiments (38, 39, 40) activation of instrumental feeding conditioned behavior was observed during electrical stimulation of the lateral hypothalamus, cortico-medial amygdala and some points in the mesencephalic reticular formation. In pre-satiated animals a similar effect may be evoked only in response to electrical stimulation of those points in lateral hypothalamus, where activation also evoked unconditioned feeding behavior. Activation of conditioned feeding behavior consisted in the following: during electrical stimulation the cats left the starting section, approached the feeders, opened the suspension door and retrieved the portion of meat. Upon cessation of stimulation, they returned to the starting section. If such a reaction was not rewarded with meat, then at repeated stimulations extinction of the activated conditioned reaction was observed. Characteristically, extinction of an activated conditioned reaction during stimulation of the lateral hypothalamus requires more non-reinforcements than during stimulation of other structures.

Activation of conditioned feeding behavior during electrical stimulation of subcortical structures proceeds typically against the background of hippocampal theta rhythm increase.

Electrical stimulation of the lateral hypothalamus was found to evoke activation also of the extinguished feeding behavior. And in this case, too, during electrical stimulation, the animal leaves the starting section, approaches the feeders and tries to get food. It is natural that during activation of both, non-extinguished (Fig. 4A, left bar) and extinguished (Fig. 4B left bar) feeding behaviors, the choice of the feeders is random. However, if electrical stimulation evoking activation of both non-extinguished (Fig. 4A right bar) and extinguished (Fig. 4B right bar) responses is applied during the directing conditioned stimulus to which the reaction had been chronically extinguished, then the choice

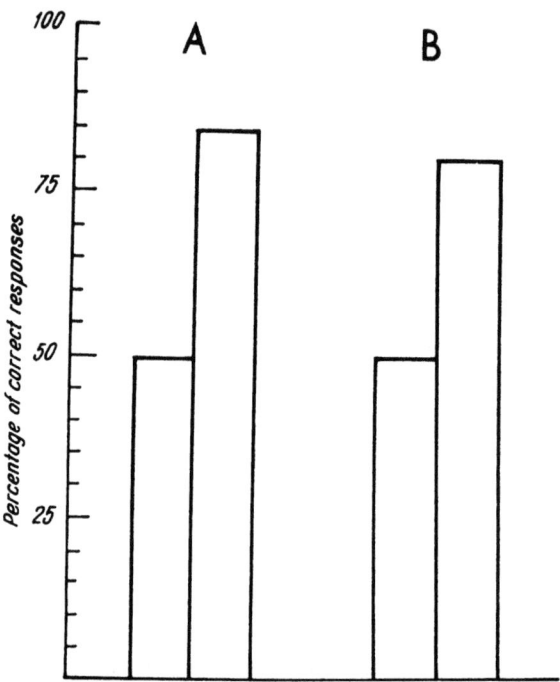

Fig. 4. Effect of triggering conditioned signal (A) and stimulation of lateral hypothalamus (B) on the extinguished conditioned reflexes. A, left bar shows the percentage of correct reactions in response to isolated application of triggering conditioned signal (opening of the door); right bar shows the percentage of correct reactions when if triggering conditioned signal (opening of the door) is delivered against the background of extinguished conditioned signals (tone and clicks). B, left bar shows the percentage of correct reactions in response to isolated application of electrical stimulation of hypothalamus, evoking activation of extinguished conditioned alimentary behavior, right bar shows the percentage of correct reactions in cases when electrical stimulation of lateral hypothalamus was applied against the background of extinguished conditioned signals.

of feeder is regular, the cats right away make for the feeder to which the sound stimulus had been a directing conditioned signal before extinction. Here the dynamics of hippocampal electrical activity is noteworthy. In response to a sound stimulus (tone or clicks) there takes place only a weak orientation reaction accompanied by neocortical and hippocampal desynchronisation (Fig. 5A), whereas during electrical stimulation activation of feeding behavior occurs against the background of increased hippocampal theta rhythm (Fig. 5B).

It should be noted that during the extinguished conditioned sound stimulus the threshold of electrical stimulation for the activation of feeding behavior is not altered. This seems to point to the lack of in-

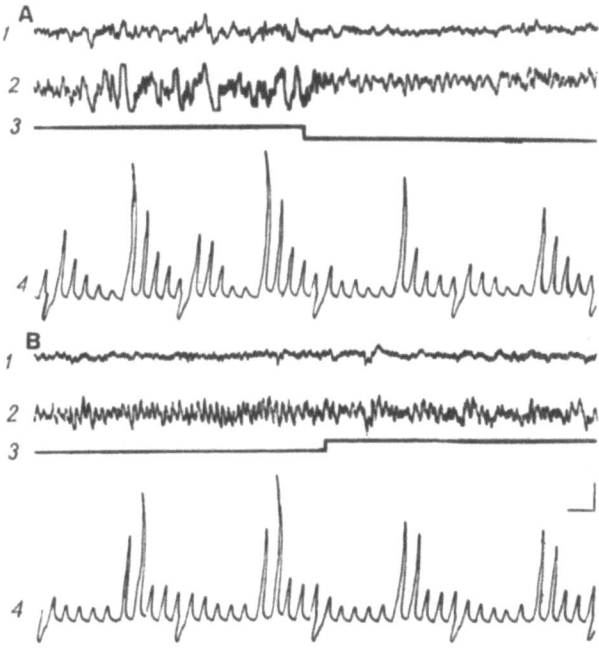

Fig. 5. Changes of electrical activity of neocortex and hippocampus in response to extinguished conditioned signal (A) and electrical stimulation of lateral hypothalamus (B), evoking activation of extinguished conditioned reflex. Leads: 1, auditory cortex; 2, dorsal hippocampus; 3, signal line (deflections downwards on A show the moment of the onset of extinguished conditioned signal and on B moment of the onset of electrical stimulation); 4, integrated values of delta, theta, alpha, beta$_1$ and beta$_2$ rhythms of neocortex (the first five deflections) and hippocampus (the following five deflections) during 5 s epoch. Calibration: 200 μV, 1 s.

hibitory influence of the signal after extinction of the conditioned reaction to it. With a view to further examining this supposition, a special series of experiments was carried out. In these experiments, after elaboration of stable discrimination of conditioned sounds to two feeders, the reaction was chronically extinguished to one of the signals and then its influence was tested on the active conditioned signal. It was found that if the active conditioned stimulus is applied against the background of a no longer active stimulus, to which the conditioned reaction had been chronically extinguished, feeding behavior is performed normally:

neither the latent period nor the time required to approach the feeder and consume the reinforcing portion of meat is altered. No considerable alterations are observed either in the EEG changes produced by the active conditioned stimulus. Figure 6 illustrates the dynamics of electrical activity in the neocortex and hippocampus during the application

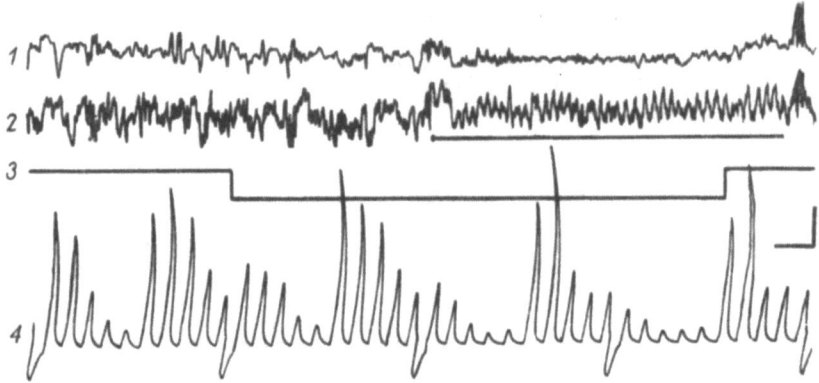

Fig. 6. Changes of electrical stimulation of neocortex and hippocampus in response to extinguished and non-extinguished conditioned signals. Leads: 1, sensorimotor cortex; 2, dorsal hippocampus; 3, signal line (deflection downwards indicates the moment of the onset of conditioned signals), the onset of an active conditioned signal is designated by horizontal line under electrohippocampogram; 4, integrated values of delta, theta, alpha, beta$_1$ and beta$_2$ rhythms of neocortex (the first five deflections) and hippocampus (the following five deflections) during 5 s epoch. Calibration: 200 μV, 1 s.

of inactive and active conditioned stimuli. The conditioned response to a tone was chronically extinguished. In the first trials it still brought about neocortical and hippocampal desynchronization (components of the orientation responses) without increasing the theta rhythm in the latter structure. However, if the tone was repeated several times, the EEG changes also disappeared. Against this background, the application of active conditioned stimulus (clicks) caused an increase of the hippocampal theta rhythm, desynchronization of neocortical electrical activity and normal triggering of feeding behavior. All this points to the lack of any considerable inhibitory influence of the inactive conditioned signal on the response brought about by the active conditioned signal.

*Influence of the functional inactivation of the hippocampus by way of induction of epileptiform discharges on the acquisition and extinction of conditioned feeding reaction*

Bekhterev (8) was the first to describe — on clinical material — a memory deficit due to bilateral hippocampal lesions. These findings were confirmed by Milner (36) and since then the role of the hippocampus in the organization of memory has attracted close attention of researchers (33, 34, 43, 50). Since acquisition and extinction of the conditioned alimentary reflex are examples of learning based on the retention of the biological significance of the stimuli, the study of the effect of functional inactivation of the hippocampus on these phenomena appeared to be promising. It is believed that during the development of epileptiform discharges there occurs a functional inactivation of the cortical structures, for their coordinated activity is disturbed (32, 42, 48). In our experiments epileptiform discharges were induced in the hippocampus by electrical stimulation. Rectangular pulses from the generator with a high-frequency output were used. Such parameters of stimulation were chosen at which epileptiform discharges were localized predominantly in the hippocampus (bilaterally) without spreading, or if they did, they spread only slightly to the neocortical structures (Fig. 7). In

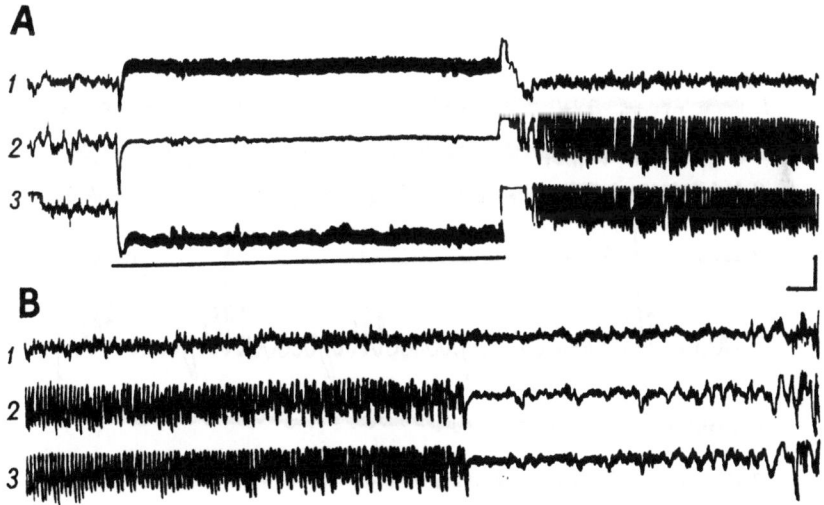

Fig. 7. A typical picture of EEG epileptiform discharges induced shortly after each combination of conditioned and non-conditioned stimuli. Leads: 1, sensorimotor cortex, 2,3, dorsal hippocampus, the moment of electrical stimulation of hippocampus is designated by a horizontal line under electrohippocampogram. B, direct continuation of A. Calibration: 200 µV, 1 s.

the course of acquisition of the conditioned alimentary reflex the hippocampal epileptiform discharges were induced each time following food reinforcement of conditioned stimuli. During extinction, electrical stimulation of the hippocampus eliciting epileptiform discharges was applied as soon as the cat opened the relevant feeder containing no reinforcing portion of meat. It was found that when the above-described procedure was strictly followed both the acquisition and extinction of feeding behavior proceeded normally and the experimental animals did not differ from the control ones. It follows from the foregoing that functional inactivation of the hippocampus by inducing epileptiform discharges does not exert any considerable influence on the acquisition and extinction of conditioned feeding behavior.

*Septal lesion effect on the acquisition and extinction of conditioned feeding behavior*

In literature there prevails a view that the hippocampo-septal system has a special role to play in the organization of the phenomenon of habituation and extinction (12, 13, 16). The septum occupies a strategic place between the rostral and diencephalic structures of the limbic system. Convincing evidence is available to prove that it is via the septum that a modulating influence of the forebrain structures is exerted on the motivational structures of the hypothalamus and mesencephalon (22, 25, 35).

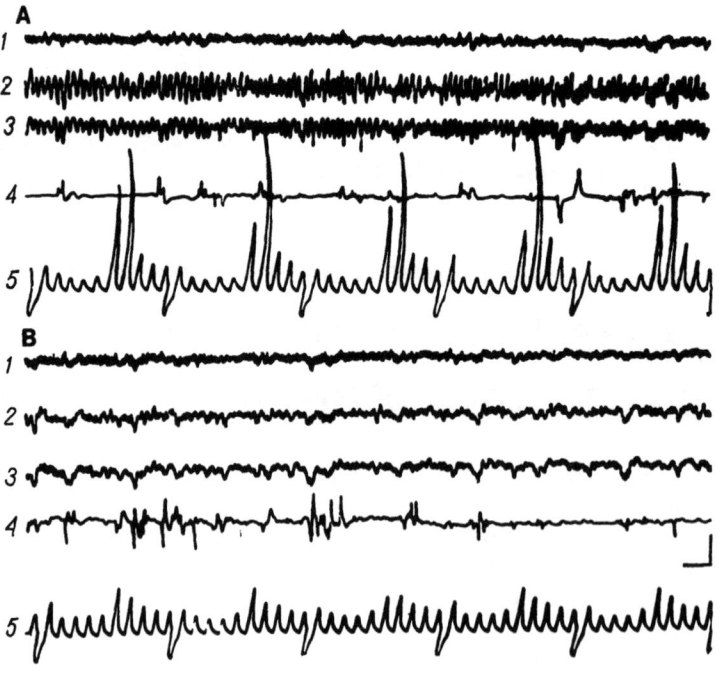

Fig. 8. Changes of electrical activity of dorsal hippocampus after septal lesion in the paradoxical phase of sleep. Leads: 1, sensorimotor cortex; 2,3, dorsal hippocampus; 4, electrookulogram; 5, integrated values of delta, theta, alpha, $beta_1$ and $beta_2$ rhythms of neocortex (the first five deflections) and hippocampus (the following five deflections) during 5 s epoch. *A*, before lesion; *B*, after lesion. Calibration: 200 μV, t. s.

In this aspect we became interested in the study of the septal lesion effect on the acquisition and extinction of the conditioned alimentary reflex. To this end, cats were chronically implanted with bipolar electrodes, bilaterally placed in the septum and the dorsal hippocampus to study the dynamics of its theta rhythm. After a detailed study of background electrical activity in the neo- and archipaleocortex in the sleep-wakefulness cycle, the septum was bilaterally lesioned by electrocoagulation. The disappearance of the hippocampal theta rhythm bilaterally served as a neurophysiological control of lesion (Fig. 8B). Then the usual procedure of acquisition and extinction of feeding behavior to the two feeders was commenced.

The cats with septal lesions did not appear to differ from the control ones in the rate of acquisition of feeding behavior and conditioned sound discrimination (Fig. 9). Yet, both acute and chronic extinction of

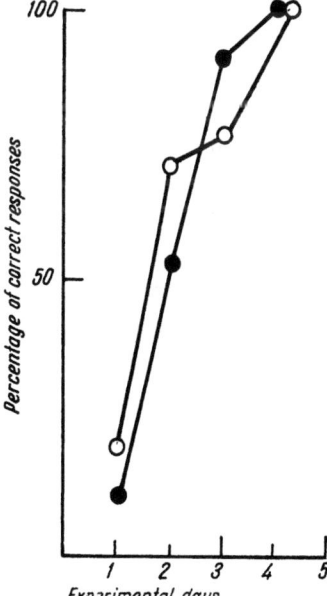

Fig. 9. Effect of septal lesion on elaboration of instrumental alimentary conditioned reflex with sound discrimination. Abscissae, days of experimentation; ordinate, percentage of correct responses. White circles show mean rate of acquisition in control animals; black circles in experimental ones.

the conditioned reflex in septumectomized cats is hindered and proceeds much slower than in the controls (Fig. 10). At the same time, we have not observed in experimental animals any marked changes in food motivation and motor activity that might be the cause of difficulty with extinction. Destruction of the septum apparently involves — in structural terms — the mechanisms responsible for extinction and not for acquisition of conditioned feeding behavior.

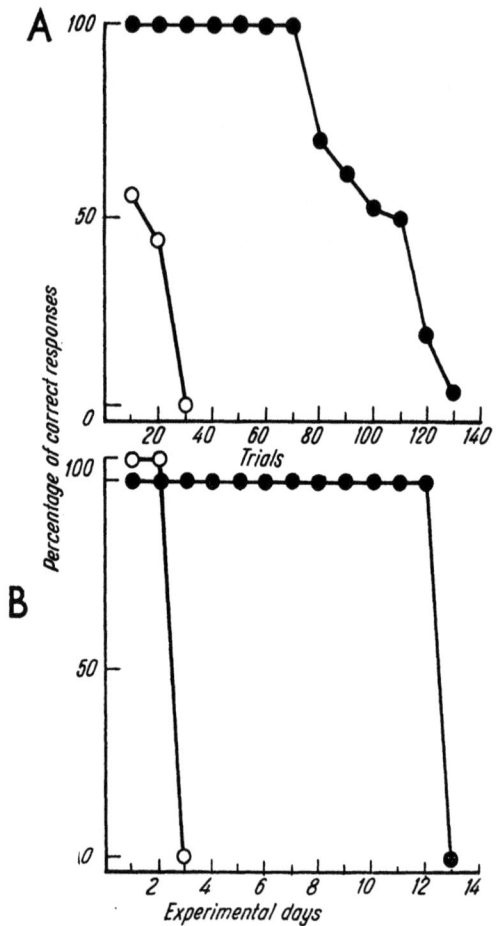

Fig. 10. Dynamics of acute (A) and chronic (B) extinction after septal lesion. A, number of trials (abscissae), the percentage of correct responses (ordinate). B, days of experiments (abscissae), the percentage of correct responses (ordinate). White circles show control animals; black circles, experimental ones.

DISCUSSION

The data described above may be analyzed from the standpoint of revealing the crucial factors of extinction of conditioned feeding behavior. Since the very inception of teaching on the conditioned reflexes to the present day (4, 6, 7, 9, 27-29, 41) non-reinforcement of a conditioned stimulus with the unconditioned one has been considered the leading cause of extinction. At the same time, it should be taken into account that non-reinforcement, especially during instrumental alimentary conditioned reactions involves multiple factors and their differentiation according to their importance is no doubt advisable. Almost all researchers share the view that among these factors the lack of food intake during a conditioned stimulus should be the decisive one (4, 28, 29). Our experiments on the extinction of the instrumental conditioned alimentary reflex involving pre-satiated animals have shown that such is not the case. In fact, hundreds of repeated conditioned stimuli applied over a number of days, without food reinforcement, failed to evoke any signs of extinction in satiated animals. After the fading of satiation they respond and discriminate conditioned stimuli perfectly well.

Non-reinforcement of a conditioned signal with unconditioned stimulation, i.e. lack of food intake, does not lead to the extinction of conditioned feeding behavior even in hungry animals, if at the time they have no opportunity to visit and explore the feeders. This indicates that non-reinforcement, even in conjunction with a high level of motivation, is not sufficient for the development of the extinction process. While considering the role of motivation in the extinction of feeding behavior, investigators (4, 29, 41) ascribe a special importance to emotional reactions resulting from non-reinforcement. It is considered that negative emotional reactions (4) or the so-called anxiety state of the animal (4, 41) are the causes which lead to the development of internal inhibition, i.e. to the extinction of a positive conditioned response. Our findings in this connection may introduce a definite correction. Emotional reactions and an anxiety state inevitably arose when hungry animals were not let out of the chamber during the onset of conditioned stimuli. They became restless, endeavored to leave the starting section; also a considerable rise in the heart rate and development of the hippocampal theta rhythm, attended by desynchronisation of neocortical activity, were observable. Repetition of conditioned stimuli during a daily session resulted in the reduction of the described behavioral, EEG and vegetative changes — to their complete disappearance (habituation), but following this, once the animals were allowed to approach the feeders, they correctly responded to the conditioned stimuli. Naturally, neither did repetition of such experiments for many days result in chronic extinction. All this shows clearly that the development of the so-called negative emotional reactions and an anxiety state per se cannot constitute crucial factors of the extinction of conditioned feeding behavior.

It is evident from our experiments that extinction of the conditioned feeding behavior develops only when, in response to a conditioned stimulus, the hungry animal is let out of the starting section and permitted to approach the feeders and explore them. This indicates that the animal must remember the absence of food in the feeders and having reassured itself, as it were, that the conditioned stimuli that had been hitherto accompanied by food reward are no longer reinforced, understands that it is useless to visit the feeders. At the same time, both acute and chronic extinction by no means lead to the forgetting of the elaborated conditioned feeding behavior. This is indicated by the well-known fact that an extinguished conditioned reflex can be readily restored by the renewal of paired stimuli (4, 5, 7, 27-29), and also by the fact that against the background of an extinguished conditioned signal, electrical stimulation of certain mesodiencephalic structures results in the recovery of response and sound discrimination. It is therefore evident that the so-called extinguished conditioned stimuli must simultaneously bring about a recollection that formerly the stimulus was accompanied by food reward, but it is no longer so.

Thus, our findings confirm I. P. Pavlov's statement, later shared by others (2, 10, 23, 26, 41, 43, 52) that extinction is a new learning that becomes a rival, as it were, of the old one. Yet, our results reveal the chief factor of extinction, attributing a crucial importance to the visual memory of the absence of reinforcement. Incidentally, the new memory is organized on the basis of and in the presence of old memory. Thus, these memories cannot differ in sign, and one should not refer to the acquisition of a conditioned reflex as a positive, and to its extinction as a negative learning.

Considering the physiological mechanisms of extinction of the conditioned reflex, I. P. Pavlov advanced the idea on the significance of the internal inhibition. According to this view, internal inhibition arises in the cortical representation of a non-reinforced conditioned stimulus, irradiating at the beginning over the entire cortex, and then again concentrating in the site of its origin. Later on, the significance of internal inhibition in the extinction of the conditioned reflex was recognized by the overwhelming majority of investigators (4, 6, 7, 28, 41 and others). Asratyan (5–7) experimentally examined Pavlov's thesis and, finding no signs of reduction of excitability during extinction of a conditioned reflex in the representation of conditioned or unconditioned stimuli, he logically assumed the possibility of localisation of internal inhibition in the temporary connections formed during the acquisition of the conditioned reflex between these points of the cortex. An original theory on the mechanisms of extinction of the conditioned reflex has been developed by Konorski (27). Originally, he assumed the possibility of establishing inhibitory conditioned connections between the centres of the conditioned and unconditioned stimuli. And what is more, in his opinion, this process occurs not as a result of reorganization of excitatory connections into inhibitory ones, but as a result of the formation of inhibitory connections with the maintenance of the earlier elaborated excitatory ones. Later on, having experimentally tested this supposition, Konorski (29) changed his views and formulated a new hypothesis on the mechanism of extinction of the conditioned reflex. According to this hypothesis it is assumed that "absence of food is a kind of unconditioned agent activating the particular centre that is in a reciprocal relationship with the centre of the presence of food in the mouth" (29, p. 247). The assumption of an unconditioned centre of food absence is in turn based on the supposition that apart from cells responding to food taste, "in the gustatory analyser there are cells which respond to the absence of food in the mouth" (29, p. 251). Therefore, when a positive conditioned stimulus not reinforced by food is extinguished, the center of this stimulus forms additional connections with the negative food center and may inhibit it. Thus, in terms of Konorski's theory, after extinction of the conditioned reflex, the center of the conditioned stimulus has excitatory connections both with the unconditioned center of the presence of food and with that of its absence. As these centers are in reciprocal relationship, the predominance of definite connections determines the occurrence or lack of a conditioned reflex response. According to Konorski's theory, the mechanism of extinction of the food-motor or instrumental conditioned reflex is almost similar. In these cases, the center of the conditioned stimulus forms excitatory connections with the center of the absence of food, which, in its turn, inhibits the center of hunger.

In the light of the above facts, at least two weak points are revealed in Konorski's hypothesis, naturally calling for definite correctives. First, during extinction of alimentary instrumental conditioned reflexes it is the visual memory of the absence of reinforcing food in the feeder rather than its absence in the mouth that forms the factor of extinction. This, no doubt, alters the situation and requires a new interpretation. Secondly, there is, apparently, no inhibition of the centre of hunger and food motivation in response to the extinguished conditioned stimuli. This is indicated by the absence of threshold rise of the lateral hypothalamic electrical stimulation to induce activation of feeding behavior against the background of an extinguished conditioned stimulus, and also the lack of the effect of extinguished conditioned stimulus on the parameters of conditioned response to a non-extinguished conditioned stimulus. Our experiments have also demonstrated that during the action of extinguished conditioned stimulus there takes place a recollection that this stimulus had earlier signalled the presence of food in a definite feeder, whereas now it signals its absence. That is why electrical stimulation of the lateral hypothalamus brings about not only activation of the extinguished feeding behavior, but also restoration of sound discrimination, as a result of which the animal, in response to the extinguished conditioned stimulus, makes for the feeder to which this stimulus had earlier been a directing one.

An important point in Konorski's hypothesis is that extinction of the conditioned reflex is explained not by transformation or replacement of excitatory processes with the inhibitory ones, but by the development of new learning, at the same time retaining the old one. Moreover, these two types of memory are competitive being realized by reciprocal relationship between the unconditioned centers of the presence and absence of food stimulus. This, naturally, militates against the recognition of the decisive role of internal inhibition in the extinction of the conditioned reflex as understood originally. Pavlov himself (41) considered extinction as a new learning based on an active process in the form of internal inhibition. Since the nature of the latter was not known, all these events observable in the course of extinction of conditioned reflex could be subsumed under it. Therefore, Konorski's hypothesis should undoubtedly be considered as a further development of the teaching on the extinction of the conditioned reflex, since it is based on concrete neurophysiological data. The principal argument here is that the elaboration and extinction of the conditioned reflexes are not due to opposite processes. This opinion is shared by a considerable number of authors (2, 10, 23, 26, 43, 51 and others) studying the extinction of the instrumental conditioned reflexes. All of them agree in that the extinction is a new learning which develops on the basis of old learning and competes with it.

When considering extinction as the development of an active competitive memory, attention is attracted by Anokhin's hypothesis (4) according to which "...in response to the stimulus being extinguished, at the very onset of extinction, sometimes even after the first non-reinforcement, two competitive unitary responses are formed in the cerebral cortex: a) the earlier elaborated conditioned food response and b) conditioned, biologically negative response arising after non-reinforcement of the stimulus being extinguished" (4, p. 130). In contrast to Konorski's view, Anokhin believed that elaboration and extinction are based on opposite neurophysiological processes. He writes: "strictly following the law of causality, the question inevitably arises: on the basis of which physiological regularities, known to us, may non-reinforcement with food, that is absence of unconditioned stimulus, transform the hitherto ongoing process of stimulation into its opposite process of inhibition?" (4, p. 111). Answering this question, Anokhin refers to I. P. Pavlov's view "that if the conditioned-stimulus-induced food excitation is not reinforced, this evokes an anxiety state in the animal, i.e. it leads to a response of biologically negative character" (4, p. 128). After discussion of this question, Anokhin comes to the following conclusion: "...after the onset of extinction, an inhibitory conditioned reflex arises to the same conditioned stimulus to which food excitation arose before extinction" (4, p. 130).

Amzel (2) also considers frustration evoked by non-reinforcement to be the factor responsible for the development of extinction through inhibition. However, he interprets this process differently. "Inhibition, he writes, is hardly a relevant and proper term to designate an event associated with the reduction of the level of responsiveness due to frustration because of non-reinforcement of the response. It would be more correct to use the term "interference" or "competition", for such inhibition cannot be thought to act directly on the excitatory process, which also forms its basis. It (frustrational inhibition) is rather a new excitatory tendency, competing with the preceding one" (2, p. 307). There can be no objection to this basically correct reasoning except for the qualification of frustration as a factor responsible for extinction.

First, we must re-emphasize that non-reinforcement with food, i.e. lack of food intake, is not the cause of extinction. This was well demonstrated in our experiments in which conditioned stimuli were not reinforced with food, but in the case where the animal had no opportunity to approach the feeders, no extinction developed. Secondly, the anxiety state, arising during non-reinforcement, does not lead to extinction or to a response of biologically negative character when the animal has no chance to approach the feeders, though the anxiety state in these conditions develops no less intensively than when the animal is allowed to visit the feeders with no reinforcement. Thirdly, calling this negative response an inhibitory conditioned reflex lacks argumentation. All evidence indicates that extinction is a new positive learning. In all probability, it has a positive rather than a negative biological significance, for the animal does not make a no longer needed, non-expedient response. The claim that extinction develops due to the anxiety state arising under non-reinforcement, i.e. due to negative emotional reactions, has not been proved experimentally and, moreover, it is not clear how these emotional reactions may account for the development of the so-called inhibitory conditioned reflex.

The views of the authors who regarded extinction of the conditioned reflex as fatigue (24, 45) or as a process of inverse structural change of the temporary connections formed during learning (9), can today be only of historical interest, for during the period of time required even for chronic extinction neither fatigue nor forgetting can occur. Experiments show that even in the case when the animal, as a result of hundreds of applications of conditioned stimuli without food reinforcement, ceases to visit the feeders, no extinction of the conditioned reflex, in the literal sense of this word, occurs. By that time, the animals remember well that each of the extinguished stimuli used to be a signal of the presence of food in a definite feeder, but then they lost that significance. The occurrence of conditioned reflexes and discrimination of directing stimuli after the so-called chronic extinction can be revealed well by electrical stimulation of mesodiencephalic motivatiogenic structures. As shown in our experiments, electrical stimulation of the lateral hypothalamus leads to restoration, as it were, of extinguished conditioned reflexes and conditioned stimulus discrimination. Actually, it is not restoration, but bringing to light of something that exists. The experimenter, closely watching the animal's behavior in the course of extinction of the alimentary reflex, may well see that the animal remembers what the given signal meant before extinction. The animal looks towards the signalled feeder, occasionally even makes some steps forward, but then, as if remembering that the signal will no longer be followed by reinforcement, returns to the starting section. There is apparently no ground to doubt that in response to the extinguished conditioned stimuli the animals simultaneously remember what the given signal meant earlier and what it means now. This, naturally, complicates the signalling significance of the extinguished stimulus and requires involvement of additional brain mechanisms in order to perform adequate behavior. It is quite possible that these mechanisms involve mainly brain structures of higher level, without affecting mesodiencephalic motivational structures. This sems to be indicated by the absence of the hippocampal theta rhythm in response to the extinguished conditioned stimuli — so characteristic of non-extinguished stimuli. After all, indifferent stimuli become conditioned signals as soon as they, as a result of combination with unconditioned stimulation, acquire a biological significance, i.e. may trigger motivational processes. Activation of motivational structures is evidenced by the development of the hippocampal theta rhythm. However, during non-reinforcement, conditioned stimuli acquire an additional, new significance. They now serve as indicators that where earlier there was reinforcement, now there is none and therefore triggering of relevant motivation leading to feeding performance is no longer advisable. Whereas activation of motivatiogenic structures and feeding behavior in response to non-extinguished stimuli electrographically manifested itself in an increase of the hippocampal theta rhythm, now, after extinction, this phenomenon was lacking, though activation of neocortical electrical activity was maintained.

It follows that a response to extinguished conditioned stimuli is a result of the resolution like go no–go discrimination. According to present-day evidence (17–19, 37), decision making must be regulated by forebrain structures, in particular by the cerebral cortex and basal ganglia. This points to the important role of the limbic system in the extinction of conditioned reflex activity. First, it has been demonstrated that the removal of the medial prefrontal cortex leads to disinhibition of inhibitory alimentary reflexes (11, 31, 47), and secondly, according to our data, lesion of the septum — this functional point of the limbic system — drastically slows down the process of extinction.

According to Kimble (26), two levels — low and high — may be discerned in the mechanisms responsible for the elaboration of the conditioned reflex. Moreover, the latter exerts a predominantly inhibitory action on the former, but inhibition is understood as a consequence of competing responses. It is not difficult to see that a similar situation is obtained during extinction of a conditioned alimentary instrumental reflex. The memory that the given signal was earlier accompanied by food reward undergoes a rival influence of the memory that recently this signal is no longer followed by food reward — hence no feeding behavior is performed. The presence of such competitive interaction is undoubtedly based on the coordinated activity of the brain mechanisms, involving the processes of excitation and inhibition, but at present it is difficult to conceive precisely how and where they are realized. It may be only assumed that this interaction is realized at the level of the forebrain, i.e. at the higher level of mechanisms responsible for learning and conditioned reflex activity.

## REFERENCES

1. AIVAZASHVILI, I. M. 1963. On the capability of food location image and elaboration of automatized feeding behavior in satiated dogs (in Russian). Soobshch. Akad. Nauk. Gruz. SSSR, 30, 1: 59–65.
2. AMZEL, A. 1973. Regularities of formation of the conditioned reflex connection (in Russian). In V. S. Rusinov, P. V. Simonov and M. N. Rusalova Mekhanizmy formirovania i tormozhenia uslovnykh refleksov. Nauka, Moscow, p. 297–316.
3. ANDERSSEN, B. and WYRWICKA, W. 1957. The elicitation of a drinking motor conditioned reaction by electrical stimulation of the hypothalamic "drinking area" in the goat. Acta Physiol. Scand. 41: 194–198.
4. ANOKHIN, P. K. 1958. Internal inhibition as a problem of physiology. Meditsina, Moscow.
5. ASRATYAN, E. A. 1965. Characteristics of formation, functioning and inhibition of a two-way conditioned reflexes (in Russian) In E. A. Asratyan (ed.), Refleksy golovnogo mozga. Izdat. Nauka, Moscow, p. 114–126.
6. ASRATYAN, E. A. 1974. I. P. Pavlov. Izdat. Nauka, Moscow.
7. ASRATYAN, E. A. 1977. Ocherki po vysshei neivrnoi deyatel'nosti. Izdat. Akad. Nauk Armyanskoi SSR, Erevan, 346 p.
8. BEKHTEREV, V. 1900. Demonstration eines Gehirns mit Zerstörung der vor-

deren und inneren Theile der Hirnrinde beider Schläfenlappen. Neurol. Zentbl. 19: 990–991.
9. BERITOV, I. S. 1969. Struktura funktsii i kory bolshogo mozga. Izdat. Nauka, Moscow, 532 p.
10. BLUMFIELD, T. M. 1973. Excitation, inhibition and contingency. (in Russian). In V. S. Rusinov, P. V. Simonov and M. N. Rusalova Mekhanizmy formirovania i tormozhenia uslovnykh refleksov. Izdat. Nauka, Moscow, p. 39–56.
11. BRUTKOWSKI, S. and DABROWSKA, J. 1966. Prefrontal cortex control of differentiation behavior in dogs. Acta Biol. Exp. 26: 425–439.
12. CARLTON, P. L. 1965. The hippocampus, brain-acetylcholine and habituation. In Brain, biochemistry and behavior. Sympos. AAS, Berkeley, p. 205–211.
13. CARLTON, P. L. 1968. Brain-acetylcholine and habituation. Prog. Brain Res. 28: 48–59.
14. COHEN, B. D., BROWN, G. W. and BROWN, M. L. 1957. Avoidance learning motivated by hypothalamic stimulation. J. Exp. Psychol. 53: 228–233.
15. DELGADO, J. M. R., ROBERTS, W. W. and MILLER, N. E. 1954. Learning motivated by electrical stimulation of the brain. Am. J. Physiol. 179: 587–593.
16. DOUGLAS, R. J. 1967. The hippocampus and behavior. Psychol. Bull. 67: 416–442.
17. EVARTS, E. V. 1968. Relation of pyramidal tract activity to force exerted during voluntary movement. J. Neurophysiol. 31: 14–27.
18. EVARTS, E. V. 1969. Activity of pyramidal tract neurons during postural fixation. J. Neurophysiol. 32: 375–385.
19. EVARTS, E. V. 1974. Precentral and postcentral cortical activity in association with visually triggered movement. J. Neurophysiol. 34: 373–381.
20. GRASTYAN, E., LISSAK, K. and KEKESI, F. 1956. Facilitation and inhibition of conditioned alimentary and defensive reflexes by stimulation of the hypothalamus and reticular formation. Acta Physiol. Hung. 9: 133–151.
21. GROSSMAN, S. P. 1962. Direct adrenergic and cholinergic stimulation of hypothalamic mechanisms. Am. J. Physiol. 202: 872–882.
22. GROSSMAN, S. P. 1967. A textbook of physiological psychology. John Wiley and Sons, New York.
23. HOLLIDAY, M. S. 1973. Inhibition and instrumental learning (in Russian). In V. S. Rusinov, P. V. Simonov and M. N. Rusalova Mekhanizmy formirovania i tormozhenia uslovnykh refleksov. Izdat. Nauka, Moscow, p. 257–280.
24. HULL, C. L. 1943. Principles of behavior. Appleton-Century Co. New York.
25. ISAACSON, R. L. 1974. The limbic system. Plenum Press, New York.
26. KIMBLE, G. A. 1973. Elaboration and inhibition of conditioned reflexes (in Russian). In V. S. Rusinov, P. V. Simonov and M. N. Rusalova. Mekhanizmy formirovania i tormozhenia uslovnykh refleksov. Izdat. Nauka, Moscow, p. 416–428.
27. KONORSKI, J. 1948. Conditioned reflexes and neuron organization. Cambridge Univ. Press, London, 267 p.
28. KONORSKI, J. 1967. Integrative activity of the brain. An interdisciplinary approach. Univ. Chicago Press, Chicago, 531 p.
29. KONORSKI, J. 1973. Some ideas, concerning physiological mechanisms of internal inhibition (in Russian). Izdat. Nauka, Moscow, p. 241-256.

30. LAGUTINA, N. I. and ROZHANSKI, N. A. 1949. On localization of subcortical food centers (in Russian). Fiziol. Zh. SSSR im. I. M. Sechenova 35: 587–593.
31. ŁAWICKA, W. 1957. The effect of the prefrontal lobectomy on the vocal conditioned reflexes in dogs. Acta Biol. Exp. 17: 317–325.
32. McDONOUGH, J. H. and KESNER, R. P. 1971. Amnesia produced by brief electrical stimulation of amygdala or dorsal hippocampus in cats. J. Comp. Physiol. 77: 171–178.
33. McGAUGH, J. L. 1972. Impairment and facilitation of memory consolidation. Activ. Nerv. Super. 14: 64–79.
34. McGAUGH, J. L., ZORNETZER, S. E., GOLD, P. E. and LANDFIELD, P. W. 1972. Modification of memory systems: Some neurobiological aspects. Q. Rev. Biophys. 5: 163–181.
35. MILNER, P. M. 1970. Physiological psychology. Holt, Rinehart and Winston Inc., Nev York.
36. MILNER, B. 1962. Les troubles de la memoire accompagnant des lesions hippocampiques bilaterales. In Physiologie de l'hippocampe. In J. Brunisseau (ed.), Centre National de la Recherche Scientifique, Paris, p. 257–272.
37. MOUNTCASTLE, V. B., LYNCH, J. C., GEORGAPOULOS, A., SAKATA, H. and ACUNA, C. 1975. Posterior parietal association cortex of the monkey: command functions of operations within extrapersonal space. J. Neurophysiol. 38: 871–908.
38. ONIANI, T. N. KORIDZE, M. G. and ABZIANIDZE, E. V. 1972. Electrical acitvity of the hippocampus during delayed responses and hypothalamic stimulation in the cat. Acta Neurobiol. Exp. 32: 799–816.
39. ONIANI, T. N., MGALOBLISHVILI, M. M. and KESHELAVA, M. V. 1976. Activation and inactivation of food motor conditioned reflex in response to electrical stimulation of subcortical structures of the brain and dynamics of delayed reactions (in Russian). In T. N. Oniani (ed.), Neirofiziologia emotsii i tsikla bodrstvovanie-son. II. Metsniereba, Tbilisi, p. 117–134.
40. ONIANI, T. N. 1978. On the nature of neurophysiological processes underlying delayed responses on discrimination of conditioned stimuli. In T. Oniani (ed.), Neurophysiology of emotion and wakefulness-sleep cycle. III. Metsniereba Publ. Tbilisi, p. 94–122.
41. PAVLOV, I. P. 1924. Dvadtsatiletnii opyt obiektivnogo izuchenia vysshei nervnoi deyatielnosti (povedeniya) zhivotnykh. Gosudarstvennoe Izdat., Leningrad.
42. POSLUNS, D. and VANDERWOLD, C. H. 1970. Amnestic and disinhibitory effects of electroconvulsive shock in the rat. Comp. Physiol. Psychol. 73: 291–306.
43. RESCORLA, P. A. 1973. A model of formation of conditioned reflex according to I. P. Pavlov (in Russian). In V. S. Rusinov, P. V. Simonow and M. N. Rusinov (ed.), Mekhanizmy formirovania i tormozhenia uslovnykh refleksov. Izdat. Nauka, Moscow, p. 25–39.
44. SIMONOV, P. V. 1975. Vysshaya nervnaya deyatel'nost cheloveka. Motivatsiya i emotsionalnyi aspekt. Izdat. Nauka, Moscow.
45. SKINNER, B. F. 1938. The behavior of organisms. Appleton-Century Co., New York.
46. SUDAKOV, K. V. 1971. Biologicheskaya motivatsiya. Meditsina, Moscow.

47. SZWEJKOWSKA, G., KREINER, J. and SYCHOWA, B. 1963. The effect of partial lesions of the prefrontal area on alimentary conditioned reflexes in dogs. Acta Biol. Exp. 23: 181–192.
48. VINOGRADOVA, O. S. 1975. Gippokamp i pamyat. Nauka, Moscow, p. 99–100.
49. VINOGRADOVA, O. S. 1975. Functional organization of the limbic system in the process of registration of information. Facts and hypothesis. In R. L. Isaacson and K. Pribram (ed.), The hippocampus. Plenum Press, New York, p. 3–69.
50. WELZEL, W., OTT, T. and MATTHEIS, H. 1977. Post-training hippocampal rhythmic slow activity ("Theta") elicited by septal stimulation improves memory consolidation in rats. Behav. Biol. 21: 32–40.
51. WYRWICKA, W. and DOTY, R. W. 1966. Feeding induced in cats by electrical stimulation of the brain stem. Exp. Brain Res. 1: 152–160.
52. ZIELIŃSKI, K. 1979. Extinction, inhibition, and differentiation learning. In A. Dickinson and R. A. Boakes (ed.), Mechanisms of learning and motivation. L. Erlbaum Ass. Hillsdale, p. 269—293.

T. N. ONIANI and N. G. VARTANOVA, I. S. Beritashvili Institute of Physiology, Georgian Academy of Sciences, 14 Gotua Str., 380060 Tbilisi, USSR.

Lecture delivered at the Warsaw Colloquium on Instrumental
Conditioning and Brain Research
May 1979

# EXTERNAL STIMULUS CONTROL OF SELF-STIMULATION BEHAVIOR

Bogdan SADOWSKI

Laboratory of Behavioral Physiology, Institute of Genetics and Animal Breeding
Mroków, Poland

*Abstract.* Dogs with electrodes implanted in the anterior forebrain region were successfully trained to leverpress for brain stimulation reward upon the presentation of a 1,000 Hz tonal conditioned stimulus and to withhold responding during its absence and upon a 2,000 Hz tonal discriminative stimulus. The conditioned response extinguished upon removal of reward and was promptly restored when reward became again available. The results are discussed in terms of a conditioned drive induced by the action of the cue. It is proposed that environmental factors play an important role in the initiation and maintenance of self-stimulation behavior. Brain stimulation reward may serve as reinforcer of conditioned instrumental reflexes in an essentially similar way as natural rewards.

Self-stimulation is commonly regarded as a kind of operant behavior in which both reinforcement and drive are produced by the same electric stimulus delivered to the brain reward system through chronically indwelling electrodes. Due to this dual action of brain stimulation, the animal's working for electrical reward differs essentially from behaviors reinforced with natural rewards since the required level of motivation for the latter is achieved by controlled food (or water) deprivation prior to each experimental session.

The lack of enduring humorally induced motivation is believed to account for several pecularities of self-stimulation. For more details see

Trowill et al. (24). First, self-stimulation, compared to behaviors reinforced with natural rewards, appears extremely sensitive to non-reinforcement. Extinction develops promptly already after a few leverpresses or other learned movements which are not followed by brain stimulation. Second, Howarth and Deutsch (8) found that rats would fail to resume pressing for brain reward if their access to the manipulandum was hindered for a period of a few seconds. This phenomenon, termed extinction without responding, gave support to the suggestion that extinction in self-stimulation is not a function of the number of non-reinforced responses but of the time elapsing from the last brain stimulation. Third, rats when put on partial reinforcement schedule cease responding for brain stimulation sonner than for a natural reward. This is clearly evidenced when the ratio of the number of responses to the number of stimulations is gradually increased. Although rats make a remarkable number of presses for one food pellet they will stop working already at a few responses for one train of electric pulses (24).

Also, according to some data, the self-stimulating animal is to some extent "inattentive" to external stimuli. In this way Brady (6) tried to explain the lack of suppression of self-stimulation by an external stimulus which was earlier paired with a foot shock. The conditioned emotional response could be easily demonstrated when water was used as reinforcer. A similar mechanism may account for the difficulty to obtain a secondary reinforcement effect with brain stimulation reward. In the case of natural rewards, this effect develops if the animal's operant, besides causing reward (which is a primary reinforcer), also activates an external signal. The reinforcing property of the latter is evidenced as a delay of extinction when the animal's responses no longer produce reward but further switch on this signal. Another method to investigate this process is a two-lever situation, in which one of the levers activates a tone earlier associated with brain stimulation, in addition to the actual reward, and the other produces reward without the tone. The animals will press more frequently on the first lever. There is some controversy about the possibility to reproduce the secondary reinforcement effect using brain stimulation as a primary reinforcer. Positive results have been reported by Stein (21), Knott and Clayton (10), Trowill and Hynek (23) and Beninger and Milner (2), but Seward et al. (20) and Mogenson (15) obtained negative data.

Many of these patterns of self-stimulation fit well into the drive-decay theory proposed by Deutsch and Howarth (7) which claims that the electric stimulus simultaneously activates motivating and reinforcing pathways. Whereas the reinforcing effect is limited to the duration of the stimulating train, the motivating effect outlasts it for a short time and instigates the animal to perform the next response. This response triggers a new train which provides reinforcement and reactivates the drive for the subsequent response, and so forth. Nevertheless, the results of some investigations suggest that another process may be also involved in the mechanism of self-stimulation. First, if self-stimulation is based only on the positive feedback as postulated by the drive-decay theory (7), it should be necessary to administer at least one brain stimulation at the onset of each experimental session in order to make the animal start pressing. This experimenter — produced stimulation, commonly termed priming, is really needed at the beginning of training, but subsequently many animals approach the lever as soon as they are placed in the experimental situation. This last phenomenon is particularly demonstrative in dogs that tend to press the lever even before being connected to the stimulator (18) or make a considerable number of responses without any reward (19). Second, self-stimulation in some placements does not proceed regularly, but is organized in bouts. Although consecutive responses within each bout are repeated at a frequency which permits to assume their dependence on the drive in terms of the Deutsch-Howarth theory, the intervals between bouts exceed the proposed decay time of this drive. Therefore, a different mechanism must be accepted for the first response in each bout. Third, incompatible with the drive-decay theory is the remarkable resistance to extinction seen in some dogs (19). In addition, extinction without responding was absent in these dogs, because despite their not being allowed to press for one min or longer, they consistently approached the manipulandum as soon as it became again available.

These observations support the incentive-motivational model of self-stimulation (24) which posits that cues associated with rewarding brain stimulation acquire motivational properties and instigate the animal to perform for further reward. Accordingly, it is the experimental environment which plays the role of a complex conditioned motivational stimulus.

More information about the conditioning mechanisms in self-stimulation comes from studies where animals learned to press for brain stimulation reward upon a specific cue. Keesey (9) and Terman and Kling (22) investigated this problem in a two-lever situation. Which of the manipulanda delivered brain stimulation reward was signalled by the intensity of light over each lever. Moreover, in order to switch on the lamps the animal had to break a photobeam at the rear of the cage, otherwise reward was not available.

Lenzer and Frommer (13, 14) demonstrated that rats will easily learn to leverpress upon presentation of a sporadic cue and to withhold their performance in the absence of the signal. Only single investigations in other species were reported. Anschel and Anschel (1) trained nine monkeys to press upon a green light and to suspend pressing during presentation of white light. Four animals acquired complete discrimination mastery whereas five others continuously explored the lever and responded as on a mixed schedule of reinforcement without external cues. A no-cue situation was not used in that study. The first investigation where dogs learned to press upon a sound was published by Perez-Cruet et al. (17). In their study, however, this behavior was established only for the practical purpose of comparing vegetative phenomena during self-stimulation and at rest. More extensive observations were reported by Pavlova et al. (16) who used a procedure resembling Type II conditioned motor reflexes with food reward described by Konorski and Miller (12). The dogs were presented with a tone signalling the availability of reward. Their pressing was reinforced only during the tone, but not during intertrial intervals which lasted several times longer than the signal. As the animals learned to approach the lever upon the tone and to withhold from pressing between the trials, another tone was introduced, and during its presentation the pressing was not effective. This paradigm resembles the procedure of differentiation in Type II motor conditioned reflexes (11) or go–no go discrimination learning. Finally the conditioned response was extinguished and subsequently restored. The results of this research will be summarized below.

Six dogs were used. All had electrodes chronically implanted into the rostral forebrain region comprising the septum, the preoptic area, the nucleus accumbens and related structures. The animals were screened for self-stimulation and the best electrodes were qualified for further study. The sessions were conducted in a sound-attenuating chamber where the animals were partially restrained on a Pavlov stand. In front of the dog was an easily accessible lever activating a stimulator which delivered 0.5 s train of 240 Hz sine wave after each leverpress. The stimulator was also connected to a key operated by the experimenter carefully observing the animal's behavior through a window.

During the first session the experimenter stimulated the dog's brain successively through all 4 to 6 implanted electrodes until the indices of positive reinforcement appeared such as closing of the mouth, sniffing and exploration of the environment (see 18). The placements where these phenomena appeared at the lowest current were selected for further self-stimulation training which proceeded according to a shaping procedure. At the beginning of training the dogs were given motivating stimulations at the time they were inactive, and all searching movements incidentally directed to the lever were reinforced with the same stimuli. Gradually the lever became a goal intensively explored by the dog and touched with his nose or foreleg. The experimenter was more and more stringent in evaluating the correctness of the movements and reinforced only definite contacts of the animal with the manipulandum. A crucial moment in the acquisition of the task was when the animal for the first time touched the lever strongly enough to activate the stimulation circuit without the aid of the experimenter. Several sessions with a continuous reinforcement schedule were needed to stabilize performance. The animals were then trained to leverpress for brain stimulation reward upon a specific cue, and to withhold responding in its absence and during presentation of a discriminative cue.

During acquisition of the conditioned response the dog obtained brain stimulation reward only when his leverpressings were contingent upon a 1,000 Hz tone termed a conditioned stimulus. In the initial stage of training the tone was activated for 15–60 s at varying intervals to resemble an intermittent reinforcement schedule. Four dogs repeatedly returned to the lever, so that their leverpresses occurring during the cue were reinforced. The two other dogs reacted to this intermittent reinforcement with suppression of responding and abandoned the lever. They failed to resume pressing unless stimulated *gratis* by the experimenter.

The pairing of a rewarding brain stimulation with a neutral stimulus prior to the execution of an instrumental act causes that this stimulus acquires conditioned motivational properties (3). Bindra and Campbell (4) associated a metronome with rewarding hypothalamic stimulation and noticed that after this procedure the sound produced a marked increase in perambulation scores. The same phenomenon was observed when a natural reward (water given to thirsty rats) was used instead of brain stimulation (5). These results led to a general conclusion that neutral stimuli associated with a reinforcer (the unconditioned stimulus) acquire, through classical conditioning, incentive motivational properties and promote the animal to respond for further reward.

Our results with dogs support this conclusion only partially, in that repeated pairing of a neutral stimulus with the electrically produced reward according to principles of classical conditioning causes this stimulus to enhance the animal's general motor activity. In our experiments this finding was manifested by the animal's restlessness or arousal. We assume that the restlessness displayed by our dogs reflected a motivational state created by the external stimulus through a conditioning process. Nevertheless, this central motivational state is by itself not sufficient to promote the animal to perform the instrumental act that earlier resulted in brain stimulation reward. This is particularly evident in dog 1 that despite about 140 pairings of the tone with experimenter — produced brain stimulation and the subsequent self-stimulation, failed to approach the lever upon the cue (see Fig. 2). Acquisition of the task was achieved only with the use of a special reinforcement strategy requiring the animal to perform a definite goal-directed movement before the experimenter activated the stimulator. At the beginning of conditioning any behavior which occurred upon the tone onsets such as an orienting response or turning of the head toward the lever, was reinforced, but later on more and more adequate movements were required to produce reward. As during initial stage of self-stimulation training, the animal's first effective leverpress, now made upon presentation of the cue, was a crucial point in the process of

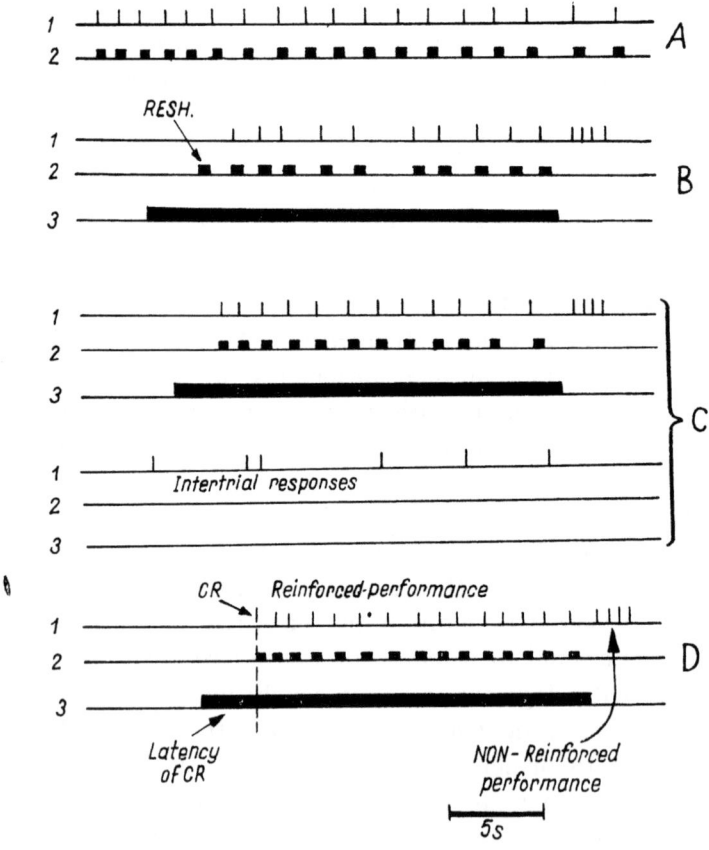

Fig. 1. Schematic representation of the training procedure. *1*, dog's responses; *2*, brain stimulations; *3*, conditioned stimulus, A, self-stimulation without external cue; B, initial stage of training; C, advanced stage of training; D, fully developed response to the conditioned stimulus. Resh., reshaping stimulation administered by the experimenter. CR, conditioned response.

acquiring the task. Brain stimulation according to this procedure was termed reshaping, in order to distinguish it from priming where no attention is paid to the animal's behavior.

Figure 1 summarizes the steps by which external stimulus control of self-stimulation was achieved. The fully developed response is shown in part D of this figure. The dog which is standing quietly on the stand and making no or negligible intertrial responses, upon tone onset approaches the lever and presses it. The first press in each trial is believed to be executed under the conditioned motivation induced by the cue, and therefore is regarded as a conditioned response. The mechanism of subsequent presses in the trial is the same as during self-stimulation not controlled externally, each press being a consequence of preceding reinforcement. Together with the end of the tone the current is cut off. Those presses appearing during the 5 s after the tone reflect the animal's reaction to a sudden removal of reinforcement and are not regarded as intertrial responses. Often, particularly in initial acquisition sessions, their rate exceeds the reinforced leverpressing rate which resembles the pattern seen at the begining of extinction.

Parts B and C of Fig. 1 show successive stages of the acquisition of the task. After having been reshaped for self-stimulation upon the tone the dog started to "examine" the lever both during the cue and in its absence. At this stage there is still no reason for regarding the approach to the lever upon the tone as a conditioned response. Rather, this behavior resembles the pattern of responding on a partial reinforcement schedule, with intermittent availability of reward. In the course of further training intertrial responses decreased or disappeared, and the latency of the animal's first press on the lever upon the cue (i.e., the latency of the conditioned response) shortened and stabilized. Behaviorally, the dogs reacted to each presentation of the cue with a closing of the mouth, turning of the head to the lever and an approach reaction. Stabilization of this pattern was achieved during several sessions where 20 s presentations of the cue were separated by variable intertrial intervals.

Figure 2 illustrates the acquisition of the task by all six dogs. Dogs 1 and 3 required reshaping. This procedure was also occasionally used in other animals. Five dogs fulfilled the acquisition criteria which consisted of high percentage of conditioned responses and substantial decrease in intertrial responses. However, in dog 5 the conditioned responses were unstable, therefore this animal was excluded from further discrimination training. Dog 6 was also eliminated due to an enormous amount of intertrial responses which did not decrease.

The stages of acquisition of the conditioning task by our dogs led us to a conclusion that besides a central motivational state created by the pairing of a neutral stimulus with reward, an instrumental strengthening of a particular movement is needed in order to transform the latter into a purposeful response emitted upon the cue. In other words, in order to acquire the task the dog had to perform the given response actively during presentation of the conditioned stimulus. The necessity of this active performance was also stressed by Konorski (12) with respect to food rewarded instrumental conditioned reflexes.

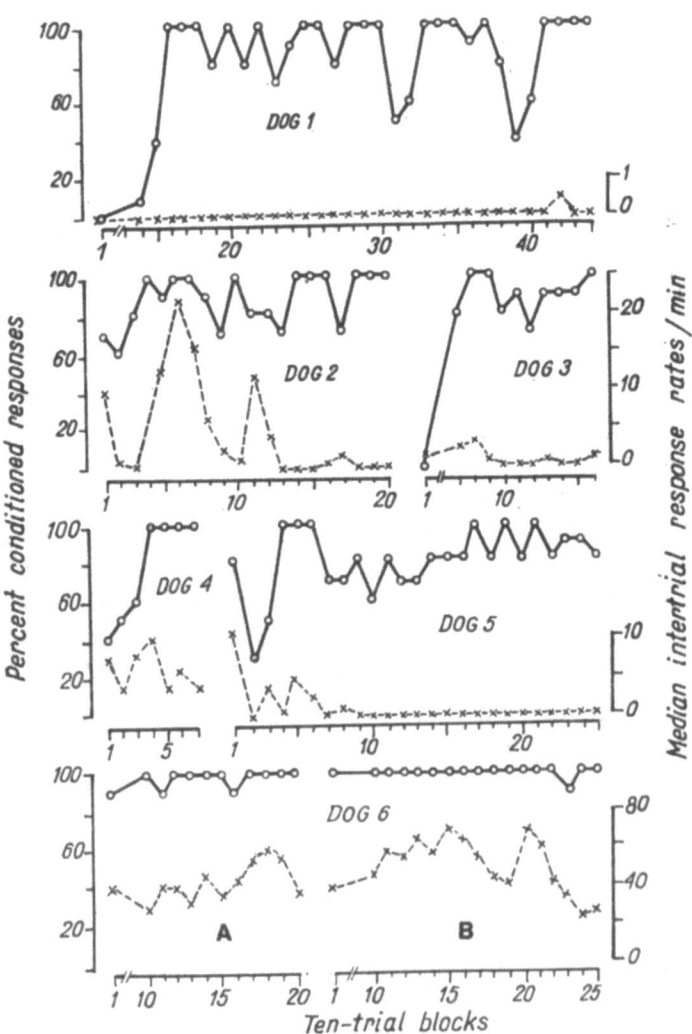

Fig. 2. Acquisition of the conditioned response. In dog 6 two electrodes (A and B) were used to sustain self-stimulation. White circles, conditioned responses; crosses intertrial responses.

Figure 3 illustrates discrimination learning in 4 dogs. The animals were presented with two cues: a 1,000 Hz tone signalled, as previously, the availability of reward, and a 2,000 Hz tone was a discriminative stimulus. Leverpressings which occurred during presentation of the latter were not rewarded. In each session the animals were presented with 20 conditioned and 10 discriminative stimuli. Each stimulus lasted 20 s and was preceded by a variable intertrial interval. As seen from

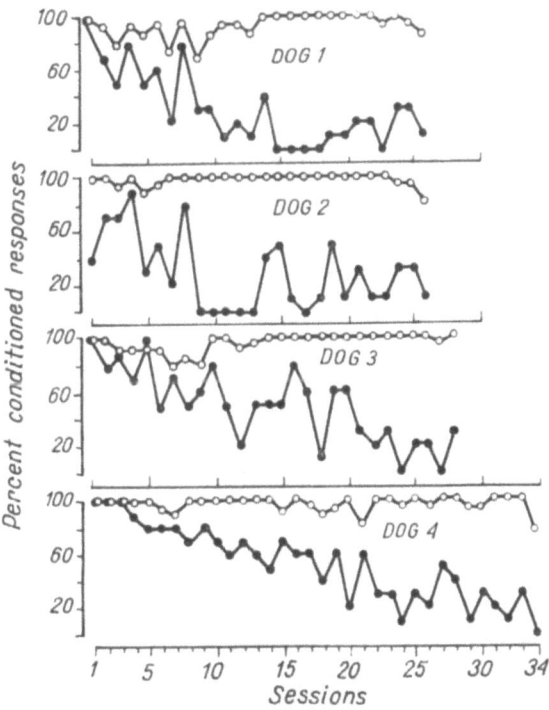

Fig. 3. Discrimination learning. White circles, percent of conditioned responses upon the conditioned stimulus; black circles, upon the discriminative stimulus.

the figure, all dogs learned to discriminate the two cues, although a persistent 100% discrimination was not reached. Intertrial responses were absent or scarce during the entire discrimination training.

Extinction of the conditioned response is shown in Fig. 4. The dogs were presented with 20 conditioned stimuli during each session, but leverpressing was not reinforced. Although the conditioned response extinguished promptly in a single session, it reappeared again in an overnight session. Therefore, several sessions were needed to achieve a chronic extinction.

At the beginning of the first restoration session the dogs were given a priming stimulation and allowed to self-stimulate for a short period without an external cue. Thereafter they were presented with the conditioned stimulus, and their pressing during the action of the latter

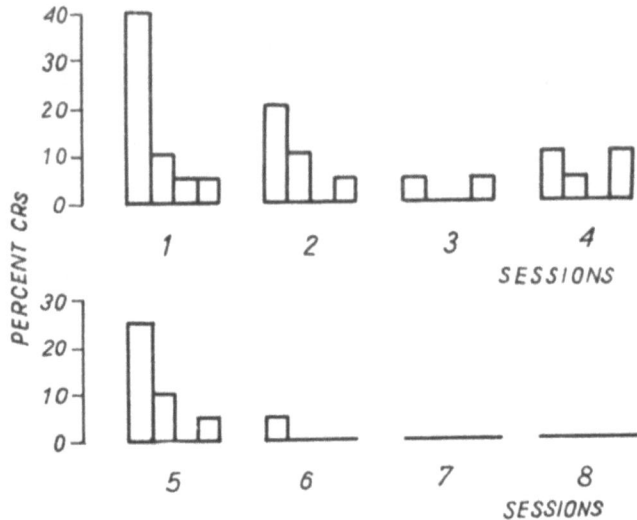

Fig. 4. Combined data showing extinction of the conditioned responses in all four dogs tested. Each session consisted of 20 presentations of the conditioned stimulus. The animals' leverpressing was not reinforced. Bars represent successive blocks of 5 trials.

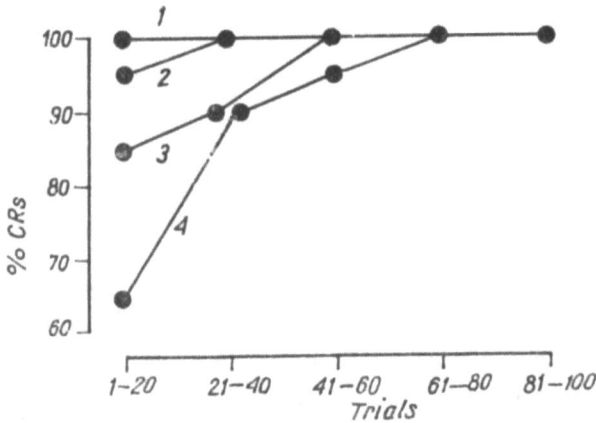

Fig. 5. Restoration of the conditioned response. At the beginning of the first session each dog obtained a priming (experimenter — produced) brain stimulation and was allowed short self-stimulation without external cue. Then the conditioned stimulus was presented 20 times in each session, and the dog's responses during its action were reinforced in the usual way. The numbers close to the solid lines refer to individual animals.

was again reinforced. As shown in Fig. 5, the conditioned response was promptly restored in all the dogs.

The intrinsic process of the conditioning mechanism in self-stimulation is far from being elucidated. There are some indications that it may be related to the function of the catecholaminergic pathways. Wauquier et al. (25) studied the action of apomorphine, a dopaminergic agonist, on conditioned reflexes with brain stimulation reward established in dogs according to the same procedure as in the present study. At a certain dose a dramatic increase in intertrial responses was noted. By this pattern the dogs behaved as our dog 6 or the five monkeys of Anschel and Anschel (1) which did not acquire the discrimination mastery. The effect of apomorphine was counteracted by haloperidol, a dopaminergic blocker.

As shown by our data, brain stimulation may be successfully used as reinforcer of instrumental learning controlled by external signals, and by this does not differ essentially from natural rewards. The lack of humorally produced drive may be easily compensated by a motivational state induced through the action of environmental factors acting as a complex conditioned stimulus.

## REFERENCES

1. ANSCHEL, S. and ANSCHEL, C. 1974. Visual stimulus control of intracranial self-stimulation in the squirrel monkey (*Saimiri sciureus*). Physiol. Behav. 12: 457–465.
2. BENINGER, R. J. and MILNER, P. M. 1977. Conditioned reinforcement based on reinforcing electrical stimulation of the brain: chain schedules. Physiol. Psychol. 5: 285–289.
3. BINDRA, D. 1968. Neuropsychological interpretation of the effects of drive and incentive-motivation on general activity and instrumental behavior. Psychol. Rev. 75: 1–22.
4. BINDRA, D. and CAMPBELL, J. F. 1967. Motivational effects of rewarding intracranial stimulation. Nature (Lond.) 215: 375–376.
5. BINDRA, D. and PALFAI, T. 1967. Nature of positive and negative incentive-motivational effects on general activity. J. Comp. Physiol. Psychol. 63: 288–297.
6. BRADY, J. V. 1958. Temporal and emotional factors related to electrical self-stimulation of the limbic system. In H. H. Jasper, L. D. Proctor, R. S. Knighton, W. C. Noshay and R. T. Costello (ed.), Henry Ford International Symposium on the Reticular Formation of the Brain. Little, Brown and Co. Boston, p. 689–703.
7. DEUTSCH, J. A. and HOWARTH, C. I. 1963. Some tests of a theory of intracranial self-stimulation. Psychol. Rev. 70: 444–460.
8. HOWARTH, C. I. and DEUTSCH, J. A. 1962. Drive decay: The cause of fast "extinction" of habits learned for brain stimulation. Science 137: 35–36.

9. KEESEY, A. E. 1966. Hypophtalamic stimulation as reinforcer of discrimination learning. J. Comp. Physiol. Psychol. 62: 231–236.
10. KNOTT, P. D. and CLAYTON, K. N. 1966. Durable secondary reinforcement using brain stimulation as the primary reinforcer. J. Comp. Physiol. Psychol. 61: 151–153.
11. KONORSKI, J. 1967. Integrative activity of the brain. An interdisciplinary approach. Univ. of Chicago Press, Chicago, 531 p.
12. KONORSKI, J. and MILLER, S. 1933. Podstawy fizjologicznej teorii ruchów nabytych. Ruchowe odruchy warunkowe. Medycyna Dośw. i Społ. 16: 1–167.
13. LENZER, I. I. and FROMMER, G. P. 1968. Successive sensory discriminative behavior maintained by intracranial self-stimulation reinforcement. Physiol. Behav. 3: 345–349.
14. LENZER, I. I. and FROMMER, G. P. 1971. Successive sensory discriminative behavior maintained by forebrain self-stimulation reinforcement. Psychon. Sci. Sect. Anim. Physiol. Psychol. 23: 88–90.
15. MOGENSON, G. J. 1965. An attempt to establish secondary reinforcement with rewarding brain stimulation. Psychol. Rep. 16: 163–167.
16. PAVLOVA, O., KOSOWSKI, S. and SADOWSKI, B. 1976. Conditioning of self-stimulation in the dog to acoustic and visual stimuli. In A. Wauquier and E. T. Rolls (ed.), Brain-stimulation reward. North-Holland Publ. Co., Amsterdam, p. 403–405.
17. PEREZ-CRUET, J., McINTIRE, R. W. and PLISKOFF, S. S. 1965. Blood-pressure and heart-rate changes in dogs during hypothalamic self-stimulation. J. Comp. Physiol. Psychol. 60: 373–381.
18. SADOWSKI, B. 1972. Intracranial self-stimulation patterns in dogs. Physiol. Behav. 8: 189–193.
19. SADOWSKI, B. and DEMBIŃSKA, M. 1973. Some characteristics of self-stimulation behavior of dogs. Physiol. Behav. 33: 757–769.
20. SEWARD, J. P., UYEDA, A. and OLDS, J. 1959. Resistance to extinction following cranial self-stimulation. J. Comp. Physiol. Psychol. 52: 294–299.
21. STEIN, L. 1958. Secondary reinforcement established with subcortical stimulation. Science 127: 466–467.
22. TERMAN, M. and KLING, J. W. 1968. Discrimination of brightness differences by rats with food or brain-stimulation reinforcement. J. Exp. Anal. Behav. 11: 29–37.
23. TROWILL, L. A. and HYNEK, K. 1970. Secondary reinforcement based on primary brain stimulation reward. Psychol. Rep. 27: 715–718.
24. TROWILL, J. A., PANKSEPP, J. and GANDELMAN, R. 1969. An incentive model of rewarding brain stimulation. Psychol. Rev. 76: 264–281.
25. WAUQUIER, A., MELIS W., NIEMEGEERS, C. J. E. and JANSSEN, P. A. J. 1978. A putative multipartite model of haloperidol interaction in apomorphine-disturbed behavior of the dog. Psychopharmacology 59: 255–258.

Bogdan SADOWSKI. Institute of Genetics and Animal Breeding, Jastrzębiec 05-551 Mroków, Poland.

# THE ROLE OF EMOTIONS IN THE FORMATION OF INSTRUMENTAL CONDITIONED REFLEX

P. V. SIMONOV

Institute of Higher Nervous Activity and Neurophysiology,
USSR Academy of Sciences Moscow, USSR

*Abstract.* The recording of vegetative parameters of emotional stress in humans and in dogs during elaboration of instrumental conditioned reflexes showed that such stress is a function of two major factors: (i) the value of the actual need (drive) and (ii) the estimation by the brain of the probability (possibility) of its satisfaction. Comparison of the need with the possibility of its satisfaction activates the mechanisms of positive (maximized by the subject) or negative (minimized) emotions; which serve as the direct reinforcement of instrumental conditioned reflexes. Experiments show that needs and emotions have an independent morphological substratum. The hypothalamus and amygdala must be intact in order to single out the need that is dominant at a given moment and is first to be satisfied. Estimation of the probability of need's satisfaction (probability of reinforcement) is the function of the frontal areas in the neocortex and hippocampus. These four brain structures play a decisive role in the genesis of emotional states and in the organization of behavior.

For the student of the physiology of emotions, the instrumental conditioned reflex has a double heuristic value. First, it is the process of the elaboration of instrumental conditioned reflex that reveals the genesis and complex inner structure of emotional stress, its dependence on the value of the actual need and the estimation by the brain of the probability (possibility) to satisfy this need on the basis of phylo- and ontogenetic experience. Second, the formation of instrumental conditioned reflexes indicates the role of emotions in the closing of conditioned connections, in the organization of goal directed behavior.

The dependence of emotions on the need and probability of its satisfaction may be easily demonstrated on the model of elaboration of

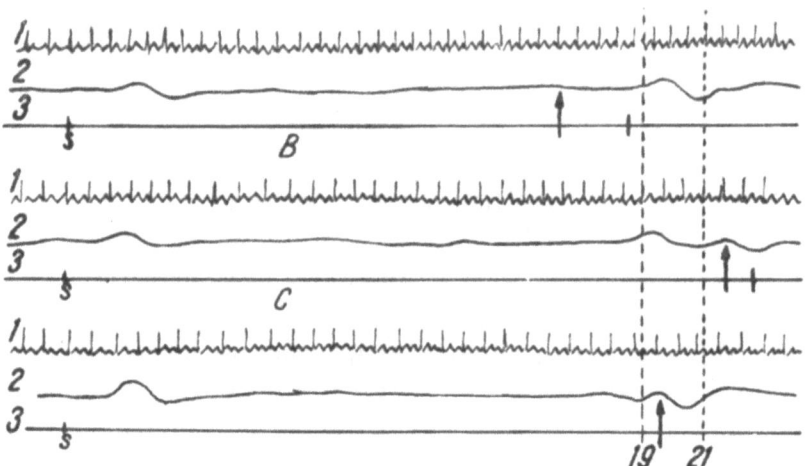

Fig. 1. Conditioned defensive reflex in humans. A and B, erroneous reactions accompanied by shock. C, correct reaction. 1, ECG; 2, GSR; 3, time. Dotted perpendicular lines denote time limits for correct reactions. S, sound signal. Arrows, reactions. Vertical lines, shocks.

defensive conditioned reflex in man (6). Subjects were instructed to press a key 20 s after a short auditory signal (Fig. 1). If the subject pressed on the key earlier than 19 s after the signal or later than 21 s, an electrical shock of 60–90 v was delivered to his forearm. After each test the subject was informed of the time of this reaction. The degree of emotional stress was assessed by the changes in heart-rate, and the galvanic skin reflex was recorded simultaneously. The increase in heart-rate was estimated by the summary duration of the first three heart beats after the sound signal and of the last three beats before the motor reaction (paper's speed = 1.5 cm/s). During the first 10 presentations of the signal the subject knew that there would be no electrical stimulation. Then electrical stimulation was used several times in isolation to determine the current intensity, exceeding the pain threshold by no less than three times. This intensity remained constant throughout the experiment.

Figure 2 (lower part) shows the time of motor reactions in successive tests. Electrostimulation was applied when the subject's errors exceeded the allowed deviation. Upper part of Fig. 2 shows the changes in heart rate. Comparison of the two stages of the experiment, containing the same number of tests (14) and of nociceptive electrostimulations (6), demonstrated that the summary deviation of heart-rate frequency from

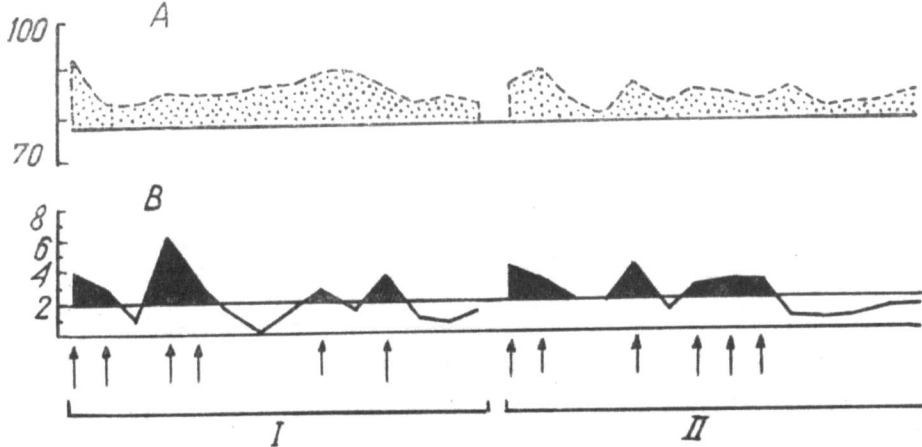

Fig. 2. Elaboration of conditioned defensive reflex. A, changes in duration of six systoles taken against the number beats per min. The horizontal line denotes the mean heart-rate before the conditioning. B, deviations of motor reaction time (in min) from the required time limits marked by two horizontal lines. I, II two parts of experiment, in the initial and final phase, matched in respect to number of shocks received by subject.

the initial background in these two experimental parts differed. Hence, it depends not only on the quantity of the punishment. An experiment on 9 subjects showed that the summary changes in heart-rate are proportional to the summary deviation of motor reactions time from the given value, i.e., proportional to the degree of perfection, precision and reliability of conditioned defensive reflexes (Table I). This rule also holds for those cases where the value of errors (but not their number, not the quantity of pain stimulations) increased along with the increase in heart-rate and where the dynamics of vegetative changes could not be explained by habituation to nociceptive stimuli throughout their repetition.

TABLE I

Relationship between total increase in heart rate and total number of mistakes

| Phase of experiment | Number of subject | | | | | | | | |
|---|---|---|---|---|---|---|---|---|---|
| | 1 | 2 | 3 | 4 | 5 | 6 | 7 | 8 | 9 |
| First | 7.5 | 1.56 | 3.28 | 5.4 | 3.08 | 0.48 | 0.66 | 0.42 | 0.73 |
| Second | 6.1 | 1.55 | 3.11 | 5.8 | 3.19 | 0.46 | 0.83 | 0.40 | 0.61 |

Experiments on animals also show that the brain forcasts the probability of punishment depending on the degree of perfection of instrumental motor reflex. Preobrazhenskaya (5) elaborated conditioned defensive reflex in dogs, where the lifting of the foreleg to a certain level and holding it at this level during 10 s prevented electrostimulation of

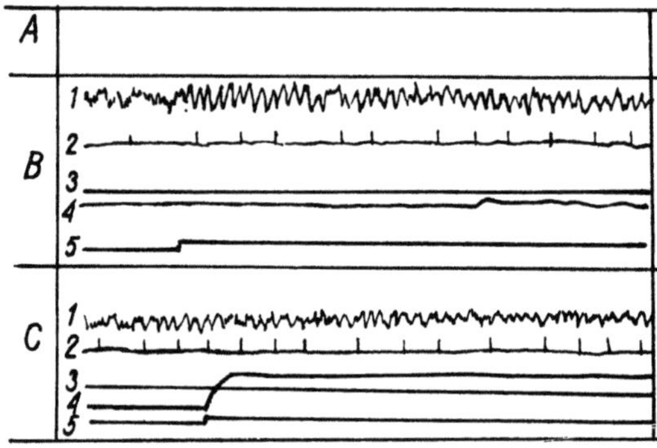

|   | | I | II | III | IV | V |
|---|---|---|---|---|---|---|
| A | | 80 | 10.4 | 0.25 | 0.75 | 0.2 |
| B | 1 2 3 4 5 | 79 | 13.9 | 0.45 | 3.0 | 0.001 |
| C | 1 2 3 4 5 | 73 | 12 | 0.21 | 1.68 | 0.1 |

Fig. 3. The changes in hippocampal theta-rhythm and in heart-rate during elaboration of instrumental conditioned defensive reflex in dog. I, EEG of the dorsal hippocampus; 2, ECG; 3, the level of switching off of the current; 4, raising of the fore-leg, switching off the current; 5, conditioned sound signal. A, before elaboration; B, experiment session No 2; C, experiment session No 19. I, mean value of integrated theta-rhythm, II, mean heart-rate, III, coefficient of correlation, IV, t, V, level of significance.

the same leg (methods of Zelenyï, Skipin-Vinnik et al.). The conditioned auditory signal was presented 10 s before electrostimulation. These pairing of sound with nociceptive stimulation before elaboration of the conditioned motor reflex led to an increase in the amplitude and percentage of theta-rhythm in the electrical activity frequency spectrum of the dorsal hippocampus (Fig. 3). Quantitative analysis revealed a positive correlation between the changes of summated hippocampal theta-rhythm (measured by the integrator's readings) and heart-rate. Both symptoms markedly weakened throughout stabilization of the motor experience, presumably saving animals from pain. Any difficulties in movements led to the repeated increase in theta-rhythm. Thus Preobrazhenskaya's experiments showed that the intensity of hippocampal theta-rhythm depends not on the motor activity itself, but on the efficiency of motor acts, on their influence on the probability of preventing pain stimulations.

In its most general form, the rule of emotions genesis in humans and in animals may be presented as a structural formula:

$$E = f\,[N, (I_n - I_a), ...],$$

where E — emotion, its degree, quality and sign; N — power and quality of the actual need (drive); $I_n - I_a$ — estimation of probability (possibility) of need's satisfaction on the basis of phylo- and ontogenetic experience; $I_n$ — information on the means, prognostically necessary for satisfaction of the need; $I_a$ — information on the means, available for the subject at the given moment.

The low probability of need's satisfaction leads to negative emotions, actively minimized by the subject. The increased probability of satisfaction, as compared to the earlier forcast, generates positive emotions, which the subject tries to maximize, i.e., to enhance, to prolong, to repeat. It is evident that the informational theory of emotions encompasses both their reflective function, the laws of their appearance, and the regulatory significance of emotions, their role in organization of behavior.

The theory is applicable not only for complex behavioral acts but also for the generation of any emotional state. For example, the positive emotion during eating occurs due to the integration of hunger excitation (need) with afferentation from the oral cavity, evidencing the increasing probability of satisfaction of this need. In another state, the same afferentation may generate the feeling of indifference or negative emotion of disgust. The estimation by the brain of the probability of need's satisfaction (i.e., integral estimation of available experiences, power resources of organism, time necessary and sufficient for realization of corresponding action) may occur in humans both on conscious and unconscious levels. The best example of unconscious forcast is intuition, where the estimation of approaching the goal or moving away from it is initially realized in the form of emotional "presentiment of decision", encouraging further analysis of the situation that elicited this emotion.

It is necessary to emphasize that the regulatory function of emotions does not come to the simple signaling of influences that are useful or harmful to the organism. For example, when an extremity is injured pain restricts motor activity, favoring reparative processes. However, the same role could be played by the mechanism, which without emotions automaticly would inhibit movements, harmful for the injured organ. Pain proves to be a more plastic mechanism: when the need in moving is very high (e.g., when the very existence of the subject is threatened) the movement is executed in spite of pain. In other words, emotions play the role of certain "brain money" — universal measure of values — and the price, as is known, is a dynamicly fluctuating variable, depending on demand (need) and supply (possibilities of satisfaction).

Experimental data gathered in our laboratory with different brain lesions and comparative analysis of the summary electrical activity showed that brain structures of the limbic system and that of the neocortex respond differently to two factors, playing the decisive role in the genesis of emotional states (Fig. 4). For example, the amygdala and the hypothalamus turned out to be important for determination of the dominant need (motivation), which at the given moment must be satisfied in first priority (4), whereas frontal parts of the neocortex and the hippocampus are necessary for the estimation of probability of conditioned signal's reinforcement. The prefrontal cortex orienting behavior to the signals of highly probable events (2) and intactness of the hippocampus being necessary for reactions to the signals with low probability of their reinforcement (4).

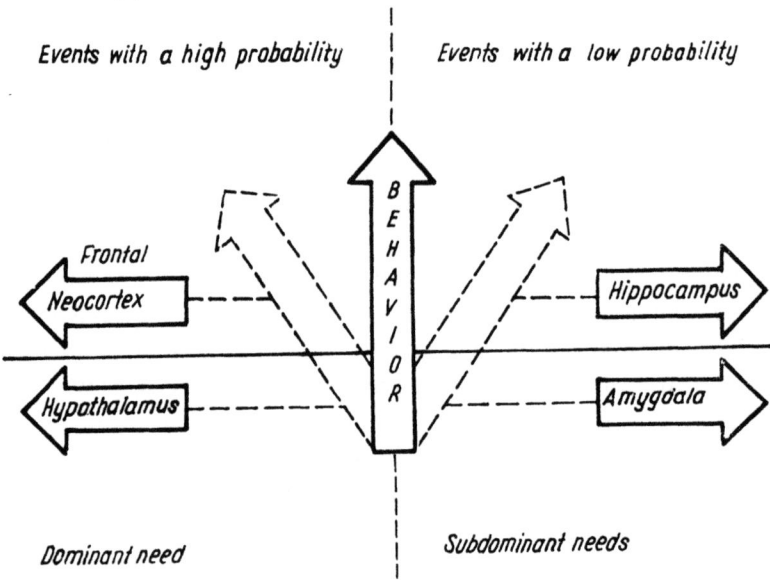

Fig. 4. The diagram of participation of different brain structures in the genesis of emotional states and in organization of behavior.

According to the informational theory of emotions, needs, motivations and emotions not only are independent, unidentified with each other phenomena of the higher nervous activity, but must also have their own morpho-physiological substratum. We believe that the consecutive involvement in behavior of brain mechanisms of needs, motivations and emotions was revealed in the experiments from our laboratory by N. G. Mikhailova and K. Yu. Sarkisova. They applied the method of gradually increasing the intensity of the stimulating current, used by A. V. Waldman, M. M. Kozlovskaya, J. P. Houston and other researchers.

In an albino rats monopolar steel electrodes with a tip of 60 µs in diameter were implanted in the lateral preoptical area and in the lateral hypothalamus (coordinates of König and Klippel: A-7.0; L-1.8; H-3.5;

and A–3.0–3.5; L–1.5; H–3.5). During stimulations, the chamber contained a pedal for self-stimulation as well as goal objects for determination of specific motivations: alimentary (sun-flower seeds or oats), drinking (water), gnawing (a piece of chalk or wood), sexual (female). The stimulation was done both with rhythmic current (right-angled monophase pulses with 100 Hz frequency, 0.1 µA duration, 0–1.0 µA intensity) and direct current of 0–70 µA. Consumatory reactions (eating, drinking, gnawing) were recorded on the myogram of chewers. Reactions latencies and the probability of their occurrence were calculated, i.e., the ratio between the number of stimulations, that caused these reactions and the general number of stimulations. The reinforcing (emotionally positive) effect of the current was estimated by the frequency of self-stimulations. Before the experiment rats had free access to food and water.

In all cases only those areas were stimulated, which are capable of responding with self-stimulation to a current with sufficient intensity. It turned out that during polarization of these areas with gradually increasing direct current as well as during their stimulation with the rhythmic current of increasing intensity, behavioral reactions always had the same sequence. Weak stimulation elicited generalized search activity without turning to the goal objects in the chamber — to food, water, female etc. Only with increased intensities of stimulation did these external stimuli become effective, the animals began to eat, sometimes to drink, gnaw etc. With further increases in rhythmic or direct current the reaction of self-stimulation occurred.

Should two different points in the hypothalamus be stimulated analogous sequence of events is observed, i.e., two stimulations of "search activity" give consumatory reactions: more often — that of eating, rarely — drinking, gnawing etc., and two stimulations of "motivating intensity" are capable of causing the reaction of self-stimulation (Fig. 5). The influence on one of the points, sufficient for eliciting self-stimulation, suppresses motivated behavior, caused by stimulation of the second point. White (9) observed the cessation of natural alimentary behavior with stimulation of the amygdala, capable of provoking self-stimulation. The experiments of N. G. Mikhailova and K. Yu. Sarkisova revealed methodical advantage of the direct current as compared to the rhythmic one. Stimulation of the two points with rhythmic current leads to enhancing (frequentation) of self-stimulation reactions. A weak cathode polarization of one point inhibits self-stimulation of the second focus, and the increased intensity of the direct current enhances self-stimulation. It should be noted that the inhibitory influence is produced by direct current intensity which in isolated use causes goal directed motivated behavior.

The observed transformation of effects is hard to explain by nonspecific additional activation of "motivational-reinforcing" structures, because stimulation of emotionally negative (eliciting avoidance) areas in the reticular formation of the midbrain inhibits the reaction of self-stimulation, whereas stimulation of emotionally neutral areas has no influence at all on self-stimulation of the hypothalamus. Only stimulation of emo-

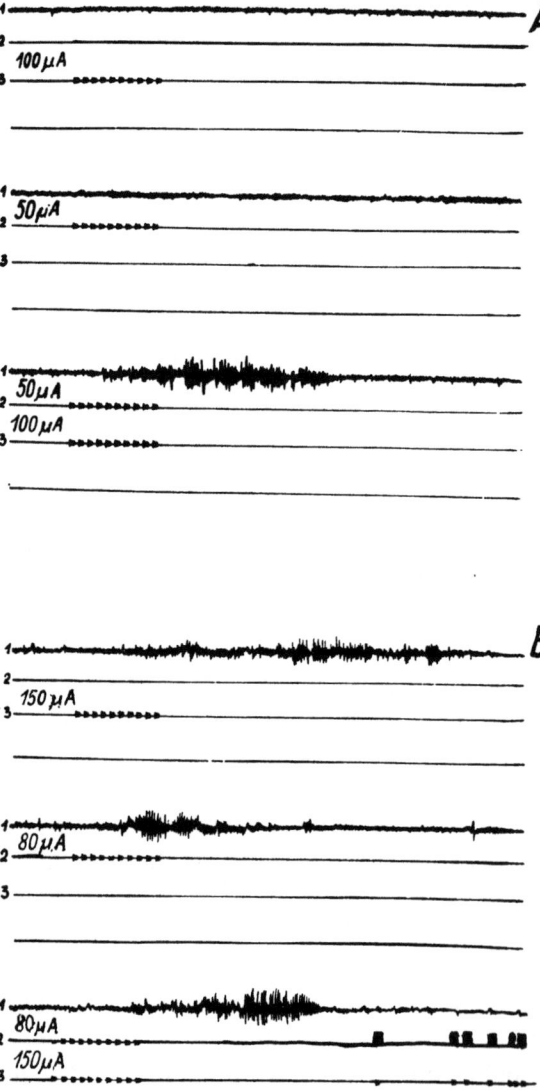

Fig. 5. Consequences of simultaneous stimulation of two points in the hypothalamus with electrical current of "searching" (A) and "motivating" (B) intensity. 1, myogram of schewers; 2 and 3, stimulation of the first and the second point. Vertical bars on the line 2, records of self-stimulation.

tionally positive structures of the reticular formation, able to cause independently at least a weak self-stimulation, are summarized with "searching". "motivating" and "reinforcing" excitation of the hypothalamic structures.

We believe, that the experiments of this model reproduce the sequence of events typical of the organization of natural behavior. Weak stimulation with the electrical current activates the system of brain structures representing a substratum of needs. The process of actualization of the need, not yet transformed into motivation, outwardly is manifested in the form of generalized searching unrest. Only an increase in excitation with the increasing intensity of direct or rhythmic current leads to activation of such structures, which hold the engrams of external objects, capable of satisfying this need. As the result external stimuli become effective, and the motivated animal starts to eat, drink, gnaw etc. However, a further increase in the current is necessary to activate the structures of emotionally positive reinforcement and to make the animal stimulate its brain in the absence of natural satisfaction of any need.

We are far from thinking that the structures of need, motivation and emotion are linearly positioned in the neighboring parts of brain tissues and that the electrical current of increasing intensity captures these structures one after another. We mean that the system of structures necessary and sufficient for actualization of a need is more simple and contains less elements than the system providing for the animal's goal directed motivation. A full complex of morpho-physiological organization of behavior (need plus motivation plus emotion) supposes additional involvement of the nervous apparatus of emotions.

Weak current stimulation seems to simulate such hunger excitation that in natural conditions comes to the hypothalamus from reticular nuclei of the pons and medulla oblongata. This excitation must be enhanced by the increasing intensity of electrical current of by stimulation of the second "alimentary" ponit (not any other) in order to activate the engrams of alimentary objects and to condition the act of eating. During eating nervous elements are excited, which generate an emotionally positive state, but this excitation still is not sufficient for the transition to artificial stimulation of emotionally positive structures by electrical current. Only further enhancing of the current or summation of two "motivating" excitations leads to the substitution of eating with self-stimulation. Let us emphasize once again that the phenomenon of artificial reinforcement is by no means the result of the enhanced motivational excitation. The transition from eating to self-stimulation occurs only after the stimulation of the points which with sufficient current intensity are individually capable of producing the reaction of self-stimulation.

The transition to self-stimulation gradually stops the alimentary behavior of a rat. This effect is an additional evidence that the direct reinforcing factor of the instrumental reflexes is not the satisfaction of any need, but the maximization of a positive (or minimization of a negative) emotional state. We believe that the results of experiments with self-stimulation agree with the data received by Oniani (3), who used the direct electrical stimulation of limbic brain structures as the reinforcement for the elaboration of the conditioned reflex. With the pairing of an external stimulus with stimulation of brain structures resulting in repleate cat eating, drinking, aggression, anger and fear, 5-50 pairings led to the elaboration of the conditioned avoidance reaction only, accompanied by fear. He failed to obtain conditioned reflexes of eating and drinking. It is of interest that the conditioned signal, reinforced by the stimulation of "aggression centers", elicited fear and avoidance.

According to Oniani, reinforcing may be the stimulation of only those brain structures, which in natural conditions are activated by external factors (fear) and not by interoceptive impulses (hunger, thirst). Oniani explains the possibility of elaboration of conditioned avoidance reflexes, reinforced by stimulation of "aggression centres", by the fact that the formation of natural states of aggression has an endogenic component (hormonal one during marital fights, hunger one etc.).

From our point of view, the results of these experiments constitute one more piece of evidence of the decisive role of *emotions* in conditioning. Fear has pronounced aggressivenes for the animal and is actively minimized by it through the avoidance reaction. Stimulation of alimentary and drinking brain systems in repleate and unthirsty animals causes stereotypical acts of eating and drinking without involvement of nervous mechanisms of emotions, which excludes the elaboration of conditioned reflexes. Stimulation of aggression centers with the given localization of electrodes and current parameters generates an emotionally negative state, which, as in case of fear, leads to minimizing avoidance reactions. If the aggressive behavior of cats is accompanied by the involvement of emotionally positive structures, then on the basis of their stimulation a conditioned reaction of self-stimulation may be elaborated, as was shown by Waldman, et al. (8).

The obligatory participation of nervous mechanisms of emotions for elaboration of instrumental conditioned reflex was shown by Wyrwicka (10) Cytawa et al. (1) and Trojniar et al. (7). Figure 6 attemps to show the role of emotions in the closing of conditioned reflex. From its inception the theory of conditioned reflexes supposed the convergence of two excitations: from the conditioned stimulus and from the stimulus, eliciting unconditioned reflex e.g., afferentation from the oral cavity with the food coming to the mouth (I in Fig. 6). At the same epoch the significance of "available functional state" was clarified, alimentary excitability, which today may be considered as the result of the excitation of brain structures, activated by the occurrence of the corresponding need, the state of hunger (II in Fig. 6). However, neither afferentation from

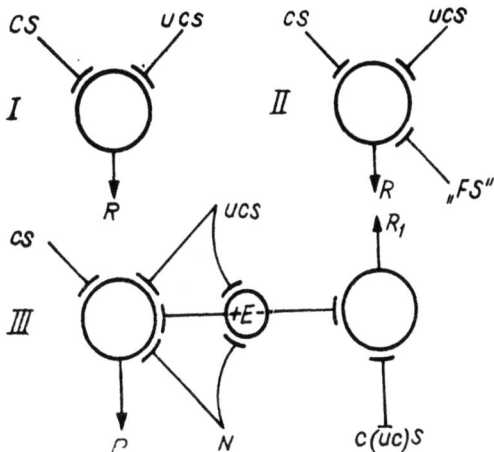

Fig. 6. Two — (I), three — (II) and four-factor (III) diagrams of the encounter of excitations in the point of convergence during formation of conditioned reflex. CS, conditioned stimulus; UCS, unconditioned stimulus; R, reaction; F, functional state; N, need; E, emotion.

the oral cavity, nor the hunger excitation itself can play the role of reinforcement, providing for the formation of the instrumental conditioned reflex. Only integration of hunger excitation with the one from the factor, capable of satisfying this need, i.e., the mechanisms, generating positive emotion, provides for the conditioning. With the other ratio between the converging excitations, for example with the food administrated into the mouth of overfed animal, activation of mechanisms of a negative emotion will lead to the defensive reaction of avoidance (III in Fig. 6).

In conclusion we would like to stress the following. If in the works of I. P. Pavlov we come across only separate mentioning of emotions, feelings and drives, separate ideas, that turned out to be extremely fruitful for further experimental studies, today the neurophysiology of motivations and emotions have become an integral part, an important sphere of the science, concerned with the higher nervous (psychic) activity of humans and animals. Suffice to say that, had not the neurophysiological mechanisms of emotions been taken into account, we would not be able to solve the central problem of the physiology of higher nervous activity: the problem of closing of conditioned reflex.

## REFERENCES

1. CYTAWA, J. and TROJNIAR, W. 1976. The state of pleasure and its role in the instrumental conditioning. Acta Nervosa Superior 18: 92–96.
2. MEKHEDOVA, A. Y. 1971. On the role of the frontal regions of the brain in the formation of conditioned responses adequate to the magnitude and probability of their reinforcement (in Russian). Zh. Vyssh. Nervn. Deyat. Im. I. P. Pavlova 21: 459–464.
3. ONIANI, T. N. 1975. On the possibility of conditioned emotional reflexes elaboration by electrical stimulation of limbic structures (in Russian). Zh. Vyssh. Nervn. Deyat. 25: 230–238.
4. PIGAREVA, M. L. 1978. Limbitcheskie mekhanizmy pereklutcheniya: gippocamp i mindalina. Izdat. Nauka, Moscow, 151 p.
5. PREOBRAZHENSKAYA, L. A. 1969. Study of the correlation between hippocampal theta-rhythm and hart rate on the initial stage of conditioned defensive reflex elaboration. In P. Simonov (ed.), Nervnoe napryazhenie i deyatel'nost' serdtsa. Izdat. Nauka, Moscow, p. 151–184.
6. SIMONOV, P. V. 1964. On the ratio of the motor and vegetative components in the conditioned defensive reflex in man (in Russian). Abstr. Comm. III Int. Symp. Central and peripheral mechanisms of motor activity in animals and man. Moscow, p. 65–66.
7. TROJNIAR, W. and CYTAWA, J. 1976. Transfer from extero- to interoceptive reinforcement in the course of instrumental conditioning in rats. Acta Neurobiol. Exp. 36: 455–462.
8. WALDMAN, A., ZVARTAU, E. and KOZLOVSKAYA, M. 1976. Psikhofarmaco logiya emotsii. Izdat. Meditsina, Moscow, 327 p.
9. WHITE, N. 1973. Self-stimulation and suppresion of feeding observed at the same site in the amygdala. Physiol. Behav. 10: 215–219.
10. WYRWICKA, W. 1975. The sensory nature of reward in instrumental behavior. Pavlovian J. Biol. Sci. 10: 23–51.

P. V. SIMONOV, Institute of Higher Nervous Activity and Neurophysiology, USSR Academy of Sciences, Butlerova 5a, 117485 Moscow, USSR.

Lecture delivered at the Warsaw Colloquium on Instrumental
Conditioning and Brain Research
May 1979

# RECENT EXPERIMENTS TESTING AN OPPONENT-PROCESS THEORY OF ACQUIRED MOTIVATION

Richard L. SOLOMON

Department of Psychology, University of Pennsylvania
Philadelphia, Pennsylvania, USA

*Abstract.* There are acquired motives of the addiction type which seem to be non-associative in nature. They all seem to involve affective phenomena caused by reinforcers, unconditioned stimuli or innate releasers. When such stimuli are repeatedly presented, at least three affective phenomena occur: (1) affective contrast effects, (2) affective habituation (tolerance), and (3) affective withdrawal syndromes. These phenomena can be precipitated either by pleasant or unpleasant events (positive or negative reinforcers). Whenever we see these three phenomena, we also see the development of an addictive cycle, a new motivational system. These phenomena are explained by an opponent-process theory of motivation which holds that there are affect control systems which oppose large departures from affective equilibrium. The control systems are strengthened by use and weakened by disuse. Current observations and experiments testing the theory are described for: (1) the growth of social attachment (imprinting) in ducklings; and (2) the growth of adjunctive behaviors. The findings so far support the theory.

The opponent-process theory of acquired motivation was suggested by the existence of three related phenomena found in many acquired motives:
1) hedonic contrast effects,
2) hedonic habituation or the growth of tolerance,
3) hedonic withdrawal syndromes or abstinence syndromes.

*Hedonic contrast effects* refer to the fact that the reinforcing properties of the presentation of, and maintenance of, many reinforcers contrast with the reinforcing effects of the removal of, or continued absence of, the same reinforcers. A mirror-image relation often exists. This relation is illustrated in Fig. 1. There we see, in the top panel, the

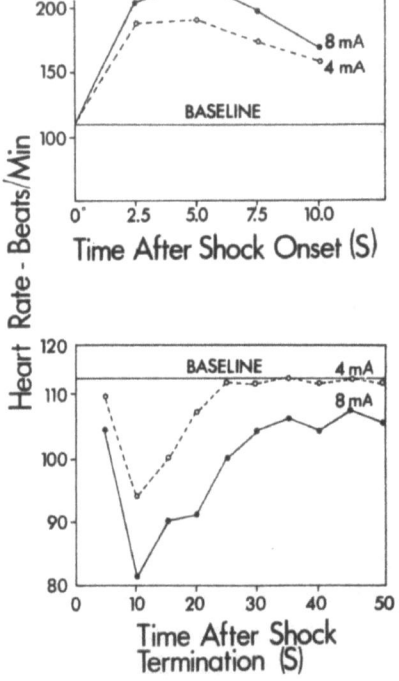

Fig. 1. The unconditioned heart-rate response in dogs. The upper panel, the response to the onset and 10-s duration of 4 mA and 8 mA shocks. The lower panel, an after-effect, after shock termination. The graphs represent the mean values from six dogs with randomized presentation of shock intensities.

dog's heart rate response to the onset, 10-s maintenance and cessation of shocks to the hind toe pads. At shock onset, the heart rate increases for about 5 s, then decreases even while the shock is still on. Then the shock is terminated after 10 s, and we see, in the bottom panel, a marked deceleration, *below* the resting baseline level, followed by a slow, sluggish recovery to baseline level. Note that the 4 mA. shocks produced less acceleration and less deceleration. The 8 mA. shocks produced more acceleration *and* more deceleration, with deceleration lasting longer than it did for the 4 mA. shocks. This is the *hedonic contrast* phenomenon. It has five distinctive features, idealized in Fig. 2, which portrays how we think the onsets and terminations of many reinforcers (unconditioned stimuli — UCSs, reinforcers — rfts, or innate releasers) affect hedonic, affective or emotional processes in the organism. This *standard pattern of affective dynamics* shows that the after-reaction involves a hedonic or affective state qualitatively different from that of the primary reaction to the presence of the reinforcer. Here, a square-wave input is transduced to a five-featured resultant. The same function can be drawn for color vision; the stimulus presentation can be red, and then the after-reaction will be green, the negative after-image. The hedonic contrast phenomenon is like a hedonic negative, affective (rather than color) after-image.

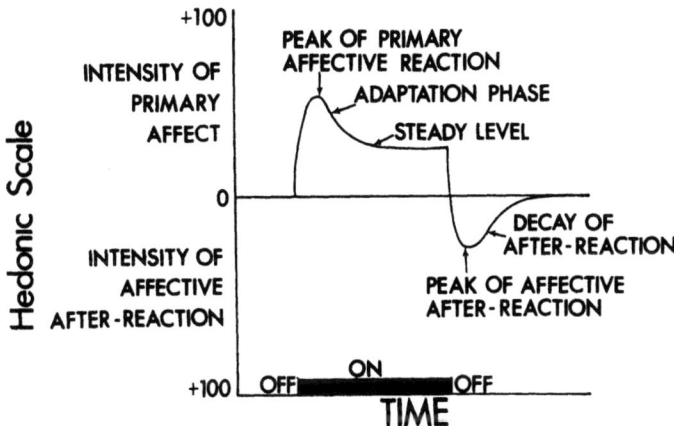

Fig. 2. An idealization of the standard pattern of affective dynamics. At reinforcer onset, there is a peak of arousal which then subsides while the reinforcer is maintained. Then, when the reinforcer is removed, there is a contrasting after-reaction which peaks quickly and then decays back to the original affective baseline.

These "after-images" have reinforcing attributes. If the onset is a positive reinforcer, then termination will function as a negative reinforcer, and vice versa. When we observe such a phenomenon, then we usually will observe the next one.

*Hedonic habituation or tolerance* effects occur when the frequent repetition of a reinforcer (UCS, rft or innate releaser) results in the decreased capacity of the reinforcer to reinforce. The reinforcer then cannot as easily influence ongoing behaviors nor form associations with CSs or with operants. The clearest examples come from repeated drug use, where we use the term *tolerance*. But other than chemical reinforcers show the same characteristic *habituation* effect. We know that many reinforcers produce smaller UCRs when repeated frequently. This phenomenon is illustrated in Fig. 3, for heart rate reactions in a "veteran" laboratory dog who has received hundreds of shocks (10-s duration, 4 mA.) over a period of several weeks. At the end of this protracted treatment, the dog's heart shows very little acceleration in response to UCS onset (see top panel). Then, when a shock is terminated, we see the emergence, shown in the bottom panel, of the *withdrawal or abstinence syndrome.*

*The hedonic withdrawal or abstinence syndrome* is characterized by a large amplification of the after-reaction, so that it is very intense and lasts a long time. Of course, we are used to observing this in connection with repeated opiate or alcohol use, when the highly tolerant drug user

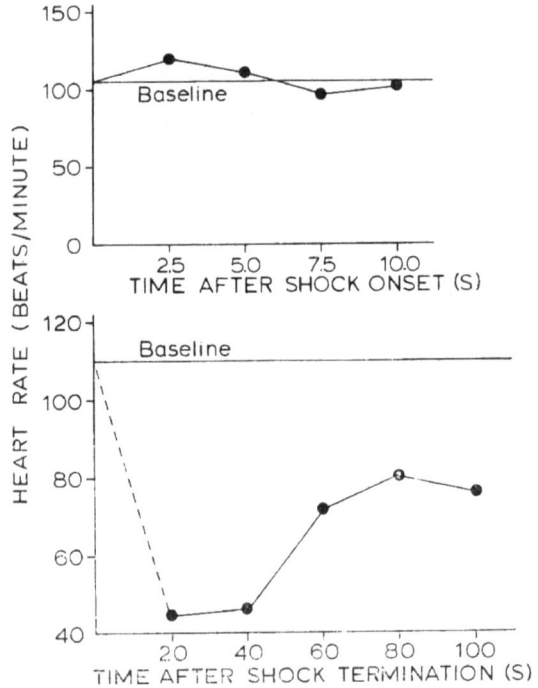

Fig. 3. The heart rate reactions of a "veteran" laboratory dog who has previously received hundreds of shocks over several weeks (means for 6 repeated shock presentations). The upper panel, response to shock onset and 10-s shock duration. The lower panel, after-effect, after shock termination. Note that the biphasic heart rate deceleration and recovery to baseline is of large amplitude and long duration relative to what it was in Fig. 1.

also shows intense, long-lasting aversive affect (negative reinforcer) when the chemical treatment is terminated. Hedonic withdrawal syndromes can be either aversive or desirable depending on the hedonic quality of the reinforcer onset and maintenance. They aren't always aversive, and so the drug case is partially misleading. For example, when military parachutists become highly tolerant of free falls (are no longer afraid), they also experience a period of exhilaration following the jump session. Or a marathon runner may experience long-lasting exhilaration or mood elevation following a long run to which the runner has already become tolerant. These are positive reinforcing effects and are sought. They are derived from constant re-exposure to originally-aversive reinforcers.

The three hedonic or affective phenomena are characteristic of many acquired motives. They are all analogous to the phenomena of opiate addiction. In all cases, the power of the substance or stimulus to reinforce a CS or an operant via onset is *reduced* by repetition. But the power of the after-reaction to reinforce becomes *enhanced* by repetition. This asymmetry of course destroys the contrast phenomenon seen in Fig. 2 and replaces it with the result seen in Fig. 3. In addition, the hedonic sign of the withdrawal syndrome is functionally *opposite* to that of the onset effect.

All three phenomena are usually found together. If you see one, you will usually see the others. All three are usually found in clear cases of *acquired,* in contrast to innate, motive systems. Acquired motives are those not built into the species without special experiences or external events. One does not have to be an opiate addict. One doesn't have to be a devotee of jogging. But these phenomena will develop if the right experiences are frequently repeated.

These hedonic phenomena are most easily explained by an opponent-process theory of motivation. The theory assumes that the brains of all mammals are, for some reason, organized to oppose or suppress many types of emotional arousals or hedonic processes, whether they are pleasurable *or* aversive, whether they have been generated by positive *or* negative reinforcers. The opposing affective or hedonic processes are *automatically* set in motion by many of those stimuli which psychologists or ethologists have shown, through defining experiments, to function as Pavlovian UCSs, operant reinforcers, or innate releasers.

All *primary* affective or hedonic processes, elicited by UCSs, rfts or innate releasers, are postulated to correlate closely in their magnitudes with the stimulus intensity, quality and duration of the reinforcer. These primary processes are phasic and sensitive to small stimulus changes. They may show some sensitization effects, but rarely do they show habituation. They are stable, unconditioned reactions. I call them *a-processes.* For example, a snake (UCS) elicits a reflex fear reaction (UCR) in a monkey. Or, the taste of chocolate syrup (UCS) elicits salivation (UCS), or excitement (UCR) and a pleasure state (UCR) in a child.

The primary process, the a-process, in turn arouses a *b-process* which functions to *oppose* and *suppress* the affective or hedonic state generated initially by the onset of the a-process. The b-process drags down the strength of an A-state. The b-process (the opponent-process) is postulated to be: (1) of sluggish latency (2) inertial, or slow to build to its asymptote and (3) slow to decay after the stimulus input (UCS) has been terminated and the a-process (UCR) has stopped. Because the b-process is an opponent-process, its affective or hedonic *quality* must be *opposite* to that of the a-process. The implications of such a simple assumption are far-reaching, as we shall see.

The affective or hedonic state of the organism at any moment is postulated to be the difference, without regard to sign, between the magnitudes of the a-process and b-process. The b-process has a negative sign because it opposes the a-process. The *state rule* is simple: (1) if /a–b/ shows a > b, then the organism is in state A, and (2) if /a–b/ shows b > a, then the organism is in state B. Furthermore, if being in state A is positively reinforcing (pleasant, desirable), then being in state B will be *negatively reinforcing* (aversive, undesirable), and vice versa.

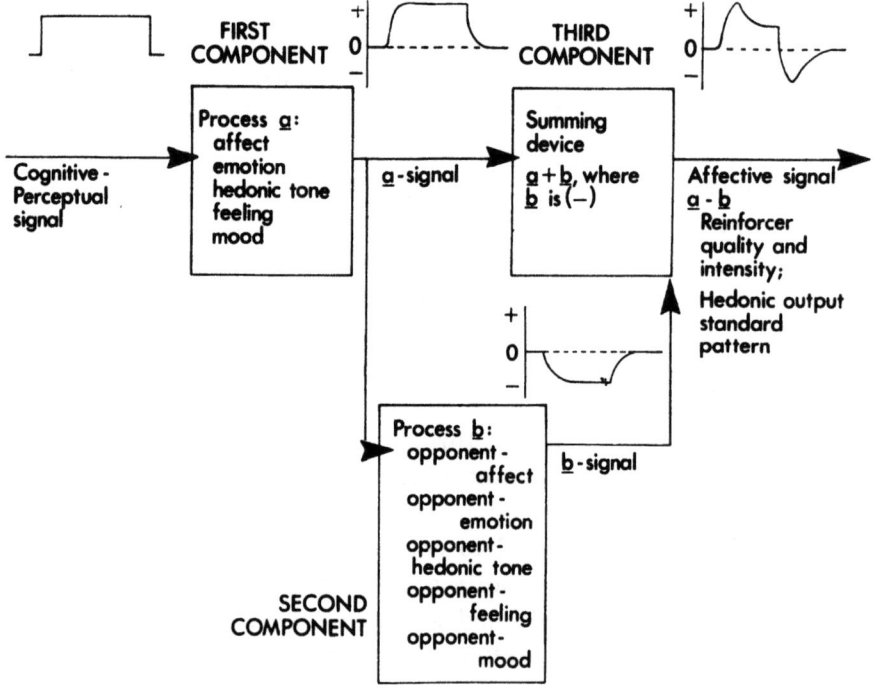

Fig. 4. The affect-processing system for the opponent-process model of acquired motivation. The presence of the reinforcer arouses the a-process, one side effect of which is to arouse the b-process, the affective opponent of the a-process. The summation of the a signal and b signal gives the current state of the organism. Because the b-process is of long latency, slow to grow and slow to decay, the sum (a–b) will produce the standard pattern of affective dynamics shown in Fig. 2.

The affect-processing system, reflecting the opponent-process assumptions made so far, is shown in Fig. 4. First, there is a cognitive-perceptual event representing the UCS, rft or innate releaser. For illustration, assume that the subject is a cat and the incoming signal is categorical, a dog! The dog can be depicted as a square-wave input. One of its side-effects is the arousal of an a-process, a primary affective or hedonic process. In this case, the UCR is a fear reaction pattern. The occurrence of this reaction pattern will then result in arousal of a b-process, the opponent-process. It will have an affective or hedonic sign *opposite* in quality to that of the a-process. At this point, we can only guess at what the quality of the opponent really is. As we shall see, its quality will only be revealed when the categorical stimulus event is terminated.

The magnitudes and qualities of the a-process and b-process are fed to a *summator* that computes /a–b/ for any moment. The summator determines whether the subject is in state A or state B, as well as the quality and intensity of those states. At UCS onset, most a-processes are more intense than their opposing b-process which, as I have indicated, has a slow build-up relative to that of the a-process. However, the slow build-up of the b-process will produce a gradual decrease in the amplitude of the A-state even while the UCS, the dog, is still present. The cat will look less fearful as time goes by. The cat will appear to be "accustomed" to the dog's presence.

When the dog goes away, there is no categorical stimulus to maintain the a-process and so it will quickly subside to zero. However, the b-process, being sluggish and slow to decay, will perseverate for a while. The peak of quality and intensity of the B-state will thus reveal itself directly after UCS termination, when the a-process goes to zero. Then the B state will slowly decay or subside. The cat may look relieved or relaxed, may show a typical feline after-reaction of pleasure, and then will slowly return to equanimity.

The processing system deduces the major facts in Figs. 1 and 2, the standard pattern of affective dynamics. However, an additional assumption is needed before the system will generate the effects of many repeated presentations, as shown in Fig. 3. The model must be able to produce the habituation effect as well as the emergence of a new, strong withdrawal syndrome. It will do so if we postulate that the b-process is strengthened by use and weakened by disuse. How this would work in the processing model is shown in Fig. 5, which compares the b-processes and resultant affective states during the first few UCS presentations and after many UCS presentations. The growth of the strength of the b-process with repeated presentations of the UCS has two consequences: (1) the sum /a–b/ during the onset and presence of the UCS is *decreased,* but the sum /a–b/ right after cessation of the UCS is *increased.* This fits our empirical generalizations quite well.

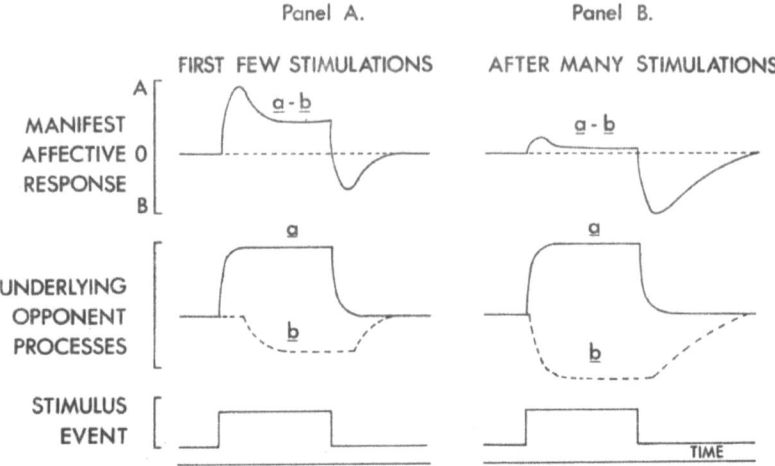

Fig. 5. A comparison of the standard patterns of affective dynamics for a relatively novel reinforcer and one that has been repeated many times. The repetition has strengthened the b-process, thus shortening its latency, augmenting its asymptote, and increasing its decay duration. These theoretical mechanisms explain the differences seen between Figs. 1 and 3. They account for habituation or tolerance, on the one hand, and the emergence of a distinct withdrawal syndrome, on the other.

### New experiments pertinent to the opponent-process theory

It is one matter to organize the known empirical generalizations about the dynamics of affect into a coherent theory of acquired motivation. It is quite another matter to test new deductions from such a theory and to find new problems and questions inspired by that theory. Such a challenge has been exciting. In the past five years, members of my research seminar have pushed experimentation in several directions, in an attempt to refute the theory, to see where revisions are needed, and to explore the generality of the theory. At the same time, in other laboratories, several new findings have been used to test the validity of deductions from the theory. In addition, several new findings, although not initially intended to test the theory, have served this purpose. It appears that the concept of the opponent-process or compensatory process (see 13) is "in the air" now.

Our experiments can be classified as follows: (1) the growth of opponent-processes in social attachment; (2) the growth of adjunctive behaviors as a function of strength of opponent-processes. I will start with the social attachment experiments, because these have taught us a great deal about opponent-processes.

*The growth of social attachement (imprinting) in ducklings.*
Imprinting has been characterized as an all-or-nothing, innate learning event of surprising suddenness (9). It occurs when a newly-hatched, precocial bird first is exposed to a moving object (or mother-surrogate, or mother). The hatchling becomes excited, looking at the moving object and often staggering toward it. Thereafter, the duckling develops more and more skilled locomotor behavior, resulting in its staying close to, or following after, the moving object.

A striking feature of imprinting is the affective reaction of the hatchling when the imprinting object is *suddenly removed*. The animal at first shows a "double-take", a perceptual startle with a very short latency. Then it becomes very active, appearing to be searching for the lost object and finally, after a 5–10 s latency, it emits high-pitched cries, or "distress calls". These distress calls can vary in the frequency with which they occur in time, and bursts of distress-calling will vary also in duration (4). They have been used as an index of degree of social attachment, much in the same way that severity and duration of opiate withdrawal symptoms have been used to index the degree of physiological and psychological dependence on heroin or morphine. If one assumes that distress-calls are an index of a b-process, an opponent caused by the presentation of a highly-reinforcing or innate releasing stimulus, then certain phenomena should be discoverable:

First, the presentation and removal of an imprinting object should have opposite reinforcing effects. This is so. Hoffman et. al. (5) showed that arbitrary operants could be shaped by presentations of an imprinting object. Furthermore, Hoffman et. al. (7) showed that removal of the imprinting object functioned effectively in a punishment contingency to weaken an arbitrary operant.

Secondly, rather than imprinting being all-or-none, or "released", it should, instead, *develop gradually* in strength as the b-process is exercised by *use*, and it should *wane* in strength should the b-process be weakened by *disuse*.

After planning sessions with members of my research seminar, Hoffman's group at Bryn Mawr designed and conducted the first experiment on the growth of an aversive opponent-process in imprinting. They showed (6) that, with 1-min exposures alternated with 1-min removals of an imprinting stimulus, the amount of distress-calling per unit of time gradually increased. Their findings are shown in Fig. 6. We can infer that the opponent-process in imprinting is strengthened by use. We are then led, as others have been, to question the all-or-none characterization of the imprinting process but based on a different type of evidence (see 12, p. 198–200). Furthermore, we can now safely assume that "following behavior" is *not* what is "released" in the imprinting process. Instead, the released behavior is an *affective* reaction, an innate a-process with positive reinforcement attributes. One can use this a-process to shape up arbitrary operants. Indeed, so-called following behavior may function as an operant. Hoffman et. al. (8) actually taught ducklings to go away from the imprinting object in order to bring about presentations of the imprinting object. If "following behavior" were released, this would not have been easy to do.

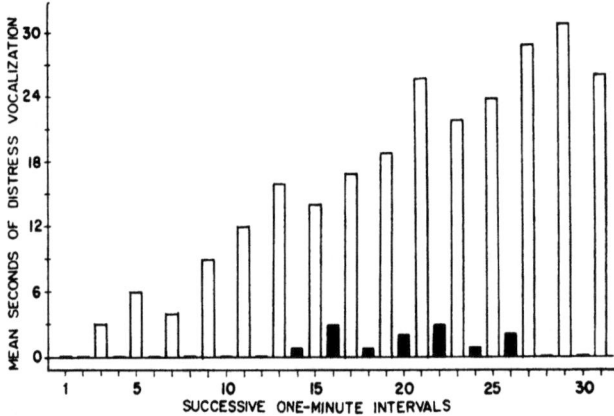

Fig. 6. The growth of magnitude of distress calling as a function of repeated 1-min presentations and removals of an imprinting stimulus. Adapted from Hoffman, et. al. (6). Imprinting, measured by distress calling, is not an all-or-none phenomenon.

At the time the Hoffman et al. (6) work was being planned, we did not know whether the opponent-process for imprinting could be weakened by disuse. Early claims suggested a negative answer. Lorenz had been impressed by the "irreversible" characteristics of imprinting. He thought it was quite different from ordinary learning in its irreversibility. In contrast, the opponent-process model deduces that the strength of social attachment, indexed by b-process magnitude, ought to decline with disuse. Starr (14) carried out the appropriate experiment to test the disuse postulate. He subjected four separate groups of ducklings to imprinting procedures. The groups were the same in their total familiarity with the imprinting object: at the end of the experiment every animal had been in the presence of a mother surrogate for a total of 6 minutes. However, the groups differed in their time intervals between exposures (their *disuse* time). Group I-1 received 12, 30-s exposures to the mother surrogate, with 1-min intervals between presentations. Group I-2 received 12, 30-s exposures with 2-min intervals between presentations. Group I-5 received 12, 30-s exposures with 5-min intervals between presentations. Finally, a control group I-0, received 6 min of continuous exposure (or 0 min between presentations).

The number of seconds of distress-calling, during a standard, 1-min observation period right after each removal of the mother surrogate, was recorded for each group. Figure 7 shows that the time interval

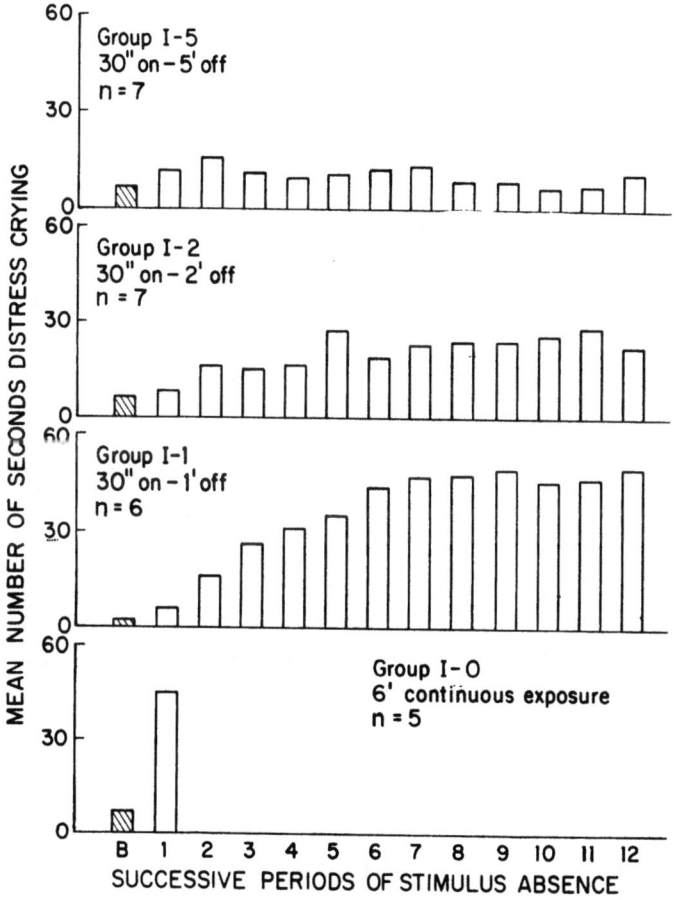

Fig. 7. The growth of magnitude of distress calling as a function of time between stimulations. Each stimulation is of 30-s duration. From the top panel, down the inter-stimulus interval is 5, 2 and 1 respectively. The bottom panel is a control condition for familiarity, a 6-min continuous exposure (zero inter-stimulus interval). Note that the distress calling magnitude stays unchanged when the inter-stimulus interval is long, but it grows when the interval is short. Adapted from Starr (14).

between presentations of the imprinting stimulus was a very powerful variable in determining the rate and amount of growth of distress-calling. The data for Group I-1 were quite similar to those recorded by

Hoffman and his colleagues at Bryn Mawr for 1-min exposures and 1-min intervals between exposures (Fig. 6). We see that distress-calling increased to an asymptote at which about 3/4 of the time was occupied by distress calls. Group I-2 showed some growth of distress-calling, but its asymptote was significantly lower than that for Group I-1. It is in Group I-5 that we see a crucial result. In this group, the repetition of exposures to the mother surrogate produced *no* growth of distress-calling over and above that level seen after the first exposure and separation. The slope of the function was zero.

Now, it could be argued that the interstimulus interval had an associative function. The repetition of "imprinting stimulus present"→ "imprinting stimulus gone" might have increased the magnitude of distress-calling by making the short presence of the imprinting object a signal for its subsequent removal. Or, it could be argued that distress-calling functioned as an operant during the separation interval and was reinforced by presentation of the imprinting object on eleven occasions. Furthermore, following this line of reasoning, the 5-min interstimulus interval group (I-5) suffered longer delays of reinforcement for distress-calling than did group I-2 or Group I-1. Association theory has a vast capacity to adapt itself to new settings! (Furthermore, I am a typical association theorist, and it is hard for me to abandon those nice old habits of thought). However, such interpretations *cannot* account for the results obtained for the control group (I-0) which received 6 min of continuous exposure to the imprinting object before the object was removed *for the first time*. This group showed the same magnitude of distress-calling (about 3/4 of the 1-min observation interval occupied by calls) during the first disappearance of the imprinting stimulus as did Group I-1 after 12 repeated presentations and disappearances. The operant contingency, though possible in Group I-1, was precluded in Group I-0, and yet the distress-calling index of attachment was the same for these two groups. The conditioning argument seems weak here. I have concluded that the interstimulus interval is a critical variable in the strengthening of the b-process. *Disuse,* or prolonged absence of the UCS or releaser, weakens the b-process *between* stimulations. The stimulations strengthen the b-process.

From this experiment, Starr induced the concept of the *critical decay duration of the opponent-process*. The critical decay duration is that *disuse time* just adequate to allow the weakening of the opponent-process to its *original,* innate reaction level. If reinforcing stimuli are presented at interstimulus intervals *greater than* the critical duration, then the opponent-process will fail to grow. In Starr's experiment, the critical decay duration must have been between 2 and 5 min.

Starr discovered something else, a *savings effect* in the already-strengthened opponent-process. When he separated ducklings from their imprinting stimulus for several days, so that distress-calling had ceased, he found that the re-strengthening of the opponent-process by repeated exposures to the imprinting object took less time and fewer exposures than had the original exposures. This phenomenon is often found in the relearning of verbal materials by human subjects. It has been called *savings*. Evidently, even though an opponent-process system has been weakened by disuse, some unique residues, or traces of past exercise of the opponent-process, remain and facilitate the restrengthening of the temporarily-dormant system. Such a phenomenon is not unexpected. For example, in alcohol addiction the abstrainer is warned that one drink may be disastrous, and the reason is the savings principle. The re-exercise of alcohol's opponent-process system strengthens the withdrawal syndrome very rapidly and sets up the special conditions for resumption of the addictive cycle. Cigarette smokers report the same phenomenon: re-addiction to nicotine takes place much more rapidly than does the initial addiction. The laws of social attachment may be identical to those for drug addiction. However we can now see the similarities extended to the fine parametric details of the opponent-process functioning.

Finally, Starr found that an enhancement of the *quality* of stimulation produced an increase in the critical decay duration of the b-process. When ducklings were exposed to an imprinting object which made honking noises, the opponent-process was strengthened rapidly, even with inter-stimulus intervals longer than the 5 min used in Starr's first experiment. A general law for the strengthening of b-processes, derived from Starr's imprinting experiments, will have to be something like this: *Opponent-processes are strengthened by use, approaching asymptotes having values directly proportional to the quality, intensity and duration of each exposure and inversely proportional to the inter-stimulus interval.*

We now have a developing science of opponent-process augmentation and weakening. It takes little imagination to see how Starr's data and concepts can be applied to drug dose frequency, quality, and size; or, for that matter, to *any* of the phenomena of acquired motivation. His ideas have considerable analytical power. We now understand some of the conditions leading either to the strengthening or weakening of opponent-processes of all types, and, consequently, to the strengthening and weakening of many experientially-acquired, new motivation systems.

It is now a matter of empirical verification to ascertain whether, in fact, the general law for strengthening b-processes does apply to many cases. Steven Seaman, working in my laboratory, has now started a series of experiments to test the application of this law to the growth of tolerance and the magnitude of the abstinence syndrome for morphine in rats. He will try to quantify the critical decay duration of the b-process for varying dose sizes and durations.

In principle, it should be possible to quantify the critical decay duration for any opponent-process system as a function of prior stimulation parameters. In such experiments, the phenomena of habituation (in opponent-process terms, /a–b/) and withdrawal syndrome intensity and duration (the B-state) would be the two major dependent variables. The analysis would be *equally applicable to positive* and negative reinforcers. Although Starr's (14) work concerned a powerful positive reinforcer, in principle it should be just as feasible to assess the strengthening of the b-process for a negative reinforcer (i.e., heat, cold, sight of an enemy predator, long-distance running, weight-lifting, shocks, free falls, etc.).

*The growth of adjunctive behaviors.* In order to stretch the boundaries of applicability of the opponent-process theory of acquired motivation, we have entered territory quite new to us, the study of adjunctive behaviors. Adjunctive behaviors are fascinating because they appear to be a reflection of the almost-senseless generalization of one acquired motivation system to a motivation system that appears to be irrelevant. The most-studied case is experientially-induced polydipsia in the white rat, usually called "schedule-induced polydipsia". When rats are hungry and put on a fixed-time or fixed-interval feeding schedule, they have lots of free time between the spaced eatings of the tiny food pellets they are given. During this free time, if sessions are frequently repeated, the adjunctive behavior of drinking will *gradually emerge* and grow in magnitude, provided, of course, that a drinking tube is available. This adjunctive drinking becomes *polydipsic*: rats will sometimes consume 5 to 10 times their normal prandial water intake during the inter-pellet intervals. Furthermore, the bouts of adjunctive drinking tend to concentrate in the early seconds of the inter-pellet interval rather than the later seconds of the interval. Thus, one reasonable view is that the adjunctive drinking is related to attributes of the pellet *just eaten* and is *not* necessarily a reflection of a learned anticipation or expectancy of the *next pellet to come*. This type of adjunctive behavior distribution, in the between-pellet free time, can be seen in other kinds of behaviors, such as running in a running wheel, if, of course, a running wheel rather than a water spout is made available. The development of adjunctive behaviors in a function of many parameters (see 3 for a fine review). However, the opponent-process theory points to parameters that have largely been ignored. It will turn out that these parameters will cast light on the essential nature of the polydipsia phenomenon.

Assume that each tasting and swallowing of a food pellet is a positive reinforcement engendering a pleasurable a-process. The termination of this a-process should make manifest its opponent-process, an *aversive* b-process. This should occur early in the inter-pellet interval, and should reach its peak early in the interval. Given that the rate of decline of the a-process stimulus after eating is a bit slow, the peak should not be right at the start of the inter-pellet interval, but ought to occur, say, between 5–15 s after the pellet has been delivered (it takes time for the pellet and its flavor to disappear). We will call this peaking b-process *taste-craving*. It is analogous to the withdrawal syndrome for opiates or to the occurrence of separation distress in ducklings. Such an aversive b-process may be quite general in its capacity to energize behaviors. Furthermore, it may be non-specific enough so that some classes of a-processes, somewhat unrelated to the b-process in question, may, nevertheless, reduce the intensity of a B-state via the summing mechanism, /a–b/. Thus, some adjunctive behaviors might be selected because of their special capacity to participate in the /a–b/ mechanism.

In commonsense terms, the rat on an FT 2' schedule is "bugged" by, or "aroused" by, a vague taste-craving experience after each pellet is tasted and swallowed. This aversive state primes an array of selected behavior classes, and if one of these happens to result in an a-process which reduces the sum /a–b/ by combining with the peaking b-process, then that behavior will be *selected* out for future action. However, the *motivation* for that behavior will depend solely on the existence of the aversive b-process, in this case, taste-craving. If that b-process $= 0$, then there will be *no* adjunctive behavior at that moment.

It takes less than great logical leaps to see the relevance of adjunctive behaviors to the opponent-process theory as elucidated in Starr's (14) work. There should be critical decay durations for the b-processes which motivate adjunctive behaviors. Each critical decay duration should be a function of the quality, intensity and duration of prior stimulation. If the critical decay duration is exceeded, there should be *no* development of adjunctive behaviors, even though stimulation is repeatedly experienced. Different adjunctive behaviors should yield differing /a–b/ summations, if we hold the b-process quality and intensity constant, and they should therefore emerge and grow at different rates. These are all matters open to empirical test. Orderly results would tend to make quite rational an area of research now clouded by mystery or fanciful, unconvincing theories (see 3). We could then think of all adjunctive behaviors as special cases of acquired motivation engendered by strengthened b-processes.

Robert Rosellini, working in my laboratory, has made a good start in exploring in the rat those aspects of experimentially-induced polydipsia of interest to an opponent-process theorist. First, he had reconfirmed the fact (see 2) that the growth of polydipsic drinking is sensitively controlled by the inter-pellet interval. There is a critical decay duration for the b-process in FT pellet schedules. Rosellini has found that, holding pellet quality and size constant, one can *prevent* the development of polydipsic drinking during the inter-pellet interval by exceeding the critical decay duration. In his research, about 200 s was the critical decay duration, using 0.45 mg pellets of chow.

The opponent-process theory requires that the quality, intensity and duration of the reinforcer must control the growth of magnitude and duration of the b-process with repetition of the reinforcer. On early trials, there should be a small contrast effect, with the b-process often being weak. On later trials, the magnitude and duration of the peak of the B-state should be greater for qualities of greater hedonic intensity. Thus, a food flavor very high in the preference hierarchy should be more potent than one low in the hierarchy in: (1) energizing a b-process, and (2) producing a longer critical decay duration of that b-process. Rosellini and Lashley (11) have tested this deduction. With a standard, fixed time, food delivery schedule known to produce polydipsia in the hungry rat (1, 2, 3) Rosellini and Lashley (11) compared the rates of development of inter-pellet drinking as a function of the flavor preference value of the particular pellets being delivered. The effects of flavor were large, indeed. In Fig. 8 are shown the adjunctive behavior growth curves for but three of many flavors Rosellini and Lashley studied: quinine-adulterated pellets, unadulterated standard pellets, and sucrose-adulterated pellets, in ascending order of taste preference, respectively. The most-preferred taste of the sucrose-adulterated pellets produced a much higher rate of development of polydipsic drinking than did the other two flavors. In addition, the asymptotic level of polydipsia was ordered by flavor preference order. Such findings are in agreement with deductions from the opponent-process model.

Because of this early success in generating and testing deductions from the opponent-process theory, I am emboldened to add a tentative hypothesis: *all adjunctive behaviors are a consequence of opponent-processes*; such behaviors are energized by the emergence of B-states correlated with sudden *termination* of a-processes, *not* the onset and maintenance of a-processes. If this is the case, then *any* clear case of adjunctive behavior should operate according to *all* of the laws of opponent-processes, to the extent that we have so far been able to discover them. Recently, Osborne (10) has shown that the magnitude of general, restless activity during the early seconds of an inter-pellet interval is directly controlled by the magnitude of the preceding food reinforcement. The principle was the same for pigeons and for rats. Figure 9 is adapted from Osborne's data. It shows the restless-activity magnitude

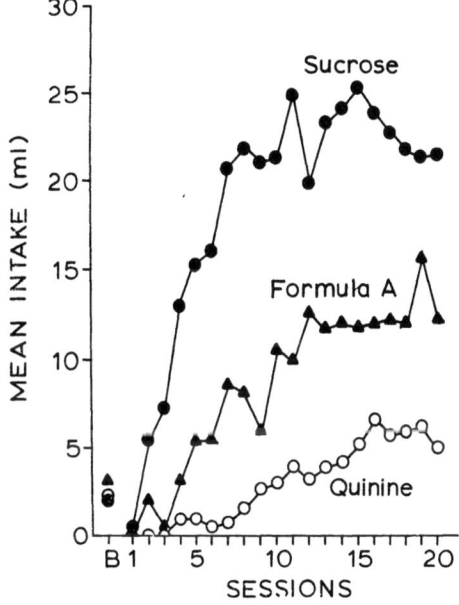

Fig. 8. The growth of adjunctive drinking as a function of the preference value of the food reinforcer. The graphs represent the mean values from 5 subjects. The greater the preference value, the more rapidly does the adjunctive drinking develop and the greater is the asymptote of polydipsia. Adapted from Rosellini and Lashley (11).

produced in pigeons when the food was available for 1.5, 4.5 and 9.0 s respectively. The inter-food time interval was 2 min during the testing period which produced the data in Fig. 9. The amount of activity rose abruptly after each occurrence of food reinforcement had terminated. The peak of post-reinforcement activity occurred early in the inter-food interval rather than late in the interval. Finally, the peak of activity was greater for the 9.0-s reinforcement duration than it was for the 4.5-s reinforcement duration; and both of these peak activity levels were greater than that following the 1.5-s duration of reinforcement.

One feature of Osborne's data is especially relevant for an opponent-process theory of motivation. Note that in Fig. 9 the peak of activity was slow in growing during the inter-food interval. Furthermore, the reaching of the peak was more delayed the greater was the peak in magnitude. Such a phenomenon could be due to the particular rate of decline of the a-process as food is ingested and its flavor stimulation *slowly* terminated. The opponent-process model ideal, as shown in Fig. 2, of course assumes a *sudden* termination of the a-process stimulus. In practice, in an eating experiment, such suddenness will rarely occur. There is, however, an alternative interpretation. A fixed *rate* of emergence of the B-state might reflect some special feature of all B-state emergences, not some property of the a-process, and if so, would re-

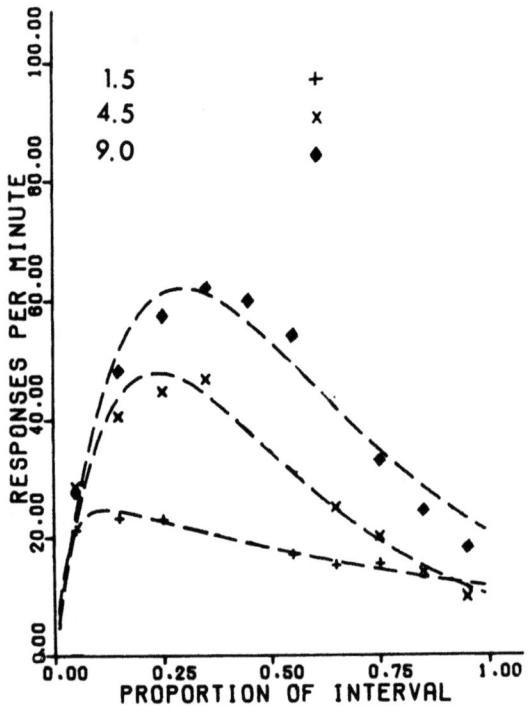

Fig. 9. The frequencies of occurrence of restless activity during different portions of a 2-min inter-pellet interval, as a function of the magnitude of food reinforcement. The 1,5; 4.5; 9.0 are the feeding times. Note that the maxima occur during the early portions of the inter-pellet interval. Note that the activity level is higher, the greater is the magnitude of each feeding. This figure is a modification from Osborne (10) p. 302.

quire a modification of the model. The strengthening of the b-process has been assumed to result in a B-state that emerges more quickly, rises to a higher asymptote, and lasts longer (see Fig. 5). The Osborne data might be reason to question this assumption, but only if we first assume that the gradual decline of taste stimulation after each reinforcement was irrelevant to the results. This possibility should be tested in appropriate experiments.

This research was supported by Grant MH-29187, from the USPHS, NIMH

# REFERENCES

1. FALK, J. L. 1961. Production of polydipsia in normal rats by an intermittent food schedule. Science 133 : 195–196.
2. FALK, J. L. 1966. Schedule-induced polydipsia as a function of fixed interval length. J. Exp. Anal. Behav. 9: 37–39.
3. FALK, J. L. 1977. The origin and functions of adjunctive behavior. Anim. Learn. Behav. 5: 325–335.
4. HOFFMAN, H. S., EISERER, L. A., RATNER, A. M., and PICKERING, V. L. 1974. Development of distress vocalization during withdrawal of an imprinting stimulus. J. Comp. Physiol. Psychol. 86: 563–568.
5. HOFFMAN, H. S., SEARLE, J. L., TOFFEY, S. and KOZMA, F., Jr. 1966. Behavioral control by an imprinted stimulus. J. Exp. Anal. Behav. 9: 177–189.
6. HOFFMAN, H. S. and SOLOMON, R. L. 1974. An opponent-process theory of motivation: III. Some affective dynamics in imprinting. Learn. Motiv. 5: 149–164.
7. HOFFMAN, H. S., STRATTON, J. W., and NEWBY, V. 1969. Punishment by response-contingent withdrawal of an imprinted stimulus. Science 163: 702–704.
8. HOFFMAN, H. S., STRATTON, J. W., NEWBY, V., and BARRETT, J. E. 1970. Development of behavioral control by an imprinting stimulus. J. Comp. Physiol. Psychol. 71: 229–236.
9. LORENZ, K. 1935. Der Kumpan in der Umwelt des Vogels. J. Ornithol. 83: 137–213.
10. OSBORNE, S. R. 1978. Analysis of the effects of amount of reinforcement on two response classes. J. Exp. Psychol. 4: 297–317.
11. ROSELLINI, R. A. and LASHLEY, R. L. 1980. An opponent-process theory of motivation: VIII. Pellet preference and adjunctive behavior. Learn. Motiv. (in press).
12. SCOTT, J. P. 1972. Animal Behavior. Univ. Chicago Press, Chicago (second edition).
13. SIEGEL, S. 1977. Morphine tolerance acquisition as an associative process. J. Exp. Psychol. 3: 1–13.
14. STARR, M. D. 1978. An opponent-process theory of motivation: VI. Time and intensity variables in the development of separation-induced distress calling in ducklings. J. Exp. Psychol. 4: 338–345.

Richard L. SOLOMON, Department of Psychology, University of Pennsylvania, 3815 Walnut Street, Philadelphia, Pa. 19104, USA.

Lecture delivered at the Warsaw Colloquium on Instrumental
Conditioning and Brain Research
May 1979

# BRAIN MECHANISMS AND HEDONIC PROCESSES

Eliot STELLAR

Institute of Neurological Sciences, University of Pennsylvania
Philadelphia, USA

*Abstract.* This paper attempts to take a broad view of investigations of brain mechanisms of motivated and emotional behavior in animals and humans. It examines the thesis that the same basic brain mechanisms are involved in physiological regulations, in various motivated behaviors and emotions, and in the hedonic experiences that can be reported by humans. It further suggests that reward and the reinforcement of learning depend on the same brain systems. Finally, it speculates on the possibility that these same brain systems play an important role in the selection of what is learned and in the consolidation, storage, and retrieval of memory. To present this conceptualization, selected experiments in thermoregulatory behavior, electrical self-stimulation of the brain, evoked approach and withdrawal behaviors, and the role of neuropeptides in thirst and hunger are reviewed. In addition, experiments will be discussed in which memory is blocked by puromycin, but in which puromycin-induced amnesia can be prevented by the administration of certain peptides such as vasopressin and some of its fragments. Speculation about the common underlying mechanism and its biological significance in the adaptation of the organism is discussed as are some of the experiments suggested by this line of thinking.

Since I have been away from the laboratory for six years as a University administrator, I cannot tell you about the results of any new experiments I have done. However, the separation from active research may have given me some perspective, so ,I would like to take advantage of the opportunity and review some investigations I have been involved in the past in the light of some of the remarkable advances that have been made in our field while I was away. Particularly, I want to review experiments of some of my closest associates, former students and colleagues at the University of Pennsylvania. My hope is that I can develop new insights of how the nervous system can work in behavior, especially in those aspects having to do with hedonic processes, including motivated behavior, reward and the reinforcement of learning, and also human subjective experience of pleasure and pain, those positive and negative hedonic experiences that we can rate and report.

This will be quite a speculative approach, for I will want to make the thesis that the same broad system within the brain is involved in all kinds of motivated behavior, reinforcement, and hedonic experience. Not only that, I will want to speculate further that the same mechanism plays a major role in the selection of which adaptive changes in the organism's behavior are learned and remembered, possibly leading us to one of the keys to the biological basis of memory. To make the speculation worthwhile, I will want, of course, to have it lead to laboratory experiments, and I will tell you about some I am involved in and have contemplated in the four months that I have been back in the laboratory.

I would like to begin by discussing some experiments in thermoregulatory behavior, for that allows us to tie together animal and human work as well as physiological regulation, motivation, reinforcement, and hedonic experience. Then I would like to go on to experiments in brain stimulation and reinforcement that might tell us something about learning and memory; then experiments on hunger and thirst that involve the peptides; and finally in a speculative leap, how study of the peptides may provide new insights into the biological basis of learning and memory.

Let me turn to thermoregulatory behavior first. Heating and cooling the brain, particularly the anterior hypothalamus, evoke heat loss and heat production physiological responses such as panting and shivering in the dog and saliva-spreading and piloerection in the rat. Using the rat, John Corbit (4) was able to show that heating or cooling the hypothalamus also leads to behavioral thermoregulation. His experimental set-up, shown in Fig. 1, allowed him to investigate the behavioral effects, not only of heating and cooling the hypothalamus, but also of heating and cooling the skin, and the interaction between the two as well.

In a nutshell, Corbit was able to show three things. First, that rats worked very hard pressing a lever for a 15-s change in temperature

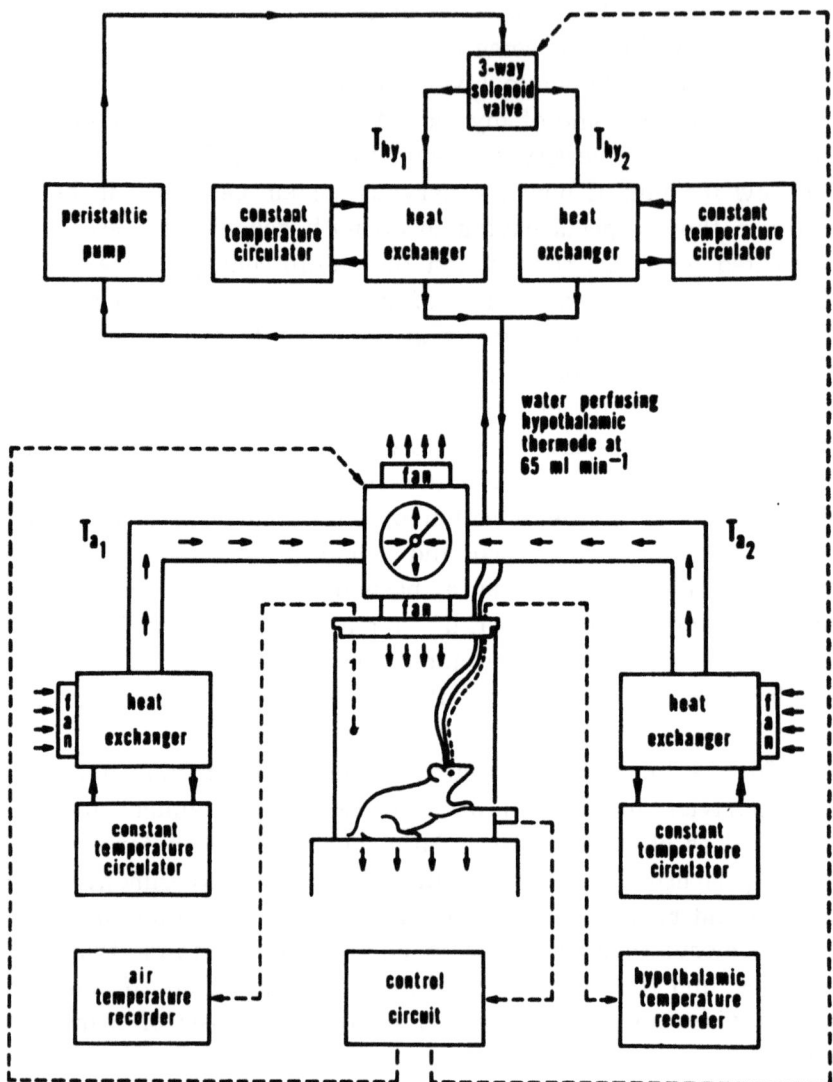

Fig. 1. Diagram of experiment in which a rat can perform instrumental conditioned responses to control the temperature of air in its chamber (bottom) or the temperature of water flowing in and out of thermodes implanted in its brain. From Corbit (4).

of the air or of the water flowing in and out of the small steel thermodes implanted in the anterior hypothalamus. The more extreme the ambient or the brain temperature, the more rapid the instrumental responding, as illustrated in Fig. 2 which shows the rate of responding for a 15-s flow of cool water through the thermode as a function of increasing brain temperature. Second, there is an interaction between the central and peripheral mechanisms, so that warming the skin will motivate the animal to cool its brain, for example. As Fig. 3 shows, the higher the skin temperature, the more the rat will work to cool a given brain temperature. Third, the animal can tell where its problem is, for when its brain is warmed, it prefers to cool its brain rather than cool

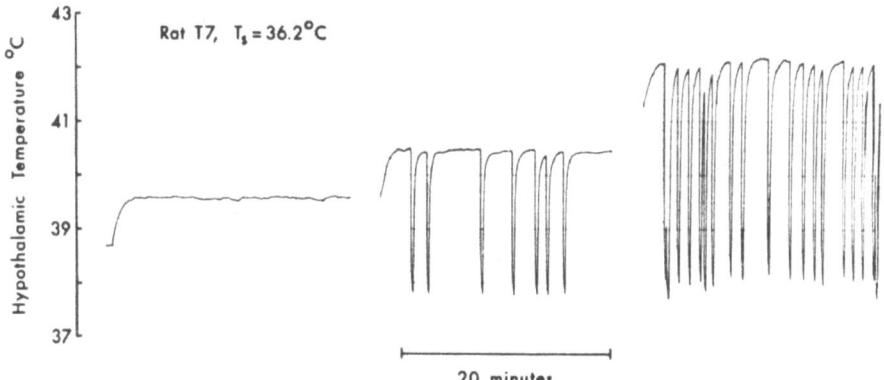

Fig. 2. Rate of instrumental responding for a 15 s lowering of brain temperature as a function of increasing basal brain temperature. From Corbit (4).

its skin if given the choice of two different instrumental responses, pulling a chain to cool the air or pressing a lever to cool its brain (5). Thus, we have clear evidence for motivated thermoregulatory behavior in which thermal change functions as an excellent reinforcement for learning instrumental responses and discriminations as well.

To investigate the hedonic experience associated with such instrumental learning under thermal reinforcement, one of my graduate stu-

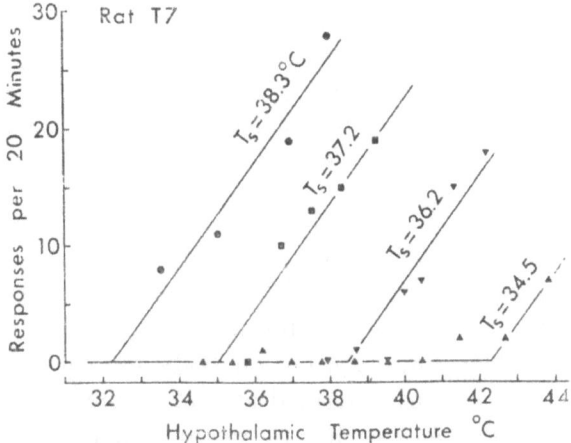

Fig. 3. Rate of instrumental responding to cool brain temperature as a function of different skin temperatures. From Corbit (4).

dents, Ray Hawkins (12), did a parallel study of human subjects. In his experiments, volunteer subjects sat in a constant-temperature tub up to their necks in warm or cool water. After a shorter or longer period of time, we had the subject stand on a stool just out of the tub water and turn on a shower which was at tub temperature. The subject was then allowed to operate a dial that controlled the shower temperature over a wide range, with the instruction to find the most pleasant shower temperature. Rectal temperature was recorded throughout the experiment.

As shown in Fig. 4, if the subject was allowed to sit in the tub long enough to raise (High $T_i$) or lower (Low $T_i$) body temperature by half

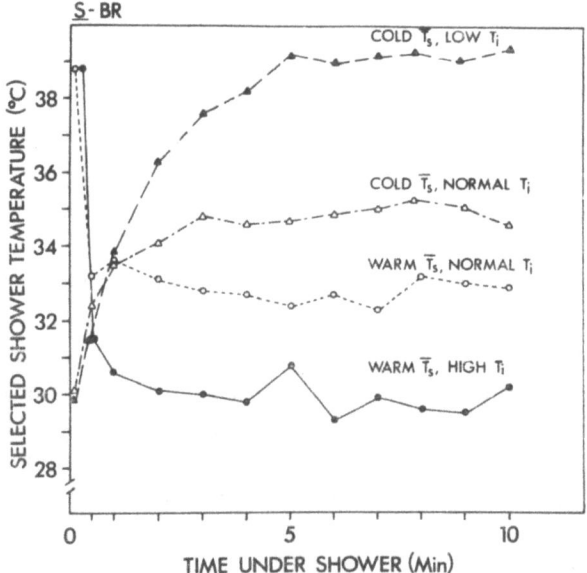

Fig. 4. Shower temperature selected as most pleasant by a human subject as a function of skin temperature ($T_S$) and internal or core temperature ($T_i$). From Hawkins (12).

a degree, he took a shower of extreme temperature. Thus, if rectal temperature was raised, he took a very cold shower for as long as ten minutes. Furthermore, he reported the experience as highly pleasurable. This is the familiar sauna effect, followed by the proverbial hedonistic roll in the snow. The same was true if body temperature was lowered, for a long, hot shower was described as just as pleasurable. If, however, the subjects took the shower before rectal temperature changed (Normal $T_i$), then shower temperature was less extreme and the hedonic experience modest.

The same result was found, following the procedure of Cabanac (3) in which the subjects rated the hedonic value of different hand temperatures. If body temperature was lowered, then dipping the hand into a water bucket of high temperature was rated most pleasant and vice versa as Fig. 5 shows. Thus, both humans and animals vigorously per-

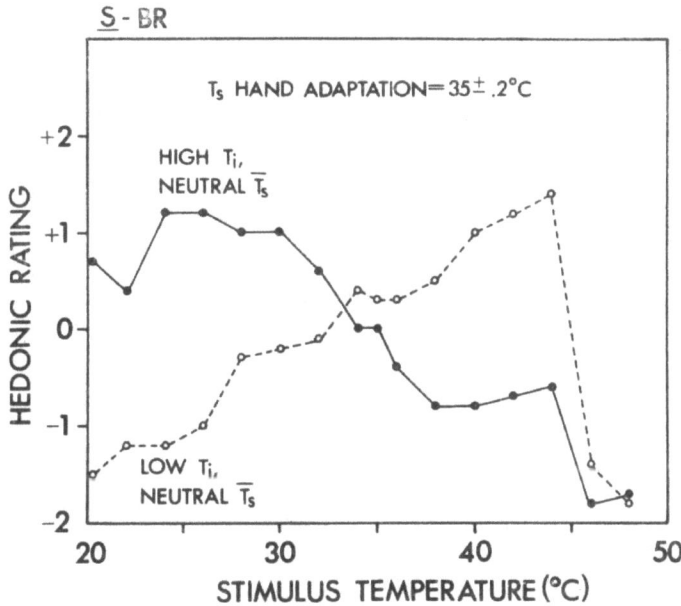

Fig. 5. Hedonic ratings of hand baths (stimulus temperature) as a function of internal temperature ($T_i$) controlled by a whole body bath. From Hawkins (12).

form instrumental responses which are motivated by changes in core temperature, and therefore, brain temperature. In humans, these motivated behaviors are accompanied by clearly reportable and strong subjective feelings or experiences of thermal comfort and discomfort, pleasantness and unpleasantness. Presumably the same brain mechanism generates the motivated behavior and the hedonic experience. This mechanism also contributes to the physiological regulation of body temperature and it participates, moreover, in the reinforcement of instrumental learning. In fact, using thermal changes in the brain as a reinforcement is another way of conducting the intracranial self-stimulation experiment, using temperature rather than electrical stimulation.

Let us turn now to electrical self-stimulation of the brain, for this remarkable discovery (15) led Olds to postulate "pleasure centers" within the brain. As he investigated the brain mechanisms underlying this phenomenon, he showed that there were extensive positive and negative systems, involving the same lateral and medial structures running through the hypothalamus and limbic system that seemed to be involved in motivated behavior.

Gallistel's experimental analysis of intracranial electrical self-stimulation into a drive component and a reinforcement component is most instructive to our thinking. In his investigations, rather than have the rat press a lever repeatedly to deliver intracranial stimulation as Olds had done, Gallistel had the rat learn the instrumental response of running down a runway to a goal-box where it could press a lever for lateral hypothalamic stimulation (11). Before the animal was placed in the runway, it was given a priming stimulation through the same electrode. Running speed was used as the measure of performance, and it was found that the more intense the priming stimulation, the faster the running. Also, the more intense the rewarding stimulation in the goal box, the faster the running. But there was a big difference. The effects of priming begin to decay in about 90 s although they sometimes last up to 10 min; the animal would run very slowly, if at all, after such a delay, even though the priming stimulation might be very intense. The effects of stimulation in the goal box, on the other hand, seemed to last a long time, at least overnight and perhaps longer, for if the animal received intense stimulation at the end of one day, it began its running at high speeds when tested the next day, for example. So it seemed as though the animal remembered what the goal-box stimulation was like from one day to the next.

Further experiments by Gallistel and his co-workers (10) have shown that there are two different populations of neurons involved in priming and reinforcement as judged by their characteristic refractory periods. One population may be described as the rapidly decaying, longrefractory-period (1.0 ms) neurons involved in priming and drive. The other population is made up of long-lasting, short-refractory-period (0.6 ms) neurons involved in reward and reinforcement and perhaps also in memory.

Using these same electrodes that produce self-stimulation, my son, Jim Stellar (19, 20), working in Gallistel's laboratory was able to make animals approach various objects and stimuli and react much more positively to them. For example, with lateral hypothalamic stimulation, rats reacted to a weak sugar solution in the mouth as though it were a strong, positive solution, and they pursued it very avidly when it was withdrawn. Lateral hypothalamic stimulation even made them react positively to a noxious odor they would normally avoid or reject and it served to attenuate or eliminate their startle response to loud noises and their escape or avoidance of electric foot shock. On the other hand, medial hypothalamic stimulation often made rats withdraw from objects they normally approached and made them unresponsive until positive stimulation was made very strong. Thus, with medial hypothalamic stimulation, it was necessary to increase the concentration of a sugar solution by a factor of ten to get the animals to respond positively to it. So it appears that electrical stimulation of the medial and lateral hypothalamic areas can bias the hedonic balance of the brain in either the positive or the negative direction, toward approach or withdrawal, acceptance or rejection.

Two other recent advances by colleagues at the University of Pennsylvania add fuel to my thinking. In one set of investigations in Alan Epstein's laboratory (7), it is becoming ever so much clearer that the octapeptide, angiotensin II, is playing an important role in motivated behavior, in both thirst and salt appetite. If angiotensin II is injected into the subfornical organ of the circumventricular system (Fig. 6) (16) then minute amounts of angiotensin, measured in femtomoles, will elicit

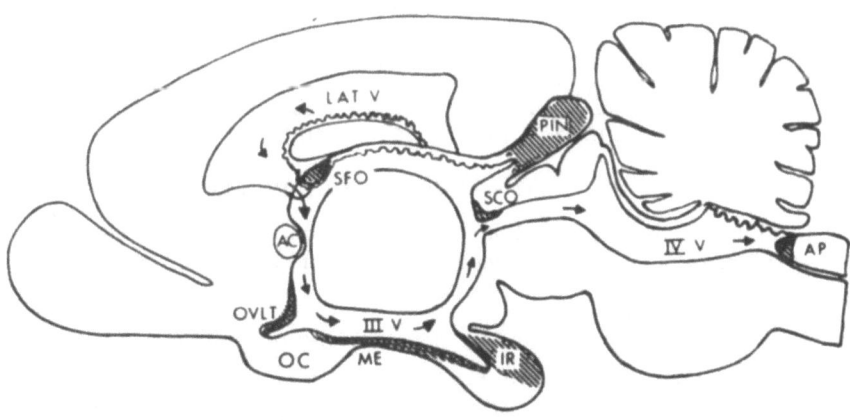

Fig. 6. Diagram of the seven circumventricular organs of the rat's brain. SFO, subfornical organ; OVLT, organum vasculosum of the lamina terminalis; ME, median eminence; IR, infundibular recess and neurohypophysis; SCO, subcommissureal organ; PIN, pineal; AP, area postrema. From Phillips (16).

drinking (Fig. 7). Angiotensin is a potent dipsogen, indeed. If it is continuously and slowly infused into the ventricles, then the rats will drink as much as their body weight in water per day. That would be the equivalent of a man drinking 50–60 l a day! In addition, rats which drink virtually no 3% sodium chloride solution, drink enormous quantities of this solution, as high as 150 ml in 24 h (2). What is more, in some animals, the salt appetite may persist for months after the infusion even though water drinking returns to normal when the angiotensis is stopped.

Angiotensin, as you know, is synthesized in two steps in the body by the kidney hormone, renin, and by a converting enzyme in the lung. It now appears that the same renin-angiotensin synthesic sequence occurs entirely within the brain, so angiotensin is a brain peptide. A si-

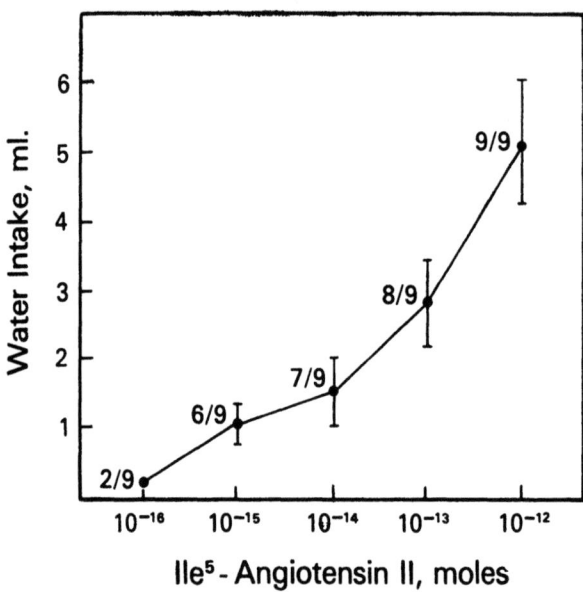

Fig. 7. Drinking elicited by injection of different concentrations of angiotensin II into the subfornical organ. Threshold is between $10^{-16}$ and $10^{-15}$ mol. From Epstein (7).

milar story is unfolding about the peptide, cholecystokinin (CCK) which is secreted by the intestine in response to food in the stomach. Experiments by Smith and his colleagues (17) in which CCK is injected systemically show that it is a satiety factor, for it causes a cessation of eating in rats and all the behavioral signs of satiation. It turns out that CCK is also made in the brain, and recent work by Yalow and her co-workers (21) has shown that the brains of genetically obese mice have about one-third the amount of CCK as their lean controls. At the University of Pennsylvania, Della-Fera and Baile are showing that injection of minute quantities of CCK-octapeptide into the ventricles of sheep causes marked reduction in food intake (6). Finally, preliminary findings by Kissileff and co-workers (14) at St. Luke's Hospital in New York indicate that human subjects, administered CCK systemically, report satiety and reduce their food intake.

Quite clearly, the brain peptides can play an important role in hunger and thirst and perhaps other kinds of motivated behavior as well.

Given what we are learning about the endorphins, peptides also seem to be important in pain perception and mood as well. It is not surprising, therefore, that de Wied and others (22, 23) have asked about the role of peptides in learning and memory, particularly vasopressin, some of its fragments, and related peptides. This possibility appeared particularly attractive to my colleague, Louis Flexner (8), when he discovered that vasopressin, given 12 h before or 24 h after a learning experience could protect mice against the amnesia induced by the antibiotic, puromycin. Some of the fractions of vasopressin, oxytocin, and other peptides were even more potent in their protection against puromycin-amnesia.

Although we originally thought that puromycin blocked memory because it inhibited protein synthesis in the brain when we first published our results on puromycin in 1963 (9), it is now clear that puromycin also has other effects, including inducing prolonged hippocampal seizures and altering catecholamine levels, especially reducing noradrenalin levels. This finding is interesting because the peptides which protect the animal against puromycin-amnesia also have effects on catecholamine levels although as far as I know, there is no evidence that one action counteracts the other.

How to understand the role of the catecholamines in learning and memory is not yet evident although Stein (18) has shown the role of noradrenalin in intracranial reinforcement. Speculation by Kety (13) suggests that the catecholamines play an important role in the brain as well as in the periphery in the adaptions of the organism. He sees the catecholamines functioning in biologically significant situations of a positive (eating, drinking, sex) or negative (pain avoidance) sort to select and reinforce synapses that may be active during the learning process. He believes that the catecholamine fibers are particularly well-suited to this function, for (i) they operate in circumstances of high motivation, reward, and reinforcement; (ii) they find their way to the cortical surfaces of the cerebrum, cerebellum, and hippocampus and wind around the large dendrites there or run along the surface in close proximity to the cerebrospinal fluid (CSF) surrounding the brain; and (iii) 95% of them end without synapses, secreting their neurotransmitter substances around the dendrites and possibly in the CSF. This conception suggests a neural network system, involving widespread structures in the brain, in which the engram is laid down in many synapses that are repeatedly active during biologically and emotionally significant events. If this is so, we must find a way to ask how it is that puromycin can block this mechanism and how the peptides can protect against the block and facilitate learning and memory.

At the present time, Dr. Flexner, Jim Sprague, and several other colleagues and I are engaged in examining the effects of depleting dopamine in the suprasylvian gyrus of split-chiasm cats by topically bathing it with 6-hydroxy-dopamine. Berlucchi, Sprague and their colleagues (1) have shown that lesion of the suprasylvian gyrus results in slow learning by the eye on that side, but good transfer of training to the other eye. In contrast, the contralateral eye learns rapidly, but there is little or no transfer to the eye ipsilateral to the lesion. Our question is whether we can find similar defects with local catecholamine depletion and whether vasopressin or other peptides will attenuate those deficits. In addition, we hope to investigate the role of the brain peptides in other learning situations in the mouse, rat, and cat, using visual discriminations, delayed alternation, and possibly delayed-matching-to-sample.

All of this is still quite speculative, but I would like to conclude now and make three points about the central mechanisms of motivation that are involved in instrumental conditioning.

1. I believe that the same brain mechanism is involved in physiological regulations and in a wide variety of motivated behavior which contribute to the regulation of the internal environment (hunger, thirst, thermal behaviors), as well as other motivated behaviors which are not concerned with physiological regulation or the survival of the individual (sexual, maternal, aggressive behaviors).

2. The same brain mechanism participates in the generation of hedonic experience in man, in approach and withdrawal behavior, and reinforcement in animals and man.

3. Are these same mechanisms also involved directly in the acquisition, consolidation, storage, and retrieval of memories? Are the peptides one of the important keys to our future experimental elucidation of the biology of memory? We shall see.

## REFERENCES

1. BERLUCCHI, G., SPRAGUE, J. M., ANTONINI, A. and SIMONI, A. 1979. Learning and interhemispheric transfer of visual form discrimination following unilateral suprasylvian lesions in split-chiasm cats. Exp. Brain Res. 34: 551–574.
2. BRYANT, R. W., FLUHARTY, S. J. and EPSTEIN, A. N. 1978. Excessive drinking of sodium solutions by rats receiving continuous intracranial infusion of angiotensin. Fed. Proc. 37: 323.
3. CABANAC, M. 1971. Physiological role of pleasure. Science 173: 1103–1107.
4. CORBIT, J. D. 1973. Voluntary control of hypothalamic temperature. J. Comp. Physiol. Psychol. 83: 394–411.
5. CORBIT, J. D. and ERNITS, T. 1974. Specific preference for hypothalamic colling. J. Comp. Physiol. Psychol. 86: 24–27.

6. DELLA-FERA, M. A. 1979. Cholecystokinin octapeptide: continuous picomole injections into the cerebral ventricles suppress feeding. Science 206: 471–473.
7. EPSTEIN, A. N. 1978. The neuroendocrinology of thirst and salt appetite. In W. F. Ganong and L. Martini (ed.), Frontiers in neuroendocrinology. Vol. 5. Raven Press, New York, p. 101–134.
8. FLEXNER, J. B. and FLEXNER, L. B. 1971. Pituitary peptides and the suppression of memory by puromycin. Proc. Soc. Natl. Acad. Sci. USA 68: 2519–2521.
9. FLEXNER, J. B., FLEXNER, L. B. and STELLAR, E. 1963. Memory in mice as affected by intracerebral puromycin. Science 141: 57–59.
10. GALLISTEL, C. R., ROLLS, E. T. and GREENE, D. 1969. Neuron function inferred from behavioral and electrophysiological estimates of refractory period. Science 166: 1028–1030.
11. GALLISTEL, C. R., STELLAR, J. R. and BUBIS, E. 1974. Parametric analysis of brain stimulation reward in the rat. I. The transient process and the memory-containing process. J. Comp. Physiol. Psychol. 87: 848–859.
12. HAWKINS, R. C. 1975. Human temperature regulation and the perception of thermal comfort. Ph. D. Thesis. University of Pennsylvania.
13. KETY, S. S. 1970. The biogenic amines in the central nervous system: Their possible roles in arousal, emotion, and learning. In F. O. Schmitt, (ed.), The Neurosciences. Second Study Program. Rockefeller University Press, New York, p. 324–336.
14. KISSILEFF, H. R., PI-SUNYER, F. X., THORNTON, J. C. and SMITH, G. P. 1979. Cholecystokinin octapeptide (CCK-8) decreases food intake in man. Amer. J. Clin. Nutrit. 32: 939.
15. OLDS, J. and MILNER, P. 1954. Positive reinforcement produced by electrical stimulation of septal area and other regions of rat brain. J. Comp. Physiol. Psychol. 47: 419–427.
16. PHILLIPS, M. I. 1978. Angiotensin in the brain. Neuroendocrinology 25: 354–377.
17. SMITH, G. P. and GIBBS, J. 1975. Cholecystokinin: A putative satiety signal. Pharmacol. Biochem. Behav. (Suppl. 1, Central neural control of eating and obesity) 3: 135–138.
18. STEIN, L. 1975. Norepinephrine reward pathways: Role in self-sitmulation, memory consolidation and schizophrenia. Neb. Symp. Motiv. 22: 113–159.
19. STELLAR, J. R. 1976. Approach-withdrawal analysis of the lateral and medial hypothalamus. Doctoral dissertation, University of Pennsylvania.
20. STELLAR, J. R., BROOKS, F. H. and MILLS, L. E. 1979. Approach and withdrawal analysis of the effects of hypothalamic stimulation and lesions in rats. J. Comp. Physiol. Psychol. 93: 446–466.
21. STRAUS, E. and YALOW, R. W. 1979. Cholecystokinin in the brains of obese and nonobese mice. Science 203: 68–69.
22. Van REE, J. M., BOHUS, B., VERSTEEG, D. H. G. and de WIED, D. 1978. Neurohypophyseal principles and memory processes. Biochem. Pharmacol. 27: 1793–1800.
23. WALTER, R., HOFFMAN, P. L., FLEXNER, J. B. and FLEXNER, L. B. 1975. Neurohypophyseal hormones, analogs and fragments: Their effect on puromycin-induced amnesia. Proc. Natl. Acad. Sci. USA 72: 4180–4184.

Eliot STELLAR, Institute of Neurological Sciences, University of Pennsylvania, Philadelphia, Pennsylvania 19104, USA.

Lecture delivered at the Warsaw Colloquium on Instrumental
Conditioning and Brain Research
May 1979

# STRUCTURES INVOLVED IN ACQUISITION AND PERFORMANCE OF VISUALLY GUIDED MOVEMENTS IN THE CAT

M. FABRE and P. BUSER

Laboratoire de Neurophysiologie Comparée, Université Pierre et Marie Curie
Paris, France

*Abstract.* Cats were trained to press on a lever either fixed or moving at different speeds from left to right or right to left at random. In the first series of animals, the following bilateral lesions were performed after reaching criterion on the moving lever: (1) deep cutting of cortico-cortical fibers connecting visual and frontal cortex; (2) subtotal electrolytic destruction of n. ventralis lateralis of the thalamus (nVL); (3) ablation of the anterior (3a) or middle (3b) suprasylvian cortex. No postoperative deficits were noticed after lesions (1), (2), (1 + 2) or (3b). On the other hand, impairment was found after lesion (3a), strongly suggesting the role of the anterior suprasylvian cortex in visual guidance of the forepaw movement in cat. The second group of animals was first trained to press the fixed lever and then underwent bilateral n. VL lesions. After lesion, performance with fixed lever remained normal while the animals displayed a deficit with the moving lever. The conclusion is that the n. VL is essential only during the acquisition phase of the tested visuomotor performance.

## INTRODUCTION

The structures and/or circuits that are essential for visual guidance of forelimb movements toward a target remain uncompletely known. Various data exist, but no complete picture of the involved mechanisms can yet be suggested. In the cat, observations were performed by Sprague et al. (63-65) on the role of the superior colliculus and pretectum. In the monkey, sections of the cortico-cortical pathways from occipital to frontal cortex were shown to impair a particular type of visually guided movements (26, 27, 49), while Trouche et al. (6, 7, 67, 68), also using a "pointing" task, studied the effects of reversible cryoblockade of dentate nucleus and pallidum. Also in the monkey, quite a different set of data was recently obtained, with emphasis placed upon the role of the parietal cortex. In connection with single unit studies in parietal areas 5 and 7, some ablation experiments were performed, indicating that these areas are involved in perception and action in the "extrapersonal" space (21, 37, 44, 46).

If we except these ablation data (irreversible or reversible), most other heuristic hypotheses on visual guidance of movements mainly derive from electrophysiological or even anatomical results: in fact any pathway carrying visual messages to some place within the motor systems may be considered "essential" or at least important for a visual-motor performance.

In this paper, we present some data regarding the role of two sets of structures in a visual guidance in the cat. Our experimental paradigm mainly implied pressing for food on a lever moving in front of the animal. Depending on the structures which were lesioned, deficits were noticed either in the performance of the preoperatively learned task, or in the acquisition itself of the movement toward the rotating lever.

The selected sites were both of a "premotor" type, i.e., situated as possible interfaces between the sensory signal and the motor discharges. One was n. ventralis lateralis (n. VL) of the thalamus, a structure which has been claimed to exert a very close control on the motor cortex (18, 43, 52, 55); moreover electrophysiological findings could show that not only somatic impulses (42), but also visual information may reach this nucleus (at least in some experimental conditions) and that these latter projections are under "gating" control of the visual cortex (36, 50, 51). Lesions which were performed here in VL were, at least in some experiments, combined with interruption of the cortico-cortical connections from occipital to frontal areas. Another investigated brain site was the suprasylvian cortex. In selecting this cortical zone, we were guided by current ideas regarding the role of areas 5 and 7 in the monkey (31, 40, 46, 47): histologically, a part of the cat anterior suprasylvian cortex is precisely occupied by the architectonic equivalent of these parietal areas (28, 29).

## METHODS

*Training apparatus (Fig. 1).* The cat was losely maintained in a hammock (a) and stood on a horizontal circular plate placed on the bottom of a cylindric aluminium "pan" (diameter 40 cm; side height 10 cm). A horizontal opaque ring plate (e) followed the upper contour of the cylinder side, thus forming a tunnel where the lever (f) (which was attached to a horizontal disk, not shown) could rotate out of the cat's

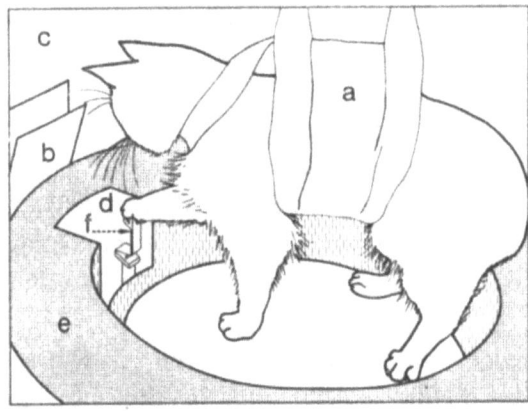

Fig. 1. Some of the major features of the training apparatus. The cat, held in a hammock (a), is just performing a grasping and adduction movement of the lever (f); e, opaque ring plate hiding pedal; d, window or action field of the animal; c, circular wall with door (b) for food-tray (Modified from 20a).

view. Side flank and horizontal annular plate were interrupted on a certain distance to form a "window" (d) that was faced by the animal ("action field", width 20 cm). The lever was a vertical brass bar (width 10 mm, height 50 mm) whose base was fixed on a microswitch, the effective movement being an adduction of the bar toward the animal, of the order of 5 mm. The lever was either left immobile in the window (for preliminary training, see below), or it crossed the action field at a given adjustable speed. Because of the circular arrangement of the system, the lever could move in the window either clock or-counterclockwise in a quasi-random way, not necessarily in a predetermined sequence, left to right, then right to left.

Whenever the lever was appropriately pulled by the animal (grasping upper edge of bar through flexion of fingers followed by adduction), solid food was delivered: a tray containing the food was moved toward the animal by a compressed-air driven piston from behind a vertical semicircular wall (c). At rest and in intertrial periods the tray was invisible and unaccessible to the animal, door (b) in wall (c) being opened only under the mechanical pressure of the food delivery system.

*Trials in trained animals.* A usual trial with a trained animal began by switching on (at time t1) the rotation of the lever (Fig. 2). In principle, information on the initial position of the lever, on its direction and speed of rotation and therefore its delay and side of appearance in the window was not accessible to the cat. However, in order to mask any possible noise artefact, which could serve as a clue, the rotation was always accompanied by a sound (1,500 to 1,700 Hz) delivered at

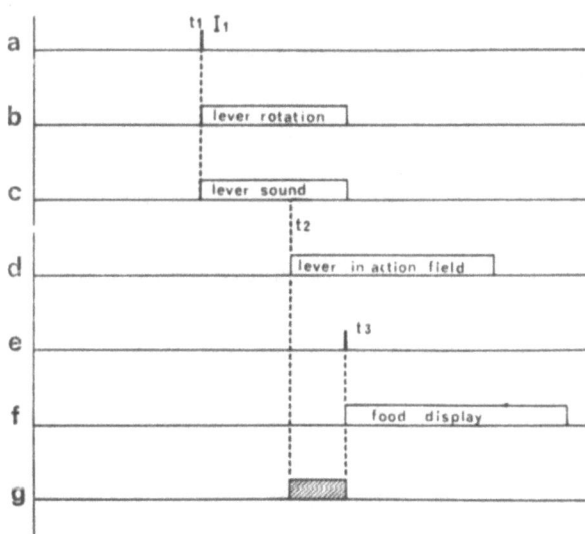

Fig. 2. Diagram of a typical trial with rotating lever. At time t1, an impulse $I_1$ is delivered (a); it triggers lever rotation (b) and lever sound (c). At time t2 (d), lever enters the window and at t3, the animal presses the lever (e), triggering the food delivery system (f). Delay t3–t2 indicates response latency (g).

time t1, and raising the cat's attentiveness. At time t2, the pedal appeared at one or the other end of the window, and if then the cat performed the movement, the lever microswitch triggered the food delivery system (and interrupted sound) at time t3. The *response time* t3–t2 was automatically computed by a digital counter and stored in memory.

*Training and scoring.* Animals were first trained with lever in fixed position in the action field. In this preliminary stage, the lever was rendered operational only during delivery of a sound (acting as a go-signal); the same sound was used later on during rotation of the lever. After this first period, the animal was either operated or immediately trained to press on the moving lever (see results, section I). In the final stage, the lever just passed the window at a speed which was gradually increased in the successive sessions until it reached a value of the order of 40°/s, corresponding to a linear speed of 27 cm/s, and a total time of accessibility of 750 ms (the width of the window being about 20 cm).

When the animal was overtrained, non-sequential histograms of the response times (t3–t2) were established for each session with 100 ms bin width, separately for all rotations of the lever from left to right, then for those from right to left and also for all movements pooled together.

In addition to these quantitative evaluations of delays at maximum speed, estimates were made by the operator, of three types of non-effective movements, using a digital pulse code: (a) trials with inadequately directed movements, i.e., movements well developed but missing the target; (b) trials with uncomplete movements, like touching the pedal without grasping or pulling, or performing an abortive movement, limited to its initial phase: those occurred e.g., in satiated animals; (c) trials which could not be taken into consideration for various reasons (animal inattentive to the passage of the lever, animal holding the lever with or without pulling).

Histograms of response times, means and standard errors, and percentages of inadequate trials of each category (with respect to total number of trials), were automatically established using a program written for a PDP 11/05 computer. In part of the experiments, failures *a* and *b* were computed together as percentage of total number of trials; later on, each category was considered separately.

*Surgery.* All cats underwent bilateral brain lesions (as symmetrical as possible). Those were performed under deep penthiobarbital anesthesia, the cat being held in a stereotaxic apparatus. To interrupt the cortico-cortical, occipito-frontal connections, a bone flap was opened on both sides to discover frontal levels A16 to A18; the dura was removed and incisions were performed with a blunt knife stereotaxically oriented (from midline to L10, down to a depth corresponding to ca. H9) [1]. Cortical ablations were performed by succion, care being taken to remove as little white matter as possible. Thalamic lesions were carried out using a monopolar steel electrode, isolated except the tip, stereotaxically oriented toward the adequate target. At each coagulation point, a current of 2 to 2.5 mA was passed during 60 s. As a rule, the experiments were resumed after 3–4 days.

At the end of the experimental period, the animal was killed with an overdose of anesthetic, and its brain was perfused with saline, then

---

[1] This part of our study was made possible by using historical data kindly provided by Drs. C. Fuentes and R. Marty (see also 22).

formalin. All lesions were histologically controlled on serial frozen-sectioned, 100 μm Nissl stained slides.

One animal underwent only anesthesia with no lesions, and served as a control subject.

RESULTS

We shall first describe the behavior of a normal animal in our system and then the results of the lesions.

*Standard performance of a normal animal*

The animals were first trained with the immobile lever, a tone being used as a go-signal, until reaching a 90% correct response score. At this time a distinction was made: if the animal had revealed particular capabilities in the speed of training to the fixed lever (i.e., less than 3 sessions to start correct pressing), it was selected to undergo lesions before starting any training to the moving bar (group II). If, on the other hand, its training had taken somewhat more time (but less then 10 sessions), it was immediately trained to the moving bar, before any surgery was performed (group I). After reaching the final chosen criterion at maximal speed (less than 30% failures of any of the 3 types, as defined above at the 40°/s speed), lesions were performed. Finally, all animals which required more than 10 sessions were eliminated.

Animals usually showed one preferential leg; however, this lateralization was not considered existing before the initial training session. In fact we discovered that the position of the operator with respect to the side of the animal in the first few sessions was determinant: the greater majority of subjects first used the forearm on the side of the operator. Later on, they kept using it throughout the experiment, whatever the direction of rotation of the lever, and although the operator was at that time invisible to the cat.

*Effect of lesioning n.VL and interrupting cortico-cortical occipito-frontal connections in group I*

Our aim in this first approach was to interrupt two of the main possible ways of visual information to the motor cortex, namely n. VL and the cortico-cortical pathways linking the occipital to the frontal cortex. In five animals (B1, B4, B5, B6, B15) the occipito-frontal pathways were cut bilaterally. In addition, two of them underwent a bilateral subtotal lesion of n. VL (B5, B15) also partially involving other thalamic areas: ventralis anterior (VA), centralis lateralis (CL), medialis dorsalis (MD), anterior medialis (AM), and anterior ventralis (AV).

In three cats (B1, B5 and B15), no significant postoperative deficits could be noticed. The response latency distribution, and number of unsuccessful trials were identical to their preoperative scores. The animals immediately performed at maximal speed of the lever. Figures 3, 4 and 5 illustrate one such case. The two other animals, B4 and B6, displayed some increase in their unsuccessful trials, but only for one direction of the lever; moreover, this deficit was not of the *a* type of error (missing target) but rather of the *b* type (uncomplete movement with no correct pulling).

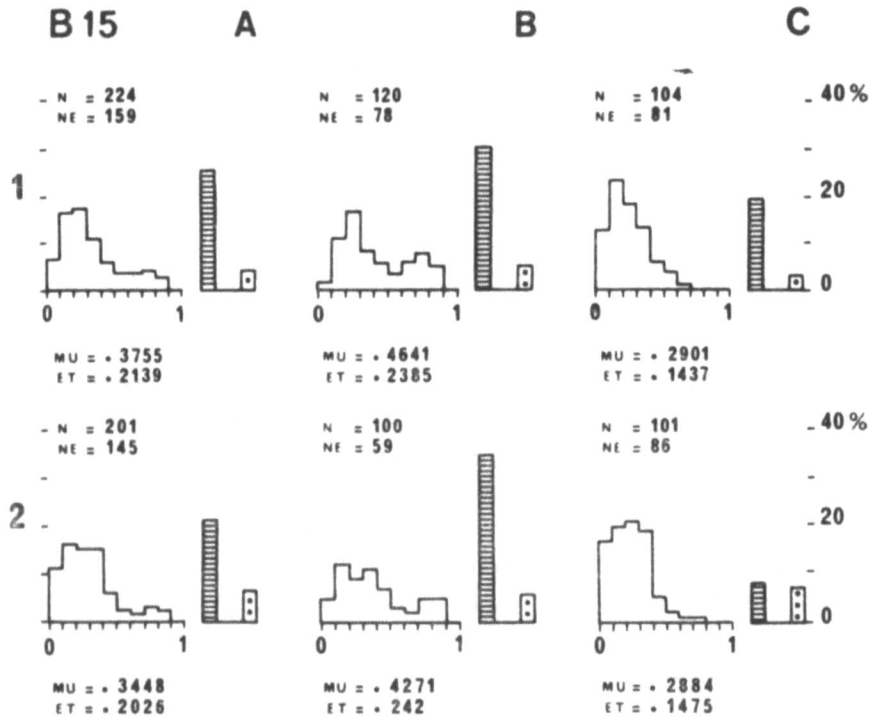

Fig. 3. Cat B15. Performances before (1) and after (2) simultaneous bilateral interruption of cortico-cortical pathways and lesion of n. ventralis lateralis. Results from all trials pooled together are given in A; in B, those with lever moving from left to right and in C, those from right to left; N, overall number of trials; NE, numer of trials with effective bar presses. Distributions of response latencies of effective movements in % of N are presented at left of A, B and C. Bin width, 100 ms. Means (MU) and standard errors (ET) are expressed in s. Ineffective trials: horizontal hatchings cumulated unsuccessful movements of the (a) and (b) types (see text), in % of N. Dots give number of (c) type trials (also in % of N.)

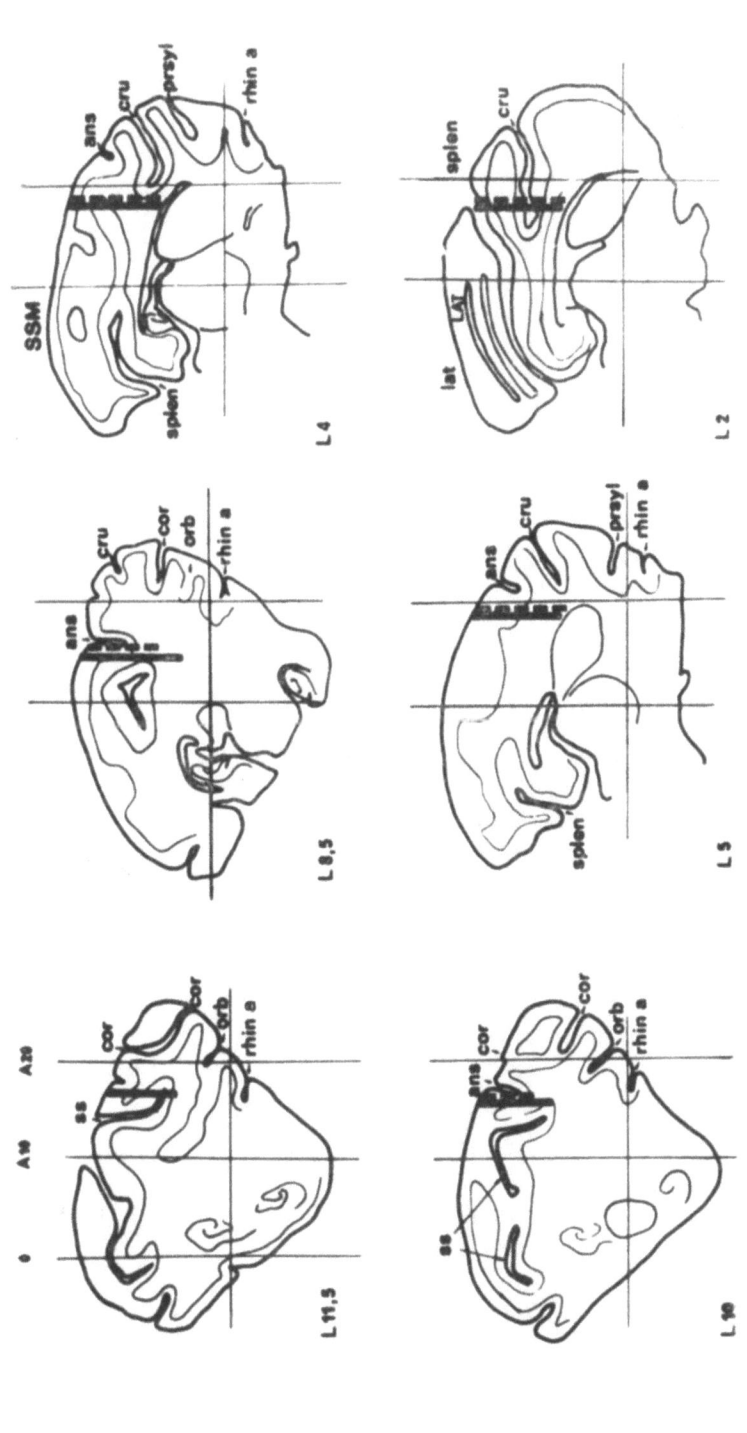

Fig. 4. Extent of cortico-cortical transections in cat B15. Full vertical line indicates left hemisphere transection; dashed heavy line, right hemisphere transection. Presented sections are parasagittal, from L 11.5 to L2. LAT, lateral gyrus and SSM, medial suprasylvian grysus. Sulci: ans, ansatus; cor, coronalis; cru, cruciatus; lat, lateralis; orb, orbitalis; prsyl, preasylvius; rhin a, rhinalis anterior; splen, splenialis; ss, suprasylvius.

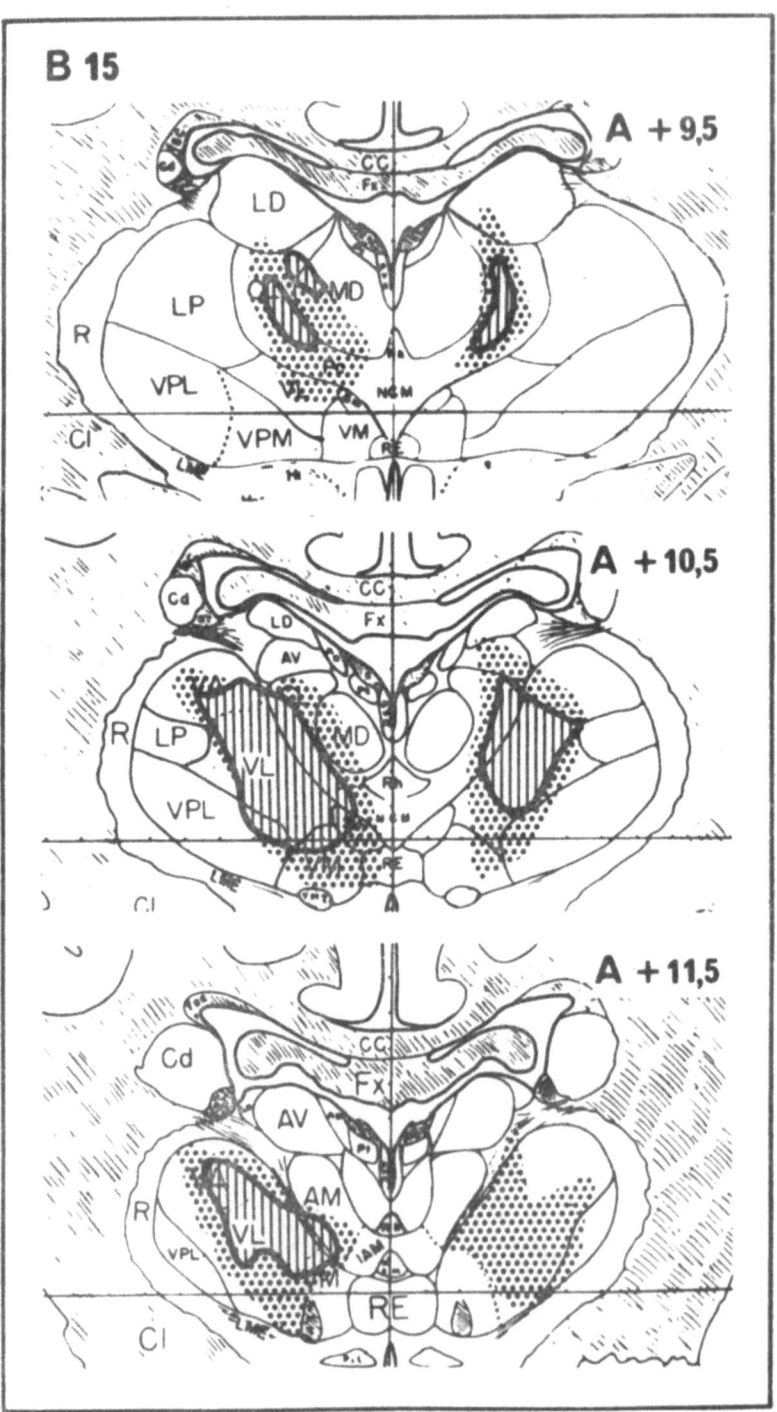

Fig. 5. Thalamic lesions performed in cat B15. Electrolytic lesions are indicated with hatching, and areas with gliosis, with dots. Frontal sections A 9.5, A 10.5 and A 11.5 were taken from Jasper and Ajmone–Marsan (32a).

## Effects of lesions of suprasylvian cortex in group I

The reasons for lesioning the suprasylvian (SS) cortex were mentioned above: at least the anterior aspect of this cortical zone contains in part areas 5 and 7, and may thus be the equivalent of the parietal cortex in primates.

Three subgroups were distinguished depending on the size and site of the lesions: animals with combined ablations of the anterior and middle suprasylvian cortex (AM), those with lesions limited to its anterior part (A), and those with lesions limited to its middle part (M).

*Subgroup AM.* Two animals (B9, B11) were used. Both were tested for the first time 2–3 days after the operation. During one week they were unable to press at maximal speed (Fig. 6), but this became possible later on. However, both subjects showed postoperatively a marked

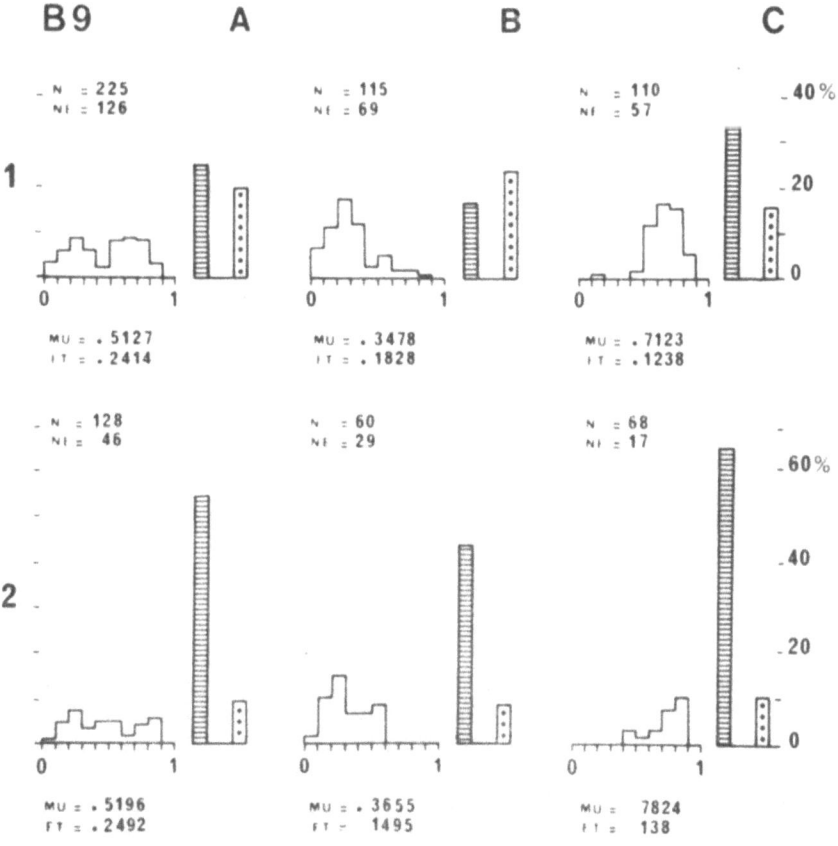

Fig. 6. Cat B9. Performance before (1) and after (2) bilateral removal of anterior and medial suprasylvian cortex. For significance of letters and symbols, see Fig. 3.

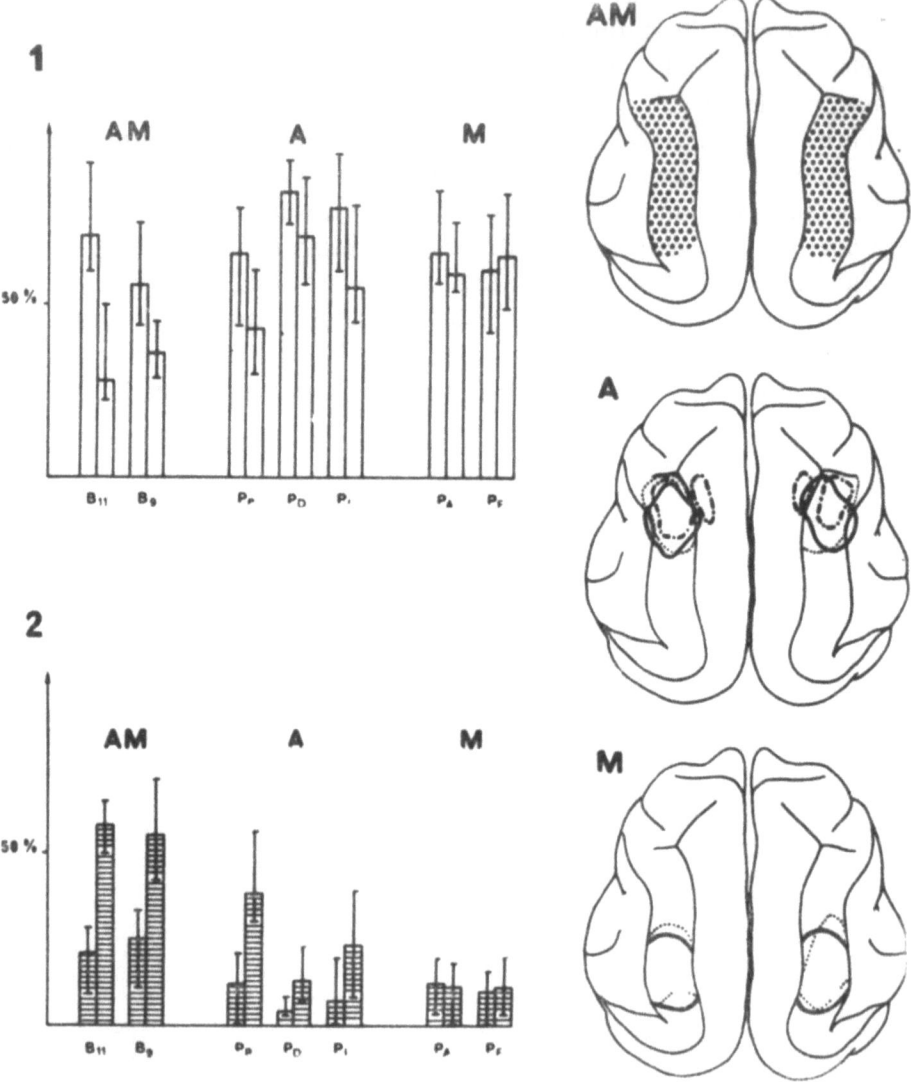

Fig. 7. Comparison between performances of animals of groups AM, A and M (see text). Percentages of effective movements (NE) with respect to total number (N) of trials (and standard errors) are given in (1) and those of type (a) ineffective trials (NI), also with respect to (N), are given in (2). For each animal (B11 and B9 for group AM; $P_B$ $P_D$ and $P_L$ for A; $P_A$ and $P_F$ for M), left column is before lesion and right column after lesion. Extents of lesions are indicated on diagrams of cortex: in A case $P_B$ is outlined with dots, $P_D$ with dashes and $P_L$ with full line. In B, $P_A$, full line and $P_F$, dots.

increase in the number of ineffective trials of type *a* defined above (i.e., movements missing the lever). Contrasting with this deficit was the fact that the distribution of latencies of effective movements (whenever those occurred) was practically unchanged.

*Subgroup A.* Ablations restricted to the anterior SS gyrus were performed in three cats (PB, PD, PL). All cats displayed a high number of ineffective presses, but to a lesser extent than those of the AM group. Whenever the animal performed an effective movement, its latency was quite within the normal range, just as in the AM subgroup.

*Subgroup M.* Ablations limited to the medial SS gyrus (SSM) were performed in two animals. From the first postoperative session, their performances were normal, and no deficit was observed during the following 5 wk. To summarize (Fig. 7), while no significant deficit was found in the M subgroup the AM subset displayed a somewhat more important deficit than A, thus suggesting the occurrence of some *mass action*.

*Effects of lesions of n. VL on the acquisition of the visuomotor guidance (group II animals)*

The lesions of n. VL (in combination with those of the cortico-cortical pathways) did not interfere with the performance of the learned task with the rotating lever, a surprising finding in view of the well-accepted hypothesis that this thalamic site plays an important role in the control of motor behavior (see above). We therefore planned another series of experiments with VL lesions, this time testing the way in which the lesioned animals would acquire their ability to direct a movement toward a moving target.

Three animals (PH, PM, PO) were first trained on the fixed lever (with a go-signal, see methods). After reaching the score, they underwent bilateral destruction of the VL.

These animals were selected because they had reached the score with the fixed lever in a particularly short time, and presumably could learn on the rotating target within 3–5 wk (if no deficit would be produced by the lesion).

Three n. VL lesioned animals displayed a severe deficit, as soon as the target speed was above a certain value. Postoperative training to the moving lever required special care, to avoid repeated non-reinforced trials in cases when the animal was not able to learn within the normal time, as a consequence of the lesion. Moreover in many cases, the maximal speed could never be reached, the animal missing the lever above a given speed (Fig. 8).

The lesions included, for one subject (PH) the subtotal extent of VL,

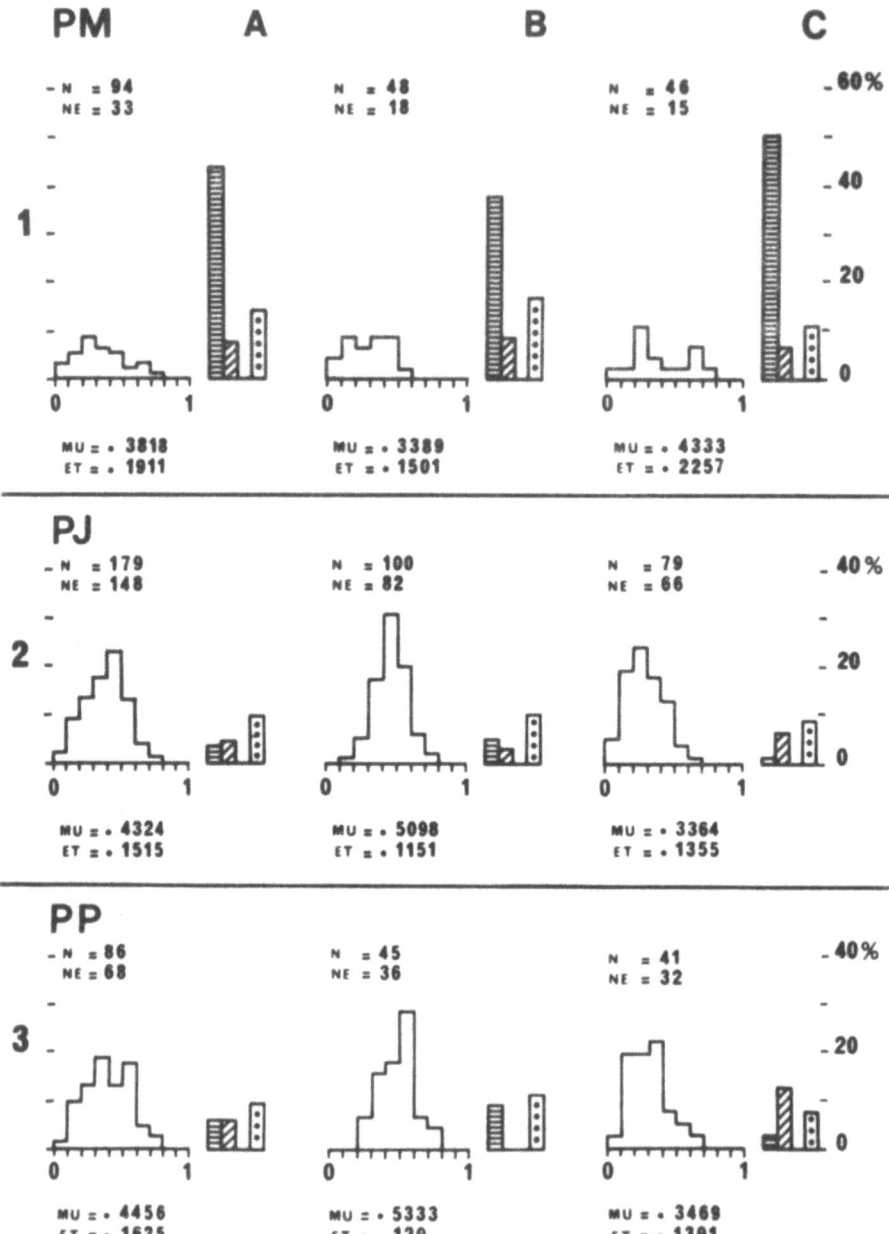

Fig. 8. Postoperative performances of cats $P_M$, $P_J$ and $P_P$ in the acquisition of the movement on the rotating lever. All three animals had been trained on fixed lever before operation (see text). $P_M$, after bilateral lesion of n. ventralis lateralis; $P_J$, after bilateral lesions of parts of n. lateralis posterior. $P_P$, after anesthesia not followed by surgery. All symbols and letters have the same meaning as in Figs. 3 and 6, except that here, horizontal hatchings only present type (a) ineffective trials, while type (b) misses are indicated by oblique hatchings (always in % of N).

encroaching on CL, MD and VA, and in the other subject (PM), the entire VL with little involvement of CL and VPL (n. Ventralis posterolateralis). In the third subject (PO), only a part of VL was lesioned. None of these animals could reach normal performance with target moving at maximal speed: PH could not reach criterion of less than 20% of incorrect movements when speed was over 20°/s (1.5 s total presentation time, 0.13 m/s linear speed). PO never succeeded at the 30°/s speed (1 s total presentation, 0.20 m/s linear speed). Finally PM could eventually perform individual trials at maximal velocity, but with a very high percentage of misses and thus did never reach the imposed score.

At any rate, we noticed in these n. VL lesioned cats two kinds of disturbances of their movement: (i) the animal would perform two successive movements, the first clearly anticipating the position of the moving lever (paw ahead of lever), and the second, being either correct or directed behind the lever; (ii) the subject would perform a hypermetric gesture, not ending with an effective pulling of the bar (for mechanical reasons); one animal also performed hypometric movements, well directed, but not reaching pedal.

Two remaining animals served as controls: one subject only underwent anesthesia with no lesion, and another (PJ) was lesioned in the posterior part of n. Lateralis posterior. In both cases, the postoperative training on the moving bar was performed within the normal delay, with score reached at maximal target velocity within 4 wk (Fig. 8).

## DISCUSSION

Our data concern two distinct types of effects of central lesions; the effects on the performance of the previously learned motor task, and the effects on the acquisition of the motor skill, with our training apparatus. Although the problems posed are numerous and fairly distinct, when considering either the deficit in performance or the deficit in acquisition, we shall briefly comment upon points which are more or less common to both situations.

### Role of the direct cortico-cortical occipitofrontal connections

Earlier data, and above all theoretical considerations on the cortical mechanisms in general, have often emphasized the potential importance of cortico-cortical connections as a means of conveying messages from a sensory cortex (visual, in our case), to the areas where motor elaboration takes place, in particular the motor cortex or structures connected to it, such as the "premotor" areas (49, 57). It is only recently, however, that data have become available, concretely illustrating this possible role of cortico-cortical connections in the monkey: contralateral visually directed finger movements were significantly impaired after a unilateral section of the occipito-frontal pathways, combined with commissurotomy to prevent interhemispheric transfer and control (26, 27, 49). In principle, these studies implied a visually directed finger movement; the difficulty is however that in this test (as in many others including ours) a final somesthetic adjustment could hardly be avoided. Keating (34) also reached the same general conclusion on the role of cortico-cortical pathways, although he stated that these pathways only play an important role in movements of distal joints. If this latter conclusion could be extended to the cat, it would not be surprising that interruption of cortico-cortical pathways were of minor importance in this latter species.

One question which arises is whether our sections did really interrupt all cortico-cortical fibers from visual to anterior cortex. Actually, our surgery was based upon histological data obtained in a joint study with Marty's Laboratory (22). Performed sections were deep enough to cut all myelinated fiber systems that had been identified anatomically with the used Marchi technique (no other histological data were available to us).

### Role of the suprasylvian cortex (SS)

Contrasting with the lack of impairment of the visuomotor adjustment on a moving target after combined cortico-cortical and n. VL lesions, is the fact that after ablations of the anterior SS cortex, the cats displayed deficits in their ability to perform the learned "pointing" movement.

1. The SS cortex has been known for more than 20 years as a structure with "polymodal" projections, from all three main modalities, visual, auditory and somatic, these projections being mainly concentrated in two foci, one anterior and the other, more posterior (1, 2, 10, 12, 19, 48, 56, 66). Although we did not explore here, in a terminal experiment, the type of sensory response that remained after the performed ablation, it is likely on the basis of a purely topographical viewpoint, that the anterior focus was affected by our lesions, while the posterior focus was only slightly encroached upon by the M type of lesions (subgroups AM and M).

2. Another point may be stressed in the same line. More anteriorly than the polymodal focus, other types of sensory projections exist; the tip of the SS cortex actually represents both the posterior limit of area SII and the most anterior part of the auditory cortex, termed "supra-

sylvian fringe" (SF) (72, see also 33). No visual projections have been observed in this area at least to our knowledge.

At any rate, it would certainly be very difficult and also too early to speculates on a relation between a possible suppression of one or an other type of sensory projection field and the observed deficits.

3. Classical data have also shown that n. Lateralis posterior (n. LP) projects to the SS cortex (25, 29). The question could thus be raised, whether lesions in n. LP would duplicate the observed deficits. In fact, no positive answer can be given, yet; in the only case when a lesion of this thalamic nucleus was performed in our present experimental series, this lesion was very uncomplete and no significant deficit could be perceived in the acquisition on the motor performance. It remains for future studies to determine the effect of a complete destruction of n. LP on acquisition and/or performance of visual pointing.

4. It is probably more interesting to stress the fact that SS cortex establishes reciprocal connections with the cerebellum, from cerebellum to SS through n. ventralis anterior of the thalamus and back from SS to the cerebellum (59, 60). It is not impossible that the long-lasting deficits after ablation of SS are in one way or another due to the interruption of this loop with the cerebellar system.

5. Little other data is available on the effects of SS ablations in the cat. Observed deficits have mainly concerned either the animal's ability to discriminate patterns or its possibilities to solve "Umweg" problems and in all cases they concerned the posterior aspect of SS ("prestriate" area) (5, 70, 71).

Now considering monkeys, recent unit studies performed in area 5 and 7, have well indicated that this parietal sector is involved in higher visual and somatic integration in relation to a purposive movement performed in the surrounding space (31, 40, 47, 58). Ablation studies have also documented these notions, by showing syndromes after unilateral ablations (21, 37, 44, 46) that recalled unilateral left apraxia in human patients (15, 16, 30, 69). If we notice that our SS ablations have largely encroached upon the architectonic areas 5 and 7, as they were delineated in the cat through architectonic criteria (28, 29), it is of course interesting — from a comparative viewpoint — to give these cat areas some sort of role in visual guidance in space.

It is also too early to speculate on the reasons of this deficit. One may suggest an impairment in the prediction that the animal makes of the position of the lever. At time $t_2$ (see Methods), the animal perceived the lever, and guessed its position and speed within the next few milliseconds. From this instant, the position of the lever was perfectly predictable to the animal who could start a program of a given ballistic movement, in a certain direction of space, after comparing its own speed of movement with that of the lever. It is this program that might be desorganized by the lesion (20).

*Effects of n. VL lesions on the acquisition of visually guided movements*

We already mentioned above that various sets of data, anatomical and electrophysiological (18, 42, 52, 55) had led to the suggestion that n. VL may exert a close control over operations at the level of the motor cortex. Therefore, our prediction was, when we initiated these experiments, that lesioning this structure would impair the execution of the learned visually guided movement. Since this was not the case, we then investigated whether a lesioned animal would learn pointing at a moving target; this time we could indeed notice significant impairments, with our paradigm. Let us briefly comment upon these data.

1. The relative unimportance of VL in performing an already acquired motor skill may be related to other recent data, also pointing to the unimportance of this structure for motor performance (45). On the other hand, our lack of positive observations may be due to the fact that the used paradigm was probably not enough "constraining" and could not allow evaluation of some deficits, like those which were observed in a reaction-time situation (8, 9, 61, 62). Therefore, there may be no real contradiction between the two categories of data. In our study, the reaction time per se was relatively small with respect to the total latency of the movement, the measured variable in our case: a small variation in RT might thus not have been perceived in this particular paradigm. The only complementary remark is that whenever the animal was performing a correct press after lesion, its response latency fell within the normal range of values, indicating that no systematic lengthening of the overall time taken to press the lever had occurred.

2. Contrasting with the first set of data is that VL appeared of essential importance for the acquisition phase of the movement. The reason of this difference remains unclear. It may be of interest to recall here the theories which have been proposed to account for the role of the cerebellum (23, 24, 32, 41) and also of the olivo-cerebellar system (39) in learning spatio-temporal motor programms. Since n. VL is a strategic site for the transfer of information between cerebellum and motor cortex (3, 4, 13, 35), such a set of hypotheses could of course be of interest to understand the importance of this thalamic site during acquisition of a motor skill. In favor of this suggestion is the fact that many visually directed movements observed here after VL lesions during the acquisition phase were hypermetric. Interestingly enough changes

in the pattern of movements occurred in the monkey during coling of the dentate nucleus (11, 14, 67).

This, however, is only one of the possible explanations of the importance of n. VL. At least in the cat, this nucleus also relays other pathways, such as those coming from the pallidum, and others from red nucleus, substantia nigra and intralaminar nuclei of the thalamus (17, 38, 52). Thus the actual system responsible for motor learning largely remains to be determined.

The technical help of Madame C. André is gratefully acknowledged. This work was supported by the following grants: CNRS, ERA 411; ATP, CNRS: 36.22; Fondation pour la Recherche médicale française.

### REFERENCES

1. ALBE-FESSARD, D. and ROUGEUL, A. 1958. Activités d'origine somesthésique évoqués sur le cortex non-spécifique du chat anesthésié au chloralose: rôle du centre médian du thalamus. Electroenceph. Clin. Neurophysiol. 10: 131-152.
2. AMASSIAN, V. E. 1954. Studies on organization of a somesthetic association area, including a single unit analysis. J. Neurophysiol. 17: 37-58.
3. ANGAUT, P. 1970. The ascending projections of the nucleus interpositus posterior of the cat cerebellum: an experimental anatomical study using silver impregnation methods. Brain Res. 24: 377-394.
4. ANGAUT, P. and BOWSHER, D. 1970. Ascending projections of the medial cerebellar (fastigial) nucleus: an experimental study in the cat. Brain Res. 24: 49-68.
5. BAUMANN, T. P. and SPEAR, P. D. 1977. Role of the lateral suprasylvian visual area in behavioral recovery from effects of visual cortex damage in cats. Brain Res. 138, 3: 445-468.
6. BEAUBATON, D., AMATO, G., TROUCHE, E. and GRANGETTO, A. 1978. Cooling of the pallidum or the putamen in monkeys performing a visuomotor task. Neurosci. Lett. Suppl. 1. S: 118.
7. BEAUBATON, D., TROUCHE, E., AMATO, G. and GRANGETTO, A. 1978. Dentate cooling in monkeys performing a visuomotor pointing task. Neurosci. Lett. 8: 225-230.
8. BENITA, M., CONDE, H., DORMONT, J. F. and SCHMIED, A. 1976. Effets du refroidissement du noyau ventral-latéral du thalamus du chat sur les temps de réaction d'un mouvement de flexion conditionné. J. Physiol. (Paris) 72; 20A.
9. BENITA, M., CONDE, H., DORMONT, J. F. and SCHMIED, A. 1979. Effects of ventrolateral cooling on initiation of forelimb belistic flexion movements by conditioned cats. Exp. Brain Res. 34: 435-452.
10. BIGNALL, K. E., SINGER, P. and HERMAN, C. 1967. Interaction of cortical and peripheral inputs to polysensory areas of the cat neocortex. Exp. Neurol. 18: 194-209.
11. BROOKS, V. B. 1974. Some examples of programmed limb movements. Brain Res. 71: 299-308.

12. BUSER, P. and BIGNALL, K. 1967. Nonprimary sensory projections on the cat neocortex. Int. Rev. Neurobiol. 10: 11–165.
13. CONDÉ, H. and ANGAUT, P. 1969. An electrophysiological study of the cerebellar projections to the VL of thalamus in the cat. II Nucleus lateralis. Brain Res. 20: 107–119.
14. CONRAD, B. and BROOKS, V. B. 1974. Effect of dentate cooling on rapid alternating arm movements. J. Neurophysiol. 37: 792–804.
15. CRITCHLEY, M. 1953. The parietal lobes. Hafnner Publ. Co., New York.
16. DENNY-BROWN, D. and CHAMBERS, R. A. 1958. The parietal lobe and behavior. Res. Publ. Assoc. Res. Nerv. Ment. Dis. 36: 35–117.
17. DORMONT, J. F. and OHYE, C. 1971. Entopedoncular projection to the thalamic ventrolateral nucleus of the cat. Exp. Brain Res. 12: 254–264.
18. DORMONT, J. F. and MASSION, J. 1967. Etude topographique des relations entre le cortex cérébral et le noyau ventrolatéral. J. Physiol. (Paris) 59: 233.
19. DOW, B. M. and DUBNER, R. 1971. Single unit responses to moving visual stimuli in middle suprasylvian gyrus of the cat. J. Neurophysiol. 34: 47–55.
20. EVARTS, E. V. and THACH, W. T. 1969. Motor mechanisms of the CNS: Cerebro-cerebellar interrelations. Ann. Rev. Physiol. 31: 451–498.
20a. FABRE, M., ANDRE, C. and BUSER, P. 1979. Testing guided forepaw movements in the cat. Training apparatus and procedure. Physiol. Behav. 23: 263–266
21. FAUGIER-GRIMAUD, S., FRENOIS, C. and STEIN, D. G. 1978. Effects of posterior parietal lesions on visually guided behavior in monkeys. Neuropsychologia 16: 151–158.
22. FUENTES, C., BUSER, P. and MARTY, R. 1977. Projections des aires visuelles corticales sur le gyrus sigmoïde postérieur et le noyau caudé chez le chat. Z. Mikrosk. Anat. Forsch. 91: 385–396.
23. GILBERT, P. F. C. 1974. A theory of memory that explains the function and structure of the cerebellum. Brain Res. 70: 1–18.
24. GILBERT, P. F. C. 1975. How the cerebellum could memorize movements. Nature 254: 688–689.
25. GRAYBIEL, A. M. 1972. Some ascending connections of the pulvinar and nucleus lateralis posterior of the thalamus in the cat. Brain Res. 44: 99–125.
26. HAAXMA, R. and KUYPERS, H. G. J. M. 1974. Role of occipitofrontal cortico-cortical connections in visual guidance of relatively independent hand and finger movements in rhesus monkeys. Brain Res. 71: 361–366.
27. HAAXMA, R. and KUYPERS, H. G. J. M. 1975. Interahemispheric cortical connections and visual guidance of hand finger movements in the rhesus monkey. Brain. 98: 239–260.
28. HASSLER, R. and MUSH-CLEMENT, K. 1964. Architektonischer Aufbau des sensomotorischen und parietalen Cortex der Katze. J. Hirnforsch. 6: 377–420.
29. HEATH, C. J. and JONES, E. G. 1971. The anatomical organization of the suprasylvian gyrus of the cat. Adv. Anat. Embryol. Cell Biol. 45: 5–64.
30. HECAEN, H. and GIMENO ALAVA, A. 1960. L'apraxie idéomotrice unilatérale gauche. Rev. Neurol. 102, 6 : 648–653.
31. HYVARINEN, J. and PORANEN, A. 1974. Function of parietal associative area

7 as revealed from cellular discharges in alert monkeys. Brain 97: 673–692.
32. ITO, M. 1972. Neural design of the cerebellar motor control system. Brain Res. 40 : 81–84.
32a. JASPER, H. H. and AJMONE-MARSAN, C. 1956. A stereotaxic atlas of the diencephalon of the cat. Natl. Res. Council. Publ. 15, p. 54.
33. JONES, E. G. and POWELL, T. P. S. 1973. Anatomical organization of the somatosensory cortex. In A. Iggo (ed.), Handbook of sensory physiology. Vol. II, Somatosensory system. Springer, New York 579–620.
34. KEATING, E. G. 1973. Loss of visual control of the forelimb after interruption of cortical pathway. Exp. Neurol. 41 : 635–648.
35. KIEVIT, J. and KUYPERS, H. G. J. M. 1971. Fastigial cerebellar projections to the VL nucleus of the thalamus in cat and rhesus monkey. Anat. Rec. 169 : 358.
36. KOSZUL, M. F., RICHARD, D. and BUSER, P. 1977. Effect of cryogenic blockade of visual and motor cortices on visual unit responses in nucleus ventralis lateralis of the cat thalamus. Brain Res. 138 : 180-184.
37. La MOTTE, R. M. and ACUNA, C. 1978. Defects in accuracy of reaching after removal of posterior parietal cortex in monkeys. Brain Res. 139 : 309–326.
38. LANOIR, J. and SCHLAG, J. 1976. Le thalamus du Chat et du Singe. Données anatomiques, hodologiques et fonctionnelles. J. Physiol. (Paris) 72b : 50–84.
39. LLINAS, R., WALTON, K. and HILLMAN, D. E. 1975. Inferior olive: its role in motor learning. Science 190 : 1230–1231.
40. LYNCH, J. C. MOUNTCASTLE, V. B., TALBOT, W. H. and YIN, T. C. T. 1977. Parietal lobe mechanisms for directed visual attention. J. Neurophysiol. 40 : 362–389.
41. MARR, D. 1969. A theory of cerebellar cortex. J. Physiol. (Lond.) 202 : 437–470.
42. MASSION, J., ANGAUT, P. and ALBE-FESSARD, D. 1965. Activités évoquées chez le chat dans la région du nucleus ventralis lateralis par diverses stimulations sensorielles. I. Etudes macrophysiologiques. II. Etudes microphysiologiques. Electroenceph. Clin. Neurophysiol. 19 : 433-441, 452-469.
43. MASSION, J. and RISPAL-PADEL, L. 1973. Activity of ventrolateral thalamic neurons related to posture and movement during contact placing responses in the cat. Brain Res. 61 : 400–406.
44. MILNER, A. D., OCKLEFORD, E. M. and DEWAR, W. 1977. Visuospatial performance following posterior parietal and lateral frontal lesions in stumptail macaques. Cortex 13 : 350–360.
45. MOLL, L. and KUYPERS, H. G. J. M. 1975. Role of premotor cortical areas and VL nucleus in visual guidance of relatively independant hand and finger movement in monkey. Exp. Brain Res. suppl. : 142.
46. MOUNTCASTLE, V. B. 1975. The view from within: pathways to the study of perception. Johns Hopkins Med. J. 136 : 109–131.
47. MOUNTCASTLE, V. B., LYNCH, J. C., GEORGOPULOS, A., SAKATA, H. and ACUMA, C. 1975. Posterior parietal association cortex of the monkey; command functions for operations within the extrapersonal space. J. Neurophysiol. 38 : 875–908.

48. NARIKASHVILI, S. P., KAJAIA, D. V. and TIMCHENKO, A. S. 1969. On the origin of cortical association responses to visual stimulation in the cat. Brain Res. 14 : 417–425.
49. PANDYA, D. N. and KUYPERS, H. G. J. M. 1969. Cortico-cortical connections in the rhesus monkey. Brain Res. 13 : 13–36.
50. RICHARD, D. 1976. Actions corticifuges à partir de l'aire visuelle primaire du chat. Analyse unitaire. Thèse de Doctorat d'Etat, Paris.
51. RICHARD, D., KOSZUL, M. F. and BUSER, P. 1977. Size and characteristics of visual receptive fields in nucleus ventralis lateralis in cat under chloralose anaesthesia. Brain Res. 138 : 175–179.
52. RISPAL-PADEL, L. 1975. Etude électrophysiologique de la voie cérébello-thalamo-corticale chez le chat. Thèse de Doctorat d'Etat, Marseille.
53. RISPAL-PADEL, L. and LATREILLE, J. 1974. The organization of projections from the cerebellar nuclei to the contralateral motor cortex in the cat. Exp. Brain Res. 19 : 36–60.
54. RISPAL-PADEL, L. and MASSION, J. 1970. Relations between the ventrolateral nucleus and the motor cortex in the cat. Exp. Brain Res. 10 : 331–339.
55. RISPAL-PADEL, L., MASSION, J. and GRANGETTO, A. 1973. Relations between the ventro-lateral nucleus and motor cortex and their possible role in the central organization of motor control. Brain Res. 60 : 1–20.
56. ROBERTSON, R. T., MAYERS, K. S., TEYLER, T. J., BETTINGER, L. A., BIRCH, H., DAVIS, J. L., PHILLIPS, D. S. and THOMPSON, R. F. 1975. Unit activity in posterior association cortex of cat. J. Neurophysiol. 38 : 780–794.
57. RONDOT, P., De RECONDO, J. and RIBADEAU-DUMAS, J. L. 1977. Visuomotor ataxia. Brain 100 : 355–376.
58. SAKATA, H., TAKAOKA, Y., KAWARASAKI, A. and SHIBUTANI, H. 1973. Somatosensory properties of neurons in the superior parietal cortex (area 5) of the rhesus monkey. Brain Res. 64 : 85–102.
59. SASAKI, K., MATSUDA, Y., KAWAGUCHI, S. and MIZUNO, N. 1972. On the cerebello-thalamo-cerebral pathway for the parietal cortex. Exp. Brain Res. 16 : 89–103.
60. SASAKI, K., OKA, H., MATSUDA, Y., SHIMONO, T. and MIZUNO, N. 1975. Electrophysiological studies of the projections from the parietal association area to the cerebellar cortex. Exp. Brain Res. 23 : 91–102.
61. SCHMIED, A. 1979. Participation du noyau ventro-latéral du thalamus à la commande d'un mouvement conditionné. Thèse de Doctorat d'Etat, Paris.
62. SCHMIED, A., BENITA, M., CONDÉ, H. and DORMONT, J. F. 1974. Corrélations entre l'activité des cellules du noyau ventrolatéral et les temps de réaction chez le chat conditionné. J. Physiol. (Paris) 69 : 296A–297A.
63. SPRAGUE, J. M. 1966. Interaction of cortex and superior colliculus in mediation of visually guided behavior in the cat. Science 153 : 1544–1547.
64. SPRAGUE, J. M., BERLUCCHI, G. and Di BERARDINO, A. 1970. The superior colliculus and pretectum in visually guided behavior and visual discrimination in the cat. Brain Behav. Evol. 3 : 285–294.
65. SPRAGUE, J. M. and MEIKLE, T. H. 1965. The role of the superior colliculus in visually guided behavior. Exp. Neurol. 11 : 115–146.
66. THOMPSON, R. F., SMITH, H. E. and BLISS, D. 1963. Auditory, somatic

sensory and visual response interactions and interrelations in association and primary cortical fields of the cat. J. Neurophysiol. 26 : 365–378.
67. TROUCHE, E., BEAUBATON, D., AMATO, G. and GRANGETTO, A. 1978. Dentate nucleus and visuo-motor control in monkeys performing a pointine task. Neurosci. Lett. Suppl. 1, S 153.
68. TROUCHE, E., BEAUBATON, D. and GRANGETTO, A. 1976. Effets du blocage reversible par refroidissement du noyau dentelé sur une performance visuo-motrice chez le Babouin; résultats préliminaires. J. Physiol. (Paris) 72 : 31A.
69. TZAVARAS, A., OZONAS, G. and CHODKIEWICZ, P. 1975. Apraxie idéomotrice à prédominance unilatérale gauche avec ataxie optique lors d'une lésion pariétale gauche. In F. Michel and B. Schott (ed.), Les syndromes de disconnexion calleuse chez l'homme. Lyon, 265–285.
70. WARREN, J. M., WARREN, H. B. and AKERT, K. 1961. "Umweg" learning by cats with lesions in the prestriate association cortex. J. Comp. Physiol. Psychol. 54 : 629-632.
71. WOOD, C. C., SPEAR, P. D. and BRAUN, J. J. 1974. Effects of sequential lesions of suprasylvian gyri and visual cortex on pattern discrimination in the cat. Brain Res. 66 : 443–466.
72. WOOLSEY, C. N. 1961. Organization of cortical auditory systems: a review and a synthesis. In Rasmussen and W. F. Windle (ed.), Neural mechanisms of the auditory and vestibular systems. Charles C. Thomas, Springfield, p. 165–180.

M. FABRE and BUSER, Laboratoire de Neurophysiologie Comparée, Université Pierre et Marie Curie, 4, Place Jussieu, 75230 Paris-Cedex 05, France.

Lecture delivered at the Warsaw Colloquium on Instrumental
Conditioning and Brain Research
May 1979

# THE POSTURAL SUPPORT OF MOVEMENT IN CAT AND DOG

Y. GAHÉRY, M. IOFFE, J. MASSION and A. POLIT

Département de Neurophysiologie générale, INP, CNRS
Marseille, France

Institute of Higher Nervous Activity and Neurophysiology, Academy of Sciences
Moscow, USSR

*Abstract.* The postural adjustment which accompanies single limb movement in the standing cat and dog was analyzed. Four trays equipped with strain gauges were used for measuring the vertical forces exerted by each limb before and during movement performance. Three types of movements were analyzed: flexion movements elicited by motor cortex stimulation, placing movements, conditioned movements of either forelimb or hindlimb (lift-off in cat, flexion with maintained final position in dog).

In both cats and dogs the postural adjustment during movement consists of a bipedal stance on two diagonally opposite limbs. Large quantitative differences were observed depending on the type of movement. Cortical stimulation elicited an adjustment where changes of forces exerted by the forelimb a hindlimb were nearly equal. During conditioned fore- and hindlimb lift-off in the cat there was a tendency to use only forelimbs for the postural adjustment associated with forelimb movement and hindlimbs for the adjustments associated with hindlimb movement. For placing in the cat and conditioned movement in the dog, the adjustment was intermediate, that is a predominant contribution of forelimb support with forelimb movements but nevertheless an associated contribution from hindlimbs. The general significance of the results with respect to the mechanism of postural adjustment associated with movement is analyzed.

## INTRODUCTION

It is well established that different kinds of movements of animals and humans are accompanied by appropriate changes in posture (1, 2, 10–12, 17).

During either forelimb or hindlimb movement the standing quadruped uses a diagonal support pattern: one forelimb and the opposite hindlimb are loaded, whereas the limb diagonally opposite to the moving limb is unloaded. These results have been obtained in the dog (3, 11) during conditioned fore- and hindlimb movements, and in the cat during movement elicited by cortical stimulation (7), during placing movements (6) and during conditioned lift-off movement of the limb (13). If the general pattern of the postural support for movement is the same, one may question whether there are quantitative differences in the way that the postural support is organized both for different species, such as cat and dog, and for different types of movement.

The aim of this paper was to compare the results obtained separately in two laboratories on the cat and the dog, with different types of movement. Common criteria for the reanalysis of the results were defined. No basic differences were found between the postural adjustment in cat and dog, but there are large quantitative differences in the postural adjustment according to the type of movement.

## METHODS

These results were collected from 3 dogs weighing from 12 to 25 kg and 7 cats weighing from 2.5 to 3.5 kg.

### FORCE RECORDING

The animals stood on four platforms equipped with strain gauges for measuring the vertical force exerted by each limb.

DC force recordings were made with inkwriting EEG. Traces from cats were digitized and stored on disk using a digital computer.

### EXPERIMENTAL PROCEDURE

*Dogs.* Two types of conditioned reflexes were elaborated. In the first type, the dog was trained to lift one limb and keep it above a certain level (about 10 cm) to avoid electrical stimulation (15 Hz, 1 ms pulse duration, 0.6–3.0 mA) of the skin of the same limb (5). The second one involved a more precise motor reaction: to avoid electrical shock (US) the dog had to lift the limb into a "safety zone" 4 cm wide located at about 10 cm above the platform and to hold it there (4). A tone 200 Hz served as conditioned stimulus (CS). The CS–US interval was 0.6 s in the first case and 5 s in the second one, and the combined presentation of both stimuli lasted for 4.5 s and 10 s respectively. In this paper the data reported concern hindlimb movement for the first kind of conditioned reflex (latency between onset of conditioned stimulus and lift-off: 0.6–0.65 s) and forelimb movement for the second one (latency between onset of conditioned stimulus and lift-off: 0.45–0.7 s).

*Cats.* Several types of movement and associated postural adjustments were compared.

1. *Movements elicited by cortical stimulation.* Electrodes (10 to 20 nickelchrome needles, insulated except at the tip) were permanently implanted in the fore- and hindlimb motor areas bilaterally, to a depth of 1.5 to 2 mm. Monopolar stimulation was used to elicit movement, the intracortical electrode being the cathode, the large silver indifferent electrode at the level of the frontal sinus, the anode. Ten to twenty stimuli were delivered to a given cortical site during each session. The on-line calculation by the computer of the projection of the center of gravity allowed for stimulation only when desired conditions were maintained for a period of one second, i.e. $25\% \pm 10\%$ of the animal's total weight supported by each leg and speed of displacement of the center of gravity less than 20 mm/s. Two experimental series were performed, one with intensity of cortical stimulation adjusted to produce a displacement of the limb of 4 to 5 cm above the supporting tray, the other being subthreshold for lift-off. The time between the onset of cortical stimulation and the lift-off was between 0.07 and 0.09 s for forelimb movements and between 0.08 and 0.11 s for hindlimb movements.

2. The *placing reaction* was elicited in the standing cat by contact of a forelimb with a moving tray. This tray was mobilized when certain conditions of weight distribution (as described for cortical stimulation) were held for 0.5 to 1 s. The stimulated limb was pushed backwards and the animal then performed a placing movement onto the moving tray. This moving tray was 35 mm above the level of the supporting tray. The time from the contact of the moving tray with the corresponding forelimb to the lift-off was between 0.18 and 0.35 s.

3. *Conditioned lift-off movements* were elicited by a discontinuous tone serving as conditioned stimulus and a milk reward was given as soon as the appropriate limb was lifted off, that is when the weight supported by that limb dropped to zero. The cat raised the limb only

a few mm to a few cm above the supporting tray. The time allowed from the beginning of the conditioned stimulus until the lift-off was 2 s. Prior to the conditioned stimulus, a continuous tone of 0.5 to 0.8 s duration was used as a preparatory signal during which the cat had to have an appropriate distribution of weight on the four legs, as explained for cortical stimulation.

During the training procedure, the cat was first required to keep a quiet posture during 0.5 to 0.8 s. A continuous sound was delivered if the quiet posture was adequate. Thereafter, a discontinuous sound (conditioned stimulus) was added during which the lift-off movement of the appropriate limb had to take place. The first movements were obtained by manually pushing the limb backwards.

Only movement of a given leg were elicited during a session. Training was then repeated for many sessions before the movement of another limb was tested. Data were obtained from both forelimbs in two cats (latency between CS and lift-off 0.5–0.6 s), from right forelimb only in a third, and from left hindlimb (mean latency of lift-off 0.7 s) only in the fourth.

## DATA ANALYSIS

The purpose of this paper was to compare the data obtained from cats and dogs under several experimental conditions. For each animal and for each type of movement, the data were collected from one or two representative sessions, each having from 10 to 30 trials. The choice of the session was made after having verified from the previous data analysis that very little variation in the results took place from one session to another. The following parameters were systematically measured:

1. *Instantaneous indices.* Antero-posterior weight distribution (AP)

$$AP = \frac{W1 + W2 - W3 - W4}{W1 + W2 + W3 + W4}.$$

Lateral weight distribution (L)

$$L = \frac{W2 + W4 - W1 - W3}{W1 + W2 + W3 + W4}.$$

Torsion (T)

$$T = \frac{W1 + W4 - W2 - W3}{W1 + W2 + W3 + W4}.$$

W1, W2, W3 and W4 are the forces measured for the left forelimb, right forelimb, left hindlimb and right hindlimb respectively (see Fig. 1).

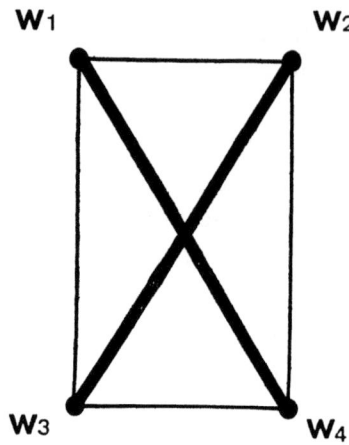

Fig. 1. The four corners of the rectangle represent the four platforms supporting the limbs. W1, W2 are the forces measured for the left and right forelimbs while W3, W4 are for the left and right hindlimbs. The difference in the force (weight) distribution along the two diagonal lines joining one forelimb and the opposite hindlimb is used for the measure of torsion.

Each of these indices was measured twice, an *initial value* calculated at the onset of cortical stimulation, contact of the moving tray with the corresponding forepaw, or onset of the conditioned stimulus and a *final value* at the time of lift-off. In the case of cortical stimulation subthreshold for lift-off, measurement of the final value was made at the time of minimum force of the "moving limb". In addition, a third measure was obtained from dogs during maintained position of the lifted limb.

2. *Differential indices.* Two differential indices were used for the analysis of the data. The reason for their choice will be explained in the result section.

Diagonal index (D)

$$D = 1 - \frac{|\Delta W1 - \Delta W4| + |\Delta W2 - \Delta W3|}{|\Delta W1| + |\Delta W2| + |\Delta W3| + |\Delta W4|}.$$

$\Delta W$ is the difference in weight between the final and the initial values.

Antero-posterior differential index (A-P)

$$A\text{-}P = \frac{|\Delta W1| + |\Delta W2| - |\Delta W3| - |\Delta W4|}{|\Delta W1| + |\Delta W2| + |\Delta W3| + |\Delta W4|}.$$

### RESULTS

#### RIGID OBJECT

For a better understanding of the biomechanical events taking place during movement of one limb in the standing quadruped, it is interesting to examine first the behavior of a rigid, four-legged object when one of the four supporting trays is dropped. Let us consider a first case in which the weight is equally distributed on the four legs (Fig. 2). Before dropping one supporting tray, initial values of the Antero-Posterior, Lateral and Torsion indices are equal to zero. After dropping one support, the values of A-P and L are still zero, because the center of gravity remains in a central position. However torsion increases to the maximum absolute value of 1, the rigid object being supported by only one pair of diagonally opposite legs.

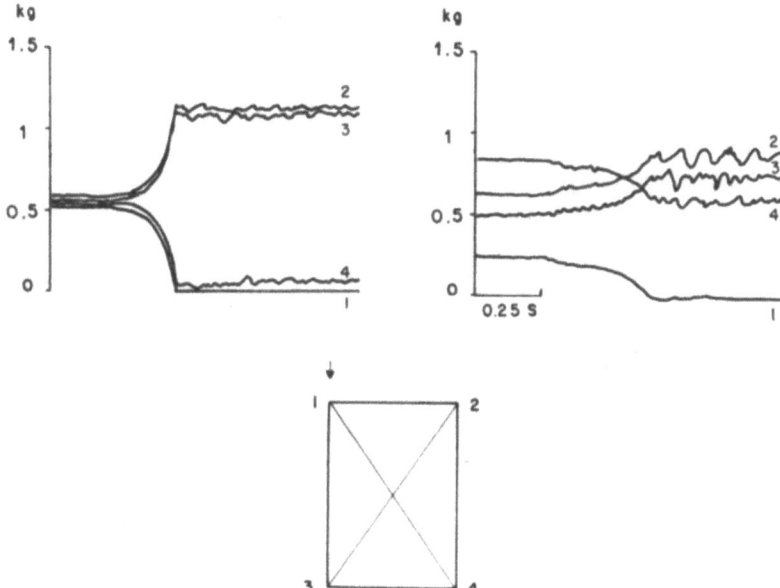

Fig. 2. Changes in weight distribution of a rigid object. The recordings were obtained with a table supported by four legs on the four platforms. The diagram at the bottom of the figure shows the position of the four supporting trays from which the vertical forces represented in the upper part of the figure were recorded. The arrow indicates the platform which is dropped (left foretray). On the left, the initial weight distribution is equal for the four trays; on the right, unequal. Notice that in both cases the same amount of weight is added or substracted to the four limbs (diagonal differential index equal to 1).

Let us now examine the case of a rigid object with unequal distribution of the weight on the four trays. Initial weights, as a percentage of the total are for limb 1 : 12%, for limb 2 : 29%, for limb 3 : 22% and for limb 4 : 37%. The initial values of A-P and L are different from 0. In the case presented in Fig. 2 right side, the following values were measured: A-P = $-0.18$, L = $+0.32$ and T = 0. The final weights are for limb 1 : 1%, for limb 2 : 40%, for limb 3 : 32% and for limb

4 : 27%. A-P and L values do not change because the center of gravity remains in the same position, but the absolute value of T increases (T = −0.44) although not to the maximum as in the previous case.

In both cases however, the weight distribution on the four supporting legs was modified in the same way. The same amount of weight was gained or lost by the four supporting legs, i.e. lost by both the one from which the support was dropped and the diagonally opposite leg, and gained by the other pair of diagonally opposing legs. Therefore a differential index was elaborated[1] to calculate quantitatively this diagonal pattern, for differing initial conditions, and/or when the limb flexion was isometric (i.e. force measured from limb did not reach zero).

This differential diagonal index (see Methods) is equal to 1 when the weight change is equal for each of the four limbs but of opposite sign for the two diagonal pairs. The index is zero when only the forelimbs (or only the hindlimbs) show a weight change (see Fig. 4). This latter situation was not encountered with the rigid object but was seen with the experimental animals where the strength of the link between forelimb and hindlimb can be changed by the nervous system.

A second differential index (antero-posterior index described in Methods) permitted evaluation of the contribution of the forelimbs and of the hindlimbs to the redistribution of weight.

With this index, a value of 0 indicates an equal contribution of fore- and hindlimbs to the weight change, a value of 1 corresponds to weight changes of forelimbs exclusively and −1 to weight changes of hindlimbs exclusively (see Fig. 5).

PHASIC MOVEMENT

Changes in instantaneous indices and values of differential indices during phasic movements will be successively analyzed.

*Instantaneous indices (Fig. 3).* The *antero-posterior index* is the one which changes the least whatever the experimental conditions or animals. The initial values of A-P index were from −0.04 to +0.16 in the cat and from +0.23 to +0.5 in the dog. This indicates a tendency for the center of gravity to be localized nearer the forelimbs than the hindlimbs as already noticed by Gray (8) in many quadrupeds. Forelimbs movements were usually accompanied by a slight backward shift of the center of gravity. Maximal change in A-P index value was of 0.2 that is 10% of the animal's weight being displaced backwards. Hindlimb movements were accompanied by the slightest A-P shift.

---

[1] The authors are indebted to Dr. A. A. Frolov who proposed this index.

291

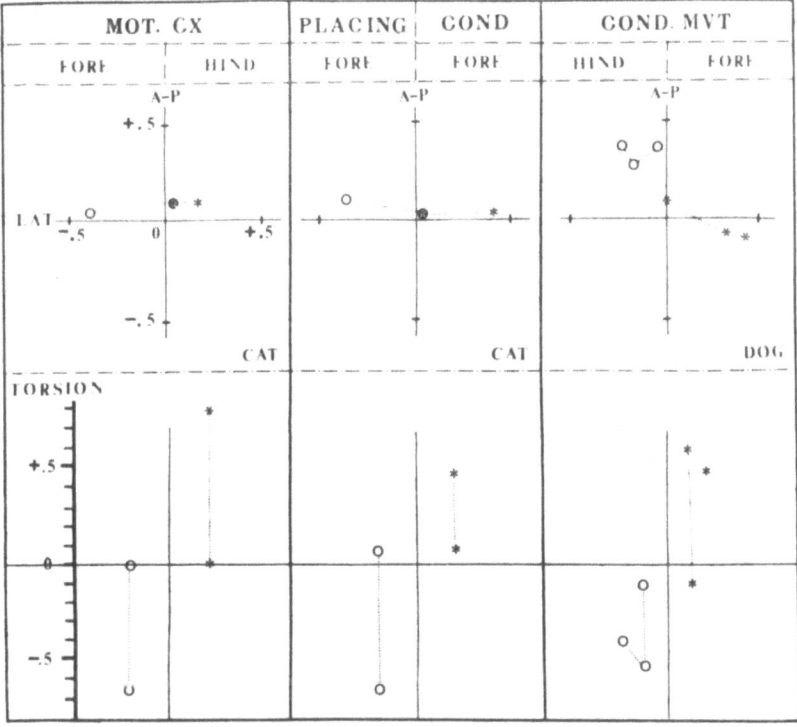

Fig. 3. Examples of anteroposterior, lateral and torsion instantaneous indices before movement (initial values closest to zero) and at the time of lift-off are represented. Three categories of samples are shown, each one corresponding to the mean values obtained during one experimental session. On the left, forelimb and hindlimb movement elicited by cortical stimulation in the cat (movements with lift-off), in the middle, forelimb placing movement and conditioned lift-off movement of forelimb in the cat, and on the right fore- and hindlimb conditioned movement in the dog. For these last movements, a third value is given which corresponds to the final maintained position of the limb. The AP and L indices are combined, and thus indicate the position of the center of pressure (projection of the center of gravity under static conditions). In each graph, results from two different sessions are represented and symbolized one by a circle, the other by an asterisk. Torsion is represented separately at the bottom. Notice the restricted AP displacement, the more important lateral displacement (except for hindlimb cortically induced movements) and torsion.

The *lateral index* was in all cases modified during movements. Movements of a left limb were accompanied by a shift of the center of pressure (resultant of the vertical forces exerted by the limbs) from a midline position towards the right. At the time of lift-off, the center of pressure was always located inside the triangle formed by the three remaining supporting limbs. The L index shifted from 0.2 to 0.5, which would correspond under static conditions to a displacement of the body weight from 10 to 25%. This lateral displacement is seen in dogs as well as in cats for all types of movements. However, hindlimb cortically elicited movements were an exception since very little lateral shift of the center of pressure was seen.

*Torsion* was the index showing the largest change during movement. Starting from an initial value near zero in most cases, the index rose 0.4 to 0.8 at lift-off time. This means that from 70% to 90% of the weight was supported by one pair of diagonally opposite limbs at the time of lift-off.

*Differential indices (Figs. 4 and 5).* Some interesting observations can be made by analyzing the differential indices, based on the differences in weights between the final values (time of lift-off) and initial values.

The *diagonal index* estimates in fact the forelimb and hindlimb contribution to the diagonal postural adjustment which takes place during movement (Fig. 4). The highest values were seen for movements produced by cortical stimulation, those for hindlimb movements (0.7–0.9) being higher than those for forelimb (0.4–0.7). No significant differences were noticed when comparing the series with cortical stimulation adjusted for a movement of 4–5 cm above the supporting tray and the series subthreshold for lift-off. The values obtained with cortical stimulation were actually the closest to those observed with a rigid object. Intermediate values were seen for placing movement, and the lowest indices were obtained for forelimb and hindlimb conditioned movements in the cat, where the weight changes are restricted almost exclusively either to forelimbs or to hindlimbs (see Fig. 4). For conditioned movements in the dog, values from 0.3 to 0.6 were observed, which are more comparable to those values from cats performing placing movements rather than conditioned lift-off movements.

The *antero-posterior index* also estimates the relative contribution of forelimb and hindlimb to the postural adjustment associated with movement (Fig. 5). The highest positive or negative values were observed for cat's conditioned forelimb or hindlimb movements (highest contribution of forelimbs to the postural adjustments with forelimb movement and from hindlimbs to adjustments of posture associated with hindlimb movement). The lowest values were seen for movements induced by cortical stimulation (almost equal contributions of fore- and hindlimbs to postural adjustment). Intermediate values were recorded for placing movement in the cat and for limb movement in the dog. The results obtained with the antero-posterior index are thus in good agreement with those furnished by the diagonal one. In addition, this index shows

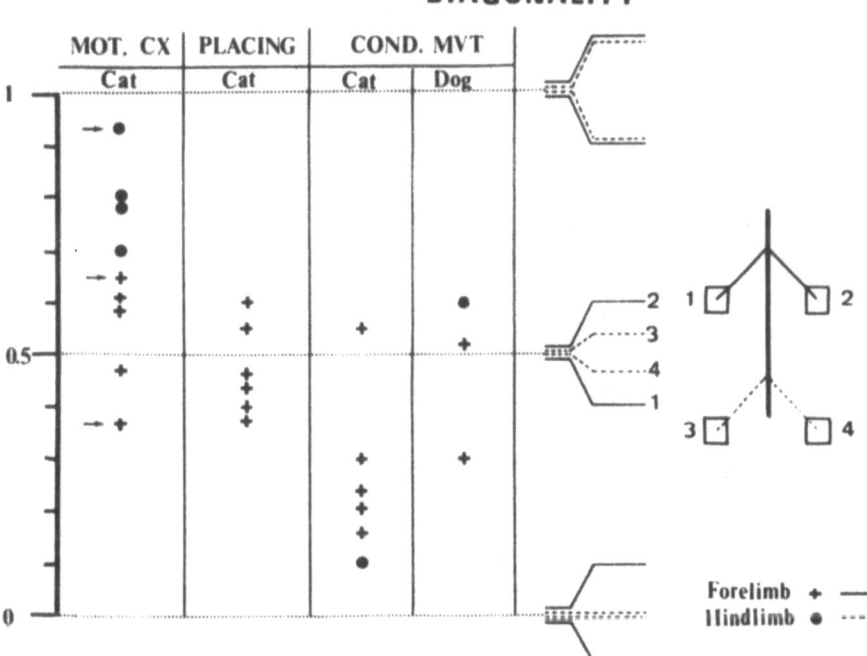

Fig. 4. Values of differential diagonal index during different types of limb movements: motor cortical stimulation in the cat (Mot Cx), placing movement in the cat (placing) and conditioned movement in cat and in dog (cond MVT). Each symbol represents the mean value obtained from one experimental session. For cortical stimulation, the results are from three cats each one having stimulating electrodes in the forelimb and in the hindlimb motor cortical area. Placing movements are obtained from 3 cats, with results from left and right placing movements. The conditioned movements in the cat were obtained from 4 animals, two trained to perform the lift-off movement of forelimb on the right and left side, one performing forelimb movement on the left side and one performing hindlimb movement of the left side. Three dogs were used, two with forelimb movement, one with hindlimb movement. On the right, diagrammatic representation of quadruped with the paws resting on four trays. The force changes which would be observed for different index values (+1, 0.5, 0) are represented. When limb 1 corresponds to the moving limb, the force exerted by it falls to zero (see for comparison Fig. 2). The lower traces, at the time of lift-off, are 1 and 4 and the upper traces are 2 and 3. The representation of force changes for the different values of the index is only an approximation based on the hypothetical case in which no shift of the projection of the center of gravity occurs during movement. On the left, values of the diagonal index. For motor cortical stimulation two sets of values are represented. Those without arrow correspond to sessions without lift-off, and those with arrow to sessions where cortical stimulation was adjusted for raising the limb 4–5 cm above the supporting tray. For the comments on the figure see text.

a result, at first sight surprising, concerning the hindlimb movement elicited by cortical stimulation. In this case, the value of the index is greater than or equal to zero (−0.02 to +0.3) which means that the postural adjustment associated with hindlimb flexion has a forelimb

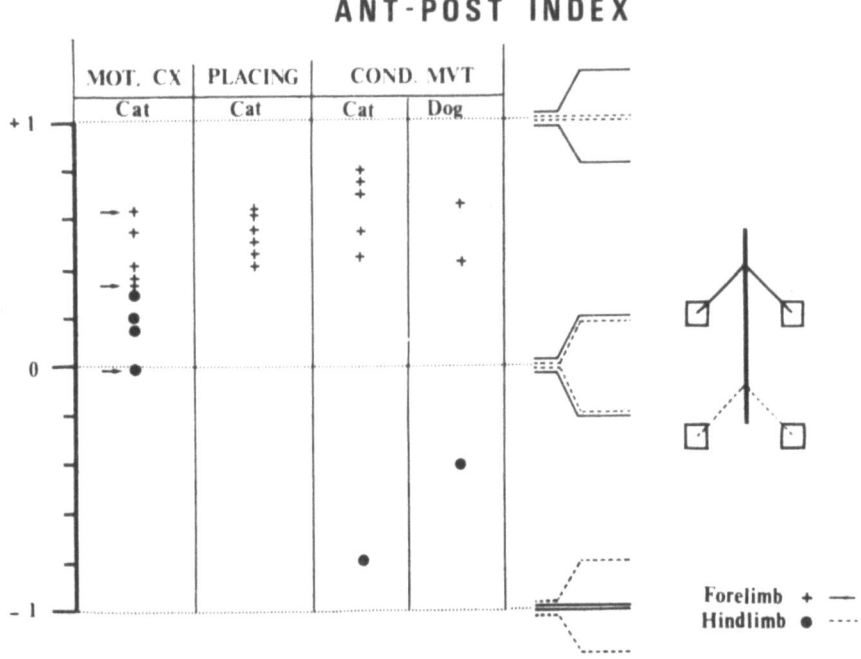

Fig. 5. Differential antero-posterior index measured from the same animals and sessions as those represented in fig. 4. On the right, diagrammatic representation of quadruped with the paws resting on four trays. The force changes which would be observed for different index values (+1, 0, −1) are illustrated. Two different sets of trials are represented for cortical stimulation as in Fig. 4.

component greater than or equal to the hindlimb contribution. This can be explained by the fact that the initial position of the center of gravity is located nearer to the forelimbs than the hindlimbs and thus a higher isometric weight change of forelimbs with respect to hindlimbs is possible.

MAINTAINED POSITION

The experimental procedure used to elicit a conditioned limb movement in the dog permitted comparison of two situations. One corresponds to the time of lift-off, that is the moment when the moving limb leaves the supporting tray. The second, a more static situation, corresponded to the maintained posture of the limb in the flexed position.

TABLE I

Comparison between diagonal differential indices during phasic movements and maintained position in the dog. The diagonal differential index was measured between lift-off and initial position and between final and initial position. Notice the lower value of the index at the time that the final position is reached

| Dog | Difference between lift-off and initial | Difference between maintained position and initial |
|---|---|---|
| LIT (Forelimb) | 0.52 | 0.31 |
| POL (Forelimb) | 0.29 | 0.1 |
| TIM (Hindlimb) | 0.57 | 0.36 |

Both instantaneous and differential indices were compared for the two situations. The main difference between the dynamic phase of the movement and the maintained posture was reflected in the diagonal differential index (Table I). This index was clearly lower during the postural fixation of the limb. At the same time, the lateral displacement of the center of pressure was increased and torsion was reduced (Fig. 3).

These results indicate that the support of the body on the two diagonal limbs was maximal during the dynamic phase of the movement, but this diagonality was less during the maintained posture of the limb in a flexed position.

DISCUSSION

One of the main objects of the experiments reported in this paper was to analyze the postural support accompanying movement in the standing quadruped. Two questions were raised. First, is this postural support qualitatively or quantitatively different in two different species of quadruped such as dogs and cats. Second, are there qualitative or quantitative differences in the postural support for movement according to the way that the movement was elicited. For this comparison, forelimb and hindlimb movements evoked by cortical stimulation, forelimb placing movements and fore- and hindlimb conditioned movements with or without maintained position were analyzed.

A first general conclusion is that qualitatively the different types of movement performed in the two species were supported by a diagonal stance using one forelimb and the opposite hindlimb. This diagonal pattern is similar to that observed when a rigid object supported by four legs loses one of its supports (Fig. 2). It is also used during locomotion, especially during trotting, where alternate supports on two diagonally opposite limbs are observed in succession (8, 9, 14, 16).

A second observation concerns the quantitative analysis of the diagonal support. It appears that the quadruped does not behave as a rigid object and that the contributions of each of the four limbs to the weight changes during movement are not equal. The analysis of the quantitative measurements of the forelimb or hindlimb contribution to the postural support has revealed that a hierarchy exists in the cat in the way that postural adjustment associated with movement is organized. The cortically evoked movements have the highest diagonal index, i.e. the closest to a rigid object where an equal contribution of fore- and hindlimb to the diagonal support are observed. An intermediate state is found with the forelimb placing movement, where the forelimb contribution to the postural support is higher than that of the hindlimb. Finally, the postural support for conditioned lift-off movement is the furthest from what is seen with a rigid table. There is a tendency that only forelimbs for forelimb movements and only hindlimbs for hindlimb movements participate in the postural adjustment associated with movement.

In attempting to explain these quantitative differences in the postural support in quadrupeds for the various types of movement, several factors may be considered. The first is the effect of the amplitude of the particular movement; two experimental series with cortical stimulation were compared, one without lift-off of the limb, the other with the leg raised 4–5 cm above the tray. The results obtained from both series were comparable. Thus, amplitude of movement does not seem to be an important factor contributing to the differences. A second possible mechanism could be the speed of force changes. In fact, for cortical stimulation where a high diagonal index is observed time to lift-off was less than 0.1 s, whereas for placing, with an intermediate diagonal index, the time was 0.18 to 0.35 s, and for conditioned movement with the lowest diagonal index 0.4 to 0.5 s. Thus the factor of speed of force changes cannot be excluded. However, it must be mentioned that the low diagonal index obtained for lift-off movements is only observed after learning is achieved, and that during learning the diagonal index is much higher, notwithstanding the fact that force changes are performed slowly. Thus differences in the central command according to the types of movement are probable. Concerning the movement elicited by cortical stimulation, the short latency of the force changes suggest that the command may have a direct access to the network which is responsible for the diagonal support. This network could be located at the spinal cord level, as suggested by experiments reported by Sherrington (15), who gave evidence for the existence at that level of a neural basis for a link between flexors and extensors of diagonally opposite limbs.

The changes in the supporting pattern according to the type of movement are clearly apparent in the cat, but the question remains open for the dog where results are available for only one type of movement, a conditioned limb movement. The measures of the diagonal index at the time of lift-off made on three dogs indicate that no values were observed as low as those in the cat. This suggests that the dog may not be able to produce postural support during movement limited to the two forelimbs or to the two hindlimbs as does the cat. The dog would thus behave more "rigidly" which is not a surprising result. However, before conclusions can be drawn, more data must be collected, using the same types of movement in both species.

The authors wish to thank Dr. J. Macpherson for her aid in translating part of this manuscript. This work was done under the scope of the "Scientific exchange program" between the USSR Academy of Sciences and the French CNRS. Part of this work was supported by the CNRS (ATP No. 05Al 3618).

## REFERENCES

1. ALEXEIEV, M. A. and NAIDEL, A. V. 1973. Rapports entre les éléments volontaires et posturaux d'un acte moteur chez l'homme. Agressologie 14B: 9–16.
2. BELENKIY, V. E., GURFINKEL, V. S. and PALTSEV, E. I. 1967. On elements of control of voluntary movements (in Russian). Biofizika 12: 135–141.
3. BROOKHART, J. M., PARMEGGIANI, W. A., PETERSEN, W. A. and STONE, S. A. 1965. Postural stability in the dog. Am. J. Physiol. 208: 1047–1057.
4. BURLACHKOVA, N. I. and IOFFE, M. E. 1978. A study of the effector structure of learned coordinations in the dog. Neuroscience 3: 125–127.
5. BURLACHKOVA, N. I. and IOFFE, M. E. 1979. The analysis of the postural adjustment accompanying a local movement. Agressologie 20B: 141–142.
6. COULMANCE, M., GAHÉRY, Y., MASSION, J. and SWETT, J. E. 1979. The placing reaction in the standing cat: A model for the study of posture and movement. Exp. Brain Res. 37: 265–281.
7. GAHÉRY, Y. and NIEOULLON, A. 1978. Postural and kinetic coordination following cortical stimuli which induce flexion movements in the cat's limbs. Brain Res. 155: 25–37.
8. GRAY, J. 1968. Animal locomotion. Weidenfeld and Nicolson, London, 479 p.
9. GRILLNER, S. 1975. Locomotion in vertebrates: central mechanisms and reflex interaction. Physiol. Rev. 55: 247–304.

10. GURFINKEL, V. S. and ELNER, A. M. 1973. On two types of static disturbances in patients with local lesions of the brain. Agressologie 14D: 64–72.
11. IOFFE, M. E. and ANDREYEV, A. E. 1969. Inter-extremities coordination in local motor conditioned reactions of dogs (in Russian). Zh. Vyssh. Nervn. Deyat. im. I. P. Pavlova 19: 557–565.
12. KORIAKIN, M. F. 1958. Contribution of postural excitations in patterning the conditioned defensive motor reflex in the dog (in Russian). Fiziol. Zh. SSSR im. I. M. Sechenova 44: 393–403.
13. POLIT, A. and MASSION, J. 1979. Patterns of postural support during limb movement. Soc. Neurosci. (Abstr.) 5: 382.
14. ROBERTS, T. D. M. 1967. Neurophysiology of postural mechanisms. Butterworths, London.
15. SHERRINGTON, Sir Charles. 1947. The integrative action of the nervous system. 2nd ed. University Press, Cambridge.
16. SHIK, M. L. and ORLOVSKY, G. N. 1976. Neurophysiology of locomotor automatism. Physiol. Rev. 56: 465–501.
17. SHUMILINA, N. I. 1945. On the duplex nature of motor excitations in the central nervous system (in Russian). Fiziol. Zh. SSSR im. I. M. Sechenova 31: 272–282.

Y. GAHÉRY and J. MASSION, Département de Neurophysiologie Générale, INP, CNRS, 31 chemin Joseph Aiguier, 13274 Marseille Cedex 2, France.
M. IOFFE, Institute of Higher Nervous Activity and Neurophysiology, Academy of Sciences, Butlerova 5a, Moscow, USSR.
A. POLIT, Department of Psychology, MIT, Cambridge, Massachusetts 02139, USA.

Lecture delivered at the Warsaw Colloquium on Instrumental
Conditioning and Brain Research
May 1979

# THE MONKEY'S PREFRONTAL CORTEX FUNCTIONS IN MOTOR PROGRAMMING

John S. STAMM

Department of Psychology, State University of New York
Stony Brook, New York, USA

*Abstract.* A new experimental approach is presented which resulted in clarification of the specific functions of the monkey's prefrontal cortex. Monkeys with chronically implanted transcortical nonpolarizable electrodes were trained on delayed response (DR) and visual delayed matching-to-sample (DMS) tasks. The onset of the trial for each group depended upon on-line computer detection of one of the specified events: FN — surface-negative steady potential shifts (SPS) from principalis cortex; MN — a similar SPS from precentral cortex; FB — near baseline SP from principalis cortex; LEM — rightward eye deviations; and YC — controls, with intertrial intervals yoked to those of other monkeys. Monkeys were trained with 1-s cue presentations on successive delays of 2 to 12 s. The DR acquisition rate by the FN group was substantially faster than that of any other group, as indicated by its mean error that was only 17.24% the YC group's error. The MN and LEM monkeys acquired the task at the same rates as the YCs, while the FB monkeys were the slowest learners. The correct DR performance transferred to testing with constant intertrial intervals (without preconditions). Subsequent on-line tests with brief (0.1 s) cue duration showed high DR performance by the FN, but not by other groups. No similar rapid learning was found with the DMS task. The findings from this, and other experiments, suggest that the major function of principalis cortex is the selection, or *programming of delayed spatial choice responses.* The view seems consonant with interpretations for the role of the human prefrontal cortex.

### I. THE ISSUE: THE FRONTALLY-ABLATED MONKEY GIVES NO SATISFACTORY EXPLANATION FOR HIS IMPAIRED DELAYED RESPONSE PERFORMANCE

The reports by Jacobson (8) that total extirpation of the monkey's prefrontal lobe (anterior to arcuate sulcus) resulted in an inability to perform the delayed response (DR) task constitute a landmark in the history of brain-behavior research. This finding initiated a stream of research, as well as the formulations of many hypotheses which, on the one hand, substantiated the strong association between prefrontal cortex and DR behavior, while, on the other hand, did not lead to any one satisfactory explanation for the underlying dysfunctions. Jacobson's interpretation that "the contributions of the frontal association areas (appear to be) the recall of a particular past event which may be only in immediate association with some of the present environment and the integration of recalled elements with the organism's stable habit system" (8 p. 55–56) was immediately challenged by several investigators. In a well designed experiment Finan (2) found that monkeys with the extensive prefrontal resections were unimpaired in a difficult temporal maze discrimination between pre-response delays of 30 s and 120 s, while they responded at chance level on DR with delays as short as 2-s. His conclusion that "temporal organization ... plays no essential role in delayed response performance and hence cannot, without further qualification, be considered the mechanism for immediate memory" (2 p. 225) has subsequently been confirmed with differing paradigms for dogs (13) and for monkeys (23). Other early investigators obtained improvements in the prefrontally ablated monkey's DR performance with modifications in the experimental conditions, such as testing the monkey in the dark (15) reduction of the environmental temperature, prolonged food deprivation, or insulin injection, (19). These findings led to two lines of explanations for the DR deficits; namely, those related to the monkey's inadequate attention to cue properties (18, 19) and those stressing his susceptibility to interference effects during the delay period (15). Furthermore, the DR deficits have been attributed to impairments in motoric control required for the choice-response. Thus, Wegener and Stamm (34), who examined the monkey's flexibility to differing response requirements, concluded that prefrontal cortex was implicated in "spatio-motor integration" and Konorski (10) considered the DR deficit as a "loss of proper balance between two behavioral tendencies: one determined by the preparatory signal — and the other determined by actual stimuli eliciting definite instrumental responses (p. 608)". He attributed this imbalance to inadequate inhibition of the tendency to the preceding response.

The considerations by these investigators of each phase of the DR trial — cue presentation, intratrial delay, choice response, and intertrial interval — led to differing interpretations for the prefrontal deficit, in terms of attentive, mnemonic, or motoric processes. The confusion was further confounded by the observations that the prefrontal monkey was impaired, not only in DR (and delayed alternation) performance, but exhibited a complex of aberrant behaviors, described by the "frontal lobe syndrome" (8).

The subsequent period of research, which started approximately after the 1962 symposium at Pennsylvania State University (33), was concerned with unraveling of the complexities of this syndrome by means of fractionations of the large prefrontal surface into functional entities. One consistent finding was the delineation of the cortical locus for DR in the monkey's principal sulcus. Monkeys with resections restricted to this zone remained unimpaired in the performance of other tasks, that were affected by different prefrontal lesions. Thus, the arcuate sulcus was delineated as the locus for conditional response tasks, in which the location of the cue signalled a concurrent choice response in a different direction (4, 5, 27, 31), Also, the inferior prefrontal convexity was found to inhibit response perseverations (7, 27). Analyses of the characteristics of the DR and other tasks led Goldman et al. (5) to conclude that the salient features of those tasks that were selectively impaired by principalis lesions were the combination of spatial and mnemonic factors, so that principalis cortex is the substrate of "memory specific for spatial information". These authors also considered the relevant cues for DR to be proprioceptive, i.e. induced by the monkey's motor acts, a view that had been advanced by several investigators (9, 26, 34). Their conclusion constitutes a considerable advance from the original interpretation by Jacobson, in that it specifies the mnemonic functions of principalis cortex. However, it sheds little light on the processes, behavioral and neuronal, that are involved in task performance. The important questions with regard to the processes that are required for spatial memory seem unanswerable with the ablation technique. DR, as well as other prefrontal tasks, require that the monkey meet a sequence of demands that involve: recognition of the cue and its spatial location; the formation, storage and retrieval of the short term memory; selection and execution of the appropriate choice response; and extinction of the memory in preparation for the next trial. The ablation method is inadequate for fractionation among these component processes. Other techniques are needed for finding answers to the persisting question of *why* the principalis-lesioned monkey cannot solve the DR task.

## II. ELECTROCORTICAL STIMULATION AND RECORDINGS

Electrocortical stimulation in the behaving animal has been considered as a technique of "reversible ablation", because it permits comparison of performance decrements during stimulus applications with nonstimulation performance. Thus, each animal may serve as its own control. A further advantage of this technique is that stimulus applications with constant parameters may be varied among different epochs of the DR trial. The resulting performance decrements would then indicate the temporal involvement of the affected cortical segment and permit spe-

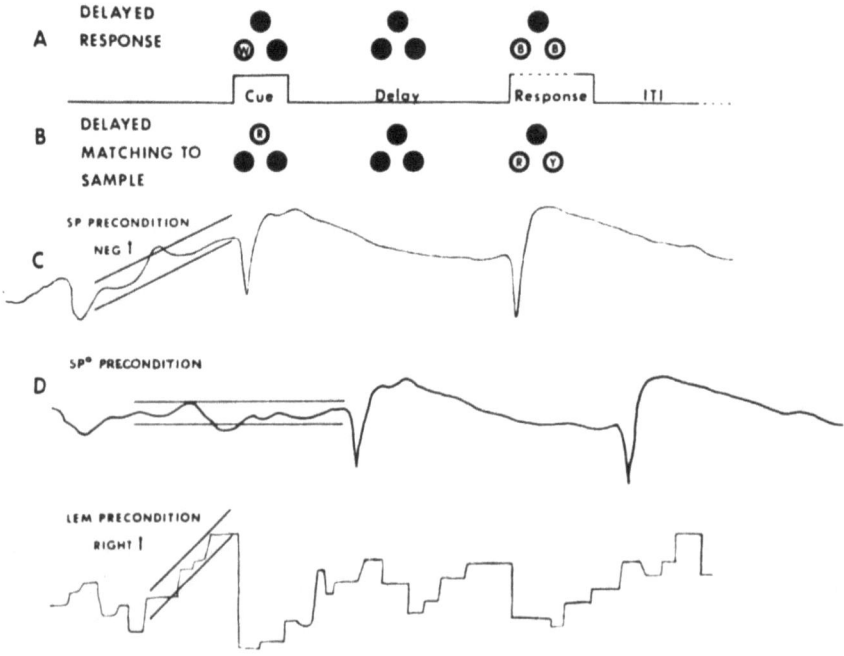

Fig. 1. Schematic representations of the (A) delayed response (DR) and (B) delayed matching-to-sample (DMS) trial, showing illumination of the display windows for the cue and choice-response epochs. The symbols indicate the illumination color as white (W), blue (B), red (R) or yellow (Y). Also, illustrations of computer sampling for the trial preconditions of: (C) cortical surface-negative steady potential shifts, (D) cortical near-baseline potentials, and (E) electro-oculograms for eye deviations to the right.

cification of the crucial period within the trial for cortical functions. Also, with implantation of multiple stimulating electrodes, the most effective cortical locus can be delineated.

Our methodology has been previously described (25, 30). During the testing session the monkey is placed in a chair which contains plastic shields for restraining of head movements. A cuff, attached to one wrist, is bolted to a shelf on the chair. The chair is placed securely in front of a vertical panel that contains three circular display windows, each 3.5 cm in diameter, located at the monkey's eye level. Two of these, which also serve as manipulanda, are in a horizontal line with 7.0 cm between their centers. The third window is located above and midway between the other two windows and a plastic foodcup is mounted midway below the bottom windows. The apparatus can be programmed for either a DR or a visual delayed matching-to-sample (DMS) task. For some of our early DR experiments the panel contained two food cups, mounted beneath each of the bottom windows. The DR trial (Fig. 1A) consists of: illumination of either the left or right bottom window with white light (the cue), non-illumination of the windows (the delay), illumination of both windows with blue light until the monkey presses lightly on one of them (the response), and non-illumination of the windows (intertrial interval). A correct response results in delivery of a sugar pellet and brief illumination of the food cup. The trial can be programmed for differing durations of cue presentations, delay, and intertrial interval. For DMS (Fig. 1B) the trial begins with presentation of the sample (cue) in the upper display window and after the intertrial delay the sample and another pattern are presented in the two bottom windows. During successive DMS trials each cue pattern and its location in the response win-

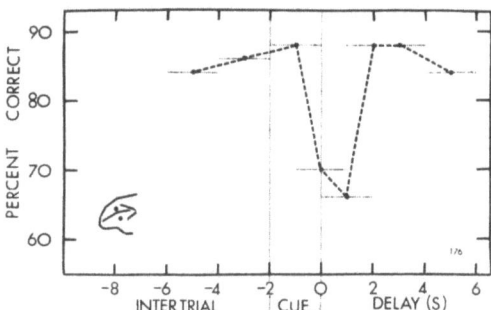

Fig. 2. Performance of a monkey on 8-s DR with 2-s stimulation (1 ms pulses at 50/s) applied across the principal sulcus (electrode locations in insert). The time scale is with reference to the start of the delay period. Horizontal lines indicate periods of stimulation. Each point on the graph represents a mean score for 50 to 110 trials. From Stamm and Rosen (30).

dows is programmed in pseudo-random order, with equal occurrence in each 10-trial block.

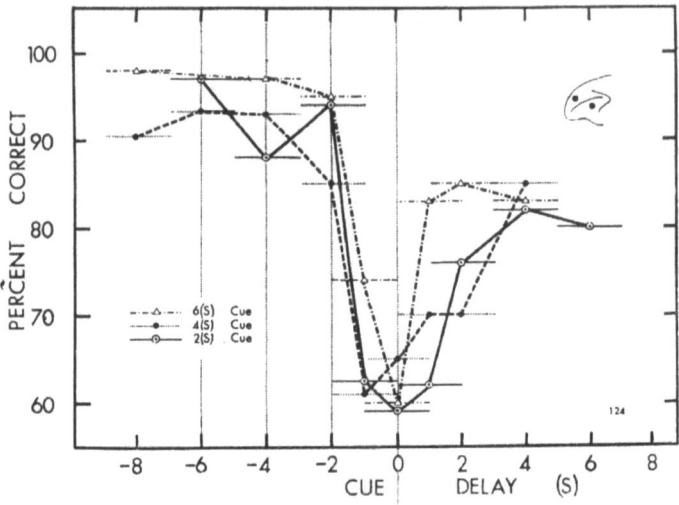

Fig. 3. Performance of a monkey on 8-s DR with cue durations of 2, 4 and 6 s. Two-s stimulation (1 ms pulses at 50/s) was applied across the principal sulcus (electrode locations in insert). The time scale is with reference to the start of the delay period. The horizontal lines indicate periods of stimulation. Each point represents a mean score for 50 to 130 trials. From Stamm and Rosen (30).

For electrical stimulation of prefrontal cortex a thin plastic sheet, which supported two arrays of electrode points, was placed on the pial surface, so that the electrodes straddled the principal sulcus. Each monkey was first trained and overtrained on the DR task without stimulation. During the following sessions the stimulus voltage was gradually increased until motoric or convulsive responses were elicited. For subsequent testing sessions the stimulating voltage was set at 80 to 90% of this threshold. During the 120-trial experimental session the stimulus train, usually of 2-s duration, was applied across one electrode pair, with the epoch for stimulus application changed for 10-trial blocks, according to a pseudo-random sequence.

Consistent results were obtained with prefrontal stimulation of 12 monkeys. The critical zone for the behavioral effects was the posterior two-thirds of the principal sulcus. Correct DR performance was severely disrupted *only* with stimulation just before and after the start of the delay, whereas stimulus applications during the remainder of the delay period had only minimal (Fig. 2), or moderate (Fig. 3) effects on correct DR responses. This time course of performance decrements was unaffected by variations of cue duration (Fig. 3), of the delay (8 to 20 s), shifts in the responding hand, or stimulation of the right or left hemisphere. These findings led to our interpretation of the major implication of principalis cortex in the formation of the short-term memory, but less in memory storage or retrieval. A similar time course for the effects of prefrontal stimulation was found with a delayed visual discrimination task (1), for which either two crosses or two Xs were presented as cues, which signalled, respectively, a right or left choice response after the intratrial delay. Maximal performance decrements with stimulus applications across the principal sulcus were found only with stimulation before and following the start of the delay, but not during other portions of the cue or delay periods. By contrast, stimulation of inferotemporal cortex during performance of this (1) or the DR task (30) disrupted correct performance only with stimulus applications following cue onset, but not during the late cue or the delay periods. The findings from these experiments were consistent with our view of the major role of principalis cortex in the *formation of the spatial short term memory*.

Different results were obtained in experiments with electrocortical stimulation during monkey's performance of a visual DMS task, with samples of either an X or O pattern (11). In addition to the usual electrode arrays across the principal sulcus, sheets supporting four electrode

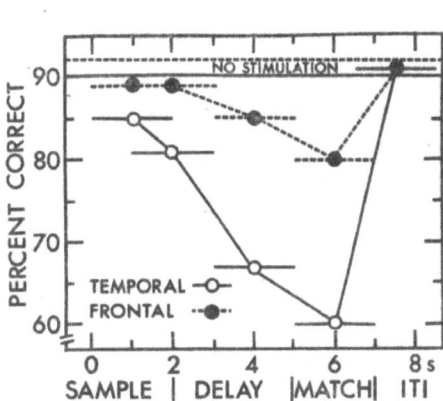

Fig. 4. Performance on 3-s DMS by four monkeys with 2-s stimulation (1 ms pulses at 50/s) applied to interotemporal or principalis cortex. The time scale is with reference to onset of the sample. The horizontal lines indicate the 2-s stimulation periods and represent the group's mean scores of correct performance. From Kovner and Stamm (11).

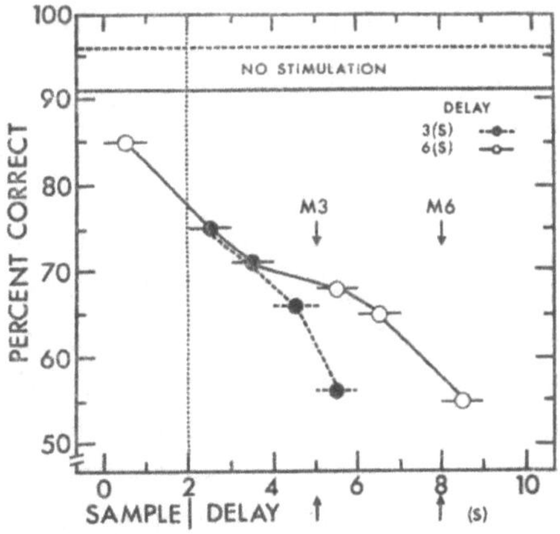

Fig. 5. Performance on DMS with 3-s and 6-s delays by two monkeys with 1-s stimulation to inferotemporal cortex. The time scale is with reference to onset of the sample and the arrows indicate the matching period following the 3-s (M3) or 6-s (M6) delay. The horizontal lines indicate the stimulation periods and represent the group's mean levels of correct performance. From Kovner and Stamm (11).

points were placed on inferotemporal cortex in each hemisphere of four monkeys. The training and stimulation procedures were similar to those of the DR experiments, except for stimulations of each cortical area during separate 120-trials sessions. The findings delineated the critical inferotemporal zone as the posterior segment of this area. Also, the results confirmed the functional dissociation between principalis and inferotemporal cortex. Stimulation of the former, even at convulsive thresholds, did not affect correct DMS responding by three monkeys and resulted in only moderate performance decrements by the fourth subject (Fig. 4). By contrast, inferotemporal stimulation had consistent and profound effects on DMS performance. The disruption of correct responses by stimulation during the matching epoch seemed consistent with the findings of ablation experiments that inferotemporal cortex is implicated in visual pattern discrimination. Stimulus applications during the delay period had variable effects on DMS performance (Figs. 4 and 5), which was minimally affected by stimulation during sample presentations or the early delay, but severely disrupted with stimulation during the late delay period. These effects seemed independent of the duration of the delay settings. The resulting relationship between performance decrement and the epoch of stimulus application appears to be almost the reverse of the one that had been found for DR impairments with prefrontal stimulation.

How can the divergent findings be explained? The common demand during the "critical epochs" of the DR and DMS tasks seems to be the monkey's selection of the choice response. For the DMS task, this requires the monkey's retrieval of the visual memory and its matching with the choice pattern, while for DR the monkey must recognize the spatial location of the cue and concurrently prepare the subsequent choice response. This analysis suggested to us that the DR memory encoding process involves, not only a passive recognition of the cue location, but the monkey's selection of the direction for his choice response. This hypothesis is of course tenuous, because prefrontal and inferotemporal cortex have quite differing functional properties and we have had no direct evidence for the latter's implication in response processes. However, the findings from these experiments led to our consideration of the regulation of response processes as an important function of principalis cortex.

The most direct evidence for neuronal involvement in DR has been obtained with recordings of single unit discharges from principalis cortex during the monkey's task performance. With this technique several classes of neurons have been identified in relation to their discharge patterns during differing epochs of the trial. Thus, Fuster (3), who used a direct DR task, found that most principalis units were activated during the cue and delay periods of the trial, with the highest overall discharge rates occurring during the early portion of the delay. These findings are consistent with those of Kubota et al. (12) who, with an indirect DR task, identified "visuokinetic" units that have similar discharge patterns. Some of these units, as well as other types, discharged shortly before and during the choice response. Furthermore, these investigators, as well as Niki (16) have identified classes of neurons that responded differentially to left and right cues and responses. In addition, for a delayed-alternation task with an apparatus that contained four response windows, Niki (17) obtained evidence for differential unit activations during the delay period for impending choice responses to the left or right manipulandum. While the majority of these units were specific to the absolute direction of the response, a minority were sensitive to the relative right or left direction, regardless of the spatial location of the response windows. The findings by the several investigators are in agreement that there is major activation of principalis units during the epoch before and after onset of the DR delay. The unit discharges also provide evidence for multiple functional processes of principalis cortex, that seem related to spatial, temporal, and response features of the task.

Our research has used techniques for recordings of transcortical steady potentials (SPs) during the monkey's DR performance (29). Pairs of nonpolarizable Ag-AgCl electrodes were chronically implanted in prefrontal, precentral and occipital cortex, with one electrode of each pair placed on the pial surface and the other in adjacent white matter, at 5-10 mm tip separation. The active prefrontal electrode was placed on cortex in the depth of the posterior principal sulcus. ECG recordings, through DC filtered amplifiers, were obtained for each electrode pair and were averaged, generally for 40-trial blocks.

The ECG recordings showed three distinct surface-negatiwe SP waves during differing epochs of the DR trial. One large SP shift (SPS) occurred after the choice response. Since this SPS became attenuated when the reward was withheld and reappeared with reinstatement of the reward, it seems to reflect cortical responses to the reinforcement. A second SPS started several seconds before cue presentation. Since this SPS did not occur with variable intertrial intervals, but developed during the course of testing with constant intervals, we consider it as an expression of the monkey's expectation of the approaching trial. The experimental procedures that affected the magnitudes of these two SPSs resulted in concurrent SP changes from all electrode locations, so that these SPSs seem to indicate general cortical arousal processes. The third SPS started during cue presentation, attained maximum negativity during the early delay, and then declined at varying rates toward baseline level. This SPS was most pronounced from prefrontal electrodes and its magnitude was found to be unaffected by changes in the duration of the cue or delay periods. Also, this SPS seemed localized to prefrontal cortex in one hemisphere and its magnitude did not change with shifts in the monkey's responding hand (28) Furthermore, the magnitude of the prefrontal SPS was significantly correlated (rs of 0.74 to 0.90) with the level of correct choise responses, whereas corresponding correlations from the other electrode locations were insignificant. These findings led to our conclusion that this prefrontal SPS reflected the specific involvement of principalis cortex in DR performance.

In summary, the findings with the techniques of electrocortical stimulation, single unit recording, and SP recording are consistent in that they delineate the epoch just before and after onset of the delay as the crucial period for involvement of principalis cortex in DR performance. Also, the striking resemblance between the time courses of the prefrontal SPS and the summated unit discharges (3) supports the interpretation cf the surface-negative SPS as an expression of neuronal activation. However, the inferences for the behavioral processes during this crucial epoch remain unresolved. While we have considered the principalis functions in terms of mnemonic and response processes, Fuster (3) interpreted his findings as indicative of the monkey's attention to the cue. This issue is more directly addressed in the following experiments.

### III. A NEW EXPERIMENTAL APPROACH TO THE OLD PROBLEM

In recent research we utilized the endogenous prefrontal SPS as a method for eliciting rapid learning of the DR task. Our interest in establishing such a phenomenon originated from an unexpected finding in an early experiment (24), in which electrocortical stimulation was applied while monkeys were trained on a delayed alternation task. An evaluation of the effects of differing stimulus parameters revealed higher levels of correct responses during stimulation at certain low-voltage settings, than was found with other voltages or no stimulation. The mean rate of task acquisition with low-voltage stimulation of six monkeys was 2.18 times greater than the acquisition rate for six monkeys that were trained without stimulation. In addition, the most pronounced stimulation effects were found with two "stupid" monkeys who could not meet the task criterion after thousands of nonstimulation trials, but subsequently attained criterion performance with low-voltage stimulation and then continued at this response level during nonstimulation testing. In our search for the cause of this facilitatory learning phenomenon we found that the home-made stimulator generated a brief negative pulse that was followed by a longer lasting positive tail of a few micro-amp current across the cortical surface. An indirect confirmation for the effectiveness of this positive current was our inability to replicate the learning effect with other stimulators that produced only negative square wave pulses. The small surface-positive currents may have established a "dominant focus of excitation" (21), which had been found to facilitate the elicitation of conditioned motor responses. Subsequent research (32) has shown that the application of anodal polarization resulted in heightened cortical excitability.

The hypothesis of anodal polarization was examined more directly in a subsequent experiment (29) in which we trained monkeys with implanted nonpolarizable electrodes on the DR task, while 8-s trains of surface-positive currents (10 to 24 μamps) were applied across the principalis cortex. Comparisons between 10-trial blocks of stimulation and nonstimulation conditions showed significantly higher performance scores for the stimulation trials. However, the facilitatory effects were small and variable. The unimpressive findings were attributed to the difficult methodological problems for eliciting adequate cortical polarization, without producing lesions.

A different experimental approach was suggested by our findings of endogenous SPSs during the monkey's DR performance. If the DR trial could be initiated during the peak of a prefrontal surface-negative wave, the affected cortex would then be in a heightened state of activation and this state might lead to faster than normal DR acquisition. This method may also serve to clarify the issue of prefrontal functions. According to the attentive interpretations, this cortical state would increase the monkey's attention to features of the cue, so that rapid acquisition would also occur with other mnemonic tasks, such as visual DMS. However, the hypotheses of modality-specific functions of prefrontal cortex would predict no corresponding facilitatory effects for non-spatial tasks. The methodology required for the detection of single endogenous SPSs has become feasible with the availability of a computer.

We trained 16 monkeys who had been implanted with pairs of nonpolarizable electrodes in principalis, precentral and occipital cortical areas. In addition, small epoxy resin electrodes were implanted subcutaneously across the eyes for EOG recordings of lateral eye movements (22). All monkeys were trained to respond with the right hand, at 98 trials per session, to the criterion of 88 correct responses in one session. They were first trained on DR with 1-s cues and 0-s delay. Monkeys were then randomly assigned among five groups, each with a different requirement for trial initiation.

For Group FN (4 monkeys) the ongoing ECG from the left principalis electrodes was monitored by a computer which generated a voltage window (Fig. 1C) for a surface-negative SPS of 50 to 100 µV amplitude over a 2.5 s epoch. The ECG voltages were sampled at 50 ms intervals and when at least 86% of the points fell within the window, the pretrial criterion was met and the appropriate cue light was initiated. During preliminary testing each monkey's criterion voltage was determined that resulted in durations of testing sessions that were about the same for all subjects. The monkey's intertrial intervals for each session were recorded and stored on magnetic tape and served for programming of the corresponding sessions for the yoked controls. Five control monkeys in Group YC were trained without pretrial contingency and with intertrial intervals that were yoked to those of three FN monkeys and one in each of the following two groups.

Three comparison groups (two monkeys in each) were trained with differing pretrial requirements. Group MN had the same pretrial SPS as that for FN, but from electrodes in left precentral cortex, located 3-6 mm anterior to the central sulcus. This requirement also permitted observation of possible behavioral mediating responses from the monkey's responding limb. For monkeys in Group LEM the trial was initiated by lateral eye deviations to the right of approximately 40° during a 1.5 s period (Fig. 1E). This motor act was selected because the pretrial requirement could be precisely controlled by the experimenter and eye movements introduced little artifacts in the ECG recordings. Also, oculomotor movements had been elicited by brief electric stimulation from the contralateral "frontal eye field" (20) which is situated in the vicinity of the principalis electrodes. For monkeys in Group FB the pretrial criterion (Fig. 1D) was a 5-s period of near-baseline ECG activity from left principalis electrodes, within a ± 25 µV voltage window. An additional monkey was trained on DR under different conditions.

The procedure consisted in training each monkey, under its respective "on-line" condition, to the DR criterion with successive delays of 2, 4, 8, and 12 s. After criterion performance was met on each delay setting, the monkey was tested for one or two sessions under the "off-line" condition, with a fixed intertrial interval that was equal to the average interval of the preceding on-line session. A monkey that did not respond at the 90% criterion during one off-line session was retrained under its on-line procedure and off-line testing was repeated. ECGs from all cortical electrodes and EOGs were recorded during all sessions and averaged.

The results for acquisition to 12-s DR are striking. The FN group met criterion on each delay setting substantially faster than did any of the other groups (Fig. 6). The total mean errors, including those for criterion sessions, were 53.2 for the FN, 308 for the YC, 205.5 for the MN, 288 for the LEM, and 485 for the FB group. Because of the few MN and LEM monkeys and their similar acquisition scores, they were combined into one comparison group. The FN group's mean error was significantly

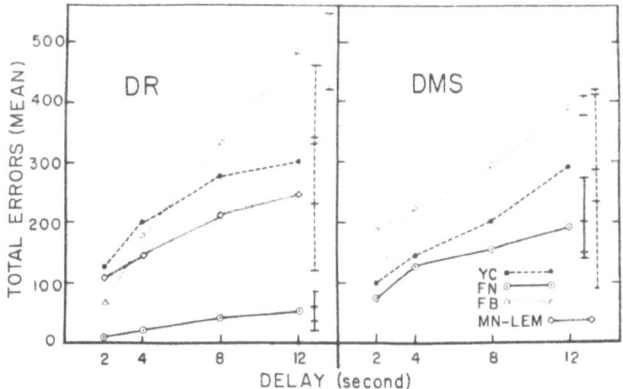

Fig. 6. Acquisition of spatial delayed response (DR) and visual delayed matching-to-sample (DMS) tasks, with successive intratrial delays of 2, 4, 8 and 12-s. The ordinate indicates mean cumulative errors on each delay setting, including errors for the 90% correct criterion sessions. The pretrial requirements for cue-onset were: FN, (N = 4); FN, a 2.5-s surface-negative steady potential (SP) shift from left principalis cortex. MN, (N = 2) a similar SP shift from left precentral cortex; LEN, (N = 2) rightward eye deviations; FB, (N = 2) 5-s baseline SP from left principalis cortex; YC, (N = 5) intertrial intervals yoked to those of monkeys in other groups. The data for the MN and LEM groups are combined for DR and not shown for DMS. The vertical lines at the right of the graphs present cumulative errors by individual monkeys (those for the MN — LEM monkeys are not shown).

below that by any of the other groups (Ps < 0.005), but no significant (Ps > 0.10) error differences were found between any two control groups. Also, there was no overlap of individual errors between FN monkeys (19 to 98 errors) and those in any of the other groups, except for the 85 errors by one MN monkey. The error range for the YC group was 121 to 480. The slowest task acquisition was found for the FB monkeys, who obtained the highest and 4th highest error scores of all subjects. Corresponding results were obtained for the number of on-line training sessions, which had medians of 6.0 for the FN, 19 for the YC, 11.5 for the MN, 13 for the LEM, and 22.5 for the FB group. One FN monkey responded above 90% correct during the first on-line session at each delay setting, while another required only five sessions. These monkey's total errors of 19 and 36 were below 10% of their total trials. The effectiveness of the training procedure is indicated by the monkeys' performance during off-line testing. The mean transfer scores from the preceding on-line session for the four delay settings were 91.3% for FN,

95.5% for YC, 82.5% for MN, 99% for LEM, and 97.4% for FB. These findings indicate that the on-line task acquisitions were not state dependent, but resulted indeed in *learning* of the DR task.

The monkeys' difficulties in meeting their pretrial requirements are expressed by their long intertrial intervals. For the first session on 2-s DR the mean intervals were 39.95 s for the FN group and 35.73 s for the combined group of six comparison monkeys. The monkeys' ability to improve their elicitation of the pretrial requirements during the course of training would be indicated by reductions of the intertrial intervals. However, the mean intervals during the final 12-s DR session differed from those of the first session by only + 18.05% for the FN and — 5.3% for the combined comparison group. There was also no evidence for the acquisition of behavioral mediating responses by the LEM monkeys,

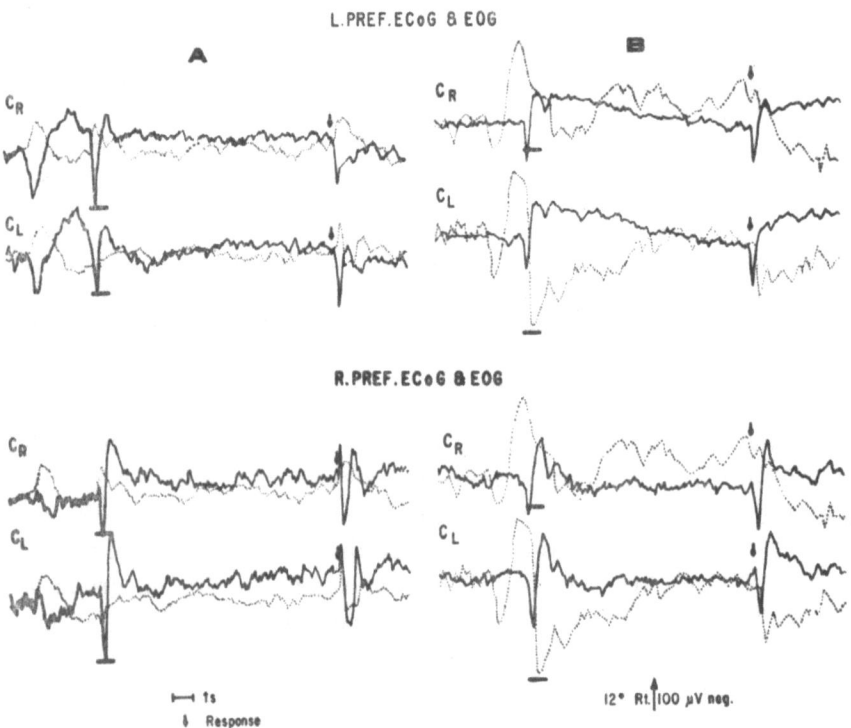

Fig. 7. Averaged left and right principalis (PREF) electrocorticograms (solid lines) and concomitant averaged electrooculograms (dotted lines) during performance of 12-s DR for (A) a FN monkey and (B) a LEM monkey. Separate averages (49 trials each) were obtained for trials with right ($C_R$) and left ($C_L$) cues. Horizontal lines denote 1-s cue presentations. Arrows signify the beginning of the choice-response period. Abbreviations as for Fig. 6. Calibrations for ECGs and EOGs as indicated. Upward deflections indicate cortical negativity or eye deviations to the right. From Sandrew et al. (22).

whose mean intertrial interval changed by only — 2.1%. These findings provide no indications that the pretrial requirements were met by the monkeys' acquisition of mediating behavioral responses.

The ECG recordings were averaged backward and forward from cue-onset. The pronounced pretrial negative SPSs from left principalis (Fig. 7A; 8 DMS) or left precentral (Fig. 8, DR) electrode locations were preceded by somewhat smaller positive waves which started about 2 sec before the 2.5-s pretrial criterion period. This precriterion positivity was found in the recordings for every FN and MN monkey. Furthermore, the pretrial SPSs appeared localized, because concomitant SPSs were not seen in the contralateral (right) prefrontal (Figs. 7 and 8) or precentral recordings nor in the ECGs from the other ipsilateral (left) electrode locations. The pretrial negative peak was followed by the evoked response to onset of the cue light and then a second negative wave could be

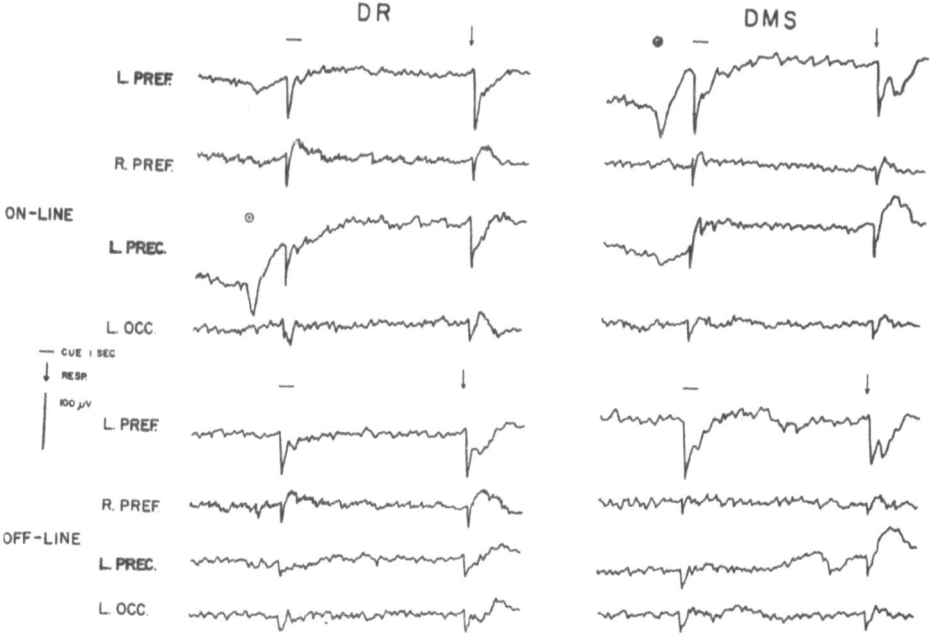

Fig. 8. Averaged electrocorticograms for monkey B316 during performance of 12-s DR (left traces) and 12-s DMS (right traces) tasks. The pretrial SPS requirement was from left precentral for DR and from left principalis cortex for DMS. The lower traces were obtained during off-line testing, with constant intertrial intervals. Each average is or 98 trials of a single testing session. Calibrations are as indicated, with upward deflections indicating surface negativity. The recordings are for left (L) and right (R) principalis (PREF), precentral, and occipital electrode locations. Horizontal lines indicate cue presentation, the circles the start of the 2.5-s pretrial criterion period, and arrows the choice response period.

identified during the delay period. This SPS generally declined toward baseline level (Fig. 7), but for a few monkeys it remained at considerable negativity until the choice response (Fig. 8). The ECGs for FB monkeys (not shown) indicate unusual steady baseline levels from left principalis electrodes for at least 10-s before cue onset, while the concurrent ECGs from the other electrode locations show small fluctuations from baseline levels. The ECG recordings obtained during off-line testing with constant intertrial intervals (Fig. 8) show near baseline SP levels at cue onset from all electrode locations.

The EOG recordings for FN monkeys (Fig. 7A) indicate no pronounced eye deviations, except for orientations toward the relevant cue and response lights. The EOGs for LEM monkeys (Fig. 7B) show the large expected pretrial rightward eye deviations, which were preceded by smaller deviations to the left. The EOGs indicate subsequent gradual eye movements toward baseline level for right DR cues and rapid large leftward eye deviations for left cues, with baseline crossing approximately 0.3 s after cue onset. Comparisons of the time courses of EOG changes and SPSs (Fig. 7) revealed no obvious relationships between these variables.

The rapid learning by the FN monkeys, as well as the brief time course of their pretrial surface-negative waves, would support the attentive interpretation for prefrontal functions. This explanation was examined more directly with the following testing procedure. The monkeys were retested under on-line conditions on 12-s DR with 1-s cue presentations. For the subsequent two sessions with on-line testing the cues were of 0.5 s, 0.2 s, or 0.1 s duration, presented in pseudo-random order,

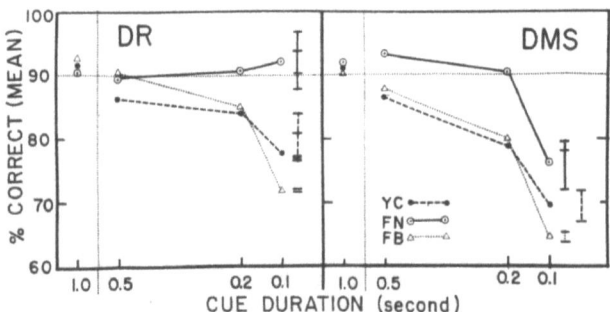

Fig. 9. Correct performance on the 12-s DR and 12-s DMS tasks with cues of 0.5 and 0.2, and 0.1 durations. The differing cue durations were presented randomly during two sessions of 98 trials each. The performance scores with 1.0-s cues were obtained during the preceding session. The vertical lines at the right of the graphs present individual scores. Designations of monkey groups are the same as for Fig. 6

so each duration occurred 16 times in the right and left display windows in a session. The results (Fig. 9, DR) show complete transfer of correct performance by the FN monkeys, whose scores with the 0.1 s cues ranged from 87.5 to 96.9% correct, with a group mean of 92.2%. By contrast, the correct performance by the YC and FB monkeys decreased with shorter cue durations and was significantly ($Ps < 0.01$) below 90% correct with 0.1 s cues. With 0.1 sec cues the means of 77.7% correct for the YC (range of 75.0 to 84.4%) and 72% for the FB groups were significantly ($Ps < 0.005$) below that of the FN group. The scores for one MN monkey (not shown) corresponded to that of the YC group, while the other MN responded below 64% correct for all cue durations. The LEM group (not shown) responded at 93.8% correct with 0.5 s cues, but performed poorly with shorter cues. This performance drop, to 62.5% correct, seems to be the consequence of these monkeys' difficulties in detecting the left cue, as indicated by the time course of their EOGs (Fig. 7B).

The attentive hypothesis for prefrontal functions was examined by subsequent training of the monkeys on a visual DMS task, with the same pretrial requirements as for DR acquisitions. The sample and choice patterns were four different colored disks (red, blue, yellow, white), presented in pseudo-random order. The training procedure was the same as that for the DR task and most monkeys were tested with their DR pretrial requirements. Because of the loss of two FN monkeys, one MN and the additional monkey were retested on 12-s DR and then added to the FN group for DMS training. Also, one YC and one LEM monkey were not trained on DMS.

The results (Fig. 6) indicate no pronounced facilitation in DMS acquisition by the FN group. The total mean errors for on-line testing to 12-s DMS were 188.5 for the FN, 289 for the YC, and 391 for the FB group. The considerable overlap of individual scores by monkeys in the FN (141 to 259 errors) and YC (86 to 419 errors), so that groups resulted in insignificant ($P > 0.01$) difference. The FB monkeys' errors (374 and 408) were significantly greater ($P < 0.01$) than those for the FN group. Corresponding results were obtained for the on-line training sessions to criterion performance, with ranges of 10 to 13 for the FN, 9 to 23 for the YC, and 17 to 22 for the FB group. The LEM monkey made 142 errors, while the MN monkey failed to attain criterion performance on 8-s DMS. Testing under the off-line conditions also showed high correct performance, with individual transfer scores from on-line sessions of 94% to 100%. The ECGs obtained during 12-s DMS testing (Fig. 8, DMS) corresponded to those obtained with the DR task. The results for DMS testing with cues of short durations differed markedly from those obtained for the DR task (Fig. 9). While the FN group responded above 90% correct with 0.5-s and 0.2-s cues, its performance dropped significantly ($P < 0.005$) with 0.1-s cues (71.9 to 79.7% correct). The YC and FB monkeys' scores were, for unexplained reasons, below those of the FN group for all cue durations, with significant group differences ($Ps < 0.05$) for 0.5-s and 0.2-s cues, but not for 0.1-s cues. The drop in correct performance between 0.5-s and 0.1-s cues was 17.66% for the FN, 17.15% for the YC, and 23.4% for the FB group.

The present results also permit comparisons of acquisition scores between the DR and DMS tasks. The two tasks seem of comparable difficulty, because the YC group's mean total errors were only 19 fewer (6.2%) for DMS than DR. By contrast, the FN group made 135 more errors on DMS than on DR, or 3.54 times as many. The somewhat faster DMS acquisition by the FN, than the other groups may be related to two variables: 1) The initial DR training by all monkeys might have affected their subsequent DMS acquisitions. This intertask transfer effect was supported by correlations of error scores between DR and DMS acquisition of $r = 0.768$ ($P < 0.005$) for the 12 monkeys. 2) In view of our previous findings of the importance of the DMS matching epoch for correct performance, we hypothesized that greater SP negativity at the end of the intratrial delay might be related to faster task acquisition. This hypothesis could not be systematically evaluated, because adequate ECGs were available only for 12-sec DMS of six monkeys. However, these recordings from left principalis cortex indicated negative SP levels at the end of the delay of 45.0 do 83.3 μV for three FN and of −6.3 to 22.0 μV for one YC and two FB monkeys. The consequent correlation between SP negativity and total DMS errors of $r = -0.772$ ($P < 0.025$) suggests a confirmation of this hypothesis.

IV. A SOLUTION TO THE RIDDLE OF FRONTAL LOBE FUNCTIONS?

First, I would like to comment on the monkeys' elicitation of their pretrial ECG requirements. The criteria for negative SPSs of 50 to 100 μV amplitude seemed difficult, because the SP fluctuations in the monkeys' ECG traces were generally of considerably smaller amplitudes. Yet, all monkeys met their pretrial requirements and the averaged ECGs showed large negative SPSs, that were preceded by smaller positive waves. Also, the pretrial criterion for near-baseline SP levels was expressed by flat ECG traces, that did not contain the usual small SP fluctuations. A second ECG finding was the cortical localization of the SP requirements, because the recordings from the other ipsilateral and contralateral electrode locations showed no concomitant SP changes. Also, we found no evidence for behavioral mediating responses, including eye deviations, that the monkeys might have used for elicitation of the SP requirements.

The most impressive finding is the rapid DR learning by the monkeys with pretrial SPSs from principalis cortex (Group FN). This group's acquisition rate, as indicated by total error scores, was 5.8 times faster than that of the yoked controls (Group YC) and each of the FN monkeys learned the task faster than did ten of the eleven comparison monkeys. A facilitatory effect on behavior of this magnitude and consistency has, to my knowledge, not been reported with either animal or human experimentation.

How can we explain the rapid DR acquisition by the FN monkeys? It seemed possible that the monkeys utilized the pretrial events as readiness signals for the approaching trial. This explanation was examined by training one group of monkeys with the motoric pretrial requirement of lateral eye deviations, that might have served as behavioral readiness responses. However, these monkeys' lack of facilitated task acquisition, as well as the independence between lateral eye movements and prefrontal SPSs, did not support this interpretation. Also, rapid DR learning occurred only with pretrial negative SPSs from prefrontal cortex, but not with similar precentral SPSs, or with pretrial baseline SPs from principalis cortex. The latter requirement seemed to have deleterious effects on learning, because it resulted in the slowest acquisition rates.

Consequently, the rapid learning effect seemed related to the time course of the pretrial SPS. This wave attained maximum negativity at cue onset and then declined gradually, but was still substantially above baseline level at the start of the delay period. This time course corresponded in general to that of the endogenous SPS that was found during DR performance by other monkeys (29) and also to discharge rates of prefrontal units. The interpretation of prefrontal cortical activation as an expression of a heightened attentive state (3) would suggest that the monkey's rapid task acquisition was the consequence of their attention to the cue. This explanation, however, seems inadequate in several respects. Most important is the lack of corresponding rapid learning of DMS, a task that certainly demands the monkey's attention to the perceptual features of the visual sample for its encoding in short-term memory. Also, this explanation does not account for the mnemonic functions of principalis cortex, because monkeys with ablations of this cortical segment have been found unimpaired in the performance of difficult tasks that involve spatial discontiguities, such as conditional positional responses (4). Finally, the attentive hypothesis provides no explanation for the specific demands that the monkeys must meet during cue presentation.

What is required of the monkey during cue presentation? Our analyses of the demands of the DR and DMS tasks, based on earlier findings with electrocortical stimulation, suggested that the crucial requirement for both tasks is the selection of the choice response. For DR this process seems to occur toward the end of cue presentation and would lead to the encoding of the spatial memory. The maintenance of this memory throughout the delay, while important and reflected by continuing neuronal discharges and generally above-baseline SP negativity, seems less demanding. Therefore, we infer that the monkey's attention during cue presentation is directed toward the selection of the subsequent choice response. Accordingly, the major function of principalis cortex would be in the selection of, or *in the programming of delayed spatial choice responses*. While emphasizing this specific function of principalis cortex, we recognize that this structure, and adjacent prefrontal zones, are also implicated in other required processes, such as inhibition of interfering response tendencies, as elicited by environmental and bodily stimuli, recognition of the spatial relations between the choice stimuli, and execution of the instrumental response.

Our interpretation for the role of principalis cortex involves processes of attention, spatial memory, and motoric interactions. Thus, each of the earlier explanations for the task impairments by prefrontally resectioned monkeys contributes to this interpretation. However, the crucial relations among these processes have only been revealed by monkeys with intact brains. The present methods for utilizing endogenous neuronal phenomena as independent variables in task acquisition are a promising technique for determinations of brain functions in behavior.

Furthermore, our view of the monkey's prefrontal functions seems consonant with interpretations that have been derived from investigations of human brain-damaged patients (6). The frontal lobes, according to Luria (14), regulate the active state of the subject, control the essential elements of his intentions, and are essential to the programming and execution of complex sequences of actions. Thus, the subject's attention must be directed and maintained toward every element of a complex sequence of task demands: anticipation and recognition of the relevant stimulus, selection and execution of the appropriate response, and control over impulsive tendencies.

In conclusion, there has been substantial progress in our understanding of the remarkable phenomenon that was described by Jacobson over 40 years ago. We have found that the prefrontal cortex is necessary for the planning and execution of specific future acts and it is indeed the substrate for the highest cortical functions in monkeys.

I want to acknowledge the substantial contributions by my collaborators, O. Gillespie, S. Rosen, and B. Sandrew, who conducted most of the experiments. My understanding of the facts and concepts of brain functions has been influenced by many colleagues, but especially by Jerzy Konorski, whose writings and discussions with me have been a major contribution to my scientific growth. This research was supported by grants from the National Science Foundation and preparation of this manuscript by NIMH Grant No. 1RO1MH3102701.

## REFERENCES

1. COHEN, S. M. 1972. Electrical stimulation of cortical-caudate pairs during delayed succesive visual discrimination in monkeys. Acta Neurobiol. Exp. 32: 211-233.
2. FINAN, J. L. 1939. Effects of frontal lobe lesions on temporally organized behavior in monkeys. J. Neurophysiol. 2: 208-226.
3. FUSTER, J. M. 1973. Unit activity in prefrontal cortex during delayed-response performance: Neuronal correlates of transient memory. J. Neurophysiol. 36: 61-78.
4. GOLDMAN, P. S. and ROSVOLD, H. E. 1970. Localization of function within the dorsolateral prefrontal cortex of the Rhesus monkey. Exp. Neurol. 27: 291-304.
5. GOLDMAN, P. S., ROSVOLD, H., VEST, B. and GLAKIN, T. 1971. Analysis of the delayed-alternation deficit produced by dorsolateral prefrontal lesions in the Rhesus monkey. J. Comp. Physiol. Psychol. 17: 212-220.
6. HÉCAEN, H. and ALBERT, M. L. 1978. Human Neuropsychology. John Wiley and Sons, New York.
7. IVERSEN, S. D. and MISHKIN, M. 1970. Perseverative interference in monkeys following selective lesions of the inferior prefrontal convexity. Exp. Brain Res. 11: 375-386.
8. JACOBSON, C. F. 1936. Studies of cerebral function in primates. Comp. Psychol. Monogr. 13: 1-68.
9. KONORSKI, J. 1967. Integrative activity of the brain. An interdisciplinary approach. Univ. Chicago Press, Chicago, 531 p.
10. KONORSKI, J. 1972. Some hypotheses concerning the functional organization of prefrontal cortex. Acta Neurobiol. Exp. 32: 595-613.
11. KOVNER, R. and STAMM, J. S. 1972. Disruption of short-term visual memory by electrical stimulation of inferotemporal cortex. J. Comp. Physiol. Psychol. 81: 163-172.
12. KUBOTA, K., IWAMOTO, I. and SUZUKI, H. 1974. Visuokinetic activities of primate prefrontal neurons during delayed-response performance. J. Neurophysiol. 37: 1197-1212.
13. ŁAWICKA, W. 1969. A proposed mechanism for delayed response impairment in prefrontal animals. Acta Biol. Exp. 29: 401-414.
14. LURIA, A. R. 1966. Higher cortical function in man. Basic Books, New York.
15. MALMO, R. B. 1942. Interference factors in delayed response in monkeys after removal of frontal lobes. J. Neurophysiol. 5: 295-308.
16. NIKI, H. 1974. Differential activity of prefrontal units during right and left delayed response trials. Brain Res. 70: 346-349.

17. NIKI, H. 1974. Prefrontal unit activity during delayed alternation in the monkey, II. Relation to absolute versus relative direction of response. Brain Res. 68: 197–204.
18. NISSEN, H. W., RIESEN, A. H. and NOWLES, V. 1938. Delayed response and discrimination learning by chimpanzees. J. Comp. Psychol. 26: 231–386.
19. PRIBRAM, K. H. 1950. Some physical and pharmacological factors affecting delayed response performance of baboons following frontal lobotomy. J. Neurophysiol. 13: 373–382.
20. ROBINSON, A. D. and FUCHS, A. F. 1969. Eye movements evoked by stimulation of the frontal eye field. J. Neurophysiol. 32: 637–648.
21. RUSINOV, V. S. 1953. An electrophysiological analysis of the connecting function in the cerebral cortex in the presence of a dominant area. Prox. XIX Int. Physiol. Congr. (Montreal), p. 719–720.
22. SANDREW, B. B., STAMM, J. S. and ROSEN, S. C. 1977. Steady potential shifts and facilitated learning of delayed response in monkeys. Exp. Neurol. 55: 43–55.
23. STAMM, J. S. 1963. Function of prefrontal cortex in timing behavior of monkeys. Exp. Neurol. 7: 87–97.
24. STAMM, J. S. 1964. Retardation and facilitation in learning by stimulation of frontal cortex in monkeys. In: J. M. Warren and K. Akert (ed.), The frontal granular cortex and behavior. McGraw Hill, New York, p. 102–125.
25. STAMM, J. S. 1969. Electrical stimulation of monkeys' prefrontal cortex during delayed-response performance. J. Comp. Physiol. Psychol. 67: 535–546.
26. STAMM, J. S. 1970. Dorsolateral frontal ablations and response processes in monkeys. J. Comp. Physiol. Psychol. 70: 437–447.
27. STAMM, J. S. 1973. Functional dissociation between the inferior and arcuate segments of dorsolateral prefrontal cortex in the monkey. Neuropsychologia 11: 181–190.
28. STAMM, J. S., GADOTTI, A. and ROSEN, S. C. 1975. Interhemispheric functional differences in prefrontal cortex of monkeys. J. Neurobiol. 6: 39–50.
29. STAMM, J. S. and ROSEN, S. C. 1972. Cortical steady potential shifts and anodal polarization during delayed response performance. Acta Neurobiol. Exp. 29: 385–399.
30. STAMM, J. S. and ROSEN, S. C., 1973. The locus and crucial time of implication of prefrontal cortex in the delayed response task. In A. R. Luria and K. H. Pribram (ed.), Frontal lobes and regulation of behavior. Academic Press, New York.
31. STĘPIEŃ I. and STAMM, J. S. 1970. Impairments on locomotor task involving spatial opposition between cue and reward in frontally ablated monkeys. Acta Neurobiol. Exp. 30: 1–12.
32. VORONIN, L. L. 1969. Action of surface polarization on intracellular unit activity in the motor cortex of waking cats. Neurosciences Translations No. 6.
33. WARREN, J. M. and AKERT, K. (ed.) 1964. The frontal granular cortex and behavior. McGraw-Hill, New York.
34. WEGENER, J. G. and STAMM, J. S. 1966. Behavior flexibility and the frontal lobes. Cortex 2: 188–201.

John S. STAMM, Department of Psychology, State University of New York, Stony Brook, N. Y. 11794, USA.

Lecture delivered at the Warsaw Colloquium on Instrumental
Conditioning and Brain Research
May 1979

# PHARMACOLOGICAL EVIDENCE ON THE SPECIALIZATION OF CNS MECHANISMS RESPONSIBLE FOR MOTOR ACT INHIBITION BY AVERSIVE EVENTS

Giorgio BIGNAMI

Section of Psychopharmacology, Laboratorio di Farmacologia
Istituto Superiore di Sanità, Rome, Italy

*Abstract.* Our use of selected pharmacological agents has now extended the range of treatment-behaviour interactions previously studied by other (e.g., lesion) approaches in the search for appropriate models of behaviour organization and underlying physiological mechanisms. The case of muscarinic blockers has special interest, because the drugs induce only one type of primary change. At present, it appears difficult to reconcile the evidence in favour of a motor (perseverative) deficit with other evidence favouring an impairment of sensory processes. However, critical experiments using several go–no go avoidance tasks show that the two deficits may be inseparable. The complex profile of cue-dependent disinhibitory effects of antimuscarinics suggests that separate sensori-motor mechanisms are employed for response suppression not simply as a function of cue type, response type, or response-reinforcement relation but as a joint function of all these factors. Sedative-tranquillizing agents with so-called anti-conflict properties add still another dimension to the problem of motor act inhibition. These agents are maximally effective in disrupting response withholding when both reward and punishment follow the emission of a particular response, less consistently effective in tests with CS paired with non-contingent shock (CER), and mostly ineffective in those go–no go avoidance tasks which show a very high sensitivity to muscarinic blockade.

## INTRODUCTION

It has become increasingly evident that the description of treatment effects and the analysis of drug mechanisms of action often fall short of their projected goals because of an insufficient knowledge of organization of behaviour and underlying physiological-biochemical mechanisms. At the same time, several model drugs with reasonably well-defined profiles can be used as "dissection tools" in the analysis of behavioural phenomena from both a neuropsychological viewpoint and a more strictly physiological-biochemical viewpoint.

In some instances the pharmacological approach has produced evidence which simply confirms (or is complementary to) the evidence obtained by other well-established methods, e.g., CNS lesions. In other instances, however, the drug-behaviour interactions observed have not allowed a direct utilization of the explanatory models or tentative hypotheses developed by other types of neurobehavioral analysis. More explicitly, several "fracture lines" created by the use of drug treatments have led many to question some important assumed processes and/or mechanisms previously handled in unitary terms.

The present discussion will be based on a limited portion of the pharmacological data which show that within a well-accepted category such as "motor act inhibition by aversive events" several sub-categories of hitherto-unexplained significance must be postulated in order to account for the interactions observed with different types of drugs. Two main chapters will be devoted, respectively, to centrally-acting muscarinic blockers and to sedative-tranquillizing agents endowed with so-called anticonflict (antipunishment) properties. Other important model drugs, for example amphetamine and related stimulants and 5-HT system agents, cannot be analyzed within the space limits of this discussion.

## THE CASE OF CENTRAL MUSCARINIC BLOCKERS

### General features of the antimuscarinic syndrome: sensory vs. motor aspects

Treatment of laboratory animals by any one of the centrally acting muscarinic antagonists — i.e., either the naturally occurring alkaloids such as atropine and scopolamine or the synthetic analogues such as benactyzine and ditran — induces a set of behavioral changes which show important analogies with those obtained by extensive septal or hippocampal lesions (for a detailed analysis and references see the recent reviews by 2, 9) [1]. The animals, particularly small rodents, show locomotor hyperactivity and hyperreactivity to stimuli, impairment of habituation, perseverative deficits in a wide variety of alternation and discrimination tasks, and an impaired ability to withhold punished responses. Furthermore, several experiments have shown some peculiar treatment-behaviour interactions which are deemed to be fairly characteristic of limbic system lesions, for example, opposite types of drug effects on one-way active avoidance (unaffected or moderately depressed) and two-way avoidance (markedly facilitated).

In this context, it should be recalled that pharmacological data have often been employed in the dispute concerning whether or not *response disinhibition* or *perseveration* (impairment of motor act inhibition) can be separated from a *motor hyperactivity* syndrome. Some of the analogies and differences between the "disinhibiting" profile of muscarinic blockers and the "hyperactivating" profile of monoaminergic (amphetamine-like) stimulants appear to be in favour of such a distinction (see, e.g., in a later section, the greater selectivity of scopolamine in enhancing responses to a CI-CS complex, relative to the more uniform response enhancements obtained by amphetamine both during CI-CS and during intertrial intervals). This point, however, cannot be analyzed further, since attention should be focussed on other more important aspects of the antimuscarinic syndrome [2].

In the first place it must be emphasized that interference with punishment suppression by antimuscarinic agents appears to be markedly affected by response factors. For example, responses which are highly

---

[1] Prior to more detailed analyses of controversial points, the present discussion makes use of brief summary statements to remind the reader of the best-known aspects of the syndromes examined. Consequently, the work of many investigators who have been active in the fields under discussion cannot be quoted here, but it can easily be located through the review papers cited.

[2] It should also be noted that, in general, the psychopharmacological literature does not attempt to draw a distinction between response enhancement (or disinhibition) which may be ascribed, respectively, to a drive enhancement or disinhibition and to a motor bias (response disinhibition or perseveration, impairment of motor act inhibition). In fact, the contingencies which have been used by neuropsychologists in the attempts to separate deficits of drive and of motor act modulation (particularly go-no go tasks with asymmetrical versus symmetrical reinforcement, i.e., "drive differentiations" versus "response differentiations under the same drive operation") have often been shown to influence the size of drug effects, but rarely to determine qualitative differences in drug action. Consequently, as exemplified by the discussions on antimuscarinic blockers and on so-called anticonflict agents which follow here, critical factors responsible for drug action and possible underlying mechanisms must often be looked for in different directions.

prepotent per se (such as stepping down from an elevated platform, or moving from a large illuminated compartment to a small dark compartment) and/or as a consequence of training which strengthens some pre-existing tendency (e.g., active avoidance conditioning of a locomotor response), once suppressed by a punishment (passive avoidance) contingency, are generally quite sensitive to antimuscarinic disinhibition. Vice-versa, manipulatory responses maintained by alimentary reinforcement and then suppressed either by a Pavlovian CS paired with shock (so-called CER paradigm), or by a CS signalling that response-contingent shock is administered simultaneously with rewards (approach-avoidance conflict), are mostly unaffected by the same treatment. (In reality, acquisition data sometimes show drug effects also in the latter types of tests. However, it is generally acknowledged that such effects are the result of complex interactions between drug-induced changes of response topographies, mild state dependence, and artifacts due to modification of unpunished baselines; for discussion see the reviews quoted above).

A second important point comes from the drug results obtained in several discrimination and alternation tasks using mainly manipulatory (bar-pressing) responses and a wide variety of interoceptive and exteroceptive stimuli and stimulus combinations. These conditions do not create room for a forceful reappearance of highly prepotent responses suppressed by training, while sophisticated analyses of the relative frequencies of omission and commission errors can show that drug effects can be interpreted mainly as a discrimination (i.e., sensory-perceptual) deficit (24–26, 35, 46, 47).

*Interactions between sensory and motor components of the antimuscarinic syndrome*

Setting aside some evidence of little relevance for the present discussion — e.g., the data suggesting that the motor component of the antimuscarinic deficit may consist of a perseveration of response tendencies (perseveration of set), rather than perseveration of responses per se (10) — one must deal now with the evidence which attempts to reconcile what appear to be separate (i.e., motor and sensory) components of the antimuscarinic syndrome. An experiment carried out several years ago in the Rome laboratory, for example, compared the effects of scopolamine on rats trained in one or the other of several active-passive avoidance discriminations in a shuttle-box with different types of cue arrangements (38). Some of the results were compatible with any one of the two interpretations of the drug profile, since tasks using simple stimuli of different modalities (light-go, noise-no go, or vice versa) were learned with little responding to the passive avoidance signal after active avoidance pretraining, and did not show a passive avoidance disruption after drug treatment.

Unexpected differences, however, were found when the highly sensitive discrimination using a simple active avoidance cue and a compound passive avoidance cue, made up of the same active avoidance signal and another stimulus of a different modality (light-go, noise/light-no go), was compared with the task using a symmetrically opposite arrangement of cues (noise-go, light/noise-no go). In fact, the latter discrimination was much easier to learn (see also 21, 39) and almost insensitive to scopolamine treatments. In other words, neither response factors per se nor cue factors per se, but an interaction between the two (depending on which signal had to be used as response activator and which one as a suppressor in the presence of the former) determined whether or not a marked disinhibition would appear after treatment.

At this point, i.e., in the early seventies, a logical step was to go back to differentiation (active avoidance-extinction) tasks to check on the interactions between treatment, cue and response factors in the absence of punishment of active avoidance responses to no go signals. In fact, several data from the Nencki Institute laboratory in Warsaw were available, pointing out (i) a separation of signalling systems used respectively for drive inhibition in differentiation tasks with asymmetrical reinforcement and for instrumental response withholding in the face of drive activation in go–no go discriminations with symmetrical reinforcement, and (ii) differential frontal lesion effects as a joint function of cue and response factors (for discussion and references see 29, 30, 48).

Both go–no go avoidance differentiations with a simple active avoidance CS and compound nonreinforced cues (i.e., the active avoidance CS preceded for 1 s and accompanied for its whole duration by a CI of the other modality, with a light CS and a noise CI in one task, and viceversa in the other) required very extensive training before a satisfactory baseline of differential responding could be established. The differences between the tasks using, respectively, light or noise as the active avoidance cue were in the same direction as in the corresponding active-passive avoidance tasks so far as acquisition of differentiation was concerned (22), but the differences in scopolamine disinhibition disappeared (20). In other words, the results were partly against the theory of separate signalling systems as a function of symmetry or asymmetry of reinforcement (acquisition data) and partly in favour of such a theory (drug data). (It could also be argued, however, that the absence of difference in sensitivity to scopolamine was an artifact. In fact, the lowered response cost in the absence of punishment for responses to no go signals may have caused a marked overall increase in the drug disinhibition, which in turn might have eliminated the differences between the two tasks *via* the so-called ceiling effect).

The data on differentiations with asymmetrical reinforcement, while incapable of showing an interaction between cue arrangement and drug treatment, nevertheless pointed out another type of treatment-cue interaction, which, however, requires a digression on a particular aspect of the antimuscarinic profile. In fact, these agents can enhance not only responses to discrete stimuli, but also intertrial responses. In two-way active avoidance acquisition this reflects itself in a much greater difference between active avoidance responses to the CS and intertrial responses in the drug state when the latter are punished, than when they are not punished (5). This is easily explained since intertrial response punishment creates an easy active-passive avoidance task with extreme differences between the go and no go cues (discrete exteroceptive stimulus as active avoidance CS1 and absence of the same stimulus as passive avoidance CS2). In the drug-sensitive active-passive avoidance task (light-go, noise/light-no go) intertrial responding varies considerably between animals, since at one extreme one can find treated rats which show only increase of commission errors during cue presentation, while at the opposite extreme there are rats which show high frequencies both of commission errors and of intertrial responses.

As concerns the tasks with asymmetrical reinforcement, intertrial responding was enhanced by scopolamine in both situations, but to a much lesser extent than responding in the presence of the no go stimulus complex. A comparison with amphetamine indicated that this should be considered a genuine cue-dependence of at least a substantial portion of the antimuscarinic disinhibition, since the differences in hyperresponding in the presence and in the absence of cues were much less marked in the case of the monoaminergic stimulant, which apparently causes a genuine overall hyperactivity (see, however, the discussion in, 20).

Several remarks can be made on the basis of the treatment-behavior interactions summarized above. For example, the situation appears to be much more complex than that revealed by the analysis of medial and lateral frontal lesion effects on go—no go avoidance differentiations and discriminations with different cues in dogs (16). In fact, the latter data suggest a single type of mechanism mediating the suppression of responses generalized during training from a go to a no go signal; and this, independently of whether reinforcement is asymmetrical (differentiation between an active avoidance CS and a CS—) or symmetrical (discrimination between active avoidance CS1 and passive avoidance CS2). In other words, response suppression in the face of nonreinforcement, presumably because of the punishing properties of CS⁻ nontermination by the avoidance response during generalization, appears to be served in dogs by a mechanism similar to that underlying response suppression by punishment. The only major source of variance remains that stemming from presence vs. absence of generalization from go to no go cue during training, independently from symmetry or asymmetry of reinforcement. (On this point, of course, dog-lesion and rat-drug data are in agreement, but the respective acquisition data appeared to be sufficient to settle the question before the introduction of treatment fracture lines).

In addition to the last-mentioned factor, scopolamine tests in rats showed two unexpected differences, namely, (i) a marked variation of the treatment effect as a function of compound cue arrangement in active-passive avoidance tasks, and (ii) a disappearance of such difference in tasks with asymmetrical reinforcement, in spite of a maintenance of the differences in task difficulty (but see above the caveat concerning a possible ceiling effect).

A somewhat tautological conclusion must be that rats make use of several different mechanisms to achieve response suppression in locomotor go-no go avoidance in the face of different combinations of cues and response-reinforcement relations. Although it is not possible to venture beyond this generic statement at the present state of knowledge, it must be emphasized that there are other lines of evidence which provide independent support of this conclusion. For example, it has been shown that scopolamine disinhibition is maintained for an extended series of treatment sessions in tasks with asymmetrical reinforcement, which excludes a genuine tolerance phenomenon (20), while it can be compensated by relearning in the drug state — although with considerable variation from one animal to another — in the face of punishment of hyperresponding in the more difficult active-passive avoidance task (8, 12.)

Furthermore, compensation for the scopolamine deficit by relearning does not prevent a frontal pole lesion from exerting its disruptive effect, consisting mainly of a passive avoidance impairment similar to that induced by the drug. When the frontal lesion effect is also compensated by relearning, however, further challenges by scopolamine show a much greater re-increase of the sensitivity to drug in the lesioned rats than in appropriate controls (sham-operated after desensitization to scopolamine). This obviously suggests not only that the drug and the frontal lesion disrupt different mechanisms, but also that the system affected by the lesion has a role in the relearning to withhold punished responses in the drug state (6).

## THE CASE OF AGENTS WITH SEDATIVE-TRANQUILLIZING (ANXIOLYTIC) PROPERTIES
### General features of the sedative-tranquillizer syndrome

A wide variety of sedative agents with quite different chemical structures and important areas of non-overlap of the respective pharmacological profiles have one important pharmacological property in common, namely, the attenuation of behavioural suppression in several conflict tasks using simultaneous reward and punishment of the same response. This applies to ethanol, to hypnotic-sedatives of the barbiturate type or related to barbiturates, and to minor tranquillizers (so-called anxiolytics) such as meprobamate and several benzodiazepine derivatives.

The impairment of a particular type of passive avoidance, which appears at doses lower than those which cause sedation and/or gross motor incoordination, contrasts with an absence of a comparable depression of active avoidance, which is affected only after neurotoxic doses of the same agents (for the non-generality of the passive avoidance deficit see later). Furthermore, this drug profile is symmetrically opposite to that observed with sedatives of the neuroleptic type employed as antipsychotics (mainly phenothiazine and butirophenone derivatives), which interfere with active avoidance fairly selectively, but do not induce a response disinhibition in conflict tasks. (For a more exhaustive description of the effects of anti-conflict agents and for references see 3, 4, 13–15, 17, 18, 23, 27, 40, 41).

With more space available, several methodological problems should be discussed at this point in order to avoid overconfidence in the simplified picture provided above. For example, it has been known for several years that the disinhibition of responses which are both rewarded and punished can become more and more marked with repeated treatment, since an initial response-depressant component of the drug action seems to be attenuated by a selective tolerance which allows a progressive magnification of the disinhibiting component (11, 13–15, 31, 40). It appears more interesting, however, to analyze here the pros and the contras of various tentative models which have been proposed in order to account for the "anticonflict" syndrome. In fact, this allows us to pose several questions about the mechanisms responsible for motor act inhibition.

In the first place, there is good agreement that general sensory and associative deficits cannot account for the antipunishment action of the drugs, since treatment consequences can vary widely in situations which should yield homogeneous results if the aforementioned deficits were responsible for the changes observed (e.g., in the reviews quoted above several comparisons can be found between tasks with positive and with negative reinforcement, and so forth). An exception should perhaps be made for certain aspects of the treatment effect in acquisition, suggesting mild associative and/or state-dependence deficits depending on the circumstances, in addition to the main "anticonflict" action of the drugs (see 1, 19, 37).

In the second place, it has already been mentioned that the drugs do not exert comparable effects in the face of different response requirements with the same reinforcement (active vs. passive avoidance). This obviously excludes a general effect on those common mechanisms responsible for fear conditioning which are assumed to underlie learning and performance of a wide range of avoidance tasks with different, or even opposite response requirements (e.g., locomotion versus withholding of locomotion).

At this point, it must also be considered that the drug profiles, although showing wide areas of non-overlap with respect to those of amphetamine-like stimulants and those of muscarinic antagonists, nevertheless include a fairly wide range of disinhibitory changes such as impairment of some types of extinction and habituation, reduction of spontaneous alternation, impairment of discriminations which require response withholding, and facilitation of certain types of bar-press and two-way locomotor avoidance (see the reviews quoted above).

Two different interpretations have been provided for this profile of drug effects. The first one postulates an overall impairment of motor act inhibition, presumably due to the well-documented effects of the drugs on limbic mechanisms, and broadly comparable to that obtained by antimuscarinics, although obviously due to a different type of primary drug effect (see particularly the discussion by Iwahara, 27). The second one favours a more selective interference with those limbic mechanisms which are assumed to trigger several behavioural changes upon exposure to stimuli previously associated with a painful experience or with stress resulting from non-reception of an anticipated reward (18, 23; this model, incidentally, excludes a drug action on unconditioned responses to painful and stressful stimuli and on immediate consequences of "frustration" and therefore relies on drug effects on classical conditioning mechanisms).

Both models, however, have considerable difficulties explaining at least an important portion of the available results. Unqualified disinhibition hypotheses, for example, cannot account for drug interference with response enhancements such as the so-called partial reinforcement extinction effect (retardation of extinction as a consequence of intermittent reinforcement during training with an alimentary reward). The second of the models mentioned above has difficulties in deciding between (i) treatment effects on *overall behavioral changes* induced by stimuli associated with painful events (or nonreward), an (ii) treatment effects limited to the *behavioural inhibition* induced by the same CS. In fact, in the former case the model could account for drug interference with the partial reinforcement extinction effect, but not with the lack of drug effects on active avoidance (at least at nontoxic doses endowed with an "anticonflict" action). In the latter case it could account for selective passive avoidance changes not paralleled by active avoidance changes, but not for the abatement of response enhancements making up the partial reinforcement extinction effect. [3]

*Non-generality of the antipunishment action of sedative-tranquillizers*

A closer analysis of the available data indicates that the contradictions are, in reality, much more serious than can be inferred from the examples provided above. In the first place, experiments using response suppression by a CS paired with unavoidable shock (CER paradigm) have yielded inconsistent results, as compared with the fairly constant drug disinhibition obtained in approach avoidance tasks (see reviews 3, 4, 28, 36), or even paradoxical results consisting of a drug enhancement of suppression (44). Attempts have been made to cope with these discrepancies by further limitations of the working models; for example, by making recourse to a separation of mechanisms responsible for general suppression by a fear CS from those responsible for suppression of a specified act previously punished in the presence of the CS itself, and by postulating a drug action limited to the latter type of mechanism. (Obviously this distinction, when translated into a conditioning model, is tantamount to making recourse to a selective action on the processing of different stimuli which can become fear CS — exteroceptive versus

---

[3] The effects of sedative-tranquillizers on nonreinforcement and partial reinforcement consequences should obviously be re-examined in the light of alternative explanations of the "frustration" syndrome, particularly changes in the relative proportions of consummatory and drive CR's (43). To the best of the author's knowledge, however, no experiments have been carried out with drugs to check specifically on this point. Therefore, since the emphasis of the present discussion is mainly on other aspects of the sedative-tranquillizer syndrome this argument cannot be pursued further. It can only be added here that the drugs enhance alimentary responses of the consummatory type without, apparently, interfering with instrumental outputs — which obviously suggests that an investigation of treatment effects on mechanisms such as those analyzed by Sołtysik and Gasanova (43) and by Sołtysik (42) may well constitute a valid alternative to the approach so far used.

response-produced — and/or at the level of the outputs which are suppressed upon activation of the fear drive by the CS — all motor behaviour versus a specified response — again excluding any effect on fear activation per se).

With this modified model the assumption should be that CER tasks leave variable room for so-called adventitious punishment (noncontingent punishment "perceived" as if it were response-contingent). This is not unreasonable, due to variable duration of CS-US intervals in different experiments and to differences in the procedures ("off the baseline" vs. "on the baseline", with much greater room for adventitious punishment in the latter condition) hence a possible source of variation, should the treatments really interfere with the suppression of specified responses, rather than with overall behavioural suppression. The available data, however, appear to be insufficient to evaluate the merits of this tentative explanation relative to others. For example, insufficient attention has been given to possible paradoxical effects of an interference with inhibition of delay in cases in which extended training with a long CS-US interval allows a recovery of responding in the early portions of the intervals prior to treatment administration (32). This phenomenon per se could account for a wide range of net effects on averaged rates during the pre-shock CS, depending on CS-US interval duration, CS modality and intensity, stage of training, and several other factors which influence the profile of the suppression obtained as baseline.

Still other data emphasize the peculiar profile of drug effects on responses suppressed by punishment, thereby leading to a more and more pessimistic view of the highly diversified mechanisms responsible for motor act inhibition. For example, it has recently been shown that two versions of a step-down passive avoidance test which are equally sensitive to antimuscarinic (atropine) treatment (which indicates that they must have at least some important element in common) are differentially sensitive to tranquillizer (chlordiazepoxide) treatment. In fact, the latter drug induced a considerable impairment of passive avoidance acquisition when rats were given only one trial per day and taken out of the apparatus as soon as they stepped down and took shock. Vice versa, chlordiazepoxide-treated rats did not differ from controls when tested continuously for 15 min in a single session in which repeated shock received for stepping down in the initial portion of the test led to a fast (within-session) acquisition of response withholding (45).

Similar problems arose in the Rome laboratory when using the two active-passive avoidance tasks which had shown, respectively, a marked and a slight disruption of passive avoidance after muscarinic blockade (see a previous section in this discussion). In fact, the experiments so far reported only in abstract form (7) showed quite limited chlordiazepoxide effects even in the task highly sensitive to scopolamine (see Fig. 1). (Incidentally, the chlordiazepoxide disinhibition was much less than that obtained with scopolamine or amphetamine in the tests on rats

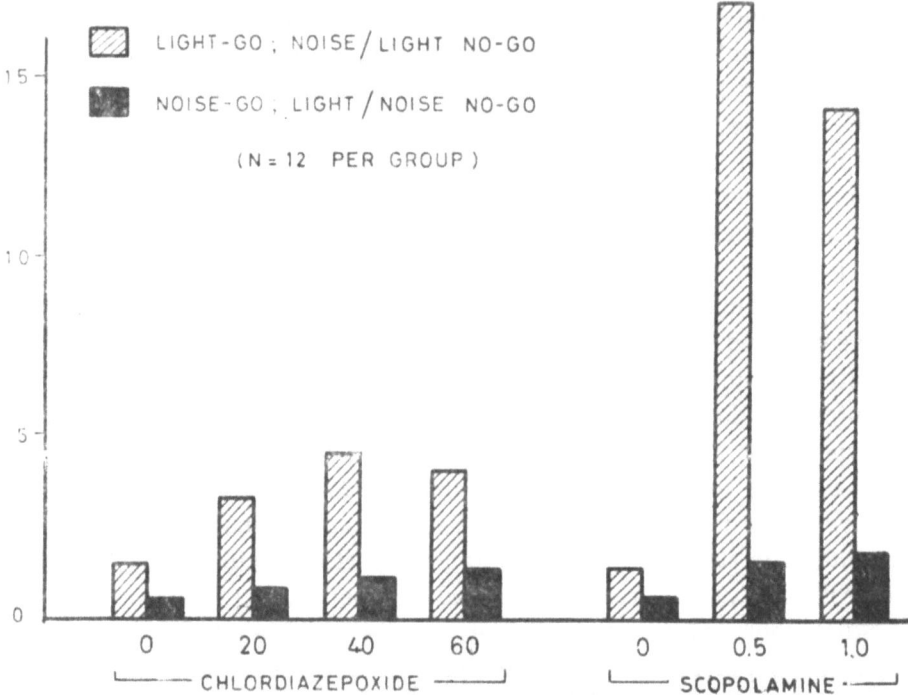

Fig. 1. Passive avoidance changes induced by a benzodiazepine tranquillizer (chlordiazepoxide) and a centrally active muscarinic blocker (scopolamine) in rats performing go–no go avoidance with different cue arrangements. The characteristics of the two tasks are described in Rosić and Bignami (38). The bars indicate increases of responses to passive avoidance signals (commission errors) above control baselines in sessions including 25 active avoidance and 25 passive avoidance trials. Changes in active avoidance performance were absent or negligible and are not indicated in the graph. Drug doses are in mg/kg given subcutaneously 30 min before testing. Data from experiments by Bignami and Chesher (7), so far reported only in abstract form.

trained with asymmetrical reinforcement contingencies; see Frontali et al. 20). This excludes the interpretation that contingent shock employed to suppress a particular prepotent response — in this case, a locomotor act originally trained as active avoidance response, instead of a locomotor or a manipulatory act trained as an alimentary response — is the main factor allowing a tranquillizer disinhibition. Experiments conducted by McKearney (34) in squirrel monkeys to compare punishment of rewarded responses and punishment of responses trained to terminate a CS associated with shock appear to have a similar meaning, since a barbiturate derivative (pentobarbital) had disinhibiting effects in the former paradigm while further enhancing punishment suppression in the latter.

The problem raised by these data, however, is more complex than can be inferred from averaged results such as those on go–no go avoidance reported above. In fact, the slight chlordiazepoxide increase of punished responses was the consequence of a marked variation between experimental animals. Most rats showed no drug disinhibition whatsoever, other showed a moderate loss of response control, while an occasional animal showed a complete disruption of passive avoidance. After the assessment of the acute effects of increasing doses of the tranquillizer, and after checking for sensitivity to scopolamine which appeared to be quite uniform several rats were also employed for repeated treatment studies, based on the aforementioned findings of a progressive enhancement of the disinhibiting effects with repeated drug exposure in conflict tasks. The picture obtained was somewhat puzzling, since some rats exhibited a progressive increase of passive avoidance disruption, others, on the contrary, tended to show desensitization, while occasionally both phenomena appeared in the same animal, with a gradual shift from an absence of treatment consequences to a marked disinhibition, and then a gradual return to control baseline in spite of continued treatment.

It should be pointed out that this marked variability of drug action cannot be ascribed to an *overall* variation in sensitivity to drug. In fact, from a certain threshold dose upwards it is quite easy to recognize the consequences of benzodiazepine treatment, consisting of a marked muscular relaxation and passivity to manipulation, without, however, a marked sedation, nor the catatonia which characterizes neuroleptic treatments. This syndrome was regularly observed in treated animals, independently of duration of exposure and of presence or absence of passive avoidance disruption; therefore, neither a genuine lack of sensitivity nor a genuine tolerance can be invoked to account for variation of drug effects on passive avoidance both within and between subjects.

The variability of the tranquillizer effect in a go–no go task which allows a fairly homogenous antimuscarinic effect indicates that even in this particular situation response suppression is served by multiple mechanisms which are partially separated either "upstream" or "downstream" relative to that (or those) influenced by muscarinic blockade. This statement has a strong tautological flavour, while at the same time the whole situation appears to pose difficult problems, since each of the sub-categories of motor act inhibition by aversive events resulting from challenges with a particular agent appears to be further split into sub-

sub-categories as a consequence of challenges with other agents. Therefore, the question is whether or not it will be possible in the future to design additional variants of the tasks allowing us to obtain more homogeneous tranquillizer effects in either direction (complete disappearance of the effect on the one side versus uniform sensitivity on the other). This appears to be essential in order to achieve a better understanding of the different mechanisms responsible for motor act inhibition, which in turn can break the ground for a more rigorous physiological analysis.

## GENERAL DISCUSSION

The data on antimuscarinics and sedative-tranquillizers discussed in previous sections are insufficient to propose any satisfactory working model of the diversity and specialization of CNS mechanisms responsible for motor act inhibition by aversive events. In other words, this evidence is barely sufficient to support a general statement which denies, much more forcefully than the comparable neuropsychological evidence, the unitary character of assumed processes and underlying physiological mechanisms which serve the response changes considered here.

A more complete picture of the situation at the present state of knowledge should consider still other types of interactions, such as those observed with amphetamine and with various types of 5-HT system agents which also induce particular response enhancement or disinhibition syndromes. Furthermore, it should be underlined that at some point different syndromes eventually require quite different experimental strategies, due to the characteristics of various agents. For example, all antimuscarinics have qualitatively similar profiles, and most antimuscarinic effects stem from a single type of primary change at the molecular level, i.e., an interaction with a well-defined receptor site. Therefore, an adequate strategy must discriminate between different consequences at the behavioural level of a single type of bias (e.g., malfunctioning of a particular type of sensory-motor interface) and different behavioural changes due to influences on different systems at different sites, although triggered by the same primary change.

Tranquillizers no doubt constitute a much more complex case. In fact, the discussion here is on a particular property which is shared by quite different compounds (e.g., ethanol, barbiturates, meprobamate and benzodiazepines). This property is not necessarily determined *via* the same physiological-biochemical changes underlying other actions which, in turn, can be found in still other types of agents (e.g., certain classes of antiepileptic drugs) in the absence of an antipunishment effect. Since the intense biochemical and physiological research carried out on benzodiazepines often does not attempt to outline areas of overlap and areas of non-overlap with respect to other agents which also induce a disinhibition of the same type, the prospect for integrating molecular and neuropsychological models in the not too distant future is quite bleak.

Finally, treatment-behaviour interactions explored in recent years have become so complex that two opposite tendencies have inevitably developed. At one extreme, research on brain-behaviour relations which remains very close to physiological-biochemical correlates of drug action is often forced to make recourse to simplified behavioural tests and simplified models of behaviour organization which cannot consider the complex interactions outlined above. From this level, there is a very short distance to the point where individual behavioural responses are reduced by the experimenter to convenient dependent variables (33). The brain-behaviour correlations observed lose any general significance for the understanding of behaviour organization, and the "models" require revision each time a modification of a test brings about a change in the results, other things being apparently equal at the physiological-biochemical level.

At the opposite extreme the analysis of complex interactions may lead the experimenter very far away from any one of the available physiological-biochemical approaches, which appears to be the case when apparently simple phenomena such as motor-act inhibition are broken down by treatment challenges into sub-categories and sub-sub-categories, each requiring at least a partially separate underlying mechanism. From this level, there is an awfully short distance to the point of abandoning any attempt to account for any but the more elementary behavioral items in terms of underlying physiological mechanisms.

It seems, therefore, that the old question of reductionism and antireductionism in the neural and behavioural sciences, far from finding a satisfactory settlement in the light of more recent and sophisticated data, behaves like a universe in expansion whose limits are not perceivable from our present perspective.

## REFERENCES

1. BERGER, B. D. and STEIN, L. 1969. Asymmetrical dissociation of learning between scopolamine and Wy 4036, a new benzodiazepine tranquilizer. Psychopharmacologia 14: 351–358.
2. BIGNAMI, G. 1976. Nonassociative explanations of behavioral changes induced by central cholinergic drugs. Acta Neurobiol. Exp. 36: 5–90.
3. BIGNAMI, G. 1976. Behavioral pharmacology and toxicology. Annu. Rev. Pharm. Toxicol. 16: 329–366.

4. BIGNAMI, G. 1978. Effects of neuroleptics, ethanol, hypnotic-hedatives, tranquilizers, narcotics, and minor stimulants in aversive paradigms. In H. Anisman and G. Bignami (ed.), Psychopharmacology of aversively motivated behavior. Plenum Press, New York, p. 385–453.
5. BIGNAMI, G., AMORICO, L., FRONTALI, M. and ROSIĆ, N. 1971. Central cholinergic blockade and two-way avoidance acquisition: The role of response disinhibition. Physiol. Behav. 7: 461–470.
6. BIGNAMI, G., CARRO-CIAMPI, G. and ALBERT, M. 1968. Effects of frontal lesions on "go-no go" avoidance behaviour in normal and scopolamine-treated rats. Physiol. Behav. 3: 487–493.
7. BIGNAMI, G. and CHESHER, G. 1978. Variability of antipunishment effects of antiemotional agents in active-passive avoidance tasks. In C.I.N.P., Abstracts of papers presented at the 11th Congress of the Collegium Internationale Neuro-Psychopharmacologicum, July 9–14 1978, Vienna, Austria, p. 38.
8. BIGNAMI, G. and GATTI, G. L. 1969. Repeated administration of central anticholinergics. Classical tolerance phenomena versus behavioural adjustments to compensate for drug-induced deficits. In S. B. de C. Baker and J. Tripod (ed.), Sensitization to drugs. Excerpta Medica (I.C.S. No. 181), Amsterdam, p. 40-46.
9. BIGNAMI, G. and MICHAŁEK, H. 1978. Cholinergic mechanisms and aversively motivated behaviors. In H. Anisman and G. Bignami (ed.), Psychopharmacology of aversively motivated behavior. Plenum Press, New York, p. 173–255.
10. BIGNAMI, G. and ROSIĆ, N. 1971. The nature of disinhibitory phenomena caused by central cholinergic (muscarinic) blockade. In O. Vinař, Z. Votava and P. B. Bradley (ed.), Advances in neuro-psychopharmacology. North-Holland, Amsterdam, p. 481–495.
11. CANNIZZARO, G., NIGITO, S., PROVENZANO, P. M. and VITIKOVA, T. 1972. Modification of depressant and disinhibitory action of flurazepam during short term treatment in the rat. Psychopharmacologia 26: 173–184.
12. CARRO-CIAMPI, G. and BIGNAMI, G. 1968. Effects of scopolamine on shuttle-box avoidance and go-no go discrimination: Response-stimulus relationship pretreatment baselines, and repeated exposure to drug. Psychopharmacologia 13: 89–105.
13. COOK, L. and SEPINWALL, J. 1975. Reinforcement schedules and extrapolations to humans from animals in behavioral pharmacology. Fed. Proc. 34: 1889–1897.
14. COOK, L. and SEPINWALL, J. 1975. Behavioral analysis of the effects and mechanisms of action of benzodiazepines. Adv. Biochem. Psychopharmacol. 14: 1–28.
15. COOK, L. and SEPINWALL, J. 1975. Psychopharmacological parameters of emotion. In L. Levi (ed.), Emotions — their parameters and measurement. Raven Press, New York, p. 379–404.
16. DĄBROWSKA, J. 1975. Prefrontal lesions and avoidance reflex differentiation in dogs. Acta Neurobiol. Exp. 35: 1–15.
17. DANTZER, R. 1977. Behavioral effects of benzodiazepines. A review. Biobehav. Rev. 1: 71–86.
18. DANTZER, R. 1978. Dissociation between suppressive and facilitating effects of aversive stimuli on behavior by benzodiazepines. A review and reinterpretation. Progr. Neuro-Psychopharmacol. 2: 33–40.

19. DANTZER, R., MORMEDE, P. and FAVRE, B. 1976. Fear-dependent variations in continuous avoidance behavior of pigs. II. Effects of diazepam on acquisition and performance of Pavlovian fear conditioning and plasma corticosteroid levels. Psychopharmacology 49: 75–78.
20. FRONTALI, M., AMORICO, L., DE ACETIS, L. and BIGNAMI, G. 1976. A pharmacological analysis of processes underlying differential responding: A review and further experiments with scopolamine, amphetamine, lysergic acid diethylamide (LSD-25), chlordiazepoxide, physostigmine, and chlorpromazine. Behav. Biol. 18: 1–74.
21. FRONTALI, M. and BIGNAMI, G. 1973. Go–no go avoidance discriminations in rats with simple "go" and compound "no go" signals: Stimulus modality and stimulus intensity. Anim. Learn. Behav. 1: 21–24.
22. FRONTALI, M. and BIGNAMI, G. 1974. Stimulus nonequivalences in go/no go avoidance discriminations: Sensory, drive, and response factors. Anim. Learn. Behav. 2: 153–160.
23. GRAY, J. A. 1977. Drug effects on fear and frustration: Possible limbic site of action of tranquilizers. In L. L. Iversen, S. Iversen and S. H. Snyder (ed.), Handbook of psychopharmacology, Section II: Behavioral pharmacology in animals. Vol. 8: Drugs, neurotransmitters, and behavior. Plenum Press, New York, p. 433–527.
24. HEISE, G. A. 1975. Discrete trial analysis of drug action. Fed. Proc. 34: 1898–1903.
25. HEISE, G. A., CONNER, R. and MARTIN, R. A. 1976. Effects of scopolamine on variable interval spatial alternation and memory in the rat. Psychopharmacology 49: 131–137.
26. HEISE, G. A., HRABRICH, B., LILIE, N. L. and MARTIN, R. A. 1975. Scopolamine effects on delayed spatial alternation in the rat. Pharm. Biochem. Behav. 3: 993–1002.
27. IWAHARA, S. 1977. Contributions from the animal laboratory — Drug and response inhibition. Mod. Probl. Pharmacopsychiatr. 12: 59–76.
28. KELLEHER, R. T. and MORSE, W. H. 1968. Determinants of the specificity of behavioral effects of drugs. Ergeb. Physiol. 60: 1–56.
29. KONORSKI, J. 1967. Integrative activity of the brain. Univ. Chicago Press, Chicago, 531 p.
30. KONORSKI, J. 1972. Some hypotheses concerning the functional organization of prefrontal cortex. Acta Neurobiol. Exp. 32: 595–613.
31. MARGULES, D. L. and STEIN, L. 1968. Increase of "anti-anxiety" activity and tolerance of behavioral depression during chronic administration of oxazepam. Psychopharmacologia 13: 74–80.
32. MASER, J. D. and HAMMOND, L. J. 1972. Disruption of temporal discrimination by the minor tranquilizer, oxazepam. Psychopharmacologia 25: 69–76.
33. McCLEARY, R. A. 1961. Response specificity in the behavioral effects of limbic system lesions in the cat. J. Comp. Physiol. Psychol. 54: 605–613.
34. McKEARNEY, J. W. 1976. Punishment of responding under schedules of stimulus-shock termination: Effects of D-amphetamine and pentobarbital. J. Exp. Anal. Behav. 26: 281–287.
35. MILAR, K. S., HALGREN, C. R. and HEISE, G. A. 1978. A reappraisal of scopolamine effects on inhibition. Pharm. Biochem. Behav. 9: 307–313.
36. MILLENSON, J. R. and LESLIE, J. 1974. The conditioned emotional response (CER) as a baseline for the study of anti-anxiety drugs. Neuropharmacology 13: 1–9.

37. PATEL, J. B., CIOFALO, V. B. and IORIO, L. C. 1979. Benzodiazepine blockade of passive-avoidance task in mice: a state-dependent phenomenon. Psychopharmacology 61: 25–28.
38. ROSIĆ, N. and BIGNAMI, G. 1970. Scopolamine effects on go-no go avoidance discriminations: Influence of stimulus factors and primacy of training Psychopharmacologia 17: 203–215.
39. ROSIĆ, N., FRONTALI, M. and BIGNAMI, G. 1969. Stimulus factors affecting go–no go avoidance discrimination learning by rats. Commun. Behav. Biol. 4: 151–156.
40. SEPINWALL, J. and COOK, L. 1978. Behavioral pharmacology of antianxiety drugs. *In* L. L. Iversen, S. D. Iversen and S. H. Snyder (ed.), Handbook of psychopharmacology, Section III: Human Psychopharmacology. Vol. 13: Biology of mood and antianxiety drugs. Plenum Press, New York, p. 345–393.
41. SIMON, P. and SOUBRIÉ, P. 1979. Behavioral studies to differentiate anxiolytic and sedative actions of the tranquilizing drugs. Mod. Probl. Pharmacopsychiatr. 14: 99–142.
42. SOŁTYSIK, S. 1975. Post-consummatory arousal of drive as a mechanism of incentive motivation. Acta Neurobiol. Exp. 35: 447–474.
43. SOŁTYSIK, S. and GASANOVA, R. 1969. The effect of "partial reinforcement" on classical and instrumental conditioned reflexes acquired under continuous reinforcement. Acta Biol. Exp. 29: 29–49.
44. STEIN, L. and BERGER, B. D. 1969. Paradoxical fear-increasing effects of tranquillizers: Evidence of repression of memory in the rat. Science 166: 253–256.
45. WADDINGTON, J. L. and OLLEY, J. E. 1977. Dissociation of the anti-punishment activities of chlordiazepoxide and atropine using two heterogenous passive avoidance tasks. Psychopharmacology 52: 93–96.
46. WARBURTON, D. M. 1972. The cholinergic control of internal inhibition. *In* R. Boakes and M. S. Halliday (ed.), Inhibition and learning. Academic Press, New York, 1972.
47. WARBURTON, D. M. 1977. Stimulus selection and behavioral inhibition. *In* L. L. Iversen, S. D. Iversen and S. H. Snyder (ed.), Handbook of psychopharmacology, Section II: Behavioral pharmacology in animals. Vol. 8: Drugs, neurotransmitters, and behavior. Plenum Press, New York, p. 384–431.
48. ZIELIŃSKI, K. 1979. Extinction, inhibition, and differentiation learning *In* A. Dickinson and R. A. Boakes (ed.), Associative mechanisms in conditioning. L. Erlbaum Ass., Hillsdale, p. 269–293.

Giorgio BIGNAMI, Istituto Superiore di Sanità, Viale Regina Elena 299, I-00161 Rome, Italy.

# PREFRONTAL CORTICAL EFFECTS ON AVERSIVELY MOTIVATED INSTRUMENTAL CONDITIONING IN RATS: SOME ONTOGENIC CONSIDERATIONS

James F. BRENNAN

Department of Psychology, University of Massachusetts Boston, Massachusetts, USA

*Abstract.* During the last 20 years, an emerging body of data has delineated critical variables controlling the acquisition and retention of aversive experiences across ages. Focusing on the rat as subject organism, the behavioral literature on task- and age-specific findings is reviewed. Response inhibitory deficits in younger subjects are related to augmentation of stimulus control through discrimination training and reinstatement of components of original learning. Somewhat parallel and complementary to studies of behavioral development, advances in the neurophysiology and neuroanatomy of cortical functions have indicated the critical role of the prefrontal cortex in acquisition and retention of aversively motivated instrumental responses. Several studies of prefrontal damage administered at varying ages reveal the importance of neural development in both performance deficits as well as recovery of function. These preliminary experiments are discussed in light of constraints from appropriate cortical influences in consideration of the ontogeny of fear.

## INTRODUCTION

The critical importance of early experience upon subsequent maturity is well documented in the psychological literature focusing upon the determinants of behavior. Within the past 20 years, attempts have been made to operationally delineate the residual effects from specific experiences that occur early in ontogeny (e.g., 16, 92). The results of many experiments have measured the significance of events that range in specificity from maternal deprivation to neonatal shock administration, and these data represent an emerging body of information that relate adult behavioral processes to their underlying developmental determinants. The efforts of researchers from several fields of behavioral science are increasingly synthesized to embody a more comprehensive core of developmental data that is well articulated within a behavioral model of experience.

In the analysis of behavioral observations, several trends in the literature have pointed to the role of cortical morphology as differentially affecting experiences administered to several age levels within a given species (e.g., 21). Specifically, such explanatory themes focus on concomitant physiological states, and changes in those states, that might underly and interact with the acquisition and the retention of specific early experiences. Although there exist historical as well as recent explications of the developmental morphology of the neonatal and weanling rat (e.g., 2, 39, 91), the relationship between physiological development and behavioral ontogeny is still evolving. Diverse studies ranging from the effects of crowding and hormonal manipulations (e.g., 38) to examinations of developmental effects of cortical damage in higher mammals (e.g., 36) contribute to the delineation of this relationship.

Experimental studies of the behavioral effects of early aversive events have typically focused upon initial acquisition comparisons among age levels and then examined their immediate and/or long-term retention. Variations of this paradigm also include measures of relearning of partial or complete components of the original task requirements after various interpolated temporal intervals. The review by Campbell (16) summarized several methodological approaches to this area as well as some major behavioral evidence. The present effort focuses on the rat as subject organism, and this selection for the study of developmental effects of aversive events is compelling for several reasons in addition to obvious ethical considerations. First, the relative immaturity of the central nervous system (CNS) in the neonatal rat is attractive, since critical postnatal changes that presumably affect behavior may be assessed. Second, anatomical studies (e.g., 29, 54) have delineated the cortical projection fields of the species, permitting phylogenetic comparisons of the frontal cortex.

Developmental studies of specific aversive events in the rat indicate a lower age limit of approximately 18 days of age, which is often dictated by various task requirements. An upper age level, after which learning deficits in young rats cease to be pronounced, has been approximated in several tasks at 35 days of age (e.g., 17, 18, 82). The period of hyperactivity (e.g., 5) overlaps this interval up to approximately 20 days of age. In their review of memory deficits and age, Campbell and Spear (21) summarized the concomitant neurological and physiological changes that mark early development. Myelin is initially observed at 2 days of age in the spinal roots, and myelinization is detected in thalamic fibers by age 12 days, reaching full extent by 40–50 days of age (39). Cortical myelinization begins by 20 days of age in most regions and is produced most rapidly between 30 and 40 days of age, with full myelinization completed by age 60 days. Campbell and Spear (21) further cited evidence of dendritic elaboration indicating that formation of synaptic junctions is relatively completed between age 10 and 30 days. Mendell, Rein, Horth-Edel and Mendell (65) showed the most rapid acceleration of DNA content in the rat to occur between ages 7 and 18 days. Accordingly, on the basis of constraints dictated by behavioral tasks, supported by parallel neurological growth, the critical period between 18 and 35 days seems appropriate.

## BEHAVIORAL PRECEDENTS: MEASURES AND PROCESSES

Much of the behavioral evidence is derived from comparisons of passive and active avoidance learning at various developmental stages in the rat. These paradigms have been defined by the relative contingency between the response and the aversive agent (43, 52). If shock is to be delivered on an acquisition trial, it is produced after the response in passive avoidance whereas shock is presented before the active avoidance response.

### Passive avoidance

An early attempt to systematically examine age effects in the retention of specific aversive early trauma was reported by Campbell and Campbell (17). They employed a passive avoidance task wherein subjects were placed in the safe side of a two compartment chamber. Upon entry into the "fear" compartment, a guillotine door was closed and subjects were exposed to a series of inescapable shocks. This procedure offers several advantages. First, as with the case in other experimental procedures using shock, a type of foot-shock (matched impedance) is used which attenuates differences in weight among age levels of subjects (see 22). Campbell (16) has reported relevant psychophysical data indicating that this type of shock source is equally aversive across ages at moderate intensities. A second advantage, unique to passive avoidance measures, is that weanling rats are not penalized for inferior motor development, insofar as the major dependent variable involves the subject simply remaining stationary in the "safe" compartment if conditioned aversiveness to the fear compartment has occurred. The results of the Campbell and Campbell (17) experiments were rather typical of subsequent findings using passive avoidance behavior. That is, among the various age groups (18, 23, 38, 54 and 100 days of age), all subjects showed comparable avoidance behavior initially, but the weanling rats yielded performance deficits as the retention intervals were increased to 42 days after original conditioning.

Further investigations have confirmed that retention deficits occur in young rats when a major component of the dependent variable presumably involves withholding or inhibiting behavior (e.g., 33, 76, 87). Additionally, studies of stimulus generalization in young and adult rats using passive avoidance (e.g., 35, 79) have reported evidence of steeper gradients in young rats. The study by Frieman et al. (35) is illustrative of this finding. Essentially, onset of a tone was paired with shock presentations upon a subject's entry into the fear compartment of the passive avoidance chamber. Subsequent variation of tonal frequency during non-shock, retention trials revealed steeper generalization gradients for the pups to values other than the original training stimulus, relative to adult gradients. Again, this evidence of less generalization found in immature subjects when the testing situation involved passive avoidance is consistent with interpretations focusing on the relative inhability to withhold behavior in young rats.

*Active avoidance*

Several studies have compared age levels in rats using a component of motor activity intrinsic to the instrumental response. While recognizing that inferior motor coordination in the young subjects may necessarily confound measure of active avoidance, these studies have been informative when the two kinds of measures, active and passive, are compared. Kirby (44) trained rats of 25, 50, or 100 days of age in a runway type of directional active avoidance situation. Although he found retention deficits in the 25 day old group after intervals of 25 and 50 days to relearning, he reported no differences in original acquisition of the task and only a statistically non-significant tendency in the adults for greater resistance to extinction. While Kirby's evidence of comparable acquisition of one-way avoidance between ages has found support in several experiment (e.g., 33, 76) a study reported by Egger and Livesey (31) suggested that the age similarity warrants qualification. Specifically, those investigators obtained an age effect in one-way active avoidance when the performance criterion was made more stringent than that used in previous (e.g., 76) studies. Their finding supports the contention that perhaps the basic incremental units of learning are somewhat different between ages, such that a given avoidance response may represent "appreciably less learning" in young rats than their adult counterparts.

More intensive studies of extinction of one-way active avoidance have confirmed the tendency for an age effect noted by Kirby (44, see also 31). Klein and Spear (45) examined extinction in terms of comparisons between ages on the relearning of a one-way avoidance task at various intervals up to 24 h. While they found similar retention curves showing a "Kamin Effect" at both ages, overall retention measures indicated a deficit in the young rats. Similarly, Riccio and Marrazo (75) found age differences in resistance to punishment of active avoidance responses during extinction.

Stimulus generalization has been examined in the context of age comparisons in active avoidance behavior. While one-way active avoidance offers the advantage of relatively rapid acquisition of the instrumental motor response at both ages, it is not ideal for studies of generalization. That is, in one-way active avoidance, the conditioned stimulus (CS) is composed of a complex of cues, such as an exteroceptive stimulus (e.g., tone or light), the opening of a door separating compartments or to a runway, and the constant direction of the required motor activity. These factors complicate generalization studies, because it is difficult to determine if all of the cues are equally salient for the subjects, a problem that may be further confounded by potentially different stimulus saliencies as a function of age (see below). Although two-way active avoidance requires a more prolonged training period, the shuttle behavior is typically controlled by the single cue that signals the avoidance contingency. Two reports (11, 78) examined age effects in this situation using the same procedure. Adult and young (21 days) rats were given 3 days of two-way active shuttle avoidance training with a CS of 3,000 Hz at a moderate intensity. Three counterbalanced generalization tests were given over subsequent days of avoidance extinction: Frequency change only, intensity variation only, and concurrent variation of both CS dimensions. Gradients obtained from both age levels under all testing conditions were highly comparable. That is, decremental slopes around the frequency value used in training were obtained when that dimension was varied alone. On the other hand, a striking dynamism was obtained, wherein response speed increased with the absolute magnitude intensity, upon variation of the CS intensity only. This dynamism effect was later shown to be highly persistent and resistant to discrimination training in rats (12) and similarly in cats (99). Interestingly, when the frequency and intensity dimensions were varied concurrently, both age groups revealed gradients that showed the effects of each dimension — i.e., frequency control was enhanced as the absolute intensity of the test stimuli increased.

Since both young and adult rats show comparable frequency control from the single stimulus training that is implicit in the acquisition procedure of two-way training, age comparisons after discrimination training are important. A study (14) of adult rats had found that the frequency dimension of a two-way avoidance CS was particularly sensitive to differentiation training administered through interpolated Pavlovian pairings ($CS^+$ = original "fear" frequency value of instrumental training; $CS^-$ = no shock, safety frequency value). Accordingly, Brennan and Riccio (13) trained young and adult rats in the instrumental shuttle procedure. Prior to the single frequency generalization test administered in extinction, subjects were given 50 presentations each of the 4000 Hz value with inescapable shock and a 3,500 Hz tone always in the absence of shock. For the adults, the frequency gradients revealed a shift in maximum responding away from $CS^+$ in the direction opposite to that of $CS^-$. However, relative to single stimulus control pups, the gradients of the pups given discrimination training showed only a distortion in the gradient of higher frequency values. This finding is of interest because it relates to the response inhibitory deficits noted earlier in young rats from passive avoidance measures. That is, the peak shift in the adult gradients is predicted from the traditional discriminative hypothesis of Spence (86) that focuses upon the summation of excitatory and inhibitory processes in post-discrimination gradients. If response inhibitory processes are not well articulated in young rats, it is reasonable to expect only a distortion, but not a clear shift in gradients obtained after discrimination training.

This interpretation was examined in a study (6) that attempted to assess inhibitory control by employing an extradimensional relationship between Pavlovian differentation training and generalization of the instrumental avoidance behavior (cf., 37, 66). Essentially, weanling and adult rats were trained in two-way active avoidance training and subsequently given Pavlovian differentiation training along the CS intensity dimension with the frequency value of both tones held constant. Separate groups at each age level were then tested for frequency generalization at either the fear or the safety intensity, or after pseudo-conditioning along the intensity dimension. The adults responded with maximum vigor to the original CS frequency when tested at the fear intensity, and they tended to respond least to the original CS frequency when tested for frequency generalization at the safety intensity value. The young subjects tested at the fear intensity also showed decremental response gradients around the original CS frequency. However, the results from the remaining pup groups were not as clear. Young subjects, given frequency tests at the $CS^-$ safety intensity or after pseudo-conditioning, retained an overall elevated level of responding, suggesting retention of some responsivity to all frequencies, while the overall gradient in the

CS⁻ group of pups was only somewhat depressed. Indeed, observations of both the CS⁻ and pseudotraining pups showed that their general activity level was increased, and they engaged in other behaviors that were antagonistic to the instrumental running response. Thus, for the young subjects tested at the CS⁻ intensity, a response inhibitory deficit emerged since both CS⁻ testing and pseudo-training conditions resulted in functionally equivalent effects.

An additional behavioral process that offers evidence of response inhibitory deficits in pups involves the examination of habituation as a function of age in rats. The experimental paradigm in such studies typically consists of nonreinforced presentations of the to-be-conditioned stimulus during pretraining and measurement of acquisition deficits (e.g., 56). It was found (32) that young rats were deficient in the habituation of a head-poking response, which was related to the inferior development of cholinergic functions involved in response inhibition. Wilson et al. (94) examined this hypothesis under aversive conditions in one-way active avoidance. Weanling and adult rats received either 0 or 10 preexposure presentations of a CS complex consisting of the opening of a guillotine door (againts a false partition) and the onset of a tone. All groups were subsequently given identical treatment in acquisition of one-way avoidance with the CS complex, followed by extinction of the instrumental response. The results indicated retardation of acquisition in the adults given 10 pre-exposures, compared with non-pre-exposed adults, while no acquisition differences were noted between the pup groups. The age difference in the habituation effect was interpreted in light of Rescorla's (74) argument that latent inhibition, accumulated as the result of nonreinforced CS presentations, contributes to a decrease in stimulus saliency. Wilson et al. (94), noting the suggestion offered in the passive avoidance generalization study by Frieman et al. (35), observed that differences in selective attention may exist between weanling and adults. Specifically, adult rats reared in a typical laboratory situation have a history of environmental changes that occur independent of reinforcement. Accordingly, the pre-exposure manipulation may have been more effective for the adults, and prepotent responses to the CS rapidly habituated. Conversely, weanlings, without this unspecified history of noncontingent events, may be more of "selective attenders", thereby requiring a greater number of nonreinforced pre-exposures to achieve a comparable habituation effect. Of further interest in the Wilson et al. (94) study was the clear evidence of an overall extinction deficit in the young rats, a tendency that was only noted in the earlier study by Kirby (44).

A study by Brennan and Barone (8) attempted to further examine properties of stimulus saliency in terms of attentional as well as general extinction differences. Groups of adults and weanlings were given acquisition and extinction treatments in one-way active avoidance with either a compound or a simple CS. While all dependent measures indicated comparable acquisition to a moderately stringent criterion, extinction differences were obtained between ages as well as the type of CS employed. Specifically, several measures of extinction consistently indicated significantly less resistance to extinction in the pups than the adults. Moreover, comparable extinction was found in the pup groups trained with the simple or compound CS, whereas adults trained with the compound CS extinguished more rapidly than adult subjects trained and extinguished with the single-cue CS. In light of temporal contingencies associated with the CS elements, the additional information contained in the compound stimulus may have been utilized by the adults to more effectively discriminate the extinction procedure from acquisition. Conversely, the lack of differential extinction performance in the pups suggests their insensitivity to the differing values of each type of CS, supporting the interpretation of differential age saliency with respect to environmental cues. An additional experiment in this study extended the question of CS complexity to a consideration of pre-exposure effects. Essentially, the results of this experiment indicated that habituation effects occurred only in the adults, and only when the pre-exposure and avoidance CSs were identical, regardless of complexity. Accordingly, the habituation deficit in the pups was repeated, with the type-of-CS employed exerting no additional effect.

Extinction of two-way active avoidance in weanling and adult rats was further examined in a recent study (7). Resistance to extinction of pups and adults was measured in the presence of two types of novel stimuli or no additional stimuli. The disruptive effect of an animate stimulus was most dramatic in the pups, and a novel light was also disruptive for the young subjects as well. The adults were seemingly unaffected by either of the extinction manipulations. In a second experiment pups and adults received two-way active avoidance training with a CS consisting of simultaneous tone and light onset. During extinction, responses terminated only one of the stimulus elements and produced continuation of the remaining element for an additional 5 s. Continuation of the tonal element resulted in greater discruption than the light in the adults, although the pups had faster, nondifferential extinction rates. Additional pups and adults were then presented with various delays of tonal CS reactivation following extinction responses, which produced increased numbers of extinction trials in the adults, but the pups failed to respond differentially to any of the delay intervals. Collectively, these experiments suggest comparable acquisition and extinction of two-way active avoidance between ages only when environmental factors and associative contingencies remain constant. Upon variation of either environmental or associative cues, differences in extinction rates between ages emerge.

The behavioral evidence indicates age differences in rats under conditions of aversive motivation when the dependent measures involve both response activation as well as witholding behavior. It is of great importance to note the deficits common to both kinds of behavioral measures. That is, young rats have difficulty in the retention of witholding motor behavior in passive avoidance and a lack of preservation of instrumental behavior during the extinction of active avoidance. Both types of deficits reflect the problem of age differential mediation of inhibitory responses that are perceptual and motor.

The review by Campbell and Spear (21) indicated that performance deficits in young rats in aversively motivated, as well as appetitive conditions, are correlated with changes in central nervous system structure and functions during early development. Citing neurological and metabolic evidence, they concluded that the neonatal rat undergoes critical periods of development that presumably affect the mediation of experience as well as the storage of memory. Relating to the memory deficits in young rats, they presented a summary of evidence suggesting improved retention of specific experiences when periodic reinstatements of part of the original learning conditions are presented during development. Campbell and Jaynes (18) reported that abbreviated re-exposures to the original learning situation, although not enough exposure to affect acquisition in naive subjects, resulted in greatly improved retention in weanlings. Further, Silvestri et al. (84) found that reinstatement experiences involving only the presentation of the CS, without shock, resulted in significant improvement of passive avoidance retention in young rats. While behavioral measures of retention may improve with the reinstatement procedure, the neurological evidence, particularly relating to cortical myelinization and EEG activity in the developing rat, emphasize the important role of cortical development in the ontogeny of memory (50).

## NEUROBEHAVIORAL EVIDENCE

*The prefrontal cortex.* The vast neurobehavioral literature contains data derived from several levels of phylogenetic development that reflect similarities and differences among animals in the functional role of the prefrontal cortex (61). The identification of various fields within the frontal cortex along with concomitant structural and functional generalizations among species are typically derived from observations of cytoarchitectural differences and/or similarities, afferent–efferent relationships and the analysis of postoperative behavioral changes. It is the former two methodological approaches that provide the empirical foundation for the latter, behavioral assessment, and presumably the examination of behavioral change reveals the functional integration of neural mechanisms or systems.

*Cytoarchitecture.* Cytoarchitectural definitions of the frontal cortex are troubled by the lack of agreement concerning topographical demarcations (1). Various maps of cortical fields clearly distinguish granular from agranular cortex (Brodmann's criterion) in primates. However, when descending the phylogenetic scale, the distinction between granular and agranular cortex is less recognizable, and the localization of subfields must include other types of justification. That is the less evolved brain of the rat does not permit separation of frontal cortices as seen in higher mammals (cf., 1) and cortical maps of the rat brain often supplement the cytoarchitectural arrangement with functional, behavioral considerations (e.g., 85). Since the lissencephalic rat cortex does not permit the topographical demarcations of cytoarchitectural areas, possible in higher mammals, this disparity in phylogenetic sophistication has resulted in the conclusion that the cortical surface in rodents is essentially agranular.

*Afferent–efferent relationships.* Within mammalian phylum the prefrontal cortex receives the essential afferent projections of nucleus medialis dorsalis (n. MD). In carnivores and primates, projections of n. MD allow the dissociation of prefrontal fields. In the rat, such dissociation was questionable which led Leonard (54) to an antereograde analysis of the thalamic nuclei. In brief, she found that n. MD was indeed dissociated into anterior and posterior components projecting respectively to the ventromedial and dorsomedial (sulcal) cortex, and there was no evidence of n. MD projections to the dorsal convexity of the hemisphere (orbital). The corroborative evidence implicating ventromedial projection sites, coextensive with anterior n. MD sites in the sulcal cortex, was found by Leonard (54) but she considered this evidence somewhat inconclusive since chromatolysis was also elicited in the ventromedial projections of the thalamus following massive lesions of the ventromedial cortex. Nevertheless she concluded that dissociation at the cortical level is conceivable for the anterior n. MD and ventromedial projections, since retrograde analysis unavoidably damages the sustaining collaterals from numerous origins. Later, Leonard (55) and Domesick (30) provided substantial evidence for a laminar organization of thalamic projections. In Leonard's study subdivisions of n. MD displayed preferential termini indicative of an organizational martix within the laminar of the cortex, while Domesick's study replicated the findings for n. MD and extended the concept to anterodorsal and anteroventral components of the anterior thalamus. The implication suggested by Domesick and formally developed by Shephard (83) is that differential laminar distributions arising from related neural substrates provide a mechanism allowing a given subsystem to interact with the same cells at different locations.

From the examination of anterograde degeneration methods, Domesick (30) concluded that there are three dissociable thalamo-cortical projections in the rat. Afferents from the anterodorsal-anteroventral components of the anterior thalamic nuclei project ventrodorsally to medial aspects of the posterior cortex. Fibers from the anterior and posterior components of n. MD project to dorsal sectors of precentral cortex along pathways ventral to anterodorsal-anteroventral and extend rostrodorsal beyond the anterodorsal-anteroventral field. A third cortical projection field was identified in the rostrocaudal aspects of the precentral cortex, and was described as coextensive with the anterior n. MD projection field. Anterograde analysis of the adjacent paratenial thalamic nuclei identified a small projection field of cortex situated distantly rostral to the genu of the corpus callosum. However, it was not possible to clearly distinguish this cortical field from the more diffuse projection of n. MD. In addition, Domesick described intrathalamic connections of the extrinsic transit fibers associated with the anterodorsal component and n. MD.

Reciprocation of thalamocortical afferents, investigated by Leonard (54, 55) and Domesick (30), was found to be somewhat incomplete. Results from retrograde analysis reported by Domesick gave no indication of cortical afferents to the anterodorsal component, although a general correspondence was suggested for the related anteroventral component. Leonard (54) and Domesick (30) agreed that both aspects of n. MD receive reciprocal connections from the anterior cortex. Cortico-thalamic projections appear to reciprocate identical thalamo-cortical connections, and overlap with cortical afferents to n. MD in supracallosal regions (30). The lack of complete correspondence between dissociated projection fields in the cortex and their thalamic dependency remains open.

More recent investigations have clarified the projections of cortical subfields, based upon evidence derived from the technique of horseradish peroxidase. Specifically, studies of the rat (29), as well as the rat relative to other species (28, 64) have identified projections to the prefrontal cortex. The prefrontal cortex receives dominant projections from n. MD, but also receives the only neocortical projections from the ventral tegmental area and the amygdala. Accordingly, Divac et al. (29) described the prefrontal cortex as containing parts that can be termed cingulate and premotor. Further, they concluded that the prefrontal cortex is distinguished into the dissociated areas indicated for higher mammals, especially monkeys.

*Functions of prefrontal cortex*

The neurobehavioral literature indicates that prefrontal cortices in primates, carnivores, and rodents may subserve similar functions. Despite differences in behavioral measures that attempt to assess prefrontal functions, performance deficits observed for experimentally lesioned subjects represent a unified symptomology. In general, somatosensory and affective alterations in the behavior of prefrontal animals are related to both the site and the size of the tissue mass removed. Dissociations of functional impairment are attributed primarily to the anatomical subdivisions of the prefrontal cortex, which are differentially sensitive to the manipulations of stimulus-response requirements inherent in experimental situations.

In rats, most of the behavioral assessments of prefrontal functions to date has been generated from studies of alimentary conditioning. Reports of aversively motivated behavior after prefrontal manipulations are relatively scarce. Nevertheless, the relationships established between the prefrontal cortex and alimentary behavior in rats do provide a basis for eventual articulation of patterns of defensive behavior. Specifically, functional similarities emerge between alimentary and classical defensive conditioning, which are not found in instrumental defensive reflexes (98).

Extensive prefrontal ablations in rats produce dramatic alimentary behavioral changes. Perseverative tendencies appear related to the difficulty of discriminating alternatives (61) and have been attributed to decreased utilization of proprioceptive cues (59, 60, 61). However, frontal lesions do not necessarily lead to perseveration if the delay factor between stimulus and response is omitted, (58) or if the response to be alternated is made visually discriminable (25, 57). Divac (26) found no impairment of spatial delayed alternation in rats with prefrontal lesions restricted to the frontopolar cortex, however, his lesions were conservative in comparison to those reported by Łukaszewska (57, 58). Wikmark et al. (93), contending that in the rat prefrontal injury is not sensitive to tests of spatial delayed alternation, also performed relatively limited prefrontal damage and, in contrast to the Łukaszewska studies, employed a non-correction, within session alternation procedure. Apparently both the extent of lesion site and the nature of the motor experience introduce a relevant variation. Łukaszewska (59) concluded that frontal rats, sometimes superior to controls in visually discriminated alternations, utilize visual cues more than normal rats to compensate for deficits associated with spatial or positional cues. The overall pattern of prefrontal deficits in rats, unlike the case observed in monkeys, may involve inadequate sensory processing. It is not entirely clear that either olfactory or visual mechanisms are left intact following extensive prefrontal removal in the rat. However, it does seem that some portion of the rat's prefrontal cortex is functionally related to delay and spatial discrimination factors, measured by delayed response alternation tests.

*Dissociated prefrontal effects.* Analysis of prefrontal functions is facilitated when the extent of damage can be localized within anatomically distinct prefrontal subregions. Recourse to such techniques has resulted in the functional dissociation between various behavioral changes characteristic of prefrontal animals. That is, in the experimental literature an attempt has been made to approportionate the relative contribution of prefrontal subregions, with regard to the more general syndrome.

Delayed response type tests, particularly sensitive to prefrontal injury in various primate species and somewhat less sensitive when lower mammalian orders are tested, may be fractioned into at least three behaviorally meaningful, components. Delay factors involve withholding a response and possibly some element of proprioceptive short-term memory. Spatial factors comprise a second functional aspect of delayed response tests, especially in the case of response reversals. Response reversal itself may represent a dissociated component of delayed response tests and may be discriminated (e.g., go, no–go alternation) or non-discriminated (e.g., conditional position reversal). As indicated by these examples, behavioral measures intended to assess prefrontal injury often involve multiple components. For example, delayed response type tests may contain delay, spatial and response reversal factors in various combinations, and experimental contingencies may be arranged to enhance or attentuate the relative contribution of each factor to solution.

Lesions in rats selectively damaging orbital, ventrolateral (VL) or dorsomedial (DM) prefrontal subregions result in functionally dissociated effects. Anatomically (54) and by retrograde techniques (29), both subdivisions have been shown to have dissociated projection sites of anterior and posterior n. MD efferents. Behaviorally this distinction is maintained in ways strikingly similar to the case observed for primates by dissociating orbital and dorsolateral prefrontal areas. Prefrontal comparisons (59) in this functional sense recognize that the frontopolar (FP) cortex in the rat does not correspond to orbital cortex in monkeys although both areas occupy the same relative position and receive similar thalamic efferents (from the medial, magnocellular region of n. MD). In tasks involving postoperative retention and performance of delayed alternation and spatial reversal tasks, DM, but not FP, lesions produced deficits (26, 93).

Consideration of DM and FP lesions in terms of their frontal, dorsomedial continuity suggests that FP lesions results in functional sparing of spatial and delay response components. In contrast to such single dissociation of effects, orbital VL and DM lesions produce a double dissociation of effect. Orbital VL lesions in rats result in perseverative response tendencies in bar press extinction, while DM lesions do not (51). Rats with DM lesions are impaired on problems involving spatial and delay factors, while orbital VL rats are not impaired with respect to spatial cues, but show slight perseverative tendencies in postoperative retention, performance and extinction under appetitively motivated, schedules involving differential reinforcement of low rates. Apparently, this latter finding depends upon the size of the orbital lesion (cf., 51, 71). Both prefrontal subregions are dissociated in qualitative respects: activity (62) hoarding behavior and habituation (51), social behavior in an open field situation (47), and emotionality (71). Finally, orbital VL and DM prefrontal areas have different relations to consummatory mechanisms (49).

*Comparisons among species.* Examination of the behavior changes associated with partial prefrontal lesions in monkeys, dogs, cats, and rats indicates more striking similarities between monkeys and rats. It should be noticed, however, that on the whole, such comparisons between rats and monkeys involving aversively motivated behavior and response disinhibitory effects are precluded. Perseverative behavior in prefrontal rats towards non-reinforced alternatives may be considered a type of disinhibition, but is not as clearly established as in the case of disinhibition of instrumentally conditioned inhibitory reflexes in dogs and cats. Functional considerations (49, 51, 71) of the rat and monkey data indicate that orbitofrontal and dorsolateral prefrontal subdivisions in the monkey are similar to orbital ventrolateral and mediodorsal subdivisions respectively in the rat.

The relation to deficits produced by cortical intervention raises the question of subcortical mechanisms. The role of cortico-subcortical interactions in mediating behavior typically imparied by prefrontal injuries in monkeys has been reviewed by Rosvold and Szwarcbart (81), Nauta (67), and Rosvold (80). Neurobehavioral data warrant implicating neostriatal (27) structures as functionally related components of a prefrontal system subserving delayed-response type behavior. Specific neural constituents of this system include the prefrontal cortex, caudate nucleus, globus pallidus, substantia nigra, subthalamic nucleus, septal nuclei, hippocampus, centromedian nucleus, and the hypothalamus (80). Double dissociation of prefrontal cortex, sustained functionally and anatomically, has suggested a dichotomous organization of the proposed neural substrate. That is, two well defined systems appear to exert differential control upon behavior: a "dorsal" system originating in the dorsolateral prefrontal cortex involving the anterodorsal sectors of the caudate, the lateral pallidum, the subthalamic nucleus, and the hippocampus; and an "orbital" ventral system, originating in the orbital prefrontal cortex and involving the ventrolateral sector of the caudate, the medial pallidum, the centromedian nucleus, the septal nuclei, and the hypothalamus. Further reevaluation or restatement of the vast amounts of data contained in those reviews (see also, 68) of cortical-subcortical integrative functions in monkeys would take us far afield. Rather, it would seem that a complimentary organizational framework delineated for non-primates merits further development. Data showing that in the rat, as in the monkey, prefrontal deficits may be obtained by lesioning subcortically justifies, to a certain degree, such an attempt. Indeed, the specific subcortical structures involved in a prefrontal system in the rat are remarkably similar to those suggested by Rosvold and Szwarcbart (81) and Rosvold (80) for the monkey.

*Subcortical studies.* In the rat (93) and similarly in the cat (27), lesions to the mediodorsal prefrontal cortex (rats), to the proreal gyri (cats) or to the caudate nucleus (rats and cats), resulted in comparable deficits on delayed alternation tasks. Bilateral caudate lesions in rats proved more debilitating than prefrontal lesions when tested by an appetitively motivated spatial reversal task (26). Avoidance behavior, on the other hand, is not similarly affected by prefrontal and caudate lesion in rats (89), while in cats the converse occurs (34). Despite the contrast between appetitive and aversive conditioning procedures and the inherent task differences obtained when testing rats and cats, both types of subjects sustaining prefrontal or caudate lesions are characterized by a marked inability in withholding, or inhibiting preoperatively established approach tendencies. The effects described after lesions to the caudate nucleus could have been produced because of disruption to overlying cortex. Divac's (27) review of this problem suggests that destruction of the caudate tissue, and not damage to the perforant fibers, is responsible for the deficits in prefrontal tasks, a problem also discussed by Kolb (48).

Inclusion of thalamic nuclei, in particular n. MD as integrated components of the prefrontal system in rats and cats is by far more readily justified on anatomical than behavioral grounds. A convincing rationalization of the discrepancy between the effects of prefrontal and thalamic lesions has yet to be developed. It should be mentioned that in rats (90) as in monkeys (80), large lesions to the dorsal thalamus failed to produce any gross behavioral changes which typically accompany prefrontal damage. The major effect of thalamic lesions (centromedian nucleus) in monkeys (80) appears to be an increase in shock escape thresholds.

Thompson (90), however, did find that medial dorsal thalamic damage in rats profoundly impaired retention of auditory or visually discriminated avoidance responses. It seems likely that lesion size rather than species is the critical factor in determining the effects of thalamic lesions on learned behavior (89).

Septal lesions in cats (34) result in passive avoidance retention deficits and seem to induce hyperactivity and hyperemotionality. Heightened emotionality following septal lesions in rats as well as passive avoidance deficits (34) have been reported. Knook (46) reported that in the albino rat fibers of critical origin reach the fornix via the septal midline, and other indirect relations are possibly established via hippocampal fibers transectioning the precommissual fornix through the precommissural septum and coursing to the medial border of the nucleus accombens (anterior septal area). Assuming hippocampal involvement in prefrontal functions, such indirect fronto-septal connections might provide an important modulatory effect on prefrontal activity. Furthermore, Powell (72) has demonstrated septal effects in the rat, which establish direct and indirect feedback circuits via the fornix body and stria medullaris connecting both the hippocampus and anterior thalamic nuclei respectively to the cingulate portions of neocortex.

Representation of hypothalamic functions mediating consummatory behavior in the prefrontal cortex of rats has been investigated by Kolb and Nonnemann (49). These investigators found that prefrontal lesions to the ventrolateral (orbital) field of n. MD resulted in transient aphagia and adipsia and therefore replicated in part the well-documented lateral hypothalamic syndrome. Related findings in cats and rhesus monkeys (49) suggest that food and water intake modified by prefrontal lesions indicates encroachment upon hypothalamic mechanisms. Raismann (73) reviewed numerous studies which showed direct connections from the pyriform cortex, located immediately beneath the rhinal sulcas in rats, to the lateral hypothalamus. Knook (46) also reported prefronto-hypothalamic efferents that course ventrally in the peripheral part of the rostral half of the globus pallidus and join the internal capsule, entering the area preoptical lateralis as well as the area hypothalamic lateralis. He concluded that from the frontal cortex, fibers distribute and terminate over the entire extent of the lateral preoptic and hypothalamic areas.

The effects of hippocampal damage in rats and cats produce patterns of impairment very similar and sometimes more exaggerated than prefrontal lesions. Lesions to the frontal cortex or hippocampus in cats produced identical deficits in discrimination reversal learning. Combined lesions to both structures produced an impairment which was not greater than the deficit arising from either type of isolated lesion (71). Hippocampal or prefrontal damage will result in similar deficits on low reinforcement tasks in rats (23) and in cats (69, 70). Reversal learning, response alternation, bar pressing extinction and passive avoidance learning are all impaired following hippocampal or prefrontal lesions in rats (42, 77, 87) and in cats (69). Alternations in social behavior and emotionality (71), elevated levels of spontaneous activity (88), and operant bar press rate (41) are also characteristic of hippocomectomized or prefrontal rats.

Anatomical relationships between prefrontal cortex and the hippocampal formation via the fasciculus cinguli have been reported for rabbits and monkeys (15), but similar connections in rats (46) have not been verified, and in cats (or dogs) remain unclear. Frontal and temporal aspects of the prepyriform cortex in the rat may receive projections from the anterior hippocampal rudiment (95), however there may also be indirect fibers of passage from rostral aspects of the fornix longus, which at that level is continuous with the hippocampal rudiment (72). The presence of subcortical structures that receive overlapping frontal and hippocampal projections, namely the lateral and medial hypothalamus, midbrain tegmentum, central gray substance, and the zona incerta (71) may serve to relate prefrontal and hippocampal functioning.

Thus, the neurobehavioral literature indicates which prefrontal areas and subcortical structures exhibit functional similarities. A promising direction for the analysis of prefrontal symptomology seems to suggest that attention be focused on what occurs to incoming sensory processing as it affects the formation of the experimentally contingent motor act. Both the execution of a motor response and its inhibition are critically dependent upon the reciprocal activation and utilization of sensory input which is interioceptive and that which arises as feedback from ongoing motor activity. Characteristically, those tasks that seem to depend most specifically on the presence of intact prefrontal areas are neither purely sensory nor purely motor, but truly sensorimotor.

## AGE COMPARISONS OF PREFRONTAL LESION EFFECTS

Given the behavioral observations of age differences in the performance of aversively motivated tasks and the demonstrated effects of the prefrontal cortex, we have been recently engaged in an examination of the integration of these two research themes. Specifically, our experiments have concentrated on the measurement of prefrontal damage effects along task dimensions known to produce developmental performance differences.

## *Dorsomedial lesion effects on avoidance acquisition, extinction, and generalization*

In a study by Brennan et al. (10), experiments were reported that compared dorsomedial prefrontal lesion effects in weanling and adult rats under active and passive avoidance contingencies. In two experiments young and adult rats acquired two-way active avoidance responses to a 3000 Hz tonal CS. Subjects were subsequently tested in avoidance extinction for generalization along the tonal frequency dimension. The two experiments differed in terms of the time of prefrontal intervention. Bilateral dorsomedial prefrontal lesions were administered to half of the subjects at each age level either prior to avoidance acquisition or following acquisition. The lesions given prior to training impaired avoidance acquisition, but by 30 days after surgery subjects at both age levels recovered to the extent permitting generalization testing. When surgery occurred after avoidance acquisition, minimal interference in retention was noted, indicating that established avoidance levels were preserved immediately following lesions. However, retesting 30 days later was characterized by a deterioration in avoidance performance of lesioned subjects relative to controls. In both experiments, frequency control was not found in lesioned subjects at either age level. Accordingly the results of dorsomedial intervention in active avoidance performance indicate that the lesion effect was the most salient determinant of behavior, and age-related differences were obscured.

Rather different results were obtained in passive avoidance behavior reported in two additioned experiments of the Brennan et al. (10) study. Rats distributed at three age levels (18–20, 30–32, and 100 + days of age) were trained to avoid the black side of a black/white chamber and subsequently tested for retention one week after original learning. One experiment examined the effects of dorsomedial lesions administered prior to initial acquisition, while lesions were given between acquisition and retention testing in the other experiment. In both experiments, passive avoidance deficits related to age per se emerged as critical, with lesion and time of lesion effects of only secondary importance. Collectively, the passive and active avoidance experiments extend the behavioral information, summarized earlier in this review, relating age to task requirements in aversively motivated behavior. That is, performance differences emerge when prepotent ontogenic differences exist, such as response inhibitory deficits in the passive avoidance behavior of weanlings. Prefrontal lesion effects are most dramatic without the overriding age effects in those contexts in which performance between age levels is comparable, such as stimulus control of active avoidance.

The generalization of passive avoidance behavior was further examined in a recently completed study modelled after the Frieman et al. (35) experiment, described above, with the addition of a surgical variable involving bilateral lesions of the dorsomedial prefrontal cortex. Five days after surgery for the lesioned and sham operated subjects, passive avoidance training began for the weanling and adult rats. After several minutes of acclimation to a double compartment chamber each rat was given five inescapable shocks randomly distributed during the 2-1/2 min continuous presentation of a 2,000 Hz tonal CS. After another 2-1/2 min interval of no tone or shock, the tone was again activated for 2-1/2 min and five more shocks were administered. Subjects were then placed in a holding cage for five min after which a generalization test was conducted. During 20 testing trials, subjects could avoid onsets of tones of 1,000-, 2,000- or 3,000 Hz or no-tone which were contingent upon the subjects entering the shock compartment.

The results of generalization testing are summarized in Table I, which shows group mean time spent in the chamber with activation of each test stimulus for subgroups at each age and surgical treatment. Considering the age variable first, the pups showed overall less passive avoidance, but better differentiation of test values — i.e., greater stimulus

TABLE I

Mean time spent in "fear" compartment with activated test stimuli for pups and adults under each surgical treatment ($n = 5$ rats per subgroup)

| Surgical treatment | Test stimuli (Hz) | Ages at surgery | | | | | | | |
|---|---|---|---|---|---|---|---|---|---|
| | | Pups (18–20 days) | | | | Adults (80–100 days) | | | |
| | | CS | | | No Tone | CS | | | No Tone |
| | | 1,000 | 2,000 | 3,000 | | 1,000 | 2,000 | 3,000 | |
| | Intact (normal) | 93.6 | 68.8 | 100.2 | 110.2 | 76.8 | 71.0 | 73.0 | 79.8 |
| | Sham operated | 92.2 | 70.2 | 95.6 | 109.4 | 71.6 | 67.4 | 68.8 | 70.8 |
| | Dorsomedial lesion | 94.2 | 62.7 | 110.0 | 117.2 | 76.2 | 74.6 | 74.2 | 85.8 |

control — than the adults. This finding is consistent with the response inhibitory deficit typically found for young subjects in such tasks. Of greater interest, however, is the somewhat paradoxical lesion effects between ages. As the respective lesioned group means indicate, the lesioned pups showed better differentiation of test stimuli, whereas the adult differentiation was worse after lesions. Consistent with the passive avoidance data of pups in the Brennan et al. (10) study, the lesioned pups in the present experiment tended to show less passive avoidance than their adult counterparts, but they also articulated this disinhibition to finer discriminations among test values.

## Blocking of instrumental behavior

Another recently completed experiment in our laboratory attempted to examine the course of one-way active avoidance extinction under several different procedures. Basically employing a response blocking procedure (cf., 3, 24) that has been shown to facilitate extinction, this experiment trained active avoidance behavior in young and adult rats that previously were either lesioned in the dorsomedial prefrontal cortex or remained intact. Subjects were then subdivided into groups and received either pre-extinction blocking (i.e., CS onset, but instrumental responses prevented by a false door), a time out period between acquisition criterion and extinction testing, or no interpolated treatment and immediate extinction testing.

The results of this experiment are presented in Table II, which shows the mean numbers of trials to acquisition and extinction criteria for all subgroups of the experiment. Within both the intact adults and pups, the blocking procedure effectively facilitated extinction, although the pups had less overall resistance to extinction and the time-out control

TABLE II

Mean trials to acquisition and extinction criteria in pups and adults of each extinction treatment subgroup ($n = 4$)

|  |  | Age at surgery | | | |
| --- | --- | --- | --- | --- | --- |
|  |  | Pups (18–20) | | Adults (80–100) | |
|  |  | Acquisition | Extinction | Acquisition | Extinction |
| Control | Blocked | 36.5 | 14 | 20.5 | 7.5 |
| | Time out | 22.75 | 21.5 | 26.0 | 35.5 |
| | Normal | 26.0 | 26.25 | 25.25 | 50.25 |
| Lesioned | Blocked | 24.5 | 42.25 | 24.25 | 8.75 |
| | Time out | 28.0 | 36.5 | 28.25 | 10 |
| | Normal | 28.25 | 39.75 | 23.75 | 26 |

was ineffective for that age level. The lesions in the adults resulted in an overall decrease in their resistance to extinction, but the distinction in extinction treatments was retained nevertheless. Conversely, not only were the extinction treatments nondistinguishable in the lesioned pups, but these subjects generated enhanced resistance to extinction compared to their intact counterparts. These data from the lesioned pups are consistent with previous findings. That is, as noted above, avoidance extinction involves a response inhibitory component. If we view the lesion effects as exaggerating that deficit in pups, the present data suggest another instance of an age and lesion interaction.

## Suppresion of open field activity

A final experiment extends the present concerns to another task and measurement of aversive conditioning. A recent experiment compared suppression of open field activity in various age levels after lesions of the dorsomedial or the orbital frontal aspects of the prefrontal cortex. At the time of surgery, 108 rats were evenly distributed at age levels of 18–20 days, 30–32 days, and 100–130 days of age. Following recovery from surgery for the lesioned subjects at each age level, all subjects were assigned to one of three Pavlovian treatment groups for two successive days of training:

1. Tone/shock pairings (T/S). Four medial, lateral and sham subjects at each age level received 20 presentations of the 3000 Hz, 78 dB tonal CS, and tonal duration on each trial was 30 sec. The inescapable shock unconditioned stimulus of 0.5 s duration was randomly distributed during either the 1st, 2nd or 3rd 10 s block of the total CS duration. Trials were separated by varying intervals averaging 30 s.

2. Tone only (TO). Subjects of the three surgical procedures at each age level were treated the same as the T/S groups, except that the shocks were omitted.

3. No tone (NT). The remaining medial, lateral and sham groups at each age were placed in the chamber used for Pavlovian conditioning for approximately 20 min with neither tones nor shocks presented.

Four hours after each of the conditioning sessions, baseline activity in each subject was measured by recording the numbers of ambulations and rearings in the open-field chamber for a period of 20 min.

After the two days of conditioning and subsequent baseline activity measures, subjects were given three consecutive days of testing for activity suppression to tonal presentations in the open-field chamber. Tests of 20 min duration in the open-field involved recording of ambulations and rearings for each 30 s period of alternating intervals of tone-on and no tone.

The results of baseline activity measures are depicted in Fig. 1 for ambulations and Fig. 2 for rearings. The statistical analysis confirmed the impression from both figures that the baseline measures showed decreasing activity with increasing age, and both surgical treatments resulted in more activity than the sham operated subjects, with lateral subjects most active. The results of the activity suppression tests are shown in Figs. 3 and 4. Briefly, these data suggest that activity in the laterals given tones and shocks was most disrupted by tone presentations and shams showed least suppression. The lateral suppression tended to be greatest in the youngest subjects.

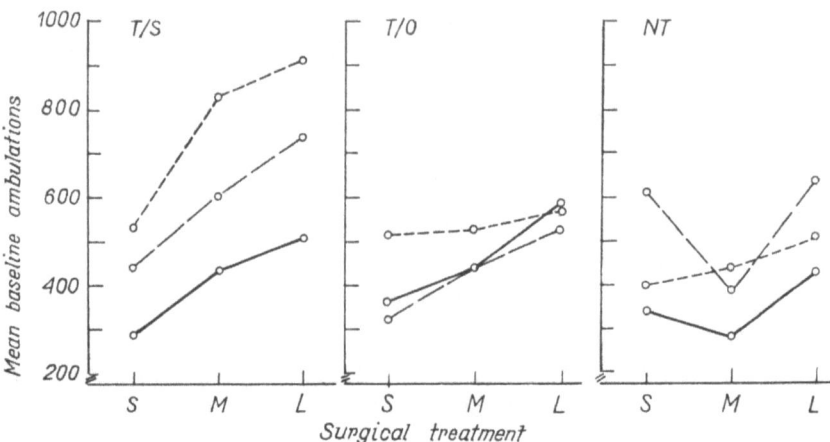

Fig. 1. Mean open field ambulations for 2 days of baseline testing following Pavlovian tone and shock pairings (T/S), tone only presetantions (T/O) or apparatus exposure for subjects of 18–20 days (dashed lines), 30–32 days (broken lines) and 100+ days (solid lines) at the time of lateral (L) or medial (M) prefrontal lesions or sham (S) control operations.

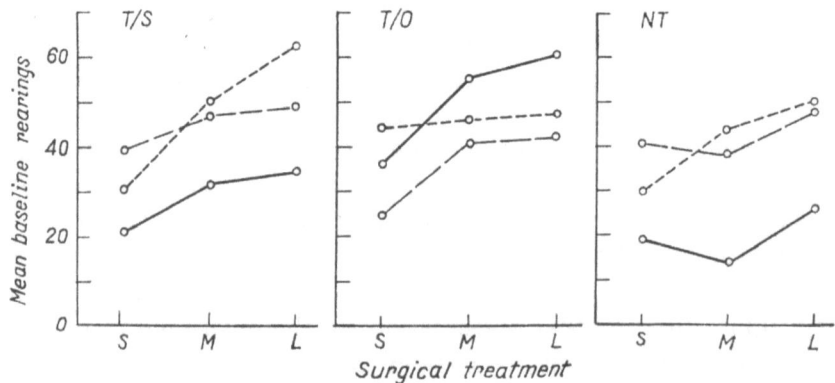

Fig. 2. Means of 2 days tests of baseline rearings after Pavlovian pairings (T/S), presentations of tone only (T/O), or no-tone (NT) exposure to apparatus cues only. Age level and surgery labels are the same as denoted in Fig. 1.

The inverse relationship between age and both activity measures during baseline measurement was most pronounced after tone and shock pairings. This evidence is consistent with ontogenic studies of behavioral arousal (19) reporting peak arousal at 15 days of age followed by a decline to an adult level by 30 days. Campbell and Raskin (20) examined this phenomenon under several conditions and concluded that fear in a novel environment is critical. The present findings of a strong transfer

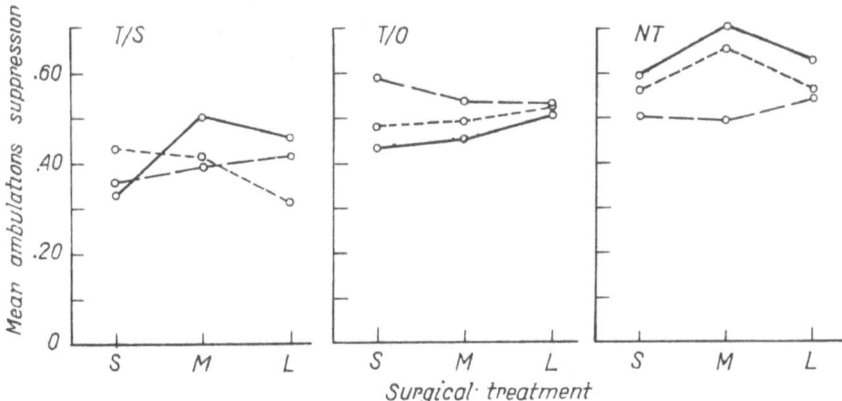

Fig. 3. Mean ambulations suppression ratios over 3 days of testing. Labels for Pavlovian experiences, age levels and surgical treatments are the same as indicated in Fig. 1.

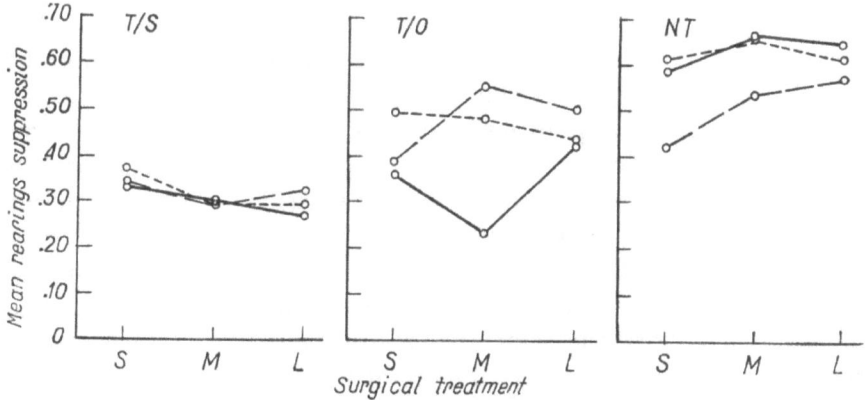

Fig. 4. Mean ratios of rearing suppression over 3 days of testing. Labels for Pavlovian experiences, age levels and surgical treatments are the same as indicated in Fig. 1.

effect from Pavlovian conditioning suggest that the circumstances for the emergence of differential arousal among ages were met. Moreover, the heightened ambulations and rearings in the lateral subjects and similar, although not as pronounced, increased activity in the medials support the results of Kolb (47).

Given the lateral performance in the tone/shock condition on the baseline activity measures, the extent of their suppression on the subsequent tests in somewhat surprising. That is, on the basis of their hyperactivity, it might be expected that they would show least suppression relative to the other groups with Pavlovian fear conditioning. However, within the tone/shock groups, ambulation suppression was greatest for the 18-day laterals and rearings suppression was most pronounced in the 18- and adult tone/shock laterals. It appears that Pavlovian extinction influences in the young tone/shock lateral subjects were minimal. This finding is consistent with a report of heightened activity and an inability to discriminate stages of appetively motivated extinction from acquisition following lateral prefrontal lesions in dogs (9). If the ratios of the laterals in the Pavlovian controls are compared, it may readily be seen that without the aversive Pavlovian experience, the laterals showed little suppression and some enhancement of responding as did subjects of the other surgical treatments that responded to the eliciting capacity of the novel tone during testing.

The question of why suppression was greatest in the youngest tone/shock lateral group remains. That is, despite the youngest group being tested at ages when they should emulate adults, some heightened activity, in contrast to suppression, should be present. However, the complexity of the present assessment of Pavlovian suppression of activity provides some insight. Specifically, the age-related deficit in response inhibitory systems may be interacting with the lesion effects to produce the persistence of suppression in the young, tone/shock lateral subjects. Consistent with our findings of lesion effects on passive and active avoidance, if spontaneous performance of the young subjects reflects characteristic response inhibiting deficits and the lateral lesions interfer with subjects' inability to extinguish, the net result of these joint processes would be a perseveration of the learned Pavlovian response. Such is the case presently, and these data are consistent with interpretations of behavioral ontogeny that stress the sequential maturation of excitatory and inhibiting centers in forebrain activity (21, 63). Moreover, the results suggest that the selectivity of cortical action (53) may be differentially affected by this sequential maturation due to age.

## CONCLUSIONS

Having reviewed the behavioral bases of ontogenic differences in aversively motivated performance and considered the neurobehavioral aspects of the prefrontal cortex in rats, data focusing on developmental differences in prefrontal lesion effects point to several generalizations. Specifically, these data indicate task-specific and age-specific implications of prefrontal damage. In the case of age, it appears that the complexities of the prefrontal cortex produce paradoxical relationships indicating the critical importance of the time at surgery. Concerning task requirements, the underlying role of response inhibitory processes results in radically different behavioral outcomes.

In the context of aversive conditioning, these experiments collectively suggest the need for creative strategies to adequately understand the interacting variables of age and task requirements relative to the prefrontal area. In this light, one interesting and potentially productive strategy involves an examination of developmental data in terms of the character of the response after prefrontal damage. The distinction between short- and long-latency responses made in studies by Zieliński (40, 97) may provide more subtle insight to the interaction of these variables. The analysis of response latencies, after prefrontal lesions in various species reveals differences between alimentary and both classical and instrumental defensive responses (96, 98). In alimentary and classically defensive response classes, prefrontal manipulations result in the shortening of response latencies and overall response disinhibition. In both active and passive instrumental defensive tasks, prefrontal lesions appear to result in a lengthening of response latencies and no disinhibition. It is intriguing to suggest that a similar analysis would differentiate the character of defensive responses within age levels as well.

Participation in the symposium and preparation of this paper were supported by a grant from the National Academy of Sciences, USA. The author is grateful to Prof. Kazimierz Zieliński and Dr Anna Kosmal for their comments on an earlier version of this paper.

## REFERENCES

1. AKERT, K. 1964. Comparative anatomy of the frontal cortex and thalmocortical connections. In J. M. Warren and K. Akert (ed.), The frontal granular cortex and behavior. McGraw Hill Book Co., New York, p. 372-396.
2. ALTMAN, J. and DAS, G. D. 1965. Autoradiographic and histological evidence of postnatal neurogenesis in rats. J. Comp. Neurol. 124: 319-336.
3. BAUM, M. 1970. Extinction of avoidance responding through response prevention (flooding). Psychol. Bull. 74: 276-284.
4. BECKSTEAD, R. H. 1976. Convergent thalamic and mesencephalic projections to the anterior medial cortex in the rat. J. Comp. Neurol. 166: 403-416.
5. BOLLES, R. C. and WOODS, P. S. 1964. The ontogeny of behavior in the albino rat. Anim. Behav. 12: 427-441.
6. BRENNAN, J. F. 1975. Differential response gradients to a Pavlovian safety signal following active avoidance training in young and adult rats. Anim. Learn. Behav. 3: 277-281.
7. BRENNAN, J. F. 1979. Differential extinction of 2-way active avoidance in young and adult rats. Dev. Psychobiol. 12: 27-37.

8. BRENNAN, J. F. and BARONE, R. J. 1976. Effects of differential cue availability in an active avoidance CS for young and adult rats. Dev. Psychobiol. 9: 237–244.
9. BRENNAN, J. F., KOWALSKA, D. and ZIELIŃSKI, K. 1976. Auditory frequency generalization with differing extinction influences in normal and prefrontal dogs trained in instrumental alimentary reflexes. Acta Neurobiol. Exp. 36: 475–516.
10. BRENNAN, J. F., POWELL, E. A. and VICEDOMINI, J. P. 1977. Differential effects of dorsomedial prefrontal lesions on active and passive avoidance in young and adult rats. Acta Neurobiol. Exp. 37: 151-177.
11. BRENNAN, J. F. and RICCIO, D. C. 1972. Stimulus generalization along dimensions of an active avoidance CS in young rats. Psychonomic Sci. 29: 170–172.
12. BRENNAN, J. F. and RICCIO, D. C. 1972. Persistance of dynamism effects in rats following auditory intensity differentiation. Learn. Motiv. 3: 259–271.
13. BRENNAN, J. F. and RICCIO, D. C. 1972. Stimulus control of shuttle avoidance in young and adult rats. Can. J. Psychol. 26: 361–373.
14. BRENNAN, J. F. and RICCIO, D. C. 1973. Stimulus control of avoidance behavior in rats following differential or non-differential Pavlovian training along dimensions of the CS. J. Comp. Physiol. Psychol. 85: 313–323.
15. BRUTKOWSKI, S. 1965. Functions of prefrontal cortex in animals. Physiol. Rev. 45: 721–746.
16. CAMPBELL, B. A. 1967. Learning in infra-primate mammals. In H. W. Stevenson, E. H. Hess and H. L. Rheingold (ed.), Early behavior: Comparative and developmental approaches. John Wiley Publ., New York, p. 43–71.
17. CAMPBELL, B. A. and CAMPBELL, E. 1962. Retention and extinction of learned fear in infant and adult rats. J. Comp. Physiol. Psychol. 55: 1–8.
18. CAMPBELL, B. A. and JAYNES, J. 1966. Reinstatement. Psychol. Rev. 73: 478–480.
19. CAMPBELL, B. A. and MABRY, P. D. 1972. Ontogeny of behavioral arousal: A comparative study. J. Comp. Physiol. Psychol. 81: 371–379.
20. CAMPBELL, B. A. and RASKIN, L. A. 1978. Ontogeny of behavioral arousal: The role of environmental stimuli. J. Comp. Physiol. Psychol. 92: 176–184.
21. CAMPBELL, B. A. and SPEAR, N. E. 1972. Ontogeny of memory. Psychol. Rev. 79: 215–236.
22. CAMPBELL, B. A. and TEGHTSOONIAN, R. 1958. Electrical and behavioral effects of different types of shock stimuli on the rat. J. Comp. Physiol. Psychol. 51: 185–192.
23. CLARK, C. and ISAACSON, R. 1965. Effect of bilateral hippocampal ablation on DRL performance. J. Comp. Physiol. Psychol. 59: 137–140.
24. CRAWFORD, M. 1977. Brief "response prevention" in a novel place can facilitate avoidance extinction. Learn. Motiv. 8: 39–53.
25. DĄBROWSKA, J. 1968. Effects of frontal lesions in black-white discrimination test in white rats. Acta Biol. Exp. 28: 197–203.
26. DIVAC, I. 1971. Frontal lobe system and spatial reversal in the rat. Neuropsychologia 9: 175–184.
27. DIVAC, I. 1972. Neostriatum and functions of prefrontal cortex. Acta Neurobiol. Exp. 32: 461–477.

28. DIVAC, I., BHORKLUND, A., LINDVELL, O. and PASSINGHAM, R. E. 1978. Converging projections from the mediodorsal thalamic nucleus and mesencephalic dopaminergic neurons to the neocortex in three species. J. Comp. Neurol. 190: 59–72.
29. DIVAC, I., KOSMAL, A., BJORKLUND, A. and LINDVALL, O. 1978. Subcortical projections to the prefrontal cortex in the rat as revealed by the horseradish peroxidase technique. Neuroscience 3: 785–796.
30. DOMESICK, V. 1972. Thalamic relationships of the medial cortex in the rat. Brain Behav. Evol. 6: 457–483.
31. EGGER, G. J. and LIVESEY, P. J. 1972. Age effects in the acquisition and retention of active and passive avoidance learning by rats. Develop. Psychobiol. 5: 343–351.
32. FEIGLEY, D. A., PARSONS, P. J. HAMILTION, L. W. and SPEAR, N. E. 1972. Development of habituation to novel environments in the rat. J. Comp. Physiol. Psychol. 79: 443–452.
33. FEIGLEY, D. A. and SPEAR, N. E. 1976. Effect of age and punishment conditions on long-term retention by the rat of active- and passive-avoidance learning. J. Comp. Physiol. Psychol. 73: 515–526.
34. FOX, S., KIMBLE, D. and LICKEY, M. 1964. Comparison of caudate nucleus and septal area lesions on two types of avoidance behavior. J. Comp. Physiol. Psychol. 58: 380–386.
35. FRIEMAN, J. P., ROHRBAUGH, M. and RICCIO, D. C. 1969. Age differences in the control of acquired fear by tone. Can. J. Psychol. 23: 230–244.
36. GOLDMAN, P. S., ROSVOLD, H. E. and MISHKIN, M. 1970. Evidence for behavioral impairment following prefrontal lobectomy in infant monkeys. J. Comp. Physiol. Psychol. 70: 454–455.
37. HEARST, E., BESLEY, S. and FARTHING, G. W. 1970. Inhibition and the stimulus control of operant behavior. J. Exp. Anal. 14: 373–409.
38. HULL, E., HAMILTON, K. L., ENGWALL, D. B. and ROSELLI, L. 1974. Effects of olfactory bulbectomy and peripheral deafferentation on reactions to crowding in gerbils (Meriones Ungiuculatus). J. Comp. Physiol. Psychol. 83: 247–254.
39. JACOBSEN, S. 1963. Sequence of myelinization in the brain of the albino rat: Cerebral cortex, thalamus and related structures. J. Comp. Neurol. 121: 5–27.
40. JAKUBOWSKA, E. and ZIELIŃSKI, K. 1979. Avoidance acquisition in cats as a function of temporal and intensity factors. Acta Neurobiol. Exp. 39: 67–86.
41. JARRARD, L. 1965. Hippocampal ablation and operant behavior in the rat. Psychon. Science 2: 115–116.
42. KIMBLE, D. 1963. The effects of bilateral hippocampal lesions in rats. J. Comp. Physiol. Psychol. 56: 273–283.
43. KIMBLE, G. A. 1961. Hilgard and Marquis' conditioning and learning. Appleton-Century Crofts Pub., New York, 590 p.
44. KIRBY, R. H. 1963. Acquisition, extinction and retention of an avoidance response in rats as a function of age. J. Comp. Physiol. Psychol. 56: 158–162.
45. KLEIN, S. B. and SPEAR, N. E. 1969. Influence of age on short-term retention of active avoidance-learning in rats. J. Comp. Physiol. Psychol. 69: 583–589.

46. KNOOK, H. L. 1965. The fibre connections of the forebrain. F. A. Davis Company, Philadelphia, 477 p.
47. KOLB, B. 1974. Social behavior of rats with chronic prefrontal lesions. J. Comp. Physiol. Psychol. 87: 466–474.
48. KOLB, B. 1977. Studies on the caudate putamen and the dorsomedial thalamic nucleus of the rat. Implications for mammalian frontolobe functions. Physiol. Behav. 18: 237–244.
49. KOLB, B. and NONNEMAN, J. 1975. Prefrontal cortex and the regulation of food intake in the rat. J. Comp. Physiol. Psychol. 88: 800–815.
50. KOLB, B. and NONNEMAN, A. J. 1976. Functional development of prefrontal cortex in rats continues into adolescence. Science 193: 335–336.
51. KOLB, B., NONNEMAN, A. and SINGH, R. 1974. Double dissociation of spatial impairments and perseveration following selective prefrontal lesions in rats. J. Comp. Physiol. Psychol. 87: 772–780.
52. KONORSKI, J. 1948. Conditioned reflexes and neuron organization. Cambridge Univ. Press, New York, 267 p.
53. LARSEN, J. K. and DIVAC, I. 1978. Selective ablations within the prefrontal cortex of the rat and performance of delayed alternation. Physiol. Psychol. 6: 15–17.
54. LEONARD, C. 1969. The prefrontal cortex of the rat. I. Cortical projection of the mediodorsal nucleus. II. Efferent connections. Brain Res. 12: 324–343.
55. LEONARD, C. 1972. The connections of the dorsomedial nuclei. Brain Behav. Evol. 6: 534–541.
56. LUBOW, R. E. 1965. Latent inhibition: Effect of frequency of non-reinforced pre-exposure of the CS. J. Comp. Physiol. Psychol. 60: 454–457.
57. ŁUKASZEWSKA, I. 1970. Frontal rats and some visual tests. Acta Neurobiol. Exp. 30: 33–42.
58. ŁUKASZEWSKA, I. 1971. Perseverative errors in normal and frontal rats in returning behavior test. Acta Neurobiol. Exp. 31: 101–109.
59. ŁUKASZEWSKA, I. 1972. Impairment of utilization of response produced cues after frontopolar lesions in rats. Acta Neurobiol. Exp. 32: 513–524.
60. ŁUKASZEWSKA, I. 1972. Distance discrimination in frontopolar rats. Acta Neurobiol. Exp. 33: 523–526.
61. ŁUKASZEWSKA, I. 1973. Frontopolar rats performance in Dashiell maze. Acta Neurobiol. Exp. 33: 491–496.
62. LYNCH, G. S. 1970. Separate frebrain systems controlling different manifestations of spontaneous activity. J. Comp. Physiol. Psychol. 70: 48–59.
63. MABRY, P. D. and CAMPBELL, B. A. 1974. Ontogeny of serotonergic inhibition of behavioral arousal in the rat. J. Comp. Physiol. Psychol. 86: 193–201.
64. MARKOWITSCH, H. J., PRITZEL, M. and DIVAC, I. 1978. The prefrontal cortex of the cat: Anatomical subdivisions based on retrograde labeling of cells in the mediodorsal thalamic nucleus. Exp. Brain Res. 32: 335–344.
65. MENDELL, P., REIN, H., HORTH-EDEL, S. and MENDELL, R. 1964. Distribution and metabolism of ribonucleic acid in the vertebrate central nervous system. In D. Richter (ed.), Comparative neurochemistry. Pergamon Press, New York.
66. MOORE, J. W. 1972. Stimulus control: studies of auditory generalization in rabbits. In A. H. Black and W. F. Prokasy (ed.), Classical conditioning:

Current theory and research. Appleton-Century-Crofts, New York, p. 206–230.
67. NAUTA, W. J. H. 1964. Some efferent connections of the prefrontal cortex in the monkey. In J. M. Warren and R. Akert (ed.), The frontal granular cortex and behavior. McGraw-Hill Book Co., New York, p. 397–409.
68. NAUTA, W. J. H. 1972. Neural associations of the frontal cortex. Acta Neurobiol. Exp. 32: 125–146.
69. NONNEMAN, A. and ISAACSON, R. 1973. Task dependent recovery after early brain damage. Behav. Biol. 8: 143–172.
70. NONNEMAN, A. and KOLB, B. 1974. Lesions of hippocampos or prefrontal cortex alter species-typical behaviors in the cat. Behav. Biol. 12: 41–54.
71. NONNEMAN, A., VOIGT, J. and KOLB, B. 1974. Comparisons of behavioral effects of hippocampal and prefrontal cortex lesions in the rat. J. Comp. Physiol. Psychol. 87: 249–260.
72. POWELL, E. 1963. Septal efferents revealed by axonal degeneration in the rat. Exp. Neurol. 8: 406–422.
73. RAISMAN, G. 1966. Neural connextions of the hypothalamus. Brit. Med. Bull. 22: 197–201.
74 RESCORLA, R. A. 1971. Summation and retardation tests of latent inhibition. J. Comp. Physiol. Psychol. 75: 77–81.
75. RICCIO, D. C. and MARRAZO, M. S. 1972. Effects of punishing active avoidance in young and adult rats. J. Comp. Physiol. Psychol. 79: 453–458.
76. RICCIO, D. C., ROHRBAUGH, M. and HODGES, L. A. 1968. Developmental aspects of passive and active avoidance learning in rats. Develop. Psychobiol. 1: 108–111.
77. ROBERTS, W., DEMBER, W. and BRODWICK, M. 1962. Alternation and exploration in rats with hippocampal lesions. J. Comp. Physiol. Psychol. 55: 655–700.
78. ROHRBAUGH, M., BRENNAN, J. F. and RICCIO, D. C. 1971. Control of two-way shuttle avoidance in rats by auditory frequency and intensity. J. Comp. Physiol. Psychol. 75: 324–330.
79. ROHRBAUGH, M. and RICCIO, D. C. 1968. Stimulus generalization of learned fear in infant and adult rats. J. Comp. Physiol. Psychol. 66: 530–533.
80. ROSVOLD, H. E. 1972. The frontal lobe system: cortical-subcortical interrelationships. Acta Neurobiol. Exp. 32: 439–460.
81. ROSVOLD. H. E. and SZWARCBART, M. 1964. Neural structures involved in delayed-response performance. In J. M. Warren and K. Akert (ed.), The frontal granular cortex and behavior. Mc Graw-Hill Book Co., New York.
82. SCHULENBURG, J., RICCIO, D. C. and STIKES, E. R. 1974. Acquisition and retention of a passive avoidance response as a function of age in rats. J. Comp. Physiol. Psychol. 74: 75–83.
83. SHEPERD, G. 1974. The synaptic organization of the brain. An introduction. Oxford Univ. Press, Toronto, 288–328.
84. SILVESTRI, R., ROHRBAUGH, M. and RICCIO, D. C. 1970. Conditions influencing the retention of learned fear in young rats. Develop. Psychol. 2: 389–395.
85. SKINNER, J. E. 1971. Neuroscience: A laboratory manual. Saunder Press, Philadelphia.
86. SPENCE, K. W. 1937. The differential response in animals to stimuli varying within a single dimension. Psychol. 44: 430–444.

87. SNYDER, D. and ISAACSON, R. 1965. Effects of large and small bilateral hippocampal lesions on two types of passive-avoidance responses. Psychol. Rep. 16: 1272–1290.
88. TEITELBAUM, H. and MILNER, P. 1963. Activity changes following partial hippocampal lesions in rats. J. Comp. Physiol. Psychol. 56: 284–289.
89. THOMPSON, R. 1964. A note on cortical and subcortical injuries and avoidance learning by rats. In J. M. Warren and K. Akert (ed.), The frontal granular cortex and behavior. McGraw-Hill Book Co., New York, p. 16–25.
90. THOMPSON, R. W. 1964. Transfer of avoidance learning between normal and functionally decorticate. J. Comp. Physiol. Psychol. 57: 321–325.
91. WATSON, J. B. 1903. Animal education: an experimental study on the physical development of the white rat correlated with the growth of its nervous system. Contr. Phil. 4: 5–122.
92. WHIMBEY, A. E. and DENENBERG, V. H. 1966. Programming life histories: creating individual differences by the experimental control of early experiences. Multivariate Behav. Res. 1: 279–286.
93. WIKMARK, R., DIVAC, I. and WEISS, R. 1973. Retention of spatial delayed alternation in rats with lesions in the frontal lobes. Brain Behav. Evol. 8: 329–339.
94. WILSON, L. M., PHINNEY, R. L. and BRENNAN, J. F. 1974. Age-related differences in avoidance behavior in rats following CS preexposure. Develop. Psychobiol. 7: 421–427.
95. ZEMAN, W. and INNES, J. R. M. 1963. Craige's Neuroanatomy of the rat. Academic Press, New York.
96. ZIELIŃSKI, K. 1966. Retention of the avoidance reflex after prefrontal lobectomy in cats. Acta Biol. Exp. 26: 167–181.
97. ZIELIŃSKI, K. 1972. Effects of prefrontal lesions on avoidance and escape reflexes. Acta Neurobiol. Exp. 32: 393–416.
98. ZIELIŃSKI, K. 1975. Effects of prefrontal lesions on defensive conditioned reflexes (in Polish). Acta Physiol. Pol. 26: Suppl. 11, 83–142.
99. ZIELIŃSKI, K. and CZARKOWSKA, J. 1974. Quality of stimuli and prefrontal lesions effects on reversal learning in go–no go avoidance reflex differentiation in cats. Acta Neurobiol. Exp. 34: 43–68.

James F. BRENNAN, Department of Psychology, University of Massachusetts-Boston, Harbor Campus, Boston, Massachusetts 02125, USA.

Lecture prepared for the Warsaw Colloquium on Instrumental
Conditioning and Brain Research
May 1979

# CENTRAL MEDIATION OF HORMONAL INFLUENCES ON INSTRUMENTAL AVOIDANCE CONDITIONING

F. R. BRUSH and S. M. FRALEY

Experimental Psychology Laboratory, Syracuse University
Syracuse, New York, USA

*Abstract.* Two behaviorally active hormones of the pituitary-adrenal system are adrenocorticotropic hormones (ACTH) and corticosterone, and their behavioral effects are facilitation and inhibition of performance of previously learned avoidance responses, respectively. Their uptake, distribution and effects on central nervous system are reviewed. Hypothalamic neurotransmitter control of corticotropin releasing hormone (CRH) is described together with hypothalamic and extra-hypothalamic (hippocampal) regulation of pituitary-adrenal activity. Extra-hypothalamic mediation of the behavioral effects of ACTH is evaluated. Recent isotopic mappings of the efferents of the hippocampal formation have identified pathways from hippocampal subiculum to hypothalamus and posterior lateral and anterior thalamic nuclei. The evidence reviewed suggests a complex circuit involving hippocampal subiculum, thalamus and hypothalamus may be involved both in regulating pituitary-adrenal responses to stress and in mediating the effects of ACTH on avoidance behavior.

## INTRODUCTION

Instrumental avoidance behavior is complexly influenced by a number of structures in the central nervous system (CNS) and by some of the hormones of the pituitary-adrenal axis. These hormones, in turn, are regulated by hypothalamic and possibly extra-hypothalamic regions of the CNS. In general, most of the extra-hypothalamic structures that have been implicated in pituitary-adrenal regulation lie within the Papez circuit (69, 70), now identified as the limbic midbrain area (65), which has long been known to have significant effects on avoidance behavior (60). Thus, the brain regions important for avoidance behavior are also important for pituitary-adrenal regulation. In this paper we wish to examine the possibility that the same structures are involved in mediating the effects of pituitary-adrenal hormones on avoidance behavior.

HORMONAL EFFECTS ON AVOIDANCE BEHAVIOR AND THE CNS

When addressing the role of the CNS in mediating the effects of pituitary-adrenal hormones on avoidance behavior the following questions arise: 1) are there demonstrable and reliable effects of the hormones on avoidance behavior, 2) to what extent and where are these hormones found in the CNS and 3) what neural events are modulated or changed as a result? These issues will be taken up in turn.

*Effects of pituitary-adrenal hormones on avoidance behavior*

Two pituitary-adrenal hormones that are important for avoidance behavior are adrenocorticotropic hormone (ACTH), a 39-amino-acid peptide from the anterior pituitary gland, and corticosterone (B), a glucocorticosteroid from the adrenal cortex. The first (N-terminal) 24 amino acids of ACTH ($ACTH_{1-24}$) are required for full steroidogenic action in the adrenal cortex, whereas there is little or no steroidogenic effect from $\alpha$-MSH (melanocyte stimulating hormone) which is identical to $ACTH_{1-13}$. By comparing the behavioral effects of purified $ACTH_{1-39}$ or synthetic $ACTH_{1-24}$, both of which stimulate the adrenal, with those of $\alpha$-MSH, which does not, it is possible to separate the behavioral effects of the pituitary peptide, ACTH, from those of the glucocorticoid, B.

The most notable effect of ACTH on avoidance behavior is to improve performance of previously learned avoidance responses, an effect which has been demonstrated in numerous experiments involving various conditions and treatments. For example, Bohus and Endroczi (7) found that exogenous ACTH improved avoidance performance early in training and reduced intertrial responding both early and late in training in both intact and adrenalectomized rats. DeWied showed that hypophysectomized and adenohypophysectomized rats were grossly deficient in learning two-way shuttle and pole-jump avoidance responses and that this deficiency could be reversed, in a dose-related fashion, by exogenous administration of ACTH (17, 19). All of the behavioral effects of ACTH appear to be independent of its steroidogenic action because essentially identical behavioral effects are seen with $ACTH_{1-13}$, $ACTH_{1-10}$, and $ACTH_{4-10}$ (35). In general, the adrenocorticosteroids are either without effect or have an effect opposite to that of ACTH, i.e., B tends to inhibit performance of previously learned avoidance responses (18).

*Uptake and distribution of pituitary-adrenal hormones in rat brain*

It is obvious that in order for these hormones to influence behavior they must be available to the CNS, and since thay are secreted into the systemic circulation, they must bridge the blood-brain barrier in order to be effective. In the case of the corticoids, McEwen et al. (58) have shown that tritiated B, injected either intraperitoneally or intravenously, is selectively concentrated in various brain structures and that brain concentrations reach their maximum about 20–30 min after administration. In general, concentration of B in brain areas, relative to that found in cortex, is best described by the following rank order: Hippocampus > pituitary > septum > amygdala > hypothalamus (2, 45, 56). In brain areas other than hippocampus clearance of the hormone parallels the decline in plasma. Thus, there appears to be a free exchange of the hormone between the brain and blood (57, 59). In contrast to the other areas, B is retained by cell nuclei in the hippocampus and septum for at least 2 h after administration (58), but hippocampal binding sites saturate at physiological doses of B, whereas septal binding sites, for example, do not. McEwen et al. (58) found evidence of a dorsal-ventral gradient of concentration of labelled B in the hippocampus, the concentration being highest in dorsal and lowest in ventral areas. Concentration of B also tended to be highest in the cells of the CA1 and CA2 fields and lowest in the cells of the CA3 field (56).

The distribution of ACTH in the CNS is a matter of controversy, in part because of measurement difficulties. However, van Riezen et al. (85) intravenously injected labelled analogues of behaviorally active ACTH fragments and were able to recover $1 \times 10^{-4}$ of the administered dose in brain tissue 15 min later. Other investigators have found evidence of endogenous ACTH in brain (37, 47, 53), although the route of entry is under discussion (3, 62) and is thought by some to be the result of leakage from the sella turcica (62) whereas others argue that brain tissues may be capable of ACTH synthesis (36, 37, 53).

*CNS modulation by pituitary-adrenal hormones*

ACTH or its absence has a number of dramatic and reliable effects on CNS activity. For example, $ACTH_{4-10}$ has been found to increase protein metabolism and the incorporation of amino acid precursors into brain protein (74), and hypopsysectomy alters brain protein chemistry in complex ways (33). Effects of ACTH or fragments of ACTH can be seen at the level of reflex activity in the spinal cord (49, 50) and at the level of whole brain EEG where, for example, it has been shown to disinhibit the final (synchronized EEG) stage of habituation (25). In contrast to these effects, corticosteroids such as cortisone and cortisol have been shown to increase the amplitude of sciatic nerve evoked response in brain stem and hypothalamus in rats (26, 28) and to produce sedation and inhibition, but these effects can be complexly related to dose (38). In some instances, hippocampal and hypothalamic recordings suggest that ACTH and B may have opposite effects on single unit activity (71, 79, 80). The safest inference may be that the electrophysiological effects of these hormones, which include increased amplitude of evoked potentials (30) and EEG activation (29), are dependent on the species and the specific structures being studied.

## Conclusion

It seems, in view of these findings, that pituitary-adrenal hormones significantly influence avoidance behavior, that they are present in CNS, ACTH in hypothalamus and perhaps elsewhere and B in extra-hypothalamic areas, and that they can significantly modify the electrophysiological activity of these structures. We now turn to the question of the regulation of the pituitary-adrenal system by hypothalamic and extra-hypothalamic regions of the CNS.

## CNS REGULATION OF PITUITARY-ADRENAL ACTIVITY

There are two aspect of CNS regulation of pituitary-adrenal activity that we wish to discuss: 1) neurotransmitter control of the hypothalamic releasing hormone for ACTH, i.e., corticotropin releasing hormon (CRH) and 2) feedback regulation by corticosteroids of ACTH release in stress. These will be taken up in turn.

## Neurotransmitter control of CRH

It is generally accepted that CRH is secreted by neurons whose cell bodies are diffusely distributed in the hypothalamus from supra-chiasmatic, paraventricular and arcuate neuclei to posterior hypothalamic sites as caudal as the mammillary bodies (39a, 40, 68). The CRH secreting neurons are presumed to terminate adjacent to portal vessels in the median eminence. Excitatory and inhibitory inputs that control the CRH neurons are integrated in the hypothalamus, and in vitro analyses of the hypothalamus of the rat have shown that CRH secretion is stimulated by acetylcholine (Ach) and serotonin (5-HT), the action of 5-HT being dependent on a cholinergic interneuron (40, 41). Inhibitory control is exerted by norepinephrine (NE), gamma aminobutyric acid (GABA) and melatonin (40, 41). In general, these stimulatory and inhibitory effects have been confirmed by others using both in vivo and other in vitro techniques (1, 9, 10, 48). The CRH is carried, via the portal circulation, to the anterior pituitary where it stimulates release, and synthesis, of ACTH, which, in turn, stimulates synthesis and subsequent release of B from the adrenal cortex. Free (unbound) plasma B then stimulates negative feedback mechanisms which regulate further secretion of CRH and ACTH.

## Feedback regulation of CRH and ACTH by corticoids

*Hypothalamic regulation.* There are two temporally distinct negative feedback mechanisms that control pituitary-adrenal activity. Both operate at both the hypothalamic and the pituitary levels (38, 41, 42), although the hypothalamic feedback tends to override the effects seen at the lower pituitary level. The first of these mechanisms is called fast feedback, and it operates over a period of minutes. It is characterized by rate sensitivity and saturation. Rate sensitivity is shown by the fact that inhibition of ACTH release is induced by a rapid increase in the plasma concentration of exogenous or endogenous B. Rates of increase in excess of 1.3 µg/100 ml/min will block further release of ACTH in response to most stressors. Rates of increase below that value are ineffective (39). Initially, fast feedback is independent of the absolute level of plasma B. However, after a sustained, but moderate, elevation of plasma B, there is no rate of increase of plasma steroid concentration that will contemporaneously inhibit ACTH release. This is the saturation phenomenon (39). This fast feedback mechanism, then, is one that can respond quickly to the dynamic changes in hormone concentration during acute stress (rate sensitivity) and that can shut itself off and allow continued ACTH secretion during prolonged stress (saturation). There is evidence to suggest that the fast feedback effect in the hypothalamus works by inhibition of CRH release which may be a surface membrane, receptor binding phenomenon (40, 41).

The second negative feedback mechanism is called slow or delayed fedback, and it operates over a period of hours or days. Exogenous administration of B can induce a period of inhibition of ACTH release one or more hours after injection, by which time the plasma B concentration has returned to basal or near basal levels. The larger the dose of exogenous B the longer delayed is the period of maximal inhibition. Thus slow feedback shows neither rate sensitivity nor saturation. The adaptive significance of this mechanism may be that it permits modulation of the overall excitability of the pituitary-adrenal axis. Slow feedback may require intracellular binding of B and probably reflects influences on CRH synthesis as well as release (39a, 40, 41, 55).

*Extra-hypothalamic regulation.* Although the hypothalamus may be the final common path for control of the pituitary-adrenal system, recent analyses suggest that the concept of a hypophysiotropic region of the hypothalamus is no longer viable. It is now thought that many remote brain regions may communicate chemically with the hypothalamus via the cerebrospinal fluid of the ventricles (75, 81, 82), and a number of structures within the Papez–Nauta limbic brain have been assumed to exert direct or indirect neural control of the pituitary-adrenal axis. One such structure, the hippocampus, has been thought to have predominantly inhibitiory control over pituitary-adrenal activity (31, 54, 88, 89).

Afferent inputs to the hippocampus arise principally in midbrain, septum and entorhinal cortex. 5-HT and NE mediated inputs to the hippocampus originate in the midbrain raphe nuclei and locus coeruleus, respectively (63, 77, 78), and projections from medial septal nuclei to the hippocampus provide Ach mediated inputs via the fornix (67). Afferents from cingulate cortex to hippocampus originate in part in the mammillary bodies and anterior thalamus.

There are three efferent pathways from hippocampus to hypothalamus: 1) From hippocampus proper, precommissural efferents traverse the septal area and reach the anterior and lateral hypothalamus via the medial forebrain bundle (MFB). Since this multisynaptic tract contains ascending and descending fibers from various limbic regions, it probably also serves to interconnect many other limbic regions (15). 2) From the ventral subiculum adjacent to the hippocampal CA1 field, efferents join the medial corticohypothalamic tract and terminate in the ventromedial hypothalamus and medial mammillary region (61, 72, 83, 84). 3) From dorsal and ventral subiculum and from pre- and para-subiculum efferents in the fimbria, fornix and posterior thalamic radiation terminate extensively in the mammillary bodies and in dorsolateral and posterolateral thalamic nuclei (61, 72, 83, 84). Thus efferents from the hippocampal formation project throughout the rostral-caudal extent of the hypothalamus. CRH secreting neurones are distributed throughout the same hypothalamic regions, so it appears that there is sufficient hippocampal input to the appropriate hypothalamic areas to support the hypothesis that the hippocampus exerts some control over pituitary-adrenal activity. Furthermore, a number of investigators have found that hippocampal units are responsive to a variety of sensory inputs that normally activate the pituitary-adrenal axis (73, 76), and stimulation of dorsal hippocampus and fornix influences unit activity in the median eminence and mammillary bodies (46, 52).

Electrical stimulation of the hippocampus has resulted in both increased and decreased pituitary-adrenal activity. The predominant effects, however, seem to be decreases. For example, early work by Endroczi and colleagues (23, 24) showed that reduced ACTH secretion resulted from low (12 Hz) frequency stimulation. And low frequency stimulation has been shown to inhibit the pituitary-adrenal response to pain or cold stress (22) or to presentation of a Pavlovian conditioned aversive stimulus (66). On the other hand, increased pituitary-adrenal activity was induced by high (120 and 250 Hz) frequency stimulation (11, 23, 24), and in one case even low (25 Hz) frequency stimulation during the a.m. trough of the circadian rhythm of plasma B concentration resulted in increased pituitary-adrenal activity (11).

As we have seen, endogenous steroids are selectively bound by hippocampal cell nuclei, and electrical stimulation, presumably some of those same cells that bind corticoids, frequently results in inhibition of pituitary-adrenal activity. Implantation of steroids might therefore be expected to activate these cells and inhibit ACTH release. In accordance with this hypothesis, Davidson and Feldman (14) found that implants of dexamethasone in posterior and lateral hippocampus and in hippocampal commissure and precommissural fornix inhibited the compensatory adrenal hypertrophy that normally follows unilateral adrenalectomy. Others found no effects of dorsal implants of cortisol (8, 27), although it should be noted that the hippocampus does not preferentially bind cortisol, so it might be expected that such implants would be ineffective (56).

The effects of electrical stimulation and steroid implants seem relatively consistent and suggest a predominantly inhibitory role for hippocampus in regulating pituitary-adrenal activity. Lesions of the hippocampus would thus be expected to release these inhibitory effects and permit supra-normal responses to stress. However, a number of studies of lesions of the hippocampus, dorsal or ventral or both, failed to find effects of these lesions on pituitary-adrenal responses to stress (15, 43, 44, 51). However, Conforti and Feldman (12) found that dorsal hippocampal ablations resulted in a significant reduction in the magnitude of the pituitary-adrenal response to sciatic nerve stimulation (under barbiturate anesthesia). There were no effects of dorsal hippocampal damage on the magnitude of the responses to flashing light or ether stress. On the other hand, Murphy et al. (64) inflicted massive ablations of dorsal and lateral hippocampus and found increased plasma concentrations in response to stress (immobilization or immobilization plus electric shock to the feet).

In contrast to the above results, Fraley (32) studied the effects of small radio frequency lesions or more extensive aspiration lesions of the dorsal hippocampus and found neither a change in the pituitary-adrenal response to foot shock nor a change in slow feedback inhibition of the pituitary-adrenal response to the same stressor. She also failed to find any evidence of dorsal hippocampal function in mediation of or slow feedback inhibition of the pituitary-adrenal response to light or ether.

*Conclusion*

On the basis of these results we can safely conclude that the hypothalamus is the final common pathway for efferent control of pituitary-adrenal activity by the CNS. Fast and slow feedback mechanisms have been observed there in the in vitro preparation of Jones (40). Although the evidence is hardly consistent, the hypothesis that extra-hypothalamic limbic structures such as hippocampus may exert some control over pituitary-adrenal activity cannot be ruled out. The anatomy and physiology of the hippocampus seem appropriate to support such a role for that structure. Certainly the neural interconnections between hippocampus and hypothalamus and selective uptake of corticosteroids in the hippocampus are adequate for the role. Electrical stimulation of dorsal hippocampus often, though not always, results in inhibition of pituitary-adrenal activation, and implantation of the appropriate steroids can induce inhibitory effects. The conflicting results of the ablation studies may be resolvable on two accounts: 1) Many of the experiments employ systemic stressors that can act directly on the pituitary and/or median eminence, so that not even hypothalamic release of CRH is required for a normal elevation of plasma B (21). 2) There may be important differences among experiments as to which tissues were destroyed and which spared. Fraley's careful analysis of her lesions and reports of some others (12, 64) suggest that a critical structure may be the subiculum and its projections via the fimbria. In Fraley's experiments these structures were spared. In Conforti and Feldman and in Murphy et al., they were not. Since the response a shock stress was not altered in Fraley's experiments but was modified in the other two, further research to identify the functional role of the subiculum and its projections seems warranted.

### HIPPOCAMPUS AS MEDIATOR OF THE BEHAVIORAL EFFECTS OF PITUITARY-ADRENAL HORMONES

There can be little doubt that hippocampal structures significantly influence avoidance learning and performance. Black et al. (4) have recently reviewed the extensive literature on this topic so we will not do so here. However, a number of recent studies have found some interesting effects of ACTH or ACTH fragments on hippocampal theta activity (20, 34) which led Bohus (5) to conclude that ACTH stimulated or increased the excitability of the theta-generating network, thus providing a possible mechanism for behavioral facilitation by ACTH. Even more recently, vanWimersma Greidanus and deWied (87) were able to eliminate the effects of $ACTH_{4-10}$ on avoidance performance in extinction with relatively discrete lesions in the dorsal hippocampus. Unfortunately, detailed histological reconstructions were not provided in that report.

After a thorough review of the existing literature, Bohus (5) in 1975 suggested that the behavioral and endocrine functions of the hippocampus might be spatially separated, the behavioral elements being dorsal, the endocrine elements ventral. Recent isotopic mappings of the efferent projections of the hippocampal formation by Swanson and Cowan (83, 84) and by Meibach and Siegel (61) provide some anatomical basis for such functional differentiation. Earlier degeneration studies indicated that the postcommissural fornix projection to the mammillary bodies was composed of axons from cell bodies located primarily in the CA1 field of the hippocampus and that the medial corticohypothalamic tract to the medial basal hypothalamus was composed of axons originating in the ventral subicular portion of the hippocampal formation. The recent data suggest, however, that the postcommissural axons that enter the mammillary region of the hypothalamus also originate in the subicular areas, either only in the dorsal (84) or in both posterior dorsal and ventral portions (61). Furthermore, there is some topographical organization of the hippocampal efferent target sites in the mammillary region and in the septum.

These efferent projections of the various levels of the hippocampal formation may provide the anatomical substrate for differential control of behavior and endocrine function. From Fraley's analysis of the lesion studies, it was suggested that whether hippocampal lesions produced a significant endocrine effect depended on whether the subiculum was destroyed or its projections disrupted. If damage to that system occurred, changes in pituitary-adrenal activity were observed, despite significant variation in the extent of damage elsewhere. Conversely, if that system were spared, no alteration of pituitary-adrenal function was found. Whether the lesions involving the subicular system increased or decreased pituitary-adrenal responses to stress probably depends on the precise locus of the lesion within that system and on the kind of stress used. This suggests that the hippocampal subiculum, with its projections to the hypothalamus may be a system that normally provides an input to regulate the pituitary-adrenal response to some neurogenic stressors. The input to the hypothalamus is mixed, both inhibitory and excitatory elements being present. But the inhibitory elements probably predominate. Lesions that destroy essential subicular elements or transect key projections to the mammillary bodies or ventral medial hypothalamus would be expected to alter the pituitary-adrenal response, most often by increasing the response, but on occasions decreasing it.

These results suggest that the hippocampal subiculum and its projections may be important in controlling avoidance behavior and pituitary-adrenal activity and in mediating the effects of hormones of that system on avoidance behavior. However, other brain areas have also been implicated in the hormone-behavior interactions. Earlier studies found that the parafascicular area of the posterior thalamus is a primary region where pituitary-adrenal hormones were exerting their effects on avoidance performance (6, 16, 86). The recent isotopic mappings, however, reported evidence of hippocampal subicular projections only to posterior lateral and anterior thalamic nuclei, not to parafascicular nuclei (83, 84). Further analyses may reveal such projections from hippocampal subiculum or intrathalamic connections between posterior lateral and parafascicular nuclei. If found, such projections or interconnections would provide the needed substrate for a complex circuit involving hippocampal subiculum, thalamus and hypothalamus that may be important both in regulating the pituitary-adrenal responses to stress and in mediating the effects of ACTH on avoidance behavior. Further research on this seems warranted by the data.

We wish to thank J. C. Froehlich for her helpful and critical comments on an earlier draft of this paper. We also appreciate the use of the facilities of the Oregon Regional Primate Research Center where the first author was a Visiting Scientist when this paper was written.

## REFERENCES

1. ABE, K. and HIROSHIGE, T. 1974. Changes in plasma coricosterone and hypothalamic CRF levels following intraventricular injection or drug-induced changes in brain biogenic amines in the rat. Neuroendocrinology 14: 195–211.
2. AXELROD, L. R. 1971. The metabolism of corticosteroids by incubated and perfused brain tissues. *In* D. H. Ford (ed.), Influence of hormones on the nervous system. S. Karger, Basel, p. 74–84.
3. BERGLAND, R. M. and PAGE, R. B. 1979. Pituitary-brain vascular relations: A new paradigm. Science 204: 18–24.
4. BLACK, A. H., NADEL, L. and O'KEEFE, J. 1977. Hippocampal function in avoidance learning and punishment. Psychol. Bull. 84: 1107–1129.
5. BOHUS, B. 1975. The hippocampus and the pituitary-adrenal system hormones.

In R. L. ISAACSON, and K. PRIBRAM (ed.), The hippocampus. Vol. 1. Plenum Press, New York, p. 323–353.

6. BOHUS, B. and deWIED, D. 1967. Failure of α-MSH to delay extinction of conditioned avoidance behavior in rats with lesions in the parafascicular nuclei of the thalamus. Physiol. Behav. 2: 221–223.
7. BOHUS, B. and ENDROCZI, E. 1965. The influence of pituitary-adrenocortical function on the avoiding conditioned reflex activity in rats. Acta Physiol. Acad. Sci. Hung. 26: 183–189.
8. BOHUS, B. and LISSAK, K. 1967. The sites of feedback action of corticosteroids at extrahypothalamic levels. Gen. Comp. Endocrinol. 9: 434–435.
9. BUCKINGHAM, J. L. and HODGES, J. R. 1977. The use of corticotrophin production by adenohypophysial tissue *in vitro* for the detection and estimation of potential corticotrophin releasing factors. J. Endocrinol. 72: 187–193.
10. BUCKINGHAM, J. L. and HODGES, J. R. 1977. Corticotrophin releasing hormone activity of rat hypothalamus *in vitro*. J. Endocrinol. 73: 30P.
11. CASADY, R. L. and TAYLOR, A. N. 1976. Effect of electrical stimulation of the hippocampus upon corticosteroid levels in the freely-behaving non-stressed rat. Neuroendocrinology 20: 68–78.
12. CONFORTI, N. and FELDMAN, S. 1976. Effects of dorsal fornix section and hippocampectomy on adrenocortical responses to sensory stimulation in the rat. Neuroendocrinology 22: 1–7.
13. COOVER, G. D. GOLDMAN, L. and LEVINE, S. 1971. Plasma corticosterone levels during extinction of a lever-press response in hippocampectomized rats. Physiol. Behav. 7: 727–732.
14. DAVIDSON, J. M. and FELDMAN, S. 1967. Effects of extrahypothalamic dexamethasone implants on the pituitary-adrenal system. Acta Endocrinol. 55: 240–246.
15. DeGROOT, J. 1966. Limbic and other neural pathways that regulate endocrine function. In L. Martini and W. F. Ganong (ed.), Neuroendocrinology. Vol. 1. Academic Press, New York, 81–104.
16. DELACOUR, J. 1970. Specific function of a medial thalamic structure in avoidance conditioning in the rat. In D. deWied and J. A. W. M. Weijnen (ed.), Pituitary, adrenal and the brain. Prog. Brain Res. Vol. 32. Elsevier, Amsterdam, p. 158–170.
17. deWIED, D. 1964. Influence of anterior pituitary on avoidance learning and escape behavior. Am. J. Physiol. 207: 255–259.
18. deWIED, D. 1967. Opposite effects of ACTH and glucocorticosteroids on extinction of conditioned avoidance behavior. Excerpta Med. Int. Congr. Ser. 132: 945–951.
19. deWIED, D. 1969. Effects of peptide hormones on behavior. In W. F. Ganong and L. Martini (ed.), Frontiers in neuroendocrinology, New York, p. 97–140.
20. deWIED, D., BOHUS, B., vanREE, J. M. and URBAN, I. 1978. Behavioral and electrophysiological efects of peptides related to lipotropin ($\beta$-LPH). J. Pharmacol. Exp. Ther. 204: 570–580.
21. DUNN, J. and CRITCHLOW, V. 1969. Pituitary-adrenal response to stress in rats with hypothalamic islands. Brain Res. 16: 395–403.
22. DUPONT, A., BASTARACHO, E., ENDROCZI, E. and FORTIER, C. 1972. Effect of hippocampal stimulation on the plasma thyrotropin (TSH) and corticosterone response to acute cold exposure in the rat. Can. J. Physiol. Pharmacol. 50: 364–367.

23. ENDROCZI, E. and LISSAK, K. 1960. The role of mesencephalon, diencephalon and archicortex in the activation and inhibition of the pituitary-adrenocortical system. Acta Physiol. Acad. Sci. Hung. 17: 39–55.
24. ENDROCZI, E., LISSAK, K., BOHUS, B. and KOVACS, S. 1959. The inhibitory influence of archicortical structures on pituitary-adrenal function. Acta Physiol. Acad. Sci. Hung. 16: 17–22.
25. ENDROCZI, E., LISSAK, K., FEKETE, T. and deWIED, D. 1970. Effects of ACTH on EEG habituation in human subjects. In D. deWied and J. A. W. M. Weijnen (ed.), Pituitary, adrenal and the brain. Prog. Brain Res. Vol. 32, Elsevier, Amsterdam, p. 254–261.
26. ENDROCZI, E., LISSAK, K., KORANYI, L. and NYAKAS, Cs. 1968. Influence of corticosteroids on the hypothalamic control of sciatic-evoked potentials in the brain stem reticular formation and the hypothalamus. Acta Physiol. Acad. Sci. Hung. 33: 375–382.
27. FELDMAN, S., CONFORTI, N. and DAVIDSON, J. M. 1972. Failure of corticosteroid implants in extrahypothalamic limbic structures to inhibit adrenocortical responses to stressful stimuli in the rat. Isr. J. Med. Sci. 8: 588–593.
28. FELDMAN, S. and DAFNY, N. 1970. Effects of adrenocortical hormones on the electrical activity of the brain. In D. deWied and J. A. W. M. Weijnen (ed.), Pituitary, adrenal and the brain, recent. Prog. Brain Res. Vol. 32. Elsevier, Amsterdam, p. 90–100.
29. FELDMAN, S. and DAVIDSON, J. M. 1966. Effects of hydrocortisone on electrical activity, arousal threshold and evoked potentials in the brains of chronically implanted rabbits. J. Neurol. Sci. 3: 462–472.
30. FELDMAN, S., TODT, J. C. and PORTER, R. W. 1961. Effect of adrenocortical hormones on evoked potentials in the brain stem. Neurology 11: 109–115.
31. FORTIER, C. 1966. Nervous control of ACTH secretion. In G. W. Harris and B. T. Donovan (ed.), The pituitary gland, Vol. 2. Univ. Calif. Press, Berkeley, p. 195–234.
32. FRALEY, S. M. 1979. The effects of dorsal hippocampal lesions on mediation and feedback inhibition of pituitary-adrenal activity in response to light, electric shock and ether stimuli in rat. Unpublished Ph. D. Dissertation, Experimental Psychology Laboratory, Department of Psychology, Syracuse University.
33. GISPEN, W. H. and SCHOTMAN, P. 1973. Pituitary-adrenal system, learning and performance: Some neurochemical aspect. In E. Zimmerman, W. H. Gispen, B. H. Marks and D. deWied (ed.), Drug effects on neuroendocrine regulation. Prog. Brain Res. Vol. 39. Elsevier, Amsterdam, p. 443–448.
34. GRAY, J. A. and BALL, G. G. 1970. Frequency-specific relation between hippocampal theta rhythm, behavior and amobarbital action. Science 168: 1246–1248.
35. GREVEN, H. M. and deWIED, D. 1967. The active sequence in the ACTH molecule responsible for inhibition of the extinction of conditioned avoidance behavior in rats. Eur. J. Pharmacol. 2: 14–16.
36. GUILLEMIN, R. 1977. The endocrinology of the neuron and the neural origin of endocrine cells. In Porter, J. C. (ed.), Hypothalamic peptide hormones and pituitary regulation. Plenum Press, New York, p. 1–12.
37. GUILLEMIN, R., SCHALLY, A. V., LIPSCOMB, H. S., ANDERSON, R. N. and LONG, C. N. H. 1962. On the presence in hog hypothalamus of $\beta$-cortico-

tropin releasing factor, α and β-MSH, ACTH, lysine vasopressin and oxytocins. Endocrinology 70: 471–477.

38. HEUSER, G. LING, G. M. and BUCHWALD, N. A. 1965. Sedation or seizures as dose-dependent effects of steroids. Arch. Neurol. 13: 195-203.

39. JONES, M. T., BRUSH, F. R. and NEAME, R. L. B. 1972. Characteristics of fast feedback control of corticotrophin release by corticosteroids. J. Endocrinol. 55: 489–497.

39a. JONES M. T. and HILLHOUSE, E. 1977. Neurotransmitter regulation of corticotropin-releasing factor in vitro. Ann. N. Y. Acad. Sci. 297: 536-558.

40. JONES, M. T., HILLHOUSE, E. and BURDEN, J. 1976. Secretion of corticotropin-releasing hormone in vitro. In L. Martini and W. F. Ganong (ed.), Frontiers in neuroendocrinology. Vol. 4. Raven Press, New York, p. 195-226.

41. JONES, M. T., HILLHOUSE, E. W. and BURDEN, J. L. 1977. Dynamics and mechanics of corticosteroid feedback at the hypothalamic and anterior pituitary gland. J. Endocrinol. 73: 405–417.

42. JONES, M. T., TIPTAFT, E. M., BRUSH, F. R., FERGUSSON, D. A. N. and NEAME, R. L. B. 1974. Evidence for dual corticosteroid-receptor mechanisms in the feedback control of acrenocorticotrophin secretion. J. Endocrinol. 60. 223–233.

43. KEARLY, R. C., vanHARTESVELDT, C. and WOODRUFF, M. L. 1974. Behavioral and hormonal effects of hippocampal lesions on male and female rats. Physiol. Psychol. 2: 187–196.

44. KNIGGE, K. M. and HAYS, M. 1964. Evidence of inhibitive role of hippocampus in neural regulation of ACTH release. Proc. Soc. Exp. Biol. Med. 114: 67–69.

45. KNIZLEY, H. Jr. 1972. The hippocampus and septal area as primary target sites for corticosterone. J. Neurochem. 19: 2737-2745.

46. KOSTOPOULOS, G. K. and PHILLIS, J. W. 1977. Mammillothalamic neurons activated antidromically and by stimulation of the fornix. Brain Res. 122: 143-149.

47. KRIEGER, D. T., LIOTTO, A. and BROWNSTEIN, M. T. 1977. Presence of corticotropin in brain of normal and hypophysectomized rats. Proc. Natl. Acad. Sci. USA 74: 648-652.

48. KRIEGER, H. P. and KRIEGER, D. T. 1971. Pituitary-adrenal activation by implanted neurotransmitters and ineffectiveness of dexamethasone in blocking this activation. In D. H. Ford (ed.), Influence of hormones on the nervous system. S. Karger, Basel, p. 98-106.

49. KRIVOY, W. A. 1970. Effects of ACTH and related peptides on spinal cord. In D. deWied and J. A. W. M. Weijnen (ed.), Pituitary, adrenal and the brain. Prog. Brain Res. Vol. 32, Elsevier, Amsterdam, p. 108-118.

50. KRIVOY, W. A. and GUILLEMIN, R. 1961. On a possible role of β-melanocyte stimulating hormone (β MSH) in the central nervous system of mammalia: an effect of β-MSH in the spinal cord of the cat. Endocrinology 69: 170–175.

51. LANIER, L. P., vanHARTESVELDT, C. WEISS, B. M. and ISAACSON, R. L. 1975. Effects of differential hippocampal damage upon rhythmic and stress-induced corticosterone secretion in the rat. Neuroendocrinology 18: 154-160.

52. MANDELBROD, I. and FELDMAN, S. 1972. Effects of sensory and hippocampal stimulation on unit activity in the median eminence of the rat hypothalamus. Physiol. Behav. 9: 565–572.

53. MARKS, N. 1978. Biotransformation and degradation of corticotropins, lipotro-

pins and hypothalamic peptides. *In* W. F. Ganong and L. Martini (ed.), Frontiers in neuroendocrinology. Vol. 5. Raven Press, New York, p. 329–377.
54. MASON, J. W. 1957. The central nervous regulation of ACTH secretion. *In* H. H. Jasper, L. D. Proctor, R. S. Knighton, W. C. Noshay and R. T. Costello (ed.), Reticular formation of the brain. Little, Brown and Co., Boston, p. 645–670.
55. McEWEN, B. S. 1977. Adrenal steroid feedback on neuroendocrine tissues. Ann. N. Y. Acad. Sci. 297: 568–579.
56. McEWEN, B. S., GERLACH, J. L. and MICCO, D. J. 1975. Putative glucocorticoid receptors in hippocampus and other regions of the rat brain. *In* R. L. Isaacson and K. H. Pribram (ed.), The hippocampus. Vol. 1. Plenum Press, New York, p. 286–314.
57. McEWEN, B. S. and WALLACH, G. 1973. Corticosterone binding to hippocampus: Nuclear and cytosol binding *in vitro*. Brain Res. 57: 373–386.
58. McEWEN, B. S., WEISS, J. M. and SCHWARTZ, L. S. 1969. Uptake of corticosterone by rat brain and its concentration by certain limbic structures. Brain Res. 16: 227–241.
59. McEWEN, B. S., WEISS, J. M. and SCHWARTZ, L. S. 1970. Retention of corticosterone by cell nuclei from brain regions of adrenalectomized rats. Brain Res. 17: 471–482.
60. McLEARY, R. A. 1961. Response specifity in the behavioral effects of limbic system lesions in the cat. J. Comp. Physiol. Psychol. 54: 605–613.
61. MEIBACH, R. C. and SIEGEL, A. 1977. Efferent connections of the hippocampal formation in the rat. Brain Res. 124: 197–224.
62. MOLDOW, R. and YALOW, R. S. 1978. Extrahypophysial distribution of corticotropin as a function of brain size. Proc. Natl. Acad. Sci. USA 75: 994–998.
63. MOORE, R. Y. and HALARIS, A. E. 1975. Hippocampal innervation by serotonin neurons of the midbrain raphe in the rat. J. Comp. Neurol. 164: 171–184.
64. MURPHY, H. M., WIDEMAN, C. H. and BROWN, T. S. 1979. Plasma corticosterone levels and ulcer formation in rats with hippocampal lesions. Neuroendocrinology 28: 123–130.
65. NAUTA, W. J. H. 1963. Central nervous organization and the endocrine motor system. *In* A. V. Nalbandov (ed.), Advances in neuroendocrinology. Univ. Illinois Press, Urbana, p. 5–28.
66. NYAKAS, C. and ENDROCZI, E. 1970. Effect of hippocampal stimulation on the establishment of conditioned fear response in rat. Acta Physiol. Acad. Sci. Hung. 37: 281–289.
67. ODERFELD-NOWAK, B., NARKIEWICZ, O., BIAŁOWĄS, J., DĄBROWSKA, J., WIERASZKO, A. and GRADKOWSKA, M. 1974. The influence of septal nuclei lesions on activity of acetylcholinesterase and choline acetyltransferase in the hippocampus of the rat. Acta Neurobiol. Exp. 34: 583–601.
68. PALKOVITS, M. 1977. Neural pathways involved in ACTH regulation. Ann. N. Y. Acad. Sci. 297: 455–476.
69. PAPEZ, J. W. 1937. A proposed mechanism of emotion. Arch. Neurol. Psychiatr. 38: 725–744.
70. PAPEZ, J. W. 1958. Visceral brain, its component parts and their connections. J. Nerv. Ment. Dis. 126: 40–56.
71. PFAFF, G., SILVA, M. T. A. and WEISS, J. M. 1971. Telemetered recording of hormone effects on hippocampal neurons. Science 172: 394–395.
72. RAISMAN, G. COWAN, W. M. and POWELL, T. P. S. 1966. An experimental analysis of the efferent projections of the hippocampus. Brain 89: 83–108.

73. RANCK, J. B. 1974. Studies on single neurons in dorsal hippocampal formation and septum in unrestrained rats. Exp. Neurol. 41: 462–555.
74. READING, H. W. and DEWAR, A. J. 1971. Effects of $ACTH_{4-10}$ on cerebral RNA and protein metabolism in the rat. Third Inter. Meet. Internatl. Soc. Neurochem. Akademiai Kiado, Budapest, p. 199.
75. RODRIGUEZ, E. M. 1976. The cerebrospinal fluids as a pathway in neuroendocrine integration. J. Endocrinol. 71: 407–443.
76. SEGAL, M. 1974. Convergence of sensory input on units in the hippocampal system of the rat. J. Comp. Physiol. Psychol. 87: 91–99.
77. SEGAL, M. 1975. Physiological and pharmacological evidence for a serotonergic projection to the hippocampus. Brain Res. 94: 115–131.
78. SEGAL, M. and BLOOM, F. E. 1976. The action of norepinephrine in the rat hippocampus, III: Stimulation of nucleus coeruleus in the awake rat. Brain Res. 107: 499–511.
79. STEINER, F. 1970. Effects of ACTH and corticosteroids on single neurons in the hypothalamus. In D. deWied and J. A. W. M. Weijnen (ed.), Pituitary, adrenal and the brain. Prog. Brain Res. Vol. 32. Elsevier, Amsterdam, p. 102–106.
80. STEINER, F. A., RUF, K. and AKERT, K. 1969. Steroid-sensitive neurons in rat brain: Anatomical localization and responses to neurohumors and ACTH. Brain Res. 12: 74–85.
81. STUMPF, W. E. and SAR, M. 1973. Hormonal inputs to releasing factor cells, feedback sites. In E. Zimmerman, W. H. Gispe, B. H. Marks and D. deWied (ed.), Drug effects on neuroendocrine regulation. Prog. Brain Res. Vol. 39. Elsevier, Amsterdam, p. 53–70.
82. STUMPF, W. E. and SAR, M. 1977. Localization of steroid hormone receptors in the central nervous system in relation to function. In V. H. T. James (ed.), Endocrinology. Vol. 1. Proc. V Internatl. Congr. Endocrinol. Hamburg, July 18–24, 1976, Excerpta Med. Amsterdam, p. 18–22.
83. SWANSON, L. W. and COWAN, W. M. 1975. Hippocampo-hypothalamic connection: origins in subicular cortex, not ammon's horn. Science 189: 303–304.
84. SWANSON, L. W. and COWAN, W. M. 1977. An autoradiographic study of the organization of the efferent connection of the hippocampal formation in the rat. J. Comp. Neurol. 172: 49–84.
85. vanRIEZEN, H., TIGTER, H. and GREVEN, H. M. 1977. Critical appraisal of peptide pharmacology. In L. H. Miller, C. A. Sandman and A. J. Kastin (ed.), Neuropeptide influences on the brain and behavior. Adv. Biochem. Psychopharmacol. Vol. 17, p. 11–27.
86. vanWIMERSMA GREIDANUS, T. B., BOHUS, B. and deWIED, D. 1974. Differential localization of the behavioral effects of lysine vasopressin and of $ACTH_{4-10}$: A study in rats bearing lesions in the parafascicular nuclei. Neuroendocrinology 14: 280–288.
87. vanWIMERSMA GREIDANUS, T. B. and deWIED, D. 1976. The dorsal hippocampus: A site of action of neuropeptides on avoidance behavior? Pharmacol. Biochem. Behav. 5: Suppl. 1, 29–34.
88. WOODBURY, D. M. 1954. Effect of hormones on brain escitability and electrolytes. Recent Prog. Horm. Res. 10: 65–107.
89. WOODBURY, D. M. 1958. Relation between the adrenal cortex and the central nervous system. Pharmacol. Rev. 10: 275–375.

F. R. BRUSH and S. M. FRALEY, Experimental Psychology Laboratory, Syracuse University, Syracuse, New York 13210, USA.

Lecture delivered at the Warsaw Colloquium on Instrumental
Conditioning and Brain Research
May 1979

# ON THE MECHANISM OF THE POST-ASYMPTOTIC CR DECREMENT PHENOMENON

J. Bruce OVERMIER, Ralph J. PAYNE, Robert M. BRACKBILL,
Bruce LINDER and Janice A. LAWRY

Department of Psychology, University of Minnesota
Minneapolis, Minnesota, USA

*Abstract.* Three experiments with 49 dogs explored the decrease in CR magnitude that sometimes occurs when CS–US pairings are continued beyond those needed to reach maximum CR magnitude. The first experiment confirmed the existence of the phenomenon, obtaining less conditioned excitation after 300 CS–US trials than after 18 CS–US trials. The second experiment demonstrated that the phenomenon is not dependent upon paired CS–US presentations because 18 CS–US pairings yielded little excitation if preceded by, followed by, or intermixed with 282 US presentations. The third experiment indicated that in contrast to the decremental effects on excitatory conditioning, inhibitory conditioning was faciliated by large numbers of prior US exposures, suggesting explanation of the post-asymptotic decrement phenomenon by opponent-process theory.

Pavlov (29, lecture 14) argued that the ultimate destiny of conditioned stimuli (CSs) — reinforced or not — was to become "inhibitory". In support of this, he described experiments by Shishlo, Petrova, and Speransky in which a stimulus (CS) was reliably reinforced and after the first few trials elicited a large stable conditioned reflex (CR); but with continued reinforced trials the CR diminished and even disappeared. The phenomenon, later reported in the American literature as well (11, 12 and others), may be described as post-asymptotic CR decrement and has been variously called "over-reinforcement", "extinction-with-reinforcement" and "inhibition-with-reinforcement". But the post-asymptotic decrement effect is not always obtained, even when sought (45). Prokasy (31), based upon a careful detailed review, suggested that not only was the phenomenon not ubiquitous but that many purported "reports" did not actually contain evidence for post-asymptotic decrementing of the CR, thus casting sufficient doubt on the phenomenon that today it is often not even discussed in major texts on learning (e.g., 7).

Nonetheless, Kimmel (16), Libby (22), Sherman and Maier (37), and Zieliński (46), among others, have confirmed that the CR does diminish in magnitude with continued CS-US presentations beyond those necessary to achieve the maximum CR. The degree of decrementing however seems to be determined by a complex interaction of CS-US interval (27), CS intensity/quality (16, 28, 47), US intensity (5), intertrial interval (31; Speransky, cited in Pavlov, 1927), and possibly the functional significance of the CR (2, 17). And often the obtained effect is quite modest, indeed (e.g., 5, 22, 46).

Not surprisingly given a complex and ephemeral phenomenon, the mechanism underlying this post-asymptotic decrementing of the CR with continued CS-US pairings is still not established. Pavlov (29) made clear that the phenomenon was not attributable to satiation or reflex exhaustion; nor was it, he argued, attributable in inhibition of delay. Instead, Pavlov argued that the underlying mechanism is the "functional exhaustion" of the cortical elements as a result of their activity in response to the CS (p. 244 f). In contrast, Hilgard and Marquis (11) and Hull (13) hypothesized that the effect resulted from fatigue of the response. The dependence of the phenomenon upon relatively short intertrial intervals is congenial to *both* of these views, but (i) the ability of the CS eliciting the diminished CR to function well as a signal for a new US and (ii) the effect of shifts in the temporal location of the US to reinstate the CR, respectively (both reported by Pavlov, 29), argue against the preceding hypotheses.

Zieliński (46) proposed a third mechanism, inhibition of delay, as the cause of the post-asymptotic CR decrementing — a proposal echoed by Kimmel (18). Zieliński noted that longer CS-US intervals conductive to the development of inhibition of delay also seemed necessary for CR diminution effect as well. His detailed analyses of trial-by-trial changes in patterns of within-CS responses in the CER paradigm (4) showed that there was a parallel between the development of inhibition of delay and the CR diminution. A strong argument against inhibition of delay as the sole *causal* factor in post-asymptotic decrements is provided by a similar CER experiment by Millenson and Hendry (25). Millenson and Hendry compared the degree of CR diminution over trials of CS–US pairings when the CS–US interval was, as in Zieliński's experiment, fixed to that when it was variable. Variable CS–US intervals typically block the development of inhibition of delay (21, 29) and did so here. Nonetheless, Millenson and Hendry (25) observed significant CR diminution under *both* experimental conditions. This indicates that some process other than, or in addition to, inhibition of delay contributes to the post-asymptotic decrementing of the CR.

Focusing on the functional value of the CRs in promoting adaptation, Anokhin (2) has suggested a two-stage model of conditioning in which initial associative bonding of CS and US becomes consolidated if the CR contributes to adaptive coping with environmental contingencies. On this view, should the CR not contribute positively to coping, the initial associative bond would be debased and the CR would diminish and eventually disappear. Kimmel has (17, 19) proposed a similar hypothesis with respect to defensive conditioning in which the subject could not modulate the aversive US in any way. [See Lykken (23) and Perkins (30) for related views of the Pavlovian CR]. To the extent that CRs can be shaped by their consequences (24), there is support for this view rooted in concepts of biological utility. Unfortunately, the hypothesis of modifiability of CRs by their functional significance is not uniformly sustained by the literature, and CRs often function to *reduce* obtained rewards (6, 41, 44).

What then is the mechanism of the post-asymptotic decrement in CRs? The present experiments focus on the possibility, originally suggested by Annau and Kamin (1), that the CR diminution arises from an habituation-like process to the US. Our strategy was (i) to demonstrate the post-asymptotic decrement of the CR phenomenon, choosing parameters so as to maximize the magnitude of the effect — even to achieving disappearance of the CR if possible, (ii) then to explore whether CR diminution is dependent upon associative relations between CS and US, and (iii) finally to determine whether the critical parameter that produces the phenomenon results in a change in the US's properties available for conditioning in accordance with Solomon and Corbit's (39) opponent-process theory of US function.

## MATERIALS AND METHODS

*Subjects.* Mongrel dogs 38-50 cm tall at the shoulder and weighing 9-12 kg were obtained from the University of Minnesota Animal Hospital for these experiments. These dogs were adapted to the laboratory regimen for several days before experimentation. Experiment I used 20 dogs divided into three groups; an additional three dogs constituted a post-hoc experimental group for Experiment I. Experiment II used 20 dogs divided into three groups. And, Experiment III used six dogs in a within-subjects experimental design. Experiments I and II were run concurrently.

*Apparatus.* Two experimental apparatus units were used in these experiments: (i) a two-way shuttlebox, and (ii) a restraining hammock. Both were housed in separate sound-reducing rooms.

The shuttlebox was used for training (Type II) the dogs in temporally paced avoidance responding (38) and for later tests of the properties of Pavlovian CSs as indexed by the avoidance response controlling properties of the CSs (3). The shuttlebox consisted of two large, brightly illuminated 113 cm long, 60 cm wide and 100 cm high black chambers separated by a barrier adjusted to the dog's shoulder height. Each chamber had a flat grid floor for the delivery of scrambled electric shocks (1/2 s at 4.5 mA on day 1, 6 mA thereafter) and speakers overhead for presenting the auditory stimuli at 10 dB above the background noise level of 75 dB (re 0.0002 dynes/cm$^2$). The stimuli were tone (200 Hz or 700 Hz — always counterbalanced within groups), clicker (20/s), and background Gaussian noise. Barrier jumping responses were detected by photocells located on both sides of the barrier. Timing and control was by automatic equipment located in another room.

The restraining hammock was used for the Pavlovian conditioning (Type I) in which CSs (tone, click, or reduced illumination) were paired with shock. The dog's body rested in the hammock with its legs extending through four holes and secured to the supporting stand; the dog's chin rested on a flat tray. The hammock was in a well illuminated white cubicle which contained speakers for the presentation of the auditory stimuli noted above. Shock USs (1 s at 6 mA) were administered to the dog through 5 $\times$ 7.5-cm metal electrodes coated with electrode paste and fastened to the dog's rear foot-pads.

*Procedures.* Each subject in each experiment received the following sequence of treatments: (i) training on the temporally paced instrumental avoidance task, (ii) aversive Pavlovian conditioning, (iii) transfer-of-control tests of the properties of the CSs.

In the first treatment phase, the dogs were trained to cross from one shuttlebox compartment to the other by jumping over the barrier to avoid unsignalled brief shocks to the feet presented according to a temporal schedule. In the absence of responding, shocks were presented every 5 s (Sh-Sh = 5 s); each barrier jumping response prevented shocks for the next 30 s (R-Sh = 30 s). Each dog received a minimum of three daily 40 min training sessions. Additional daily sessions were provided if necessary until the dog met the criterion of avoiding 99% of the possible shocks. No subject required more than 5 days of training.

The second treatment phase of Pavlovian conditioning was different for each group in each experiment. The details of the parameters will therefore be specified for each group as each experiment is presented. The common features of this treatment were as follows. On the day after achieving the avoidance criterion, the dog was placed in the hammock and electrodes attached for a single conditioning session which varied across groups. This session lasted approximately 6 h beginning and ending with 5-min periods without any conditioning trials.

The third treatment phase was to test the degree of conditioning to the CSs as indexed by the power of the CSs to modulate instrumental avoidance behavior. This test occurred on the day after Pavlovian conditioning. For testing, the dog was returned to the shuttlebox; then the dog was given 15-min of re-training on the instrumental avoidance task before the first test presentation of a CS. The purpose of this brief retraining was to insure that all dogs were responding in as reliable and stable manner during testing as prior the Pavlovian conditioning. The mean barrier-jumping rate on the test day was 9.3 responses per minute (SE = 1.61) and did not differ significanty among any of the groups from all three experiments.

Immediately following the retraining, a series of 12 test trials with the CS were presented; CS presentations were independent of instrumental responding. Test trials were separated by an average of 80 s. The CSs were never reinforced during this testing phase. Moreover, the avoidance schedule of shocks was suspended during each CS test presentation (5 s in Experiments I and II and 10 s in Experiment III) and for equal periods of time immediately before and after the CS.

The data of primary interest were the rates of instrumental responding during the CS relative to those prior to the CS. When superimposed on a previously established avoidance baseline, an excitatory aversive CS (a fear signal) typically causes an increase in the response rate relative to the pre-CS baseline rate while an inhibitory aversive CS has the opposite effect (e.g., 3). These effects in the shuttle box are dependent upon the Pavlovian signal values of the CSs and independent of any motor-acts during the Pavlovian conditioning phase (28, 40, 43).

*Experiment I*

The goal was to determine whether or not a large number of Pavlovian conditioning trials would result in *less* excitatory conditioning than a small number of trials. This effect, when previously obtained, has typically been the product of relatively large numbers of pairings when small numbers of CS-shock pairings are quite sufficient to yield substantial excitatory conditioning (10, 20, 29, 46). Therefore, we compared the degree of conditioning obtained with 300 or 18 CS–US pairings, relative to a nonassociative control group.

*Conditioning phase procedures.* The six dogs composing Group I received 300 pairings tone CS and shock US. The post-asymptotic diminution of CR strength proceeds rapidly with CS–US intervals of 30 s, according to Pavlov (29, p. 236). Therefore CS duration was 30 s. Because the post-asymptotic CR decrement phenomenon is supposed to occur more readily with massed trials (29, 31), a 30-s average intertrial interval was used. The seven dogs in Group II were treated identically except that they received only 18 pairings of CS–US, after which the dogs simply rested in the hammock until the end of the session. The seven dogs in Group III constituted a control group and received only US presentations; they received 300 US delivered according to a schedule (i.e., VI 60-s) that matched the shock distribution of Group I. Group III provided the baseline against which to assess associative effects. Behavioral testing for degree of conditioning occurred on the following day for all groups.

*Results and discussion.* The test day percentage change from baseline responding [(Σ During CS — Σ Pre CS baseline)/Σ Pre CS baseline] was computer for each dog from its 12 test trials. The mean test day percentage changes are shown in Fig. 1. Eighteen CS–US pairings in Group II were sufficient to establish an excitatory CS indicated by the 74% increase in rate of responding that the CS evoked. In contrast, after 300 pairing in Group I, the CS only evoked a 7% increase in rate of responding. This is very close to the 6% increase in baseline evoked by the novel tone CS in Group III which received only US presentations during the conditioning phase.

Statistical analyses confirmed that the groups differed significantly in the excitatory power of the CSs ($F_{2/17} = 6.57$, $P < 0.01$) [1]. Pairwise comparisons of the groups indicated that the group that received only 18 CS-US pairings responded significantly faster during the CS than did either the group that received 300 CS–US pairings or the group that received only the 300 US presentations ($t_{17} = 2.25$ and 2.30 both $P \leqslant 0.05$). Moreover, these latter two groups did not differ from one another ($t_{17} = 0.03$, $P > 0.5$). This result clearly confirms that the conditioned excitatory state established after a modest number of trials is decremented if a large number of additional conditioning trials are carried out.

The decremental effect obtained here was so complete, and to us, so

---

[1] All comparisons reported as significant in this paper were confirmed using non-parametric statistical tests as well.

surprising that we worried that the low level of conditioning persisting after 300 pairings might be an artifact and unique to those particular conditioning and testing parameters. Therefore we ran a post-hoc group of three additional dogs to check on this possibility. This group received 300 CS–US pairings using a shorter 10-s CS–US interval and a longer 80-s average intertrial interval. Additionally in the test phase, longer 10-s CS test durations were used. The results obtained from this post-hoc group are also shown in Fig. 1. The relative increase in responding

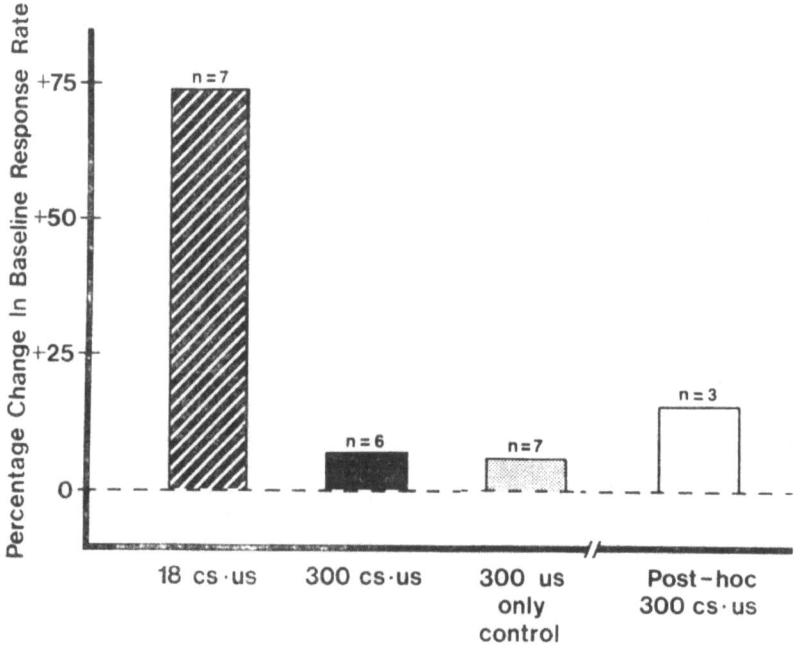

Fig. 1. Percentage changes in rate of free-operant, temporally paced avoidance responding produced in the test phase of Experiment I by presentations of CSs established with different numbers by aversive Pavlovian conditioning trials. (Number of dogs in each group is indicated above each bar.)

obtained during the test CS presentations was 16%. Although this degree of facilitation was slightly larger than that originally obtained in Group I, the two 300 pairings groups did not differ statistically ($t_7 = 0.3$; $P > 0.5$). Thus, the dramatic diminution of the excitatory conditioning obtained in our Experiment I is not an artifact of the particular conditioning or testing parameters but is the product of some systematic process correlated with number of trials.

The post-hoc group that received 300 CS–US pairings provided confirmation of the post-asymptotic decrementing of the CR phenomenon in a second way. During the Pavlovian conditioning treatment, heart rate was recorded during successive 5 second periods before the CS, during the CS, and immediately following the US. Changes in heart rate during the CS and after the US, relative to the pre-CS heart rate provide indices, respectively, of CR and UR development over trials. All three subjects showed the same pattern of CRs increasing over the early trials and

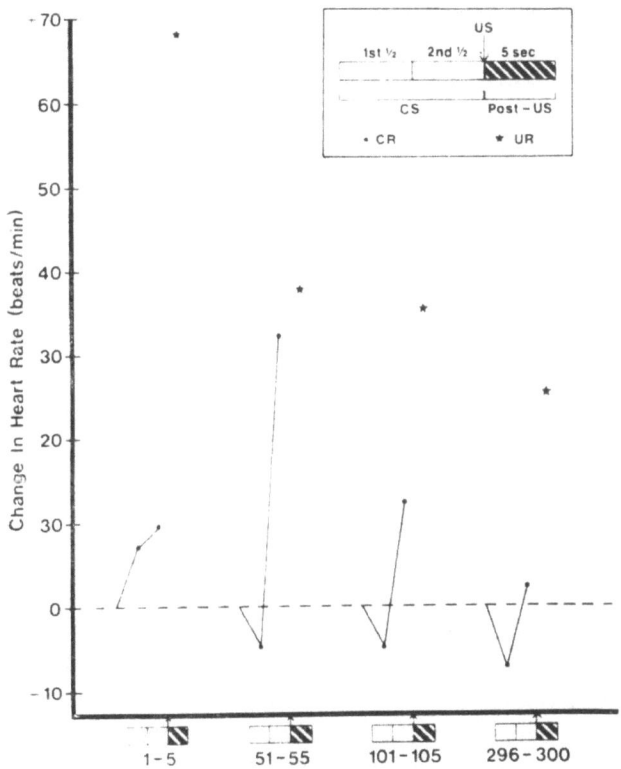

Fig. 2. Changes in heart rate, relative to pre-CS rate, produced by CS–US presentations in selected blocks of trials during the 300 CS-US pairings given to a representative dog from the post-hoc group of Experiment I. Reflected are (i) the increasing CR magnitude on early trials, (ii) the decrementing of the CR over later trials, and (iii) the monotonic diminution of the UR across all trials.

decreasing on later trials, although the trials of maximum CR differed across the subjects. Moreover, the subjects also showed decreasing UR magnitudes across trials as well. This pattern is shown for one subject in Fig. 2 which presents mean heart rate changes on early, mid, and late blocks of trials; the trials 51 to 55 are included because for this subject that was the point of maximum CR. These data supplement the behavioral data from the test sessions and show a consistency between the two measures of conditioning.

## Experiment II

Given that we had successfully demonstrated that 18 CS–US pairings result in significant conditioning while 300 CS–US pairings did not, we turned our attention to identifying the mechanism underlying this post-asymptotic decrementing of the CR. Most hypotheses offered argue that the CR diminution arises from some associative feature of the conditioning process (e.g., inhibition of delay). Here we wished to explore the possibility that some non-associative mechanism may be responsible. Annau and Kamin (1) suggested that the CR diminution observed over long sequences of pairing trials might well be due to some habituational process. An habituated US evokes a lesser UR and support conditioning of a lesser CR. On this view, subjects become habituated to the US over trials intervening between the 18th and the 300th pairing and thus the US cannot sustain the maximum CR. The observation that a series of exposures to US alone prior to conditioning impairs later development of the CR is now a well established phenomenon (e.g., 14, 26, 32 for review).

To test whether some habituation-like mechanism was the source of the decrementing in the group which received 300 CS–US pairings relative to the group which received only 18 CS–US pairings, Experiment II held the number of pairings constant at 18 and the total number of the US shocks constant at 300 by presenting 282 "extra" USs unsignalled by any CS. If the same low level of conditioning obtains, this would be evidence that the decrementing mechanism is not dependent upon some feature of the associative process arising from extensive CS–US pairings.

*Conditioning procedures.* All features of Experiment II were identical to those of Experiment I, except as noted here.

Twenty dogs were assigned to three groups which differed only with respect to the treatment phase 2 Pavlovian conditioning session. Group A ($n = 7$) received a series of 282 USs delivered on a VI 60 s schedule followed by 18 CS–US pairings like those for Group II in Experiment I. Group B ($n = 7$) received exactly the same except in reverse order; thus, Group B received 18 CS–US pairings followed by 282 USs. Group C ($n = 6$) also received 18 CS–US pairings and 282 USs. However, for Group C these two kinds of trials were randomly intermixed with one another with the average inter-shock interval of 60; thus successive CS–US trials were separated by the standard average of 30 s and successive US trials by the standard average 60 s.

Testing of Groups A, B, and C proceeded on the next day just an in Experiment I.

*Results and discussion.* All three groups showed very small percentage changes in baseline rates of responding when their CSs were superimposed upon the free operant avoidance behavior. These changes ranged between —3% and +8%. These data are presented in Fig. 3 with the levels of conditioning obtained in Experiment I indicated for ready comparison.

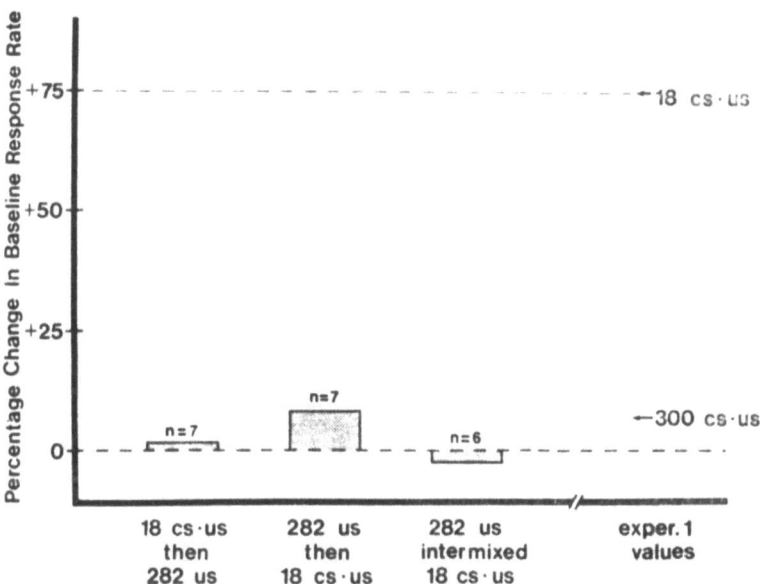

Fig. 3. Percentage changes in rate of free-operant, temporally paced avoidance responding produced in the test phase of Experiment II by presentations of CSs established with different patterns of distribution of 18 CS–US pairings and 282 unsignalled USs during aversive Pavlovian conditioning.

Statistical analysis confirmed that the CSs in Group A, B, and C did not yield rates that differed from one another ($F_{2/17} = 0.06$, NS) nor from the pre-CS baseline (largest $t_{17} = 0.3$, NS). Because the six groups I, II, II, A, B, and C were in fact run concurrently, an additional analysis of variance was computed incorporating them all to provide a basis for additional pairwise comparisons. These comparisons confirmed the impression provided by Figure 2 that Group II which received only the 18 CS–US pairings differed from each of the other five groups (smallest $t_{34} = 2.25$, $P \leqslant 0.05$) and that these latter groups did not differ among themselves (largest $t_{34} = 0.38$, NS).

Thus, while 18 CS–US pairings alone resulted in significant excitatory conditioning, 18 CS–US pairings *combined* with 282 USs resulted in a level of excitation that did not differ from either that of the 300 pairings group or the nonassociative 300 USs control group. This pattern of results indicates that the decrementing process is *not* dependent upon pairings of CS and US. The decrementing process seems to be related primarily to US presentations.

Accounts might well be offered for the specific outcomes obtained in Group A, or B, or C. For example, Kamin (15) provides a "blocking" account of way 18 CS-US pairs *preceded* by 282 USs should result in less excitatory conditioning to the tone CSs than 18 pairings. The "blocking" argument suggests that during the first 282 US presentations, the background stimuli come to reliably predict the US so that when the CS is later introduced, the CS is a redundant predictor thus resulting in no conditioning. Tomie (42) has shown that background cues *may* function in exactly this way, at least in alimentary conditioning.

However, "blocking" cannot account for why 18 CS–US pairings *followed* by 282 USs should result in a low level of conditioning. Rescorla's theorizing that the magnitude of the CR is a function of the vividness of the memorial representation of the US (35) could account for the low level of excitatory conditioning obtained if one assumed that the postconditioning US presentations reduced the vividness of this memorial representation. Habituation to the US during the 282 USs might so modify the representation and thus decrement the CR. Indeed, Rescorla (35) has shown that habituation to a noise US following conditioning does reduce the elicitable CR.

A still different account might be offered for Group C in which the 18 CS–US pairings were randomly intermixed with 282 USs. According to a contingency analysis of Pavlovian conditioning (33), the degree of excitatory conditioning is a function of the magnitude of the inequality between two conditional probabilities: $P(US/CS) > P(US/no\ CS)$. When the dog's total conditioning experience is 18 CS–US pairings, this inequality takes the form $1 > 0$ and is maximal; when exposed to 18 CS-US pairings intermixed with 282 USs this inequality takes the form $1 > 0.94$ and is very small. On this view, the former treatment should result in larger CRs than the latter. Supporting this analysis, Rescorla (34) has shown that the amount of conditioning decreases systematically as degree of contingency decreases. Thus, according to this view, Group C should have shown little conditioning — as indeed it did. But difficult for this contingency hypothesis to account for is why the 300 pairings group showed little excitatory conditioning. This difficulty arises because the degree of contingency is the same in both the 300 pairings group (I) and the 18 pairings group (II), but which Experiment I showed to differ markedly in behavior during the CSs.

While each of the above accounts fits well with the results of one group, none seem sufficiently facile to account for the full pattern of results obtained in Experiments I and II.

Most promising seems the view suggested early (1) that directly invoked some habituation-like mechanism with respect to the US. This is because the only parameter common to all the experimental groups which showed low conditioning was that of 300 USs, whether paired or unpaired and independent of the distribution of pairings. Because all shared this extensive US experience, an "habituation" account can be applied to all, whereas exhaustion of the CS center, inhibition of delay, blocking, memory and contingency accounts cannot.

Recently, Solomon and Corbit (39) offered a theory which details the dynamics for reduced affective impact of USs over trials, as reflected by the UR values in our Fig. 2. According to this theory the affective impact is of a given US presentation is a function of the interaction of two opposing processes. One, the a-process, is the primary excitatory affect to the US and is a fixed function of the strength of the US showing no change over successive presentations of the US. The opponent second process, the b-process, is thought to be a slow one evoked in turn by the occurrence of the a-process and to function to inhibit the excitatory state. In contrast to the a-process, the b-process in hypothesized as a consequence of repeated evocations to decrease in latency and increase in intensity and duration — coming to substantially outlast the a-process — when these evocations occur *sufficiently close* together in time. The net affective impact of a US, then, is the result of the algebraic summation of the two opposing processes, "a" and "b". Important for our purposes is that this theory predicts that the affective value of a US, and thus its effectiveness in excitatory conditioning, should diminish over massed trials as the b-process grows.

## *Experiment III*

The opponent-process theory finds the post-asymptotic decrementing of the CR, seen in the present experimens, concordant. These data do not, however, provide a direct test of the opponent-process account because they do not differentiate between simple loss of an a-process over trials from growth of a b-process. But having identified effective parameters (number of USs and the interval between successive USs) for producing decrementing, we can use these parameters to carry out a test of the opponent-process account. Because, the b-process grows over successive US presentations (appropriately distributed) and substantially outlas the US-locked a-process, then the *post*-US state of the animal should change over trials, becoming increasingly inhibitory. If so, then establishing a conditioned inhibitor through the backward conditioning paradigm should be facilitated when the backward conditioning trials have been preceded by numerous prior US exposures. This is because the CS-would be paired with a stronger post-US inhibitory state than in the absence of numerous prior US presentations. Experiment III sought to provide a test of this prediction.

*Conditioning procedures.* The apparatuses, baseline avoidance training, and testing procedures were exactly as described for Experiments I and II. Six dogs were used in Experiment III. During the second treatment phase, these dogs received 300 backward conditioning trials in which CS onset coincided with US termination; the CS duration was 10 s. The inter-US interval averaged 60 s. Three CSs were used: tone, clicker, and dimmed illumination. These CSs differed as to whether they served as a CS- on early, middle, or late trials in the 300 trial series. $CS_1$ was the backward CS on the first 40 trials; $CS_2$ was the backward CS on the middle 220 trials; $CS_3$ was the backward CS on the last 40 trials. For one half the dogs $CS_1$ was tone and $CS_3$ was clicker; for the other half, $CS_1$ was clicker while $CS_3$ was tone. $CS_2$ was always dimmed illumination.

On the day after conditioning these CSs were superimposed on the avoidance baseline to test for their relative conditioned properties.

*Results and discussion.* Our primary interest is in comparison of the percentage changes in baseline responding produced by $CS_1$ and $CS_3$. Although $CS_2$ is of some interest, it is not directly comparable to either $CS_1$ or $CS_3$ because of differences in CS quality (visual vs. auditory) and number of backward pairings (220 vs. 40). On the other hand, $CS_1$ and $CS_3$ are directly comparable.

The degree of inhibitory conditioning was directly related to the number of prior US exposures. $CS_1$, the CS-conditioned on the first 40 backward conditioning trials did not suppress the baseline responding as an inhibitory stimulus typically does (36). In contrast, $CS_3$, the CS-conditioned on the last 40 backward trials, produced marked suppression of the baseline avoidance responding: responding during $CS_3$ was 36% below the baseline rate of avoidance. The results of Experiment III are presented in Fig. 4.

The prediction derived from Solomon and Corbit's (39) opponent-process theory was that degree of inhibitory conditioning should be greater when the backward conditioning trials are preceded by numerous US presentations than when not. The pattern of data obtained is consistent with this prediction. Statistical comparison of responding to $CS_1$ and $CS_3$ detected a significant difference ($t_5 = 2.46$, $P < 0.05$) confirming the prediction.

The small degree of excitatory conditioning that was apparent to $CS_1$, while unexpected, is not without precedent (9) and Heth (8) has shown than such excitatory conditioning diminishes when larger numbers of trials are used. Thus, this observation should not detract from the trend of increasing inhibitory conditioning as the number of US exposures increases. Because $CS_1$ and $CS_3$ were paired with US termination on

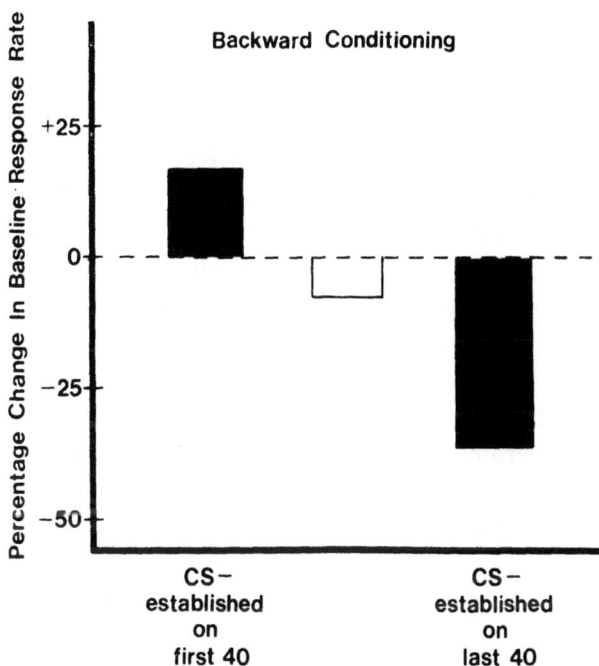

Fig. 4. Percentage changes in rate of free-operant, temporally paced avoidance responding produced in the test phase of Experiment III by presentations of backward conditioned CSs as a function of which USs in a series of 300 USs the CSs were paired with. (The unfilled bar represents the effect of visual $CS_2$ backward paired with the middle 220 USs.)

equal numbers of trials, the differential conditioning may be attributed to the changing consequences of those US terminations. Solomon and Corbit's theory predicted just such changes in the consequences of US termination.

## GENERAL DISCUSSION

The post-asymptotic decrementing of CR magnitude with continued CS-US pairings was our initial point of interest because it seemed counterintuitive and the mechanism underlying the phenomenon not broadly agreed upon. But it is just one of a set of CR diminishing operations for which no unitary account exists; the others are the pre-conditioning US exposure effect, the post-conditioning US exposure effect, and the intermixed US exposures effect. These four phenomena are usually given disparate theoretical treatment. But, they have the common feature of being dependent upon many US exposures and are enhanced when such exposures are massed. This suggested that Solomon and Corbit's (39) opponent-process theory of affective dynamics which focuses upon just these parameters might provide a basis for systematic, coherent treatment.

Our series of there experiments demonstrated a very large post-asymptotic decrement effect — larger than that contained in most reports in the literature. The evidence was obtained from both the motor and autonomic response systems, with the latter also revealing UR diminution (cf., 18). The second experiment made clear that the CR decrementing is not dependent upon CS-US pairings. This is not to say that some product of the CS–US associative process cannot contribute to the decrementing (cf., 37 Exp. III), but that US presentations alone are sufficient for the decrementing to occur. Although the opponent-process theory was consonant with the full pattern of data on both CRs and URs in Experiments I and II, so too was any theory of simple reduced responsiveness to the US (1, 14, 45). To differentiate these views, a unique prediction was derived from opponent-process theory. This prediction was that while a long series of closely spaced USs impairs excitatory conditioning, the same series of USs should *facilitate* inhibitory conditioning of CSs presented according to the backward conditioning paradigm. Our third experiment used a within subject comparison of the degree of inhibition obtained through backward conditioning when CSs were paired with USs early vs. USs late in the series. The results of greater inhibitory conditioning in the latter case confirmed the theory's prediction.

Thus, the reduction in the functional effectiveness of the US for *excitatory* conditioning over a long series of US presentations — as in extended CS-US pairings — seems to be attributable at least in part to the dynamic development of an inhibitory process opposing the excitatory process. While the present experiments are not definitive and further tests are called for, the viability of the opponent-process theory in accounting for the post-asymptotic decrement phenomenon is illustrated through the present experiments.

This research supported by National Science Foundation Grants BNS 77-28171, BNS 77-22075, and BNS 79-17338 to JBO.

# REFERENCES

1. ANNAU, Z. and KAMIN, L. J. 1961. The conditioned emotional response as a function of intensity of the US. J. Comp. Physiol. Psychol. 54: 428–432.
2. ANOKHIN, P. K. 1974. Biology and neurophysiology of the conditioned reflex and its role in adaptive behavior. Pergamon Press, Oxford.
3. BULL, J. A. and OVERMIER, J. B. 1968. The additive and subtractive properties of excitation and inhibition. J. Comp. Physiol. Psychol. 66: 511–514.
4. ESTES, W. K. and SKINNER, B. F. 1941. Some quantitative properties of anxiety. J. Exp. Psychol. 29: 390–400.
5. GOLDSTEIN, M. L. 1960. Acquired drive strength as a joint function of shock intensity and number of acquisition trials. J. Exp. Psychol. 60: 349–358.
6. GRASTYÁN, E. and VERECZKEI, L. 1974. Effects of spatial separation of the conditioned signal from reinforcement: A demonstration of the conditioned character of the orienting response or the orientational character of conditioning. Behav. Biol. 10: 121–146.
7. HALL, J. F. 1976. Classical conditioning and instrumental learning: A contemporary approach. Lippincott, Philadelphia, 518 p.
8. HETH, D. C. 1976. Simultaneous and backward fear conditioning as a function of number of CS-US pairings. J. Exp. Psychol. Anim. Behav. Proc. 2: 117–129.
9. HETH, D. C. and RESCORLA, R. A. 1973. Simultaneous backward fear conditioning in the rat. J. Comp. Physiol. Psychol. 82: 434–443.
10. HILGARD, E. R. 1933. Modification of reflexes and conditioned reactions. J. Gen. Psychol. 9: 210–215.
11. HILGARD, E. R. and MARQUIS, D. G. 1935. Acquisition, extinction, and retention of conditioned lid responses to light in dogs. J. Comp. Psychol. 19: 29–58.
12. HOVLAND, C. I. 1936. "Inhibition of reinforcement" and phenomena of experimental extinction. Proc. Natl. Acad. Sci. USA, 22: 430–433.
13. HULL, C. L. 1943. Principles of behavior. Appleton-Century-Crofts, New York.
14. KAMIN, L. J. 1961. Apparent adaptation effects in the acquisition of a conditioned emotional response. Can. J. Psychol. 15: 176–182.
15. KAMIN, L. J. 1968. Attention-like processes in classical conditioning. In M. R. Jones (ed.), Miami Symposium on the Prediction of Behavior: Aversive Stimulation. University Press, Miami, p. 9-31.
16. KIMMEL, H. D. 1959. Amount of conditioning and intensity of conditioned stimulus. J. Exp. Psychol. 58: 283–288.
17. KIMMEL, H. D. 1963. Management of conditioned fear. Psychol. Rep. 12: 313–314.
18. KIMMEL, H. D. 1966. Inhibition of the unconditioned response in classical conditioning. Psychol. Rev. 73: 232–240.
19. KIMMEL, H. D. and BURNS, R. A. 1975. Adaptational aspects of conditioning. In W. K. Estes (ed.), Handbook of learning and cognitive processes. Vol. 2. Conditioning and behavior theory. Lawrence Hillsdale, N. J. Erlbaum Assoc. p. 99–142.
20. KIMMEL, H. D. and GARRIGAN, H. A. 1973. Resistance to extinction in planaria. J. Exp. Psychol. 101: 343–347.
21. KONORSKI, J. 1948. Conditioned reflexes and neuron organization. Univ. Chicago Press, Cambridge.

22. LIBBY, A. 1951. Two variables in the acquisition of depressant properties by a stimulus. J. Exp. Psychol. 42: 100–107.
23. LYKKEN, D. T. 1962. Preception in the rat: Autonomic response to shock as a function of length of warning interval. Science 137: 665–666.
24. MARTIN, I. and LEVEY, A. B. 1969. The Genesis of the Classical Conditioned Responses: International Series of Monographs in Experimental Psychology, No. 8. Pergamon, Oxford.
25. MILLENSON, J. R. and HENDRY, D. P. 1967. Quantification of response suppression in conditioned anxiety training. Can. J. Psychol. 21: 242–252.
26. MIS, F. W. and MOORE, J. W. 1973. Effect of preacquisition UCS exposure on classical conditioning of the rabbit's nictitating membrane response. Learn. Motiv. 4: 108–114.
27. MORROW, M. C. and KEOUGH, T. E. 1968. GSR conditioning with long interstimulus intervals. J. Exp. Psychol. 77: 460–467.
28. OVERMIER, J. B., BULL, J. A. and PACK, K. 1971. On instrumental response interaction as axplaining the influences of Pavlovian CS+s upon avoidance behavior. Learn. Motiv. 2: 103–112.
29. PAVLOV, I. P. 1927. Conditioned Reflexes (translated by G. V. Anrep). Dover, New York.
30. PERKINS, C. C., Jr. 1971. Reinforcement in classical conditioning. In H. H. Kendler and J. T. Spence (ed.), Essays in neobehaviorism. Appleton-Century-Crofts, New York, p. 113–136.
31. PROKASY, W. F. 1960. Postasymptotic performance decrements during massed reinforcements. Psychol. Bull. 57: 237–247.
32. RANDICH, A. and LoLORDO, V. M. 1979. Associative and non-associative theories of the UCS pre-exposure phenomenon: Implications for Pavlovian conditioning. Psychol. Bull. 86: 523-548.
33. RESCORLA, R. A. 1967. Pavlovian conditioning and its proper control procedures. Psychol. Rev. 74: 71–80.
34. RESCORLA, R. A. 1973. Effect of US habituation following conditioning. J. Comp. Physiol. Psychol. 82: 137–143.
35. RESCORLA, R. A. 1974. A model of Pavlovian conditioning. In V. S. Rusinov (ed.), Mechanisms of formation and inhibition of conditional reflex. Academy of Sciences, Moscow, USSR.
36. RESCORLA, R. A., and LoLORDO, V. M. 1965. Inhibition of avoidance behavior. J. Comp. Physiol. Psychol. 59: 406–412.
37. SHERMAN, J. E. and MAIER, S. F. 1978. The decrement in conditioned fear with increased trials of simultaneous conditioning is not specific to the simultaneous procedure. Learn. Motiv. 9: 31–53.
38. SIDMAN, M. 1953. Avoidance conditioning with a brief shock and no exteroceptive warning signal. Science 118: 157–158.
39. SOLOMON, R. L., CORBIT, J. D. 1974. An opponent-process theory of motivation: I. Temporal dynamics of affect. Psychol. Rev. 81: 119–145.
40. SOLOMON, R. L. and TURNER, L. H. 1962. Discriminative classical conditioning in dogs paralyzed by curare can later control discriminative avoidance responding in the normal state. Psychol. Rev. 69: 202–219.
41. STĘPIEŃ, I. 1974. The magnet reaction: A symptom of prefrontal ablation. Acta Neurobiol. Exp. 34: 145–160.
42. TOMIE, A. 1976. Interference with autoshaping by prior context conditioning. J. Exp. Psychol. Anim. Behav. Proc. 2: 323–334.
43. TRAPOLD, M. A. and OVERMIER, J. B. 1972. The second learning process

in instrumental behavior. In A. H. Black and W. F. Prokasy (ed.), Classical conditioning II: Current theory and research. Appleton Century Croft, New York, p. 427-452.
44. WILLIAMS, D. R. and WILLIAMS, H. 1969. Auto-maintenance in the pigeon: Sustained pecking despite contingent nonreinforcement. J. Exp. Anal. Behav. 12: 511-520.
45. ŻERNICKI, B. and KONORSKI, J. 1959. Fatique of acid conditioned reflex. Acta Biol. Exp. 19: 327-337.
46. ZIELIŃSKI, K. 1966. "Inhibition of delay" as a mechanism of the gradual weakening of the conditioned emotional response. Acta Biol. Exp. 26: 407-418.
47. ZIELIŃSKI, K. and WALASEK, G. 1977. Stimulus intensity and conditioned suppression magnitude: Dependence upon the type of comparison and stage of training. Acta Neurobiol. Exp. 37: 299-309.

J. Bruce OVERMIER, Center for Research in Human Learning, Elliot Hall, 75 East Road, University of Minnesota, Minneapolis, Minnesota 55455, USA.
Ralph, J. PAYNE, Shippensburg State College, Shippensburg, Pennsylvania, USA.
Robert, M. BRACKBILL, University of Arkansas, Monticello, Arkansas, USA.
Bruce LINDER, McMaster University, Hamilton, Ontario, Canada.
Janice A. LAWRY, University of Minnesota, Minneapolis, Minnesota, USA.

Lecture delivered at the Warsaw Colloquium on Instrumental
Conditioning and Brain Research
May 1979

# MATHEMATICAL MODELLING OF REACTION LATENCY: THE STRUCTURE OF THE MODELS AND ITS MOTIVATION

A. PACUT

Institute of Automatic Control, Technical University of Warsaw
Warsaw, Poland

*Abstract.* The basic structural assumptions concerning the dynamic models of the reaction latency are presented. The linear dynamic stochastic model of the reaction latency is considered as a special case of dynamic model. The biological motivation for using these models are outlined. These models express the reaction latency as a first access time of the random threshold of a certain stochastic process. This approach was used in modelling the reaction latency in escape and avoidance experiments (the results will be presented in subsequent papers).

## THE ROLE OF MODELLING IN BIOLOGY

It is a frequently met question: is mathematical modelling truly required in biology? To answer this question one must think about the advantages of mathematical modelling which are not accesible with other methods.

The first advantage of modelling is the possibility of presenting otherwise known results in an exact, precise form. Therefore the modelled results should be capable of having the required degree of precision. However, the biological facts often have only a qualitative nature, which must be recognized at the very earliest stages of model building. But just this qualitative nature of biological description requires development of adequate mathematical tools to handle and analyze such phenomena.

The second, frequently noted advantage of mathematical models is that they make it possible to handle large sets of data and facts. This is useful in analyses of results of experiments. One cannot handle results of neuron stimulation or of behavior experiments without even simple models. The most popular models treat these results as a sample of a number of random variables. Because of this assumption, statistical analyzes may be applied. Initially a distribution function is computed, but this can be computed without using a probabilistic (modelling) framework (it is then referred to as cumulative frequency function). In such cases one cannot treat the cummulative frequency function as a distribution function and to interpret the results statistically. Thus, one cannot use the results of this analysis without adding a probabilistic framework. Even frequently used averaging of sequences of data of an experiment may, in fact, be used only when related to a mathematical model. In fact, averages as they are normally computed, do not have the interpretations usually given to them unless the averaged elements are considered as random variables with appropriate mathematical properties which permit the use of the limit theorems of probability theory. Often averaging is done without checking the assumptions about the model, but this situation becomes dangerous only when one is using the model without bearing this hazard in mind.

The third role of mathematical models is the most important, because mathematical models of composite phenomena make it possible to handle and analyze such phenomena. In consequence, models may lead to discovering new features of these phenomena which were not earlier noticed because of their complicated nature. Properly chosen models can handle the observed facts in a simpler way than one could do without them. In effect, mathematical models allow one to explain known facts or to present them in simpler form. Models may also lead to new experiments which verify these models and also uncover new facts.

The goal of modelling has a strong impact on the shape of the model. In the modelling of technical systems, many types of models can be distinguished on the basis of their purpose. For example, there are models for control applications; on the basis of these models the rules of control of a technical system are developed. Another type constitute models for prediction. These models are used to predict future behavior of the system under analysis. We also meet with "cognitive models". In these models it is required to have all behavior of the model very close to the behavior of the real system. Depending on the goal, we would choose different models of the same system. Models should describe only those features of the modelled system, which are essential from the point of view of the goal of modelling. All unessential features should be neglected. In effect, elements of the model do not necessarily have to be identical with the elements of the modelled event. If, for instance, a random variable is included in the model, then this random variable would not exist as a physical object in the modelled system. Elements of mathematical models are mathematical abstracts. They should be interpretable in the sense that the behavior of elements of the model should have the same behavior as elements of the modelled system.

In all cases there exist a number of differences between the model and the reality. These differences provide the basis of one important way for the classification of the models. Deterministic models interprete these differences in terms of numbers. Stochastic (probabilistic) models treat them as random variables or stochastic processes. There are other interesting types of models, for instance fuzzy models.

Frequently in modelling one can assume only the structure of the model and the parameters must be determined on the basis of experimental data. Therefore, an essential role is played by the theory of identification together with the connected branches of mathematics such as estimation theory.

The last stage of modelling should be verification of the model. The verification procedure has its mathematical framework: theory of hypothesis testing and other branches of statistics can be used. But one should also perform a second, no less important, stage of verification, that is client verification. In biology, only the biologist can accept or reject the model on the basis of biological behavior of the model. This phase of modelling which will be called *acceptation* is often skipped in practice but, in the author's opinion, it is the most important stage of verification and, therefore, of modelling. When either the verification or the acceptation have not been

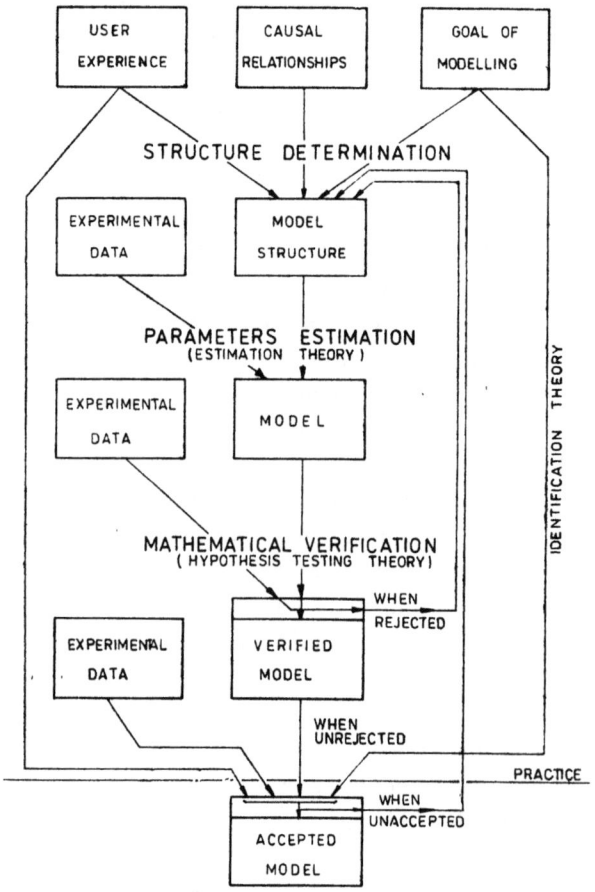

**Fig. 1. Stages of mathematical modelling.**

uccessful, one must return to the stage of structure determination (or to the stage of estimation but with the use of larger — or better — data sets). The whole procedure of model building is presented schematically in Fig. 1.

## THE MODELLED PHENOMENON

The subject of modelling presented in this paper is the latency of reaction in the experiments on cats which were performed in the Nencki Institute of Experimental Biology. Each cat was placed in a cage and subjected to a sequence of stimulations. The cats were trained to press the bar placed on the wall of their cage. Two experimental procedures will be discussed. Both experiments consist of series of trials.

The first procedure called escape experiment (12) consists in applying an electric shock to the paws of the cat in each trial. In this procedure the cats were trained to perform an escape response, that is, pressing the bar, in order to terminate the painful stimulation.

The second procedure called avoidance experiment (13, 14) consits in possibility of applying two stimuli in each trial. Each trial begins with applying the conditioned stimulus (an acoustic white noise with constant applitude). If the animal performs the proper reaction during given period (that is, bar-pressing response) than the trial terminates at the moment of pressing. If it doesn't answer properly then after this given period the unconditioned stimulus is applied together with the conditioned one. The proper reaction causes termination of the trial. A lack of such reaction after sufficiently long time forces experimenter to make some extraordinary activity which is beyond the scape of the model. It is worth pointing out that it also happens that the animal performs the reaction without stimulation (so called intertrial responses).

The latency of the reaction was measured in both the experiments. The results of the experiments are — from the numerical point of view — a data series expressing latencies of reactions in consecutive trials. The models presented in the paper treat this data series as a series of random variables, that is, as a discrete-time stochastic process generated by the model. In the course of the experiment, the cats underwent a learning process, which caused changes in the model parameters. These changes were the basis of the model verification. This problem will be treated later (Pacut; Pacut and Tych, in preparation). In the following, the structure of the model will be specified and some remarks concerning the model behavior and identification problems will be made.

## GEN ERAL STRUCTURE OF THE MODEL: DYNAMIC STOCHASTIC MODEL

The model under consideration, called the dynamic stochastic model, needed to be useful in analysis of the experimental data so it had to be relatively simple to make it posssible to identify its coefficients, but at the same time complicated enough to be able to handle all the substantial features which were contained in the data. Therefore the model had to have a structure that would be consistent with the biological principles associated with the experiment.

The structure of the dynamic stochastic model can be divided into three functional parts. These are the transformation system, the decision system, and the execution system (Fig. 2). The model aims to describe the dependence of reaction latency on

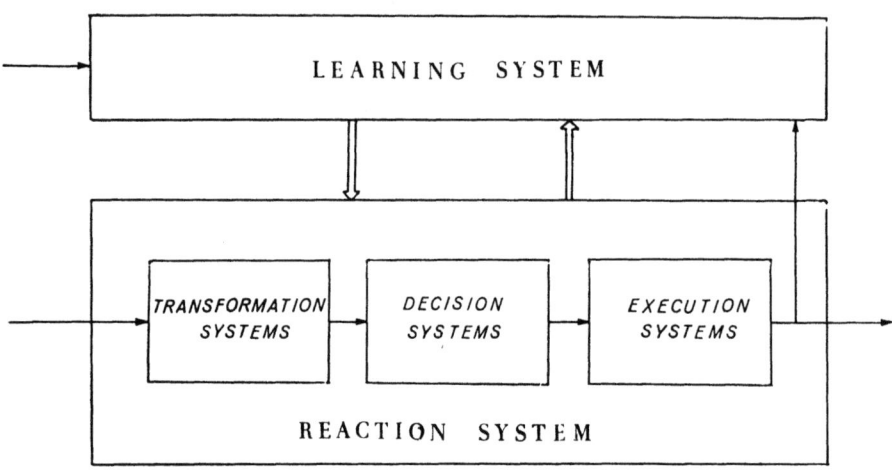

Fig. 2. General structure of the model.

the stimulus. Therefore the stimulus is the input signal to the model (it does not matter what we mean by the term "stimulus"). The activity of the transformation system is connected to the sensory systems of the animal, among other things. It generates a hypothetical time function called the excitatory potential. This function is a development of ideas of Hull (6), Spence (11) and Grice (3, 4). The transformation system represents all the transformations of the stimulus before it reaches the next link — the decision system. This decision part of the model compares the excitatory potential with a certain threshold value and may initiate execution of the reaction which, in turn, is performed by the execution system. One may imagine the composite net of transformation, decision and execution systems (Fig. 3) which connects all the sensory inputs through many decision systems with many reactions. There are pathways in this net which describe nonspecific reactions or performance of the learned reaction to nonexperimental stimuli. Nevertheless, only one small portion of this net will be considered more deeply, that is the pathway(s) which connects the experimental stimulus (or stimuli) with the reaction desired by the experimenter (called the experimental reaction). The paths that have to be used in the "active" part of the model depend on the type of experiment.

In the escape experiment (12) the animals were trained to respond to the unconditioned stimulus of electric shock by pressing the bar. The input used by the model

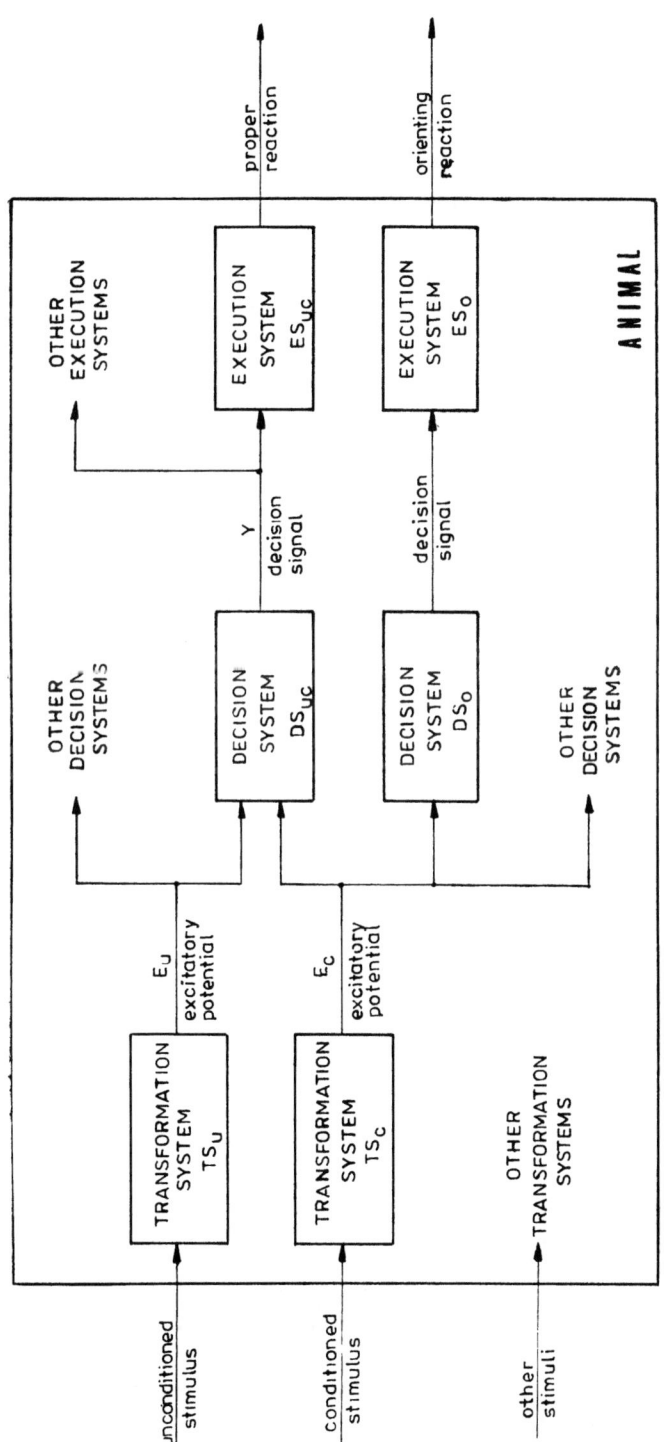

Fig. 3. Composite net of model elements.

is the intensity of the electric shock. But there are other influences, such as fear or the whole experimental setting, which affect the animals behavior, and these may be treated as other inputs in the model, which were not measured. In our models we will extract only one input, the intensity of electric shock, and the remaining influences will be treated as stochastic disturbances in the model.

In avoidance experiment (13) there are two basic stimuli: an electric shock and an accoustic white noise. Therefore at least two inputs must be considered in the model: electric noise intensity and acoustic noise intensity. The remainder of the stimuli which act on the animal will be treated as the model noise.

It is worth noting that the extraction of only these one or two inputs does not imply that the rest of the cues is beyond treatment by modelling methods. If the experimental procedure were changed even without changing these one or two controlled stimuli, the model would ha˙ to be modified because the structure of the stochastic inputs to the model would be changed.

There is a similar problem with the output from the model. The controlled reaction is (in the experiments described in (12) and (13)) pressing the bar placed on the wall of the cage. But there are of course many other reactions to the same experimental stimulation. Only a small portion of these reactions is dealt with instrumental procedure. The instrumental procedure can be treated as negative feedback — say "instrumental feedback" — which connects the desired reaction with the experimental stimulus (Fig. 4). Let us underline that the reaction which is fed back by feedback

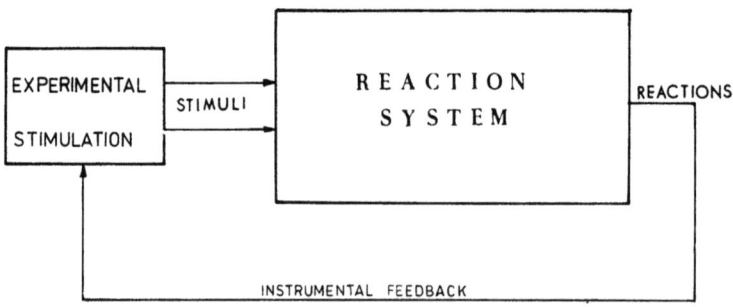

Fig. 4. Instrumental conditioning as feedback.

loop is not only pressing the bar but also all the other activity which is naturally connected with this pressing. Therefore from the point of view of identification there is no possibility of differentiating between elements of the different patterns of actions which might be performed during desired response to the stimulus. All o- them will be treated together and called the output of the model.

The logic of the model is as follows. The external stimulation changes the level of the hypothetical excitatory potential. Values of this potential are compared with the threshold and decisions concerning if and how to react are then taken. Several models of reaction latency previously considered by other authors can be considered as special cases of the dynamic model discussed here. Spence (11) postulated that the overt reaction may be considered as one of unobserved reactions in which oscillations are generated. The observed reaction occurs when the oscillation attains an amplitude greater than some level. In consequence Spence's model may be included in the class described above (Fig. 5). Its transformation system transforms input

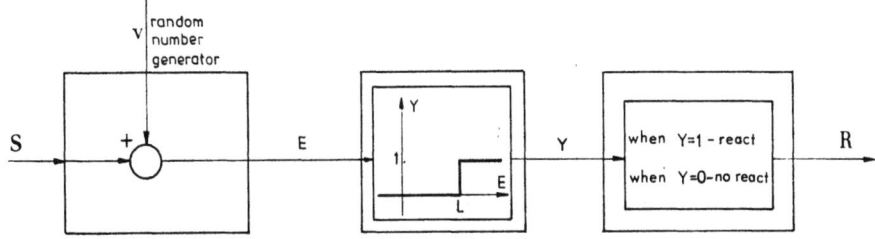

Fig. 5. Spence's model as special case of dynamic model.

signals to a white Gaussian sequence[1] and its decision system is given by a threshold element. A modification of Spence's model which leads to a better behavior of this model is given in another work (7).

A very impressive model has been proposed by Grice (3). He supposed that the stimulus consists of series of impulses which are counted by some probabilistic counter in the organism. The reaction occurs at the first moment that the cumulative

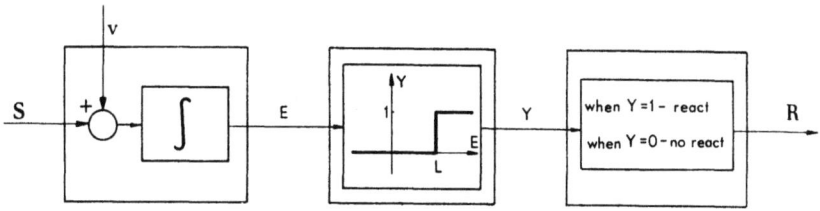

Fig. 6. Grice's model as special case of dynamic model.

count reaches some given level (random, in general). Some generalizations are given in the later work (4). Remarks on modified versions of these models are given in another work (7). The model of Grice may also be included in the class of general dynamic models (Fig. 6). Grice's transformation system is an integrating (counting) element and the decision system is a threshold element with a possibly random threshold.

---

[1] i.e. the sequence of independent random variables which have Gaussian distribution.

In consequence Spence's and Grice's models have the structure of the dynamic model which has been proposed. This structure is too general to be useful without beeing specified further, and we now proceed to do this.

FURTHER STRUCTURAL ASSUMPTIONS: LINEAR DYNAMIC STOCHASTIC MODELS

The special structure which we call the linear dynamic stochastic (LDS) model, is described as follows. Let the transformation system be described by a linear first-order dynamic system driven by a stochastic noise. Let the decision system be based on a threshold mechanism and the execution system acts without delay. The above assumptions, which are specified below in detail for the escape and avoidance experiment, completely describe structure of the LDS model.

For the escape experiment we consider one stimulus (one input) and one reaction (one output). Therefore, one transformation system, one decision system and one execution system, coupled in series, will be considered. The execution system will be modelled by the first order stochastically disturbed differential equation. This type of equation may be considered as a local approximation of more complicated relationships and was successfuly used in various applications. Namely,

$$dE(t) = (ks(t)-aE(t))dt + dw(t), \quad E(0) = E_0, \quad (1)$$

where:
$t$ is time which is measured from the moment of the initiation of the stimulation,
$E(t)$ is the excitatory potential created by the stimulus,
$w(t)$ is a stochastic Wiener process with $\mathscr{E}w(t) = 0, \mathscr{V}dw(t) = \sigma_w^2 dt$, where $\mathscr{E}$ denotes expectation, $\mathscr{V}$ denotes variance and $dw$ is a stochastic differential (see, for instance 2),
$E_0$ is a Gaussian random variable independent of the process $w(t)$ with $\mathscr{E}E_0 = m_0$, $\mathscr{V}E_0 = \sigma_0^2$,
$(st)$ represents the stimulus strength for the stimuli which will be considered, $s(t) = 0$ for $t < 0$ and $s(t) = S$ for $t \geqslant 0$,
$k$ denotes the static gain
$a$ denotes the dynamic gain.

The decision system will be modelled by a random threshold element whose output we denote by $y(t)$. This means that

$$y(t) = \begin{cases} 0 & \text{if } E(t) < L, \\ Y & \text{if } E(t) \geqslant L, \end{cases} \quad (2)$$

when $y(t) = Y$ the reaction occurs and there is no reaction if the output is 0. $L$ is a Gaussian random variable representing the random threshold, independent of $w(t)$ and $E_0$ with parameters

$$\mathscr{E}L = m_L \text{ and } \mathscr{V}L = \sigma_L^2.$$

In other words

$$y(t) = \begin{cases} 0 & \text{if } E(t) \notin R, \\ Y & \text{if } E(t) \in R, \end{cases} \quad (3)$$

where

$$R = \langle L, \infty) \quad (4)$$

is a region on a line called reaction region (see Fig. 8a). This notion will be more useful in the modelling of avoidance reaction. The execution system will be modelled by simple nondynamic transducer. Therefore the whole model structure is as shown in Fig. 7.

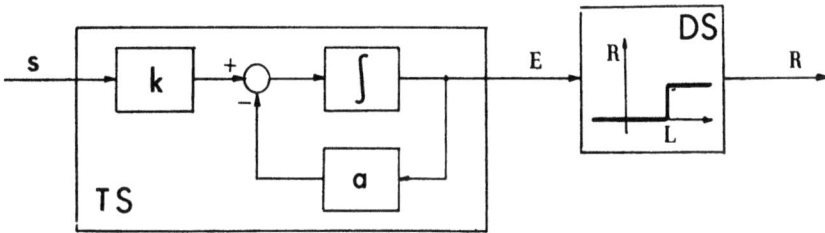

Fig. 7. Model of the escape experiment.

For the avoidance experiment we consider two stimuli (two inputs) and one reaction (one output). Therefore two transformation system will be assumed, each of them having the same form as described above for the excitatory potential. In other words,

$$dE_c(t) = (k_c s_c(t) - a_c E_c) dt + dw_c(t), \quad E_c(t) = E_{c0}, \quad (5)$$

where the subscript $c$ denotes functions and parameters connected with conditioned stimulus, and

$$dE_u(t) = (k_u s_u(t) - a_u E_u) dt + dw_u(t), \quad E_u(0) = E_{u0}, \quad (6)$$

where the subscript $u$ denotes functions and parameters connected with the unconditioned stimulus. It may be assumed that $s_u(t) = 0$ for $t < T$, where $T$ is the moment of initiation of the unconditioned stimulus. Nevertheless it cannot be assumed that the random variables which are connected with the $u$-subscripted (unconditioned) part are independent of the corresponding random variables which are connected with the $c$-subscripted (conditioned) part.

The decision system is more complicated because it makes a decision based on a comparison two excitatory potentials (to be called conditioned and unconditioned, respectively) with some associated thresholds. It will be assumed that the decision system consists of two thresholds, each of them being a random variable, and correlated with the other. The execution signal which causes the activation of the execution system will be sent when any one of the excitatory potentials reaches its threshold value.

This means that

$$y(t) = \begin{cases} 0 & \text{if} \quad E_c(t) < L_c \text{ and } E_u(t) < L_u, \\ Y & \text{if} \quad E_c(t) \geq L_c \text{ or } E_u(t) \geq L_u, \end{cases} \tag{7}$$

where $L = \begin{bmatrix} L_c \\ L_u \end{bmatrix}$ is a random Gaussian vector with expectation vector $\mathscr{E}L = m_L = \begin{bmatrix} m_{Lc} \\ m_{Lu} \end{bmatrix}$ and covariance matrix

$$\mathscr{V}L = \begin{bmatrix} \sigma_{Lc}^2 & \varrho \sigma_{Lc} \sigma_{Lu} \\ \varrho \sigma_{Lc} \sigma_{Lu} & \sigma_{Lu}^2 \end{bmatrix},$$

where $\varrho$ denotes the correlation coefficient between the two thresholds: one connected with the conditioned stimuli and the other connected with the unconditioned one. It may be more fruitfull to write relation (5) in the form

$$y(t) = \begin{cases} 0 & \text{if} \quad E(t) \notin R, \\ Y & \text{if} \quad E(t) \in R, \end{cases} \tag{8}$$

where

$$E(t) = \begin{bmatrix} E_c(t) \\ E_u(t) \end{bmatrix}$$

is called the excitatory vector and the region $R$, to be called the reaction region, is given by

$$R = \{(e_c, e_u) : e_c > L_c \quad \text{or} \quad e_u > L_u\}. \tag{9}$$

In other words, $R$ is a region on the plane in which at least one of the coordinates is greater than the corresponding threshold value. Because $L_c$ and $L_u$ are random variables, $R$ is a random region.

In consequence, for both models (of the escape reaction and the avoidance reaction), the decision system has the common form (3) or (8). The decision regions are given in Fig. 8 for both models. The execution system will be assumed to have the same form as for the escape model. The whole structure of the avoidance model is given in Fig. 9.

Therefore the LDS models for escape and avoidance have a common structure. As we will see, the parameters and the functions of this structure have a biological interpretation.

Consider the transformation systems. There are two parameters for each of them: the static gain $k$ and the dynamic gain $a$. The static gain may be interpreted as a parameter connected with unit and scale changes as well as with changes of carrier of information, for instance from the acoustic signal to the electric one. The dynamic gain is the coefficient in the negative feedback. Therefore the rate parameter (time constant) of the excitatory potential is equal to $1/a$. This means that the average speed of excitatory potential increase depends on this dynamic gain.

One of the inputs to the transformation system is the stimulus and the second is a stochastic white noise which is a convenient description of the derivative of a Wiener

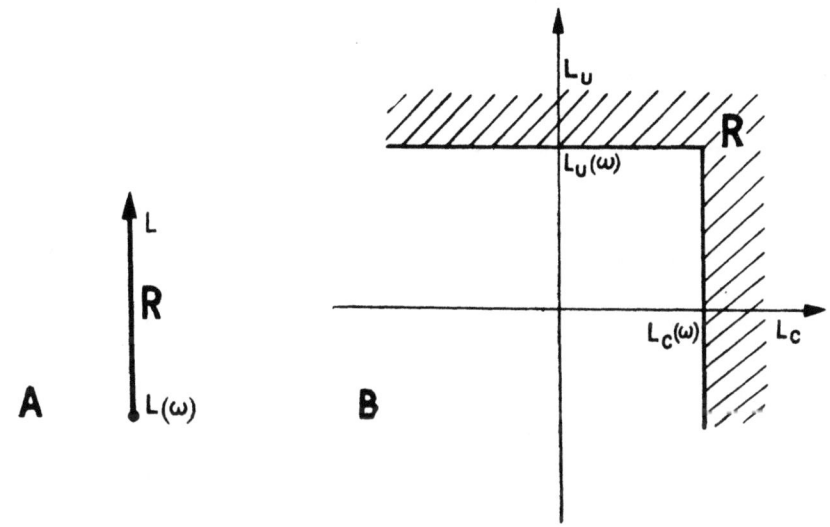

Fig. 8. Decision regions R for escape A, and avoidance B, model.

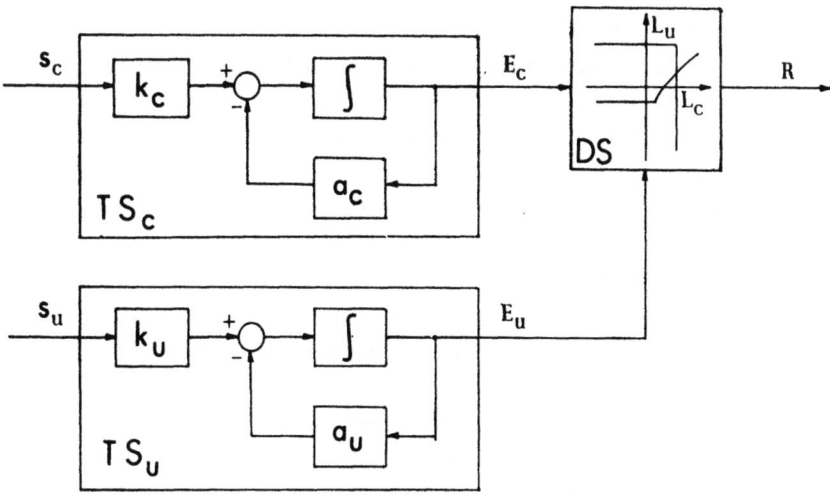

Fig. 9. Model of the avoidance experiment.

process with respect to time. This noise may be interpreted as existence of uncertainty in nervous system which causes changes of its activity even without stimulation. The integrator in the transformation system is responsible for accumulating the impressions caused by the stimulus.

Next consider the decision system. For both models the elementary actions of the decision system consist in comparing the excitatory potential with a given value. Such a comparison is easily interpreted on the basis of the threshold-like activity of neurons. The randomness of the threshold represents random variations of the threshold from trial to trial.

The situation may be compared to that in the theory of signal detection, where the noise is compared with the signal corrupted by noise. The decision what is observed: noise or signal plus noise may be taken on the base of distribution function analysis (5).

A very simple model has been chosen for the execution system. It was assumed that the delay caused by this system is small in comparison to other sources of delay. Some effects of this delay has been modelled as random disturbances. A direct inclusion of this delay into the model would make the model identification much more complicated. Special attention should be devoted to random elements in the model. There are at least two main sources of randomness. The first one is connected with fluctuations of the state of the system even without stimulation. This source is expressed by the Wiener process at the input of the model and by assuming random initial conditions. It may be noted that when there exists no stimulation and $s(t) = 0$, then the excitatory potential $E(t)$ is given by the equation $dE(t) = -aE(t)dt + dw(t)$ and therefore $E(t)$ has an asymptotically (for $t \to \infty$) Gaussian distribution with parameters $\mathscr{E}E(t) \to 0$, $\mathscr{V}E(t) \to \sigma^2_w/2a$ (for $a > 0$). Thus if one assumes that the time distance between two consecutive trials is much greater than the latency, then the parameters of the random variable expressing the initial conditions may be expressed through input noise parameters.

The second source of randomness, which is assumed to be independent of the above one is connected with the threshold. For the case of the escape experiment this is a one-dimensional Gaussian variable and for the case of the avoidance experiment this is a two-dimensional Gaussian variable.

It may be noted that the zero and the unit of excitatory potential is not subject to identification. The excitatory potential is hypothetical function and therefore the hypothetical zero and unit may be chosen.

It will be reasonable to assume that zero of this scale is connected with the asymptotic situation when no stimulation exists. In consequence, it leads to the assumption $m_0 = 0$.

The choosing of the unit of the excitatory potential will be considered when analyzing the properties of random threshold models.

It is easy to check that the models of Spence and Grice are also the special cases of LDS models of the escape-type reactions (for this problem see 7).

## ANALYSIS OF LDS MODELS

It is not a simple matter to fully analyze the behavior of the LDS models. It is a simple task to know the behavior of the excitatory potential but it is a rather involved one to solve the so-called first access problem for the stochastic process $E(t)$. We consider both problems briefly.

It can be shown (8) that the excitatory potentials have an exponential shape "in the average", namely

$$E(t) = ks/a + (E_0 - ks/at)\exp(-at) + \int_0^t \exp(a\vartheta)\exp(-at)\,dw(\vartheta) \qquad (10)$$

for the escape experiment and, in a slightly modified form, for the avoidance experiment. Therefore the individual excitatory potential may be illustrated as in Fig. 10 for the escape model and in Fig. 11 for avoidance model.

For both models the latency time $\tau$ can be written as

$$\tau = \begin{cases} \inf_{t \in \mathcal{S}} \{t : E(t) \in R\} & \text{if } \{t : E(t) \in R\} \neq \emptyset, \\ \infty & \text{for the opposite case,} \end{cases} \qquad (11)$$

where $\mathcal{S}$ is the stimulation period and $\mathcal{S} = \langle 0, t_{\max} \rangle$. Because of the randomness of $E(t)$ and $R$ the latency $\tau$ is a random variable. The problem of determining the latency distribution is, therefore, the first access time problem[2] for the process

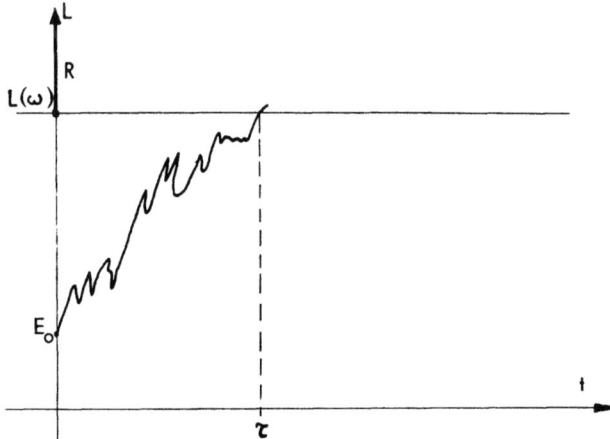

Fig. 10. Reaction evocation for escape experiment.

$E(t)$ and the region $R$. This problem has been effectively solved only for a small class of stochastic processes (1, 10). A very useful survey of existing techniques is given in Ricciardi (9).

---

[2] First access problem. For a given stochastic process and a region R find the distribution of random variable $\tau$ given by (11).

The first access problem can be resolved for both models. A method for the escape model is given in (8). A method for the avoidance model can be based on the same approach. However, only the Laplace transformation of the distribution function of

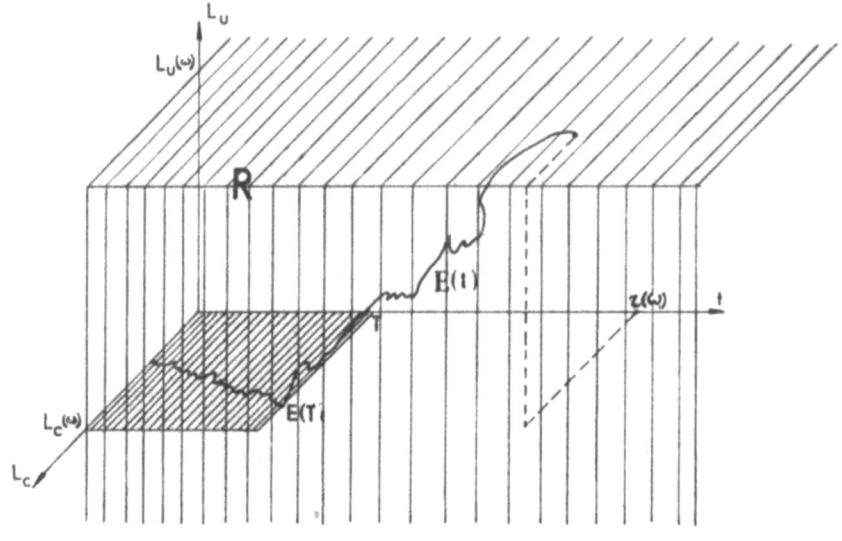

till the moment T  E(t) lies on the plane $L_c \times t$
from the moment T  E(t) lies in the space $L_c \times L_u \times t$

Fig. 11. Reaction evocation for avoidance experiment.

the first access time is known. This solution is not useful for solving the identification problem and a numerical method of solving the problem in terms of a time-domain analysis must be used. For a discussion of these problems the reader is referred to (8).

## IDENTIFICATION PROBLEMS

The only observations which can be used for the identification of the model parameters is the first access time. However this time is a random variable and it cannot be identified on the basis of a single observation. The sample distribution of this time must be derived from experimental data.

A problem arises of how to compute the sample distribution. The problem consists of the fact that there is learning during the experiment and therefore the latency data cannot be treated as a stationary series of random variables. There are several ways to handle this problem. One good approach is to model the reason for the nonstationarity. This approach cannot be directly used. The reason is that the structure of the nonstationarity is unknown. Knowledge of this structure is equivalent to knowledge of the structure of the learning process and this in its turn may be known after examination of the described model. Therefore another approach must be taken. It is assumed that the learning process behaves like a stationary one when considering only a small number of consecutive trials. If the latencies in these trials can be treated as independent then the averaging procedures can be used and sample distributions may be computed. An analogous approach may be used in examination of the structure of learning. Therefore this approach may be also treated as the initial step in modelling the learning process.

There is a substantial constraint on the quality of the identification. This depends on the number of latency measurements which can be used in the computation of the sample distribution computation. The data taken together will be called stage of learning. For the considered experiment it may never be assumed that the stage lasts longer than 50–200 trails. Moreover, it changes from animal to animal. It seems to be appropriate to assume that the time-lenght of the stage depends on the animal and is given by dividing the total number of trials by the number of stages which was assumed to be constant. In consequence, the number of parameters which may be identified on the basis of stage data is drastically constrained.

The detailed development of the concerning identification problem for the escape and avoidance models will be presented in subsequent papers (Pacut; Pacut and Tyoh, in preparation).

## CONCLUSIONS

In the paper the basic structural assumptions concerning the dynamic models and then the linear dynamic stochastic models of the reaction latency have been presented. The models have a uniform structure. Moreover, the same structure can be applied to other experiments, like differentiation conditioning. The biological motivation for using these models has been outlined. The models express the latency time as the first access time of the random threshold of a rather simple stochastic process, but even with such simple models identification is very complicated.

In the result the model must be identified in simpler form and then verification must be performed. In subsequent papers an analysis of the identified models will be presented as well as the results of the identification procedure.

I thank Prof. H. J. Kushner of Brown University, Providence, Rhode Island, Prof. F. R. Brush of Syracuse University, Syracuse, and Prof. K. Zieliński for helpful discussion on previous version of this paper.

## REFERENCES

1. DARLING, D. A. and SIEGERT, A. J. F. 1953. The first passage problem for a continuous Markov process. Ann. Math. Statist. 24: 625–639.

2. GIKHMAN, I. I. and SKOROKHOD, A. V. 1969. Introduction to the theory of random processes. W. B. Saunders Co., Philadelphia (translation of 1966 Russian edition).
3. GRICE, R. G. 1968. Stimulus intensity and response evocation. Psychol. Rev. 575: 359–373.
4. GRICE, R. G. 1971. Conditioning and a decision theory of response evocation. *In* G. H.Bower (ed.), Psychology of learning and motivation. Vol. 5, Academic Press, New York.
5. HELSTROM, C. W. 1960. Statistical theory of signal detection. Pergamon Press, Oxford
6. HULL, C. L. 1943. Principles of behaviour. Appleton-Century Comp., New York.
7. PACUT, A. 1977. Some properties of threshold models of reaction latency. Biol. Cybernetics 28: 63–72.
8. PACUT, A. 1978. Stochastic model of the latency of the conditioned escape response. *In* Progress in cybernetics and system research. Vol. 3. Advance Publ., London, p. 633–640.
9. RICCIARDI, K. M. 1977. Diffusion processes and related topics in biology. Springer-Verlag, Berlin.
10. SIEGERT, A. J. F. 1952. On the first passage time probability problems. Physical. Rev. 81: 617–623.
11. SPENCE, K. W. 1954. The relation of response latency and speed to the intervening variables and N in S-R theory. Psychol. Rev. 61: 209–216.
12. ZIELIŃSKI, K. 1970. Retention of the escape reflex after prefrontal lobectomy in cats. Acta Neurobiol. Exp. 30: 43–57.
13. ZIELIŃSKI, K. 1971. Increase versus decrease in noise intensity as a cue in avoidance conditioning. Acta Neurobiol. Exp. 31: 331–340.
14. ZIELIŃSKI, K. 1972. Stimulus intensity and prefrontal lesion effects on latencies of the bar-pressing avoidance response in cats. Bull. Pol. Sci. Ser. Biol. 20: 821–826.

A. PACUT, Institute of Automatic Control, Technical University of Warsaw, Nowowiejska 15/19, 00-665 Warsaw, Poland.

Lecture delivered at the Warsaw Colloquium on Instrumental
Conditioning and Brain Research
May 1979

# LOCUS COERULEUS LESIONS AND AVOIDANCE BEHAVIOR IN RATS

Adam PŁAŹNIK and Wojciech KOSTOWSKI

Department of Pharmacology and Physiology of the Nervous System,
Psychoneurological Institute
Warsaw, Poland

*Abstract.* Bilateral electrolytic lesions of the locus coeruleus were performed in rats. In comparison with controls, lesioned rats showed decreased acquisition of two-way avoidance response, less intertrial responses, longer latency of avoidance responses on the first day of training, and needed more trials to reach the extinction criterion. It is suggested that locus coeruleus, contrary to the ventral noradrenergic bundle, facilitates avoidance behavior in rats.

## INTRODUCTION

In recent years, outstanding interest has been directed toward the possible role of brain noradrenergic (NA) neurons in conditioned behavior and self stimulation mechanisms (2, 4, 15, 24, 25). According to the Stein and Wise theory of two opposite systems (26, 27), catecholamines such as dopamine (DA) and NA are involved in a reward system, while acetylcholine and serotonin (5-HT) are involved in punishment system.

Noradrenergic neurons form at least two main fiber systems. The dorsal bundle (DB) originates mainly in the area corresponding to the nucleus locus coeruleus (LC), and the ventral bundle (VB) originates more heterogenously from NA cell groups in the ventral tegmentum (5).

The DB projects to the cerebral, hippocampal and cerebellar cortical areas as well as to the amygdala, thalamic and hypothalamic regions. The VB innervates a variety of basal brain structures associated with hypothalamus and limbic system (5, 16).

It has been suggested that Na released from neurons belonging to the LC is critically involved in the learning processes (10). According to Crow and co-workers (3, 4) NA released in the cortex stimulates synaptic processes which are necessary for initiating the transmission from a short- to a long-term memory system. This hypothesis is supported by some papers reporting impairment of acquisition of a food rewarded runway response after electrolytic lesion to the LC (2, 11, 23).

However, some observations seem to argue against the role of the LC and DB in learning mechanisms. It was reported that animals with chemical or electrolytic lesions to the LC or the DB, when showing an impairment in some types of conditioned responses, can still learn a wide variety of other conditioned tasks (18, 20, 21). Moreover, lesions to the DB produced significant resistance to extinction i.e., perseveration in responding to no longer reinforced stimuli (18–20). On the basis of similar results, Gray (8) suggested that DB is involved in the coding of "frustration" produced by omission of an expected reward. This hypothesis is supported by findings that omission of reward is correlated with the occurrence of hippocampal theta rhythm at the frequency of 7.7 Hz, and that lesions to the DB raise the threshold for this frequency (6, 7). Another explanation for the resistance to extinction in DB lesioned animals was proposed by Mason et al. (19, 21). This explanation is based on the observation that the number of stimuli connected with learned responses during acquisition training may determine the rate of extinction (17). In the brain there are mechanisms which select irrelevant stimuli and are responsible for habituation (17). Mason et al. (19, 21) suggested that DB may play a role in these processes and that lesions of this structure result in resistance to extinction due to an increase in the number of stimuli that the animal has sampled during training and have become attached to reward.

The present study proposed to determine the effect of bilateral lesions to the LC on avoidance acquisition, extinction and retention. The role of this structure in avoidance behavior is unclear and some contradictory results have been reported (11, 23). Recently we found (13) that electrolytic lesion of the second main NA system in the brain, the VB, surprisingly **facilitated** two-way avoidance acquisition in rats. We emphasized that brain NA neurons are not functionally homogenous and supposed that VB seems to play an opposite role to the DB in behavioral processes (9).

METHODS

*Subjects.* Twenty seven naive male Wistar rats, weighing 190–200 g at the start of the experiment, were used. Rats were housed in groups of 3–4 in macrolon cages (20 × 24 × 43 cm) with food and water freely available. The holding room was kept constant at 20 ∓ 2°C and a 12 h reversed light/dark cycle was maintained (light 7 : 00 p.m. to 7 : 00 a.m.). Then rats were randomly divided into 2 groups: 15 for LC lesions and 12 for sham operated control.

*Surgical and histological procedures.* Electrolytic lesions were performed under chloral hydrate anaesthesia (400 mg/kg i.p.) using a stereotaxic Stoelting instrument and stainless steel electrode (0.25 mm in diameter insulated except for the tip). The coordinates for LC were: P 1.6 mm, L 1.1 mm and H 2.5 mm above the interaural line (according to König and Klippel, 14). Lesions were made by passing d.c. current of 1.0 mA for 5 s. Operated control subjects (sham lesioned) were treated in the same manner, except that the electrode was not lowered into the brain. The lesions were made one week before the experiment started. At the end of the experiment the animals were killed by decapitation, their brain stems were placed in 10% formalin solution, and 15 µm sections were stained with haematoxylin and eosin. The placement and size of lesions were checked microscopically without knowledge of individual behavioral results.

*Apparatus.* Avoidance testing was performed in a shuttle box consisting of two identical compartments (24 × 24 × 76 cm) separated by a 8 cm high hurdle. The grid floor consisted of stainless steel rods 2.0 mm in diameter, spaced 1.5 cm apart. Illumination was provided by 25 W bulbs centered 30 cm above each compartment. An electric shock of approximately 2.0 mA was delivered through the grid floor of the compartments.

*Behavioral procedure.* On the first day of two-way avoidance training the animals were habituated to the shuttle box apparatus for 5 min. The conditioned stimulus (CS) was a light of 5 s duration, and the unconditioned stimulus was the electric footshock. The CS–UCS interval was 5 s, the intertrial interval averaged 30 s. Rats received 10 trials daily during 10 consecutive days or until the criterion 9 avoidance responses in two consecutive days was reached. When the avoidance criterion was established, extinction testing began, during which the CS was not reinforced. Initially the animals were subjected to 6 CS presentations without any possibility of making avoidance responses and than spontaneous extinction of avoidance was observed until the extinction criterion of 1 avoidance response on 10 trials was obtained. During the next 2–3 days, rats were retrained to the previous avoidance criterion. Two weeks after the last retraining session, subjects that reached the retraining criterion level received a retention test. During avoidance conditioning and retention the following parameters were recorded: the number of avoidance responses for each session, the number of non-punished intertrial crossings between both compartments, and the latencies of avoidance and escape responses.

*Statistical analysis.* The latency data were analyzed by the Kolmogorov-Smirnov test. The remaining data were analyzed using the Mann and Whitney two tailed test.

## RESULTS

*Location of lesions.* Histological examination showed that lesions bilaterally involved the LC. Usually lesions also partially destroyed the mesencephalic root nucleus of trigeminal nerve (Fig. 1). In 5 rats, lesions were not accurately positioned in the LC and these animals were excluded from the analysis of the results.

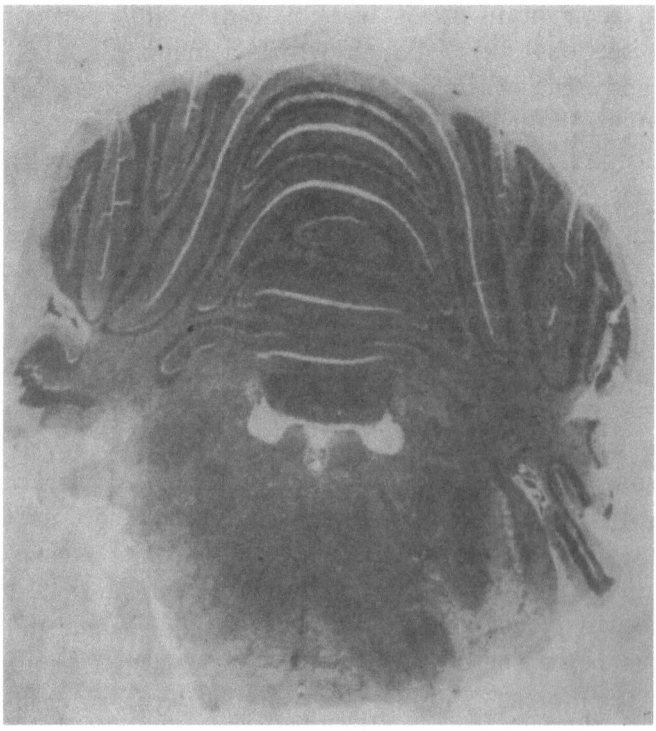

Fig. 1. Frontal section showing typical bilateral lesion involving the locus coeruleus.

*Two-way avoidance.* Lesions involving the LC significantly decreased the rate of acquisition of this task during first 5 days of training, (1st day: $U = 90.0$, $P = 0{,}05$; 2nd day: $U = 77.0$, $P = 0.005$; 3rd day: $U = 84.0$, $P < 0.025$; 4th day: $U = 78.0$, $P < 0.01$, 5th day: $U = 90.0$, $P = 0.05$; 6th day: $U = 91.0$, $P > 0.05$; 7th day: $U = 99.0$, $P > 0.05$). The number of intertrial responses was smaller in LC lesioned animals thant in control group, but the difference was not significant (Fig. 2).

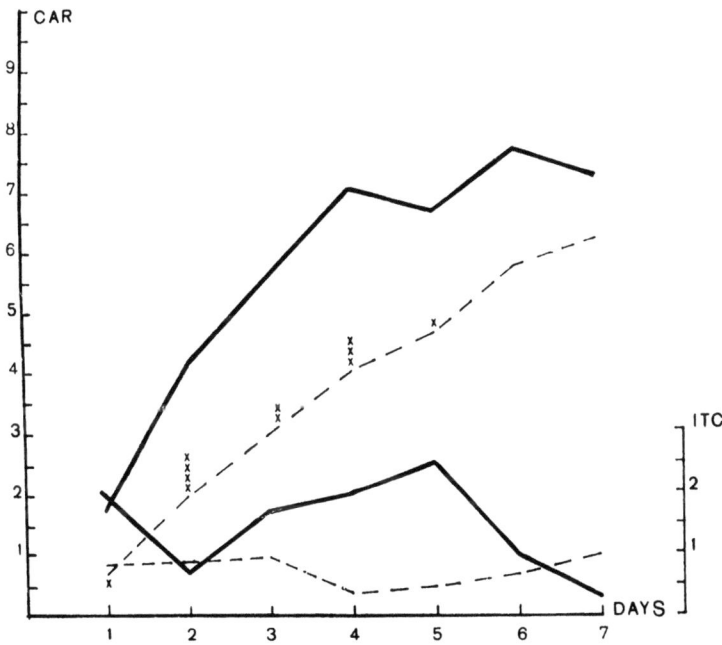

Fig. 2. Acquisition of two way conditioned avoidance response (CAR) and number of intertrial crossings (ITC) during avoidance sessions. Solid lines, sham operated rats; dashed lines, rats with lesioned locus coeruleus. Differences from sham operated rats: $x = P < 0.05$; $xx = P < 0.025$, $xxx = P < 0.01$, $xxxx = P < 0.005$. For number of rats see text.

The latencies of avoidance responses were prolonged in animals with lesioned LC. This effect was significant on the very first day of training, ($k = 1.418$ $P < 0.05$). The latency of escape reaction was also prolonged but this effect was not significant (Fig. 3).

Four rats with LC lesions and 3 sham operated control, did not obtain avoidance criterion and were rejected from futher experiment. It was left: 6 rats with lesions to the LC and 9 sham operated control animals. Rats with lesioned LC needed significantly more trials to reach the extinction criterion then sham operated animals, ($U = 64.5$ $P < 0.025$), (Fig. 4). One sham operated rat did not reach avoidance criterion during retraining and was rejected from the retention test. There was, however, no differences in performance between both groups during retention test, $(U = 41.5\ P > 0.05)$, (Fig. 4).

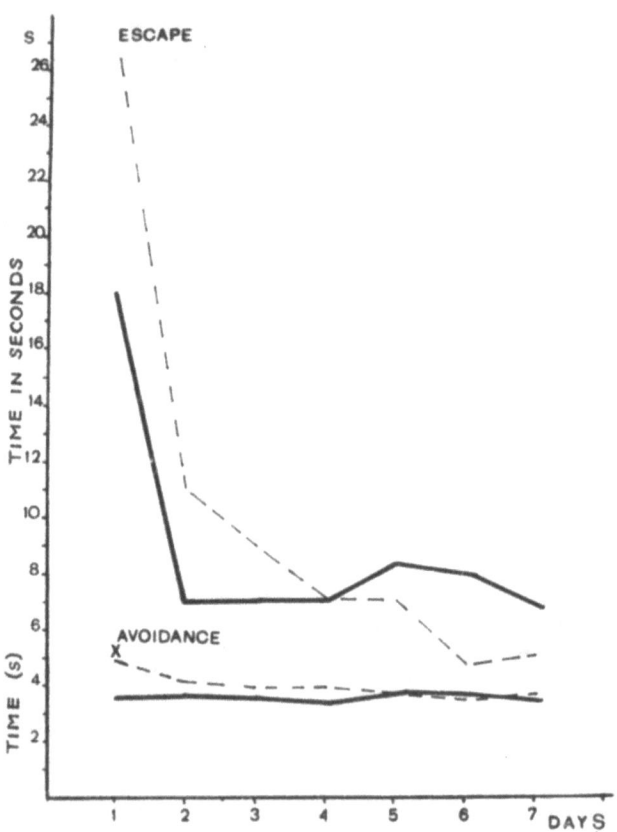

Fig. 3. Latencies of avoidance and escape responses during training of two way avoidance in s. Other denotations as in Fig. 2.

## DISCUSSION

Our results support the hypothesis of an involvement of the LC and DB in learning processes. The performance decrements on avoidance and appetitive tasks after electrolytic lesions to the LC have been observed by numerous investigators (2, 11, 23). There are, however, studies which do not confirm these findings (1, 19, 20). The most intriguing question arising from these conflicting data is of the essence of action of NA on the learning processes. Gray (6–8) suggested that this amine is involved in the conding of nonreward and that this mechanism is responsible for retardation of extinction after destruction of the LC. Mason and Fibiger (19) proposed that in LC lesioned animals, the changes in affectional processes might lead to the retardation of extinction. According to these authors NA is important for the regulation of activity of brain structures selecting the incoming informations from the environment. Lesions of NA neurons should, therefore, lead to the disinhibition of these processes resulting in resistance to extinction.

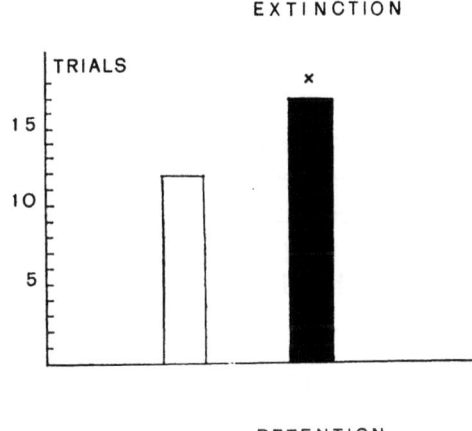

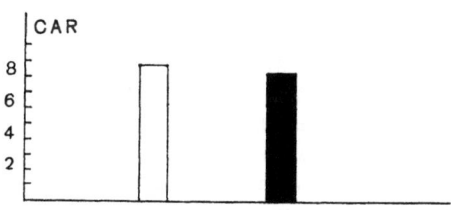

Fig. 4. Number of trials to the extinction criterion (upper) and performance of rats during the retention test (lower). White columns, sham operated rats; black columns, rats with lesioned locus coeruleus. Difference from sham operated rats: $x = P < 0.05$.

The possible involvement of NA in emotional processes may also explain some effects observed in LC lesioned animals. This problem may be discussed in term of reward and punishment theory, according to which NA neurones are involved in positive reinforcement and 5-HT neurones in punishment — i.e., negative reinforcement. It should be mentioned that the decrease in activity of NA neurones leads to increased activity in 5-HT neurones (12). This mechanism may be involved in increased fear and unpleasant feelings that may disturb avoidance acquisition and performance. The resistance to extinction may be also explained by the increase in fear after lesion of NA neurons. Such an interpretation is compatible with finding of Mason et al (21) who observed increase in "neophobia" to a number of novel tasks in rats with lesioned DB by intracerebral injections of 6-OH dopamine. Our result demonstrating that resistance to extinction following electrolytic lesion to the LC can be obtained in a two-way avoidance situation has important implications for theories which explain the resistance to extinction in terms of "frustration". It is difficult to suppose that rats are "frustrated" when they not receive a punishment, (i.e., an electric footshock), which they expect (as might take place in the case of positive reinforcement). This conclusion is compatible with that of Mason and Fibiger (19) who demonstrated that resistance to extinction following injection of 6-OH dopamine to the DB can be demonstrated in passive avoidance situation.

It is proper to add that recently we found that lesions of the second NA pathway, the VB, facilitated avoidance acquisition in rats (13). Such a result was unexpected, since up to that time lesions of the brain NA system either reduced or failed to change the acquisition of a variety of tasks. According to this finding, the VB seems to play an opposite role in learning processes than the DB does. It should be therefore concluded that NA neurones in the brain are not functionally homogenous.

We are grateful to Mrs Barbara Kucharska and Mr Jan Borszewski for helpful technical assistance.

## REFERENCES

1. AMAREL, D. G. and FOSS, J. A. 1975. Locus coeruleus lesions and learning. Science 188: 377–378.
2. ANLEZARK, G. M. CROW, T. J. and GREENWAY, A. P. 1973. Impaired learning and decreased cortical norepinephrine after bilateral locus coeruleus lesions. Science 181: 682–684.
3. CROW, T. J. and ARBUTHNOTT, G. W. 1972. Function of catecholamine — containing neurons in mammalian central nervous system. Nature 238: 245–246.
4. CROW, T. J. and WENDLANDT, S. 1976. Impaired acquisition of a passive avoidance response after lesions induced in the locus coeruleus by 6-OH dopamine. Nature 259: 42–44.
5. DAHLSTRÖM, A. and FUXE, K. 1964. Evidence for the existence of monoamine containing neurons in the central nervous system. I. Demonstrations of monoamines in the cell bodies of the brain stem neurons. Acta Physiol. Scand. Suppl. 232: 1–55.
6. GRAY, J. A. 1972. Effects of septal driving of the hippocampal theta rhythm on resistance to extinction. Physiol. Behav. 8: 481–490.
7. GRAY, J. A., McNAUGHTON, N., JAMES, D. F. and KELLY, P. H. 1975. Effect of minor tranquillisers on hippocampal theta rhythm mimicked by depletion of forebrain noradrenaline. Nature 258: 424–425.
8. GRAY, J. A. 1977. Drug effects on fear and frustration: possible limbic site of action of minor tranquilizers. In L. L. Iversen, S. D. Iversen and S. H. Snyder (ed.), Handbook of psychopharmacology. Vol. 8, Plenum Press, New York.
9. JERLICZ, M., KOSTOWSKI, W., BIDZINSKI, A. and HAUPTMANN, M. 1978. Effects of lesions in the ventral noradrenergic bundle on behavior and res-

ponse to psychotropic drugs in rats. Pharmacol. Biochem. Behav. 9: 721–724.
10. KETY, S. S. 1970. The biogenic amines in the central nervous system, their possible roles in arousal, emotions and learning. In F. O. Schmitt (ed.), The neuroscience. Rockefeller Univ. Press, New York, p. 324–336.
11. KOOB, G. F., KELLEY, A. E. and MASON, S. T. 1978. Locus coeruleus lesions: learning and extinction. Physiol. Behav. 20: 709–716.
12. KOSTOWSKI, W. 1975. Interactions between serotonergic and catecholaminergic systems in the brain. Polish J. Pharmacol. Pharm. 27: Suppl. 15–24.
13. KOSTOWSKI, W. and PŁAŻNIK, A. 1978. Effect of lesions of the ventral noradrenergic bundle on the two-way avoidance behavior in rats. Acta Physiol. Pol. 29, 6: 509–514.
14. KÖNIG, J. T. and KLIPPEL, R. A. 1963. The rat brain: a stereotaxic atlas of the forebrain and the lower part of the brain stem. Williams and Wilkins, Baltimore.
15. LENARD, L. G. and BEER, B. 1975. Modification of avoidance behavior in 6-OH dopamine treated rats by stimulation of central noradrenergic and dopaminergic receptors. Pharmacol. Biochem. Behav. 3: 887–983.
16. LINDVALL, O. and BJÖRKLUND, A. 1974. The organization of the ascending catecholamine neuron systems in the rat brain as revealed by the glyoxylic acid fluorescence method. Acta Physiol. Scand., Suppl. 412: 1–48.
17. MACKINTOSH, N. J. and HOLGATE, V. 1968. Effects of inconsistent reinforcement on reversal and nonreversal shifts. J. Exp. Psychol. 76: 154–159.
18. MASON, S. T. 1978. Parameters of the dorsal bundle extinction effect: pre vious extinction experience. Pharmacol. Biochem. Behav. 8: 655–659.
19. MASON, S. T. and FIBIGER, H. C. 1978. 6-OH dopamine lesion to the dorsal noradrenergic bundle alters extinction of passive avoidance. Brain Res. 152: 209–214.
20. MASON, S. T. and IVERSEN, S. D. 1977. Effects of selective forebrain noradrenaline loss on behavioral inhibition in the rat. J. Comp. Physiol. Psychol. 91: 165–173.
21. MASON, S. T. and IVERSEN, S. D. 1978. Reward, attention and the dorsal noradrenergic bundle. Brain Res. 150: 135–148.
22. MASON, S. T., ROBERTS, D. C. S. and FIBIGER, H. C. 1978. Noradrenaline and neophobia. Physiol. Behav. 21: 353–361.
23. SESSIONS, G. R., KANT, G. J. and KOBB, G. F. 1977. Locus coeruleus lesions and learning in the rat. Physiol. Behav. 17: 853–859.
24. STEIN, L. C., BELLUZZI, J. D., RITTER, S. and WISE, C. D. 1974. Self-stimulation reward pathways: norepinephrine vs. dopamine. J. Psychiatr. Res. 11: 115–119.
25. STEIN, L. C. and WISE, C. 1973. Amphetamine and noradrenergic reward pathways. In Frontiers in catecholamine research. Pergamon Press, p. 963.
26. STEIN, L. C. and WISE, C. D. 1974. Serotonin and behavioral inhibition. Adv. Biochem. Psychopharmacol. 11: 281–286.
27. WISE, C. D., BERGER, B. D. and STEIN, L. C. 1973. Evidence of noradrenergic reward receptors and serotonergic punishment receptors in the rat brain. Biol. Psychiatr. 6: 3–8.
28. WISE, R. A. 1978. Catecholamine theories of reward: a critical review. Brain Res. 152: 215–247.

Adam PŁAŻNIK and Wojciech KOSTOWSKI, Psychoneurological Institute, Sobieskiego 1/9, 02-957 Warsaw, Poland.

Lecture prepared for the Warsaw Colloquium on Instrumental
Conditioning and Brain Research
May 1979

## PROTECTION FROM EXTINCTION BY A CONDITIONED INHIBITOR

Stefan SOŁTYSIK and George WOLFE

Mental Retardation Research Center, School of Medicine
University of California at Los Angeles
Los Angeles, California, USA

*Abstract.* The phenomenon of protection from extinction (PFE) of a conditioned stimulus (CS) by a conditioned inhibitor (CI) has not been yet unequivocally demonstrated for the CS–CI compound in which the CS precedes the onset of the CI. Preliminary data from a project addressed to this problem strongly indicate that PFE is a real and robust phenomenon. Moreover, the protection is demonstrated not only for the CS duration overlapping with the CI but also for the early part of the CS which is not prevented by the CI from eliciting a conditioned response. The review of a few theories of conditioning suggests that the phenomenon of PFE is theoretically acceptable and predicted within the framework of any hypothetical mechanism which allows for post-trial "processing" or "consolidation" of information acquired during the trial.

### INTRODUCTION

Jerzy Konorski was first to propose a causal link between a conditioned inhibitor (CI, a signal of nonreinforcement in a Pavlovian conditioning paradigm) and the protection from extinction (PFE) of a nonreinforced conditioned stimulus (CS). In 1948 he posed the following question: "So why is it that the character of this stimulus (a CS previously paired with the injection of acid into the mouth and presently eliciting a leg flexion avoidance response, S.S.) as a conditioned defensive one is somehow 'preserved' when it is accompanied by a movement constituting a conditioned inhibitor? One has the impression that this very movement somehow protects the stimulus against extinction, and thus the conditioned reflex second type (i.e., instrumental, S. S.) is maintained perpetually. But we are unable to explain the causes of this phenomenon" (10, p. 231). Since then, there were a few attempts to verify the idea that there really is a prevention or retardation of the extinction of the conditioned responses to the nonreinforced CS, if it is presented in compound with the CI.

In dogs trained to press a bar for food in response to a CS, Chorążyna (3, 4) observed that after multiple presentation of a *simultaneous* compound of a CS and CI, the CS retained its capacity to elicit a conditioned response. This finding was not applicable for the avoidance situation where the CS precedes the CI and *elicits,* presumably, at least some suprathreshold fear CR necessary to motivate the avoidance instrumental response.

Sołtysik (17) described an experiment on one dog which had been trained to salivate and press a bar for food in response to a 12 s CS, but not to respond to this same CS if it was preceded by a CI. To simulate more closely the situation of avoidance learning, the author presented 120 times (in 10 daily sessions) a CS–CI compound in which the CS onset preceded for 3 s the onset of a CI and both stimuli coterminated after 12 s from the onset of the CS. During the first 3 s of the CS acting alone the initial stages of the CR were observed: orienting towards the CS, acceleration of heart rate, occasionally the onset of salivation or raising of the paw. After 120 trials of nonreinforced CS–CI compound, the CS alone elicited the full conditioned response. The evidence of PFE was unquestionable, though somewhat limited to the first few trials of CS alone, because the testing trials were reinforced with food and some rapid reinstatement of a CR could be assumed for the consecutive trials. In concordance with the view, prevalent at that time, that there is considerable generality of learning laws, the author confidently assumed that his finding applied also to defensive conditioning, the more so, that food CRs extinguish more easily than defensive CRs.

The next experimental work to address the question of protection against extinction was provided by LoLordo and Rescorla (12). Ten dogs were trained in a Sidman avoidance procedure and in a Pavlovian paradigm, on alternate days. The Pavlovian paradigm included aversive CSs and CIs; CIs, when presented, followed the CSs to simulate the avoidance response. The testing compared the course of extinction of two CSs, one of which was presented nonreinforced but protected by a CI, while the other was simply nonreinforced. No evidence for protection from extinction was found in this study.

Similarly, no evidence for a protective role of a CI was obtained in the study of Johnston, Clayton and Seligman (unpublished 1972, quoted by Seligman and Johnston, 16, p. 83) on rats. Seligman and Johnston felt justified to make a strong statement: "To summarize, our results and those of LoLordo and Rescorla suggest that protection from extinction is nonexistent when the inhibitor follows the conditioned fear stimulus as it must in avoidance paradigms." and later: "There was no reason to have hoped that protection could occur under these circumstances" (16, p. 83).

There are a few reasons, mostly procedural, why these two studies are not very convincing. The first reason for the failure in obtaining protection from extinction is the short training of a conditioned inhibitor in both LoLordo and Rescorla's and Johnston et al. papers. This brief training period probably did not allow the stimulus intended as a CI to become an inhibitor. No independent evidence of the inhibitory properties of the "CI" was presented. The second reason is the relative salience of CSs and the CI. It is generally believed that the CI has to be a strong or salient stimulus. It is in a double disadvantage in matching the CSs: firstly, it elicits an opposing (inhibitory in respect to the CR) process or response, which is known to be a weaker, more slowly trained, and easily disturbed one; and secondly, it is trained after the CR to the CSs is acquired, so it is a later addition to subject's behavioral repertoire. The strength of acquired responses is believed to reflect, besides other factors, their order of acquisition. In both studies which yielded evidence against the protection from extinction, the CSs were auditory stimuli while the conditioned inhibitor was a visual cue: turning off the light in LoLordo and Rescorl's study, and a light stimulus in Jonston's et al. In the latter experiment the "CI" was introduced only after the CS–US pairing was completed, so the "CI" was more like a cue for extinction than a real CI, which normally is acquired by long training including both CS–US and CS–CI trials mixed in each daily session.

The recent study of Hendersen and Harris (7) provides weak support for PFE, limited to the first trial of testing. This result is furthermore undermined by the use of a compound of fully overlapping CS and CI, so that no resemblance to the signalled avoidance paradigm was attempted. The authors were able, however, to provide an independent estimate of the inhibitory role of the CI. It was rather mediocre and the weak protective effect might simply reflect the weak inhibitory potential of their CI.

In summary, the interesting idea of Konorski that the conditioned inhibitor may *somehow* interfere with the extinction process, has produced very little experimental effort so far. Supporting evidence of Chorążyna (3, 4) and Sołtysik (17) are in the realm of food conditioning and need not apply to aversive situations. Hendersen and Harris' (7) data provided some evidence of PFE in aversive conditioning, but the effect was very weak and short lived, probably because the CI was not sufficiently established in its role as an inhibitor. The negative data of LoLordo and Rescorla (12) and of Johnston et al. (cited in Seligman and Johnston 16) should not be accepted because of the lack of evidence that an actual CI was used in their studies.

This dearth of experimental data on the PFE gives us an excuse to report the preliminary results from the first subject in an ongoing study on cats.

THE GENERAL STRATEGY AND THE EXPERIMENTAL DESIGN

Forewarned by the difficulties encountered in the aforementioned studies, the experiment, started a few months ago, was designed in such a way that only well trained and behaviorally clearly defined conditioned stimuli and conditioned inhibitor were used. Moreover, in contrast to all previous studies, all stages of our experiment were carried out without change in the density of shock-reinforced trials. Thus any sudden transitions from an overall situation of shock to no-shock were avoided. As was postulated by Capaldi (2) discrimination between US and noUS situations, and by extrapolation, between different densities of US trials, may contribute to the extinction phenomenon. In this study we wanted the extinction to be cued to particular stimuli and not to the entire situation. Besides, we expected the extinction, on such an "excitatory background", to procede more slowly enabling us to make more precise comparisons between conditioned responses elicited by differently treated CSs.

All stages of the experiment were performed with the animal secured in a modified Pavlovian stand. Although its head was fixed through a cranial implant (to enable physiological recording of heart rate, respiratory movements, etc.) it could walk or run freely on a treadbelt and the distances of locomotor responses were recorded. The unconditioned stimulus (electric shock) was given through two electrodes: one on the left foot (hindleg) and the other on the tail, about 4 cm from its base. The US reliably elicited high leg flexion, a vocal response, changes in respiration and heart rate, and occasionally a locomotor response. The movements of both hindlegs were recorded using light attachements connected to potentiometers, so that raising (flexing) the leg produced a change in electric resistence allowing the onset and the amplitude of flexion to be recorded.

The subject was first trained using Pavlovian leg flexion procedure to respond to three CSs: $L_{CS}$ = continuous light from the panel placed 20 cm in front of the cat's head; $A_{CS}$ = a continuous air blow to the sacral region about 5 cm forward from the tail base; and $R_{CS}$ = a rotating cylinder painted in black and white stripes, placed above the light panel in front of the subject. The duration of these CSs was 5,200 ms. The shock US (2.5 mA, 60 Hz) was delivered 5,000 ms after the onset of a CS and lasted 300 ms, i.e., it terminated 100 ms after the termination of the CS, overlapping with it for 200 ms.

Following the establishment of the CRs to these CSs a conditioned inhibitor, a clicking sound (10 clicks/s) delivered from a laudspeaker situated behind the animal was added. In CS–CI compound trials, the CI was presented 2,000 ms after the onset of the CS and both stimuli were terminated simultaneously 3.2 s later; thus the duration of the CS was the same as in the CS–US trials, and the CI was always presented for 3.2 s.

There were three stages of the experiment. In the first stage (CONTROL) each session consisted of 18 trials: each of the CSs were paired with a shock US twice and the remaining 12 trials consisted of six $L_{CS}$–CI and six $A_{CS}$–CI presentations. The trials were randomly mixed, with one constraint, that the same trials did not occur three times in a row. The second stage (PROTECTION/EXTINCTION) started when the subject performed nearly 100% correctly, exhibiting the vocal and leg flexion CRs on the positive trials but not on inhibitory trials. During the Stage 2 the sessions consisted also of 18 trials, six of which were $R_{CS}$–US trials and the remaining 12 trials were nonreinforcement trials. Six times per session the $A_{CS}$ was presented in compound with the clicker CI: this was the "protected from extinction" CS. And, six times the $L_{CS}$ was presented alone. The selection of the protected and unprotected stimuli was done by a coin flip. The CRs to $L_{CS}$ and $A_{CS}$ were of comparable intensity, with slightly stronger vocal response and shorter latencies of the CRs to the $L_{CS}$. But this would work against the protection hypothesis. Also the location of the stimuli ($L_{CS}$ in the proximity of always reinforced $R_{CS}$, and $A_{CS}$ close to the CI) was such that could only work against the evidence for PFE. Still, one more adjustment was made to make the comparison even fairer (or more demanding for the protection hypothesis): the unprotected $L_{CS}$ was presented during stage 2 of the experiment for only 2 s. This was done to control for the possibility that only the first 2 s of the protected $A_{CS}$ really underwent extinction, because the remaining three seconds were "masked" by the CI. Had the unprotected $L_{CS}$ not have been shortened, a differential rate of extinction migth have resulted from such uneven treatment. Of course, there was a risk that the subject could learn to discriminate between short nonreinforced and long reinforced CS (using the early termination of a CS as a CI) but luckily this was not so in this first experiment[1]. The PROTECTION/EXTINCTION stage lasted 25 days and each of the CSs was presented 150 times: $R_{CS}$ always reinforced, $A_{CS}$ always with the CI, and $L_{CS}$ always alone for 2 s. The third stage (TEST) of the experiment consisted of 5 sessions which were made up of 6 $R_{CS}$–US trials, six $L_{CS}$ alone and six $A_{CS}$ alone trials. Both $L_{CS}$ and $A_{CS}$ were presented for 5 s and nonreinforced, so this was extinction of the full duration CSs. The Table I summarizes the duration and trial composition of each stage for this experiment.

TABLE I

General outline of the experiment

| Stage 1: 10 sessions Control | | Stage 2: 25 sessions Protection/Extinction | | Stage 3: 5 sessions Test (extinction) | |
|---|---|---|---|---|---|
| Type of trial | No of trials in a session | Type of trial | No of trials in a session | Type of trial | No of trials in a session |
| R – US | 2 | R – US | 6 | R – US | 6 |
| L – US | 2 | L (2 s) | 6 | L (5s) | 6 |
| A – US | 2 | A – CI | 6 | A (5s) | 6 |
| L – CI | 6 | | | | |
| A – CI | 6 | | | | |

*Leg flexion data.* Table II shows the conditioned leg flexion responses in each type of trial during all stages of the experiment. Percentages of leg flexion occurrence within 5 s after the onset of the CS are computed for 5 session blocks. Note the following facts. There is 100% correct responding to all three CSs during the last 5 control sessions. Responding to $L_{CS}$–CI and $A_{CS}$–CI is nearly absent during this period. There is no responding to $A_{CS}$–CI compound during the stage 2 of the experiment, but the $L_{CS}$ (alone for 2 s) elicits 60% responses in the first 5 day block of the stage 2. In the stage 3, $L_{CS}$ presented for

---

[1] It should be mentioned, however, that when the subject was later retrained and again exposed to nonreinforced CS of short duration, he was able to learn such a discrimination, and also a PFE was found with the 2 s CS alone, without additional CI.

TABLE II

Percent of conditioned leg flexions in 5 session blocks

| Types of trials | Control | | Protection/extinction | | | | | TEST |
|---|---|---|---|---|---|---|---|---|
| | Blocks of five sessions | | | | | | | |
| | 1 | 2 | 3 | 4 | 5 | 6 | 7 | 8 |
| R – US | 100 | 100 | 97 | 100 | 100 | 97 | 97 | 100 |
| L – US | 100 | 100 | — | — | — | — | — | — |
| A – US | 90 | 100 | — | — | — | — | — | — |
| L – CI | 13 | 0 | — | — | — | — | — | — |
| A – CI | 17 | 3 | 0 | 0 | 0 | 0 | 3 | — |
| L (2) | — | — | 60 | 7 | 3 | 3 | 0 | — |
| L (5) | — | — | — | — | — | — | — | 0 |
| A (5) | — | — | — | — | — | — | — | 43 |

5 s does not elicit any responses while $A_{CS}$ produces a considerable 43% CRs over the 5 days of testing. Since the table disregards the amplitude of responses and the distribution of responding over the 5 days of the TEST stage, Fig. 1 presents the record of the left hindleg flexions from the first 4 days of testing. The first responses to the $A_{CS}$ are full size flexions. Even on the fourth day of testing (i.e., extinction) there is a noticeable tendency for the $A_{CS}$ to elicit a CR.

Impressive as this result is, it has one potential weakness. It refers only to the response which comes late during the CS–US interval and, being of "consummatory" nature, need not characterize the emotional and motivating conditioned processes which are of greater interest for the theory of PFE as conceived for explaining avoidance behavior. Therefore the remaining data will be presented in such a way as to facilitate the discussion of the PFE of the initial part of the CS and/or CR.

*Heart rate data.* There is no point in rehearsing now the arguments for and against the notion that heart rate changes during the CS–US period reflect some central processes related to attentional and motivational machinery of the brain. But it seems worthwhile to present the data which show that in the cat, the heart rate change regularly accompanies the aversive CS and that this change is suppressed by the CI, protected from extinction by the CI, and, extinguished when the CS is presented without reinforcement. Figure 2 presents computer plots of averaged heart rate curves for the three sets of trials. The two top plots are responses to $L_{CS}$–US and $A_{CS}$–US during the first stage (CONTROL) of the experiment. Note the triphasic shape of the heart rate (HR henceforth) response: initial acceleration followed by a bradycardic wave that again reverts into late tachycardia. The US elicits a large tachycardic response which, in this well trained animal, looks like a natural continuation of the late "conditioned" acceleratory response. We postpone the discussion on what these three waves [2] may represent to the next paper (Sołtysik and Wolfe, in preparation) and draw the reader's attention to the middle plots. These are averaged $L_{CS}$–CI and $A_{CS}$–CI trials from the control sessions. The dotted line

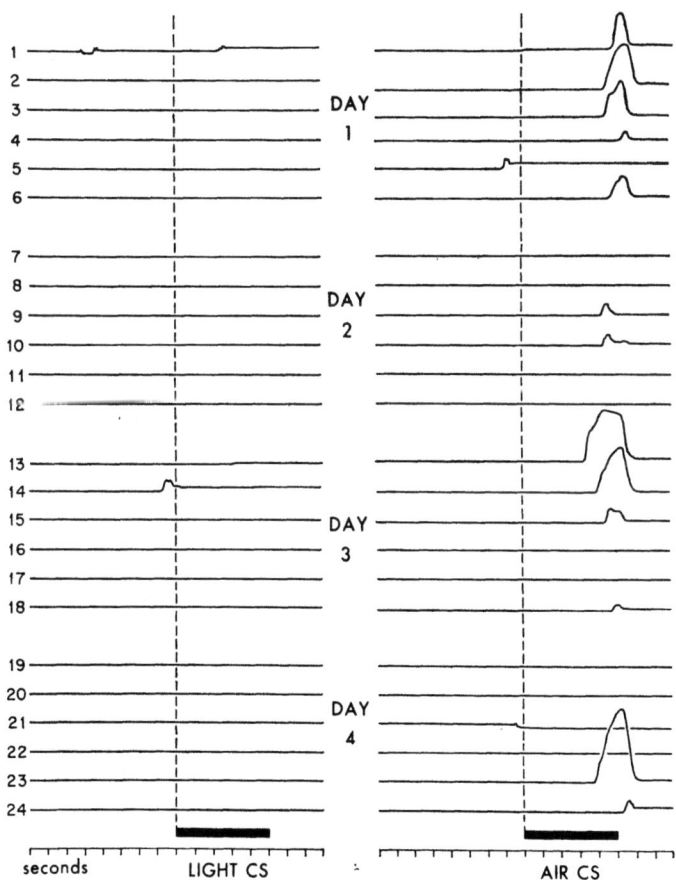

Fig. 1. Extinction of the conditioned leg flexion during first four test sessions. Each horizontal line is the new flexion record from an individual trial. Abscissae, time in seconds from the beginning of the trial to 3 s after CS termination.

---

[2] We refer to these deflections from the baseline as *waves* because we consider them as analogus to the evoked potential. If our recording was at the level of the heart's pacemaker membrane, the record (after substracting the steady shift of the pacemaker potential) would be very similar to our averaged HR plot, with the depolarization corresponding to tachycardia and hyperpolarization reflecting and, in fact, causing, bradycardia.

repeats the top plot from the CS–US trials for comparison. The initial acceleratory response is preserved on CS–CI trials, simply because for the first 2 s the CS is presented alone. But the addition of the CI erases the bradycardic wave and the late tachycardic part of the HR response. The bottom plots represent averaged HR responses to $L_{CS}$ and

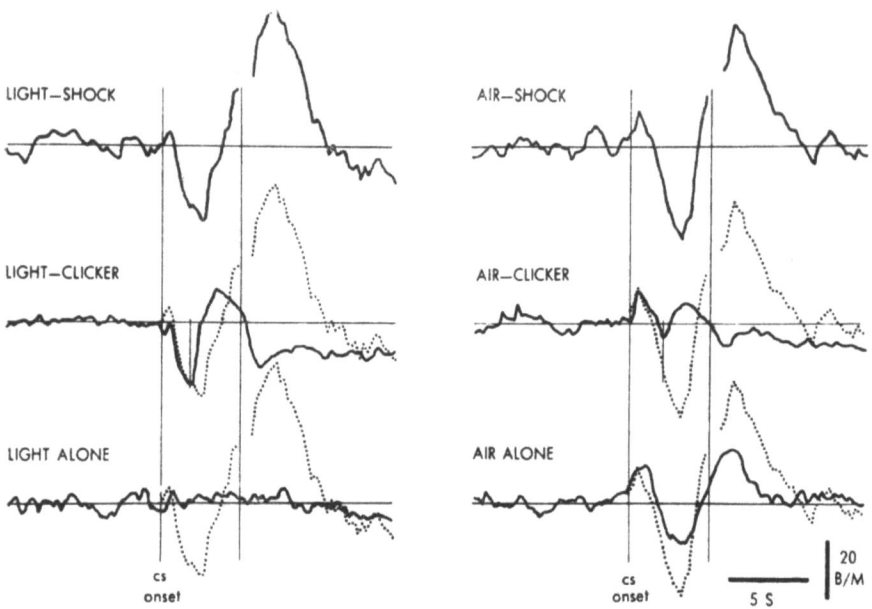

Fig. 2. Comparison of heart rate CRs during various light (left) and air (right) CSs. Each plot is the average instantaneous heart rate (four samples/s) beginning 10 s before CS onset and ending 10 s after CS termination for a number of trials of the type indicated. All plots have the same vertical and horizontal scale. The vertical lines indicate the CS onset and termination; the short vertical lines indicate CI onset. Top plots are from the stage 1 of the experiment (CS–US trials). Middle plots are CS–CI trials from the same stage. Bottom plots are from the first 2 days of the Test stage. The $L_{CS}$–US, and $A_{CS}$–US plots (above) have been replotted on the middle and bottom plots to facilitate comparison (dotted lines). $L_{CS}$–US, $n = 10$; $L_{CS}$–CI, $n = 27$; $L_{CS}$ alone, $n = 12$. $A_{CS}$–US, $n = 9$; $A_{CS}$–CI, $n = 30$; $A_{CS}$ alone, $n = 12$.

$A_{CS}$ alone during the first two days of testing in the third stage of the experiment. The response to $L_{CS}$ is extinguished whereas the $A_{CS}$ which was protected in the stage 2, still elicits the entire triphasic HR response. Of particular interest is the fact that not only the middle bradycardic and lae tachycardic parts of the response are preserved, but that also the initial tachycardic response survived well 150 nonreinforced (but assumedly "protected") trials of the stage 2 of the experiment. While the reason for the non-extinction of the middle and the late waves could be ascribed to the fact that they were not elicited on the

CS–CI trials, the first acceleratory wave was preserved in spite of the fact that it *was elicited* and not reinforced by the US for 150 times during 25 days.

*Respiration data.* Equally interesting are our finding on the changes in respiration. Briefly, the recording was made from a thermistor placed in front of the external nares. Changes in both frequency and amplitude were recorded. A more extensive discussion on the method and analysis of data will be presented elsewhere (Sołtysik and Wolfe; Wolfe, in preparation). Frequency changes yielded interesting data but they seemed to be related more to attentional processes and will not be discussed now. On the other hand, the changes of the amplitude were found to be of great interest. The typical response to a CS signalling shock is a gradual reduction of the amplitude which drops to zero at the delivery of the shock. After we checked that this reduction is not due to redirecting of the breathing from the nose to the mouth,

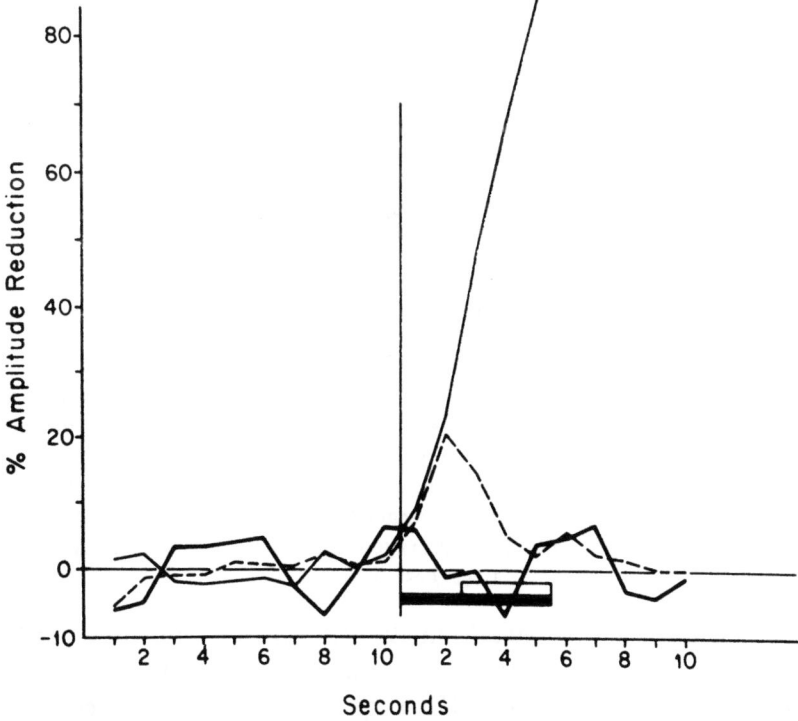

Fig. 3. Comparison of respiratory CRs during $L_{CS}$–US, $L_{CS}$–CI, and $L_{CS}$ alone trials. Ordinate, percent reduction from the mean respiratory amplitude of the 10 s pre-CS period. Abscissae, time in seconds from the beginning of the trial to 10 s after the CS onset period. Thin solid line, averaged for 18 $L_{CS}$–US trials during the 10 control sessions. Dashed line, averaged CR for 45 $L_{CS}$–CI trials during the 10 control sessions. Thick solid line, averaged CR for 6 $L_{CS}$ alone trials during the first Test session.

the reduction was assumed to be a real phenomenon, probably reflecting the increasing tension of the body musculature in anticipation of the upcoming noxious US. This, of course, "degrades" the amplitude reduction response as being extrinsic to the respiratory function, i.e., a sort of peripheral interference at the level of the "final common

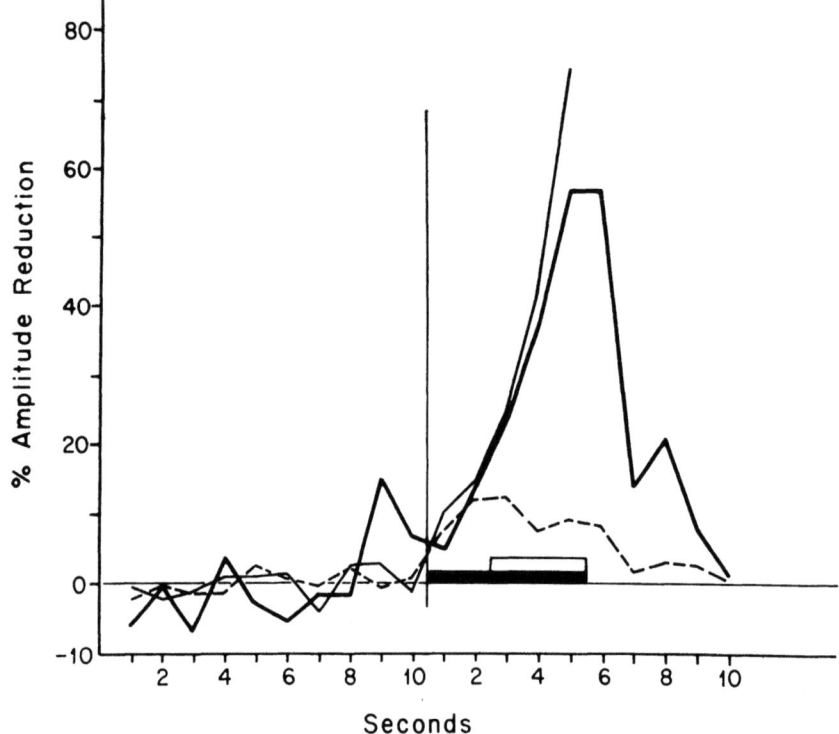

Fig. 4. Comparison of respiratory CRs during $A_{CS}$–US, $A_{CS}$–CI, and $A_{CS}$ alone trials. Ordinate, percent reduction from the mean respiratory amplitude of the 10 s pre-CS period. Abscissae, time in seconds from the beginning of the trial to 10 s after the CS onset period. Thin solid line, averaged CR for 17 $A_{CS}$–US trials during the 10 control sessions. Dashed line, averaged CR for 50 $A_{CS}$–CI trials during the 10 control sessions. Thick solid line, averaged CR for 6 $A_{CS}$ alone trials during the first Test session.

path", but it does not detract from its usefulness as an integrated index of emotional state, or preparatory readiness for the impending aversive US. In contrast to the complex HR response, this physiological concomitant of the CS was a monotonic and almost linear change from the baseline with the peak precisely timed at the moment of the delivery of the US. Figures 3 and 4 compare computer plots of respiratory amplitude changes to $L_{CS}$ and $A_{CS}$ from three sets of trials: CS–US during the CONTROL stage of the experiment (thin line), CS–CI trials during the same period (dashed line), and CS alone during the first day of

TEST stage (heavy line). The reduction in amplitude is expressed as upward deflection and scored as percentage change in relation to the mean amplitude during the 10 s prior to the CS onset [3]. The 100% response in our plot corresponds to zero amplitude, while the negative values denote an increase in amplitude. An example of the reliability and stability of these responses is provided by the next Fig. 5, in which the responses to a nonreinforced $L_{CS}$ of 2 s duration are shown for the first 12 sessions of the stage 2 of the experiment. Each plot is an average of only six responses, but the shape of each plot is an inverted V

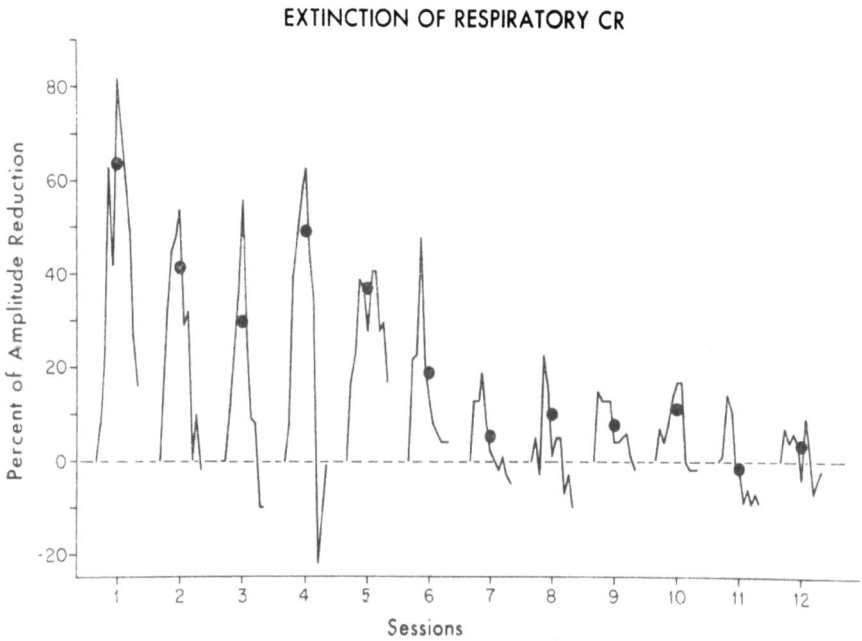

Fig. 5. Respiratory CRs during unreinforced 2 s light CS trials in PROTECTION/EXTINCTION sessions one through twelve. Each curve, the average of six respiratory CRs, begins from a baseline of the average mean-pre-CS-amplitude and continues with ten values, for the 10 s after the CS onset, expressed as percent of the reduction relative to the pre-CS mean. The solid circles represent the average reduction throughout a five second period starting three seconds after the CS onset. Abscissa, percent amplitude reduction from mean pre-CS level.

[3] Expressing it as percentage of the pre-CS amplitude was indicated by the following considerations. Although the position of the thermistor should not change within a session, it varies between sessions, as does the gain setting of the amplifier. Occasionally the adjustment in gain has to be made between trials, so that the recorded quiet breathing in the intertrial intervals be of comparable size: 10–20 cm on the XY plotter or 1–2 V on the FM tape. Thus, the amplitude is arbitrary and the best expression of the change is in relation to some standard period, as in our case, the 10 s pre-CS period.

with the peak at about 5 s after the onset of the CS, i.e., precisely when the shock was presented on the CS–US trials in the earlier training. Even more astonishing is this timing of the peak when one realizes that the CS duration was only 2 s in this stage of training, so the timing of the peak could not be cued to the termination of the CS.

Returning to the Figs. 3 and 4, we find that on the CS–CI trials this respiratory response, already initiated in the first 2 s of CS, was promptly suppressed by the CI. The third, and most revealing plot, from the first day of the TEST stage, when the $L_{CS}$ and $A_{CS}$ were both presented six times nonreinforced for 5 s, fully supports the findings on the leg flexion and HR responses. The response to $L_{CS}$ is extinguished, whereas the $A_{CS}$ elicits a practically full size response. Again, the

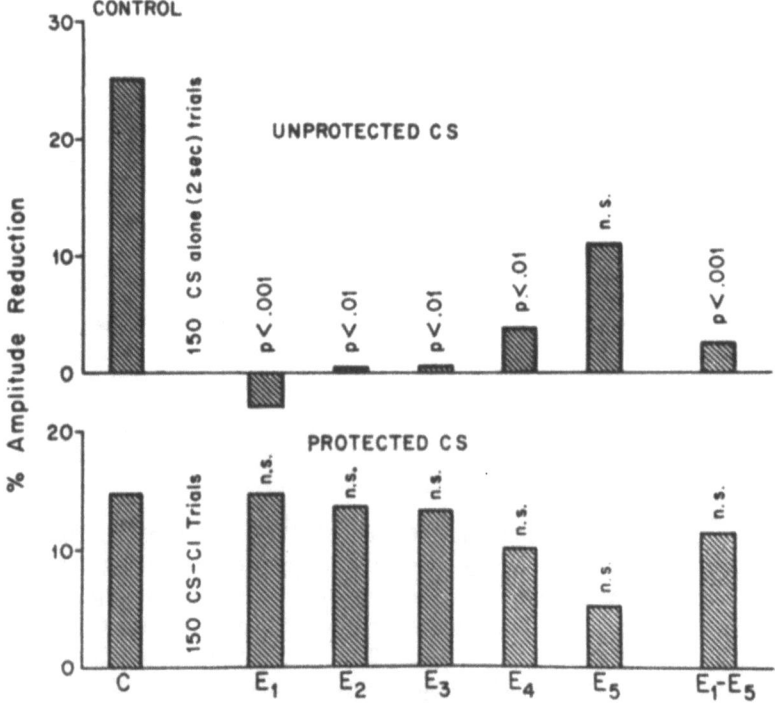

Fig. 6. Comparison of average respiratory CRs during the five Test sessions and the control sessions. Responses to Light-CS, above, and to Air CS, below. The columns represent averaged percent reduction computed in the following way: Respiration was sampled 1.5 s after the CS onset, the percent reduction of this sample from the mean pre-CS amplitude of that trial was determined, and the average percent reduction at 1.5 s was calculated. The columns represent these averages for all $L_{CS}$–US and $A_{CS}$–US trials in the control stage ($n = 12$), for the $L_{CS}$ alone and $A_{CS}$ alone trials in each of the five extinction sessions ($n = 6$), and for all five Test sessions combined ($n = 30$).

fact that the response was not extinguished during the last 3 s of the CS may be attributed to the non-elicitation when the CS–CI compound was presented 150 times in the second stage of the experiment. But, similarly to the HR response, the early part of the respiratory response, which *was* elicited on the CS–CI trials, also was preserved.

Figure 6 adds a little quantitative information about this early response. The scores from the 1.5 s point after the CS onset are presented from the CONTROL stage (from trials CS–US only) and compared by a t-test with the same response on each of the five TEST sessions, as well as with the average of all TEST sessions. In the case of the unprotected $L_{CS}$ there was complete extinction on the first three days and only on the fifth day the difference is not statistically significant. No trace of extinction was found for the protected $A_{CS}$ and only on the 4th and 5th day of testing (i.e., extinction) there was noticeable decrease in response, not attaining, however, statistical significance.

### GENERAL DISCUSSION

The results presented in this admittedly preliminary report indicate that the phenomenon of PFE is fully demonstrable and powerful when a well trained CI (predictor of no US) is used. The protection from extinction was observed for all categories of conditioned responses studied in this experiment: leg flexion, heart rate change and change in the amplitude of respiration [4].

Thus, assuming tentatively that the PFE does in fact exist, let us briefly review a few theories of learning in order to find out if the PFE is a theoretically acceptable phenomenon.

*Pavlov's hypothesis of "protective inhibition"* (14, lecture 22; 1). This old-fashioned theory of the maintenance of conditioned "reflexes" satisfactorily predicts and "explains" the PFE. The CS–US pairing originally creates a new connection between CS and US cortical "centers". Afterwards, however, the conditioned reflex is maintained by the following mechanism. The CS activates the CS center. This activation, responsible for the CR, would linger and exhaust the neurons of this center if the US was not presented. The activation of the US center inhibits the CS center, or middle link of CS–US connection in Asratian's version of this hypothesis, and thus protects it from over-exhaustion; repeated omission of the US results in lowering reactivity (or "working capability" in Pavlov's words) which in overt behavior is ob-

---

[4] The PFE was also observed in vocal CRs, not discussed in this paper.

served as extinction. The logical extension of this theory is that any other stimulus exerting inhibitory effects upon the center of the CS (or the CS–US bond) might also prevent extinction. Obviously the well-trained and specific CI is a prime candidate for this role. Unfortunately, this theory of Pavlov has many weaknesses, both empirical (it has never been verified) and logical; e.g., the inhibitory process is both a result of exhaustion and an active process preventing the exhaustion.

*Consolidation hypothesis of associative learning.* There are many versions of the hypothetical consolidation process which is initiated during the trial but takes place in the posttrial period. The simple version could be exemplified by the notion (6) of the coexistence of the traces of the stimuli which could interact and cause permanent changes in the common structural elements involved in the trace processes (e.g., neurons which participated in the reverberation of impulses, etc.). More elaborate accounts (post-trial "backward scanning" (9) or "rehearsal" (19) belong rather to the information processing and cognitive theories of learning. At any rate, the extension of the consolidation account to extinction learning (18) assumes the interaction between the trace of the CS and the trace of no US (assuming that the absence of the expected US is a perceived event equally capable of leaving a "trace") [5]. This interaction should lead to new learning, with the former CS becoming "extinguished", i.e., a predictor of no US. However, if the trace of a non-reinforced CS is terminated before it enters the post-trial phase of consolidation, no consolidation of a CS-no US association will occur. The CI might be such a terminator of a CS trace because it was trained to suppress the response elicited by the CS. Even more importanty, the CI, being a predictor of no US, may prevent also the initiation of the trace of the no US, which no longer comes as a surprise. This aspect of the presenting CI, is even better articulated in terms of Wagner's rehersal hypothesis of Mackintosh's attentional theory (13).

*Rescorla-Wagner model of classical conditioning.* As pointed out by Henderson and Harris (7) the Rescorla-Wagner Model of conditioning (15) predicts PFE because there is no discrepancy between the $V_A$ (current associative strength) and $\lambda$ (limit of associative strength for a given reinforcer) on a CS/CI-noUS trial, when there is no response, and the value of $\lambda$ for no US is assumed to be zero. However, the theory is not elaborated for evaluating the "net associative strength" for a compound of CS–CI when there is no full overlap of the eliciting and suppressing stimuli and the compound is arranged sequentially. It seems, however,

---

[5] Konorski (11) proposed elaborate arguments for such representation of no-stimuli in the brain.

that the "net" should refer either to the time instant when the reinforcer would normally be presented, or better to the onset of the posttrial period, when the "processing" of the, acquired during the trial, information begins. This would be in accord with Wagner's process-oriented development (19) of the original model. With such an amendment, the model is predictive of PFE, and, as will be discussed in the next section, allows also for the occurrence of the initial part of the CR to the CS prior to the CI onset.

*Cognitive theory of avoidance conditioning (16)*. This theory includes a protection from extinction notion but applies it only to the expectancy of US, while allowing for the fear CR to be extinguished. In other words, CS in a CS–CI trial retains it signalling (predicting) informational role, because only CS alone would constitute a disconfirmation of a CS–US bond. On the other hand, the CS as an elicitor of emotional CR undergoes extinction on the CS–CI trial, because it is not reinforced by the US. Such a view of two levels of learning (cognitive and emotional) following different courses, preservation for expectancies, and extinction for emotions, was prompted by the data which seemingly disconfirmed the PFE and left unanswered questions of how the avoidance response is motivated in the lack of drive. If, however, the existence of PFE of the fear CR could be ascertained, the theory need not be abandoned. One solution would be to assume more interdependence between expectancies and CRs; it is rather displeasing in the present version of the theory that the elicitation of the expectation of shock which causes the subject to respond instrumentally to provide the expectancy of no US, could occur without an emotional CR, acquired originally by pairing the CS with the US. If any such discrepancy between the cognitive and conative levels of responding is possible, it should be rather in the opposite direction, when the subject is emotionally responding contrary to the better knowledge that nothing important is going to happen. It is the persistence of emotional CRs and not of expectancies that sends people to psychotherapists. And, contrarywise, expectancies alone seldom determine behavior if the emotions are absent or directed elsewhere. The other solution would come in the form of accepting both sources of motivation, emotional CRs and expectancies (in presence of preferences), interacting and summating. This adds to the complexity of the theory, but there is no reason to assume that a complex behavioral machinery would not be selected for in the evolution if it increases survival fitness.

This brief and, admittedly, very superficial review of a few theories of behavior, makes clear that PFE does not pose any threat or require major revision.

*Retrograde protection from extinction of the CR elicited by the early portion of the CS, prior to the onset of the CI.* If the PFE phenomenon is to be of help in explaining the resistance to extinction of avoidance responses the mechanism of protection must not depend upon prevention of the elicitation of the CR. As Seligman and Johnston (16) correctly point out, "protection of CS from extinction must occur while this very CS is eliciting a CR". Their theoretical objection to a role for PFE in avoidance centers on this issue. They argue that the reason a "conditioned inhibitor provides protection for a CS may be precisely because the CR is inhibited", and therefore that there is no cause to expect that the conditioned response elicited by the early portion of the CS, prior to the onset of the CI, could be protected. Our data fully support the possibility that PFE also applies to the initial part of the CR elicited prior to the onset of the CI, at least for the heart rate and respiratory responses.

Two issues are now raised: in what theoretical context can this retrograde protection from extinction (RPFE) be placed, and what might the mechanism of RPFE be, if it does not (as it logically must not) involve inhibition of the initial part of the CR? Neither Pavlov's nor consolidation accounts encounter any difficulties, because in both cases it is posttrial hypothetical aftereffects that determine the outcome of the trial, and as shown above, the CI may be assumed to prevent the normal course of the aftereffects of nonreinforcement.

The Rescorla-Wagner model encounters a problem, unless amended along Wagner's rehersal hypothesis. The original version of the model could be restated in the following way: What is elicited undergoes extinction on the nonreinforced trial. Therefore, the initial part of the CR elicited by the CS prior to the onset of the CI should be extinguished in the course of 150 CS–CI trials. There are even ways to explain RPFE without any amendment invoking posttrial processing of information. Assuming that the similarity and therefore generalization between the early and later portions of the CS is strong, one could hypothesize that protection of the $CS_1$ (late part of the CS) somehow extends to the unprotected $CS_e$ (early part of the CS). In other words, as long as the $CS_1$ retains the capacity to elicit the CR, also the $CS_e$ will remain functional elicitor of the CR. The experiment with a two-compound CS, where $CS_e$ and $CS_1$ are very different, may verify this hypothesis. Such a bipartite CS should exhibit much less or no RPFE if it is due to the generalization between $CS_e$ and $CS_1$.

Another mechanism for RPFE might be derived from the ethological concept of a "preparatory-consummatory" sequencing of behavior.

If consummatory CR *must* be preceded by the preparatory response, then PFE of the former should automatically preserve the anteceding preparatory CR. In our subject, the consummatory CR was a leg flexion and the fear CR might be considered a preparatory CR. If HR and respiratory changes are in some way peripheral concomitant indices (not necessarily perfect correlates) of this preparatory CR, then their RPFE could be explained by the notion of the sequential structure of the behavior conditioned with the aversive US.

An interesting modification of the Rescorla-Wagner model by Frey and Sears (5) endows the CI with a property of a dynamic, acquired control of salience. If this property of reducing attention could be assumed as also reducing posttrial processing (according either to "consolidation of traces" or "rehearsal" accounts), the improved model would be predictive of the RPFE.

The hybrid theory of Seligman and Johnston (16) does not offer any solution for the RPFE. In the course of long training with CS-CI trials only, there should be a build-up of an expectancy that $CS_e$ will be followed by the $CS_i + CI$. So even the informational content (that of predicting US) of the cognitive response to $CS_e$ should undergo diminution during the series of CS-CI trials. The only hope lies in the separation of cognitive response from the conditioned one. Even if the expectancies tend to change, due to the absence of CS-US trials, the PFE mechanism, operating on the CR level, may still prompt the organism to assume a conservative attitude and to respond, contrary to the "knowledge", with the fear CR. But in such a case the cognitive part of the Seligman theory becomes redundant, at least as an explanation of the resistence to extinction of the avoidance responses.

Finally, one can adopt a neuroethological inclination (i.e., a state of mind rather than theory), which, superficially trivial, may be most practical for the student of behaving organisms (as opposed to the study of behavior or learning as such). Firstly, within this broad orientation one may assume that if RPFE has any survival value for the organism, for instance, by preserving the avoidance response without requiring that the organism be periodically exposed to the aversive events (8), then there is a good chance that the mechanism providing the RPFE will be invented (by chance variability of neurobehavioral machinery) and selected for in the evolution of the species. And secondly, the final understanding of the PFE according to this theoretical bias should come from the studies on the neurobehavioral machinery, with or without the help of any formal theory of learning.

## CONCLUDING REMARKS

In the final remarks, we would like to admit that we are aware of some weaknesses of our experiment, besides its preliminary character. Presenting a short CS in the stage 2 (Protection/Extinction) and a long CS in stages 1 (Control) and 3 (Test) makes it possible for the animal to learn to discriminate between the short and long CSs. This has not happened in our subject (although he learned that discrimination later) but it may happen in others and this may reduce the difference in responding to protected and unprotected CSs in the stage 3. One remedy would be using long CSs in all three stages of experiment, but this would put the unprotected CS in a disadvantaged position relative to the protected one, as discussed in the Introduction. The difference in the CR eliciting properties would be probably very large in the stage 3 but it would not be so convincing, because one could argue that the CI is masking the protected CS and therefore, the comparison is made between 5 s $CS_1$ and 2 s $CS_2$ being extinguished in stage 2. Of course, the extinction of a short duration CS should be less complete than extinction of the longer duration CS, especially if we compare their response eliciting properties in the stage 3 using 5 s duration for both of them. Another solution, using a neutral stimulus instead of CI in a $CS_2$-CI compound (instead of a short, non-reinforced CS) is also not a good solution because with our prolonged training, such a stimulus will acquire the role of a CI. A between-subject design would be more satisfactory, but that would require much larger number of subjects, which in a study lasting many months is a costly proposition.

A better solution for these paradigmatical and theoretical problems would be a trace conditioning design with all CSs of 2 s duration and the CS–US interval of 5 s. The 3 s CI would be presented in the "gap" between the termination of the CS and the time point where the US occurs on the CS–US trials. The CS would then remain unchanged in all stages of the experiment. This, however, would increase the generalization between the CSs (the traces of the CSs are characterized by stronger generalization than the actual stimuli) so the differences between the protected and unprotected CSs in the stage 3 might be diminished. And, more importantly, this would be analogous to using a two-component CS, with the $CS_e$ being actual stimulus and $CS_1$ its trace, so the discrimination between them might be easier than in the case of a 5 s CS, and as a result of this, the RPFE might be diminished if, as argued before, it depends on the generalization along the CS duration. Still it seemed to us that the trace conditioning is possibly as good a design as the one described in this paper and we are currently training one subject using only short CSs [6].

Thus, although each has it flaws, the present design and the trace conditioning design will be used in our study to add more weight to the evidence for both PFE and RPFE which we consider a real and important phenomenon in the realm of behavioral plasticity.

Finally, we should confess our belief that the PFE phenomena are as important in the maladaptive as they are in normal behavior. Many neurotic or persistent maladaptive responses may be explained by the operation of the PFE mechanism. Better understanding of this phenomenon may, therefore, be of considerable diagnostic and therapeutic value.

We wish to thank W. J. Wilson for reading the manuscript, and G. LeNae Boddie for helping prepare the manuscript for publication. This investigation was supported by USPHS HD-05958.

## REFERENCES

1. ASRATYAN, E. A. 1969. Mechanism and localization of conditioned inhibition. Acta Biol. Exp. 29: 271–291.
2. CAPALDI, E. J. 1967. A sequential hypothesis of instrumental learning. In K. W. Spence and J. T. Spence (ed.), The psychology of learning and motivation, Vol. 1. Academic Press, New York, p. 67–156.
3. CHORĄŻYNA, H. 1957. Some data concerning the mechanism of conditioned inhibition. Bull. Acad. Pol. Sci. 5: 387–392.
4. CHORĄŻYNA, H. 1962. Some properties of conditioned inhibition. Acta Biol. Exp. 22: 5–13.
5. FREY, P. W. and SEARS, R. J. 1978. Model of conditioning incorporating the Rescorla-Wagner associative axiom, a dynamic attention process, and a catastrophe rule. Psychol. Rev. 85: 321–340.
6. HEBB, D. O. 1958. A textbook of psychology. W. B. Sounders Co., Philadelphia.
7. HENDERSEN, R. W. and HARRIS, K. 1979. Inhibitory protection of conditioned fear extinction. Acta Neurobiol. Exp. 39: in print.
8. HULL, C. L. 1929. A functional interpretation of the conditioned reflex. Psychol. Rev. 36: 495–511.
9. KAMIN, L. J. 1969. Selective association and conditioning. In N. J. Mackintosh and W. K. Honig (ed.), Fundamental issues in associative learning. Dalhousie University Press, Halifax.
10. KONORSKI, J. 1948. Conditioned reflexes and neuron organization. Cambridge University Press, London, 267 p.
11. KONORSKI, J. 1967. Integrative activity of the brain. An interdisciplinary approach. University of Chicago Press, Chicago, 531 p.

---

[6] The results of this experiment fully replicated the findings of this paper.

12. LOLORDO, V. M. and RESCORLA, R. A. 1966. Protection of the fear-eliciting capacity of a stimulus from extinction. Acta Biol. Exp. 26: 251–258.
13. MACKINTOSH, N. J. 1975. A theory of attention: variations in the associability of stimuli with reinforcement. Psychol. Rev. 82: 276–298.
14. PAVLOV, I. P. 1928. Lectures on conditioned reflexes. International Publishers, New York.
15. RESCORLA, R. A. and WAGNER, A. R. 1972. A theory of Pavlovian conditioning: Variations in the effectiveness of reinforcement and nonreinforcement. In A. H. Black and W. F. Prokasy (ed.), Classical conditioning II: Current research and theory. Appleton-Century-Crofts, New York.
16. SELIGMAN, M. E. P. and JOHNSTON, J. C. 1973. A cognitive theory of avoidance learning. In F. J. McGuigan and D. B. Lumsden (ed.), Contemporary Approaches to conditioning and learning. John Wiley and Sons, New York, p. 69–110.
17. SOŁTYSIK, S. 1960. Studies on the avoidance conditioning: 3. Alimentary conditioned reflex model of the avoidance reflex. Acta Biol. Exp. 20: 183–192.
18. SOŁTYSIK, S. and ZIELIŃSKI, K. 1963. The role of afferent feedback in conditioned avoidance reflex. In Gutmann and P. Hnik (ed.), Central and peripheral mechanisms of motor functions. Publishing House of the Czechoslovak Academy of Sciences, Prague, p. 215–221.
19. WAGNER, A. R. 1976. Priming in STM: An Information processing mechanism for self-generated depression in performance. In T. J. Tighe and R. N. Leaton (ed.), Habituation: Perspectives from child development, animal behavior, and neurophysiology. Erlbaum, Hillsdale N. Y.

Stefan S. SOŁTYSIK and George WOLFE, Mental Retardation Research Center, School of Medicine, University of California at Los Angeles, Los Angeles, California 90024, USA.

Lecture prepared for the Warsaw Colloquium on Instrumental
Conditioning and Brain Research
May 1979

# MECHANISMS OF MOTIVATION IN AVOIDANCE BEHAVIOR

Wanda WYRWICKA

Department of Anatomy and the Brain Research Institute
University of California, School of Medicine
Los Angeles, California, USA

*Abstract.* The mechanisms involved in avoidance behavior are discussed. It is assumed that the conditioned stimulus (CS) activates the memory pattern of associations related to the former applications of the unconditioned stimulus (US) and, as a result, produces an undesirable sensory state. This activates another memory pattern of associations related to the avoidance response and the postponement of the US. The performance of the avoidance response discontinues the CS, resulting in inactivation of the first memory pattern; this leads to a removal of the undesirable sensations, i.e., to an improvement in the sensory state. It is suggested that avoidance behavior obeys the same general rules which apply to approach (appetitive) behavior. In both approach and avoidance behavior the instrumental response provides a desirable sensory change (due to obtaining of the desired US in approach behavior and the postponement of the undesired US together with the discontinuation of CS in avoidance behavior). In both cases the response gradually extinguishes when its performance no longer provides the sensory "better-being".

Avoidance conditioning is one of most popular tools in studying behavioral problems related to sciences such as functional anatomy of the brain, neuroendocrinology, neuropharmacology or developmental neurobiology. In spite of the popularity of the avoidance method, the nature of avoidance conditioning is not yet fully understood. The problem of resistance to extinction and the problem of motivation in avoidance behavior are still the topic of much discussion and controversy. This article represents a further attempt to help solve these problems.

## THE PROBLEM OF RESISTANCE TO EXTINCTION IN AVOIDANCE BEHAVIOR

It is generally known that in order to establish a firm avoidance behavior it is necessary to start with pairing a conditioned stimulus (CS; e.g., a tone) with a strong unconditioned stimulus (US; e.g., a strong shock). The instrumental avoidance response (AvR; e.g., pressing a lever) trained afterwards to avoid the US, is very stable and highly resistant to extinction (1, 2, 5, 9, 14–18, 25, 26, 29, 33–37). This situation is contrary to that observed in classical conditioning or in instrumental conditioning related to food. As is well known, in classical conditioning, the withholding of electric shock to the animal's leg leads to the extinction of the defensive flexion of that leg in a few sessions (10, 19); similarly, the withholding of food results in extinction of salivary conditioned reaction in several trials (28). Also, the withholding of food in instrumental conditioning inevitably leads to extinction of the motor response (15–18).

The phenomenon of resistance of avoidance behavior to extinction had been studied by a number of investigators. It was found that the degree of resistance to extinction depends on such factors as the intensity of the US used in the initial phase of the training and the length of the interval between the CS and the US. The stronger the US (an electric shock) and the shorter the CS–US interval, the higher the resistance to extinction (12, 13, 33). In the cases where the shock was weak and the CS–US interval long, the extinction of the AvR occurred after several repetitions of the CS without the US (12, 13, 32, 33). On the other hand, when the shock was strong and given a few seconds after the CS was switched on, the AvR was always stable and observed for months of experimentation (5, 36, 37).

Mowrer and Lamoreaux (27) found that the establishment of avoidance behavior was facilitated when the response not only prevented the occurrence of the shock but also terminated the conditioned stimulus. The importance of the cessation of the CS immediately after the performance of the AvR for the establishment of the avoidance behavior and its resistance to extinction, was also pointed out by Bregadze (2), Fonberg (5), Sołtysik (36) and others.

According to Konorski's interpretation (16, p. 416–417), the performance of the AvR becomes a "barrier" preventing the formation of associations between the CS and the state of relief resulting from the postponement of the US. That way, the CS continues to be a "fear CS" as originally trained and, consequently, the extinction of the AvR is not possible.

Solomon and Wynne (35), on the other hand, based their explanation on the theory of "two-process learning" as proposed by Mowrer (25). According to that theory, pairing the CS and the US leads to the establishment of fear in the first stage of the training; the performance of the AvR enables the animal to escape from the feared stimulus in the second stage of the training. Solomon and Wynne (35) pointed out, however, that a quick performance of the avoidance response to the CS prevents not only the occurrence of the US, but also the development of fear; as a result, fear is "conserved" and, that way, protected from extinction. In addition, these authors assumed that classical conditioning is partially irreversible; this assumption was based on observations reported by Gantt (7, 8) that the classical cardiac reaction to CS was present in dogs for months and even years in spite that the US (which was responsible for the establishment of that reaction) was not used any more. The principles of "anxiety conservation" and the "partial irreversibility" therefore, are the factors responsible for the resistance of avoidance behavior to extinction.

It turned out, however, that the extinction of the avoidance response was actually possible in some conditions (besides those mentioned above). One of such conditions was making the AvR impossible to perform. In experiments of Solomon and his colleagues (34) a glass barrier was used to prevent the animal from jumping into the safety area; this procedure resulted in the extinction of the AvR in the situation with the barrier, but not in the situation without the barrier. In a study of Baum (1) the animal was forced to remain on the grid floor (where shock had previously been given) for a few minutes during which no shock was applied. This led to a rapid extinction of the AvR. More recently Prado-Alcala (29) found, however, that this method was not always effective unless combined with a "counterconditioning" through an intracranial stimulation which produced a behavior antagonistic to the avoidance response.

The most effective method of achieving the extinction of the AvR appeared to be the continuation of the CS after the performance of the AvR. This was obtained by Bregadze (2) and, independently, by Fonberg (5) in dogs. In Fonberg's study, the CS was continued after the repetitive performance of the AvR until the response did not appear for 10 s since its last performance. With the use of this procedure, a full extinction of the avoidance response occurred after only four sessions (46 trials altogether). This procedure was also successfully applied by Sołtysik (36) in his studies on differentiation and extinction of avoidance behavior. According to Sołtysik (36) the extinction of the AvR occurs in two stages. In the first stage, the CS, by being continued after the performance of the AvR, becomes a secondary reinforcement substituting for the US; this leads to the suppression of the AvR. In the second stage, after the removal of the protective (from extinction) role of the AvR, extinction takes place according to the rules governing the classical conditioning. Another explanation of this problem will be discussed below.

### THE PROBLEM OF MOTIVATION IN AVOIDANCE BEHAVIOR

The fact that a simple change in the procedure such as extension of the CS beyond the AvR, disrupts avoidance behavior, suggests that the principles of "anxiety conservation" and "partial irreversibility" (35) are not sufficient to explain avoidance behavior and that still another factor may play a critical role in the behavior. What is that factor? According to Mowrer (25, 26) the action of the CS initially related to the US (e.g., a shock) produces a state of anxiety (fear); fear is removed after the performance of the AvR and discontinuation of the CS; therefore, the reduction of fear is the factor motivating AvR. Miller (23) claimed that fear produced by the CS is a drive and feardrive reduction is the critical factor in sustaining avoidance behavior. According to these authors, the fear reduction in the Mowrer's concept, and the feardrive reduction in Miller's concept, is the reinforcement in the avoidance behavior.

Before proceeding with the present discussion, let us briefly consider the term "reinforcement". As is well known, the word "reinforcement" was first used by Pavlov (28) to describe the power of the US (food or acid solution placed in the dog's mouth following the action of the CS) to strengthen the connections between the CS and the US. In studies on instrumental behavior this term has been widely used in its empirical meaning and, frequently, as a substitute for the US (e.g., food in appetitive behavior or shock in aversive behavior). In appetitive behavior the term "reinforcement" refers to the obtaining of the US for the performance of the instrumental act. In avoidance behavior, however, the US is not used and, therefore, it cannot be called "reinforcement". In fact, the theoretical approach to the term "reinforcement" varies from author to author (14, 20, 23, 26, 30, 31, 38). Some authors expressed their doubts as to usefullness of the traditional concept of reinforcement in explaining behavior, especially avoidance behavior (30, 31). In view of this multiplicity of approaches the term "reinforcement" will not be used further in this article.

Let us now examine the meaning of the "anxiety reduction" (26) and of "fear-drive reduction" (23). Both these expressions refer to a change in the sensory state of the organism: due to the performance of the AvR a particular sensory state (anxiety, fear) disappears. If we assume that the state of fear is undesirable for the organism (as the animal acts to remove it by performing the AvR), we can say that the reduction of fear is a positive change in the sensory state (see 26, p. 129). This positive change obtained due to a performance of the AvR can also be considered an "improvement in the sensory state" or an achievement of sensory "better-being", according to the terminology previously used by this writer elsewhere (39, 40). Such an approach deals with an *induction*, i.e., a process leading to satisfying sensory state, rather than with a reduction, i.e., a process leading away from dissatisfying sensory state. Let us try to explain aovidance behavior from this point of view.

## THE ORGANIZATION OF AVOIDANCE BEHAVIOR

It is generally assumed that the phenomenon of conditioning is a result of the associations formed between the neural representations of stimuli. There is an evidence that such associations can be formed and stored at the neuronal level (9, 11, 24). For example, in experiments of Morrell (24), the unitary activity of the polysensory parastriate neurons was recorded in conscious cats during the pairing of a visual stimulus (L) with an electric shock (S). Before the pairing, each of these stimuli evoked a different pattern of discharges. A combined application of L + S evoked a combined pattern of discharges, L.S. After a number of combined applications of L + S, L alone evoked the combined pattern L.S. This combined pattern, as a response to L, was observed throughout the approximately 60 min testing period after the withholding of the shock.

Morrell's experiments reveals the basis of classical conditioning, which may be used to explain the events occurring in the first stage of avoidance training. At that phase, the CS (e.g., a tone) and the US (e.g., a shock) are paired many times; as a result, a combined pattern of sensory traces of the CS and the US may be formed and stored in the polysensory neurons. Since then, the previous traces of neuronal activation related to the CS have become "contaminated" with the traces related to the US; as a result, the CS can now evoke *only* the combined pattern, as if the US was still present with the CS.

In the first stage of the avoidance learning procedure, classical conditioning takes place, i.e., the neuronal pattern CS.US is formed. In the second stage, when the AvR is introduced to prevent the shock, a new pattern of combined sensory traces is formed against the background of the old CS.US pattern. The new pattern consists of sensory changes related to the CS, AvR and noUS. The "noUS" denotes the sensory changes related to the absence of the US. The supposition that the US is not represented in the new pattern, is supported by the results of the experiments of Fonberg (6) in which this author examined the relationship between the CS and the US in avoidance behavior. After the establishment of the AvR in dogs, a test was performed in which the US, an electric shock, was given alone. A question was asked as to whether the AvR would occur during the shock or not. It turned out that the shock itseld did not evoke the AvR.

Nevertheless, there is an evidence that after the establishment of the AvR the old sensory pattern CS.US still remain active. When studying the neuronal changes during defensive conditioning in cats, Halas and Beardsley (9) observed that repeated applications of the CS alone continued to evoke the neuronal response in spite of the fact that the behavioral avoidance response was already extinguished. Sołtysik and Kowalska (37), in their study on changes in heart beat during classical and instrumental conditioning, demonstrated that new, neutral acoustic stimuli did not evoke any significant cardiac reaction, but an acoustic stimulus which had previously been paired with a shock produced an accelaration in the heart beat. This cardiac reaction disappeared as soon as the AvR was performed and the CS was discontinued, to appear again to the next application of the CS. This perseverance of the classical cardiac reaction, earlier described by Gantt (7, 8), seems to be a result of the "partial irreversibility" of classical conditioning, as proposed by Solomon and Wynne (35). The appearance of the neuronal changes and the cardiac reaction to the CS suggests that the sensory state of the organism might rapidly become undesirable. The return of the heart beat to normal after the performance of the AvR and discontinuation of the CS suggests that the sensory state improved.

The performance of the AvR, therefore, leading to the discontinuation of the CS, leads at the same time to the inactivation of the "old" memory pattern, CS.US, and consequently, to the disappearance of the undesirable sensory state, i.e., to an "improvement in the sensory state" and achievement of "better-being" (see 39, 40). This is the sensory reward which the animal receives each time it performs the avoidance response. Consequently, each application of the CS alone producing the AvR, strengthens the avoidance response instead of making it weaker. This is why avoidance behavior is so resistant to extinction.

As described above, the extinction of the AvR is possible when the CS is continued after the performance of the AvR. In that case, the memory pattern CS.US evoked by the CS remains active as long as the CS continues, sustaining the undesirable sensory state. With the

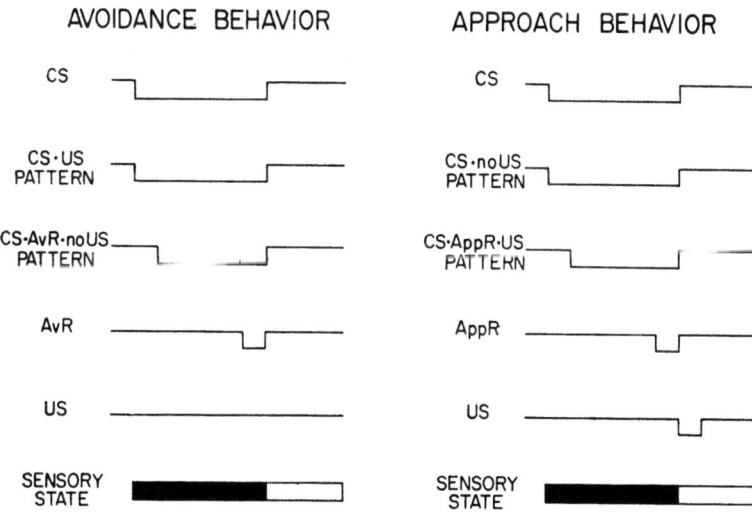

Fig. 1. A scheme showing a sequence of events occurring during and after the action of the conditioned stimulus in avoidance behavior and in approach (appetitive) behavior. CS, conditioned stimulus; US, unconditioned stimulus; AvR, avoidance response; AppR, approach (instrumental) response; CS.US, memory pattern of associations related to the presence of the undesired US during the action of CS in avoidance behavior; CS.noUS, memory pattern of associations related to the absence of the desired US during the action of CS alone in approach behavior; CS.AvR.noUS, memory pattern of associations related to CS, the performance of AvR and the postponement of the undesired US in avoidance behavior; CS.AppR.US, memory pattern of associations related to CS, the performance of AppR and the presence of the desired US; darkened field, undesirable sensory state of the organism; white field, sensory "better-being". Periods of action of CS and US, performance of AvR and AppR, and activation of memory patterns are shown by a drop in the corresponding lines. The scheme shows that in spite of the procedural differences between avoidance behavior (postponement of US after the performance of AvR) and approach (appetitive) behavior (occurrence of US after the performance of AppR), the final effect is, in general, the same: an achievement of sensory "better-being".

repetition of such procedure, the AvR loses its ability to improve the sensory state; in other words, the AvR is no longer rewarded by procuring sensory "better-being". Consequently, an extinction of the AvR takes place, similarly as it happens in approach (appetitive) instrumental behavior.

The view presented above makes it possible to extend the rules governing approach instrumental responses to avoidance behavior. Despite the procedural differences between these two behaviors, the performance of the instrumental response in either approach behavior or avoidance behavior leads to the same general result: an improvement in the sensory state (Fig. 1).

It is obvious that the terms used above such as "undesirable sensory state", "improvement in the sensory state" or "better-being" describe hypothetic, subjective feelings which cannot be strictly and objectively measured. However, these sensory states can be indirectly measured through recording the changes in autonomic responses, electrical activity of the brain, etc., accompanying the avoidance response. For example, changes in cardiac responses as recorded by Gantt (8) during classical defensive conditioning or by Sołtysik and Kowalska (37) during avoidance behavior, may serve as an indicator of the sensory state of the animal. The EEG recording may also prove useful in this respect. Some EEG correlates of reward in alimentary instrumental behavior were reported by several authors (3, 4, 21, 22). It is not excluded that an objective evidence of changes in the subjective sensory state during avoidance behavior will be found.

## REFERENCES

1. BAUM, M. 1966. Rapid extinction of an avoidance response following a period of response prevention in the avoidance apparatus. Psychol. Rep. 18: 59–64.
2. BREGADZE, A. N. 1953. The problem of establishment of the defensive conditioned reflex in dogs (in Russian). Tr. Inst. Fiziol. Akad. Nauk. Gruz. SSR 9: 43–59.
3. BUCHWALD, N. A., HORVATH, F. E., WYERS, E. J. and WAKEFIELD, C. 1964. Electroencephalic rhythms correlated with milk reinforcement in cats. Nature (Lond.) 201: 830–831.
4. CLEMENTE, C. D., STERMAN, H. B. and WYRWICKA, W. 1964. Post-reinforcement EEG synchronization during alimentary behavior. Electroencephalogr. Clin. Neurophysiol. 16: 355–365.
5. FONBERG, E. 1960. On some peculiarities of defensive conditioned reflexes, type II. In E. A. Asratyan (ed.), Central and peripheral mechanisms of motor activity in animals. Collected papers of an International Symposium in Poland, 1958. Moscow, Acad. Nauk SSSR. p. 124–134.

6. FONBERG, E. 1962. Transfer of the conditioned avoidance reaction to the unconditioned noxious stimuli. Acta Biol. Exp. 22: 251–258.
7. GANTT, W. H. 1944. Experimental basis for neurotic behavior. Hoeber, New York, 165 p.
8. GANTT, W. H. 1960. Cardiovascular component of the conditioned reflex to pain, food and other stimuli. Physiol. Rev. 40, suppl. 4: 266–291.
9. HALAS, E. S. and BEARDSLEY, J. V. 1970. Changes in neuronal activity in the cochlear nucleus as a function of classical and instrumental conditioning. Psychon. Sci. 18: 161–163.
10. JAWORSKA, K., KOWALSKA, M. and SOŁTYSIK, S. 1962. Studies on the aversive classical conditioning. 1. Acquisition and differentiation of motor and cardiac conditioned classical defensive reflexes in dog. Acta Biol. Exp. 22: 23–34.
11. JOHN, E. R., SHIMOKOCHI, M. and BARTLETT, F. 1969. Neural readout from memory during generalization. Science 164: 1534–1536.
12. KAMIN, L. J. 1954. Traumatic avoidance learning: the effects of CS–US interval with a trace conditioning procedure. J. Comp. Physiol. Psychol. 47: 65–72.
13. KIMBLE, G. A. 1955. Shock intensity and avoidance learning. J. Comp. Physiol. Psychol. 48: 281–284.
14. KIMBLE, G. A. 1961. Hilgard and Marquis' conditioning and learning, Appleton-Century-Crofts, New York.
15. KONORSKI, J. 1948. Conditioned reflexes and neuron organization. Cambridge University Press, London, 267 p.
16. KONORSKI, J. 1967. Integrative activity of the brain. An interdisciplinary approach. University Chicago Press, Chicago, 531 p.
17. KONORSKI, J. and MILLER, S. 1933. Podstawy fizjologicznej teorii ruchów nabytych. Ruchowe odruchy warunkowe. Książnica-Atlas TNSW, Warsaw, 168 p.
18. KONORSKI, J. and MILLER, S. 1936. Conditioned reflexes of the motor analyzer (in Russian). Tr. Fiziol. Lab. I. P. Pavlova 6: 119–278.
19 KONORSKI, J. and SZWEJKOWSKA, G. 1952. Chronic extinction and restoration of conditioned reflexes. III. Defensive motor reflexes. Acta Biol. Exp. 16: 91–94.
20. LIVINGSTON, R. B. 1967. Reinforcement. In G. C. Quarton, T. Melnechuk and F. O. Schmitt (ed.), The neurosciences. A study Program. Rockefeller University Press, New York, p. 568–577.
21. MARCZYŃSKI, T. J., ROSEN, A. J. and HACKETT, J. T. 1968. Post-reinforcement electrocortical synchronization and facilitation of cortical auditory evoked potentials in appetitive instrumental conditioning. Electroencephalogr. Clin. Neurophysiol. 24: 227–241.
22. MAULSBY, R. L. 1971. An illustration of emotionally evoked theta rhythm in infancy: Hedonic hypersynchrony. Electroencephalogr. Clin. Neurophysiol. 31: 157–165.
23. MILLER, N. E. 1959. Liberalization of basic S-R concepts: extensions to conflict behavior, motivation and social learning. In S. Koch (ed.), Psychology: A study of a science. Study I, Vol. 2. McGraw-Hill, New York, p. 196–292.
24. MORRELL, F. 1967. Electrical signs of sensory coding. In G. C. Quarton,

T. Melnechuk and F. O. Schmitt (ed.), The neurosciences. A study program. Rockefeller University Press, New York, p. 452–467.
25. MOWRER, O. H. 1947. On the dual nature of learning — a reinterpretation of "conditioning" and "problem solving". Harv. Educ. Rev. Spring. 102–148.
26. MOWRER, O. H. 1960. Learning theory and behavior. John Wiley and Sons, New York.
27. MOWRER, O. H. and LAMOREAUX, R. R. 1942. Avoidance conditioning and signal duration — a study of secondary motivation and reward. Psychol. Monogr. 54, No. 247.
28. PAVLOV, I. P. 1927. Conditioned reflexes. Transl. from Russian by G. V. Anrep. London, Oxford University Press.
29. PRADO-ALCALA, R., RUSH, H., STEELE, D. and REID, L. 1973. Brief flooding and counter-conditioning as treatments for persisting avoidance. Physiol. Psychol. 1: 389–393.
30. PREMACK, D. 1965. Reinforcement theory. Nebraska Symposium on Motivation. 13: 123–180.
31. SCHOENFELD, W. N. 1978. "Reinforcement" in behavior theory. Pavlovian J. Biol. Sci. 13: 135–144.
32. SHEFFIELD, F. D. 1948. Avoidance training and the contiguity principle. J. Comp. Physiol. Psychol. 41: 165–177.
33. SIDMAN, M. 1955. On the persistence of avoidance behavior. J. Abnorm. Soc. Psychol. 50: 217–220.
34. SOLOMON, R. L., KAMIN, L. J. and WYNNE, L. C. 1953. Traumatic avoidance learning: the outcomes of several extinction procedures with dogs. J. Abnorm. Soc. Psychol. 48: 291–302.
35. SOLOMON, R. L. and WYNNE, L. C. 1954. Traumatic avoidance learning: the principles of anxiety conservation and partial irreversibility. Psychol. Rev. 61: 353–385.
36. SOŁTYSIK, S. 1960. Studies on avoidance conditioning. II. Differentiation and extinction of avoidance reflexes. Acta Biol. Exp. 20: 171–182.
37. SOŁTYSIK, S. and KOWALSKA, M. 1960. Studies on avoidance conditioning. I. Relations between cardiac (type I) and motor (type II) effects in the avoidance reflex. Acta Biol. Exp. 20: 157–170.
38. TAPP, J. T. (ed.) 1969. Reinforcement and behavior. Collected papers by several authors. Acad. Press, New York, XXI, 429 p.
39. WYRWICKA, W. 1972. The mechanisms of conditioned behavior. Charles C. Thomas, Springfield, 179 p.
40. WYRWICKA, W. 1975. The sensory nature of reward in instrumental behavior. Pavlovian J. Biol. Sci. 10: 23–51.

Wanda WYRWICKA, Department of Anatomy and the Brain Research Institute, University of California, School of Medicine, Los Angeles, California 90024, USA.

Lecture delivered at the Warsaw Colloquium on Instrumental
Conditioning and Brain Research
May 1979

# CORTICAL MECHANISMS OF GOAL-DIRECTED MOTOR ACTS IN THE RHESUS MONKEY

A. S. BATUEV, A. A. PIROGOV, A. A. ORLOV and V. I. SHEAFER

Department of Higher Nervous Activity, Leningrad State University
Leningrad, USSR

*Abstract.* Multineuronal activity from the frontal cortex (s. principalis) and the motor cortex (g. precentralis) in the Rhesus monkey during the goal-directed performance was recorded. The experimental sequence comprised a chain of signals securing adequate behavioral performance. The sequence began with an anticipatory signal, then the conditioned signal was switched on, a lamp in either the right or the left key, and after a 5–10 s delay, a trigger signal opening a screen followed. Then the monkey had to press the key indicated by light and grasp the food from a box. Differences in the average firing rate and in the patterns of response were observed by comparing the unit activity during the CS expectancy with the trigger signal expectancy (delay). A mathematical treatment of the unit firing pattern allowed us to select specific firing pattern rearrangements likely to be involved in the short-term memory mechanisms, i.e. intracortical reverberation and successive recruitment of neuronal populations. Unit activity of s. principalis in one hemisphere depended on the CS right vs. left location. Similar differences were found for the delays. During the delay, information about the CS spatial location is probably stored as short-term traces in the units of s. principalis. The frontal cortex unitary responses elicited by the trigger signal depended on the direction of the subsequent movement. The motor cortex units provided less pronounced differences in response to the trigger. In contrast to the motor cortex units, some frontal cortex units reacted only to the presence of food reward.

## INTRODUCTION

Konarski (14) considered the elaboration of the conditioned instrumental reflex to be the basis of developing association between the kinesthetic neurons involved in a given movement and neurons activated by a certain drive. An immediate drive satisfaction is critical for the formation of such an association.

In connection with this idea, Konorski (16), as well as other investigators, have focused their attention on the neocortical frontal lobe functions (4, 5, 23, 29, 30). A wide range of behavioral deficits in dogs and monkeys were demonstrated mainly by lesion and local cooling of frontal lobes (2, 3, 13, 29). Disturbances in the motivational and emotional functions, in memory storage processes, in voluntary movement control, in spatial orientation, etc. have been described as consequences of the frontal lobe damage (4, 5, 16, 23, 29, 30). The above-mentioned facts determined the search for the neuronal correlates of the frontal lobe functions, which has become possible due to the development of new techniques for unit activity recording in awake animals (11, 12, 28). The pioneer works in this field appeared only some years ago. The authors aimed at finding the specific reaction of the monkey's frontal lobe units correlated with the specific behavioral elements (12, 17–19, 21, 27, 31, 32). Since the complexity of the frontal lobe morphological pathways (24–26) determines their polyfunctional nature, it allows one to consider frontal lobes in general as an associative structure (4). Its specificity lies not so much in the integration of heterosensory inputs, but rather in the possibility to control this integration by activation of the limbic system (10, 25, 30).

In order to study the cortical mechanisms related to the programming of the goal-oriented motor act, we devised a complex behavioral sequence including attention, spatial choice and short-term memory based on food reinforcement. The main objectives in this work were to investigate the neuronal correlates of the behavioral sequence stages, interneuronal relationships, plastic rearrangements of unit activity during simultaneous recording from the frontal and motor cortices. We therefore faced the problem of developing a new technique for a prolonged unit activity recording in awake monkeys.

## METHODS

Five (4–5 kg) adult Macaca rhesus monkeys, were used in the experiments. During the recording sessions the monkey was seated in a primate chair (Fig. 1, I), with its head restrained by the board. In front of the monkey a panel was situated within an arm's reach. The panel contained two keys, with a lamp in each, a moveable screen, a device for automatic reward, and a photic flash (Fig. 1, I). To determine

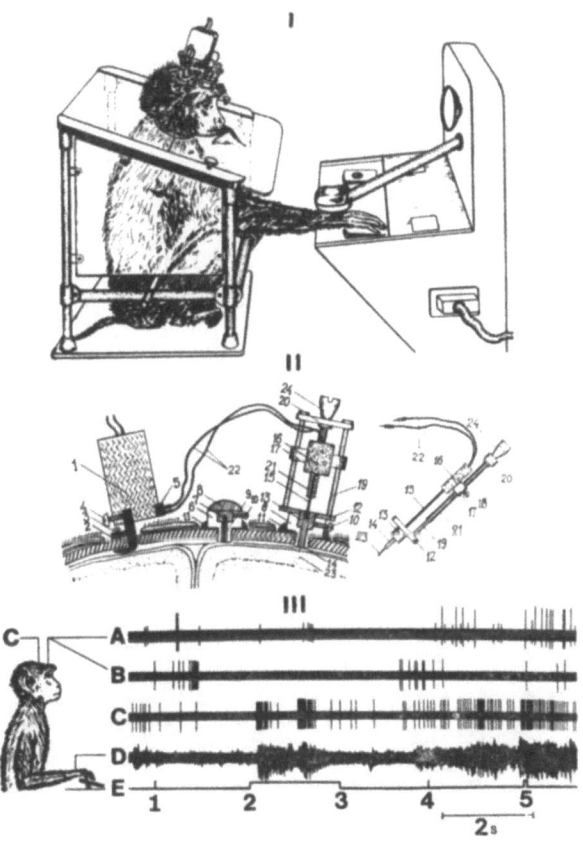

Fig. 1. General experimental setup. II: The microelectrode bundle and its location on the skull. Numbers indicate various components of the electrode system. III: Examples of unitary spike activity and of EMG recorods in the awake monkey. The unit activity was recorded from s. principalis (A, B); s. precentralis (C); EMG from elbow muscles (D); types of stimulation (E) were: 1. anticipatory signal; 2–3, conditioned signal; 4. trigger signal; 5. key-press. Further explanations in the text.

the neuronal correlates of a motor performance, we used the sequence of automatically presented signals in the following order (Fig. 2). With the screen in closed position the anticipatory signal flashed (Fig. 2b). After a 2 to 5 s expectancy interval the conditioned signal was presented for 2 s by one of the lamps in the right or in the left key (Fig. 2c). The lamp brightnesses were equal. The left or right keys were lit randomly. The monkey could see the light through the glass windows

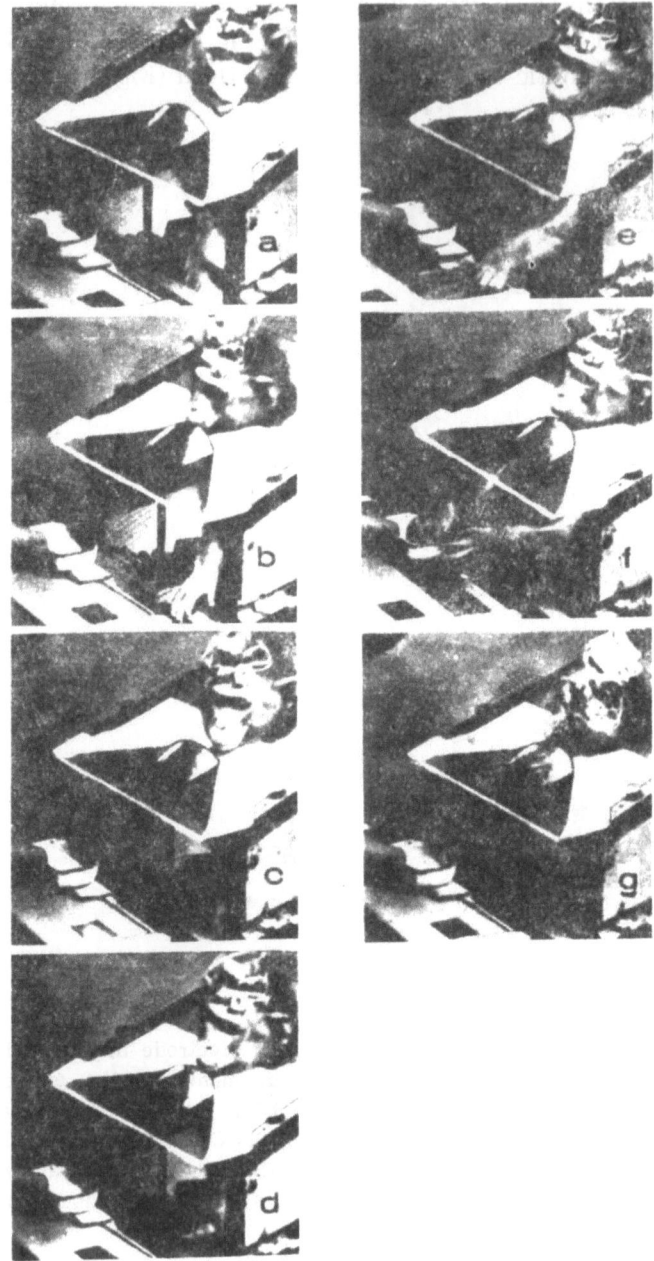

Fig. 2. Successive stages of the behavioral sequence. (Explanations in the text).

in the screen above each key. After termination of the conditioned signal, a delay of a maximum 10 s duration ensued (Fig. 2d). During the delay access to the keys was blocked by the screen. After the delay the screen was opened automatically and this served as a trigger signal (5 ms, 60–70 dB noise) for a choice of the correct key. Then the screen was closed to prevent another pressing or correction of errors. If the monkey pressed the key which was indicated earlier by the lamp (correct choice) (Fig. 2e), food reward (pieces of bread, carrot, hazel-nut) was delivered to the foodbox located equidistantly from the two keys (Fig. 2f). The monkey could perform hand-to-mouth movement through the opening in the neckband (Fig. 2g). All signal sequences and durations of the signal and delay were automatically programmed. The same sequence was presented at least 20 times, with intertrial intervals of 2 min or less.

After training, the monkey underwent surgery to install the microelectrode system on the skull. The skull was trephinned above the investigated cortical areas (middle part of s. principalis and rostral part of g. precentralis corresponding to the hand area) and the plugs were implanted (Fig. 1, II, 6–11). The microelectrode piston was fastened to the plugs and remained there for 2–8 days. One electrode bundle, consisting of 4 silver glass-insulated wires (outer diameter 38–50μm, silver tip diameter 2–10 μm), was inserted into an injection needle. The bundle diameter was approximately 100 μm. The dura mater was perforated through the plug hole before penetration by the electode bundle. A multi-channel preamplifier was attached to a previously implanted metal clamp (Fig. 1, II, 1–5), which served as an indifferent electrode.

Thus this new technique allowed us to study single cortical neurons, to change the arrangement of the electrodes in the bundle, to increase the number of simultaneously recorded units from one or more cortical areas and, finally, to avoid injury following screw implantation for restraint attachment. The recording technique is illustrated in Fig. 1, II. Simultaneously with the unit activity, the EMG of a shoulder muscle (m. deltoideus) was recorded with bipolar steel electrodes (50 μm in teflon insulation) connected by wires running under the skin to the skull.

After completion of the experiments, histological analysis was done in order to determine the location of each electrode tip, the state of cortical tissue surrounding the electrode bundles, and the electrode bundle divergence. As shown in Fig. 3, there were no significant injuries in the recording sites. Reconstructions of the electrode bundle divergence based on serial sections of s. principalis shows (Fig. 4) that the electrode divergence is about 200 μm to 250 μm at the depth of 1 mm and reaches 600–1,200 μm at 5 mm depth.

Unit activity was recorded by means of a standard electrophysiological equipment. The mini-preamplifier on the head was connected to a multi-channel amplifier, then to a tape recorder and via an amplitude discriminator, to a cathode-ray oscilloscope with a camera or to an ink-writing oscillograph. During the experimental session the animal was placed in a sound proof room. A closed-circuit TV-equipment enabled an observation of the animal. The data were processed using the Neuron-1 and Nairi-2 computers. Dot displays and peri-stimulus time histograms were plotted. Unit responses were considered present when a stable pattern could be recorded in 20 presentations of the same sequence. A correlation between the activity of several units at different stages of the sequence and their reliability were determined by correlation and reliability coefficients. Significant differences in firing rate were calculated by finding the coefficients of reliability for the difference variations.

## RESULTS

Of 238 units recorded in s. principalis, 230 displayed a background activity. Six other units yielded very low background activity (0.01 imp/s) and did not respond to any signals. These silent units were not taken into account when evaluating quantitative relationships of the sampled units. Other units provided mean firing rates ranging from 0.1 to 16 imp/s (01–10 imp/s for 91% of units). Our observations demonstrated a dependence of the background firing rate upon the levels of food motivation, attention and task performance.

*Unitary responses during the expectancy period of the CS signal.* The anticipatory signal fixed the monkey's attention to the entire experimental situation. One hundred forty eight units (64%) reacted to it, while the remaining 82 units (36%) showed indefinite responses. The responses of most units were inhibited for 0.5 to 1.5 s (62 units. 27%) and in 53 neurons firing was suppressed for 200-300 ms. A few units (19.8%) increased their firing for 0.4–1.5 s ($P < 0.05$).

Distribution of unit changes during the expectancy period was as follows. The firing rate of 74 units did not differ from the background. A majority of units (57%; Fig. 5, 2) decreased their firing at the moment the CS was turned on. Within this period 25 units (11%, Fig. 5, type 1, 3, 4) produced one or two bursts lasting for 1.5–2 s.

Since the anticipatory signal was meant to warn the animal about the time interval preceding the CS, irrespectively of its spatial location, the period of CS expectancy is labelled "unspecific". When the above period has a fixed duration, tonic rearrangements of impulse activity might be responsible for short-term memory processes underlying realization of a conditioned reflex to time.

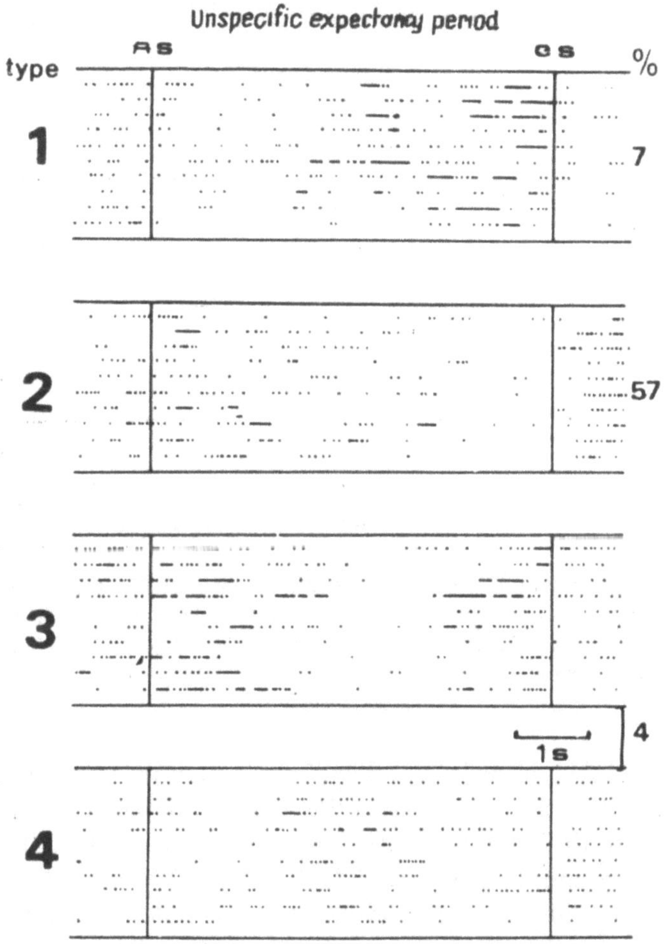

Fig. 5. Types of neuronal responses (1–4) during the conditioned signal expectancy (unspecific expectancy). AS, onset of the anticipatory signal; CS, onset of the conditioned signal. Numbers on the right, percentages of units with the given response type. Dot displays are plotted for ten presentations of the same sequence.

*Neuronal reactions to the CS presentation.* A total of 158 units (69%) gave reliable responses to the CS. The unit response falls into 6 groups. Forty units (17%) increased firing rate during the entire period (2 s) of CS presentation; 34 units (15%) provided inhibitory on-off responses, the duration of each of them being 350 ± 80 ms; 24 units (10%) showed a partial or complete suppression. On the whole, inhibitory reactions were predominant (75 units, 34%).

Our attention was directed to the dependence of the unit activity produced by the CS presentation upon its spatial location. A comparison of responses to the right or left CS presentation allowed us to find marked distinctions either in their mean firing rate (Fig. 6; type 1), or in their response pattern (Fig. 6, type 2). These were 57 units (25%)

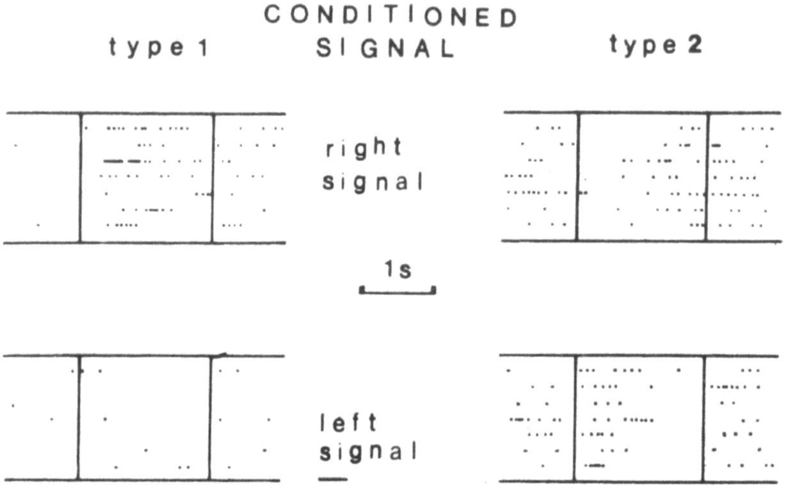

Fig. 6 Unit responses to the conditioned signal presented at varying spatial locations. Vertical lines, the onset and termination of the CS presentation. Type 1: The responses of the same unit differ by their total discharge rates during the CS presentation. Type 2: The responses differ by their patterns of activity within the same period.

and 29 units (14%), respectively. The following type of data-processing was performed for the units which produced different firing rates to the right or left CS. Two groups of units were analysed: 26 from the right hemisphere and 31 from the left. Then the rows of average firing rates during the right and left CS presentation, the average frequency

TABLE I

Effect of spatial location of the CS upon the responses of the frontal cortical neurons during the CS presentation

| Number of units | $n = 26$ | | $n = 31$ | |
|---|---|---|---|---|
| Hemisphere | Right | | Left | |
| Location of CS | Left | Right | Left | Right |
| Average firing rate (imp/s) | $5.82 \pm 0.64$ | $2.64 \pm 0.58$ | $3.97 \pm 0.82$ | $7.14 \pm 0.71$ |
| Significance of differences | $P < 0.01$ | | $P < 0.01$ | |

and the coefficient of difference reliability were calculated (Table I). It was seen that the impulses discharged by the right hemisphere units have a 2.2 times greater frequency to the left-hand CS than to the right-hand ipsilateral CS. In contrast, the left hemisphere unitary responses to the right CS are 1.8 times larger than those to the left CS.

To sum up, the mean firing rate in most frontal units reflects the spatial location of the CS which carries the main information as to the location of the key to be pressed.

*Unit responses during the delay period.* Trigger signal expectancy (delay or specific expectancy) elicited pronounced responses in 84% of the sampled units. The types of responses during a 10 s delay are illu-

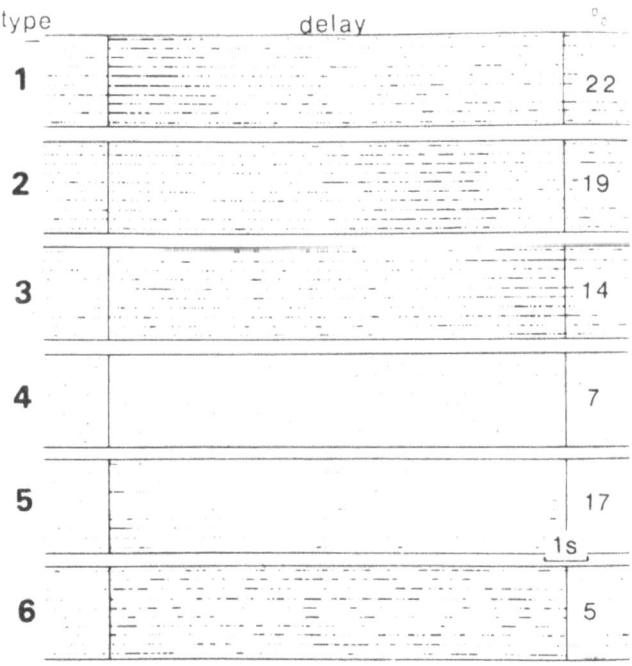

Fig. 7. Types of unit responses during the 10 s delay (trigger signal expectancy). 1-6, dot displays of the activity of various units for ten presentations of the same sequence. Numbers on the right, percent distribution of neurons according to their activity types during the delay (limits marked by vertical lines).

strated in Fig. 7. Type 1 are the units activated during 1.0–2.5 s at the beginning of the delay. Type 2 provided the same activation at the mid-delay ($\pm$ 2 s following its onset). Type 3 units showed similar activation during 1.5–2 s at the delay termination before the trigger signal. Certain types of wave-like unit activity were recorded from 72% of the units. Sometimes such types of activity could be observed simultaneously as a result of parallel recording from 2–3 units by one microelectrode bundle. Figure 7 showes such types of reactions in units silent to the anticipatory and conditioned signals. The average firing rate during a 1.0 to 2.5 s activation increased 2.3 times in comparison to the background impulsation ($P < 0.01$). In addition miscellaneous units (Fig. 7 type 4 and 5) with 1 to 3 activity bursts on the low frequency background were observed. Thus the entire delay period comprises successive response enhancements of different units, either at the beginning (type 1), middle (type 2) or end (type 3) of the delay, or within various parts of delay (type 5). It should be noted that only 12 neurons (5%) maintained a high level of activation throughout the delay (type 6).

The described pattern is comparable to a successive "relay-race" involvement of different neuronal populations in order to sustain a prolonged activation of a small group of "summator" units (type 6).

As stated above, impulse responses of most s. principalis units during the CS presentation depended on its lateralization. It was therefore important to see whether the same differences held for the delay. It

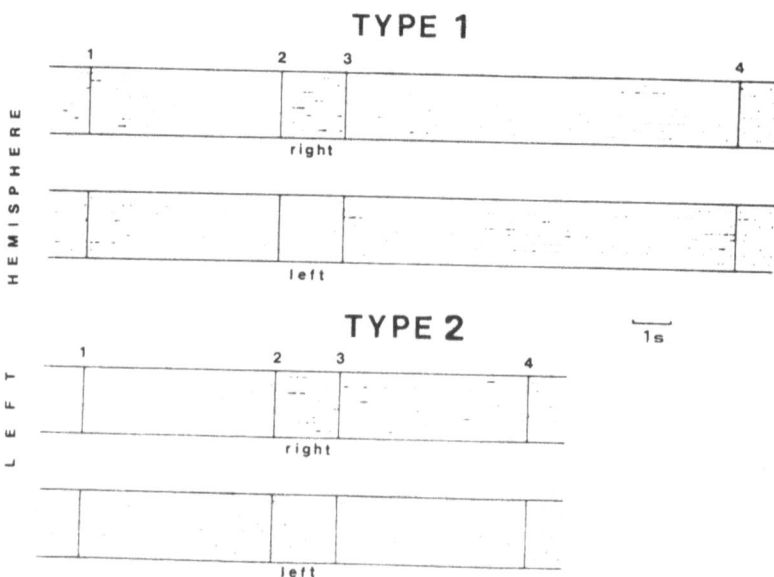

Fig. 8. Effect of the CS location on differences in activity during the delay. Left hemisphere units: type 1: differences in activity patterns; type 2: differences in the total discharge rates within the delay. 1, onset of the anticipatory signal; 2–3, the conditioned signal; 4, trigger signal. "Right" and "left", corresponds to the right and left-hand presentation of the CS. The specific unit activity related to the CS spatial location extends throughout the delay (spatially specific units). Dot displays are based on 10 presentations of the same sequence.

turned out that 96 units (42%) retained the difference in activity related to the right or left CS location. Of this group, 58 units (25%) displayed distinctions in the firing rate, while 38 units (17%) maintained their patterns of activity. Figure 8, type 2 illustrates modifications of average firing rate (9 to 10 times). Figure 8, type 1 shows changes in activity patterns. As can be seen, the delay periods are distinguished by temporal sequences of maximum discharge rate to the right or left CS location.

The data-processing showed that, in the right hemisphere, the average firing rate during the delay was 2.1 times higher in response to the left (contralateral) CS location than to the right (ipsilateral) ($P < 0.01$). Correlations for the left hemisphere were reverse: discharge rate during the delay in the case of the right CS location was 2.0 times that to the left ($P < 0.01$; Table II). Significance of the differences was higher than for similar correlations in the same units when the CS was on.

TABLE II

Effect of spatial location of the CS upon the responses of the frontal cortical neurons during the delay period

| Number of units | $n = 26$ | | $n = 26$ | |
|---|---|---|---|---|
| Hemisphere | Right | | Left | |
| Location of CS | Left | Right | Left | Right |
| Average firing rate (imp/s) | 5.19±0.32 | 2.46±0.28 | 3.12±0.21 | 6.22±0.30 |
| Significance of differences | $P < 0.01$ | | $P < 0.01$ | |

Thus, there is an interrelation between the activity of the frontal lobe units during the delay and the processes of CS spatial location storing in the short-term memory. This is evident not only from the differences in the average firing rates, but also from different response patterns within the same period.

*Unit responses accompanying rearrangements of spatio-temporal relations.* Prolonged recording from the same unit made it possible to trace plastic rearrangements of the unitary activity. In one series of experiments the monkey was trained to press only the right key (Fig. 9). Then the CS location was abruptly changed to the left key. Activity patterns were studied till the new conditioned reflex to the left key press became stable. Of 12 units sampled, 6 increased their firing, 3 decreased, and in 3 units the background activity did not chage significantly. Figure 9 demonstrates the paradigm of unit activity rearrangements from the left frontal lobe. It is obvious that the conditioned reflex alteration is concurrent with a pronounced increase in the background activity. This increase may correspond to an increase of the central excitatory state during the conditioned reflex generalization. Correspondingly, the number of erroneous trials increases, probably due to the preceding stable training. After the new conditioned reflex had become

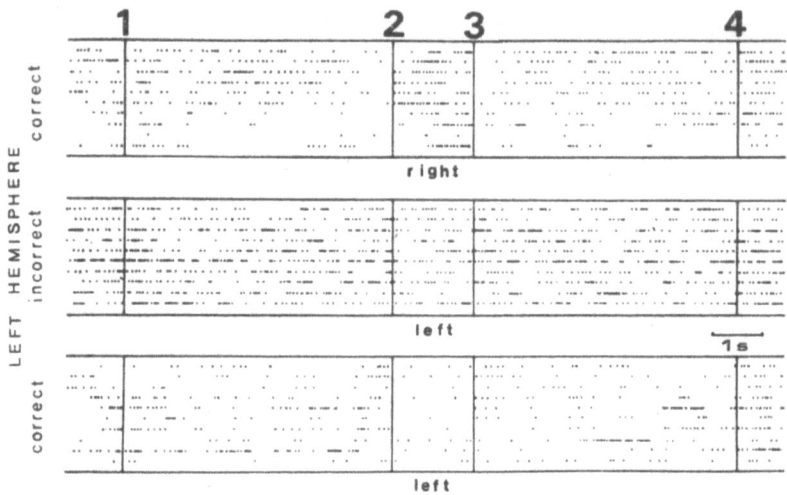

Fig. 9. Unit activity rearrangements following a sharp change in the CS location. 1, anticipatory; 2–3 conditioned; 4, trigger signals. The unit from s. principalis, left hemisphere. Top oscillogram, unit activity during performance of the conditioned reflex to the right-hand CS. Middle, the activity of the same unit accompanying a reflex reshaping to the left-hand CS. The number of incorrect trials is high, the general level of unit activity increases. Bottom oscillogram, the activity of the same unit after returning to the original right-hand CS. Adequate left key-presses are performed. Impulse activity is observed to decrease and its pattern differs from that of the original right-hand key-press reflex.

stable, the same neuron revealed quite different activity patterns: firing decreased during CS presentation on the left side. The same differences existed for the delays. Other experiments were designed to study the rearrangements of impulse activity in 34 units when the delay was suddenly lengthened. The procedure was as follows. The initial duration of the delay was 5 s. Figure 10A shows that the firing of the sampled units was enhanced only by the CS and for a certain time after the trigger. First, the monkey was trained in this situation for some time. Then the delay was abruptly prolonged to 10 s. Four neurograms correspond to the 1st, 4th, 7th and 10th presentation of the new sequence. Transition to the new sequence is seen to be accompanied by impulse enhancement from the start of the delay. Gradually maximum activity shifts to the middle, then closer to the end, and finally, the most powerful discharge is generated at 700 ms prior to the trigger signal.

The external manifestation of such dynamic rearrangements is similar to the regularity found for the Pavlovian conditioned inhibition elaborated when the reinforcement is delayed after the onset of the conditioned signal. This type of inhibition is commonly held to form the basis of conditioning to time. This is evident from the peristimulus histogram (Fig. 10C), which coincides with the stabilization of a new sequence including a 10 s delay. By comparing it to the above histogram (Fig. 10B) we can see that the maximum discharge rate occurs in the second part of the 10 s delay.

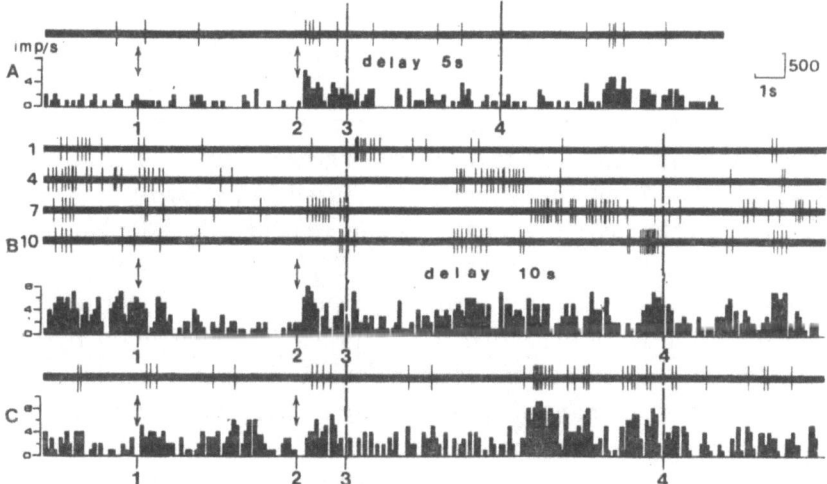

Fig. 10. Unit activity rearrangement related to delay duration. I: anticipatory; 2-3 conditioned; 4, trigger signals. A, neurogram (upper); histogram (lower) of the neuron compiled for ten repetitions when the delay duration was 5 s. B, upper — neurograms of the same unit at the 1st, 4th, 7th and 10th presentation of the sequence with a 10-s delay; lower-histogram for all 10 presentations. C, upper — sample neurogram after establishing of the 10-s delay; lower-histogram of ten repetitions at that stage. General level of unit activity decreases and the maximum unit activity shifts to the last 4 s of the delay. Only neuronal responses corresponding to corret alternation were analyzed and compared. Calibration: 1-s, amplitude, 500µV.

The background activity of a given unit becames three to four times greater than during the period of delay alteration (Fig. 10 B). Subsequently it declines to the original level following stabilization (Fig. 10C).

Thus an alteration of the spatial and temporal relationships between the elements of a sequence is accompanied by an increase in the central excitatory state and also affects the firing rate of the frontal lobe units.

*Unit reactions during the conditioned motor act.* By means of multichannel recording, we studied simultaneous activity of single units in the frontal and motor cortex as well as EMG from deltoideus muscle immediately before and during the motor act. As noted above, the delay was terminated by the screen opening (trigger signal). One hundred forty eight units from s. principalis showed reliable responses to the trigger, 92 units (40%) provided phasic or tonic inhibitory responses, and 56 units (24%) were enhanced. Of 100 motor cortex units sampled, 58% generated tonic activity: the numbers of units increasing their firing (latency approximately 107 ms) and decreasing (latency approximately 60 ms) were equal.

We were able to calculate the latencies of the EMG activity and that of the motor acts by averaging of the EMG s accompanying 20 correct responses. The onset of the EMG occurred 300–400 ms after the screen opening and 400–500 ms before the key-press. The EMG accompanied the movement. The interval between the trigger signal and the key-press was initially 0.7 to 2 s and increased up to 5–7 s following gradual satiation, (sometimes monkeys refused to perform).

From our calculations we may assume that the duration of the unitary responses to the trigger was 200 to 300 ms after its onset, because all subsequent unit activity rearrangements coincided with the key-press. The latter were classified into 6 types for the frontal cortex and 4 types for the motor cortex. Most sampled units (68, i.e. 29.6%) from both regions were enhanced or suppressed after the key-press.

Of particular interest was the analysis of differences in unit activity after the trigger, as related to the direction of the forthcoming movements to the left or right key (Fig. 11). It turned out that 31.7% of the

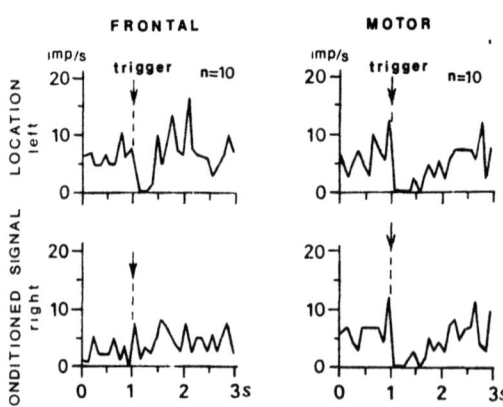

Fig. 11. Comparison of impulse activity of simultaneously recorded units from the frontal and the motor cortices (right hemisphere) elicited by the trigger signal. Histograms show that the frontal units retain the spatial properties of the CS even when the trigger works, whereas in the motor cortex such units were not found.

frontal lobe units generated different discharge patterns after the trigger signal depending on lateralization of coming movements. These distinctions were less pronounced or quite unreliable for sampled motor cortex units (29%) (Fig. 11). The responses of the frontal and motor cortical units to the trigger signal and during the movement performance are presumably of different origin. This assumption is supported by a digital analysis of unit activity rearrangements in relation to the correct and incorrect key-presses. Figure 12 demonstrates correct key-pres-

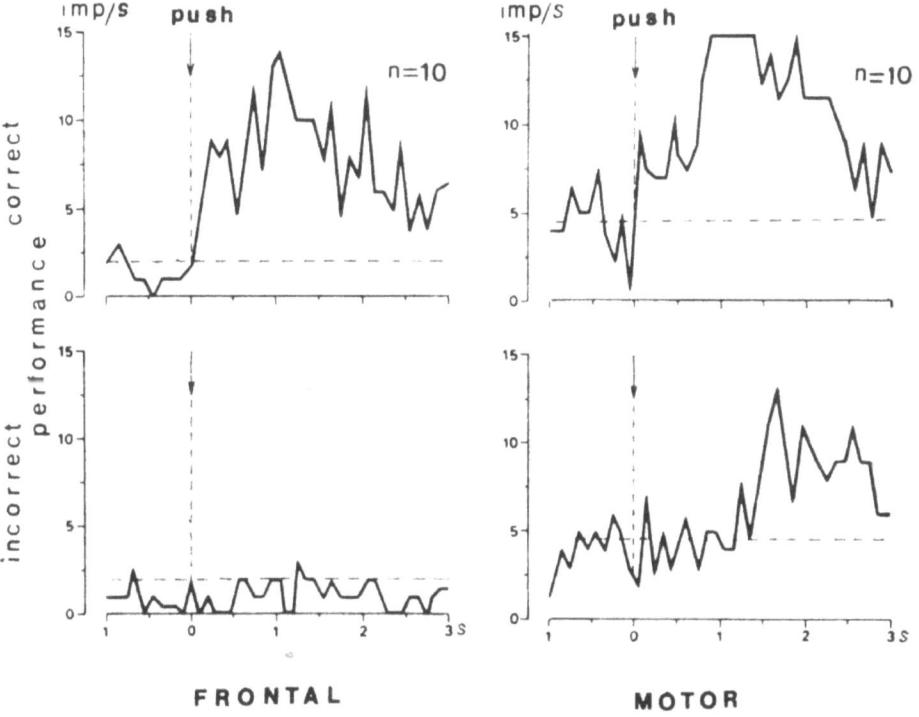

Fig. 12. Comparison of impulse activity of units simultaneously recorded from the frontal and motor cortices in response to the key-press. Correct performance coupled with food reward is concurrent with an increase in the frontal unit activity. Incorrect choice (not-rewarded) does not affect the activity of the same unit ($n = 10$). The activities of the motor cortex unit during correct or incorrect performances do not differ significantly.

ses coupled with food reward, which are characterized by response enhancement both in the frontal and motor cortices. However, in the cases of erroneous key-presses (non-rewarded), the impulse activity of the same frontal cortex units (8.7%) did not change in relation to the background level. In the same cases the unitary responses of the motor cortex did not differ from those accompanying the correct choices. Thus, in contrast to the motor cortex, the unit activity of s. principalis correlates with the direction of the coming movement and, if the choice is correct, with the presence of the food reward.

The latter prompted us to study the effects of the food motivation level (satiation) and of the adequate choice performance. In all cases analogous behavioral effects and changes in unit activity from s. principalis were observed. Figure 13 illustrates one paradigm for units res-

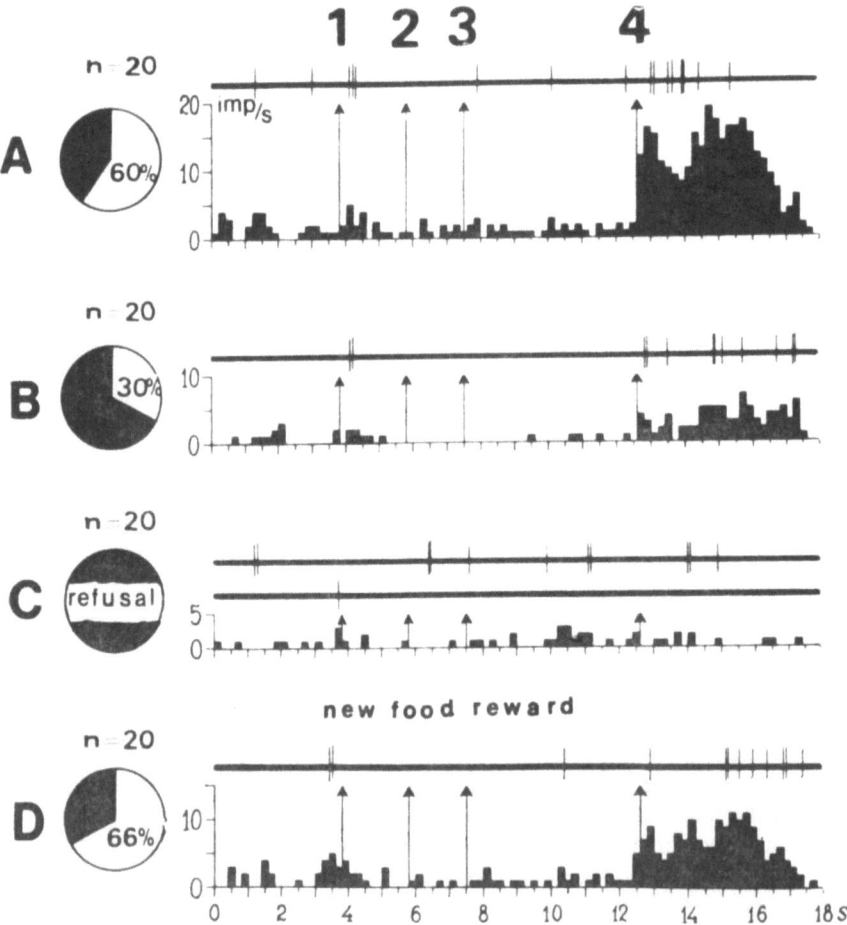

Fig. 13. Effect of food motivation level in the frontal unit activity and alternation performance. I: anticipatory; 2–3, conditioned; 4, trigger signals. Left, diagrams show the percentage of correct alternations ($n = 20$); right, neurogram and peristimulus histogram ($n = 20$). A, the initial level of motivation (carrots as food-reward); B, gradual satiation; C, refusal to perform the sequence; D, new type of food reward (hazel-nut). The conditioned reflex activity and impulse unit activity rise to the original level.

ponding by a phasic increase in their firing rate after the onset of the trigger signal. With gradual satiation, the number of incorrect responses grew until the refusal to perform the trained movement followed (Fig. 13 A–C). Parallel to this, a complete disturbance of the discharge pattern was observed (Fig. 13C). The original discharge pattern and the level of behavioral response was however restored as a consequence of a modification of the food motivation, i.e. a change in the kind of food reward (Fig. 13D). The motivation level is probably an essential factor of the frontal lobe activity and the adequate programming of the goal motor act.

## DISCUSSION

A polysensory nature of units is known to be typical of the associative structures of the brain. A wide convergence of heterosensory inflows was demonstrated for the prefrontal cortex in monkeys (6, 7). Considerable facilitation of responses during concurrent heterosensory stimulation and absence of discharge blocking in the prefrontal units provide evidence for a high responsiveness of the structures, on which specific heterosensory impulsation converges via the independent, spatially separate afferent inputs (2, 6). The latter might be viewed as a structural basis enabling the frontal lobe involvement in the analysis of all current modifications of the environment. Response inhibition in awake monkeys was induced only by repetition of indifferent stimuli. However, it was not observed in the case of repeated application of biologically significant signals. The most likely explanation is that environmental modifications are reflected in the s. principalis unit firing, whereas the transfer of information concerning indifferent stimuli is simultaneously blocked.

The experimental sequence used in our experiments was a complex chain, consisting of various signals, each enabling the animal to perform the oriented goal-motor acts adaptively. Since the interval between the anticipatory signal and the CS (unspecific expectancy) was equal for all animals, we assume that a conditioned reflex to time was elaborated. Indeed, the anticipatory signal warned the animal about the onset of the 5-s interval until the CS was presented. Therefore, tonic rearrangements in the unit firing rate within such a period can be considered as neuronal correlates of marking off time. As already described, some units had 1 to 2 periods of maximum firing (1.5–2 s) during the entire unspecific expectancy period, but a majority of the units revealed a sharp decrease in the impulse activity up to the moment when the CS was switched on. The discharge pattern for each sampled unit was stable and, therefore, we assume that such units are involved in various neuronal populations.

The CS signalling the key location is a determining factor in the adequate performance of the sequence. The significance of the CS is, first of all, manifest in a high reactivity of most frontal units, which were as a rule, multisensory. We can agree with the authors who consider that the number of units recruited depends upon the level of significance of a certain stimulus (31–33). The second important point is that the spatial features of the CS are reflected in the unitary activity of s. principalis. It should be pointed out that the CS lateralization affects the unit firing differently, not only in the right or left hemisphere, but in the same hemisphere as well, where units possessing spatial selectivity were also found. They showed stable distinctions in frequency of firing and in dicharge patterns. Such spatial selective sensitivity was retained in cases of a random CS presentation and during alteration of the conditioned reflex. It allows one to postulate an existence of units responsible for encoding spatial features of the visual stimuli.

After the CS was switched off, the delay which had a fixed duration for each animal, incorporated elements of conditioned reflex to time. Experiments employing a sharp prolongation of this delay support the idea of close correlations between the temporal properties of the delay and the plastic rearrangements of the activity in s. principalis. In contrast to the unspecific expectancy period, a 10-s delay elicited a wavelike activity in most units (72%). However, only a small number of units sustained a high level of activity during the entire delay. From these observations we can deduce a neuronal model employing closed neuronal circuits. Neurons activated within the initial phase of the delay might be considered as "input neurons", detecting information as to the CS attributes (spatial location). Furthermore, the "relay-race" phenomenon may involve various neuronal populations, each sending its axon collaterals to the "summator neurons". The latter convey their delayed impulsation toward the trigger structures of the motor cortex. Existence of this type of neuronal circuits in s. principalis was described by Demianenko (8). Figure 14 shows assemblies of pyramidal cells at layers III and IV, the axons of which branche off U-like and run to the cells of the same group. Stellate neurons can also be observed and their axons send collaterals to the large pyramidal dendrite area in layers V–VI, where these axons contact the basal dendrites. Thus, such a neuronal assembly is an example of a "trap". The inflow entering it can reverberate until relayed onto the efferent column of the large pyramidal neurons. There are closed neuronal circuits in s. principalis which may account for a short-term inflow storage and "relay-race" activation by recruiting ajacent neuronal circuits during the delay. The above-described phenomenon is related to the short-term memory and we consider it to be divided into some intervals, each associated with a certain circuit activation.

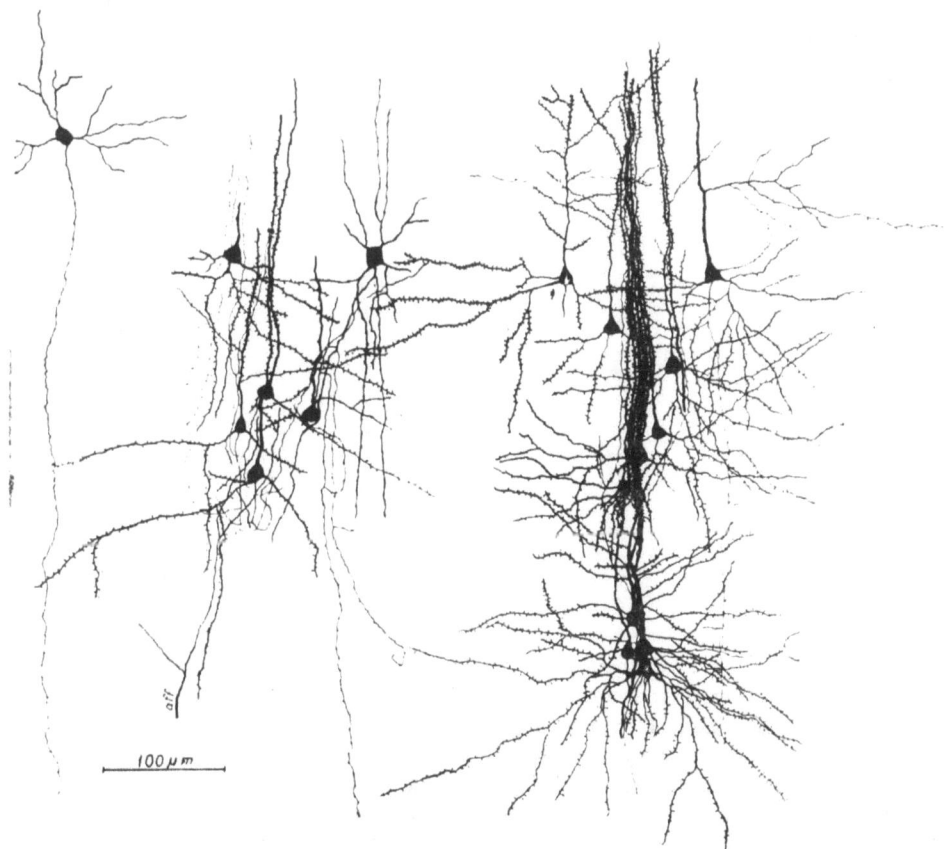

Fig. 14. Main types of neuronal assemblies in s. principalis (8). A fragment of microelectrode track was studied. Left, pyramidal cells connected by axons of 'the "trap" type layers III–IV); right, assembly of pyramidal cells in layers V–VI with nests of soma location and apical dendrite bundles.

Additional information was obtained through investigation of the interneuronal correlations. During recording from the same neuron, the correlation coefficient for a 1-s interval was calculated. Correlation coefficients were measured for 37 neuronal pairs by means of impulse quality comparison (t = 100 ms) during 5–10 applications of the sequence.

Figure 15 demonstrates general results when the correlation sign during CS presentation was used as a basis for calculation. Neurons with positive or negative correlation were divided into two groups. It allowed us to reveal the general tendency in the correlation variations for the unspecific expectancy period and the delay. As can be seen from Fig. 15, this tendency is characterized by an increase of the positive correlation within 2 to 3 s following the anticipatory signal. The correlation sign reversed 1 s before the CS onset. On the contrary, the delay is characterized by interneuronal correlation modifying its sign randomly. This might imply a "relay-race" involvement of various neuronal circuits.

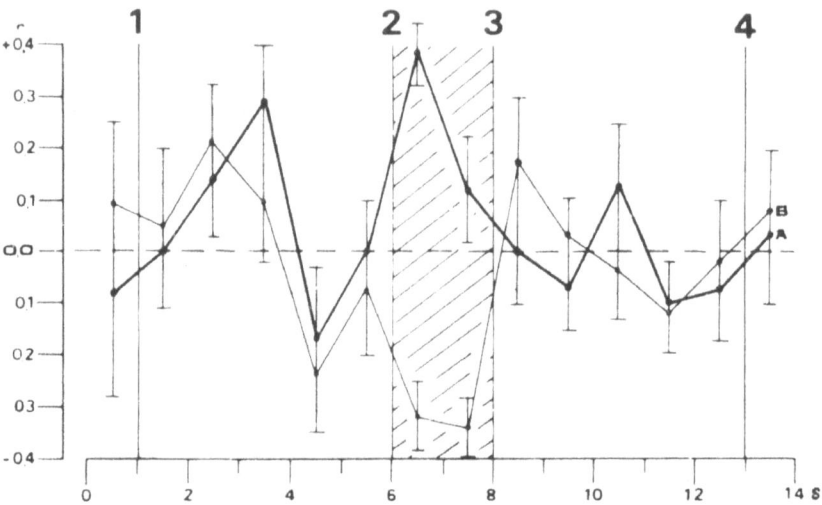

Fig. 15. Changes in the interneuronal correlations during the program. The diagram is plotted in relation to the correlation type during the CS presentation. I: anticipatory; 2–3, conditioned; 4. trigger signals. A, an averaged curve showing changes in the interneuronal correlation observed in units with a positive sign of the correlation coefficient during the CS presentation (14 pairs of units); B, the same for the units with a negative sign of the correlation coefficient (10 pairs of units). Abscissa: time in s; Ordinate: the correlation coefficient values (r) for 1 s intervals. Zero, is marked by a broken line.

Wave-like fluctuations of unit activity within the unspecific expectance and delay might be associated with intracortical reverberation (15, 22), viewed as the basic mechanism for relevant information storage (in our case, CS location) in short-term memory. Similar effects can be found in average firing rate differences an in discharge patterns for the frontal lobe units within the delay. The distinctions are associated with the location of the units sampled (left or right hemisphere). It was shown that spatially selective units responding to CS retained their characteristic discharge patterns during the delay. They might be labeled as "input units" which initiate the "relay-race" phenomenon.

The results of spatial selectivity of neuronal activation can be traced in the activity patterns after the trigger signal. Thus the differences in s. principalis unit activity determined by the CS location and, consequently, by the direction of the oncoming movements are continuously maintained in neuronal circuits up to the moment of realization of the motor act. But at that moment the given differences are absent or less clear-cut.

When the key choice was erroneous, the firing rate of the frontal lobe units following it did not change significantly for lack of food reward. This was observed in the group of units which did not respond to any event except the food reward. Special analysis showed that there exists no direct connection between this and adequate alternation performance. This connection, however can be mediated by the level of food motivation. Recent findings (10) established effects of stimulation of g. cinguli on the unit activity in the dorsolateral part of the monkey prefrontal cortex. The current viewpoint is that the prefrontal cortex, as a mediodorsal thalamic nuclei projection zone (1, 9, 20, 25) is under control from the higher limbic system (26, 29, 30). Moreover, some authors believe the frontal lobes to be the neocortical projection of the limbic system (30). Besides, the region of s. principalis receives inputs from practically all sensory systems (6) via the direct pathways from corresponding sensory cortical areas (25, 26).

As already shown, all modifications in the spatiotemporal structure of the surroundings and in the level of biological motivation affect the rearrangements of the unit activity in s. principalis. We suppose that the frontal lobes possess the necessary structural and functional properties for storing such changes in short-term memory. The latter might be achieved by means of the intracortical reverberation mechanisms, which can explain the participation of the frontal lobes in the organization of cerebral processes. They also ensure urgent rearrangements in the behavioral program according to the varying significance of external factors and dominant motivation.

### REFERENCES

1. AKERT, K. 1964. Comparative anatomy of the frontal cortex and thalamocortical connections. *In* I. M. Warren and K. Akert (ed), The frontal granular cortex and hehavior. McGraw-Hill, New York, p. 372-397.
2. ALBE-FESSARD, D. and FESSARD, A. 1954. Thalamic integration and their consequence at the telencephalic level. *In* Brain mechanisms and consciousness. Blackwell Sci. Publ., Oxford, p. 115–149.
3. BAURE, R. H. and FUSTER, M. 1976. Delayed-matching and delayed-response deficit from cooling dorsolateral prefrontal cortex in monkeys. J. Comp. Physiol. Psychol. 90: 293–302.
4. BATUEV, A. S. 1969. The frontal lobes and the processes of synthesis in brain. Brain Behav. Evol. 2: 202–218.

5. BATUEV, A. S., MALYUKOVA, I. V. and HOHRYAKOVA, I. M. 1974. Structure and functional basis for frontal lobe participation in the organization of complex behavior in cats. Brain Behav. Evol. 10: 290-306.
6. BATUEV, A. S., PIROGOV, A. A. and ORLOV, A. A. 1977. Characteristics of neuronal activity in the monkey prefrontal cortex (in Russian). Neurophysiologiya 9: 437-439.
7. BATUEV, A. S., PIROGOV, A. A. and ORLOV, A. A. 1978. Participation of frontal lobes in integrative activity of brain of *Macaca mulatta* (in Russian). J. Evol. Biochem. Physiol. 14: 144-150.
8. DEMIANENKO, G. P. 1977. Axonal system of the monkey frontal lobe neurons (in Russian). Reports fo the USSR Acad. Sci. 234: 191-194.
9. DESIRAJU, T. 1973. Electrophysiology of frontal granular cortex. I. Patterns of focal field potentials evoked by simultaneous stimulations of dorsomedial thalamus in conscious monkey. Brain Res. 58: 401-414.
10. DESIRAJU, T. 1976. Electrophysiology of the frontal granular cortex. III. The cingulate prefrontal relation in primate. Brain Res. 109: 473-485.
11. EVARTS, E. V. 1966. Methods for recording activity of individual neurons in moving animals. In R. F. Rushmer (ed.). Methods in medical research. Year Book Chicago, 2: 241-250.
12. FUSTER, Y. M. 1973. Unit activity in prefrontal cortex during delayed-response performance: neuronal correlates of transinet memory. J. Neurophysiol. 56: 61-78.
13. FUSTER, Y. M., and BAUER, R. H. 1974. Visual short-term memory deficit from hypothermia of frontal cortex. Brain Res. 81: 393-400.
14. KONORSKI, J. 1967. Integrative activity of the brain. An interdisciplinary approach. Univ. Chicago Press, Chicago, 531 p.
15. KONORSKI, J. 1972. Memory problem in physiological aspect (in Russian). In J. Beritashvili, (ed.). Gagrskie besedy. Vol. 6. Izdat. Mesnierba, Tbilisi, p. 37-56.
16. KCNORSKI, J. 1972. Some hypotheses concerning the functional organization of prefrontal cortex. Acta Neurobiol. Exp. 32: 595-613.
17. KUBOTA, K. 1975. Prefrontal unit activity during delayed-response and delayed-alternation performance. Jpn. J. Physiol. 25: 481-493.
18. KUBOTA, K. and KOJIMA, S. 1975. Prefrontal unit activity and delayed-response in over-trained and under-trained monkeys. J. Postgrad. Med. 21: 2-11.
19. KUBOTA, K. and NIKI, H. 1971. Prefrontal cortical unit activity and delayed alternation performance in monkeys. J. Neurophysiol. 34: 337-347.
20. KUBOTA, K., NIKI, H. and GOTO, A. 1972. Thalamic unit activity and delayed alternation performance in the monkey. Acta Neurobiol. Exp. 32: 177-192.
21. KUBOTA, K., TWANOTO, P. and SUZUKI, H. 1974. Visuokinetic activities of primate prefrontal neurons during delayed response performance. J. Neurophysiol. 34: 1197-1212.
22. LORENTE de NO, R. 1943. Cerebral cortex: architecture, intracortical connections, motor projections. In I. T. Fulton (ed.), Physiology of the nervous system. Oxford Univ. Press, Oxford, p. 277-301.
23. LURIA, A. R. 1962. Vysshie korkovie funktsii cheloveka. Moscow Univ. Press, Moscow.
24. NAUTA, W. J. H. 1964. Some efferent connections of the prefrontal cortex in

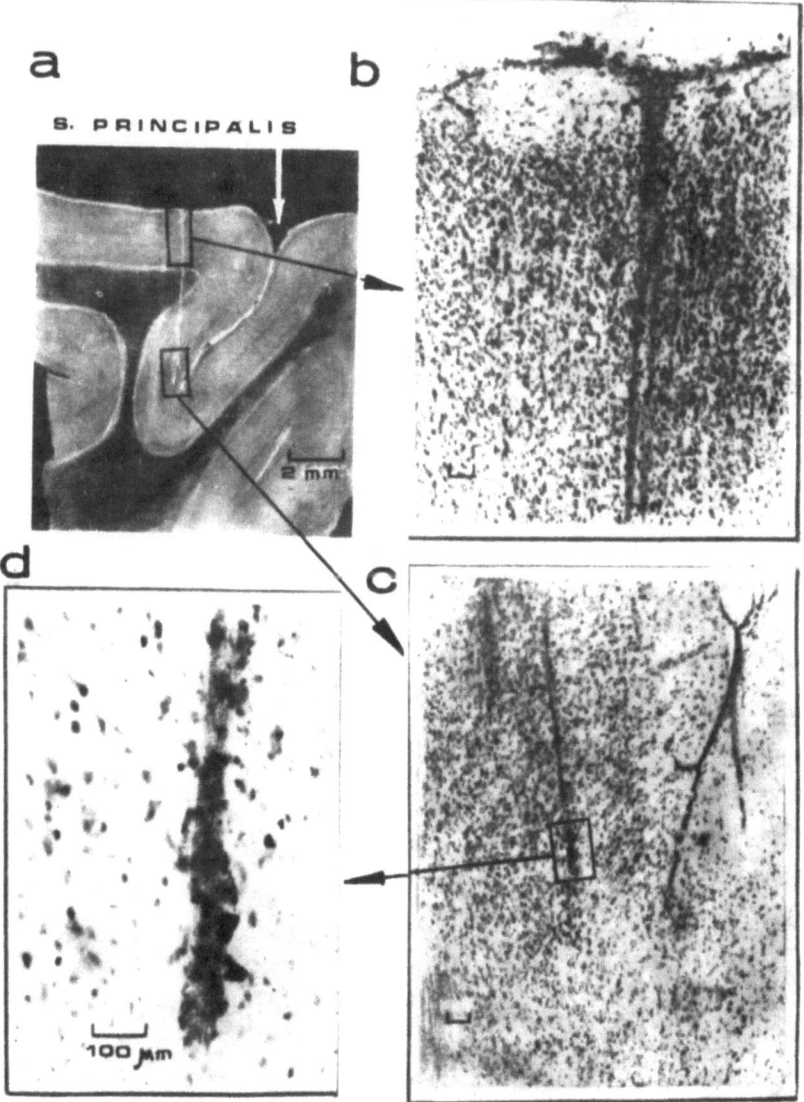

Fig. 3. An example of histological localization of the microelectrode bundle penetration, a, brain section in the area of the microelectrode bundle penetration; b, c, magnification of fragments, merked by rectangles in a; d, magnification of the fragment marked by the rectangle in c. The microelectrode bundle track is visible in a and b. The track of a particular microelectrode is shown in c and d. Cresyl-violet staining.

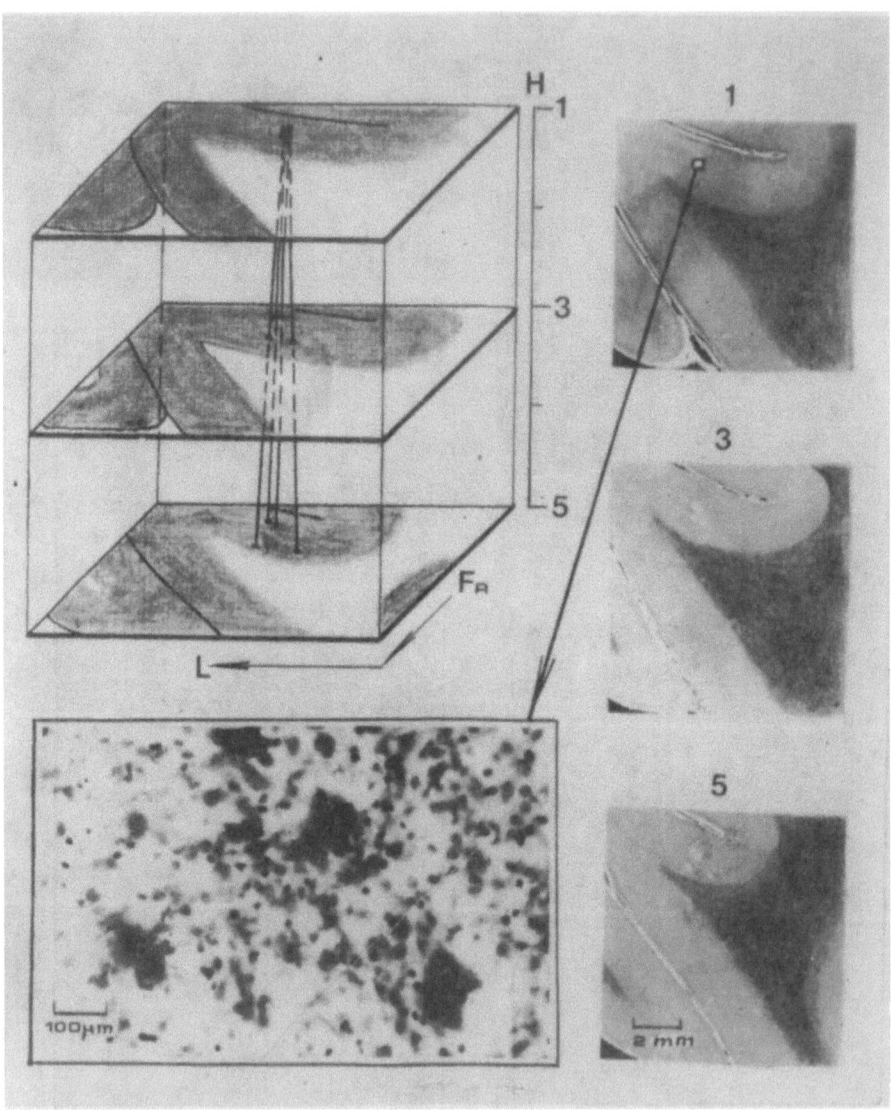

Fig. 4. Reconstruction of the microelectrode bundle divergence in the brain tissues. 1, 3, 5, three sections of the neocortex at 1, 3 and 5 mm depth from the surface, respectively. Left bottom: magnified area of section 1; traces of four microelectrode tips are seen (cresyl-violet staining). Left top: reconstruction of the microelectrode bundle divergence in three horizontal sections, whose locations are marked on the ordinate H. L, lateral direction; Fr, frontal.

the monkeys. In M. Warren and K. Akert (ed.). The frontal granular cortex and behavior. McGraw-Hill, New York, p. 397–410.
25. NAUTA, W. J. H. 1972. Neuronal associations of the frontal cortex. Acta Neurobiol. Exp. 32: 125–141.
26. NAUTA, W. J. H. 1975. Convergent sensory processing in the mammalian forebrain. In J. M. Finizi (ed.), Society for neuroscience. 4th Annual Meeting BIS. Univ. California, Los Angeles, p. 1-4.
27. NIKI, H. 1974. Prefrontal unit activity during delayed alternation in monkey. II. Relation to absolute versus relative direction of response. Brain Res. 68: 197–204.
28. PIROGOV, A. A., and ORLOV, A. A. 1977. New technique for multineuronal activity recording in conscious monkeys (in Russian). Fiziol. Zh. SSSR im. I. M. Sechenova 63: 600–602.
29. PRIBRAM, K. 1972. The primate frontal cortex — executive of the brain. In A. R. Luria and K. Pribram (ed.). The behavioral physiology of the frontal lobes. Academic Press, New York, p. 112-131.
30. PRIBRAM, K. 1975. The primate frontal cortex: progress report 1975. Acta Neurobiol. Exp. 35: 609–625.
31. SAKAI, M. 1974. Prefrontal unit activity during visually guided lever pressing reaction in the monkey. Brain Res. 81: 297–309.
32. SUZUKI, H., and ASUMA, M. 1976. Prefrontal neuronal activity during gazing at light spot in the monkey. Brain Res. 126: 497–508.
33. WOODY, C. D., VASSILEVSKY, N. N. and ENGEL J. 1070. Conditioned eye blink: unit activity at coronal-pericruciate cortex of the cat. J. Neurophysiol. 33: 851-864.

A. S. BATUEV, A. A. PIROGOV, A. A. ORLOV and V. I. SHEAFER, Department of Higher Nervous Activity, Leningrad State University, 7/9 University Embankment, Leningrad 199164, USSR.

Lecture delivered at the Warsaw Colloquium on Instrumental
Conditioning and Brain Research
May 1979

# SPECIFIC AND UNSPECIFIC NEURONAL MECHANISMS OF CELLULAR OPERANT CONDITIONING

Urfan GASSANOV and Alexandra GALASHINA

Department of Neuronal and Synaptic Mechanisms of Conditioned Reflexes,
Institute of Higher Nervous Activity and Neurophysiology, USSR
Academy of Sciences
Moscow, USSR

*Abstract.* An analysis was carried out on spontaneous multiple-unit activity recorded through one chronically implanted electrode in the auditory cortex of awake cats. Statistical relations between the neurons were evaluated by crossinterval histograms of two cells. After determination of the pattern of background interaction between three neighboring units, a cat was stimulated with clicks or electrical shocks triggered by spontaneous spikes of the selected neuron, which is considered a variant of cellular operant conditioning. Shock was applied to muscles of the eyelid or ear. Rhythmical and random stimulations with cliks were used in control experiment. A neuronal microsystem is supposed to be an elementary cell network, which allows detection of new qualitative properties of cortical activity. At the level of such a micronet it is possible to describe the specificity of systemic neuronal activity in adaptive reactions. The analysis of crossinterval histograms reveals diverse participation in network function of neurons, differing in amplitudes of spikes. Two neuronal mechanisms of learning are suggested.

## INTRODUCTION

The generally adopted theory of learning rests upon the hypothesis of specific integrative or systematic neuronal activity. Some ideas about its mechanisms have been gained with the help of microelectrode investigation of a single neuron. Observations of single cell activity support the hypothesis that basic mechanisms of learning lie in functional cell interactions. From the Pavlovian concept (8) of dynamic functional connections between cortical cells, Asratyan (1) deduced the idea of local conditioned reflexes, or local conditioned states elaborated in cortical projections of meaningful stimuli. Konorski (7) assigned to interneuronal connections an essential role in the transfer of gnostic neurons from an inactive to an active state.

The available neurophysiological data, however, are inadequate for the assessment of functional interconnections of cortical cells because they concern the activity of a single units. At present a number of investigators, principally Dickson (3), Gerstein (5), Bryant (2) and others, believe that the most productive approach to examining neuronal networks is a statistical analysis of mutual dependencies in neuronal pairs. Cross-correlations between extracellularly recorded firing patterns of two neurons are used as a statistical technique for detecting connectivity between them. In our experiment we have employed the statistical method for studying systemic neuronal activity in the auditory cortex during cellular operant conditioning caused by feedback stimulation.

## METHODS

*Analytical.* Our statistical analyses were carried out on records of spontaneous multiunit activity in the auditory cortex. These records were obtained from freely moving awake cats by using chronically implanted extracellular wire electrodes. During the recording session the cat sat in a box and could hear all the common laboratory noises. To record the multiple unit activity a plug with a miniature cathode follower was inserted into the head-socket and a lead, connected at one end to the plug, was fed both onto an oscilloscope and to a tape recorder. Subsequent analysis consisted of separating three spike trains from the multiple unit activity with specially constructed window gates. In accordance with the amplitude of selected spikes, the trains were regarded as impulses of neuron 1 (large spikes), neuron 2 (middle-size spikes) and neuron 3 (small spikes). The selected spike amplitudes were in a ration of approximate to $4:2:1$ (Fig. 1). Interval histograms (IH) and crossinterval histograms (CIH) were carried out by an electronic counter. Proceeding from IH data, a correlation epoch was established equal to 20 ms (10 bins of 2 ms each). This permitted driving the beam of the counter by each spike of the triggering train. Absence of any prominent bin counts (confidence level $\pm 3\Omega$) was interpreted as independance of the two neurons. CIH between two trains of action potentials was obtained for 30 s recording by triggering the count with the action potentials of one train and averaging the other one with respect to time. A backward CIH was performed by reversing the process and triggering with the action potential of the second train.

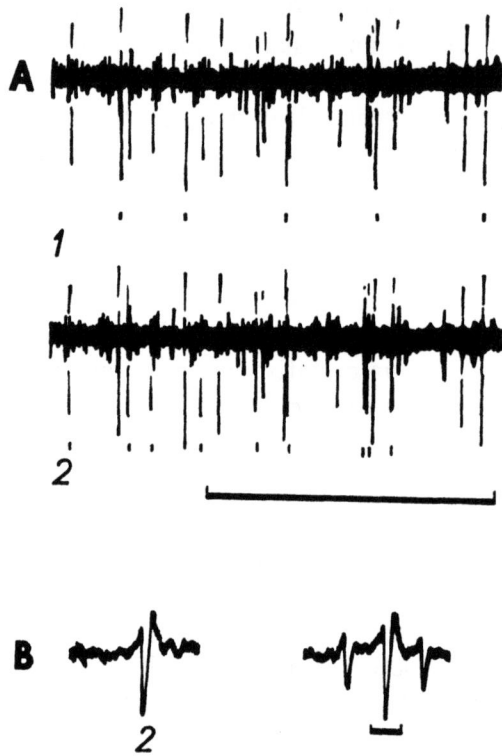

Fig. 1. Multi-unit activity recorded from a chronically implanted electrode in the auditory cortex of an awake cat. A, selection of neuron 1 with large amplitude of spikes (1) and neuron 2 with middle amplitude of spikes (2) by means of window discriminators. Time, 500 ms; B, the action potential of the neuron 2 at the beginning and the end of the experimental session. In the second oscillogram the neuron 3 can be seen. Time, 1 ms.

It was possible to identify three types of interneuronal dependence: inhibitory, excitatory and mixed (inhibitory excitatory or vice versa). The type of CIH was usually interpreted as being a consequence of direct interaction, or of shared input (2, 3, 5). Accumulated experience allows interpretation of our CIH as representing interneuronal connections because of the following characteristics: (i) the short temporal period of the statistical dependence and (ii) the unequal (asymmetrical) patterns of relationships of the two spike trains in forward and backward interactions (3). All our dependent relations between two neurons correspond to these indices.

*Experimental.* Two operant conditioning and three control experimental series were used to asses changes in cortical neuronal interactions. Operant conditioning consisted of shock or sound stimulation triggered by spikes of the neuron 1. The transformed spike train was led to a generator of rectangular impulses which were applied in one series of the experiment to eyelid or ear muscles, and in the other series fed a laudspeaker to generate cliks. The automatized stimulation lasted 3–10 min. For shock, the stimulus intensity was defined by slight movement of eyelid or ear, and for cliks the intensity was about 50 dB above the threshold of human hearing. This variant of feedback stimulation (spike dependent stimulation) was chosen to avoid the possible tendency for adjacent neurons to discharge in synchrony (4).

The control series were performed only with cliks. In one series the sound was presented at random times. In the second control series, rhythmical sound stimulation was given for 10 min at the rate of 10/s that approximately corresponded to the mean frequency of spontaneous unit activity. The last control series completely reproduced the operant conditions but was carried out on the cells of the inferior colliculi.

For simplicity of data description the following terminology was used: automatized stimulation by shock (AS1), automatized stimulation by click (AS2), control automatized stimulation by click (AS3), rhythmical click stimulation (RS), control automatized stimulation by click triggered by large spike of the collicular neurons (ASC).

Of the 312 CIHs obtained, 72 belonged to AS1, 84 to AS2, 72 to AS3, 36 do RS and 48 to ASC. The mean frequency of each neuron was defined by the number of spikes for 2 min. IH was constructed of 500 spikes and the width of a bin was 8 ms. Each CIH had two definitions: the type of dependence (inhibitory, excitatory etc.) and the direction of the connection (from one neuron to the other). The designation "connection 1–2, 3–2" means the connection from the neuron 1 to the neuron 2, and so forth.

RESULTS

In most cases the stimuli used had no clear effects on unit activity. Some neurons, however, responded to clicks in phasic form. Figure 2 presents the results of AS2 and RS obtained on such neurons. Irregular stimulation coinciding with spikes did not change the patterns of the responses, while rhythmical one depressed the responses. This could be seen in responses both to a click and to a short noise.

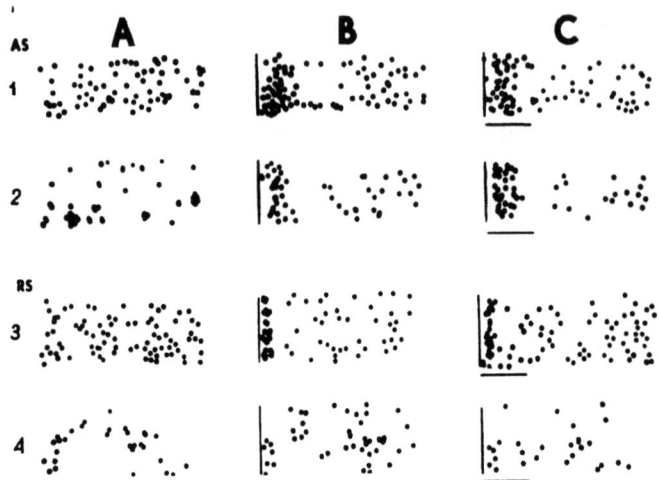

Fig. 2. Spontaneous and evoked activity of an auditory cell. AS, before (the 1st row) and after (the 2nd row) automatized stimulation with click; RS, before (the 3rd row) and after (the 4th row) rhythmical stimulation with click. A, spontaneous activity; B, responses to click; C, responses to noise. In the rasters each dot represents the discharge of a cell. Each raster represents twenty succesive trials. Time, 160 ms.

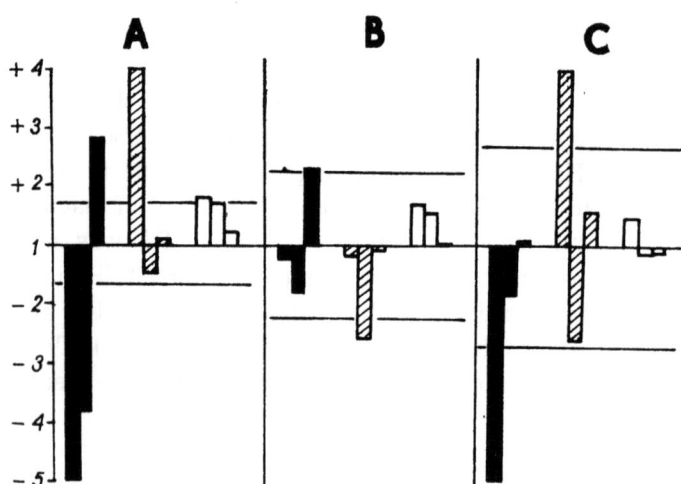

Fig. 3. Spontaneous activity of three cortical neighboring neurons in different experimental situations. Two minute periods of spontaneous activity were evaluated. The vertical line indicates the relations to mean background frequency (1). The horizontal lines above and below line 1 indicate the ranges of background activity. Each bar represents the ratio of spike frequency after shock (dark bars) or click (striped bars) automatized and rhythmical (open bars) stimulation to mean background frequency. Twenty seven neurons are divided into three groups: neuron 1 (A), neuron 2 (B) and neuron 3 (C).

487

Quantitative analysis of the spontaneous activity of 27 neurons revealed the effectivness of AS1 more that AS2 (Fig. 3). After AS1 (shock stimulation) the frequency of triggering neuron 1 firing changed

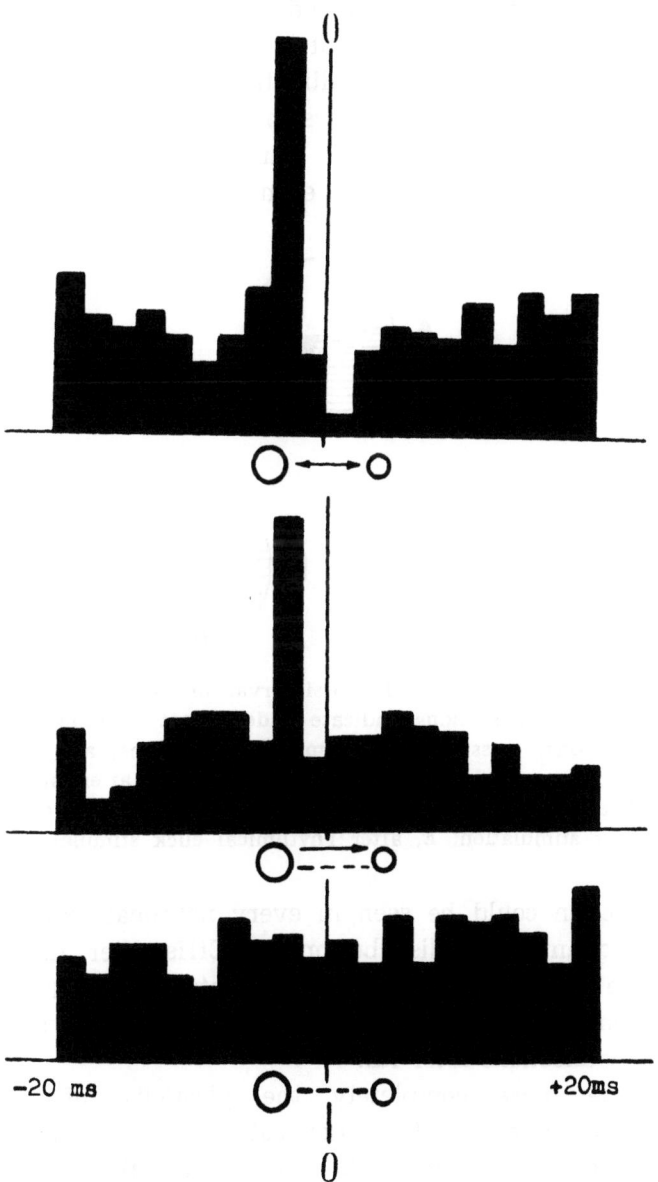

Fig. 4. Examples of mutual statistical relationships between two neighboring cortical cells. The upper crossinterval histogram shows mutual (excitatory in one direction and inhibitory in the other) dependence, in the middle histogram — one way excitatory dependence is shown and the lower histogram shows complete independence.

more dramatically than frequencies of the other two units. RS was ineffective for all of the three selected neurons.

We were particularly interested in interneuronal relations. All the analyzed CIHs were divided into four groups according to statistical signs: excitatory, inhibitory, mixed and independent. Examples of CIHs are given on Fig. 4 in the form of a combined presentation of two separately obtained CIHs in both directions. We made use of several quantitative methods for the expression of cell network plasticity. It was impossible to determine definite transformations of neuronal interaction after stimulations because in each case different changes of the

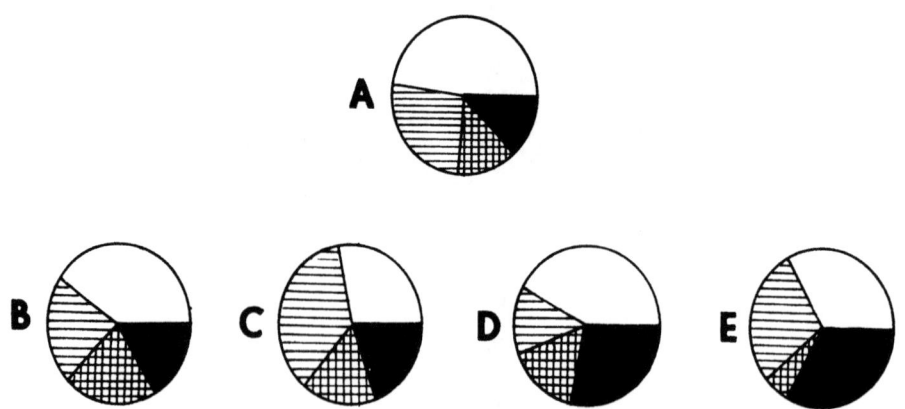

Fig. 5. Quantitative distributions of crossinterval histogram in different experimental situations. Open sections indicate independent relations; dark sections, excitatory connections; crossed sections, mixed connections; and striped sections, inhibitory connections. A, background relationships; B, after automatized click stimulation; C, after automatized shock stimulation; D, after control irregular click stimulation; E, after rhythmical click stimulation.

type and direction could be seen in every neuronal pair. Figure 5 illustrates the quantitative distributions of CIHs after AS1, AS2, AS3 and RS. There are no obvious differences between the diagrams. Some slight enhancement of inhibitory connections can be seen after AS1 and of excitatory connections after RS.

To examine more completely the plasticity of interneuronal connections, we estimated the changeability or sensitivity of each interneuronal connection within the three neuronal microsystems. The changeability meant any kind of transformation of a given connection (1-2, 2-1, 1-3, 3-1, 2-3, 3-2); i.e., the transition from one type of connection to the other, appearance or disappearance of the connections. Analysis of the plasticity of connections by the criterion of sensitivity needs a definition of the probability of sensitivity because in the awake state, the CIHs often spontaneously showed some modification in the course of 10–20 min time. We found that in intact animals, connections could change with a probability of 0.18 to 0.54. Therefore a criterion for definition of plasticity was taken as high as 0.7–1.0. The results of such an analysis are presented in Fig. 6 where only meaningful (in the sense of changeability) connections are indicated.

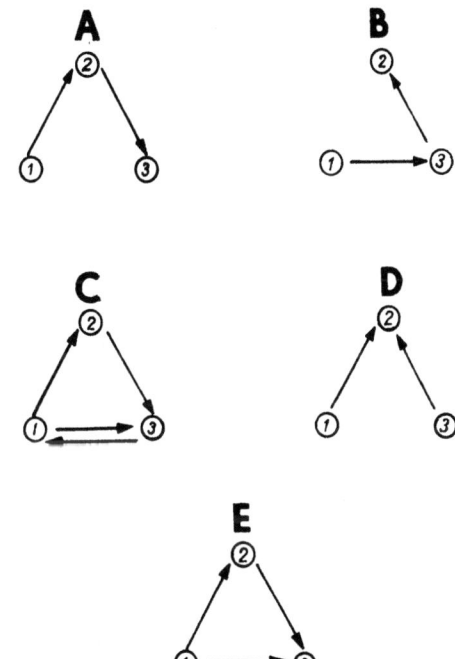

Fig. 6. Selective plasticity of interneuronal connections in three neuronal microsystems after different stimulations by click (AS2, AS3, RS, ASC) and shock (AS1). Four microsystems are cortical (AS1, AS2, AS3, RS) and one is collicular (ACS). The figures in a circle indicate a neuron according to its spike amplitude (1, high; 2, middle; 3, low). A, B, C, D, E — AS2, AS1, AS3, RS, ACS, respectively.

The schematic representation of microsystem plasticity are of considerable interest for describing some dynamic properties of the cortical cell networks. There is a high selectivity of the affected connections which is closely related to the experimental situation. Both click and shock AS triggered by spikes of neuron 1 typically led to changes in the connections placed in succession and beginning from neuron 1. At the same time these connections had different locations and directions. For AS1 they were connections 1–2–3, for AS2 — 1–3–2. With RS there were two changed connections as well, but they were 1–2 and 3–2. None of the other control stimulations had any similarity to AS1 and AS2. It should be noticed that ASC identical to AS2 resulted in completely different patterns of interneuronal relationships in the collicular microsystem, than those in the cortical one. Another control session, AS3, gave the same result as ASC.

The next point in connection with Fig. 6 concerns the level of systemic organization, at which the peculiarities of cell network plasticity become distinctive. Thus, considering neuronal pairs separately it is easy to see their similarity in different experimental situations. For instance, the plasticity of the connection 1-2 takes place both in AS1 and in RS. It is the three neuronal system level that reveals the specific property of systemic neuronal activity.

It should be noticed that the specificity with AS1, AS2 and RS was not only a quantitative but a qualitative manifestation. In all these experimental situations only two connections were modified, but in each case the modification had a different pattern. As for the AS3 it is possible that an "irrelevant" stimulation, having no internal or external regularity, tended to disorder the neuronal interactions. The same effect may be inherent for ASC.

The analysis of connections in the three neuronal microsystem suggests different participation in network activity of neurons distinguishing by amplitudes of spikes. It seems reasonable to consider the amplitude of spikes recorded by one electrode as a criterion of the size of a neuron with definite functional properties (6). In any case our experiment showed that the behavior of a neuron in the network is to a great extent defined by spike amplitude.

## DISCUSSION

All neuronal-synaptic mechanisms of adaptive reactions may be reduced to two clases: the cellular class, based on experimental investigation of single unit activity, and the class of systemic neuronal organization based on theoretical or model considerations. It is known that the integration of neurons in a system is determined by synaptic and membrane mechanisms of single cells. At the same time, the united function of several cells due to their subordination to a system of connections, is qualitatively different from the single cell's activity. Suffice to say that the uniform activity of neighboring cortical neurons may proceed with various functional interrelations between them. The method of cellular learning via feedback (the method of automatized stimulation) is an experimental approach for the study of the neurons that takes direct part in the organization of elaborated adaptive reactions. The analysis of three simultaneously functioning cortical neurons shows that the specificity of processes involved in adaptive reactions consists of interneuronal relations. It is noteworthy that the system of connections between neurons is formed depending on the modality of the stimulus and the specificity of the stimulation does not manifest itself in activity of single neurons. In any case, this was found with the microsystem of three cortical neurons under simple forms of learning caused by automatized stimulation with click or shock. Neither evoked activity nor spontaneous discharge frequency of single neurons had correlations with changing of CIH. The plasticity of single neurons had correlations with changing of CIH. The plasticity of connections in neuronal pair can be seen in different experimental situations, and at the same time frequencies of the neuron may be changed in one situation and remain unchanged in the other. From this point of views, it is important to recognize two categories of indices of neuronal plasticity: the unspecific cellular, that which accompanies any adaptive reflexes, and the systemic neuronal activity, that which is specific for the elaborated reaction. In the first case plastic changes are manifested in the activity of single cells or synapses. In the second case, they are observed in terms of neuronal interactions. It is exciting to suggest that a specifically organized neuronal microsystem reflects a memory process evaluating quality and significance of the situation. Each experimental situation exerted selective influences on interneuronal connections. Uncertain situations with irregular stimulation brought about a diffuse reconstruction of the neuronal microsystems as compared with regular stimulation.

All neurons differing in amplitude of spikes equally participated in the systemic activity, but they interacted through different inputs and in different directions. Observations of the behavior of neurons with high, middle or small spikes suggest a functional meaning of spike amplitude. Very likely, multiunit recording gives us an additional possibility of morphological and functional differentiation of cells.

We thank Dr. N. Alexeenko for her valuable advice and help during the preparation of this paper.

## REFERENCES

1. ASRATYAN, E. A. 1965. Conditioned reflex and modern neurophysiology (in Russian). Zh. Vyssh. Nervn. Deyat. im. I. P. Pavlova, 15: 202–216.
2. BRYLANT, L. H., MARCOS, R. A. and SEQUNDO, J. P. 1973. Correlation of neuronal spike discharges produced by monosynaptic connections and by common inputs. J. Neurophysiol. 36: 205–225.
3. DICKSON, J. N. and GERSTEIN, G. L. 1974. Neuronal interactions in auditory cortex. J. Neurophysiol. 37: 1239–1261.
4. FETZ, E. E. and BAKER, M. A. 1973. Operantly conditioned patterns of precentral unit activity and correlated responses in adjacent cells and contralateral muscles. J. Neurophysiol. 36: 179–204.
5. GERSTEIN, G. L. 1970. Functional association of neurons: Detection and inter-

pretation. *In* F. O. Schmitt (ed.), the neurosciences: Second study Program. Rockefeller Univ. Press, New York, p. 648–661.
6. HENNEMAN, E., SOMJEN, G. and CARPENTER, D. O. 1965. Functional significance of cell size in spinal motoneurons. J. Neurophysiol. 28: 560–580.
7. KONORSKI, J. 1967. Integrative activity of the brain. An interdisciplinary approach. Univ. Chicago Press, Chicago, 531 p.
8. PAVLOV, I. P. 1951. Polnoe sobranie sochinenii. USSR Academy of Sciences, Vol. 3, 2: 179, 308.

Urfan GASSANOV and Alexandra GALASHINA, Institute of Higher Nervous Activity and Neurophysiology, Butlerova 5a, Moscow 117485, USSR.

Lecture delivered at the Warsaw Colloquium on Instrumental
Conditioning and Brain Research
May 1979

# TWO BEHAVIORAL PARADIGMS FOR STUDY OF RAPID CHANGES IN FUNCTIONAL GROUPING OF NEURONS

George L. GERSTEIN, Dwight H. RENFREW, and Benjamin H. PUBOLS, Jr.

Department of Physiology, University of Pennsylvania
Philadelphia, Pennsylvania, USA

*Abstract.* Recent progress in extracellular recording technology and in analytic methodology for the resulting spike trains are making practical the simultaneous registration of twenty or more neurons. This begins to make possible the direct experimental observation of functional neuronal assemblies, particularly as they dynamically change in membership or properties during a behavioral task. Behavioral paradigms appropriate for such experiments have very stringent requirements in order, with high likelihood, to produce changes in neuronal assembly structure during the time that stable recording can be maintained. We describe two motor system paradigms that seem to be appropriate, the first with crayfish claw, the second with monkey paw. The crayfish task involves a rapid learning of claw position. The monkey task involves a preset state which determines the responses to a subsequent somatosensory discrimination.

## INTRODUCTION

Electrical activity of single nerve cells has been accessible to experimental investigation for approximately the past 25 years. At first such measurements were obtained from various sensory systems on deeply anesthetized preparations, but more recently neurons from many different portions of unanesthetized and even behaving animals have come under observation. Many interesting relationships between single neuron firing pattern and either stimulus or response have been described. At the same time it has become clear that changes in the firing of a single neuron usually represent only a small portion of the sensory information being processed or the behavior being produced by the nervous system.

Most theories of neural coding and function in central nervous systems envisage assemblies of neurons which are active in some coordinated way. A given neuron, it is thought, may participate in different assemblies at different times, thus giving a dynamic characteristic to the functional grouping (3, 7, 10). The neurophysiologist is also accustomed (at least implicity) to think in terms of neuronal assemblies or functional groupings, viz. Sherrington's motor pools, or cortical columns (12, 13, 17). Unfortunately such assemblies are defined only in terms of neurons that have some shared stimulus preference, rather than in terms of a shared information task.

Direct experimental observation of neural assemblies and their dynamics is possible only if many neurons are simultaneously (and separately) recorded. Sequential recording is useless for determination of connectivity and information flow. A qualitative glimpse of ensemble phenomena is seen in the pioneering experiments of Verzeano (20, 28); the formidable difficulties of multiple neuron experiments have kept subsequent, more quantitative progress slow, and usually limited to simultaneous observation of pairs or triples of neurons (4, 8, 14, 15, 23).

The difficulties associated with multiple neuron experiments and direct observation of neural assemblies are in three separate areas: (i) the recording technology, (ii) the data analysis, and (iii) the choice of an appropriate behaving animal preparation. Progress has recently been made in recording methods either by multiple fine wire electrode bundles (8, 19) or by optical methods in which a special dye produces either fluorescence change or absorption change during action potential (21). Thus it is becoming possible to put some 20 neurons under simultaneous observation. The resulting massive data flow would be prohibitive if it were necessary to analyze all possible neuron pairs with cross correlation of spike trains. Fortunately, a new way of analyzing data from many spike trains has recently been described (9). This method allows rapid and easy identification of functional groups of neurons among the recorded set, and can be used as a screening technique to select smaller groups of neurons for more detailed analysis. Alternatively, because of its great sensitivity, the method may be used to describe dynamically the structure and hierarchy within neuronal assemblies even when they are not distinguishable by previous statistical procedures.

For maximum interest we should choose a behavioral task that is likely to involve rapid change of functional neuronal groupings within the time scale of a multiple neuron recording session. Thus either we require a very fast learning task, or alternatively some sort of chained stimulus so that the significance of an particular stimulus (and the subsequent behavioral response) can rapidly be altered.

In this paper we describe two behavioral preparations that meet many of the needs for experimental studies of neuronal groupings and their dynamics.

## CRAYFISH CLAW

*Methods and results.* Under appropriate conditions, crayfish (Procambarus clarkii) can rapidly learn to maintain a particular range of claw positions. This material has largely been described (22) and will only be reviewed briefly here. The claw movement system is relatively simple, consisting of an opener and a closer muscle. The opener is supplied by a single excitatory and a single inhibitory neuron. The closer is supplied by a fast exciter, a slow exciter and an inhibitor neuron. Thus only five motor neurons are involved in a system which also includes many skin receptors, a stretch receptor in the joint (PD organ), and an unknown number of interneurons.

The experimental paradigm is related to the Horridge leg raising procedure first carried out with cockroach (11). A train of identical electric shocks is applied to *both* claws of the crayfish, conditional on the position of only *one* claw (positional claw). The other claw, which receives shocks independent of its position is used as control. Initial experiments required that the positional claw be more *closed* than some criterion value to avoid shock. Most crayfish were able to acquire the desired position criterion within 20 min, and kept it up to 2 h without further shock. The control claw continued to move normally and randomly throughout the experiment. Subsequent experiments moved the location of the shock to the animal's tail, although it still was the positional claw that determined when there would be shock. This is a much better paradigm since the site of electric shock is far removed from any of the synapses and neurons directly involved in the claw control system. Again, most crayfish can rapidly discover which claw is positional, and attain the desired criterion. More recent work by Stern-Tomlinson and Natale (*personal communication*) has extended the tail shock paradigm, and has also demonstrated that the crayfish can learn to keep the positional claw more open than some criterion value. Forman and Hoyle (6) have reported that locust leg could be trained into a "window" criterion, i.e. A<position<B by a somewhat similar leg shock paradigm. Attempts to achieve this with crayfish using the tail shock would seem appropriate.

*Discussion.* A large amount of data about the crayfish claw and other muscle control systems has accumulated over the last 40 years, starting perhaps with work of Van Harreveld and Wiersma (27). The field is reviewed by Wilson and Davis (30) and more recently by Atwood (1).

It is relatively easy to record from several neurons simultaneously in the crustacean nervous system. Such experiments together with cross correlation analysis of the spike trains have been used to define further the connectivity of the efferent and afferent portions of the claw control system (16, 29). It would seem extremely promising to extend such measurements of neuron activity with the maximum possible multiplicity of simultaneous registration to the behavioral paradigms we have described above. The entire time course of "learning", "retention" and "forgetting" fall within the possible length of stable recording, so that dynamic changes of functional connectivity or functional grouping could be assessed. It should be possible to identify as yet unknown interneurons that are involved in these changes. The behavioral paradigm is relatively easy to set up, and is largely automatic, so that most attention can be concentrated on maintaining adequate recording quality.

## MONKEY PAW

*Methods and results.* A monkey (Maccaca speciosa) sits in a behavioral chair with the right arm loosely restrained by rings above and below the elbow and more tightly restrained by a ring at the wrist. The animal holds a vertical lever which moves freely transversely to the arm; the monkey must flex or extend at the wrist to move the lever. Attached to the lever are a torque motor for mechanical stimulus, and instrumentation for measurements of lever position and acceleration. A small visual stimulus projector with red or green light is placed approximately 30 cm in front of the monkey's head; a feeding tube for liquid reinforcement is placed convenient to the monkey's mouth. The monkey is run in a liquid deprived state, so that most of his daily requirements must be earned in the apparatus. The final, complex paradigm is shown in Fig. 1. Various early and intermediate shaping procedures are of course necessary to enable the animal to solve this task. After a minimum of 5 s intertrial interval, a trial begins when the monkey brings the lever to a small center zone. Either the red or the green light goes on. A random time later, in the range 2–6 s, the torque motor applies force to the lever, either to the left or to the right, corresponding respectively to flexion or extension at the monkeys (right) wrist. If the light was red (upper portion of Fig. 1) then a motor driven flexion of the wrist is to be answered with an extension movement, while a motor driven extention is to be answered by holding the lever in the center zone. If the light was green (lower portion of Fig. 1) the required responses to the same two motor driven stimuli are reversed.

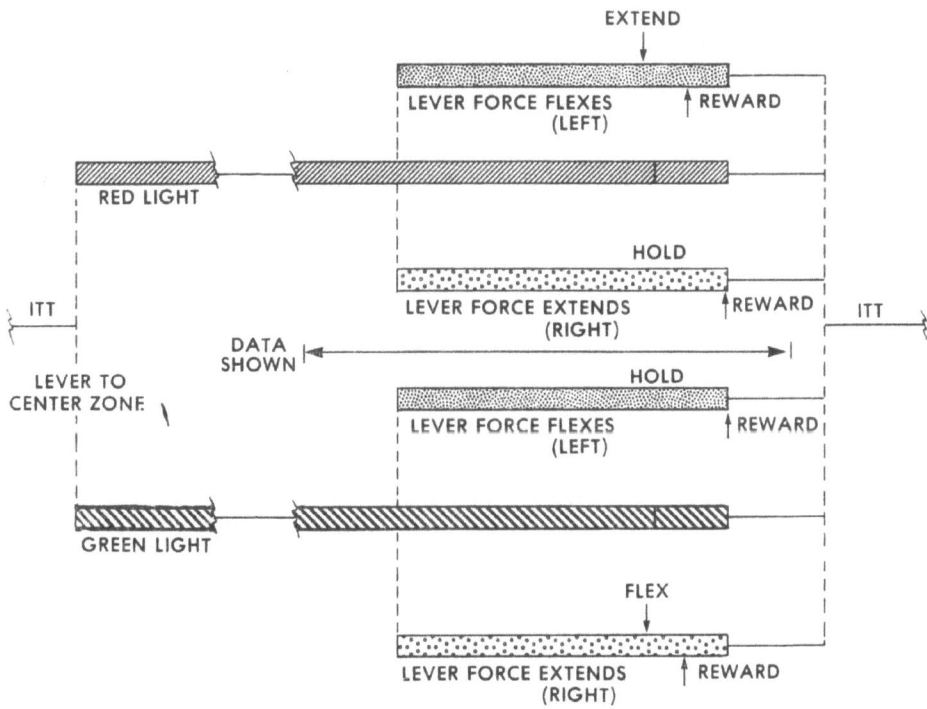

Fig. 1. The behavioral paradigm used with monkey paw. The four possible combinations of light color and lever twitch are shown in parallel. Correct behavior is indicated. Times are only approximate. See text.

For those stimulus combinations (light color, motor force direction) which required an answering movement, the maximum time allowed was progressively reduced during the training period. The monkey could easily attain a lever movement latency of about 200 ms; however performance diminished rapidly in attempts to further force a reduction of response latency. For those stimulus combinations which required holding the lever in the center zone, a minimum hold time of 400 ms was required.

Note that the light and lever twitch stimuli serve very different purposes in this complex paradigm. The light stimulus acts as an instruction to define the subsequent somatosensory discrimination problem set by the twitch direction. Thus the light stimulus sets up one of two possible preset states (red or green) in which context the somatosensory discrimination must be solved. In these circumstances the *same* somatosensory lever twitch input must be answered by *different* movements, depending on whether the preset state is red or green.

The results of an electromyographic study during this behavioral paradigm are shown in Figs. 2, 3 and 4. EMG signals were obtained with surface electrodes taped to the monkey's cleaned skin directly over the wrist flexor and wrist extensor muscle groups. Conventional differential amplifiers with pass band 100–5,000 Hz were used to record EMG signals on magnetic tape, along with lever acceleration and position signals. All behavioral equipment was under control of a computer program (in a PDP-8); a digital stimulus marking signal produced by the program was also recorded on one track of the tape. Subsequently,

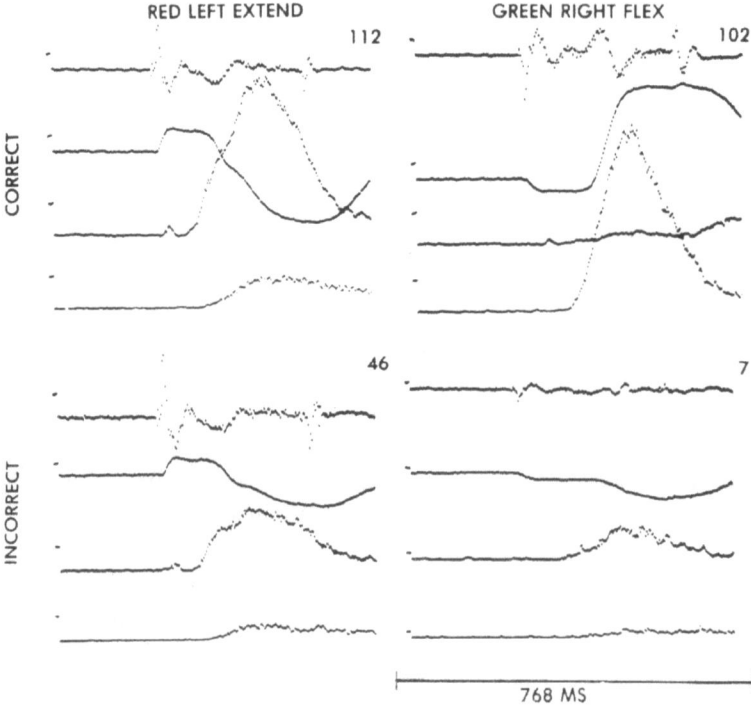

Fig. 2. Electromyographic study in a two task discrimination. Tasks are described in terms a, light color; b, twitch direction; c, correct response movement. Number of trials in each category at upper right of each group of traces. Correct trials in upper row, incorrect trials in lower row. In each group of four traces: top, lever acceleration; second, lever position; third, extensor EMG; lower, flexor EMG. All signals were averaged over the indicated number of trials. EMG signals were rectified and filtered before averaging.

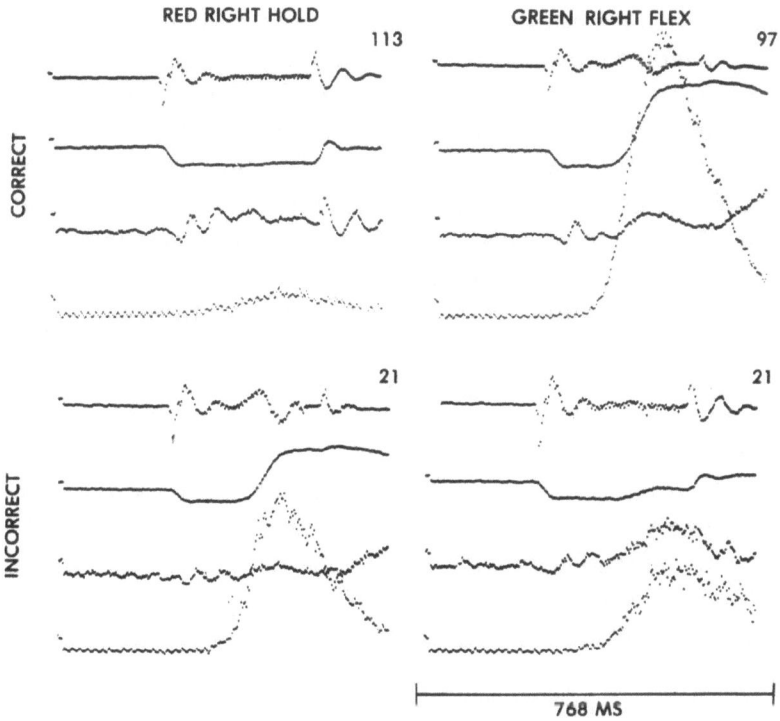

Fig. 3. **Electromyographic** study of a second two task discrimination involving the same direction of lever twitch. Denotations as in Fig. 2.

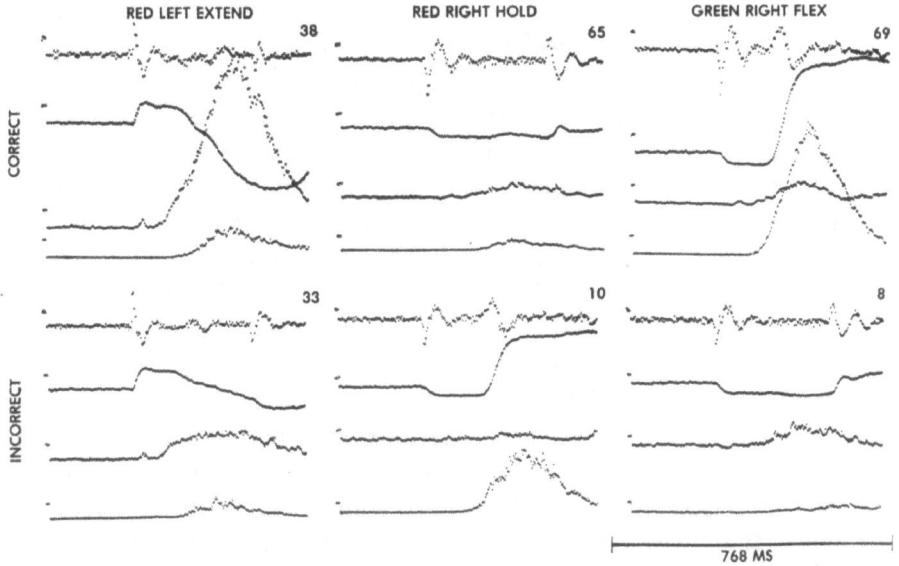

Fig. 4. **Electromyographic** study of a three task discrimination involving both light and twitch discrimination. Denotations as in Fig. 2.

data were sorted according to stimulus and response, and within each category averaged by a program in a PDP-12 computer. EMG signals were rectified before averaging. For each stimulus-response category the averaged signals are shown as four traces, which represent acceleration of the lever system, position of the lever system, the extensor EMG and the flexor EMG. The small rectangular elevation at the beginning of each trace was introduced to allow matching the vertical size of all corresponding traces within each figure. Thus the vertical scales for each stimulus response category are different, in correspondence to the number of trials falling into that category. The number of trials averaged in each category is indicated by the number at the upper right of each group of four traces. Time calibration is given in each figure.

Results from sessions where only two of the possible four stimulus combinations were used are shown in Figs. 2 and 3. Figure 2 contrasts the red-left, extend task with the green-right, flex task. Lever accelerations and movements in the correct executions of the two tasks are approximately inversions. An extend response is accompanied by a large extensor EMG, while a flex response is accompanied by a large flexor EMG. However, note that in each case the antagonist muscle is to some extent also activated. The extensor EMG shows a number of successive components with various latencies. The first such component represents the stretch reflex to the stimulus force, and is followed by a period of relative inactivity of the muscle. Subsequent activity components probably represent various longer "reflex" loops involving somatosensory to motor cortex, or cerebellum (see discussion). Note that separate components are not evident in the flexor EMG, although in some degree always present in the extensor EMG in each of the experiments shown in Figs. 2, 3 and 4. Such difference between flexor and extensor is unlikely to be physiological, and may be the result of different muscle mass, electrode position, or other recording condition.

Figure 3 contrasts the red-right, hold task with the green-right, flex task. In this situation the stimulus twitch of the lever is always in the same direction, but the correct response movement is determined by whether the red or green light was lit. Lever accelerations and movements are appropriate during the correct execution of the two tasks. We again note that in each case there is coactivation of the two antagonistic muscles. Again separate components are visible only in the extensor EMG, although they look somewhat different than in Fig. 2, since this was a different recording session with somewhat different electrode placement. The early components of the EMG after the stimulus twitch are identical for the two tasks; differences arise at the time of the second component of the extensor EMG, long before the earliest parts of the response movement.

Figure 4 shows results from a session where three of the four possible stimulus combinations were used. The monkey's performance of the red-left, extend task is rather poor in comparison to the same task in Fig. 2. Inspection of the corresponding "incorrect" traces shows that these incorrect extend responses were weaker and slower than required, but clearly differed from a hold or a flex response. "Incorrect" traces of the other tasks invariably show the wrong type of movement.

All the previous observations we have made about the data of Figs. 2 and 3 can be made about Fig. 4, which looks simply like a superposition of the other two figures. The task discriminations in each of Figs. 2, 3 and 4 could in principle have been carried out by means of different strategies. Thus the discrimination in Fig. 2 could be made on the basis of twitch direction alone, the color of the light simply acting as a preparatory "instruction". The discrimination in Fig. 3 could be made on the basis of the light color alone, with the twitch simply acting as a releasing stimulus. The three tasks of Fig. 4 however need some more complex strategy. The observation that EMG records look the same for a given task irrespective of which of the other tasks are given simultaneously suggests either that the same complex strategy is always used here by the experienced monkey, or alternatively that different strategies can not be detected at the level of the EMG.

*Discussion.* Why do we use such a complex behavioral task involving so much training time and effort? This has been an attempt to create two relatively limited states in the nervous system; during each of these states the same somatosensory information is processed, but with a different outcome. In some sense we are "switching" the animal between two preset states (signalled by the light stimulus), and this eventually leads to the appropriate stimulus twitch-response movement combination.

A preset state has been used by Tanji and Evarts (24) in a paradigm somewhat similar to that reported here. In that work a coloured light stimulus was used to indicate the direction of the desired future response movement. The actual movement was triggered (or released) by a somatosensory stimulus twitch in *either* direction; no discrimination was required. In contrast, the coloured light in our work determined stimulus twitch-response movement association to be used after discrimination of the twitch direction. Thus our preset state had to do with the subsequent discrimination rather than simply with the subsequent movement.

In all our extensor EMG records we have observed several temporal components. Similar components have been reported by several laboratories, and can be associated with several different somatosensory to motor loops. The fastest such loop is the stretch reflex via the spinal cord. A second component has been identified (25) as coming through Somatosensory I cortex. A third component has been identified (2) as coming through cerebellum. Our second and third EMG components have somewhat longer latencies than in the above references, so that direct comparisons are risky. The interesting observation in our data is, however, that all components are to some extent expressed in all three response movements (flex, hold, extend) involved in the various tasks. When different response movements follow a particular direction of stimulus twitch (Figs. 3 and 4) there is no difference between the first (stretch reflex) response components, but there are increasing differences between subsequent components. Thus the "switching" action that this paradigm is designed to produce is effective in several different pathways between the stimulus twitch and the motor response. We have also observed that each of the three possible response movements involves simultaneous activation of the flexors and extensors. The significance of the resulting increase in joint rigidity is not known.

These EMG measurements cast no light on the nature of the central preset state. In order to discover how the activity from a particular twitch stimulus can be "switched" into different motor output it will be necessary to record simultaneously from many neurons in several central structures. When the future movement can be predicted by the animal, the preset state is expressed as "preparatory" potentials (18, 26) as much as half a second before the movement, or by appropriate early activation of cortical flexor or extensor neurons (5, 24). In our situation the future movement can not be predicted by the monkey. We would hope to find an expression of our two preset states as two distinct rearrangements of functional neuronal groupings rather than in changes of firing patterns of single neurons. Dynamic reorganization of neuronal groupings is a likely physiological candidate for the type of "switching" we have observed with this behavioral paradigm. The time course of the "switching" action is (with a trained animal) the time between successive trials, well within stable recording time limitation. It would thus seem appropriate to apply the new recording and analysis methods mentioned in the introduction to this type of behavioral paradigm, in spite of its great additional experimental complexity.

This work was supported by grant number NS 05606 from the National Institutes of Health.

# REFERENCES

1. ATWOOD, H. L. 1976. Organization and synaptic physiology of crustacean neuromuscular systems. Prog. Neurobiol. (Oxford) 7: 292–391.
2. BROOKS, V. B., KOZLOVSKAYA, I. B., ATKIN, A., HORVATT, F. E. and UNO, M. 1973. Effects of cooling dentate nucleus on tracking task performance in monkeys. J. Neurophysiol. 36: 974–995.
3. CRAGG, B. and TEMPERLEY, H. 1954. The organization of neurons; a cooperative analogy. EEG Clin. Neurophysiol. 6: 85–92.
4. DICKSON, J. W. and GERSTEIN, G. L. 1974. Interactions between neurons in the auditory cortex of the cat. J. Neurophysiol. 37: 1239–1261.
5. EVARTS, E. K. and TANJI, J. 1976. Reflex and intended responses in motor cortex PT Neurons of monkey. J. Neurophysiol. 39: 1069–1080.
6. FORMAN, R. and HOYLE, G. 1978. 8th Ann. Meet. Soc. Neurosci. Abstr. 591.
7. FREEMAN, W. J. 1972. Linear analysis of the dynamics of neural masses. Annu. Rev. Biophys. Bioeng. 1: 225–256.
8. GASSANOV, U. G. and GALASHINA, A. G. 1976. Plasticity of cortical interneuronal connections. Zh. Vyssh. Nervn. Deyat. im. I. P. Pavlova 26: 820–827.
9. GERSTEIN, G. L., PERKEL, D. H. and SUBRAMANIAN, K. N. 1978. Identification of functionally related neural assemblies. Brain Res. 140: 43–62.
10. HEBB, D. O. 1949. The Organization of behavior. John Wiley and Sons, New York.
11. HORRIDGE, G. A. 1962. Learning of leg position by the ventral nerve cord in headless insects. Proc. Roy. Soc. B. 157: 33–52.
12. HUBEL, D. H. and WIESEL, T. N. 1962. Receptive fields, binocular interaction and functional architecture in the cat's visual cortex. J. Physiol. (Lond). 160: 106–154.
13. HUBEL, D. H. and WIESEL, T. N. 1974. Sequence regularity and geometry of orientation columns in monkey striate cortex. J. Comp. Neurol. 158: 295–306.
14. KOGAN, A. B. 1971. Structure and function of ensembles of neurons in rat visual cortex. Dokl. Akad. Nauk. SSSR 200: 1242–1245.
15. KRISTAN, W. B. and GERSTEIN, G. L. 1970. Plasticity of synchronous activity in a small neural net. Science 169: 1336–1339.
16. LINDSEY, B. G. and GERSTEIN, G. L. 1979. Interactions among an ensemble of chordotonal organ receptors and motor neurons of the crayfish claw. J. Neurophysiol. 42: 383–399.
17. MOUNTCASTLE, V. B. 1957. Modality and topographic properties of single neurons of cat's somatic sensory cortex. J. Neurophysiol. 20: 408–434.
18. NEAFSEY, E. J., HULL, C. D. and BUCHWALD, N. A. 1978. Preparation for movement in the cat I. Cerebral cortex. II. Basal ganglia and thalamus. EEG Clin. Neurophysiol. 44: 714–734.
19. OLDS, J., MINK, W. D. and BEST, P. J. 1969. Single unit patterns during anticipatory behavior. EEG Clin. Neurophysiol. 26: 144–158.
20. RINALDI, P., JUHASZ, G. and VERZEANO, M. 1977. Circulation of cortical and thalamic neuronal activity in wakefulness and in sleep. EEG Clin. Neurophysiol. 43: 248–259.

21. SALZBERG, B. M., GRINVALD, A., COHEN, L. B., DAVILA, H. V. and ROSE, W. N. 1977. Optical recording of neuronal activity. J. Neurophysiol. 40: 1281–1291.
22. STAFSTROM, C. E. and GERSTEIN, G. L. 1977. A paradigm for position learning in the crayfish. Brain Res. 134: 185–190.
23. STEVENS, J. and GERSTEIN, G. L. 1976. Interactions between cat lateral geniculate neurons. J. Neurophysiol. 39: 239–256.
24. TANJI, J. and EVARTS, E. V. 1976. Anticipatory activity of motor cortex neurons in relation to direction of an intended movement. J. Neurophysiol. 39: 1062–1068.
25. TATTON, W. G., FORNER, S. D., GERSTEIN, G. L., CHAMBERS, W. W. and LIU, C. W. 1975. Effects of postcentral cortical lesions on motor responses to sudden upper limb displacement in monkeys. Brain Res. 96: 108–113.
26. TAYLOR, M. J. 1978. Bereitschafts potential during the acquisition of a skilled motor task. EEG Clin. Neurophysiol. 45: 568–576.
27. VAN HARREVELD, A. and WIERSMA, C. A. G. 1937. The triple innervation of crayfish muscle and its function in contraction and inhibition. J. Exp. Biol. 14: 448–461.
28. VERZEANO, M. and NEGISHI, K. 1960. Neuronal activity in cortical and thalamic networks. J. Gen. Physiol. 43: 177–195.
29. WIENS, T. J. and GERSTEIN, G. L. 1975. Cross connections among crayfish claw efferents. J. Neurophysiol. 38: 909–921.
30. WILSON, D. M. and DAVIS, W. J. 1965. Nerve impulse patterns and reflex control in the motor system of the crayfish claw. J. Exp. Biol. 23: 193–210.

George L. GERSTEIN and Dwight H. RENFREW, Department of Physiology, University of Pennsylvania, Philadelphia, Pa. 19104, USA.
Benjamin H. PUBLOS, Jr., Department of Anatomy, Hershey Medical Center, Pennsylvania State University, Hershey, Pa. 17033, USA.

Lecture delivered at the Warsaw Colloquium on Instrumental
Conditioning and Brain Research
May 1979

# NEURONAL MECHANISMS OF CONDITIONED PLACING REACTIONS IN CATS

Boris KOTLYAR, Vladimir MAIOROV and Elena SAVCHENKO

Chair of Physiology of Higher Nervous Activity of Moscow University
Moscow, USSR

*Abstract.* Neuronal correlates of conditioned placing reaction of cat's forepaw were studied. The conditioned reaction evoking by tactile stimulation of the paw's ventral side had the same motor pattern, consisting mainly in successive flexion and extension of elbow joint, as the placing reaction evoked by paw's dorsal side stimulation in naive animals. The activity of single neurons from the m. biceps representation area in pericruciate motor cortex and VL thalamic nucleus was recorded. As a result of learning the excitatory response of cortical neurons to paw's ventral side stimulation in 20–50 ms post-stimulus interval was 2–2.5 times more than the response to the same stimulation in naive cats. This short-latency increase of response was not accompanied by modifications of sensory inflow to the motor cortex or movement related afferentation changes arriving from VL nucleus. There were no marked differences in excitability of cortical neurons of naive and trained animals as well. The results suggest a functional plasticity of the neuronal net in the motor cortex, consisting of a change in the efficiency of connections between neurons receiving sensory determined afferent excitation and the functional groups of neurons controlling the contraction of different muscles.

## INTRODUCTION

Plastic changes and their initial localization related to the neuronal mechanisms of conditioned reflexes are not yet known. The present research is devoted to the neurophysiological investigation of a model of learning based on the neuronal reorganization in the placing reaction of cats. The placing reaction in naive animals is evoked by tactile stimulation of the dorsal side of the forepaw. The conditioned reflex procedure results in the elaboration of the placing reaction to tactile stimulation of the ventral surface of the paw, which in naive animals initially blocks the dorsal placing reaction.

The extirpation of the motor cortex results in the irreversible disappearance of the placing reaction (4). The transcortical sensorimotor connections participating in placing reaction performance have been recently investigated (3), and are a part of the complicated system of transcortical reflex arcs participating in the initiation and control of movements (12). Therefore, the investigation of changes in the activity of motor cortical neurons occurring in parallel with plastic modification in the placing reaction is important for estimating the role of the motor cortex in the formation and alteration of motor habits.

## METHODS

The experiment was conducted out on adult unanesthetized cats, weighing 3.0 to 4.5 kg, that had their heads fixed in a frame with implanted screws. The forepaws remained free, but could not be seen by the cat. The conditioned placing reaction to ventral stimulation was reinforced with milk given automatically by a tube placed in the cat's mouth (8). Tactile stimulation of the paw was made by a special device ("toucher") with a photocell and light source used for the recording of tactile stimulation. The light beam crossed by the paw was recorded, and errors of measurement were less than 10 ms.

The silver electrodes implanted in m. biceps and m. triceps for chronical registration of EMG were connected by a female connector located on the animal's head with nichrome wire introduced subcutaneously. Extracellular recording of cortical neurons was done by tungsten microelectrodes introduced into the brain by means of a mechanical driver. The dura mater was not removed. The recording of nerve cells activity was made in that area of the pericruciate motor cortex where superficial electrical stimulation evoked elbow flexion.

The signals from the nerve cell were directed to a preamplifier placed on the animal's head and recorded on film and magnetic tape. The analysis of the experimental data was performed by a 512 channel analyzer (LP — 4050, NOKIA), and poststimulus time histograms (PST) were recorded on a plotter and printed. Spikes of several neurons having equal amplitudes and reactions to the stimuli presented were registered together. Microstimulation of the cortex was done through the recording electrode with trains of rectangular negative pulses. The duration of the pulse trains was 25 ms, the interpulse interval was 2.5 ms, and the duration of a single pulse was from 0.15 to 0.3 ms. The amplitude of the stimulating current ranged from 5 to 50 μA (16). Antidromic identification of pyramidal tract neurons was done by bipolar electrodes introduced into the pyramidal tract at the P11 or P3 level.

## RESULTS

### Plasticity of the cat's placing reaction

The placing reaction in naive, awake cats is elicited by touching the dorsal side of the forepaw, and consists of a sequence of movements resulting in the ventral surface of the paw contacting an object (support). The shoulder, elbow and wrist joints muscles of the forepaw all participate in this movement (Fig. 1A). The movement of the elbow joint, however, is the core of placing reaction. The placing reaction evoked by touching the dorsal side of the paw (Sd) consists of a succession of m. biceps and m. triceps contractions. Tactile stimulation of the ventral side of paw (Sv) of the naive animal evokes extention (action of m.

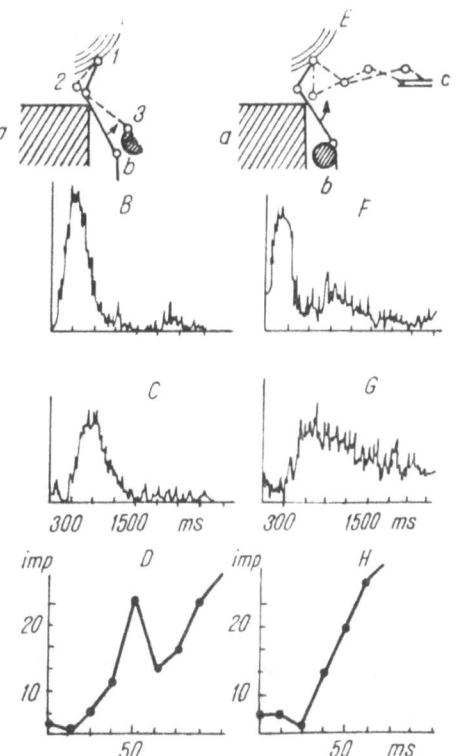

Fig. 1. The cat's placing reaction evoked by tactile stimulation of the dorsal (A, unconditioned reaction) in naive animals and ventral (E, conditioned reaction) side of the forepaw. 1–3, shoulder, elbow and wrist joints respectively; a, stand under the paw; b, tactile stimulus; c, platform (support) for placing the paw. Averaged PST of EMG m. biceps (B, F, D, H) and m. triceps (C, G) of unconditioned (B–D) and conditioned (F–H) reactions. Abscissa, time (in ms) starting from touching the paw. Ordinate, number of crossings of the pre-set level by EMG potentials. Number of trials: B, C, F, G = 20; D = 129; H = 152.

triceps) and blocks the placing reaction to the dorsal stimulation. Thus, the initial sensorimotor relations for the elbow joint may be briefly described as follows: Sd — BIC, Sv — TRI, SvN (Sd — BIC) (N — sign of negation).

The placing reaction is characterized by a number of plastic properties. Repeated stimulation of the dorsal side results in the extinction of the placing reaction. The reinforcement of the placing reaction with milk resulted in an increase in its probability of occurrence and shortening of its latency. In animals trained for 1–3 wk to raise the paw on the platform in response to touching of the ventral side the functional connection, Sv — TRI becomes ineffective, being replaced by a newly formed connection Sv — BIC (Fig. 1E). In addition the functional connection Sd — BIC of 3 animals was extinguished, which was achieved with great difficulty and required some thousands of administrations of dorsal stimulation of the paw without reinforcement. New functional relations for these animals may be briefly presented as follows: Sd — O, Sv — BIC.

*EMG of the biceps and triceps during the performance of placing reactions by naive and trained animals*

The data presented in Fig. 1B–D show the succession of m. biceps and m. triceps contractions during the performance of the reaction evoked by touching the dorsal side of the paw. The earliest activation in m. biceps occurs with about 30 ms latency. Figure 1F–H presents examples

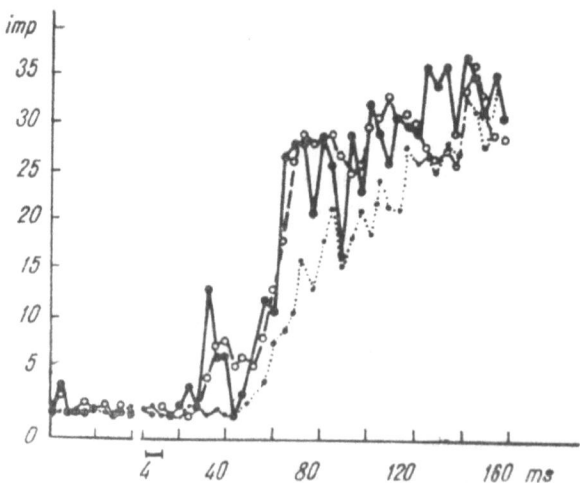

Fig. 2. Two components of biceps EMG during the placing reaction to stimulation of paw's ventral surface. PST histograms of biceps EMG of 3 different sessions in one animal. Each PST is the sum of 50 single trials. The horizontal line below abscissa shows the mark of tactile stimulation. Abscissa, time (ms); ordinate, number of impulses per 4 ms.

of the EMG of biceps and triceps in trained animals, that performed the placing reaction in response to stimulation of the ventral side of the paw (Fig. 1E). Comparison of the biceps and triceps EMG during performance of dorsal placing (naive animals) and conditioned placing reaction to ventral tactile stimulation (trained animals) show similar dynamics of m. biceps and m. triceps activity in both cases (Fig. 1B, F, C and G). The extended time scale shows the average latency of m. biceps activation, was about 30 ms (cf. Fig. 1D and H). The latency of the paw movement was measured as the time between the moment of touching and the moment of the paw's lifting from the toucher surface. It lasted not less than 40–50 ms in both cases.

On some PSTs there are two components in the EMG changes of the biceps: a first small component of about the 30 ms latency and a main component having 40–60 ms of latency (Fig. 2). These components join and become indistinguishable, especially on the PST if the level of motor readiness is high and the performance of the placing reaction is fast.

*General functional properties of the neurons in the motor cortex participating in the performance of placing reactions*

The activity of 430 neurons of the pericruciate motor cortex in 7 cats related to the representation of m. biceps and m. triceps was recorded during the placing reaction to tactile stimulation of the dorsal surface of the paw and during the conditioned placing reaction to ventral stimulation. Figure 3A shows the area of pericruciate cortex where superficial stimulation evoked contraction of the m. biceps. The same figure presents examples of neuronal activity recorded in this area during the performance of the placing reaction (Fig. 3C) and the EMG of the m. biceps reaction (Fig. 3B) during microstimulation of the cortex through the recording electrode.

The activity of neurons in the motor cortex was considered as related to m. biceps contraction if: (i) during the placing reaction the increase of neuronal spike frequency coincided with the EMG (Fig. 4); and (ii) stimulation of the cortex through the recording electrode elicited the EMG response (Fig. 3B). The majority of neurons were recorded in the deep layers of the cortex, in the region most effective for EMG activation by cortical microstimulation. A number of them were identified as neurons of the pyramidal tract with conduction velocity ranging from 10 to 60 m/s. The threshold for the cortical microstimulation in this region did not exceed 5 µA. The modifications of cortical neuronal activity occurring simultaneously and in parallel with the contraction of the respective muscles and the animal's movement will be henceforth referred to as changes of the motor type.

Figure 10A shows the summarized PST of neuronal activity in the motor cortex representation of the forepaw flexion of two animals (solid line) during performance of the placing reaction in response to stimulation of the paw's ventral surface. The curve shown is divided into two components. Besides the main motor type component associating with EMG changes, the first component also can be distinguished. The latter increased for 50 ms after tactile stimulation, that is before the initiation of the paw's removal from the surface of toucher. Such a component

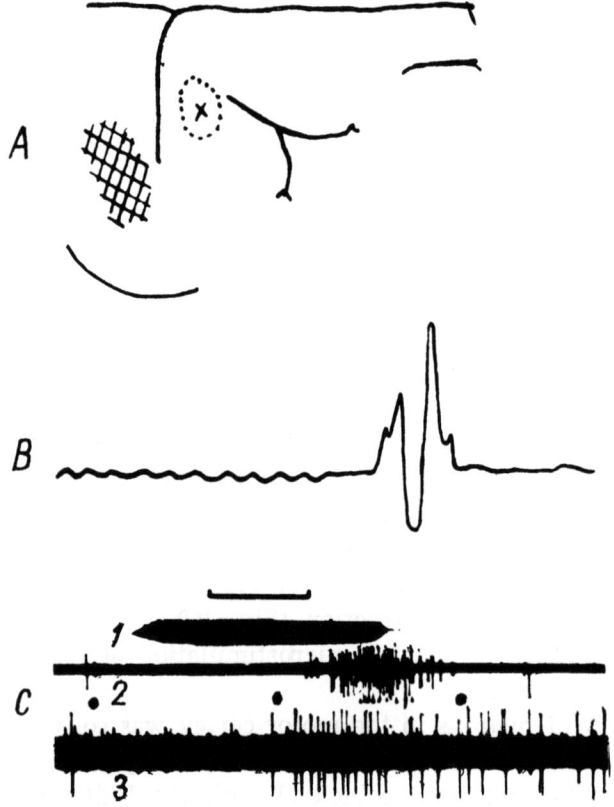

Fig. 3. *A*, The recording area of neuronal activity (dashed) in the cat's cortex. *B*, m. biceps EMG reaction to microstimulation of the motor cortex (12 µA) through the recording electrode. Time mark, 14 ms. *C*, the activity changes of motor cortex neurons and m. biceps EMG during the placing reaction. Recording of neuronal activity was done at the same cortical point stimulation of which resulted in the effect presented on Fig. 3*B*: 1, tactile stimulation; left extreme point of the dark stripe corresponds to beginning of tactile stimulation; the decrease in the dark stripe corresponds to the moment of the paw withdrawal from the tactile stimulus; 2, m. biceps EMG; 3, neuron's activity; time calibration, 0.25 s.

is also distinguished in the reactions of single neurons (Fig. 5). Temporal parameters of the first component in neuronal activity correspond to parameters of the first component observed in the EMG of the biceps (Fig. 2). However this component is more noticeable in the activity of cortical neurons (cf. Fig. 10A and Fig. 2).

The first component in the activity of some neurons remained almost without change in the absence of the paw's motor reaction (for example after acute extinction of the placing reaction, Figs. 4–3, and 5D). However, in the activity of other neurons of the motor cortex, the same changes in time were observed only in conjunction

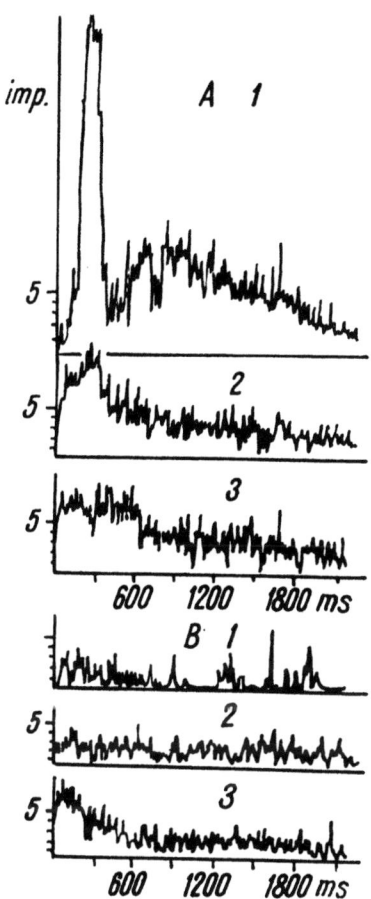

Fig. 4. PST histograms of neuronal reactions (2, 3) and m. biceps EMG (1) during the placing reaction (A) and after acute extinction of the placing reaction (B). Abscissa, time (ms) after beginning of tactile stimulation; ordinate, number of impulses per 10 ms. Each PST is a sum of 10 realizations.

with forepaw movement (Figs. 4–2 and 5A–C). The latencies of reaction of cortical neurons in both cases did not exceed 20 ms (Fig. 6), although it is difficult to determine the exact latencies since the method of measurement was not precise enough. The above-mentioned properties of the first component in neuronal activity during the performance of placing reactions allows one to consider it to be the sensory determined response evoked by incoming tactile afferentation to the motor cortex.

Since the latency of this sensory type of reaction in motor cortical neurons is about 20 ms and the time necessary for the transmission of the cortical signal to muscles, in accordance with the data of microstimulation, is about 10 ms, the minimal time of signal transmission from skin receptors to respective muscles through the motor cortex must be about 30 ms. This period coincides with the minimal latency of the biceps EMG reaction to a tactile stimulus observed in the experiment.

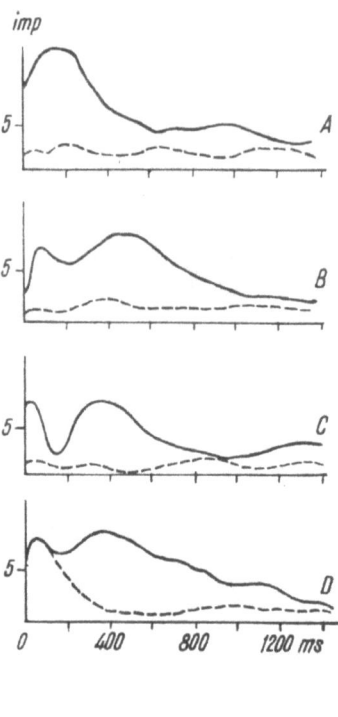

Fig. 5. Smoothed PST of activity changes of motor cortical neurons during the performance (solid lines) and after acute extinction (broken lines) of the conditioned placing reaction to stimulation of the paw's ventral side. Abscissa, time (ms) after beginning of tactile stimulation; ordinate, the number of impulses per 10 ms.

## Activity of neurons in the ventrolateral nucleus (VL) of the thalamus during performance of a placing reaction

The introduction of microelectrodes was made in accordance with the stereotaxic coordinates of the ventrolateral nucleus (15). The main functional criteria of electrode location in the nucleus were the characteristic effects of microstimulation through the recording microelectrode. Recording was performed in the area in which stimulation evoked a local flexion of the forepaw in the elbow joint. Characteristic rhythmic twitches continuing after switching off the current were observed if the strength of the stimulating current was increased (Fig. 7E). In this area neurons were recorded with increasing activity during the active flexion of the forelimb at the elbow joint. The activity of 51 neurons in VL was

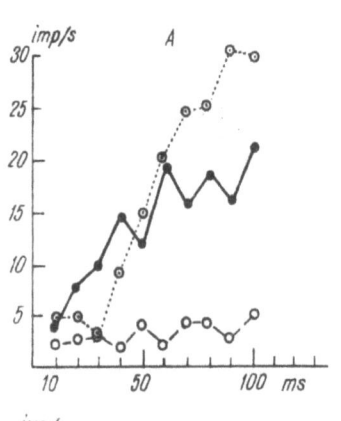

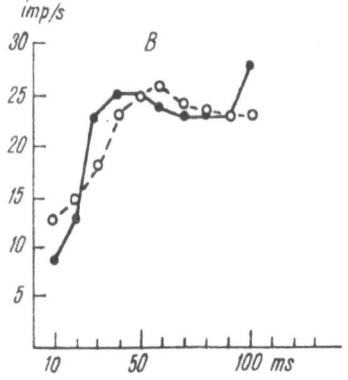

Fig. 6. PST histograms of reactions of the cortical neurons. A, averaged PST histogram of the activity of 6 neurons; B, averaged PST histogram of the activity of 5 neurons. During the investigation of each neuron 10 trials were summarized; the number of EMG trials corresponds to number of trials used for analysis of neuronal activity. Solid lines, neuronal activity during the placing reaction; broken lines, in the absence of movement after acute extinction of the placing reaction; dotted lines, changes of EMG m. bicep activity during the movement. Abscissa, time (ms) following the beginning of tactile stimulation; ordinate, averaged number of impulses per 1 s.

recorded altogether. The changes of neurons in VL during performance of the placing reaction are similar to the changes of neurons in the motor cortex (Fig. 7 A–D). However, the first component is not observed. None of the recorded neurons in VL reacted to tactile stimulation in the absence of movement. Figure 8 shows the histograms of reaction latency distribution in the motor cortical neurons, VL neurons and biceps EMG. The latency was considered as the time to attain half of the reaction in a given PST. Figure 8 A shows the reaction latency distribution in motor cortical neurons to tactile stimulation of the paw in the absence of mo-

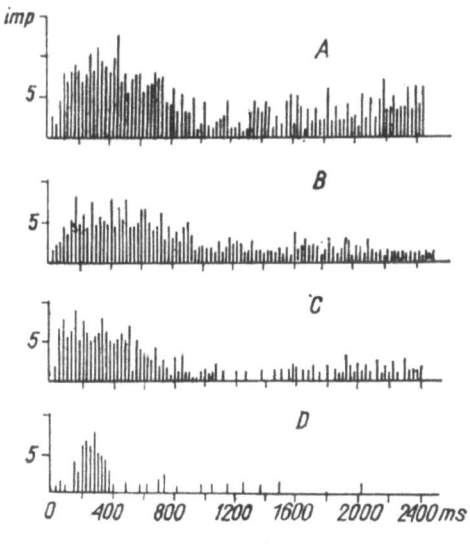
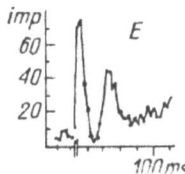
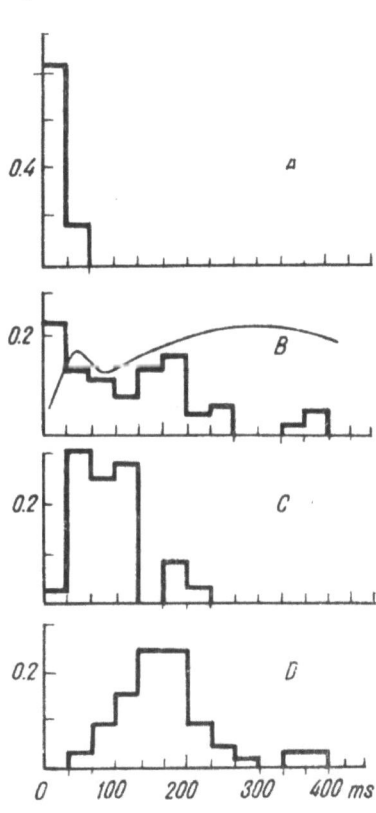

Fig. 7. Activity changes of neurons in the ventrolateral thalamic nucleus during the placing reaction (A–D). Abscissa, time (ms) after beginning of tactile stimulation; ordinate, number of impulses per 25 ms. Each PST histogram, the sum of 10 trials. E, PST histogram of biceps EMG in response to electrical stimulation of the ventrolateral nucleus (showed by arrows). Abscissa, time (ms); ordinate, the number of impulses per 4 ms.

Fig. 8. Histograms of the distribution of latencies of neuronal reactions in motor cortex (A, B), the ventrolateral thalamic nucleus (C) and biceps EMG (D) to tactile stimulation of the paw which evoked the placing reaction (B–D) and did not evoke a motor reaction after acute extinction of the placing reaction (A). Abscissa, latency (ms); ordinate, relative frequency of occurrence at a given latency. For explanation see text.

vement (after acute extinction of the conditioned motor reaction); Figure 8 B shows the reaction latency distribution in the case of the same tactile stimulation evoking the conditioned motor reaction of the paw. Also presented here is the summarized curve of the activity changes in neurons in the area of cortical representation of forelimb flexions during movement (see Fig. 10 A). Figure 8 panels C and D show the distributions of reaction latencies for neurons of VL and motoneurons (EMG) under the same conditions. It is necessary to point out that the changes of neuronal activity in VL with a latency less than 30 ms are practically missing, that is those changes forming the reaction of the sensory type in neurons of the motor cortex (cf. Fig. 8 A). Therefore, the activity changes in neurons of the motor cortex with a latency of less than 30 ms and respective changes in EMG do not connect with afferentation from VL to motor cortex, which is another argument in favor of their sensory determined origin. Comparison of the mutual disposition of maxima on histograms of latency distributions allows one to supose that the activity changes in VL neurons coincided with the development of movement arise earlier than the main motor component of activity changes in the motor cortex and the EMG of biceps.

*Relation between activity of neurons in the motor cortex and the dynamics of muscle contraction*

Analysis of firing rate changes in the neurons of the forelimb flexion area in the motor cortex, developing in association with movement, showed that specific neurons are active in a definite phase of movement (11, 13). Figure 9 shows the periods of maximal activity of 50 neurons in relation to the beginning of biceps contraction. The latter was considered as the moment of reaching a half of the maximal level of EMG frequency in a given hystogram. The length of each segment on the figure corresponds to the period when the neuronal frequency of discharges was increasing from a level equal to half of the maximal (beginning of segment) to the maximal (end of segment). As may be seen, the periods of maximal increase in the neuronal discharge frequencies are distributed equally enough in relation to the start of flexion. This distribution does not depend on the dispersion of EMG maxima positions (vertical marks on Fig. 9). For neurons of the motor cortex, the time of switching on correlated positively with the moment of achieving the maximal reaction ($r = 0.78$; $P < 0.005$) and with the time of the switching off of neuronal activity ($r = 0.29$; $P < 0.025$). The difference in periods of maximal activity, was characteristic for different neurons, since the temporal parameters of the reactions of every single neuron in successive recordings remained approximately permanent.

Figure 10 A shows the summarized and averaged reactions of cortical neurons connected functionally with the biceps and the EMG of this muscles during performance of a motor task. Reactions were averaged with steps of 10 ms (0–100 ms) and 100 ms (100–900 ms). As seen on Fig. 10 A, B the summarized neuronal activity of the motor cortex and EMG changes of the respective muscles are proportional to each other during the whole period of the reaction performance with the exception

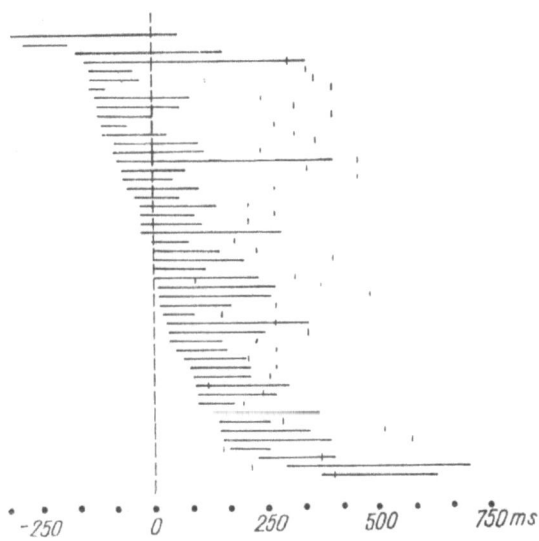

Fig. 9. The distribution of periods of highest activity of 50 neurons during the placing reaction in relation to the beginning of m. bicep contraction. Vertical broken line, the moment of reaching half of the maximal activity of biceps EMG in given PST histogram. The beginning of each horizontal line, the moment of reaching half of the maximal level of the neuron's reaction; the end, maximum of reaction. Vertical marks, maxima of EMG. Each horizontal line is based on an averaged PST histogram (10 trials). Time is shown in ms.

of the short interval of 50–100 ms from the start of stimulation. During the reaction, there is the obvious change of the proportion for neuronal and EMG activity; it increased during the interval of 100 to 700 ms in comparison with the interval of 0 to 50 ms (Fig. 10 B). As has already been mentioned, the changes of neuronal and EMG activity in the 0 to 50 ms interval were predominantly sensory determined. These changes in the activity of motor cortical neurons were expressed much better than in the EMG of the corresponding muscle.

Simultaneous with the increase of neuronal activity in the cortical area of biceps, the initial phase of the conditioned placing reaction was characterized by reciprocal supression of neuronal activity in the cortical area of triceps (Fig. 11).

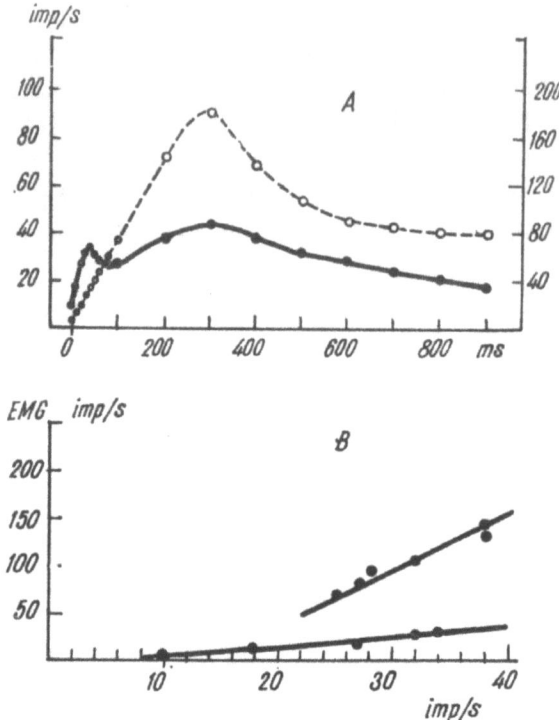

Fig. 10. A, PST histograms of neuronal reactions in the m. bicep cortical area (dark circles, averaged data from 34 neurons in two animals) and of m. biceps EMG (open circles) during conditioned placing reaction to ventral stimulation of the paw. Other denotations as in Fig. 6. The right scale, EMG; left neuronal activity. B, Relation of averaged neuronal activity in the m. bicep cortical area and biceps EMG according to data presented in Fig. 10 A. Abscissa, the average frequency of neuronal activity; ordinate, EMG activity in corresponding moments after tactile stimulation. The lines are drawn according to the regression equations. 1, regression line for 0–50 ms interval: $y_1 = -3.5 + x$, $r = 0.97$; 2, regression line for 100–700 ms interval: $y_2 = 6x - 85.5$, $r = 0.99$

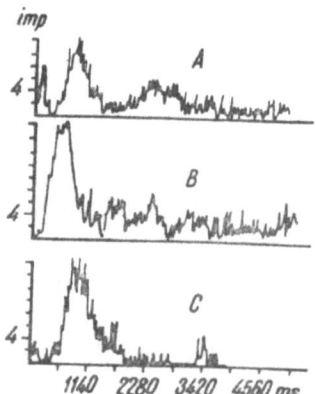

Fig. 11. PST histograms for neuronal activity (A) and EMG of m. biceps (B) and EMG of m. triceps (C) during the placing reaction. Other denotations as in Fig. 4.

*The sensory connections of the neurons in the cat's motor cortex*

The localization of the skin receptive fields of cortical neurons having projections to m. biceps and m. triceps was studied by tactile stimulation of the dorsal and ventral sides of distal parts of the paw. We selected neurons with a stable sensory component of responses to tactile stimulation, remaining even in the absence of the motor reaction of the paw. The reactivity of cortical neurons to the tactile stimulation coming from the paw skin surface was investigated in 4 naive and 5 trained animals (Table I). The neurons of the biceps cortical area are excited mainly from the dorsal side of the paw representing the receptive field for the initial phase of the placing reaction. 14 of 65 neurons (21%) in the m. biceps cortical area of the animals, performing placing reaction to stimulation of the paw's dorsal side, responded to that kind of stimulation in the absence of paw movement with stable reactions of the sensory type (Fig. 12). Only about 8% (11 of 136) of the neurons of the biceps

TABLE I

The distribution of neurons activated by tactile stimulation in the absence of forepaw movemet among the neurons in m. biceps and m. triceps cortical areas. $S_d$, dorsal surface; $S_v$, ventral surface

| Groups of neurons | Type of tactile stimulation of paw | Naive animals | Trained animals | |
|---|---|---|---|---|
| | | $n = 4$<br>Sd–BIC<br>Sv–TRI | $n = 2$<br>Sd–BIC<br>Sv–BIC | $n = 3$<br>Sd–O<br>Sv–BIC |
| Neurons of m. biceps cortical area | Sd | 14 of 65 (21%) | 5 of 70 (7%) | |
| | Sv | 11 of 136 (8%) | | |
| | | 4 of 41 (10%) | 7 of 95 (7.4%) | |
| Neurons of m. triceps cortical area | Sd | 0 of 14 | | |
| | Sv | 14 of 14 | | |

cortical area reacted to stimulation of the paw's ventral side in naive ($S_d$ — BIC) and trained ($S_v$ — BIC) animals.

The neurons of the m. triceps cortical area associated with the extension of the forelimb at elbow joint (Fig. 13) on the contrary are excited during tactile stimulation in the absence of movement exclusively from the ventral side of the paw (Fig. 13 C), that is, from the receptive field of the extension reflex. The stimulation of the dorsal surface of the paw was ineffective (Fig. 13 B) or decreased the activity of these neurons.

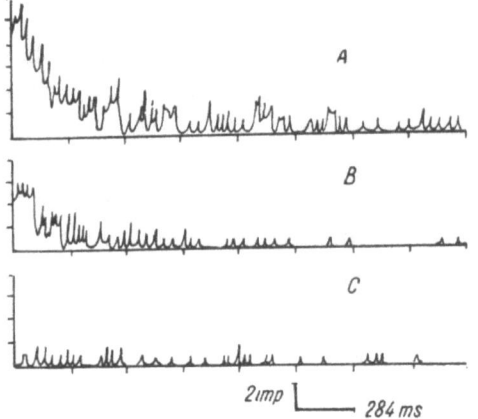

Fig. 12. PST histograms of neuronal activity in the motor cortex of a naive animal. A, during the placing reaction; B, in response to dorsal stimulation after the acute extinction of the placing reaction; C, in response to ventral stimulation. Other denotations as in Fig. 4.

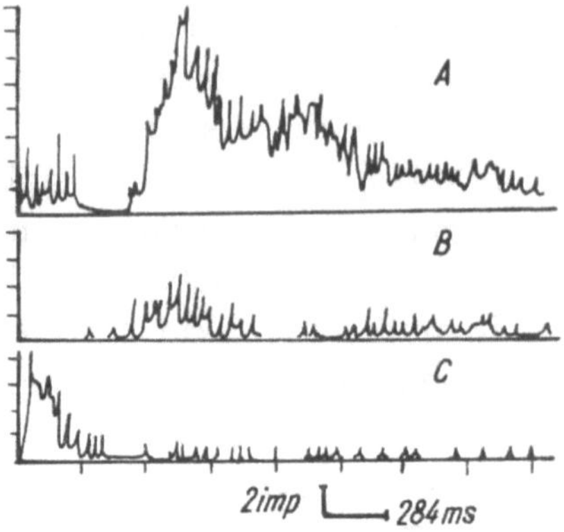

Fig. 13. PST histograms of neuronal activity in the motor cortex and EMG of the m. triceps of a naive animal. PST of EMG of the m. triceps (A) and neuronal activity (B) during the placing reaction evoked by dorsal stimulation. C, reaction to ventral stimulation in the absence of movement. Other denotations as in Fig. 4.

*Functional connections of neurons in the motor cortex after elaboration of the conditioned placing reaction*

As was seen from the above data, the process of learning consisted in a change in the functional significance of tactile stimulation of the paw's ventral surface. In naive animals touching the area of the pads on the paw's ventral surface evokes the extension reflex and blocks the placing reaction. After training touching the paw's ventral side evoked the flexion and placing reaction on the support as shown on the Fig. 1. The conditioned reaction to ventral stimulation as far as the motor pattern is concerned, was similar to the initial placing reaction evoked by touching the paw's dorsal side in naive animals.

The changes that occurred in the properties of activity in neurons of the motor cortex as a result of learning may be described as following. Tactile stimulation of the paw's ventral side resulted in conditioned placing reaction evoked the same average response of neurons in the biceps cortical area in the 20–50 ms post-stimulus interval as stimulation of the paw's dorsal side resulted in the placing reaction in naive animals (Fig. 14 A). Figure 14 A shows the averaged reaction of neurons in the biceps cortical area to tactile stimulation of the paw's dorsal side resulted in the placing movement in naive animals (Sd — BIC, open circles) and the averaged reaction to tactile stimulation of the paw's ventral side resulted in the placing movement in trained animals (Sv — TRI, close circles). The third curve on this figure (open circles with point inside) shows the averaged reaction of neurons in the biceps cortical area of naive animals to

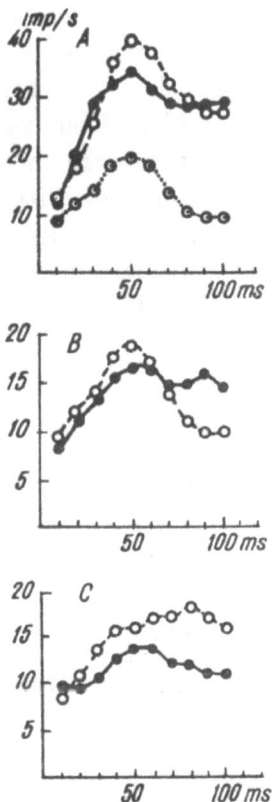

Fig. 14. Averaged PST histograms of neuronal reactions in the m. biceps cortical representation area. A, during placing reaction to ventral (dark circles, averaged data from 42 neurons) and dorsal (open circles, 36 neurons) stimulation; open circles with dot inside shows response to ventral stimulation which did not evoke placing reactions in naive animals (36 neurons). B, reactions to ventral stimulation in the absence of forepaw movement; dark circles, trained animals after acute extinction of the conditioned placing reaction (24 neurons); open circles, naive control animals (36 neurons); C, reactions to dorsal stimulation in the absence of forepaw movement; open circles, naive control animals after acute extinction of the dorsal placing reaction; dark circles, trained animals (24 neurons). Other denotations in Fig. 6.

stimulation of the ventral surface, which was ineffective in producing a placing reaction. The effective stimulation resulted in placing movement (Sd in naive animals, Sv — in trained ones) evoked an approximately equal increase of discharge frequency in the cortical pool of neurons related to m. biceps contraction. The figure shows the identity of the reaction latency in both cases as well.

Elaboration of the conditioned placing reaction does not change the relative number of neurons in the m. biceps cortical area being steadily excited during tactile stimulation of the paw's ventral side, idependent of presence or absence of the movement reaction.

From 41 neurons in naive animals (Sv — TRI; Sd — BIC) activated during the m. biceps contraction only 4 neurons (10%) were excited by stimulation of the paw's ventral side in the absence of movement. Among the trained animals (Sv — BIC) 7 neurons of 95 (about 7%) were excited by the same stimulus. Figure 14 B shows the averaged neuronal reactions in the m. biceps cortical area to stimulation of the paw's ventral side in the absence of movement in naive animals (Sv — TRI, open circles) and to ventral stimulation in trained animals (Sv — BIC) after acute extinction of the conditioned placing reaction (close circles). The comparison of these curves does not reveal any difference in averaged neuronal reactions to ventral stimulation in the absence of forepaw movement. This conclusion is illustrated also by figure 15. It can be seen that after acute extinction of the conditioned placing reaction the neuronal activity increased during the tactile stimulation of paw's dorsal side (Fig. 15 D) and was inhibited during stimulation of the paw's ventral side (Fig. 15 C). However, the stimulation of the paw's ventral side evoked an increase of neuronal activity in association with m. biceps contraction (cf. Fig. 15 B and Fig. 15 C).

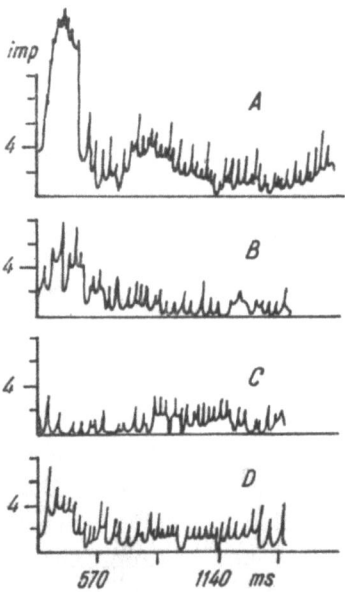

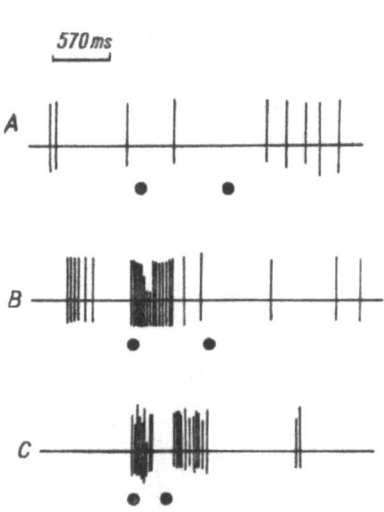

Fig. 15. PST histograms of EMG of the m. biceps (A) and neuronal activity (B–D) in the m. biceps cortical area of a trained animal. B, neuronal reaction during the placing reaction evoked by ventral stimulation. C, D, reaction to ventral and dorsal stimulation respectively with no movement. Other denotations in Fig. 4.

Fig. 16. Reactions of a neuron in the m. triceps cortical area of a trained animal to the touching of the dorsal (A) and ventral (B) side of the paw without movement and during performance of the placing reaction to ventral side stimulation (C). Recordings of neuronal activity on paper tape. The first point corresponds to beginning of tactile stimulation; the second, to the end of tactile stimulation (A, B) and paw withdrawal from the toucher surface during the placing reaction (C).

The retention of stable sensory connections to the motor cortical neurons after learning was supported by the investigation of neuronal reactions in m. triceps cortical area (Fig. 16). The neurons of this area of trained animals were excited by tactile stimulation of the paw's ventral side, although the ventral stimulation became ineffective as referred to m. triceps contraction.

Other results were obtained concerning the inputs to m. biceps cortical area after chronic extinction of the initial placing reaction to dorsal side stimulation in animals which performed the conditioned placing reaction to ventral side stimulation. A decrease in the number of neurons in the m. biceps cortical area responding to dorsal stimulation up to 7% (5 of 70 neurons) was observed in this case. All of these neurons were recorded in the motor cortex of the animal with incomplete extinction of the reaction to stimulation of paw's dorsal side.

Figure 14, C shows the reaction of the neurons in naive ($S_d$ — BIC, open circles) and trained ($S_v$ — BIC, $S_d$ — 0, close circles) animals to stimulation of the paw's dorsal side in the absence of movement. Chronic extinction of the dorsal placing reaction under condition of retention of responses to stimulation of ventral side decreased the efficiency of dorsal tactile stimulation with respect to neurons of the m. biceps cortical area.

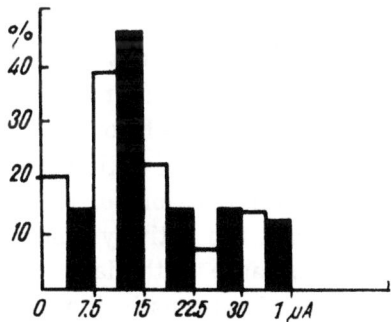

Fig. 17. Histograms of threshold currents evoking the m. biceps contraction. Microstimulation of the motor cortex. Dashed columns, data from trained animals (50 measurements); open, data from naive animals (36 measurements). Abscissa, current (µA); ordinate, percent of stimulations at a given threshold.

A threshold current evoking the electrical response in the m. biceps of naive and trained animals (ranging from 5 to 50 µA) was passed through the microelectrode at the recording points of the neurons excited in parallel with the m. biceps contraction. The comparison of histograms (Fig. 17) of the threshold current distribution for m. bicep EMG responses in trained (dashed columns) and naive animals (open columns) shows that there is no essential difference in excitability between the populations of cortical neurons controlling the forepaw flexions.

## DISCUSSION

In the activity of recorded neurons we distinguished the excitatory reactions of two types: 1) changes related to the active contraction of different muscles and to movement and 2) a sensory-determined component of reaction to tactile stimulation of the paw. This division, initially based on the observations of temporal dynamics in the activity of neurons in the motor cortex during performance of the placing reaction, are confirmed by other findings:

1. There are different relations of neuronal activity and EMG biceps during temporal intervals corresponding to the first sensory-determined component and the main motor component;

2. The changes of neuronal activity recorded in the ventrolateral nucleus of the thalamus occur with a latency corresponding to the main component of activity in neurons of the motor cortex;

3. The short-latency responses of some neurons of the motor cortex to tactile stimulation were retained in the absence of paw movement.

These data suggest a different origin of the first and the main components in neuronal activity of the motor cortex (2). Stability and short latency of neuronal reactions of the sensory type (about 20 ms) are comparable with respective characteristics of the reactions in neurons of the sensory cortical area and ventralis posterolateralis nucleus of the thalamus. These reactions appear to be determined by the entrance of specific tactile afferentation into the motor cortex through the lemniscal path. The stable component of the sensory response is likely a direct response to the afferentation from the sensorimotor cortex and ventralis posterolateral nucleus and not mediated through interneurons.

The changes of activity in the neurons of the motor cortex associated with motor reactions of the forepaw develop in parallel and partly arise later than the activity changes in neurons of the ventrolateral nucleus. In accordance with data in the literature is seems that the motor discharge of neurons in the motor cortex is partly determined by incoming excitation from the cerebellar nuclei through ventrolateral nucleus (2, 17).

The placing reaction is characterized by some plastic properties worth studying in order to investigate the neuronal mechanisms of formation and extinction of motor habits. First, the placing reaction disappears when there is no constant reinforcement. Second, a number of instrumental motor reflexes can be elaborated using the placing reaction as an unconditioned reflex. The motor pattern of these instrumental reflexes is similar to the motor pattern of the placing reaction. We elaborated the conditioned placing reaction in response to the stimulation of the paw's ventral side. This stimulation in naive cats initially blocks and inhibits the placing reaction to stimulation of the paw's dorsal side.

As a result of learning, the ventral stimulus evoking the conditioned placing reaction evokes a significantly more intense increase of short-latency excitatory activity changes in the neurons of m. biceps cortical area. This increase might be supposed to be determined by the increase of sensory afferentation coming to the neurons in the biceps cortical area from the sensory cortex (SmI) and the nucleus ventralis posterolateral (1, 5, 18), i.e., by the redistribution in "addresses" of specific afferent excitation on the motor cortical neurons. This is not the case. It was found that the relative number of neurons in m. bicep cortical area responding to stimulation of the paw's ventral side in the absence of the placing reaction does not changes. A distinct specificity of reactions to ventral stimulation in m. triceps cortical neurons is preserved despite the depression of the reaction, Sv — TRI, and the elaboration of a new opposite reaction, Sv — BIC. The reactions of neurons in SM1 sensory cortex to tactile and other kinds of specific stimulation are characterized by stability, being independent of previous learning, motor readiness and functional state of the animal (6, 19). So it seems that elaboration of the conditioned placing reaction does not result in efficiency changes in direct pathways connecting tactile receptors with the motor cortex or ending points of SM1 and VPL afferents on motor cortical neurons.

The lack of excitability changes in cortical output elements points to motor cortical interneurons as a probable place of plastic changes providing an increase of responses of biceps cortical area neurons to stimulation of the paw's ventral side, which evokes the conditioned placing reaction. The comparison of the data obtained with other data (6, 7, 9-11, 14, 20, 21) allows us to conclude that there is a high plasticity of the neuronal net of the motor cortex.

The investigation of the strict organization and physiological mechanism of plasticity of interneuronal connections in the motor cortex is the task of a further study. It probably will bring us closer to an understanding of some of the mechanisms of instrumental conditioned reflexes.

We thank Prof. E. N. Sokolov for valuable comments on the manuscript.

## REFERENCES

1. ALBE-FESSARD, D. 1975. The motor cortex as the reflectory center. In A. S. Batuev (ed.), Sensornaya organizatsiya dvizhenii. Izdat. Nauka.
2. AMASSIAN, V. E., WEINER, H. A. and ROSENBLUM, M. 1972. Neural system subserving the tactile placing reaction: a model for the study of higher level control of movement. Brain Res. 40: 171-178.
3. ASANUMA, H., STONEY, S. D. A. Jr., ABZUG, C. 1968. Relationship between afferent input and motor outflow in cat motor sensory cortex. J. Neurophysiol., 31: 670-681.
4. BARD, P. 1933 Studies on the cerebral cortex. I. Localized control of placing and hopping reactions in the cat and their normal management by small cortical remnants. Arch. Neurol. Psychiatr. 30: 40-74.

5. EVARTS, E. V. 1974. Precentral and postcentral cortical activity in association with visually triggered movement. J. Neurophysiol. 37: 373–381.
6. EVARTS, E. V. and TANJI, J. 1974. Gating of motor cortex reflexes by prior instruction. Brain Res. 71: 479–494.
7. FETZ, E. E. and FINOCCHIO, D. V. 1975. Correlation between activity of motor cortex cells and arm muscles during operantly conditioned response patterns. Exp. Brain Res. 23: 217–240.
8. FOX, S. S. and RUDELL, A. P. 1970. Operant controlled neural events: functional independance in behavioral coding by early and late components of visual cortical evoked response in cats. J. Neurophysiol. 33: 548–561.
9. KOTLYAR, B. I., MAIOROV, V. I. and SAVCHENKO, E. I. 1975. Models of learning based on plastic properties of placing reaction in the cats (in Russian). Zh. Vyssh. Nervn. Deyat. Im. I. P. Pavlova, 25: 967–973.
10. MAIOROV, V. I., KOTLYAR, B. I. and SAVCHENKO, E. I. 1977. The participation of cat's motor cortex neurons in afferent reorganization of placing reaction (in Russian). Zh. Vyssh. Nervn. Deyat. Im. I. P. Pavlova 27: 931–940.
11. MAIOROV, V. I., SAVCHENKO, E. I. and KOTLYAR, B. I. 1977. The transformation of the afferent tactile signal to the motor command in the motor cortex of the cat. Neurophysiologia 9: 115–123.
12. MURPHY, J. T., WONG, V. C. and KWAN, H. C. 1975. Afferent-efferent linkages in motor cortex for single forelimb muscles. J. Neurophysiol. 38: 990–1016.
13. PORTER, R. and LEWIS, M. 1975. Relationship of neuronal discharges in the precentral gyrus of monkeys to the performance of arm movements. Brain Res. 103: 201–213.
14. SAVCHENKO, E. I., MAIOROV, V. I. and KOTLYAR, B. I. 1976. The activity of cat's motor cortex neurons in the course of realization of the conditional placing reaction (in Russian). Zh. Vyssh. Nervn. Deyat. Im. I. P. Pavlova 26: 65–72.
15. SNIDER, R. S. and NIEMER, W. T. 1961. A stereotaxic atlas of the cat brain. Univ. Chicago Press, Chicago.
16. STONEY, S. D., THOMPSON, W. D. and ASANUMA, H. 1968. Excitation of pyramidal tract cells by intracortical microstimulation: effective extend of stimulating current. J. Neurophysiol. 31: 656–669.
17. THACH, W. T. 1978. Correlation of neural discharge with pattern and force of muscular activity, joint position and direction of intended next movement in motor cortex and cerebellum. J. Neurophysiol. 41: 654–676.
18. THOMPSON, W. D., STONEY, S. D. and ASANUMA, H. 1970. Characteristics of projections from primary sensory cortex to motosensory cortex in cats. Brain Res. 22: 15–27.
19. TOWE, A. L., TYNER, C. F. and NYQUIST, J. K. 1976. Facilitatory and inhibitory modulation of wide field neuron activity in postcruciate cerebral cortex of the domestic cat. Exp. Neurol. 50: 734–756.
20. VORONIN, L. L. 1976. Cellular mechanisms of conditioned activity. Zh. Vyssh. Nervn. Deyat. Im. I. P. Pavlova 26: 705–719.
21. WOODY, C. D., VASSILEVSKY, N. N. and ENGEL, J. 1970. Conditioned eye blink: unit activity at coronal-precruciate cortex of the cat. J. Neurophysiol. 33: 851–864.

Boris KOTLYAR, Vladimir MAIOROV and Elena SAVCHENKO, Chair of Physiology of Higher Nervous Activity, Moscow University, Lenin Hills, Moscow 117234, USSR.

Lecture delivered at the Warsaw Colloquium on Instrumental
Conditioning and Brain Research
May 1979

# SINGLE UNIT ACTIVITY IN THE VISUAL CORTEX DURING CONDITIONING IN CATS

G. MERZHANOVA

Laboratory of Conditioned Reflexes, Institute of Higher Nervous
Activity and Neurophysiology, USSR Academy of Sciences
Moscow, USSR

*Abstract.* Multiple neuronal activity, recorded through chronically implanted electrodes, was analyzed. During acquisition and consolidation of alimentary conditioned reflexes to electrostimulation of the lateral geniculate body or optical tract, the patterns of neuronal activity in the visual cortex and sensory motor cortex became organized in a way different from that observed during pseudoconditioning. The majority of the neurons showed change in activity during both the isolated action of the stimuli and their simultaneous presentation. In stabilized conditioned reflexes, the activity of neurons in the sensory motor and visual cortex was interdependent. Neuronal indices of backward conditioned connections during activation of the reinforcing structures were analogous to the reactions of visual cortical neurons during the conditioned stimulus action and were manifested by an increase of discharge activity.

## INTRODUCTION

The concept of backward temporary connections was primarily formulated by Pavlov (15) to explain mechanisms underlaying the acquisition of instrumental conditioned reflexes in dogs. For many years this idea has been experimentally developed in the laboratory of Asratyan by means of various behavioral and electrophysiological techniques. Special attention to backward connections has been also given by Beritov whose works were primarily focused an theoretical aspects of the problem.

It was shown (1–3, 13) that the pairing of two stimuli, which evoke their own reactions, leads to the formation of forward temporary connections, when the first stimulus is followed by the effect of the second. In addition backward temporary connections are formal when separate presentations of the second, reinforcing stimulus start to elicit the proper reflex of the first, signalling stimulus. On the basis of these data, Asratyan (1, 2) formulated the idea that during conditioning there occur two-way connections. With the accumulation of experimental data the two-way temporary connection grew from a phenomenological observation to a successful instrument for the study of the nervous mechanisms of conditioned activity. Should local transformations of electrographic parameters in the cortical projection of the conditioned stimulus be considered as the effect of backward influences, then, according to Asratyan (2), this influence plays a double physiological role. First, it participates in the occurrence of the local conditioned state, which is characterized by an increase in the excitability of the cortical projection of the conditioned stimulus. Second, it is manifested in the formation of the local conditioned reflex, representing a newly acquired form of interneuronal functional connections.

It is known (5, 6, 12, 14) that conditioning may be accompanied by both the enhancement and suppression of neuronal impulse responses in the cortical projection area of the conditioned stimulus. In addition, it may lead to involvement in the activity of new neurons. A question arises whether all of these listed forms of impulse transformations are related to the feedback influence or only some of them, and secondly, whether the similarity between conditioned and unconditioned neuronal responses is the manifestation of the feedback influences?

We showed previously (14) that in animals with stable food-seeking conditioned reflexes to the electrostimulation of the lateral geniculate body (LGB), the presentation of the reinforcing stimulus elicits a neuronal response at the projection of the signalling stimulus (visual cortex) analogous to that characterizing the conditioned reflex. Since in these experiments visual perception of the reinforcing stimulus, meat, was not excluded, it was possible that the pattern of neuronal activity of the visual cortex was related both by the modulating specific influences of the reinforcing stimulus, along the backward conditioned connections from the alimentary to the visual center, and by the direct influence of the reinforcing stimulus on visual reception. In order to explore the specific influences of the structures of the reinforcing stimulus on the structures of the conditioned stimulus along backward conditioned connections, in this study we used a method with milk administration into the oral cavity which excluded sight of the alimentary stimulus. Our main aim was to study neuronal indices of two-way conditioned connections on the model of conditioned reflexes with milk reinforcement, trained in cats to electrostimulation of the brain structure of the visual analyzer.

## METHODS

*Experimental.* Seven chronic cats were used. The alimentary conditioned reflex was elaborated to electrical stimulation of the LGB or the optical tract (OT). Recording of the neuronal activity in the visual and sensory motor cortical areas (representing the tongue and chewers) was done with a set of nichrome electrodes, each of 50 μm in diameter (5). Electrostimulation of the visual pathways consisted of stimuli 2 Hz in frequency and 5 s in duration. For 3 cats the CS-US interval of 2 s and for other 4 cats of 3 s was used. 2 cc of milk was administered directly into the oral cavity through in indwelling cannula. Each experimental session consisted of 15–20 trials.

*Statistical.* Neuronal activity of the visual and sensory motor cortical areas was measured before training, during pseudoconditioning, during consecutive stages of conditioning, and in special tests consisting in the presentation of the reinforcing stimulus applied before and after conditioning. The machonogram of licking was recorded simultaneously with the recording of neuronal activity. Recording of all electrical parameters was done on a general purpose electrophysiological apparatus UEFI-3, designed at the Central Design Office of USSR Academy of Sciences, and on magnetic tape of a Nihon Kohden tape-recorder. The analysis and statistical processing of multi-neuronal activity was done on Plurimat S-100 computer. The selection of spikes was carried out by considering the spike's amplitude and shape, and for this purpose the levels were set and the spike's shape was described by definite algorhythms (Fig. 1 $A, B$).

Post-stimulation histograms and cross-correlation histograms were made at the analysis epoch of 100 ms with 1 ms bin. Additionally, in order to obtain a clearer picture of spike distribution density, histograms were made at the given analysis epoch, post-stimulation density histograms (PSDH) and density histograms of cross-correlation (CCDH) (Fig. 1C) using the formula:

$$P_i = \frac{1/4\ A_{i-1} + 1/2\ A_i + 1/4\ A_{i+1}}{A_{max}} ; \ P_i \leqslant 1.$$

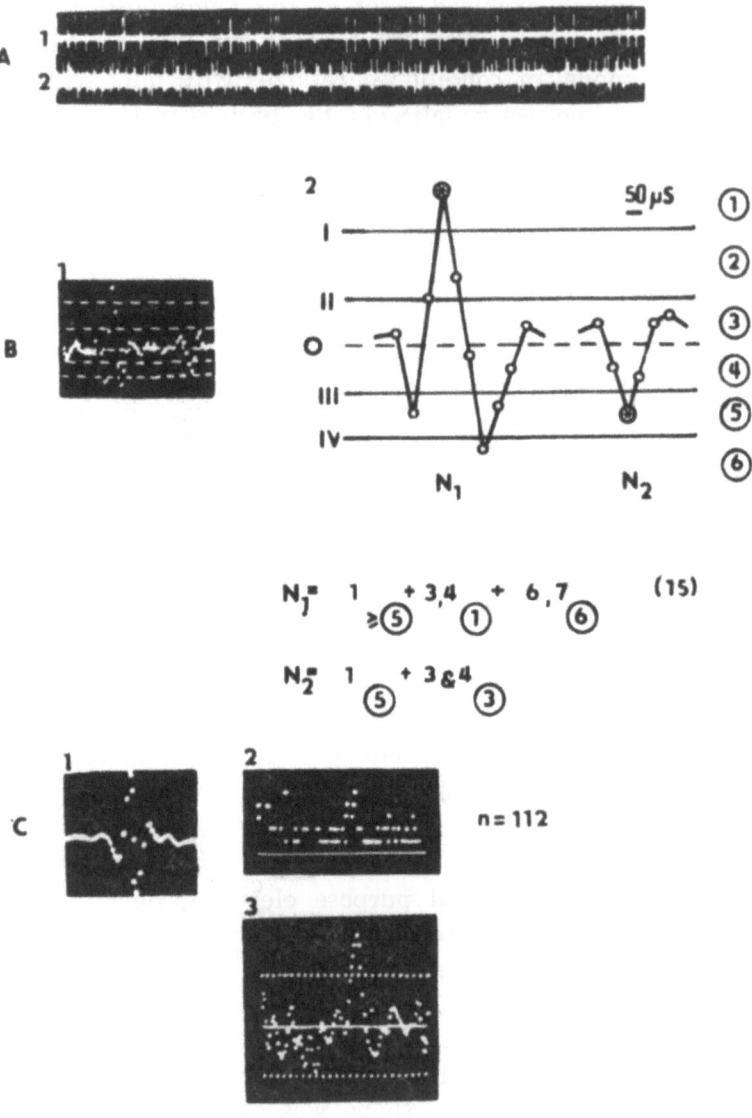

Fig. 1. Statistical analysis of chronically recorded multiple unit activity in the sensory motor (A, 1) and visual (A, 2) cortex of an unrestrained cat. Single action potentials with large and small amplitudes (B, 1); and diagramatic recording of the same action potentials (B, 2) illustrating the method of separating the large spike (71) from the discharges of its neighbors (N2), according to the algorithms shown below. Numbers in circles denote amplitude windows. For N1 point 1 occupied window 5; point 3 or 4, window 1; points 6 or 7 occupied window 6 and the next 15 points were excluded after the last point. For spike N2 point 1 occupied window 5, whereas points 3 and 4, window 3.

## RESULTS

In five cats the neuronal activity was studied throughout the entire experiment and in two others only after stabilization of conditioned reflexes. The conditioned reflex was established after 30 to 50 pairings and manifested in the reaction of licking (Fig. 2).

Figure 3 shows three simultaneously recorded neurons from the visual cortex, (projection area of the conditioned stimulus) and sensory motor cortex (representing the tongue and chewers) after stabilization of conditioned reflexes. In the impulse flows of both areas, neuronal responses similiar in shape were singled out and their reactions were studied on density post-stimulation histograms. It is seen that in the

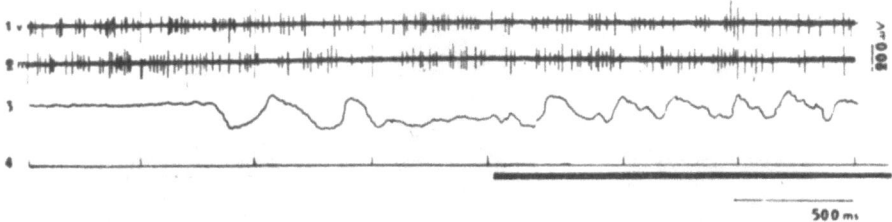

Fig. 2. Simultaneous recording of the electrophysiological and effectory indices of the conditioned milk reflex. 1, in the visual cortex; 2, unit activity in the sensory motor cortex; 3, machonogram of licking; 4, conditioned stimuli (vertical thin line) and reinforcement (heavy horizontal line). Trial 123.

visual cortex two neurons with different shapes and amplitudes (N1 and N3) responded to the conditioned stimulation with short-latency reactions. The first neuron of the motor cortex produced a cycle of excitatory responses with 40 Hz frequency, accompanied by periods of inhibition. Such forms of neuronal responses were typical of the motor cortical area and resulted from conditioning, because in the same animals they were absent before training. Thus, stable conditioned reflexes with milk reinforcement were characterized by a specific pattern of neuronal activity in the visual and sensory motor cortex.

We conducted special tests with activation of backward conditioned connections, which consisted of an isolated presentation of milk, and thus increased alimentary excitability. Figure 4 shows the original oscillograms of the stable conditioned reflex (A), the test with activation of the backward connection (B), and the intertrial licking after stabilization of conditioned reflexes (C). It turned out that in a number of cases, neurons of the visual cortex during isolated presentation of the reinforcing stimulus increased their frequency of discharges. The same was found for the intertrial licking movements after stabilization of conditioned reflexes (C), which was also considered as an electrophysiological manifestation of the backward conditioned connection.

In order to prove the functioning of the backward conditioned connection, we determined the statistical dependence of impulse neuronal responses of the sensory motor and visual cortical areas from which the cross-correlograms of neuronal activity pairs were drawn. Such analyses were performed for the background activity (30 pairs) and for the tests with activation of the backward conditioned connections (25 pairs) in the same animals before and after acquisition of the conditioned reflexes.

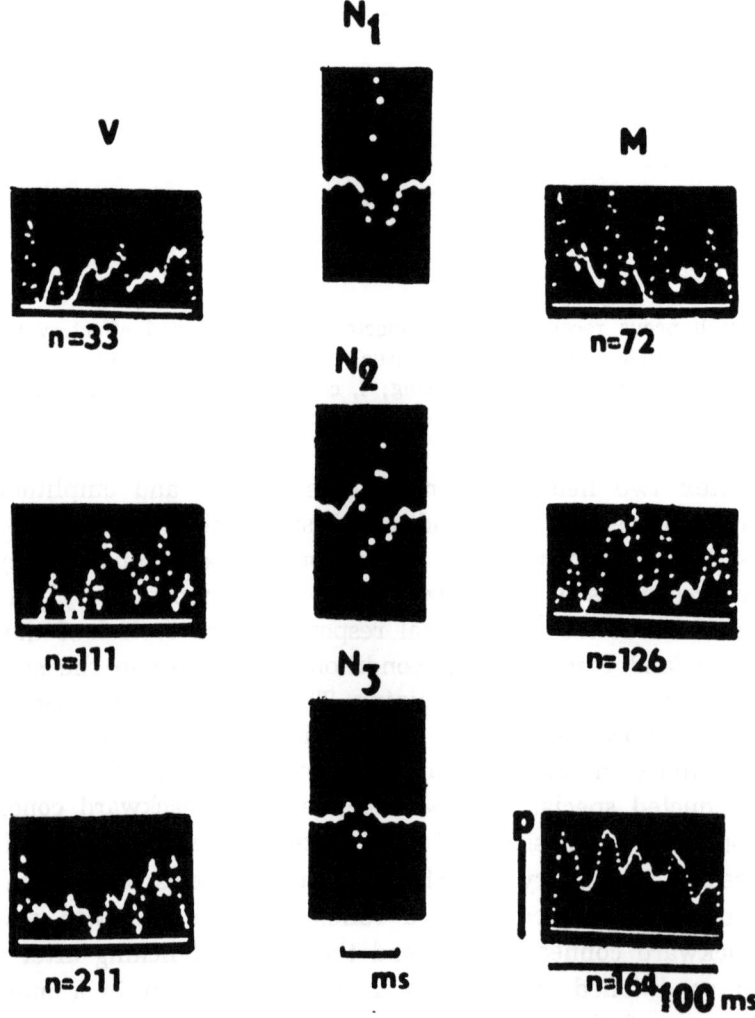

Fig. 3. The shapes (N1, N2, N3) of the similar spikes singled out from multiunit activity and their post-stimulus density histograms after conditioning. Visual cortex neurons (V), sensory motor (M).

Figure 5 shows density crosscorrelation histograms (CCDH) of cellular activity in the visual and sensory motor cortex. From the impulse row of the sensory motor cortex, a spike of maximal amplitude (M1) was singled out and intervals were studied, with which three different neurons in the visual cortex responded (V1; V2; V3). In constructing the

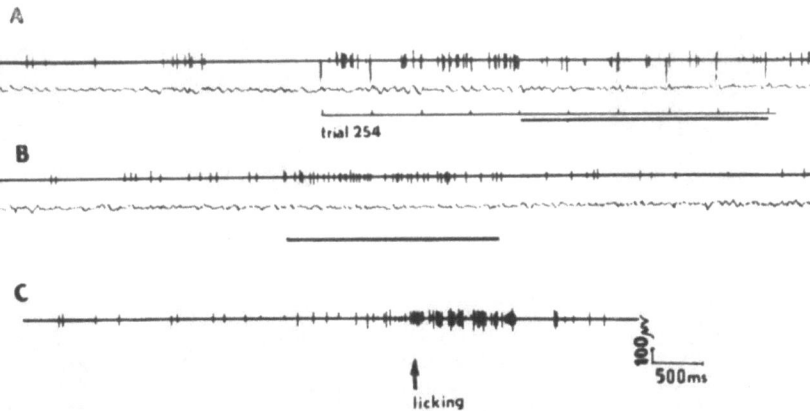

Fig. 4. The responses of visual units to the conditioned stimulus (A), to the isolated presentation of the milk test (B), and during intertrial licking (C) after stabilization of the conditioned reflex. The upper trace of each oscillogram shows visual unit activity; the middle trace represents EEG activity from the same point (both recordings were made by means of one electrode), the low trace marks the presentation of the conditioned stimulus and reinforcement. Increasing frequencies of spikes in each of three cases are shown.

CCDH, the impulses of neuron M1 (50) were presented to start the count, and impulses of neurons V1, V2, or V3, were counted for construction of corresponding CCDH shown in B and C. Some impulses of neuron M1 in the fixed analysis epoch were missing, but they were small in number because the averaged mean frequency of M1 discharges was 6–8/s. The symmetrical branch of CCDH was not investigated in the present study. It turned out that all three neurons of the visual cortex are connected differently with the motor neuron. On the CCDH it is evident that the maximal spike density of neuron V1 falls at 15 ms, that of neuron V2 at 30 ms, while that of V3 has an even distribution along the whole analysis epoch. The comparison of the CCDH obtained after acquisition of the conditioned reflex (C) and before conditioning (B) reveals that the neurons of the visual cortex V1 and V2 established clear excitatory connections with motor neurons, which was not observed in untrained animals.

In the analysis of 24 neuronal pairs during spontaneous activity before conditioning, 35% showed a statistical dependence of the visual neurons on the motor neurons (Fig. 6A, 1). Examination of the same 24 pairs after conditioning during intertrial periods indicated that 74% showed dependent relations, which was manifested predominantly in the form of excitatory connections (Fig. 6B, 1). A similar picture persisted in the special tests with activation of the backward conditioned

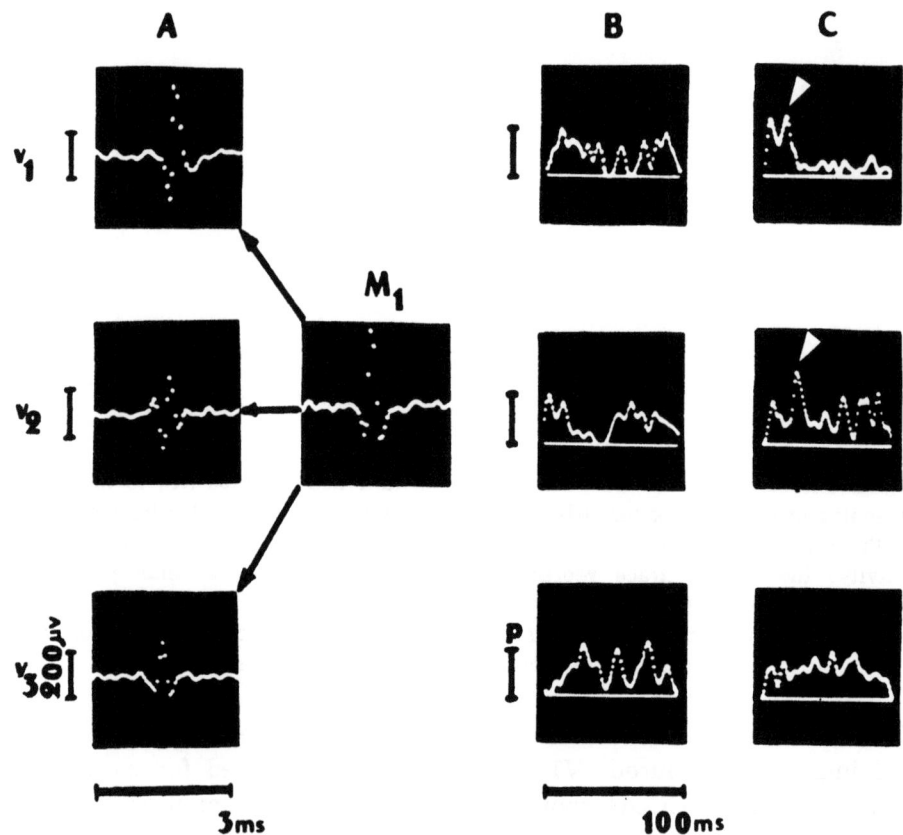

Fig. 5. Typical interactions between visual and sensory motor neurons before and after conditioning. A; V1, V2, and V3 represent three action potentials from each visual neuron to demonstrate their distinguishability. M1, a single action potential selected from sensory motor, multiple-unit activity. B and C: cross-corelation density histograms obtained from the selected neurons in the visual and sensory motor cortex of untrained (B) and trained (C) cats. The appearance the excitatory relation in the connection M1 — V1 and M1 — V2 can be seen in C.

connections when the reinforcing stimulus was presented alone. The analysis of 18 pairs showed that in 25% of the cases before (6A, 2) and 80% after conditioning (6B, 2) statistical dependence of visual neurons on the motor ones were found. Figure 7 shows the influences of different functional states on the pattern of relations between neurons of the visual and sensorimotor cortex. The increase of excitatory connections

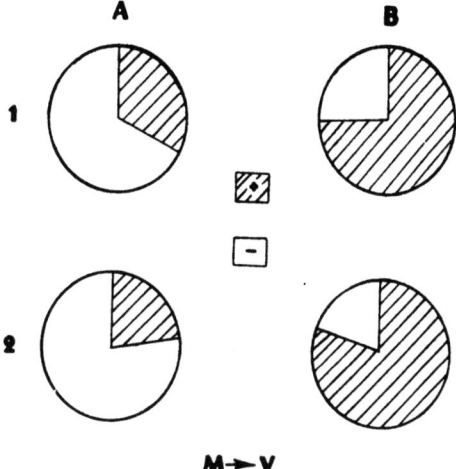

Fig. 6. Quantitative distribution of crossinterval histograms before (A) and after (B) conditioning. 1, sessions without testing milk presentation. 2, sessions with milk testing. The diagrams are drawn on the basis of data from 24 neuronal pairs concerning one type connections (from M to V).

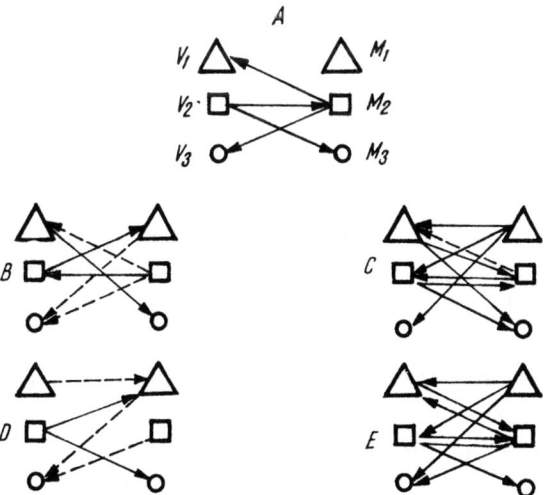

Fig. 7. The influence of different experimental situations on patterns of interneuronal connections drawn in the form diagrams. A, before conditioning; B, after pseudoconditioning; C, after conditioning; D, to the test presentations of milk before conditioning; E, to milk presentation after conditioning. The solid line indicates excitatory connection, the dotted line indicates inhibitory connections. Note the enhancement of connections between the visual (V1, V2, V3) and the sensory motor neurons (M1, M2, M3) as well as the appearance of mutual interrelations after conditioning. Considerable increase of connections may be seen when the milk was presented after conditioning (E). All the diagrams are based on the analysis of 69 neuronal pairs.

after acquisition of the conditioned reflex and the appearance of mutual forms of interrelations did occur. The neuron V1 was supposed to form connections with the sensory motor neurons (C), which did not exist before conditioning (A). The neuron M1, which was free of excitatory connections during pseudoconditioning (B) formed such connections with all visual neurons under study (C). The two neuronal group interrelations taking place during pseudoconditioning were less extensive and complex than those found during conditioning.

The difference of interneuronal organization during the milk presentation (D) was considerably more than without milk (A). As seen from Fig. 7, the visual-motor relations to milk testing were quite poor before conditioning (D) and extremely intensive after conditioning (E). The resemblance in the patterns of neuronal organization during conditioning in situations C and E, that is in sessions with and without the special milk testing, should be stressed.

## DISCUSSION

We have shown the neuronal correlates of forward and backward conditioned connections in the process of acquisition and specialization of alimentary conditioned reflexes to electrostimulation of the visual analyzer. The elements of brain structures that perceive the reinforcing stimulus were activated along the forward connections. In the present model of conditioned reflexes these are the structures of alimentary and motor centers. Both the selective increase in activity of the neurons of the visual cortex and the statistical dependence between neurons of the sensory motor and visual cortex, were considered as the effect of afferentiation along the backward connections.

The existence of two-way conditioned connections supported the hypothesis put forward by Asratyan (2) who pointed out that "it is logic to assume that unconditioned reflexes, elicited by stimuli, paired in order to elaborate the conditioned reflex, are mutually reinforcing, that the process of reinforcement is also two-way by nature". In the cortical projections of each of the paired stimuli, proper interfocal functional connections were formed, which created active functional units. The latter were able to participate in the constellation of these foci, activity of which Asratyan called a "local conditioned reflex". It seems possible to consider all of the electrophysiological data, relating to changes in the discharge activity of neurons in the foci of paired stimuli, from the view of two-way reinforcement, recalling, however, that the power of these influences may be different.

The electrophysiological indices of the backward conditioned connection have been reported (2, 13, 14). Initially it has been reported that during performance of the intertrial classical motor conditioned reflex response, when only non-signalling repetitive cliks were presented, elicited in the auditory cortex evoked potentials, analogous to that evoked by the sporadic CS. It was suggested that only the backward conditioned connection between the sensorimotor and auditory cortical areas could provide a structural basis for this phenomenon. Our presented previous data (13, 14), also showed an increase in late evoked potential components, as well as neuronal correlates of backward conditioned connections in the form of activation of neuronal responses to the reinforcing stimulus in the visual cortex.

In the present study the influence of the reinforcing stimulus along the backward conditioned connections on the structures of the conditioned stimulus projection, were revealed both in special tests with activation of the backward conditioned connection and in the forming of a specific pattern in visual neuronal discharge activity in response to the conditioned stimulus. There are some data (8, 11) that in acquired reactions the changes of evoked potentials in the projections of the conditioned stimulus may reflect the influence along the backward conditioned connections. We assume, that the pattern of discharge neuronal activity in the visual cortex, shown in the response to the sporadic conditioned stimulus, is formed under the influence of spike activity, coming along the backward conditioned connections from the structures of the reinforcing stimulus. The methodological characteristics of our studies permit us to exclude the effect of alimentary stimuli on visual perception. We think that the pattern of neuronal activation in the visual cortex reflects the result of a functional neuronal interaction, formed exclusively by the acquisition of the conditioned reflex, in contrast to other studies, where animals always saw the food, which in itself might contribute to the formation of the conditioned neuronal reactions.

After conditioned reflex acquisition to the isolated presentation of the reinforcing stimulus, neurons of the visual cortex responded with reactions analogous to that found in the period of isolated activity of the conditioned stimulus. Thus we tested the functional state and readiness of neurons in the visual cortex, the cortical projection of the signal stimulus.

Our data suggest that after stabilization of conditioned reflex the pattern of neuronal responses in the visual cortex is a result of a new functional interaction between the neurons, rather than the effect of diffuse non-specific influences on the neurons, suggested by some authors (4, 16, 17). We probably saw a reflection of non-specific influences in neuronal responses of the visual cortex during pseudoconditioning, when electrostimulation of the visual analyzer and administration of milk were done at random. The organization of the neuronal activity in the visual cortex in these cases differed from that found during conditioning.

Our data also suggest that two-way conditioned connections function not only between the structures of the alimentary and visual analyzers, but also between the motor and visual cortical areas. The appearance of specific neuronal reactions of the visual cortex during intertrial licking after stabilization of the conditioned reflexes, and their absence in untrained animals support this suggestion. Moreover, we obtained an increase in dependent relations between the neurons of the sensory motor and visual areas after conditioning.

According to the literature (6, 9, 10), in the study of spike activity of three or more neurons on cross-interval histograms of each neuronal pair, the dependent relations are treated as the indices of functional interneuronal connections of the microsystem. Different types of influences (automated stimulation with clicks, or with pulse current, rhytmic auditory stimulation) cause plastic changes in neuronal connections. Each type of influence has its own distribution of connections (6). The formation of new interneuronal relations and the changing of the existing ones may have occurred when the activity of separate cells remained within the limits of their initial values. Thus the study of interneuronal relations on CCDH is a more sensitive parameter than the study of the changes in activity patterns of single cells.

The analysis of statistical dependence in the activity of simultaneously recorded cortical neurons is an indirect method of studying synaptic processes. The presence of such dependence of neuronal pair spike activity may indicate both direct synaptic connections between these neurons and the existence of an outside general source. To date there are several criteria for the estimation of crossinterval histograms. Correlations occurring on a cross-interval histogram within an interval of several milliseconds suggest direct synaptic connections between neurons (9, 10). It was shown (12) that the most common time interval, characterizing neuronal interaction of the visual and motor cortical areas is equal to 10 ms. Neuronal responses of the sensory motor cortex to electrical stimulation of the visual cortex with a latency of 10 ms were recorded in alert rabbits in both intact and isolated cortical strips. The excitation caused by direct stimulation of the cortex seems to use approximately the same pathway for its proliferation, and along this pathway the interaction between the neurons in the state of spontaneous activity occurs.

We showed the dependence of neuronal activity in the visual cortex on the neurons of the sensory motor cortex with different latency, which may be considered as the manifestation of the functioning backward conditioned connection. Further studies of interneuronal connections, both of an intrafocal and interanalyzer character, will permit the delineation of the physiological composition of two-way conditioned connections.

I thank Mr S Varashkevich and Mr V. Dorochov for comments and assistance in programming. I also thank Mrs. N. Chugunova for technical assistance.

## REFERENCES

1. ASRATYAN E. A., 1967. Some peculiarities of formation, functioning and inhibition of conditioned reflexes with two-way connections. In A. Fessard and H. H. Jasper (ed.), Brain reflexes. Elsevier, Amsterdam, p. 8–20.
2. ASRATYAN E. A., 1977. Ocherki po fiziologii uslovnych refleksov. Nauka, Moscow, p. 348.
3. DAVIDOVA E. K., 1978. Formation of forward and backward connections during elaboration of classical and instrumental conditioned reflex (in Russian). Zh. Vyssh. Nervn. Deyat. Im. I. P. Pavlova 28, 3: 484–489.
4. FEENNY D. and OREM, J., 1972. Modulation of visual cortex inhibition during reticular evoked arousal. Physiol. Behav. 9: 805–808.
5. GASSANOV U. G. and GALASHINA A. G., 1975. Analysis of interneuronal connections in the auditory cortex of alert cats (in Russian). Zh. Vyssh. Nervn. Deyat. Im. I. P. Pavlova 25, 5: 1053–1060.
6. GASSANOV U. G. and GALASHINA A. G., 1976. Study of plastic changes in cortical interneuronal connections. Zh. Vyssh. Nervn. Deyat. Im. I. P. Pavlova 26, 4: 820–827.
7. GASSANOVA R. L. and GASSANOV U. G., 1975. Electrographic expression of conditioned feedback connections during internal inhibition. In T. N. Oniani (ed.), Mechanizmy deyatelnosti golovnogo mozga. Mestsniereba Publishers, Tbilisi, p. 126–133.
8. GASSANOV U. G. and VANETSIAN G. L., 1971. Method of chronic investigation of unit activity in alert and anaesthetized cats (in Russian). Zh. Vyssh. Nervn. Deyat. Im. I. P. Pavlova 21, 4: 820–826.
9. GERSTEIN G. L., 1970. Functional association of neurons: detection and interpretation. In F. O. Schmitt (ed.), The neurosciences. Second study program. Rockefeller Univ. Press, New York, p. 648–661.
10. GERSTEIN G. and PERKEL D., 1978. Simultaneously recorded trains of action potentials: analyze and functional interpretation. Science 164: 828–830.
11. KOSTANDOV E. A., 1975. The role of forward and backward temporary connection in the formation of the cortical evoked potentials. In The brain mechanisms, Metsniereba Publishers, Tbilisi, p. 157–164.
12. LIVANOV M. N., 1975. Neuronal mechanisms of memory (in Russian). Uspechi Physiol. Nauk. 6, 3: 66–78.
13. MERZHANOVA G. Ch. and SERDJUCHENKO V. M., 1973. Changes of potentials evoked stimulation of the Red nucleus in the case of direct and backward conditioned connections. Zh. Vyssh. Nervn. Deyat. Im. I. P. Pavlova, 23, 3: 632–635.
14. MERZHANOVA G. Ch. and SERDJUCHENKO V. M., 1977. Neuronal correlates

of forward and backward conditioned connections in food-procuring reflex to electrical stimulation of the lateral geniculate body (in Russian). Zh. Vyssh. Nervn. Deyat. Im. I. P. Pavlova 27, 3: 479–487.
15. PAVLOV, I. P., 1936. Foreword to the paper by J. Konorski and S. Miller (in Russian). Tr. Fiziol. Lab. Akad. I. P. Pavlova, 6: 115–118.
16. VENEGAS H., FOOTE W. and FLYNN J., 1970. Hypothalamic influences upon activity of units of the visual cortex. Yale J. Biol. Med. 42: 121–201.
17. YUNG R., 1957. Coordination of specific and unspecific afferent impulses at single neurons of the visual cortex. *In* H. Jasper (ed.), Reticular formation of the brain. Little, Broon and Co., Boston, p. 423–434.

G. MERZHANOVA, Institute of Higher Nervous Activity and Neurophysiology, USSR Academy of Sciences, Butlerova 5a, 117485 Moscow, USSR.

Lecture delivered at the Warsaw Colloquium on Instrumental
Conditioning and Brain Research
May 1979

# PACING OF BEHAVIORAL AND ELECTROENCEPHALOGRAPHIC EVENTS

T. RADIL-WEISS, Z. BOHDANECKÝ and P. LÁNSKÝ

Institute of Physiology, Czechoslovak Academy of Sciences
Prague, Czechoslovakia

*Abstract.* Time characteristics of lateral-hypothalamic self-stimulation were analyzed in rats using point process statistics. Inter-bar-pressing interval distributions were usually unimodal and asymmetrical and resembled shifted exponential distribution, i.e. bar-pressings seem to be performed more or less randomly over time. When self-stimulation was made ineffective for several seconds after stimulation, the probability of bar-pressing increased gradually, i.e. the brain seems to posses a system generating time intervals, which tend to reach certain predetermined values. When fixed rations (1 : 2–1 : 5) between rewarded and unrewarded bar-pressing were introduced, the animals followed certain patterns of alternation of inter-bar-pressing intervals of different duration. The mean inter-bar-pressing intervals remained at ratio 1 : 2 similar to those, when all reactions were rewarded, they decreased gradually at rations 1 : 3 to 1 : 5, i.e. the decrease of rewarding rate under one third to one half of the initial value is compensated by an increase of bar-pressing rate. Alternation of EEG alpha and non-alpha periods in human subjects with closed eyes was analyzed as a realization of an alternating random point process without and with feedback. The distribution of alpha periods was in the majority of cases of shifted exponential type and their incidence increased when feedback was used. The results are explained by means of a hypothesis about two alternating discrete states of the brain that correspond to particular levels of consciousness.

## INTRODUCTION

Behavioral events of the instrumental reflex type pioneered by Konorski (5, 8) and put into general theoretical frame work by Żernicki (15) and Zieliński (16) are suitable models for the study of complex brain phenomena among other reasons due to the circumstance that both stimuli and reactions are well "dated" with respect to the time of their appearance. Although the analysis of intrinsic brain mechanisms underlying such behavior is still difficult their timing can be studied objectively under certain conditions. Irrespective of the descriptive nature of the results gained in such a way, they might contribute to the explanation of the mechanisms of the processes studied.

This paper concentrates upon phenomena of all or none (binary) nature with incidence stationary (in statistical sense) over time, i.e., not influenced considerably by fatigue, satiation, adaptation etc. For describing and evaluating such phenomena the so called "point process statistics" (3) used in neurophysiology mainly for computer processing of impulse activity of neurons (10) and dealing with the time instance of the occurrence of a certain event only, irrespectively of its other possible features, may be adopted.

We deal with two types of such phenomena: self-stimulation of the brain with electric current in rat and alternation of EEG alpha and non-alpha periods without and with feedback signalization in human.

## ANALYSIS OF SELF-STIMULATION

The type of behavior studied is self-stimulation of the lateral hypothalamus by bar-pressing in rats (11, 12). A preliminary report of some of these experiments was published elsewhere (13). Present explanations of the intrinsic brain processes playing a role in self-stimulation are complex, their explanation more or less hypothetical (9) and we shall not deal with them here, limiting ourselves to an operational description of the experimental procedure and animal behavior only.

The naive animal placed in a small box with the lever only presses it randomly during the orienting and searching behavior in the new environment. Due to the electrode localization, however, brain stimulation triggers a so called "motivational" process, which influences the behavior of the animal in the sense of repeating the bar-pressing reactions. Thus some type of learning develops and its important aspect seems to be the formation of a feedback link between the "motivational" process and the specific behavioral reaction. Under favorable conditions (suitable electrode localization and stimulation parameters) "learning" occurs very fast, within several minutes, the behavioral activity itself being considerably stable (stationary) over a long time.

In this report we analyze: (i) the distribution of moments of self-stimulation over time, and (ii) the influence of some reward schedules on the distribution of bar-pressing reactions. The instrumentation used has been described elsewhere (14). When the lever is pressed, the brain is stimulated by 50 Hz AC of a controlled intensity of several tens of microamps for 0.5 s approximately. It is necessary to release the lever and to push it again to receive a new stimulation. During stimulation itself further lever pressings are ineffective (the circumstances related to changing the reward schedules will be explained later). The time instants of all bar-pressings, which are rewarded by brain stimulation, are recorded as uniform pulses on a two-channell tape recorder and later processed by means of a computer.

The first question we tried to answer was about the distribution of self-stimulating bar-pressings over time. Three unidimensional histograms of the inter-bar-pressing intervals in a rat each one containing 1024 intervals (Fig. 1, No. 1, 2, 3) are presented. In this case the histograms are taken from successive time periods during the same experi-

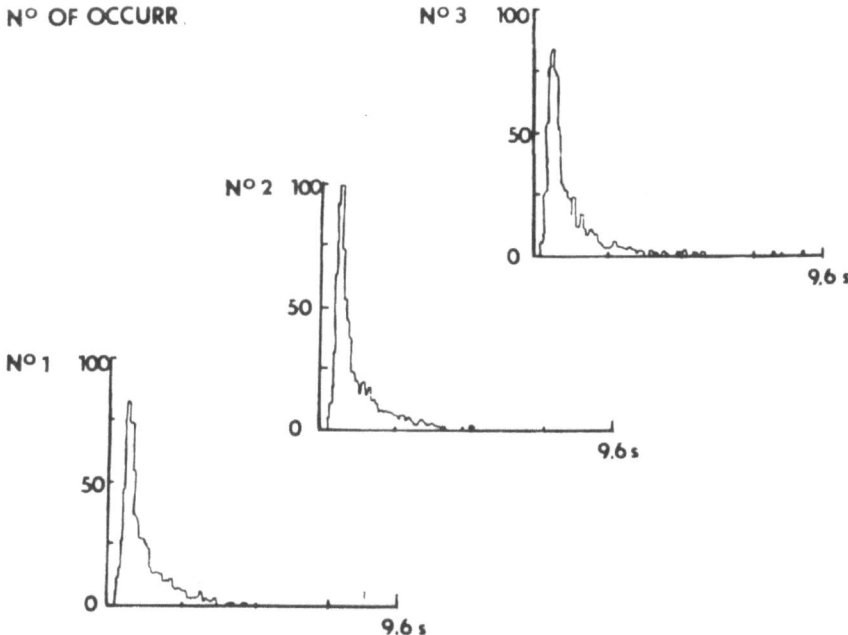

Fig. 1. The inter-bar pressing interval histograms computed from successive time periods. The abscissa corresponds to the classes of inter-bar-pressing intervals of different lenght (the whole scale being about 10 s), the incidence of intervals of certain classes in shown on the ordinate.

ment and they are remarkably similar, demonstrating the relative stability of the distribution of self-stimulation behavior over time. On the basis of many similar histograms it is possible to assume that self-stimulation neither occurs regularly, nor that alternation of certain typical length intervals takes place. The unidimensional inter-bar-pressing interval histograms are in the great majority of cases unimodal and assymetrical, the statistical behavior of the system analyzed could be at the best approximated by the Poissonian process. However, a rigorous statistical analysis of the empirical time series (for instance using a $\chi^2$) gives a positive result only in some case. In spite of that it seems to be probable that the occurrence of a self-stimulation reaction at a certain moment has a very small predictive value with respect to the occurrence of the reactions following after. In other words the bar-pressing reaction seems to be performed more or less randomly in time, as it would be determined by a kind of random generator triggering this type of behavior localized somewhere in the brain.

Manipulating the reward schedule, however, the type of distribution markedly changes. Two types of such experiments have been performed. In the first, a certain time interval has been introduced after each brain stimulation lasting either 1.3 or 4.4 s during which bar-pressing is ineffective (is not rewarded). The unidimensional interval histograms become quite different under this condition (Fig. 2). The distribution of intervals shows almost always two different peaks. The localization of the second peak of the bimodal histograms depends on the length of the artificial "refractory period". It corresponds to shorter intervals in the case the "refractory period" is shorter (1.3 s in Fig. 2, histogram 1)

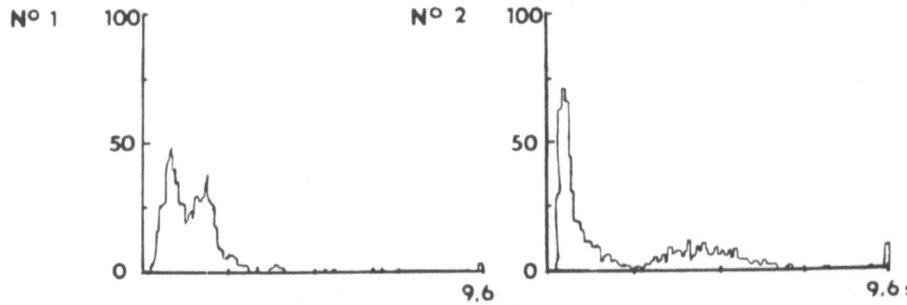

Fig. 2. The inter-bar-pressing interval histograms in experiments with artificial "refractory period". Denotations as in Fig. 1.

and to longer intervals when the "refractory period" is longer (4.4 s in Fig. 2, histogram 2). Instead of the more or less random distribution of bar-pressings over time two definitely different groups of inter-bar-pressing intervals appear. Pilot experiments in which the length of the consecutive intervals before and after changing the reward schedule have been measured, show that the adaptation to the new condition develops very rapidly (during a few minutes). In addition we computed crosscorrelation histograms (Fig. 3) reflecting the conditional probability of the occurrence of unrewarded bar-pressing during the artificially induced "refractory periods" following brain stimulation.

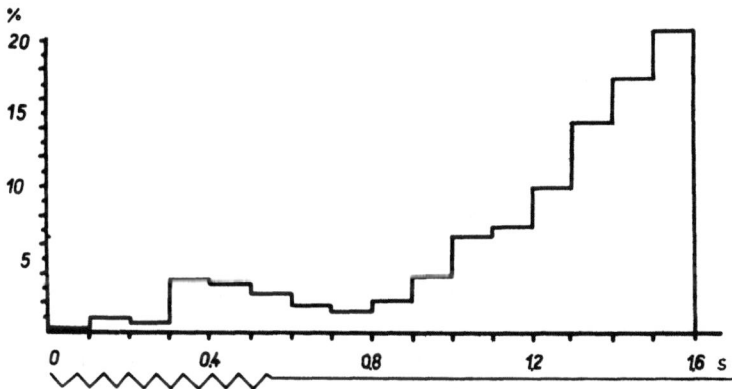

Fig. 3. The crosscorrelation histogram demonstrating the occurrence of unrewarded bar-pressing after brain stimulation. The abscissa corresponds to time after brain stimulation and the ordinate to the incidence of the unrewarded bar-pressings. (Zig-zag line indicates the time of self-stimulation, the straight line the artificial "refractory period").

These histograms help to understand how intervals of different length appear (i.e., unrewarded bar-pressings are distributed in time) with respect to the instant of the brain stimulation (i.e., the rewarded bar-pressing). The bar-pressings during the artificial "refractory periods" are not distributed randomly. The probability of performing the reaction gradually increases toward the end of this period. The basic shape is similar when the "refractory period" is short (1.3 s) or long (4.4 s). It seems to be clear that in this case the animal rapidly during a few minutes (according our orienting experiment) learns to predict with a certain probability how long the bar-pressing will be ineffective. The brain seems to posses a system capable of generating time intervals between triggering a certain behavioral event; such intervals tend to reach (with some dispersion) certain predetermined values.

In the second experiment the reward schedule has been manipulated in such a way that brain stimulation did follow bar-pressing not in all cases but in a fixed ratio (1 : 2, 1 : 3, 1 : 4, 1 : 5), the length of the intervals between the successive reactions not being important.

The most prominent consequence of this procedure in general is, similarly like in the previous case, a tendency of changing the interval histograms from unimodal (in the control experiment with ratios between bar-pressing and brain stimulation 1 : 1) to bimodal (the histograms look like in Fig. 2, histogram No. 1). Bimodal histograms often appear also when higher fixed ratios have been used. We did not observe in these cases histograms with more than two modes.

An example of the analysis of the distribution with respect to time of the unrewarded bar-pressings in an experiment with fixed ratio 1 : 5 is shown in Fig. 4. Unidimensional inter-bar-pressing interval histograms have been computed in these cases in a way shown in the upper part of the figure. It is clear that the interval corresponding to the instant in time between the peak of histogram A and B is longer than the intervals between B and C, resp. C and D. The same is true with respect to the interval corresponding to the time from the moment of rewarded stimulation to the peak in the histogram (A) of the first unrewarded reaction.

In other words in the case of fixed ratios the animal makes a longer inter-bar-pressing interval between the first ineffective bar-pressing after brain stimulation and the following ones, introducing a certain specific pattern of alternation of intervals of different length. It is worth mentioning that the mean interval length (bar-pressing rate) remains in the case of fixed ratio 1 : 2 practically the same in spite of the fact that the amount of rewarded brain stimulation decreases to 50%. The bar-pressing rate increases, however, in cases of ratio 1 : 3 and higher (13).

These experiments show that the hypothetical motivational process possesses the feature of generating repeated signals triggering self-stimulation behavior. The whole feedback system seems to work at the same repetition rate, even when a considerable reduction is introduced between the specific type of bar-pressing behavior and the hypothetical motivational process. However, a decrease of the reward rate under one third to one half of the initial value is compensated by an increase of the behavior rate.

The experimental approach mentioned may bring some new information on temporal organization of the motivational brain mechanism underlying self-stimulation and we believe it could be adopted for other cases of motivated behavior and instrumental learning.

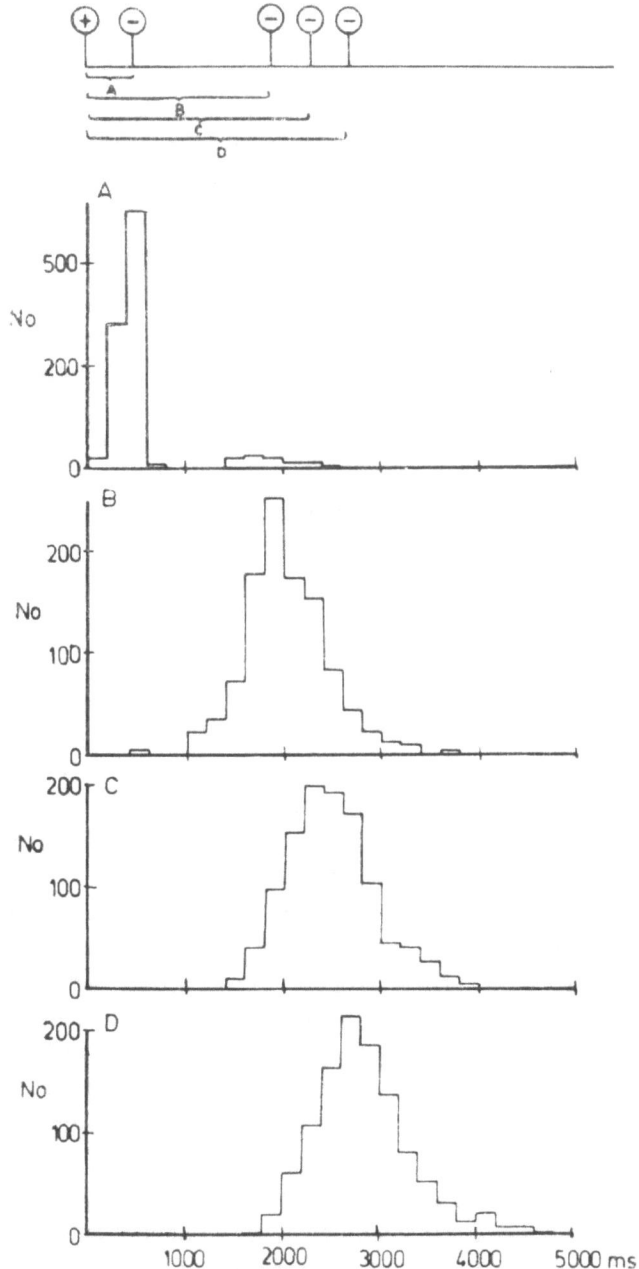

Fig. 4. Inter-bar-pressing histograms between the instant of the rewarded reaction and the following unrewarded reaction (fixed ratio 1 : 5). The histogram construction principle is pictured in the upper part. A, is the histogram of intervals between the rewarded and the next, i.e., first unrewarded bar-pressing; B, between the rewarded and the second following bar-pressing etc.

## ANALYSIS OF EEG ALPHA AND NON-ALPHA PERIODS ALTERNATION

The second phenomenon studied is the spontaneous alternation of EEG alpha and non-alpha periods in humans relaxing with their eyes closed in a dark sound-proof and electrically shielded chamber. The origin and functional significance of EEG in general and of alpha activity in the first place are complicated and contraversal matters (4, 7) and we shall not deal with them here, limiting ourselves again to the description of the experimental procedure and some of the results.

Electrodes were fixed and EEG recorded from $O_z$ position referenced to $A_1$. After filtering the activity with 8–12 Hz bandpass filter, histograms of local EEG extremes have been constructed and their median estimated. The treshold corresponding to that median enables to differentiate between alpha and non-alpha; computerized statistical analysis has been performed based on the duration of consecutive alpha and non-alpha time periods. The resulting process represents from the statistical point of view a special point process characterized by two alternating states with instantaneous transitions between them in both directions (1, 6).

Under the described conditions psychic relaxation is combined with the appearance of alpha rhythm, whereas mental activity with non-alpha intervals. We have found previously that the process of spontaneous alternation of both types of intervals fullfils in majority of cases the condition of statistical stationarity. It is important to mention that the subjects are not aware of their actual EEG.

Three 15 min consecutive different recordings, divided by short interruptions, were run in a single session. Equally remunerated students served as subjects. Another session was repeated within several days. During the first recording the spontaneous alternation of alpha and non-alpha periods was recorded. During the second one a low intensity tone signaled (as an external feedback signal) the presence of alpha activity and the task of the subjects was to maximize its duration. During the third trial no external feedback signal was given, but the subjects had to try to increase the amount of alpha activity by means of any type of internal psychological procedure they were capable.

The basic finding in all such experiments is that the histograms of alpha periods correspond in the majority of cases to a shifted exponential distribution (Fig. 5). The statistical distributions of non-alpha periods is often more complicated.

A mathematical model based upon the quequeing theory (1) — we shall not deal with in detail — has been adopted to explain this finding. Alpha intervals correspond to "non-busy" and non-alpha intervals to

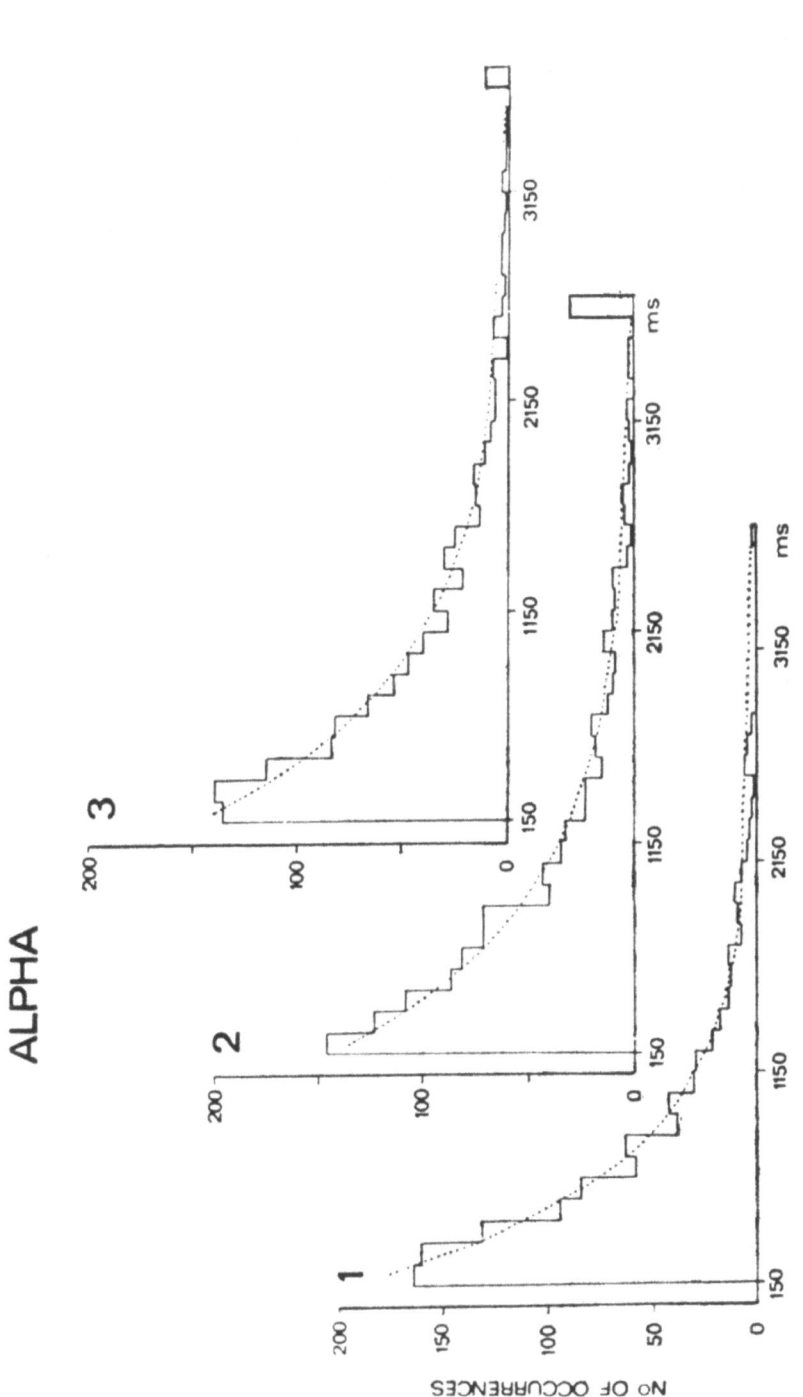

Fig. 5. An example of theoretical (dotted line) and empirical (solid line) distribution of alpha intervals duration for 3 different sessions in one subject.

"busy" periods of the brain during which mental operations are performed in a sequential way (2), the instance of their appearance being random and independent (Poisson process).

The question whether the "instrumentation" of the alpha and non-alpha alternation phenomenon by means of an external or internal feedback loop reinforcing the presence of alpha causes definite changes in the time schedule of alternation of both intervals could not be answered

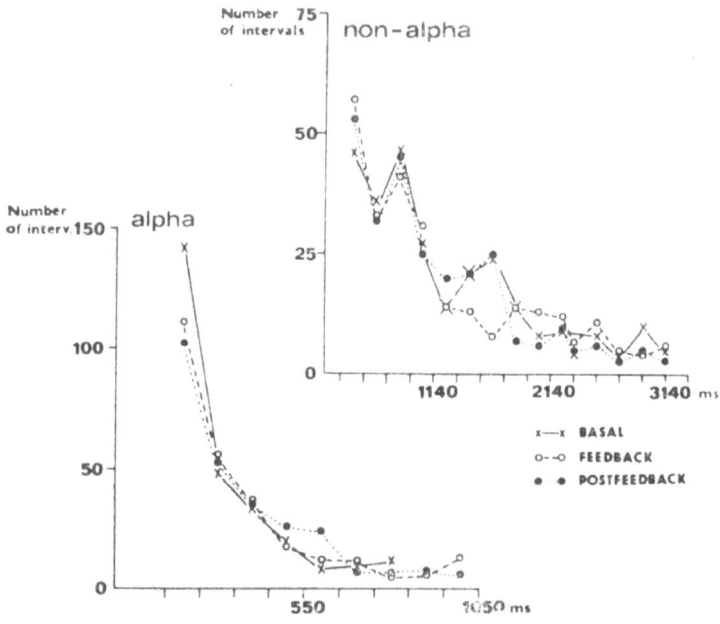

Fig. 6. An example of the distribution of alpha and non-alpha intervals duration for three different situations in one subject.

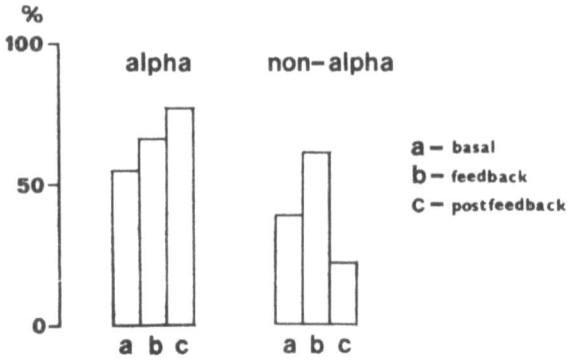

Fig. 7. The **number of sessions** (expressed in %) where alpha and non-alpha intervals are exponentially distributed.

definitely yet. The training interval was probably too short. It has been found, however, that fullfilling the criterion of shifted exponential distribution (Fig. 6) increases in the whole group of subjects in cases when feedback was used.

Further experiments are required to test the possibility whether different nonrandom time schedule generation mechanism controlling the flow of brain processes responsive for alternation of alpha and non-alpha periods as well as mental states linked to them can be built up by certain types of instrumental conditioning procedures and other similar approaches (Fig. 7).

## REFERENCES

1. BOHDANECKÝ, Z., LÁNSKÝ, P., INDRA, M., and RADIL-WEISS, T. 1978. EEG alpha and non-alpha intervals alternation. Biol. Cybern. 30: 109–113.
2. BROADBENT, D. E. 1954. The role of auditory localization in attention and memory span. J. Exp. Psychol. 47: 191–196.
3. COX, D. R. and LEWIS, P. A. W. 1966. The statistical analysis of series of events. Methuen, London, 287 p.
4. ELUL, R. 1967. Statistical mechanisms in generation of the EEG. In L. J. Fogel and F. W. George (ed.), Progress in Biomedical engineering. Vol. 1 Spartan Books, Washington, p. 131–150.
5. KONORSKI, J. and MILLER, S. 1936. Conditioned reflexes of the motor analyzer (in Russian). Tr. Fiziol. Lab. I. P. Pavlova 6: 119–278.
6. LÁNSKÝ, P., BOHDANECKÝ, Z., INDRA, M. and RADIL-WEISS, T. 1979. To alpha detection (Some comments on Hardt's and Kamiya's: Conflicting results in EEG alpha feedback studies). Biofeedback and selfregulation 4-127-131.
7. LINDSLEY, D. B. and WICKE, J. D. 1974. The electroencephalogram: autonomous electrical activity in man and animals. In R. F. Thompson and M. M. Patterson (ed.), Bioelectrical recording techniques. Part B. Academic Press, New York, p. 4–83.
8. MILLER, S. and KONORSKI, J. 1928. Sur une forme particuliere des reflexes conditionnels. C. R. Soc. Biol. 99: 1155–1158.
9. MILNER, P. M. 1976. Models of motivation and reinforcement. In A. Wanquirer and E. T. Rolls (ed.), Brain-stimulation reward. North Holland Publ. Comp. p. 543–556.
10. MOORE, G. P., PERKEL, D. H. and SEGUNDO, J. P. 1966. Statistical analysis and functional interpretation of neuronal spike data. Annu. Rev. Physiol. 28: 493–522.
11. OLDS, J. and MILNER, P. 1954. Positive reinforcement produced by electrical stimulation of septal area and other regions of rat brain. J. Comp. Physiol. Psychol. 47: 419–427.
12. OLDS, J., TRAVIS, R. P. and SCHWING, R. C. 1960. Topographic organisation of hypothalamic self-stimulation functions. J. Comp. Physiol. Psychol. 53: 23–32.
13. RADIL-WEISS, T. and BOHDANECKÝ, Z. 1979. Probabilistic analysis of

selfstimulation in rats. Proceedings of the Conference on Motivation. Pécs, Hungary (in press).
14. VALÁŠEK, J. and RADIL-WEISS, T. 1968. Simple instrumentation for generation of reward schedules in autostimulation experiments. Act. Nerv. Super. 10: 404–405.
15. ŻERNICKI, B. 1975. Drive-controlled reflexes: a theory. Acta Neurobiol. Exp. 35: 475–490.
16. ZIELIŃSKI, K. 1979. Extinction, inhibition and differential learning. In A. Dickson and R. A. Boakes (ed.), Associative mechanisms in conditioning. Erlbam Assoc. Hillsdale, p. 269–293.

T. RADIL-WEISS, Z. BOHDANECKÝ and P. LÁNSKÝ, Institute of Physiology, Czechoslovak Academy of Sciences, Budějovická 1083, 14220 Prague 4-KRČ, Czechoslovakia.

Lecture delivered at the Warsaw Colloquium on Instrumental
Conditioning and Brain Research
May 1979

# A STUDY OF SYNAPTIC PLASTICITY IN HIPPOCAMPAL SLICES

V. G. SKREBITSKY and V. S. VOROBYEV

Brain Institute, USSR Academy of Medical Sciences
Moscow, USSR

*Abstract.* Long-lasting potentiation in the hippocampal pathways is used at present as a model for long-term plasticity in the nervous system. In this study post-tetanic potentiation was investigated in the dentate gyrus-area $CA_3$ pathway by extra- and intracellular recordings from transverse slices of the mouse hippocampus. Tetanization of the dentate gyrus led to a reduction in the latency of action potentials (APs) and EPSPs recorded from area $CA_3$, to an increase in the amplitude of EPSPs and in the steepness of their ascending slope, and to an augmented probability of APs. These changes persisted for a period of several seconds to 30 min after tetanization. Of special interest were records from cells responding with EPSPs of a short latency (2–3 ms) which was not changed by an increase in the frequency and strength of stimulation. We assume that such EPSPs are monosynaptic. Our results suggest that monosynaptic EPSPs can undergo long-lasting (up to 30 min) post-tetanic potentiation.

## INTRODUCTION

The discovery of post-tetanic potentiation in the hippocampus which lasts for hours or even days (3–5) stimulated studies of synaptic activity in hippocampal pathways. Some to the studies examining various features of post-tetanic neuronal responses in the dentate gyrus and in areas $CA_1$ and $CA_3$ were performed in vitro on brain slices (1, 6, 9, 11, 14). The latter appear to be a convenient model for investigation of synaptic plasticity at least for two reasons: (i) one may be sure that changes in responses occurring during and after tetanization are localized in a given lamina; (ii) powerful IPSPs determined to some extent by between hippocampal laminar inhibitory connections and masking potentiation of EPSPs are less expressed in slices than in the intact hippocampus (10).

To obtain further information on neuronal correlates of post-tetanic potentiation, we carried out several series of experiments on slices with stimulation of the dentate gyrus and extra- and intracellular registration in area $CA_3$. Special attention was focused on changes in short-latency, presumably monosynaptic responses.

## METHODS

The methods for preparing hippocampal slices and for recording were essentially identical to those described by Yamamoto in 1972 (13). Mice aged 1–2 mo were used for the experiments. After decapitation, the brain was exposed, the neocortex removed and transverse hippocampal slices (about 300 μm thick) were cut with a razor blade under a binocular microscope. The slices were immediately transferred onto a nylon net fixed in the recording chamber so that its interior side was bathed in a stream of Simms' balanced salt solution (heated to 36–37° and gassed with 95% $O_2$ and 5% $CO_2$). Stimulating electrodes (a pair of glass micropipettes filled with Simm's solution, resistance about 100 mΩ, interelectrode distance 0.2 mm) were placed into the dentate gyrus. Stimuli (duration, 0.2 ms; strength, 1–50 V) were delivered at a frequency ranging from 0.3 to 100 Hz. Recording microelectrodes (glass micropipettes filled with 2 M potassium citrate; resistance, 50–100 mΩ) were advanced with the aid of a low-power microscope. The distance between the stimulating and recording electrodes was, as a rule, 1–3 mm.

Electrical activity was registered in 30 slices. In 50 cells the recording was extracellular. For an analysis of synaptic events, intracellular records were made from 30 cells for a period of at least 30 min. In 25 of these cells membrane potentials (MPs) and action potentials (APs) were 40–60 mV, while in the rest, MPs were less than 40 mV and generation of APs ceased during the registration procedure.

## RESULTS

### Extracellular recording

Stimulation with weak (1–10 V) single shocks often failed to evoke APs. If and when they occurred, their latency was usually 3–5 ms.

Increasing the strength of stimuli or tetanizing (5–20 Hz, 10–15 s) was accompanied by a reduction in AP latency.

After tetanization, responses appeared more often and their latencies were reduced at least for one minute. Although we did not systematically explore long-lasting potentiation (LLP) through extracellular recording, it was possible to observe that in some instances these changes persisted for a long time.

An example of LLP is given in Fig. 1. It may be seen that during tetanization the latencies of evoked APs were reduced from 2.5 ms (A, 0.1 Hz) to 1.5 ms (B, 20 Hz) and their probability [1] rose from 0.1–0.3

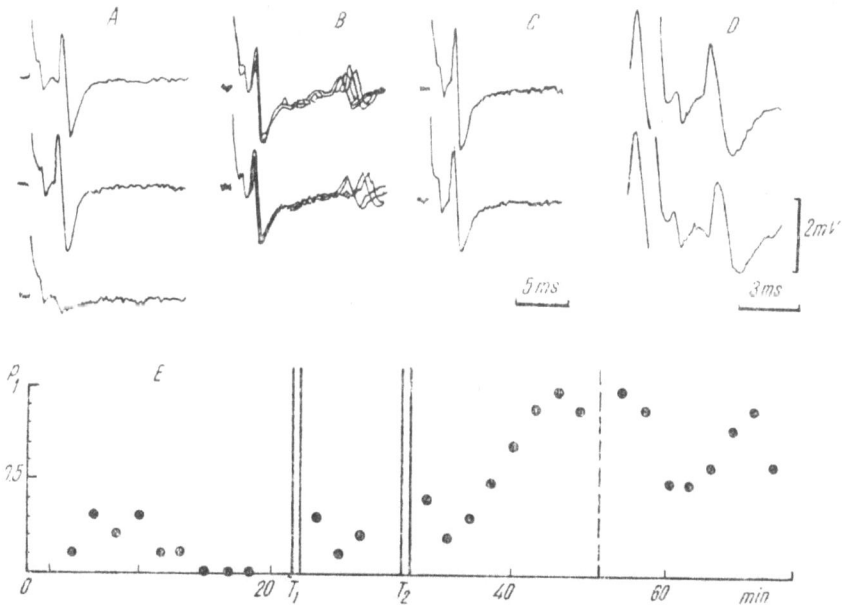

Fig. 1. Long-lasting post-tetanic potentiation of neuronal responses in area $CA_3$ to dentate gyrus (DG) stimulation. A, neuronal responses to low-frequency (0.1 Hz) stimulation; B responses during tetanization (20 Hz, 5 s); C, after tetanization; D, the collision test indicated that the responses were orthodromic; stimulation was triggered by a spontaneous spike; E, plot of probability of responses (P) before and after the first ($T_1$) and the second ($T_2$) tetanization. The probability of responses (filled circles) was calculated after presentation of 10 stimuli (0.1 Hz). Interrupted vertical lines: 10-min pauses in recording. Here and in all other records stimuli trigger the sweep. The spikes have been retouched.

to 1 (E). These changes in AP latencies (C) and probability (E) persisted for at least 40 min after tetanization was over. Record D demonstrating the collision test (2) indicates that the responses were orthodromic.

---

[1] Number of times of appearance of a AP after 10 successive stimuli.

## Intracellular recording

In many intracellular records the responses to stimulation of the dentate gyrus consisted of EPSPs followed by protracted IPSPs. The latter were especially prominent in damaged cells (with MPs less than 40 mV) and often were hardly noticeable in well-preserved cells. When the MP was about 60 mV the IPSPs were obscured, but when the membrane was depolarized in the course of registration they became clearly expressed in the same cell.

An increase in the strength of a stimulus reduced both the latency of the initial EPSP and the interval between its beginning and rise to the firing level (rise time); thus its ascending slope became steeper. Figure 2A shows the variation in latency and in the rise time of EPSPs

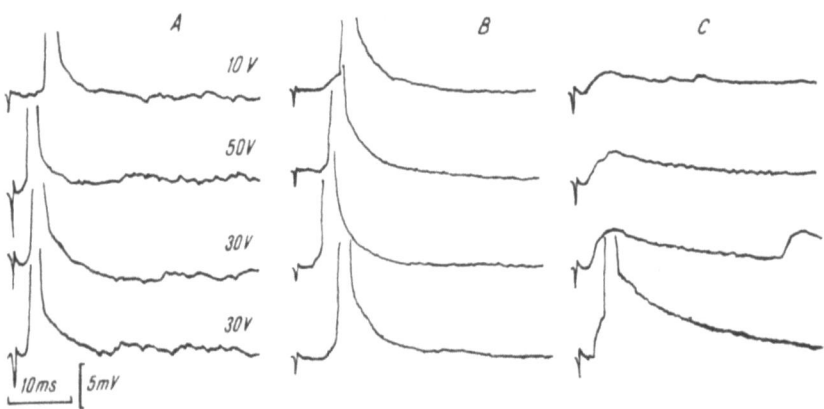

Fig. 2. Intracellular records of neuronal responses in area $CA_3$ to DG stimulation. The strength of stimulation was changed in A from 10 V to 50 V then to 30 V and was constant in B and C. The membrane potential (about 50 mV) rose by several mV from B to C for an unknown reason. Here and in the following figures (except Fig. 5) the spikes are cut.

evoked by stimuli of different strengths. It may be seen that the latency fell from 5 to 1.5 ms and the rise time was shortened when the strength of stimulation was augmented from 10 to 50 V. A decrease in strength to 30 V resulted in a latency of 2 ms.

Still both parameters fluctuated to some extent even if the strength of stimulation was unchanged. Figure 2B demonstrates spontaneous changes in EPSP latency and rise time. Depolarizing afterpotentials (DAP) described in hippocampal neurons in vivo (8) and in vitro (10) are quite prominent in this cell.

When stimuli were applied with a frequency of 0.3 Hz sequential responses did not, as a rule, affect each other. However, some records reveal potentiation of EPSPs even at such a low repetition rate (Fig. 2C).

High frequencies of stimulation also reduced latency (especially in long-latency EPSPs) and the rise time. The probability of responses was increased. These changes persisted after tetanization.

Figure 3 shows cell responses to low-frequency (0.3 Hz) stimulation before (A) and after (C) tetanization (50 Hz). During tetanization (B) short-latency (2.5 ms) EPSPs rose steeply and always triggered APs. When tetanization was stopped the slope of the EPSPs gradually decreased during the next 12 s.

Of special interest were records from cells responding with EPSPs of a short (2-3 ms) latency which was not changed significantly by an increase in the frequency and strength of stimulation. We assume that such EPSPs are monosynaptic.

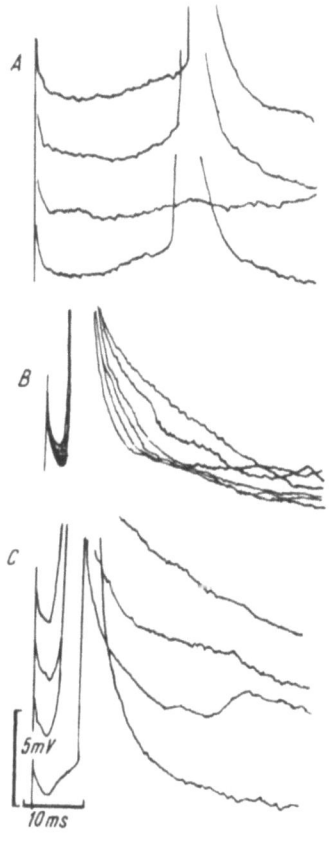

Fig. 3. Intracellular records of neuronal responses before (A), during (B) and immediately after (C) tetanization (50 Hz, 4 s). Frequency of stimulation in A and C, 0.3 Hz.

Figure 4 demonstrates the distribution of amplitudes of presumably monosynaptic EPSPs induced in a neuron of area $CA_3$ by low-frequency (0.1 Hz) stimulation of the dentate gyrus. Some examples of responses are presented in the left panel; on the right is a histogram showing the number of observations (N) as a function of the amplitude of EPSPs for class intervals of 0.1 mV.

Two examples of EPSPs induced by single stimuli which were below the threshold of AP triggering are given in Fig. 5A. Figure 5B shows responses to tetanization (100 Hz). Since high-frequency stimulation was switched on gradually, the first few shocks had a frequency of about 20 Hz. It may be seen that the amplitude of the first three EPSPs went on rising, whereas their latency remained constant. Subsequently each stimulus evoked APs, but their amplitude declined in the course of stimulation due to inactivation of the spike generator mechanism. Two potentiated responses registered 10 and 60 s after tetanization are presented in Fig. 5C. In both cases steep EPSPs (compare with the control record in Fig. 5A) reached the firing level and triggered two or three APs. It

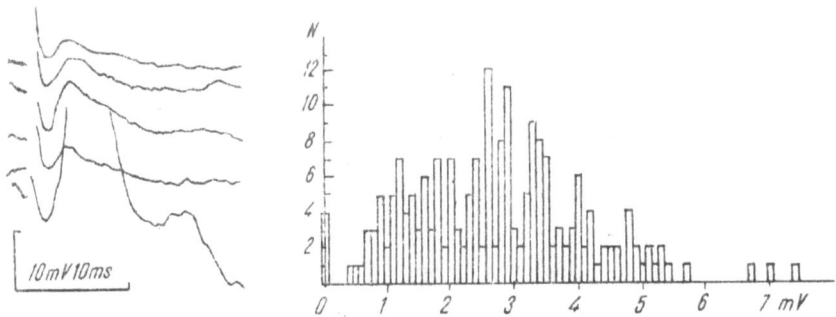

Fig. 4. Presumably monosynaptic EPSPs induced in a neuron of area $CA_3$ by low-frequency (0.1 Hz) stimulation of DG. Left, examples of synaptic and spike (lower record) responses. Right, a histogram showing the number of observations (N) as a function of the amplitude of EPSPs for class intervals of 0.1 mV. Total number of observations, 200.

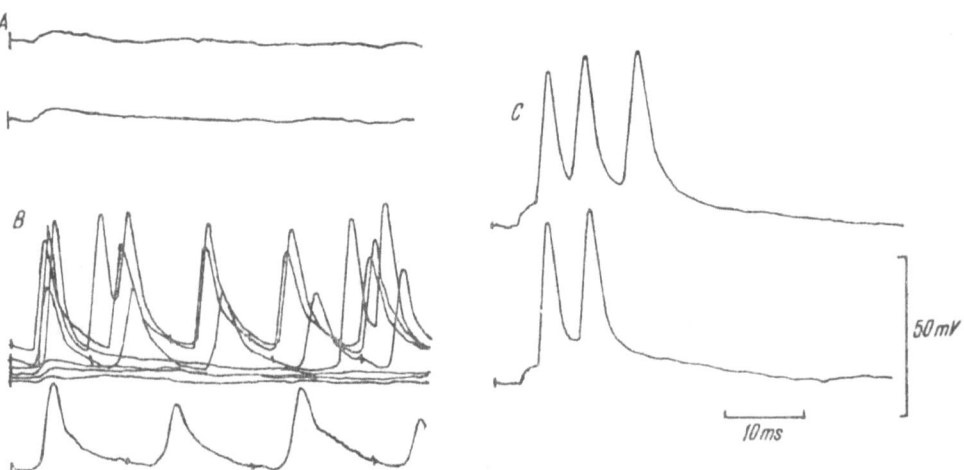

Fig. 5. Post-tetanic potentiation of presumably monosynaptic EPSPs. In all records presentation of stimuli is designated by vertical bars. A, control responses; B, frequency potentiation during tetanization (20-100 Hz, 5 s); C, responses registered 10 and 60 s after tetanization. For details see text.

should be noted that the MP was almost the same in records A and C, whereas during tetanization (B) it was lower by about 10 mV.

In records made from three neurons, post-tetanic potentiation, as described above, persisted for about 30 min. A comparison of the distribution of EPSPs during the first and second 10-min periods after tetanization (Fig. 6, central and lower histograms respectively) and in the control period (Fig. 6, upper histogram) shows that the proportion of low-amplitude (less than 1 mV) EPSPs decreased. Thus, LLP was accompanied by a rise in the amplitude of EPSPs in presumably monosynaptic pathways.

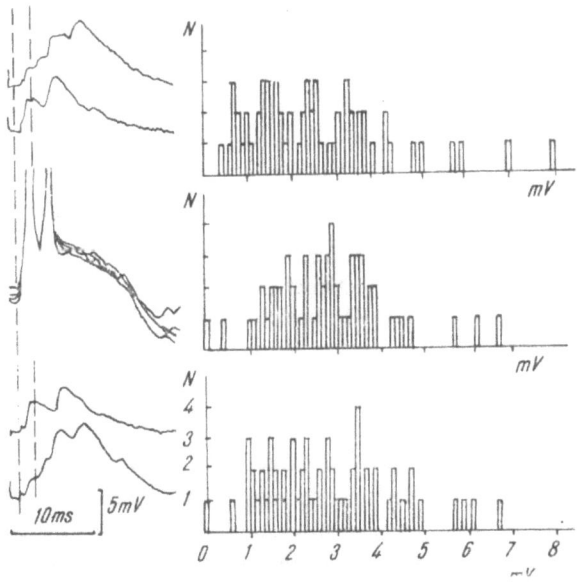

Fig. 6. Long-lasting potentiation of presumably monosynaptic EPSPs. Left, examples of responses before (top), during (center) and 10 min after (bottom) tetanization. The amplitudes of EPSPs were measured between two points indicated by interrupted lines. Right, distribution of EPSP amplitudes (see Fig. 4) before (top), within 10 min after tetanization (center) and during the following 10 min (bottom). Total number of observations for each histogram, 62.

The following synaptic events could be seen in records obtained from one cell. The responses to stimuli applied in the control period (Fig. 7A) looked like "dendritic spikes" (10). After tetanization, their amplitudes were greatly augmented (Fig. $7B_{1,2}$) and in some cases they triggered APs (Fig. $7B_3$). During this period "dendritic spikes" were clearly expressed also in spontaneous activity (Fig. 7$B_4$). Within 15 min after tetanization the amplitudes of "dendritic spikes" returned to the control level (Fig. 7$C_{1,2}$) but the probability of APs was still increased (Fig. 7$C_3$), possibly as a result of general activation of the cell (Fig. 7$C_4$).

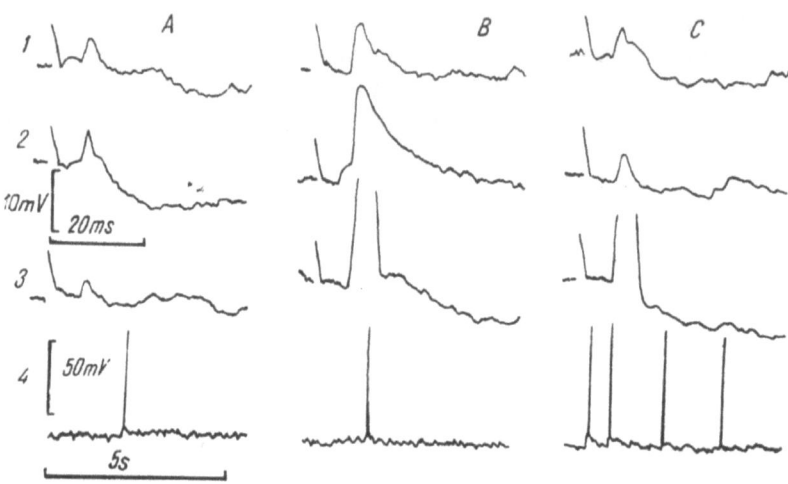

Fig. 7. Post-tetanic potentiation of "dendritic spikes". $A_{1-3}$: control responses: $B_{1-3}$ and $C_{1-3}$: responses recorded respectively 3 and 15 min after tetanization (20 Hz, 5 s); $A_4-C_4$: background activity.

## DISCUSSION

The results of this study confirm the data obtained previously in hippocampal slices (14) about neuronal correlates of long-lasting potentiation in the endings of mossy fibers on the $CA_3$ pyramids. In extracellular records potentiation of synaptic transmission manifests itself by a reduced latency and an augmented probability of APs. The corresponding synaptic events are a reduction in latency and an increase in the amplitude and in the steepness of the ascending slope of EPSPs.

We were especially interested in the problem of monosynaptic pathways. Two criteria were used for their identification: (i) short latency (2–3 ms) of the initial EPSP; (ii) absence of significant changes in latency upon increasing the frequency and strength of stimulation (some insignificant errors might have arisen from the difficulty in estimating the beginning of EPSPs). Responses satisfying both criteria were considered monosynaptic, although convincing proof for this supposition was lacking.

If these criteria are valid, the records presented here testify that long-lasting (up to 30 min) potentiation of synaptic transmission can occur in the monosynaptic pathway from the dentate gyrus to area $CA_3$. The amplitudes of presumably monosynaptic EPSPs do not have the typical Gaussian or Poissonian distribution. Therefore they are probably unsuitable for a quantal analysis of transmitter release.

It is still unclear whether LLP is mediated by pre- or postsynaptic events. There is practically no information on ultrastructural correlates of LLP in mossy fiber synapses. Some morphological evidence was obtained in studies of dentate granular cells and $CA_3$ pyramids following tetanic stimulation of the entorhinal cortex (7) and the septum (12). The increase in postsynaptic surface area ("swelling of dendritic spines") and post-synaptic density found in these studies suggests a post-synaptic origin for LLP.

## REFERENCES

1. ANDERSEN, P., SUNDBERG, S. H., SVEEN, O. and WINGSTRÖM, H. 1977. Specific long-term potentiation of synaptic transmission in hippocampal slices. Nature 266: 736–737.
2. BISHOP, P. O., BURKE, W. and DAVIS, R. 1962. The interpretation of extracellular responses of single lateral geniculate cells. J. Physiol. (Lond.) 162: 452–473.
3. BLISS, T. V. P. and LÖMO, T. 1973. Long-lasting potentiation of synaptic transmission in the dentate area of the anaesthetized rabbit following stimulation of the perforant path. J. Physiol. (Lond.) 223: 331–356.
4. BRAGIN, A. G. and VINOGRADOVA, O. S. 1973. "Chronic" potentiation of mossy fiber synapses (cortical input) on the $CA_3$ neurons in the hippocampus. In Fiziologicheskie mechanizmy pamyati. Puschino-na-Oke, p. 8–24.
5. DOUGLAS, R. M. and GODDARD, G. V. 1975. Long-term potentiation of the perforant path-granule cell synapse in the rat hippocampus. Brain Res. 86: 205–215.
6. DUDEK, F. E., DEADWYLER, S. A., COTMAN, C. W. and LINCH, G. S. 1976. Intracellular responses from granule cell layer in slices of rat hippocampus: perforant path synapse. J. Neurophysiol. 39: 384–393.
7. FIFKOVA, E. and Van HARREVELD, A. 1977. Long-lasting morphological changes in dendritic spines of dentate granular cells following stimulation of entorhinal area. J. Neurocytol. 6: 211–30.
8. KANDEL, E. R. and SPENCER, W. A. 1961. Electrophysiology of hippocampal neurons. II. After-potentials and repetitive firing. J. Neurophysiol. 24: 243–259.
9. LINCH, G. S., DUNWIDDIE, T. and GRIBKOFF, V. 1977. Heterosynaptic depression: a postsynaptic correlate of long-term potentiation. Nature 266: 737–739.
10. SCHWARTZKROIN, P. A. 1975. Characteristics of $CA_1$ neurons recorded intracellularly in the hippocampal in vitro slice preparation. Brain Res. 85: 423–436.
11. SCHWARTZKROIN, P. A. and WESTER, K. 1975. Long-lasting facilitation of a synaptic potential following tetanization in the in vitro hippocampal slice. Brain Res. 89: 107–119.

12. WENZEL, J., BOGOLEPOV, N. N. and SKREBITSKY, V. G. 1978. Strukturelle Aspekte der Plastizität in Synapsen des Hippocampus. *In* O. S. Adrianov (ed.), Lokalizatsya i organizatsya mozgovych funktsi. Moscow, p. 211–212.
13. YAMAMOTO, C. 1972. Activation of hippocampal neurons bo mossy fiber stimulation in the brain section in vitro. Exp. Brain Res. 14: 423–435.
14. YAMAMOTO, C. and CHUJO, T. 1978. Long-term potentiation in thin hippocampal sections studied by intracellular and extracellular recordings. Exp. Neurol. 58: 242–250.

V. G. SKREBITSKY and V. S. VOROBYEV, Brain Research Institute, Academy of Sciences of the USSR, Per. Obukha 5, Moscow, USSR.

Lecture delivered at the Warsaw Colloquium on Instrumental
Conditioning and Brain Research
May 1979

# CENTRAL EXCITATION AND INHIBITION IN CONDITIONED REFLEXES

## V. M. STOROZHUK

A. A. Bogomoletz Institute of Physiology, Kiev, USSR

*Abstract.* Cortical neuronal reactions were studied in cats during the conditioning of a defensive reflex to auditory stimuli. Conditioned cortical neuronal reactions were polyphasic, in which the initial responses, the temporal depression of activity, and the early and late afterdischarges could be distinguished in. In many neurons during conditioning, a strengthening of initial impulse responses was found, as was the appearance and augmentation of afterdischarges. Comparison between the latency of conditioned movements and phases of neuronal reactions suggested that the conditioned reactions represented primarily the modified afterdischarges evoked by conditioned stimuli. Elimination of the reinforcement was accompanied by the gradual recovery of neuronal reactions to the initial level. Early afterdischarges were most stable during extinction. In contrast to usual extinction, differentiated inhibition was found to be a more active process: in some neurons ordinary extinction occurred, while in the others the impulse reactions arose and were augmented. It is concluded that changes of excitatory processes underly central inhibition.

## INTRODUCTION

The neurophysiological mechanisms of conditioned reflexes are a central problem of brain research. Many investigators consider the study of neuronal activity of cerebral structures under classically and instrumentally conditioned reflexes (CRs) as a way to solve this problem (6–8, 11, 12). Neuronal activity more often has been observed in animals after preliminary elaboration of conditioned reflexes, allowing the observation of comparatively stable reactions (3, 4). But it is not always clear if such stable reactions have a specific conditioned nature. At the same time, the process of the development of conditioned neuronal reactions remains vague. On the other hand, the analysis of the development of conditioned neuronal reactions in the course of the elaboration of the classical CR has been accompanied by the estimation of the neuronal reactions for large time intervals of 0.5 to 1 s. Moreover the CR has been evaluated only as an increase or decrease of neuronal impulse activity (5, 6, 11).

It is known that neuronal activity appears earlier and extinguishes later than the corresponding conditioned movement. Neuronal reactions have often an unstable, fluctuating character during the stable conditioned stimulus (CS). Therefore, sometimes investigators do not compare neuronal reactions with the movement during the elaboration of the CR, and some investigators even artificially remove movements for better microelectrode registration of neuronal activity (5). The purpose of this investigation was to observe the dynamics of neuronal reactions during the training of classically conditioned defensive reflexes to auditory stimulation as well as to examine some forms of conditioned inhibition.

## METHODS

The experiment was conducted with male cats that were first surgically treated under nembutal anesthesia. Access to the cortex was made and a metallic curved plate was attached to the skull by dental acrylic. The plate permitted the fixing of the cat's head during the experiment. Training of the CR began on the third day after surgery. A series of 100 clicks per second during the CS–US interval of 800–1,000 ms comprised the CS, while a series of 50 clicks per second in the same interval constituted the differential stimulus. The reinforcement consisted of the stimulation of the paw by a single electric shock at the end of the auditory stimulus. Paired stimuli were presented after irregular time intervals lasting between 1.5 and 5 min. The appearance of local paw movement to the single auditory CS was the evidence of the CR acquisition.

The neuronal impulse activity was conducted by ordinary glass microelectrodes and recorded by a taperecorder and by an inkwriter.

The movements were conducted by piezodevise and were also recorded by the inkwriter. The estimation of neuronal reactions was made by peristimulus histograms of the impulse activity between 1 s before and 3 s after the beginning of the auditory stimulus; one bin of histogram 50 ms.

## RESULTS

Conditioned paw movements typically appeared after 20 to 40 pairings of stimuli with usual latencies in the range of 80 to 400 ms. In single cases the first movements arose later, approximately at the moment of reinforcement. Sometimes the first movements during the elaboration of the CR were startle reactions with latencies of 25 to 50 ms, and conditioned local paw movements appeared only with the continuation of paired stimuli.

The neuronal responses of cortical areas 3 and 4 to electrical stimuli were usually phasic reactions consisting of the initial response, the short-term depression of activity and the period of the afterdischarges. It is known that the basis of the initial responses are excitatory postsynaptic potentials (EPSPs) evoking afferent vollies coming to the cortex. A cause of the depression is disynaptic afferent and polysynaptic feedback IPSP. Afterdischarges are the result of the rebound and the additional synaptic activity of intracortical and subcortical origin (9).

Only some somatic cortical neurons, about 30% in the motor cortex, were active during the auditory stimuli before acquisition of the CR to the stimulus. Their reactions had unstable and illegible character.

The reactions of the somatosensory and motor cortical neurons to the auditory stimuli became clearer after CR acquisition (10). It appeared possible to differentiate some phases of these reactions resembling phases of the response to the electrical stimulus: the initial response of one or several impulses with latencies of 20 to 30 ms after CS onset (Fig. 1: 2, 6, 7), a phase of depression of impulse activity (50 to 100 ms) and a phase of afterdischarges, some of which developed without preliminary initial responses (Fig. 1: 3, 7) and had latencies of 80 to 200 ms. Some late afterdischarges occurred toward the end of the CS (Fig. 1: 4, 9). Neurons with long temporal depression of impulse activity during the CR were observed rarely (5, 10). Parallel neuronal reactions developed to the conditional stimuli in areas 2 and 5.

The most safe criterion for true neuron participation in the CR is parallel to its dynamics during acquisition and extinction of conditional movements. Accordingly, we emphasized the dynamics of neuronal reactions at early stage of the CR acquisition (Fig. 2A). Neurons from area 2 had rare background activity and had weak reactions to afferent stimuli. Initial impulse responses to the auditory stimulation increased and achieved a maximum at 21–25 pairings. Early afterdischarges in-

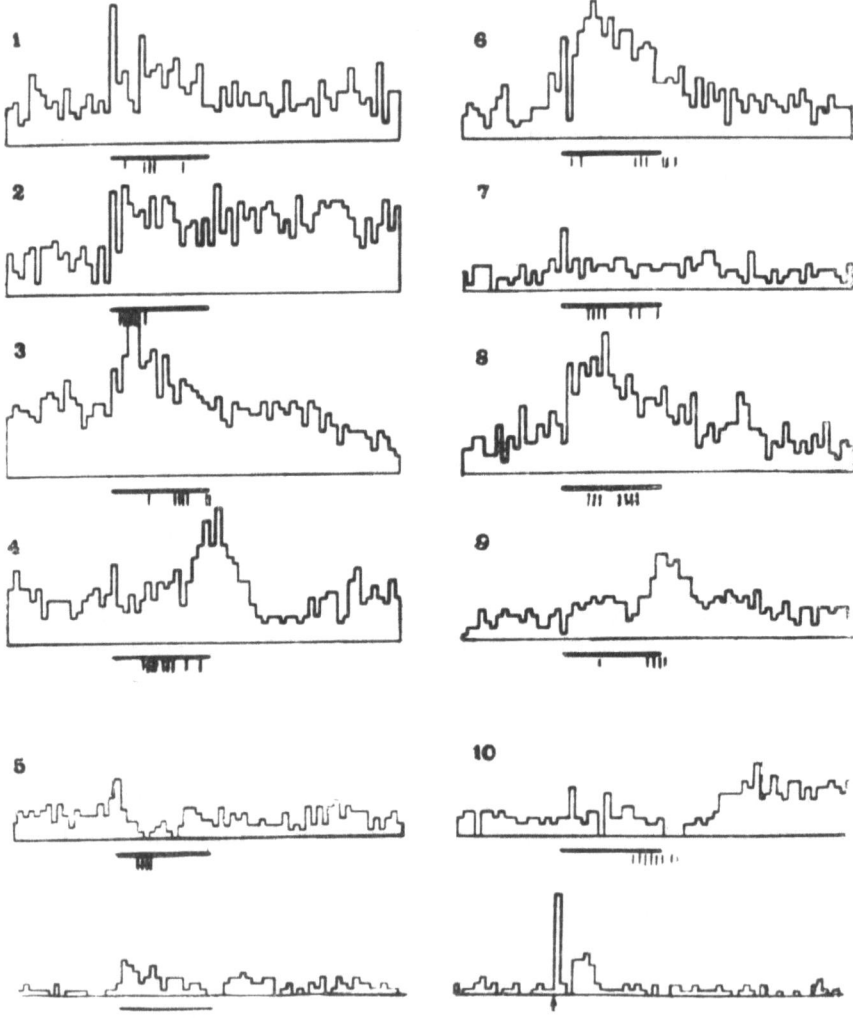

Fig. 1. Types of the neuronal reactions to the conditioned auditory stimulus. Left panel: neuronal reactions of area 3 initial response and afterdischarges (1); the initial response, short-term inhibition and afterdischarges (2); early (3) and late (4) afterdischarges; prolonged depression of impulse activity (5). Right panel: neuronal reactions of area 4 the initial response, short-term inhibition and afterdischarges (6); the initial response (7); early (8) and late (9) afterdischarges; prolonged depression (10). Below: response to unconditioned sound stimulus (left) and electrical shock (right). Each bar is the sum of 10 trials, bin width = 50 ms; horizontal line indicates the auditory stimulus duration (1 s), the vertical bar shows the begining of the conditioned movement.

creased the independence, and during the course of association, they appeared earlear then the initial response (Fig. 2A, 5–7). Late afterdischarges appeared after 20 pairings (7, 10).

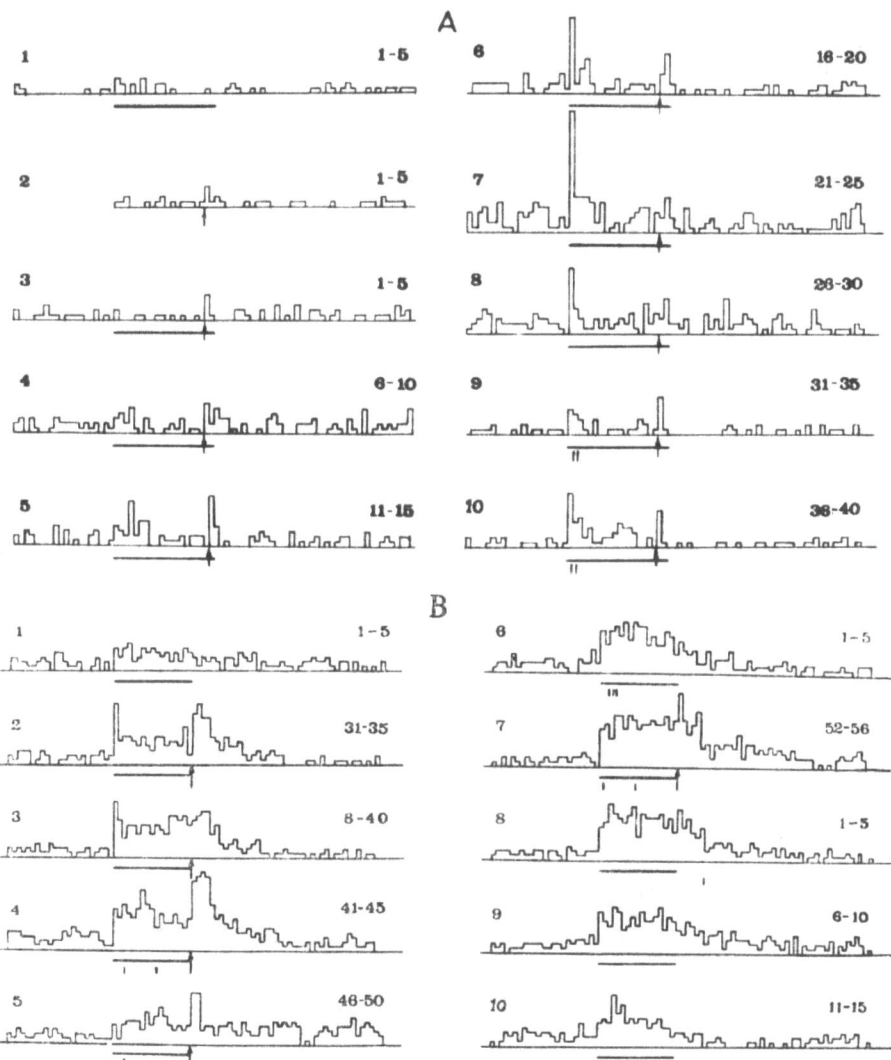

Fig. 2. The dynamics of area 3 (A) and 4 (B) neuronal responses during acquisition of the CR. The numbers on the right side of each histogram denote the numbers of trials. For further explanations see text.

A second neuron from the motor cortex, recorded after 30 pairings, responded to CSs with impulse activity only slightly exceeding the background activity (2B, 1). The initial response increased sharply, while the afterdischarges increased slowly over the course of pairings. The initial response and afterdischarges appeared only in this neuron after 30 pairings. Conditioned neuronal reactions after 50 pairings revealed intensive growth in both the initial response and afterdischarges. Abolishment of the reinforcement made the reaction shorter and decreased the initial response. Early afterdischarges over 150 ms following the beginning of the auditory stimulus appeared more stable (Fig. 2B, 10). The neuronal reactions were acquired and extinguished simultaneously with acquisition and extinction of the movement, which is why neuron reactions can be accepted as a true conditional reactions.

The neuron reactions were extinguished during 10–15 isolated CSs when the CR was unstable. Extinction developed slowly when the CR was more stable (Fig. 3). The neuron reacted distinctly to conditioned

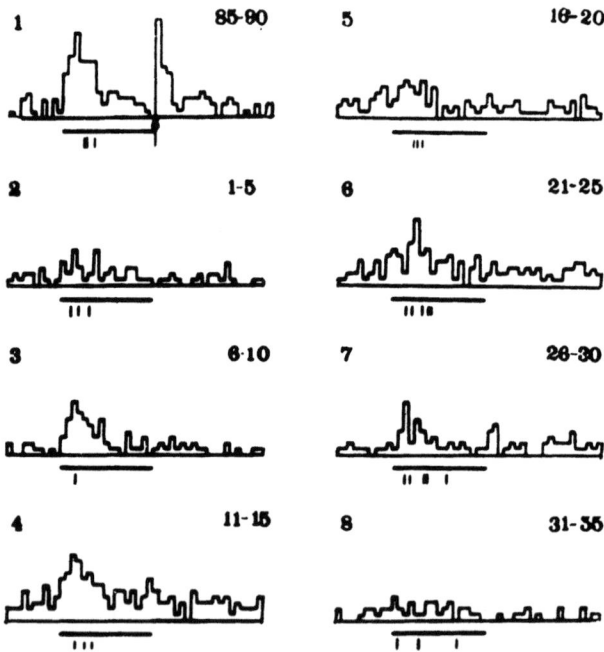

Fig. 3. Extinction of the neuronal reaction. Denotations as in Fig. 1.

and reinforcing stimuli on 85–90 of their combinations. The conditioned reactions reached the maximum level by 11 to 15 CS presentations after removal of the reinforcement. The probability of the movement stabilized during the next 20 to 30 stimuli (Fig. 3: 4, 6, 7). Distinct partial extinction of the neuronal response came by 31 to 35 CS presentations. The most stable portion of the reaction was early afterdischarges (Fig. 3: 7). It is interesting that in both cases, an extinction-related inhibition developed as neuronal activity returned to the initial background level.

Extinction during reinforcement was the other observed form of cortical inhibition (Fig. 4). The neuron with low background activity did not response initially to auditory stimuli, but clearly responded to electrical ones (Fig. 4: 1, 2). The acquisition of the CR and the appearance of movements coincided with the increase of background activity and the appearance of early and late afterdischarges (Fig. 4; 3, 4). Further conditioning led to the disappearance of the CR. As the beckground acitivity decreased, the neuronal reaction to sound vanished, but the reaction to the electrical stimulation remained (Fig. 4; 7, 8), which indicates that the excitability of the neuron did not decrease.

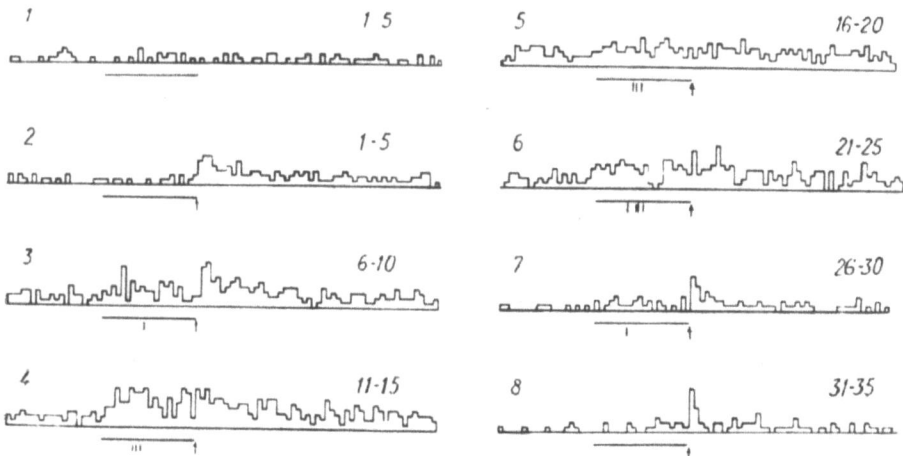

Fig. 4. Extinction during reinforcement. Denotations as in Fig. 1.

Neuronal reactions during the development of differential inhibition had certain distinctive features (Fig. 5). Unstable conditioned reflexes were acquired after 50 pairings. The neuron reacted to the conditional stimulus with initial response and afterdischarges. The auditory stimulus of 50 clicks per second did not evoke visible neuronal reactions (Fig. 5A: 1, 2). Neuronal reactions developed after 5 to 15 presentations of differential stimuli accompanied by intermittent stimulation of positive stimuli and its reinforcement, which included the initial responses and the early afterdischarges. Clear neuronal reactions were lost, accompanied with movement, only after 16 to 20 stimuli presentations (Fig. 5A: 6, 9). Thus, the neuron did not react intially to the differential stimulus and began to participate in differential reaction and preserved its participation in CR. Some motor cortical neurons and all observed neurons from area 2 had another pattern during differentiation (Fig. 5B). Intermittent differential stimuli were accompanied by the lowering of background activity and reduction of impulse responses to positive conditional stimuli. The response to electrical reinforcement was inhibited as well. The reactions to differential stimuli were somewhat similar to the reactions to conditional stimuli, and disappeared completely by 25 to 29 stimuli presentations.

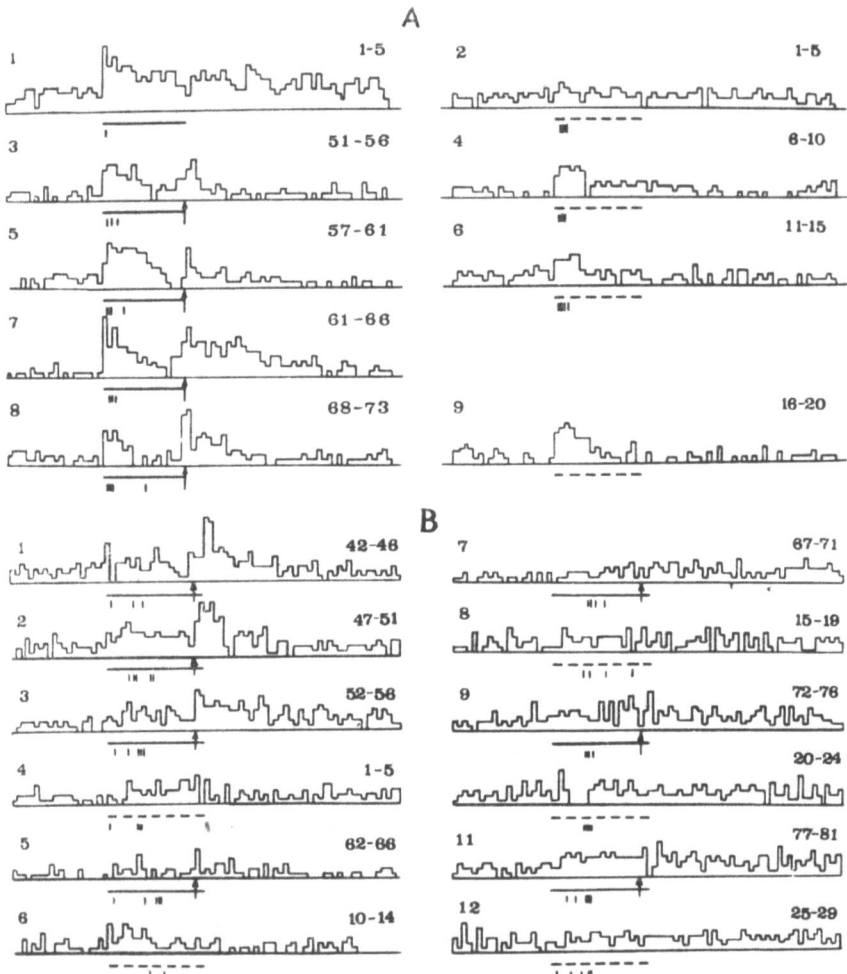

Fig. 5. Neuronal reactions during differentiation in motor (A) and somatosensory (B) cortex. The broken line indicates the differentiated stimulus. Other denotations as in Fig. 1.

DISCUSSION

Many factors do not allow us to carry out a strict mathematical, quantitative analysis of the neuronal activity during acquisition of conditioned reflexes: (i) the different speeds and individual peculiarities of CR acquisition among cats (ii) the unsimultaneous involvement of different neurons in CRs, (iii) the wave participation and mobility of neuronal reactions during the CR, (iv) the varing duration of observations of conditioned neuronal reaction, (v) the neuronal recording began from different stages of CR acquisition, (vi) the absence of special quantitative methods for evaluating nonstationary processes.

Nevertheless, that our facts reflect regularity of cortical processes, is evidenced by the following observations. For example, 45 of 105 neurons recorded in the motor cortex during CR acquisition changed their impulse activity. Changes took place mostly at the expense of afterdischarges; an initial response was observed in only 8 neurons. Extinction was observed in 30 neurons and extinction with reinforcement in 7. Six out of 12 neurons in which differentiation was observed also acquired new responses; 5 neurons (as in area 2) decreased reactions to positive auditory stimuli and reinforcement, and gradually eliminated responses to the differential stimuli. The quality of estimation of neuron reactions during and after CR acquisition shows that neuronal reactions had a phasic character. The appearance and perhaps the origin of these phases were of different character. The initial responses appeared in a small group of neurons at the beginning of CR acquisition and during pseudoconditioning, which reflected a general increase of excitement. The phase of the short-term decrease of impulse activity was expressed in responses to USs, and appeared sometimes among responses to CSs during the stage of unstable CR's. Most prominent during elaboration of CR were early afterdischarges, and less prominent were late afterdischarges in neurons that resisted extinction better and appeared before conditioned movements. We think this phase may be considered as the basis of the temporary connection. It is important to emphasize this suggestion because many investigators have concentrated on the learning of initial monosynaptic neuronal reactions as models for the mechanisms of conditioning and plasticity, instead of investigating membrane and synaptic processes during afterdischarges (4, 12, 13).

An afferent synthesis includes the triggering stimulus, environment afferentation, motivation and memory (1, 2). The triggering and environment afferentations during conditioning of clasical reflexes do not change. Impulse activity connected with motivation must become reflected in background activity before stimuli can elicit reactions. Therefore, the dynamics of neuronal reactions to conditional stimuli reflect mainly the mechanism of short-term memory, the stage of extracting the conditional movement program from it.

Extinction and extinction with reinforcement as forms of CR inhibition are significantly different neurophysiologically from ordinary forms of pre- and postsynaptic inhibition. Outer it takes place ordinary returning of impulse reactions to initial level, before elaboration of CR. During differentiation some neurons acquired to differential stimuli the same impulse reactions as to positive ones, but some had decreasing responses to differential stimuli and temporary decreasing responses to conditioned and reinforcement stimuli. Thus conditioned inhibition at the somatic cortical level is express first of all in the reorganization of excitement process.

REFERENCES

1. ANOKHIN P. K. 1968. Biologia i neurophisiologia uslovnogo refleksa. Meditsina, Moscow, 546 p.
2. ASRATYAN E. A. 1977. Ocherki po fisiologii uslovnykh refleksov. AN Armyanskoĭ SSR, Erevan, 346 p.
3. EVARTS E. V. 1974. Sensorimotor cortex activity associated with movement triggered by visual as compared to sensimotor inputs. In F. O. Schmitt and F. G. Worden The neuroscience. Third study program. The MIT Press, Cambridge, p. 327–335.
4. KOTLYAR B. M. 1977. Mekhanizmy formirovania vremennoi svyazi. Moscow University, Moscow, 208 p.
5. O'BRIEN J. H. and FOX S. S. 1964. Single-cell activity in cat motor cortex. I. Modification during classical conditioning procedures. J. Neurophysiol. 32: 267–284.
6. RABINOVICH M. Ya. 1975. Zamykatelnaya funktsya mozga. Meditsina, Moscow, 248 p.
7. SHULGINA G. I. 1978. Bioelectricheskaya aktivnosti golovnogo mozga i uslovnyĭ refleks. Izdat. Nauka, Moscow, 282 p.
8. SHVYRKOV V. B. Neurophysiologicheskoe izuchenie systemnykh mekhanizmov povedeniya. Izdat. Nauka, Moscow, 240 p.
9. STOROZHUK V. M. 1974. Funkcionalnaya organizatsya neuronov somaticheskoĭ kory. Naukova Dumka, Kiev, 270 p.
10. STOROZHUK V. M., and SEMENYUK E. F. 1978. Dynamics of the neuron reactions during elaboration of conditioned defensive reflex to sound (in Russian). Neurophysiologiya, 10, 4: 339–347.
11. VASSILEVSKIĬ N. N. 1968. Neironalnye mekhanizmy kory bolskikh polusharii. Meditsina, Leningrad, 220 p.
12. VORONIN L. L. 1976. Microelectrode study of neurophysiological mechanisms of conditioning. C. D. Woody, Los Angeles, 64 p.
13. WOODY C. G., VASSILEVSKIĬ N. N. and ENGEL J. Jr. 1970. Conditioned eye blink: unit activity of coronal preructate cortex of the cat. J. Neurophysiol. 33: 851–864.

V. M. STOROZHUK, A. A. Bogomoletz Institute of Physiology, 4 Bogomoletz Str., 252 601 Kiev 24, USSR.

Lecture delivered at the Warsaw Colloquium on Instrumental
Conditioning and Brain Research
May 1979

# MICROELECTRODE ANALYSIS OF THE CELLULAR MECHANISMS OF CONDITIONED REFLEX IN RABBITS

L. L. VORONIN

Brain Research Institute, Academy of Medical Sciences of the USSR
Moscow, USSR

*Abstract.* Spikes and postsynaptic potentials (PSPs) were recorded in the sensorimotor cortex of awake rabbits. (1) A cellular analog of cortico-cortical conditioned reflex (CR) was studied. Direct cortical stimulation of a remote point (conditional stimulus — CS) was paired with the stimulation near the microelectrode insertion (unconditional stimulus — UCS). The most neurons showed response facilitation when short intertrial intervals and strong UCS were used. These changes were short-lasting (up to 30 s) and were explained by posttetanic potentiation. Of 21 units tested with long (7 to 120 s) intertrial intervals, 9 showed significant response modification of longer duration. (2) The analog of cortico-cortical CR was modified using stimulation of homolateral cortex as the CS. The UCS was complimented by lateral hypothalamic reinforcement (LHR) in instrumental or classical paradigm. Of 37 units tested, 5 showed an increase in the probability of short-latency ($<25$ ms) spike. (3) Extracellular activity of 17 neurons was followed through the entire cycle of elaboration and extinction of a "local conditioned startle response" established by pairing click CS with cortical UCS and LHR. Responses of 7 neurons showed significant modifications. The latency distributions of averaged PSPs were compared for naive (N), conditioned (C) and extinguished (E) states. Latencies in N and E groups were similar but were different from those in C group. Of 41 C group neurons, 26 responded at a latency of $<17$

ms and 9 responded after ⩽ 7 ms. A pathway for this simple CR may pass through the cortex.

Thus PSP changes were observed in all models. Augmentation of short-latency EPSPs support the idea that an increase in efficacy of excitatory synapses underlies the conditioning. Interneuronal excitatory connections within the sensorimotor cortex are presumably changed during elaboration of simplest motor CRs. These connections can be analyzed with intracellular recordings triggered by spikes of a neighbouring neuron.

## INTRODUCTION

The changes in neuronal interconnections have long been suggested as a mechanism of memory and learning on the basis of morphological (16) and behavioral (53) experiments. The "synaptic" hypothesis of learning was further elaborated in a great number of theoretical models (11, 13, 14, 20–22, 31–33, 36, 37, 41, 42, 57, 58, 64, 70, 82). Theoretical models of Konorski (41, 42) rank high among others. He specified the "synaptic" hypothesis as a transformation of "potential" connections established between two groups of neurons into actual connections. Moreover, Konorski anticipated other cellular hypothesis of learning (e.g. 33) by suggesting some specific conditions necessary for synaptic modification such as simultaneous (or near simultaneous) activation of "emitting" and "receiving" neurons. It is worth to emphasis that other proposed conditions — preliminary activation of "emotive" system or activation of "antidrive" neurons were not pure hypothetical constructs but were tested in extensive experimental work of Konorski and his collaborators. It is also worthy of note that the discoverer of the instrumental CR who was convinced that there were many distinctions between classical and instrumental conditioning, postulated an identical mechanism at the synaptic level for both types of CRs. Although many of his suggestions were supported by skillfully planned behavioral experiments, Konorski (42) emphasized the necessity of more direct electrophysiological tests.

In recent years electrophysiological methods of single neuron recordings have come into extensive use for the study of various behavioral CRs as well as so-called "cellular analogs" of the CR as defined by Kandel and Spencer (37). Extracellular recordings were commonly used at the level of mammalian central nervous system (see reviews 12, 37, 43, 56, 67, 70). However more direct tests of "synaptic" hypothesis of learning can be done with intracellular recording and stimulation. Intracellular recordings have been used in studies of learning mechanisms mainly in invertebrates. The "synaptic" hypothesis has been strongly supported but only for the most simple type of "learning" — i.e. the habituation of defensive reflexes in invertebrates (36, 99). Moreover, in "cellular analogs" of the CR created in the molluscan CNS, changes in synaptic effectiveness have also been revealed (7, 8, 36–38). A special mechanism of heterosynaptic facilitation was postulated to account for these changes (7, 36). Heterosynaptic facilitation has been shown to be mainly unspecific to pairing and lacks many other important features of the conditioning paradigm (e.g. long term changes, "savings", spontaneous recovery after extinction, etc.). Neither participation of this mechanisms in formation of any behavioral CR nor its existence in the vertebrate CNS has been demonstrated.

During the last ten years we have been trying to find useful experimental approaches with which to attack the problem of neurophysiological mechanisms of the CR in mammalian neocortex. Different tipes of "cellular analogs" of classical and instrumental conditioning as well as a simple type of behavioral CR have been used in our experiments (73, 77, 79–88). This report will present our main data in this line.

## METHODS

Experiments were performed on unanesthetized rabbits. The head was loosely restrained in an untraumatic headholder or was held rigidly fixed to a metal frame by means of screws previously attached to the skull under Nembutal anesthesia. Methods of animal preparation and stimulation varied at different stages of this work. The details are described in the original papers (84–88) and will also be mentioned in the appropriate sections of this report. Conditioned stimuli (CS) or stimulation analogous to the CS varied (see Results) but unconditioned stimulus (UCS) was the same throughout all our study. Direct cortical stimulation delivered with bipolar silver epidural or epicortical electrodes served as the UCS. The stimuli were either single impulses (usually 0.3–1 ms, 0.3–1.5 mA) or trains of different frequencies (25 to 600 Hz) and durations (up to 0.9 s). A hole with 1–2 mm diameter was drilled above the foreleg motor projection in the sensorimotor cortex. The UCS evoked contralateral foreleg movement. In some experiments "lateral hypothalamic reinforcement" (LHR) was used in instrumental CR paradigm or as an addition to the cortical UCS in a classical paradigm. Parameters of the LHR (0.5–2 s of 0.3–1 ms, 0.1–0.3 mA pulses at 40–

80 Hz) were those which produced self-stimulation or a "feeding reaction" in the same animal as tested 1–3 days before the first recording session (85).

Recording glass microelectrodes were filled with 2 M potassium citrate and had resistances in the range of 30 to 150 megohms. They were inserted in the immediate vicinity of the UCS application.

Three kinds of recording situations were used for analysis of the postsynaptic potentials (PSPs). The first was intracellular registration of activity of neurons with a relatively high resting membrane potential of 40 to 70 mV and action potentials of 30 to 80 mV. The second was intracellular registration of neurons with low but stable membrane potential of 15 to 30 mV and with deteriorated mechanism of spike generation. The third type was called "partially intracellular" (66) or "quasi-intracellular" (49) with spike amplitude of 7 to 28 mV. "Quasi-intracellular" recordings were most commonly used as a basis for investigation because they allowed the study of PSPs and spikes in the same cortical neuron in waking animals for long periods (up to 2 or 3 h). Two needles placed subcutaneously upon a muscle with clear cut reaction to the cortical UCS were used in some experiments for recording the electromyogram (EMG). Averaging from 10 to 20 CS presentation of neuronal and EMG responses with ART–1000 (Saip, France) or PDP–8 (Digital, USA) computers was used at later stages of our experiments. Data analysis was performed using mainly nonparametric statistical methods. Differences were considered "significant" when $P$ value was less than 0.05.

## RESULTS

### A cellular analog of cortical-cortical CR

At the first stage of our investigation it seemed useful to study not true behavioral CRs but more simple electrophysiological models "cellular analogs" of the CR (37). As a first step, electrical stimulation was used as both CS and UCS (73, 77, 87, 88). Thus the experimental situation represented a cellular analog of behavioral cortico-cortical CRs (18, 27, 47). The use of the direct cortical stimulation as the CS and UCS had some additional appeal to us because a number of effects of such stimulation including various aftereffects had been previously analyzed with intracellular recordings from awake rabbits (72, 75, 76, 78).

Bipolar stimulation of the cortical surface as near as possible (0.2–1 mm) to the location of the microelectrode was used as an analog of the UCS. Stimulation of a remote caudal cortical point (from 3 to 12 mm away) was used as the CS. The parameters were selected so that the CS initially produced only some inhibition of spike discharges (Fig. 1A, 1) or no significant effects (Fig. 2A, 1) and the UCS produced spike discharges (Figs. 1A, 2, and 2A, 2) or profound inhibition of cell activity (Fig. 5A, 2). Experimental procedure consisted of control CS presentations (6 to 15 presentations in each series) (Fig. 1A, 1), 10 to 30 pairings of the CS and UCS with 0.4–1.3 s delay (Fig. 1A, 2–6) and "extinction" (Fig. 1B) — presentations of CSs alone similar to the control procedure. Pairing and extinction series were repeated untill the cell was lost. The stimulus parameters and interstimulus intervals were varied in different experiments in search of suitable means for securing reproducible plastic changes in unit activity. Different effects were found with short (1 to 5 s) intertrial intervals in comparison with long (7 to 120 s) ones. These effects will be considered separately.

In 22 of 30 cells (including 28 cells recorded quasi-intracellulary and 2 cells recorded extracellularly) tested with short intertrial intervals,

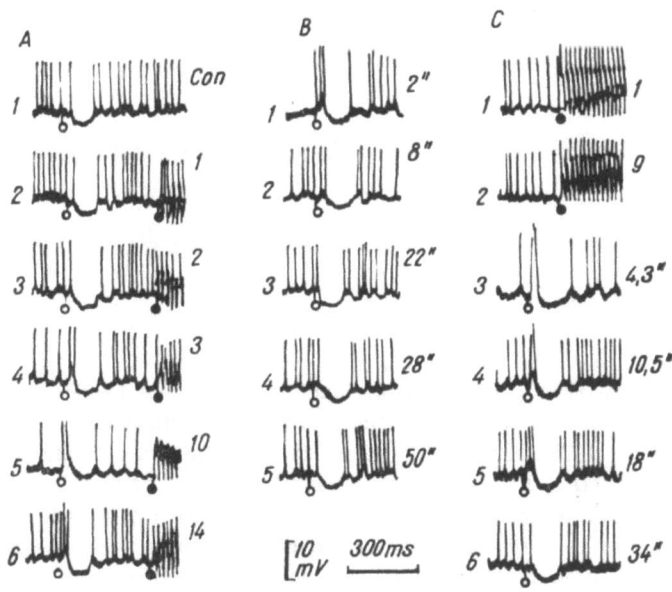

Fig. 1. Short-lasting facilitation of EPSP evoked by direct cortical stimulation. A1: control response to remote cortical CS before pairings; 2–6, paired trials. B: responses to the CS after pairings. C: pseudoconditioning procedure, 1, 2 presentations of the UCS alone; 3–6, responses to the CS after 9 unpaired UCSs. The numbers to the right of the traces are ordinal numbers of the pairings (A), or the presentations of the CS (C), or the time in seconds after the last UCS (B, C). Opened circle, the CS; filled circles, the beginning of the UCS train. From Voronin and Kozhedub (88).

significant modifications of the unit response to the CS were observed provided that a sufficiently intense and prolonged UCS was used. The principal change consisted of an increase in short latency (less than 25 ms) excitatory (spike and PSP) response to the CS independent of increase or decrease in neuronal activity to the UCS.

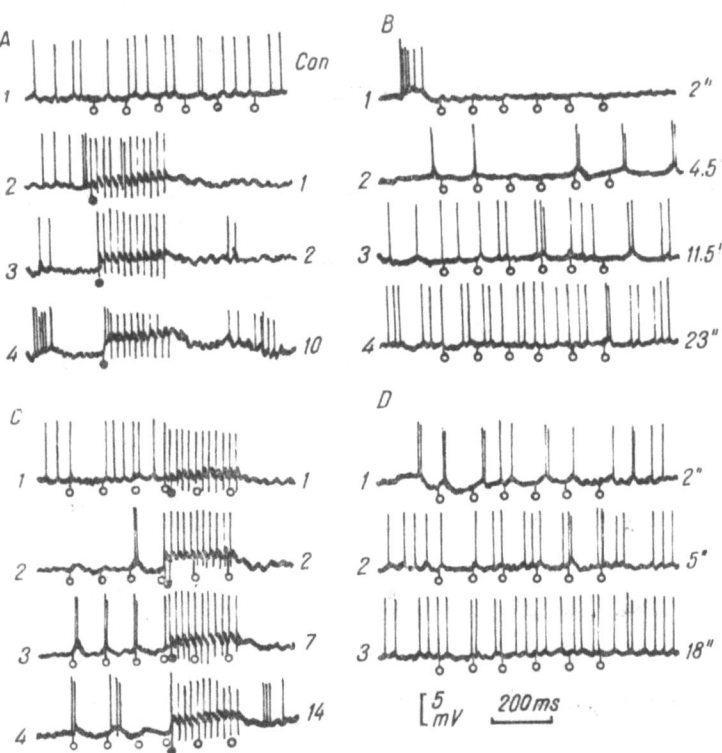

Fig. 2. A specific type of short-lasting facilitation of the EPSP evoked by direct cortical stimulation. A1: control responses to remote cortical CS; 2–4, unpaired presentations of the UCS. Most of the pulses in the UCS train evoke full-sized (2, 3) or partial (4) spikes. B: Failure to demonstrate PSP facilitation either during residual inhibition after UCS presentation (1) or during afterdischarges (2). C: paired stimulations. D: responses to the CS after pairings. Note the EPSP facilitation during inhibition of the spontaneous activity (C2–4) as well as during afterdischarges (D1). See Fig. 1 and the text for further explanation. From Voronin and Kozhedub (88).

In Fig. 1A, 1 the control response to the CS consists mainly of an inhibitory postsynaptic potential (IPSP) with only a small excitatory postsynaptic potential (EPSP). The first pairing of the single pulse CS with the UCS is shown in Fig. 1A, 2. The UCS was 0.9 s train at a frequency of about 50 s. The beginning of the train is shown in Fig. 1A, 2.

Each pulse of the UCS evoked a prominent EPSP with a spike discharge. Just after the first pairing (Fig. 1A, 3), an increase in amplitude and duration of the EPSP produced by the CS was obvious. The facilitation became more prominent in subsequent trials, when the EPSP produced bursts of spikes (Fig. 1A, 4–6). The IPSP evoked by the CS was diminished in certain cases (Fig. 1A, 5) or greatly enhanced in others, but an increase in EPSP amplitude was the most characteristic feature of these modifications.

However, as one can see in the second column (Fig. 1B), the facilitation was of short duration, lasting at most up to 30 s. The responses obtained in 28 s (Fig. 1B, 4) and 50 s (Fig. 1B, 5) after the last pairing were similar to the control record (Fig. 1A, 1). A special "extinction" procedure was not necessary to produce the diminution of the response to the CS. A testing CS applied within several seconds after the last pairing revealed the diminution of the response by itself.

The second feature of this short-lasting modification was non-specificity to pairing for the majority of the cells. For example, a very large facilitation was produced in the response of the neuron illustrated (Fig. 1C, 3) after 9 unpaired presentations of the UCS (Fig. 1C, 1, 2). The facilitation decayed after this pseudoconditioning procedure (Fig. 1C, 4–6) with a time course which was similar to that after pairing (Fig. 1B). Controls for specificity to pairing were carried out on 15 cells. Figure 2A, 1 illustrates a control presentation of the CS. The low-frequency remote cortical stimulation failed to produce distinct stable PSPs. There were no significant responses (Fig. 2B) after 10 unpaired UCSs (Fig. 2A, 2–4) as well. In contrast, a prominent EPSP was evoked by each remote cortical stimulus even after a single pairing (Fig. 2C, 2). The facilitation became stronger in subsequent trials (Fig. 2C, 3, 4) and persisted for several seconds afterwards (Fig. 2D, 1, 2). However, of 15 units tested with short intertrial intervals only 2 showed a significantly greater facilitation after pairing than after unpaired repetition of the UCS.

To eliminate short lasting, mainly unspecific posttetanic changes in the next series of the experiments, the trials were presented with longer intervals (as a rule more than 12 s and up to 2 min). A complete experiment was performed on 21 neurons consisting of control applications of the CS and at least one series of pairings (at least 12 pairings), followed by extinction trials. Of 21 units, 5 were recorded extracellularly and 16 quasi-intracellularly. 5 units showed a statistically significant modification of spike responses to the CS during the pairings. The changes in spike responses and PSPs were much weaker than the facilitation in the previous experiments with short intertrial intervals. Only one cell showed reproducible changes in the short latency (less than 10 ms) PSP (Fig. 3). The initial response of this unit to the CS (Fig. 3A, 13) was an inhibition of spike discharges and a low-amplitude IPSP. After 18 paired stimuli the CS became effective in producing an early EPSP distinguishable from the background "synaptic noise" (Fig. 3A, 19). EPSP amplitude rose in the subsequent

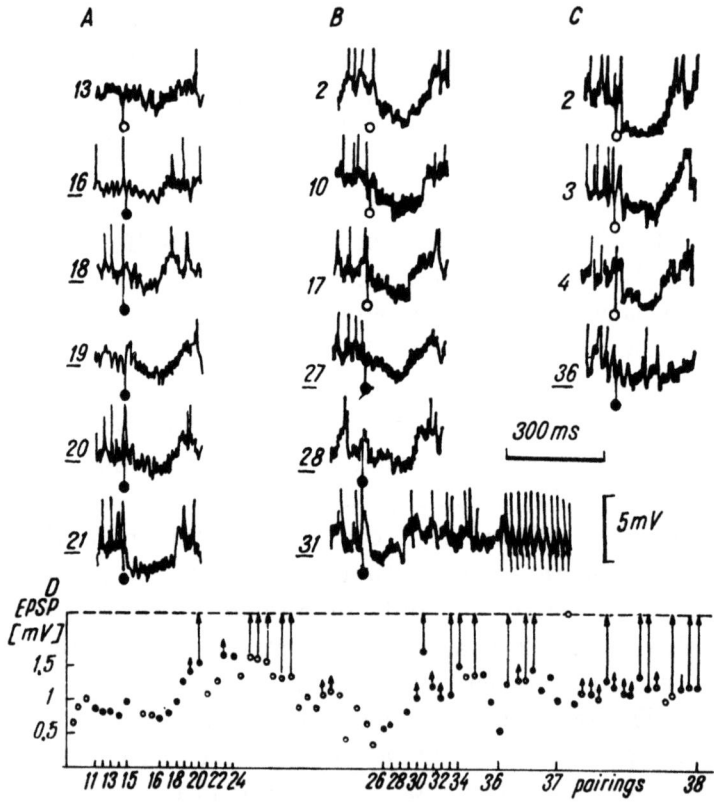

Fig. 3. „Conditioned" changes in the short-latency EPSP. A-C responses to the CS during extinction (the CS is marked by circle) and paired trials (the CS is marked by a dot). The beginning of the UCS is shown at the bottom of the middle column (B31). Only the response to the CS is shown for the rest of the records. The digits here and in Figs. 4 and 5 are the ordinal numbers of the extinction trials (independent numeration for each extinction sequence). The underlined digits are the ordinal numbers of the paired trials (continuous numeration throughout the entire experiment). Peaks of full-sized spikes are off the screen. D, EPSP changes during the whole experiment. Ordinate: amplitude of the early depolarization evoked by the CS during extinction (circles) and during paired trials (dots). Abscissa: numbers of pairings. After the 15th pairing 11 extinction trials are omitted. Small arrows indicate the generation of a partial spike; the arrows pointed to the stippled line indicate full-sized spike generation. From Voronin and Kozhedub (88).

trials and became suprathreshold for spike generation (Fig. 3A, 20, 21). The second column (Fig. 3B) illustrates an extinction series with a diminution of the EPSP and a reconditioning series in which the spike response developed again (Fig. 3B, 31). The third column (Fig. 3C) illustrates rapid second extinction of the EPSP by presenting the CS without the UCS. Changes in the EPSP amplitude during the whole experiment are shown in the bottom diagram (Fig. 3D). The intervals between pairings were varied from 20 to 120 s in this experiment.

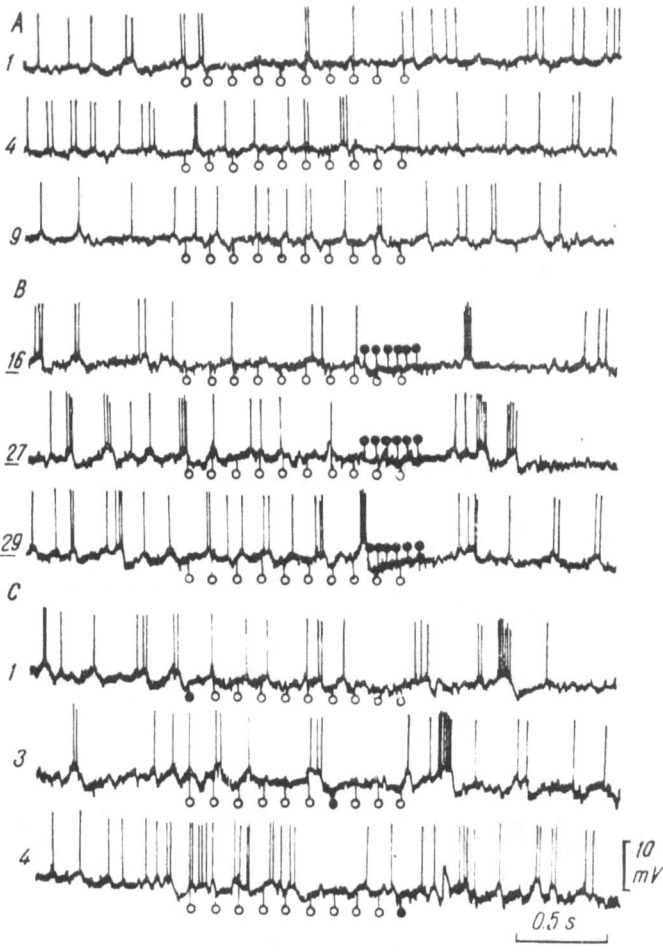

Fig. 4. Cellular "conditioned response" as revealed during extinction trials. A: control responses to the CS. B: pairings of two cortical stimuli. C: extinction trials. Note persistence of inhibition and afterdischarges (C1, 3) at the time when the UCS would have been delivered. Circles: CS; dots: UCS. See Fig. 3 for an explanation of the meaning of the digits. From Voronin (77).

A testing CS applied about one second after the 37th pairing (Fig. 3D) also revealed a powerful short-lasting facilitation.

In addition to 5 units with conditioned changes during pairings, in 4 neurons response modifications appeared only during extinction trials. These cellular "conditioned reactions" occurred at a time generally corresponding to the omission of the UCS. This type of a time-locked response is illustrated in Fig. 4. The CS was a low-frequency train of ten pulses which evoked no significant response (Fig. 4A). The UCS was a shorter train of a higher frequency which produced a clear-cut inhibition of spike discharges with a small hyperpolarization followed by a variable afterdischarge (Fig. 4B). In the early extinction trials, late inhibition and afterdischarge were evident (Fig. 4C) similar to these evoked by the UCS.

An interesting type of time-lacked inhibition was found in an other cell (Fig. 5). In this neuron each "conditional" pulse, from the second to the sixth one evoked a spike (Fig. 5A, 1), and the UCS evoked hyperpolarization and inhibition of both spontaneous activity (Fig. 5A, 2) and spike discharges evoked by the CS (Fig. 5A, 3, 4). An inhibitory "conditioned reaction" consisting of disappearance of some spike responses to the CS was elaborated as a result of the pairing (Fig. 5A, 4).

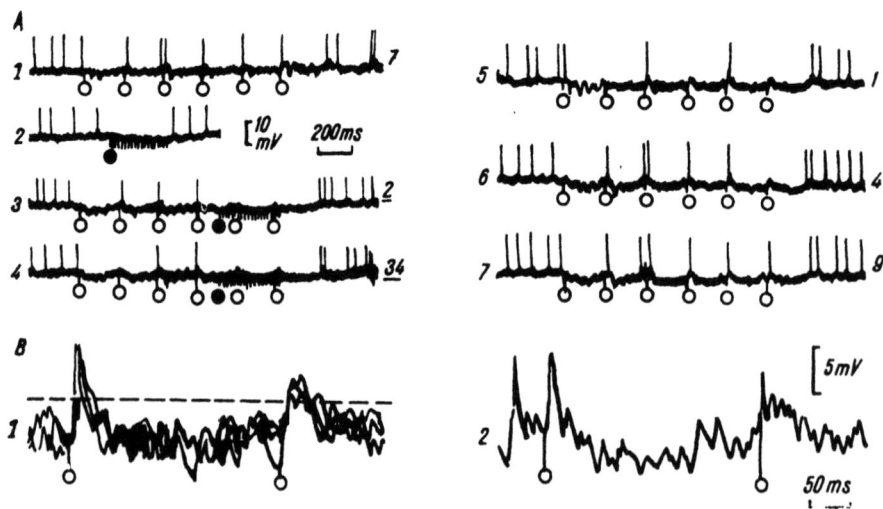

Fig. 5. Inhibitory "conditioned" changes. A1: a response to the CS before pairings: 2, the same to the UCS; 3, 4. paired CS+UCS trials; 5-7, extinction trials after the pairings. B1: superimposed responses to the 5th and 6th conditional pulses during the extinction; 2, an example of a partial spike response to the CS during an extinction trial (the 3rd and 4th conditional pulses are shown). See Fig. 3 for explanation of the meaning of the digits to the right of the traces. From Voronin and Kozhedub (88).

The inhibition of the response to the sixth "conditional" pulse was especially stable as had been revealed during early extinction trials (Fig. 5A, 5, 6). This pulse was delivered at a time corresponding to that of the UCS application. Measurements did not reveal any significant hyperpolarization or changes in the amplitude of EPSPs evoked by the CS. Figure 5B illustrates that postsynaptic depolarization in response to the sixth "conditional" pulse overcomes the firing level attained by the previous EPSPs.

Thus of 21 neurons tested with sufficiently long intervals between the pairings, 9 showed statistically significant modification of response to the CS. "Conditioned reactions" occurring at a time corresponding to the omission of the UCS (Figs. 4 and 5) indicated the specificity of these modifications. In addition, random presentations of the CS and the UCS or unpaired repetitions of the UCS alone with intervals of 5 to 30 s had no significant effect in 12 control units. In four cases two remote cortical stimuli were tested. One stimulus (CS) was paired with the UCS, while the other ("differentiated stimulus") was not. There were no statistically significant changes in spike responses to the "differentiated stimulus", whereas the responses to the CS were modified in two of these four units.

### A cellular analog of the CR using interhemispheric stimulation and hypothalamic reinforcement

Some changes in neuronal responses to the CS (Fig. 3) are exactly of that kind which could be expected to occure on the basis of the "synaptic" hypothesis of conditioning. However some drawbacks of the cellular analog used in these experiments may be noted. These drawbacks precluded direct tests of the "synaptic" hypothesis, accurate description of the conditions necessary for synaptic modification and more profound analysis of the recorded PSP changes. Firstly, the changes were mostly feeble and variable. Only exceptionally were they seen during the experiment itself, being as a rule revealed only later after statistical treatments. Practical experimentation with this analog is therefore difficult. Two methods may be used to overcome these difficulties. The first one is more precise measurements, for example with averaging techniques. The second is modification of the analog to make the changes more prominent and reproducible. It is known from behavioral experiments (42) that a movement evoked by direct cortical stimulation becomes easily conditionable with food reinforcement. An effective substitute for food reinforcement can be stimulation of locations which sustain self-stimulation (51). This is particularly convenient in experiments with animals that are confined during observation. The second drawback of cortico-cortical conditioning is the polysynaptic nature of the responses under study, mediated through complex and uncertain pathways. Crucial changes may occur not in the recorded neuron but rather in previous elements.

Thus, in the next series of the experiments, stimulation of the sensorimotor cortex contralateral to the site of registration was used as the CS. Oligosynaptic connections including monosynaptic ones are described from the contralateral hemisphere to pyramidal neurons (28).

In addition after the cortical UCS which was similar to that in the previous experiments, LHR was applied. In some experiments a cellular analog of the instrumental CR was used. In these cases the LHR alone without the cortical UCS was applied after CSs which selectively evoked short-latency (less than 25 ms) spike discharges.

Thirty seven neurons recorded extra or intracellularly have been tested with one or both of these paradigms. Probability of the short-latency spike discharges ("firing indicies") were calculated in these experiments. Nine neurons showed statistically significant changes. Of these, five neurons increased and four neurons decreased firing index during conditioning procedure.

It is shown in an example (Fig. 6H) that on the average 10% of the CSs evoked a short-latency spike early in the experiment (up to the 50th trial of the CS). Intracellularly recorded response consisted mainly of a hyperpolarization (Fig. 6A). After two sets of paired trails (filled circles in Fig. 6H) nearly every CS began to evoke a spike (Fig. 6C). After 30 extinction trials the firing index decreased again (Fig. 6H) and hyperpolarizing responses similar to the control ones reappeared (Fig. 6D). At the moment indicated by the arrow on the abscissa (Fig. 6H), the microelectrode went out of the neuron. However the cell presumably was not lost because spike responses with similar characteristics of reconditioning and extinction were recorded during the second half of this experiment (Fig. 6E–H). The presentations of the CS were interrupted for 4–10 min several times during this experiment as shown by stippled lines in Fig. 6H. When this rest period coincided with the "conditioned state" (the firing index as high as 0.7–1) there was no significant decrease in responsiveness (stippled lines a, b, c, and h in Fig. 6H). When the interruption took place during "extinguished state" (no response to the last 7 to 10 presentations of the CS), a restoration of responsiveness was found such that the first two or three CSs after rest period (d, e and f in Fig. 6H) consistently evoked spike discharge. A classical associational paradigm with

CS + UCS + LHR pairings was used in the first stage of this experiment. One could see also that the "instrumental" procedure (with every spike response followed by the LHR) was able to restore the extinguished response (Fig. 6H, 181st trough 200th presentations of the CS).

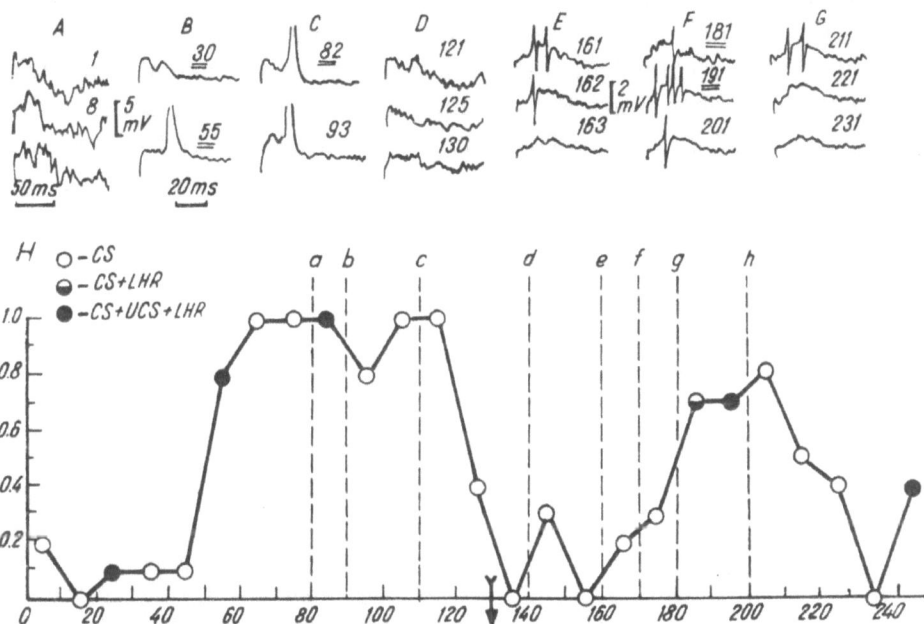

Fig. 6. "Conditioned" modifications of the short-latency interhemispheric discharge. A–G: neuronal responses to stimulation of the contralateral cortex. A–D: intracellular recordings early in an experiment: A, before pairings; B and C (upper trace) — during the pairings; C (bottom trace) and D, during extinction procedure. E–G, extracellular recordings late in the experiment: E, extinction after spontaneous restoration; F, paired trials (two upper traces) and the first extinction trial (bottom trace); G, examples of additional extinction trials. The ordinal number of the CS presentation is given at the right of each record. Underlined digits are the paired presentations. Time scale: 50 ms for A; 20 ms for B–G. Calibration: 5 mV for the intracellular records (A–D); 2 mV for the extracellular ones (E–G). H, a graph showing the firing index in every 10 presentations of the CS alone (opened circles), or followed by the LHR (a half-filled circle), or by UCS + LHR (filled circles). Ordinate: the number of CS trials evoking short-latency (20 ms) spike discharges divided by ten. Abscissa: CS presentations. Stippled lines (a-h) mark intervals of 4–10 min interrupting the CS presentations (and do not correspond to the capitalized letters above). See the text for further explanation. From Voronin (81).

One-line averaging of PSPs was used at the later stage of these experiments. Of ten neurons recorded intracellularly and quasi-intracellularly during and after ten or more pairings in a classical associational paradigm with complex cortical and hypothalamic stimulation, six showed changes in amplitude of EPSP or IPSP evoked by interhemispheric stimuli. However, the predominantly polysynaptic PSP component with latency more than 10 ms (stippled line b in Fig. 7) underwent the most prominent changes (no control for specificity of these changes to pairing have been done up to now). Only exceptionally was an early, presumably monosynaptic component of the transcortical EPSP (Fig. 7, line a) also increased. Accordingly the latency of the spike discharge recorded in the previous series of the experiments was ordinarily more than 6 ms. For example, the latency of the spike shown in Fig. 6 was 12 to 16 ms. The values and variability of the recorded latencies indicate the predominantly polysynaptic nature of the responses.

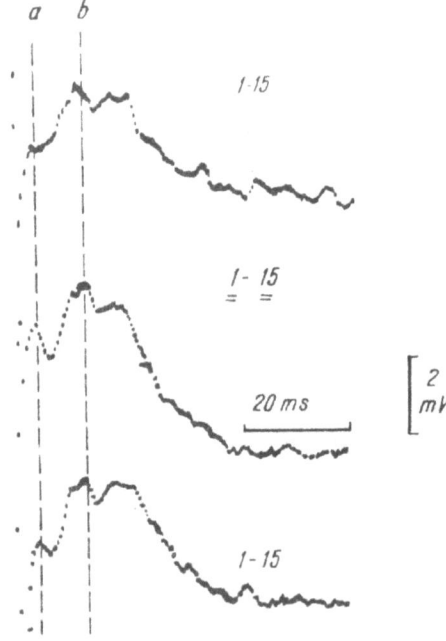

Fig. 7. Changes in interhemispheric intracellular response. A: control response to the contralateral cortical CS; B: The response during the pairing of the CS with the ipsilateral cortical and hypothalamic stimulation; C: the response to 15 unpaired CS presentations (extinction). A–C: averages of 15 presentations of the CS. The CS was applied 2 ms before the begining of the sweep. Stippled lines (a and b) mark early (peak time 5 ms) and late (peak time 15 ms) components of the response. From Voronin (81).

Thus our preliminary experiments demonstrate the possibility of creating a cellular CR analog with electrical stimulation of the contralateral cortex. Increase in EPSP amplitude was again found with conditioning procedures (Fig. 6, 7). However the changes in early polysynaptic responses were poorly reproducible.

*Elaboration of a simple behavioral CR in a single experiment*

In all previous experiments it has been not true CRs, but rather cellular analogs which have been under study. It is far from clear whether the mechanisms which one tries to study at a cellular analog level would be identical (or even similar) to those of behavioral learning paradigm. From the methodological point of view, the answer to another question is also of great importance. The question in whether the changes in cortical neurons can be more prominent during behavioral conditioning than they are during simple pairings of two stimuli in CR analogs. It may be expected that it is much easier to find appropriate conditions for prominent changes and to control the level of these changes by using a simple behavioral index. Thus we made an attempt to develop a model with simple and reproducible behavioral changes suitable for analysis with extra- and intracellular recording (84, 85). We used peripheral stimuli, clicks or flashes as the CS in these experiments. The behavioral CR was evaluated by recording the EMG of muscles in the forepaw contralateral to the applied cortical UCS. In some experiments the EMG was also recorded from other muscles (85). The CR was elaborated in a semi-automated experiment,

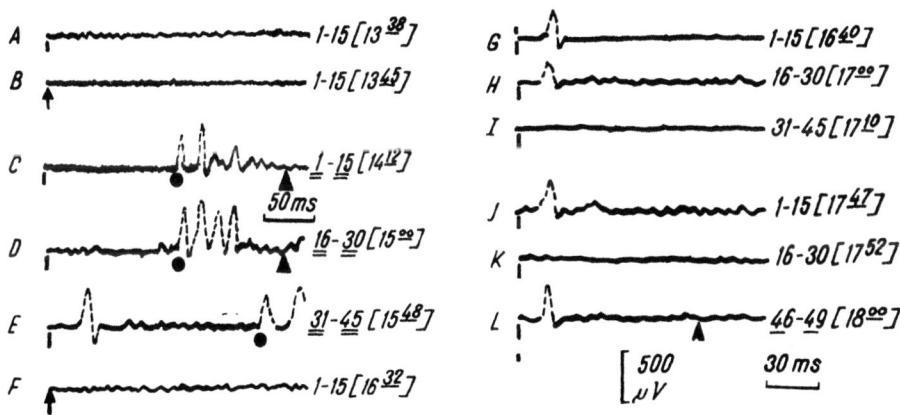

Fig. 8. One-session conditioning of EMG response. *A*: control response to click before pairings; *B*: the same to light flash; *C–E*: three series of pairings; *F*: control presentations of unpaired flashes; *G-I*: extinction series; *J*: spontaneous recovery after a pause of about 30 min. *K*: subsequent extinction series; *L*: reconditioning with the LHR. *A–L*: averages of rectified and integrated EMGs of the foreleg. 50 ms calibration for *A–D*; 30 ms for the rest of the traces. Digits in brackets are time (hour, min). Here and in the following figures, digits to the right of the traces are ordinal numbers of extinction trials; underlined digits are ordinal numbers of pairings; click, flash, UCS and LHR presentations are indicated by bars, arrows, dots and triangles respectively. From Voronin et al. (84).

using amplitude and time logic on the filtered and rectified EMG. At the beginning of an experiment isolated CSs were given (control, Fig. 8*A*, *B*), subsequently the CS, UCS and LHR were combined. The interval between CS and UCS was 120–300 ms, between UCS and LHR it was from 0 to 100 ms. Intertrial intervals during conjoint stimulation were from 15 to 180 s (usually about 2 min). During control and extinction (presentation of isolated CS subsequent to the elaboration of a CR), intertrial interval was 10–50 s (usually about 20 s). In most experiments, a scheme of instrumental CR analogous to that of Konorski and Miller (see 42) was used. Cortical UCS stimulus was switched off after several presentations of CS + UCS + LHR while the LHR was given only if a sufficiently intense short-latency (less than 0.2–0.6 s) EMG reaction was detected. The logic circuits allowed the investigators to set the minimum amplitude and maximum latency for which an EMG-reaction would lead to delivery of the LHR. If the rabbit did not receive the LHR after several trials with the CS alone, then the UCS would again be turned on. In other (later) experiments a scheme of "classical" CR was used with all stimuli (CS, UCS and LHR) presentations at all times.

In most of our experiments (in 18 of 20 with click CS, and in 5 of 10 with flash CS) EMG appeared mainly in the foreleg contralateral to the side of UCS application. The responses were found to possess the following parametric features of the CR (53): (1) elaboration with pairing and extinction with unpaired presentations of the CS; (2) spontaneous recovery after extinction; (3) long-term "saving" as revealed by quick restoration even one to three days later; (4) more rapid extinction after repeated restoration and extinction procedures; (5) specificity to pairing and (6) limited afferent and effector generalization.

Extremely short latency (12 to 17 ms) "phasic" EMG discharge to click CS was most common as the CR. An example is provided by Fig. 8. Neither click (Fig. 8A) nor light flash (Fig. 8B) evoked any significant averaged response before pairings. Results of the first series of pairings are shown in Fig. 8C. There is no response to the click CS, but the UCS evokes a foreleg EMG response (the first wave marked by a dot represents an integrated stimulus artifact; the beginning of the LHR is marked by a triangle). A small EMG response appeared in the second series (Fig. 8D). In the third series (Fig. 8E), a prominent short-latency response was evident (note the change in sweep gain). There was no generalization resulting in a response to the light stimulus (Fig. 8F), which had not been paired with the UCS and the LHR. An extinction produced by presentation of the CS alone is shown in Fig. 8G trough 8I. Spontaneous recovery after more than 30 min without stimulus presentations is shown in Fig. 8J and extinction of this recovered response is shown to be quicker than the previous extinction (compare Fig. 8K with Fig. 8G–I). In this experiment it was found that only a few CSs followed by the LHR were required to cause reconditioning according to "instrumental" CR paradigm (Fig. 8L). In special experiments, tests for pseudoconditioning were carried out (85). For this purpose we used 30 to 60 presentations of UCS and LHR. Either this complex was followed by the CS, or all 3 stimuli were given in random order. No stable pseudoconditioned responses to the CS developed in any of 8 experiments.

A striking feature of the CR under study was its extremely short-latency from 12 to 17 ms. This latency is much shorter than that of usual CRs or minimal reaction time of the human being (17). Wave form and amplitude of the EMG response indicate a synchronized discharge of a great many muscle units (see single responses in Fig. 9H,

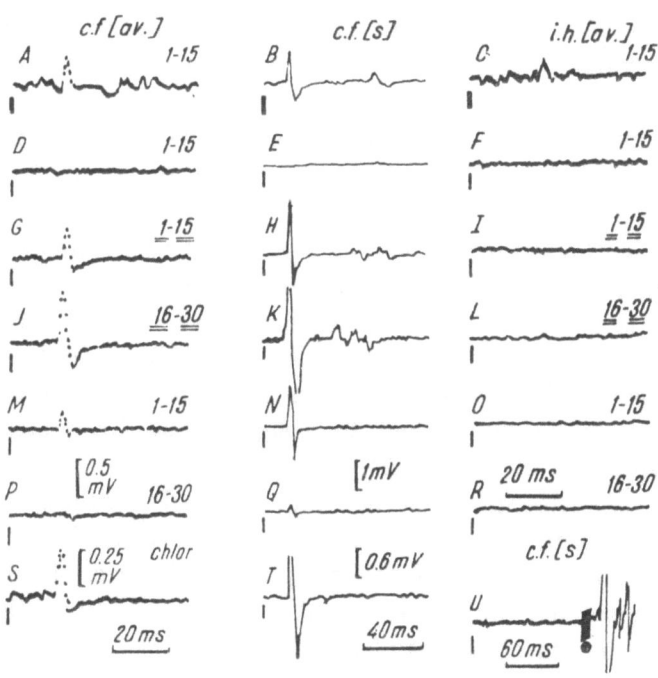

Fig. 9. Averaged (av.) and single (s) EMG responses recorded simultaneously from the foreleg contralateral (c.f.) to the side of the UCS application and from the ipsilateral hindleg (i.h.). A–C: responses to a loud "startling" click; D–F: responses to a neutral click before pairing; G–L: pairings; M–R: extiction; S, T: responses to the same low intensity click after chloralose injection; U: an example of the response to the UCS (the second pairing). Calibration: 0.25 mV for A, C, and S; 0.5 mV for the rest of the averaged records; 0.6 mV for T; 1 mV for the rest of the single records. Presentation of the loud startling click is indicated by a thick bar. All EMGs in band 10–1,000 Hz. Remaining notation as in Fig. 8. From Voronin et al. (85).

K, N). An abrupt jerk in the corresponding foreleg was observed during the CR performance.

Similar wave form and latency are known for the generalized unconditioned startle reaction (SR) produced by an intense stimulus, especially a shot-like sound (54, 69). Similar short latency responses were found also under chloralose anesthesia (see 68, 98 for further references). The similarity of the CR under study with the unconditioned SR was confirmed in the following experiments. The intensity of the click CS was 10 dB above the level of the masking background "white" noise (which was about 60 dB). It was found that a louder click (20–40 dB above the noise level) produced unconditioned SR in the majority of the animals. In Fig. 9, examples of averaged (Fig. 9A) and single (Fig. 9B) SRs are shown. The low intensity click evoked no EMG response (Fig. 9D, E). After several pairings of the low intensity click with the UCS and LHR, the click began to evoke EMG responses (Fig. 9G, J) similar to the unconditioned SR (Fig. 9A). Similar responses (Fig. 9S, T) were evoked by the low intensity click in animals after chloralose administration. The chloralose (80–100 mg/kg, i.p.) was injected 1–3 days before elaboration of the CR or after its extinction (Fig. 9S, T).

However, an important difference was found between the conditioned and unconditioned SR (or "chloralose jerk"). One can see in Fig. 9 that the SR is generalized and could be recorded in the foreleg (Fig. 9A, B) as well as in the hindleg muscles (Fig. 9C). In contrast, the CR was local, being practically absent in the hindleg muscles (Fig. 9I, L, O). Thus it seems convenient to designate the CR under study as a "local conditioned startle reaction" (LCSR).

Thus an experimental situation that allowed a rapidly elaborated and extinguished movement CR was found. In view of the extremely short and constant latency of the peripheral response, the LCRS promised to become a simple convenient paradigm for investigating the neurophysiological mechanisms of conditioning.

### Changes in cortical neuronal activity during elaboration and extinction of the LCSR

In spite of many studies of neuronal activity during elaboration of the behavioral CRs (see Introduction for References), only a limited number of works included the registration of the same cortical neuron during all phases of CR elaboration (9, 25, 52, 63a). We tried to follow the activity of the same single neuron during elaboration of the LCSR (84, 85).

Three-electrode micromanipulator and tungsten microelectrodes were used in these experiments. From 2 to 5 units were recorded simultaneously. More than 50 units had been recorded in 15 experiments however most of them were lost during control click presentations or during early paired trials. Seventeen neurons from 7 different animals

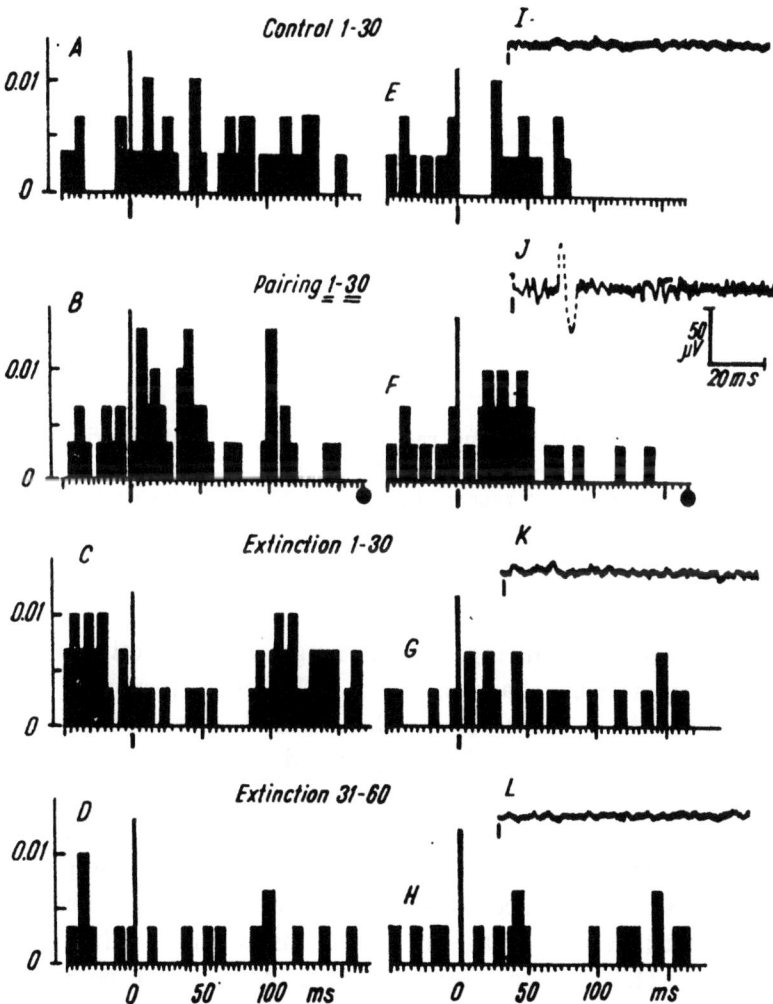

Fig. 10. Simultaneous recordings of two single cortical neurons by two separate microelectrodes during conditioning and extinction. A-D: the first unit; E-G: the second one; H: the third neuron recorded by the second microelectrode after the unit in E-G had been lost; I-L: EMG of the foreleg. A, E, I: control before pairings; B, F, J: responses during the pairings; C, G, K: the first extinction series; D, H, L: the second extinction series. Histograms ordinate: average number of spikes per bin (10 ms), per stimulus; abscissa: time (ms). Clicks were presented at 0.
See Fig. 9 for further explanation. From Voronin et al. (84).

were held from the period before the occurrence of any EMG conditioned response through its elaboration and extinction (Fig. 10). In the majority of neurons, evoked activity was very low (Fig. 10B, F) even during well established LCSR (Fig. 10J). These findings are similar to those of Woody et al. (97) for cortical neurons in blink conditioned cats. Only one neuron recorded here from a conditioned animal produced a consistent burst discharge to every CS (Fig. 11D). Of 17 cells

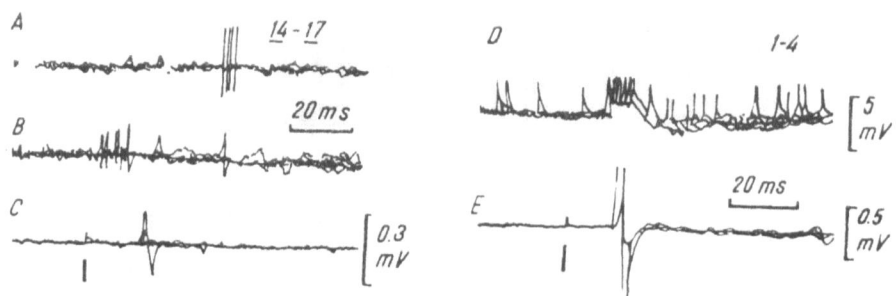

Fig. 11. Examples of single-neurone responses (A, B, D) and superimposed single EMGs (C, E) recorded from two conditioned rabbits. A–C: during the 14th to 17th pairings from the experiment represented in Fig. 10; D–E: during early extinction trials from another experiment. See Fig. 9 for further explanation. From Voronin et al. (85).

recorded continuously during conditioning and extinction, 7 showed statistically significant changes. As a rule, click responses were increased during the pairings (Fig. 10B, F) and decreased with extinction of the LCSR (Fig. 10C, G). Response decrement as a result of the conditioning procedure was found only in one neuron.

Latencies of the "conditioned" neuronal responses varied from 4–6 ms to 140 ms. The neurons with the conditioned changes can be divided into two groups according to response latencies. Three neurons gave significant responses as late as 4 ms or more after the EMG reaction (Fig. 11A). These responses, as a rule, could be readily explained by the proprioceptive feedback. Response latencies of 4 cells were similar to those of the related EMG response or even shorter (Fig. 11D). In some units occasional discharges after an extremely short latency ($\leq 7$ ms) were evident (Fig. 10B, 11B). In one experiment (see 81, 82) activity of three neurons recorded by the same microelectrode in the period of 4 to 6 ms after the click significantly exceeded the range of spontaneous activity preceeding the click. This increased activity disappeared with extinction of the EMG response.

For more detailed analysis, a statistical approach used successfully by other authors (96, 97) in extracellular studies of the blink CR was applied here to analysis of PSP changes related to the elaboration of the LCSR (81–83, 86). Toward this goal, single and averaged intracellular and quasi-intracellular records of the activity of sensorimotor cortical neurons together with EMG responses were compared in naive, conditioned and extinguished states. In addition, a group of responses to the loud "startling" click was recorded in naive animals.

The click to be used as the CS did not evoke any EMG response (Fig. 12B, F) in the naive animals (the control); no significant responses were found in averaged recordings of about half of the neurons in the period of 0 to 100 ms after the click presentation (Fig. 14A). An

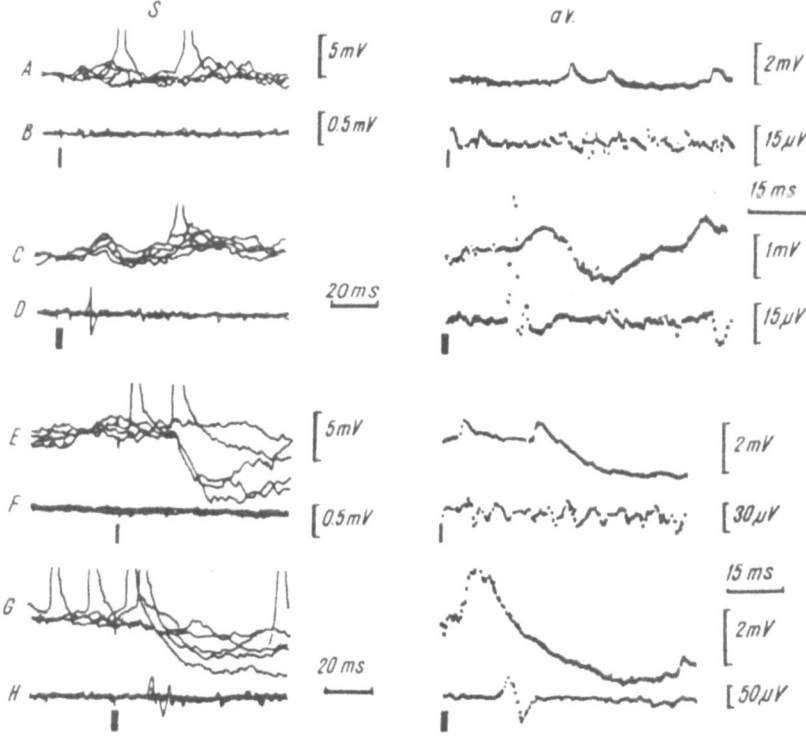

Fig. 12. Cellular (A, C, E, G) and EMG (B, D, F, H) responses to clicks of two different intensities in two naive animals (A–D and E–H). s: five superimposed single traces. av: average of 10 responses. A, B, E and F responses to a click of low intensity (marked by a thin bar). C, D, G and H responses to a loud click (marked by a thick bar) which evoked the unconditioned startle response (D, H). Peaks of action potentials (here and in Fig. 13) are off the screen. Note the latency of 4–5 ms in unit response (G, av) preceding the EMG response (H) for loud clicks. From Voronin and Ioffe (86).

example is shown in Fig. 12A. The two humps in Fig. 12A (av.) represent two random spikes (not PSPs). In the rest of the cells of this group, PSPs evoked by clicks occurred at latencies of 8 to 43 ms. IPSPs were the most prominent components of the responses (Fig. 12E). The IPSPs were sometimes preceded by EPSPs or by spike discharges.

The second group of responses was recorded in the conditioned state, with LCSR latencies of 12 to 17 ms (Fig. 13B, E). The relative number of responsive neurons was significantly greater in this group than in the control group (compare Fig. 14A and B). The averaged neuronal responses were more prominent. Of special significance seems to be the appearance of EPSPs (Fig. 13A) and spike discharges (Fig.

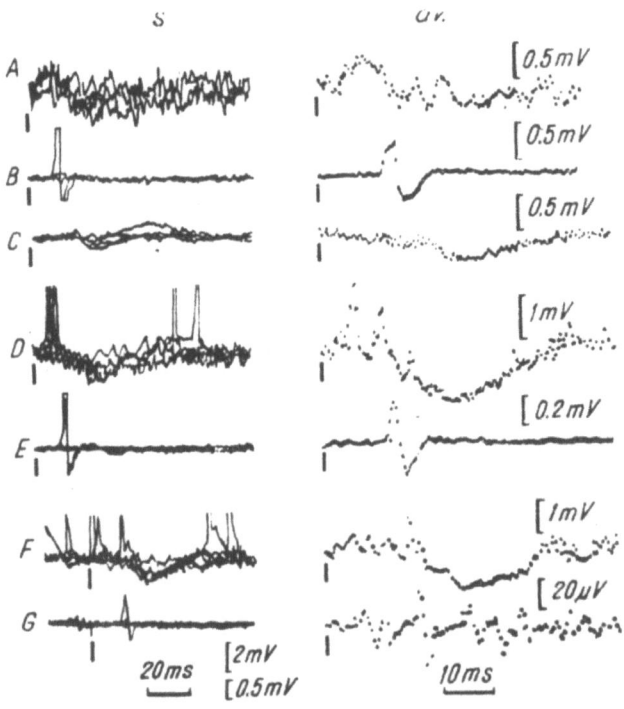

Fig. 13. Responses of three different neurons (A, D, F), successively recorded in a single penetration in the sensorimotor cortex of a conditioned animal, together with EMG recordings (B, E, G respectively), C: extracellular field potential recorded just after the death of the first neuron (A) and before obtaining response from the second cell (D). Some of the EMG responses (B, E) are limited by the amplifier (note changes in gain). Clicks were delivered alone, without succeeding cortical and hypothalamic stimulation. Extinction can be seen by comparison of the EMG averaged responses. av; average of 6 (A–C) and 15 (D–G) responses. See Fig. 12 for further explanation. From Voronin and Ioffe (86).

13D) with latency of less than 7 s (Fig. 14B). Such short latency neurons were never found in the control group (Fig. 14A).

The responses in the third group (Fig. 14C) were recorded after extinction of the LCSR by presenting the click alone. Comparatively low amplitude EMG responses were evoked only by 10–20% of clicks in the extinguished state; otherwise there were no EMG responses at

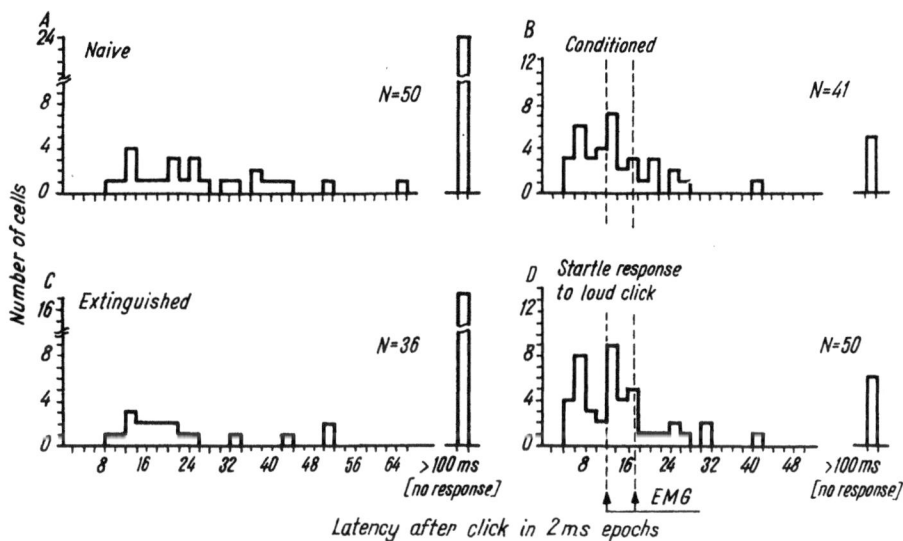

Fig. 14. Distribution of postsynaptic latencies of cortical neuronal responses in naive (A), conditioned (B) and extinguished (C) states as well as latency of responses to loud startling click in naive animals (D). Ordinate: number of cells; abscissa: latency after click in 2 ms epochs. N, the number of neurons in each group. Right columns (> 100 ms) — the number of neurons in each group with no significant response within the time interval of the averaging. Dotted lines (B, D) denote minimal and maximal latency of EMG response. From Voronin (83)

all. As shown in Fig. 13G (s), only the first click was followed by a clear EMG response. The corresponding average (Fig. 13G, av.) revealed only a poor reaction (if any) compared to the spontaneous EMG level. The amplitude characteristics (Fig. 13F) and latency distributions (Fig. 14C) of the postsynaptic responses in this group were comparable to those in the naive animals.

As had been shown previously, the latencies of the unconditioned SR evoked by the loud click did not differ significantly from those of the LCSR. The postsynaptic responses evoked by the startling click were significantly larger than those to the neutral click in the same neuron (compare IPSP in Fig. 12E, av. with that in Fig. 12G, av.). Most of the unresponsive neurons (Fig. 12A) were found to show significant responses to the loud click (Fig. 12C) and some were comparable to those in the conditioned group in responding with a latency of less than 7 ms (Fig. 12G, av.). The latency distributions were similar in the groups of responses recorded during the performance of the two types of motor reaction: the local, conditioned movement to previously neutral clicks (Fig. 14B) or the generalized, unconditioned SR to loud clicks (Fig. 14D).

Field potential responses to the same stimuli (clicks) were recorded routinely to check the influence of gross activity on the single neuron recordings. The amplitude of the field potential was always found to be significantly smaller than that of the single unit response (compare Fig. 13C and D, av.). As with the surface evoked potential in similar conditions (81) the most prominent component of the field response was a negative wave (Fig. 13C). The latency and the time course of the negative field potential were similar to those of the IPSP in most of the impaled neurons (compare Fig. 13C with D and F). It may be suggested therefore that previously described changes in the amplitude of the negative wave of the surface evoked potential during elaboration of the LCSR (81) are reflections of an increased IPSP in many cortical neurons.

## DISCUSSION

Our experiments demonstrate the possibility of investigating changes related to conditioning procedures at synaptic and membrane level in cortical neurons of awake mammalian animals. Two classes of experimental situation have been studied. The first class is "cellular analogs" of the CR, and the second one is a behavioral CR of a simple type (LCSR).

Two types of PSP changes have been found in cellular analog of cortico-cortical conditioning. The first type was a short-lasting (up to 30 s) PSP facilitation. In most neurons tested, this short-lasting facilitation was unspecific to pairing and therefore was analogous not to conditioning, but to pseudoconditioning (37) or "dominant focus" effect (3, 59). The time course and character of EPSP and IPSP changes are similar to corresponding characteristics of posttetanic potentiation (PTP) of the cellular response produced by a testing stimulus delivered via the same cortical surface electrodes used for tetanization (72, 75, 76). The short-lasting EPSP facilitation may be accounted for by the mechanism of PTP in polysynaptic cortical pathways. It is well known that PTP can produce heterosynaptic effects in polysynaptic nets by potentiation of interneuronal synapses shared by priming ("unconditional") and by testing ("conditional") pathways. Another possibility is the direct potentiation of synaptic endings formed by a "conditional" pathway.

Some other explanation based on changes in membrane potential level, cell excitability, nerve impulse reverberation or posttetanic afterdischarges have been discussed and considered to be unlikely (78, 81, 82, 87, 88).

Additionaly, short-lasting facilitation specific to pairing was found in two neurons. The nature of this phenomenon seems worthy of more detailed study because such a specificity cannot be readily explained on the PTP basis (11) and certain additional assumption must be made (26). One may suggest a participation of a special mechanism of heterosynaptic facilitation similar to that in molluscan CNS (36) which is partially specific to pairings in some cases (37). However the time course of the short-lasting EPSP facilitation is different from the time course of the molluscan heterosynaptic facilitation.

The second type of the PSP changes in the cellular analog of cortico-cortical CR was revealed with longer (up to 2 min) intertrial intervals. These PSP changes were far less pronounced than the short-lasting facilitation described previously. However, these changes seem much more interesting from the standpoint of investigating learning mechanisms. The changes were more long-lasting (more than several minutes, at least), specific to pairings and showed effects time-locked to the UCS presentation. It is interesting to note that the direction of the cellular "conditioned" response (initial activation or inhibition) was always the same as the direction of the response to the UCS in these experiments.

Similar neuronal models of "cortico-cortical" CRs were created independently (with extracellular recordings only) in the rabbit sensorimotor cortex (4), in the neuronal isolated cortex of the cat (39) and in an isolated slab of rabbit cerebral cortex (40). All these data demonstrate reproducible modifications of cellular activity during the pairings of two cortical stimuli. It is of interest that the percentage of "conditionable" neurons and the adequate number of pairings are similar in "cortico-cortical" cellular analogs to those reported for the other experimental situations (see Introduction for References).

Similar, changes in PSP have been found in a cellular analog of conditioning with interhemispheric CS. Although specificity to pairing have not been adequatly tested, some CR properties have been demonstrated including "extinction" and self-restoration (Fig. 6). Preliminary analysis of the averaged PSPs indicated changes predominantly in polysynaptic components. These data correspond to changes in gross interhemispheric delayed response encountered by Rutledge (60) after elaboration of a behavioral CR in cats with electrical stimulation of the suprasylvian cortex as the CS.

Changes in short-latency components of thalamo-cortical PSPs have been reported recently in various experimental situations, some of them being close to cellular analogs of the CR (5). The changes were similar to short-lasting facilitation described in our experiments but in several cases appeared to be more prolonged and specific to pairings.

The most prominent and reproducible modifications of cellular responses (appearance of short-latency EPSP in conditioned group) were found in our experiments with elaboration of a simple EMG response. The elaborated behavioral response satisfied most of the required characteristics of a CR on the one hand and it was similar in some respects to the unconditioned SR on the other hand. Facilitation of the SR has been demonstrated during various conditioning procedures (30, 54). Moreover, conditioning procedures enable the elaboration of a conditioned SR to initially inefficient (subthreshold) stimuli (29, 45, 46, 55). It was suggested by earlier workers (45, 55) that this stable and short latency response represents pseudoconditioning rather than a "true" CR. However specificity to pairing was demonstrated later (29, 46) and it was also found for the particular type of the conditioned SR (LCSR) described here. The short latency blink reaction (20 ms) investigated by Woody et al. (97), which is another variant of the SR was also found to be specific for pairings. The short latency of the unconditioned SR as well as lesion experiments (see 68, 98) indicate the subcortical nature of the SR, the main pathway being presumably the reticulo-spinal tract. On the other hand, gross (15) and single-neuron (68) recordings of pyramidal tract discharges in cats anesthetized with chloralose suggest participation of the cortico-spinal pathway in mediating the SR.

It is evident from our data that the occurrence of the majority of neuronal responses in the sensorimotor cortex cannot be explained on the basis of proprioceptive feedback after unconditioned SR or LCSR. This is clear from the comparison of the 17 ms latency or neuronal discharges with the 12–17 ms latency of the LCSR (or the SR) and the minimal 7 ms latency of PSPs to forelimb stimulation (71). Moreover the appearance of extremely short-latency (less than 7 ms) responses suggests that at least one path of the LCSR (or the SR) passes through the sensorimotor cortex. The difference between the latency of the unit responses (4–7 ms for the shortest latencies) and that of the EMG response is 5–13 ms. This difference corresponds to a 4–7 ms minimal delay between direct cortical stimulation and the EMG found in special experiments (85).

Of course we do not suggest that information related to the performance of the LCSR is transmitted exclusively through the cortex, if account is taken of the essentially subcortical nature of the unconditioned SR and the LCSR. Thus the organization of the conditioned and unconditioned SR seems to be similar to organization of many other types of movement which can have several parallel pathways including that through the motor cortex.

We concluded from our data that at least one of the crucial changes during this type of conditioning takes place at the level of sensorimotor cortex. Similar conclusions were made by other authors on the basis of extracellular recordings for blink CR (97) in cats and more recently for some other types of short latency conditioned movements (23, 42, 61).

The extremely short minimum latencies of the click evoked synaptic responses (4 to 7 ms) suggest fast afferent conduction through pathways with few synapses. Similar pathways must be postulated to account for the less than 6 ms latency of neuronal discharge in the rat sensorimotor cortex after conditioning (52). Olds et al. (52) data are presumably relevant not to comparatively long-latency CR described by these authors (52) but to the SR which is facilitated during conditioning procedures (30, 54). In a special series of experiments (83) we tried to analyze possible pathways mediating such short-latency EPSPs by stimulating various subcortical auditory structures. It was found that about 25% of sensorimotor cortical neurons responded with PSPs and spikes within 4 ms after the stimulation of the colliculus inferior. However most of these responses did not appear to be monosynaptic. These data as well as some other indirect evidence provided with extracellular recordings (23, 43) indicate intracortical interneuronal connections as a probable locus of plastic changes underlying the elaboration of simple motor CRs.

The most direct modern method of studying synaptic connections is recording of so-called "unitary" PSPs, e.g. PSPs evoked in "receiving" neuron after a spike discharge produce by a single "emitting" neuron. The properties of afferent (34, 94) and interneuronal (35, 62) unitary PSPs have been studied in spinal motoneurons with this method but only a few records of unitary cortical EPSPs have been published up to now (2). We are developing methods of unitary PSP recording and analysis in sensorimotor cortex of awake rabbit in collaboration with Dr. A. G. Gusev (Fig. 15). Figure 15 illustrates examples of single (Fig. 15A, 1–5) and averaged (Fig. 15B, C) intracellular recordings triggered by extracellular spikes (Fig. 15A, 6) in a neighbouring "emitting" neuron. It is interesting to note the similarity between unitary (most probably monosynaptic) EPSP shown in Fig. 15B, 1, 5 and short latency EPSP evoked by click in conditioned animal (Fig. 13A, av.). Preliminary statistical treatment within the frame of "quan-

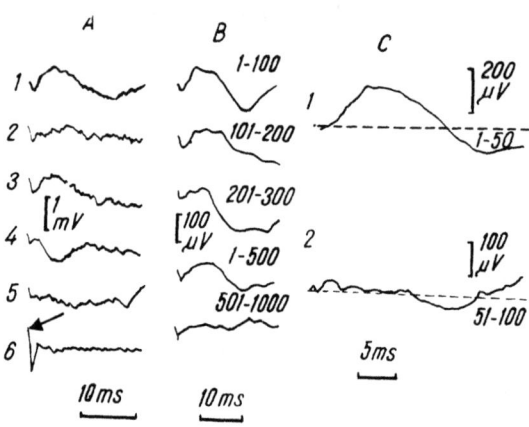

Fig. 15. Postsynaptic potentials found in a cortical neuron by triggering from spikes of an adjacent neuron. A: examples of intracellular single records (1–5) triggered by extracellular spike similar to that shown by an arrow in A6. B: intracellular spike-triggered averages from the same neuron. C: average of the first (C1) and last (C2) 50 sweeps from 100 sweeps used to compute the record B1. The numbers to the right of the traces are ordinal numbers of averaged sweeps. The presynaptic neuron was de polarized through the extracellular microelectrode to produce frequent spiking. The target neuron was hyperpolarised through the intracellular microelectrode to abolish spontaneous discharges.

tum" hypothesis of Katz (see 44) was made for several neurons. The data indicate very low mean quantum content (less than 1) and quantum size of the order of 0.1–0.2 mV. The latter values are similar to those reported for other neuronal synapses (44) including hippocampal synapses (6). The averaged unitary EPSP were sometimes repeatable in successive summations (compare Fig. 15B, 1 with 15B, 2) but great variability was found in other cases (15B, 4, 5; C, 1, 2). This variability is presumably a characteristic of unitary intracortical connections in unanesthetized brain, but one needs more data to support this notion and to reveal the underlying mechanisms.

In general, PSPs changes including increase in short-latency EPSPs and longer latency IPSPs were observed in all experimental models studies. Augmentation of EPSP in cellular analogs of conditioning and especially the appearance of short latency EPSPs after elaboration of behavioral CR support the idea that an increase in efficiency of excitatory synapses underlies the mechanisms of conditioning (see Introduction). The synaptic hypothesis seems more attractive also for explanation of our studies of the cellular analog of time conditioning (77, 80, 81). On the other hand, some studies in simple analogs of learning in molluscian CNS indicate participation of intracellular of "membrane" mechanisms (36, 67). Some authors (96, 97) found that thresholds for direct electrical stimulation in motor cortex were lower in conditioned than in naive animals. These data were denied for other type of CR (43). Nevertheless some measurements, especially those of Woody and Black-Cleworth (95) with intracellular recording may be interpreted as an indication of a decrease in membrane thresholds as a postsynaptic mechanism underlying the elaboration of the CR. At first glance one of our records (Fig. 5) seems to support such a "membrane" hypothesis suggesting selective modification (decrease in our case) of neuronal threshold. However at least two "synaptic" explanations can be considered in our case. The first possibility is that of a simultaneous increase in excitatory and inhibitory synaptic inputs, with the latter resulting in an increase in membrane permeability which prevents spike generation. In fact, this is the simplest way to explain the lack of the spike generation by high-amplitude EPSPs, which can often be seen in response to direct cortical stimulation (74). The second possible explanation is based on blockage of a spike in a zone remote from the recording microelectrode. The possibility of remote spike generation is supported by occurrence of partial spikes in these cells (Fig. 5B, 2, see also Fig. 2A, 4 and 3). It is woth noting that, in fact, two discussed mechanisms ("synaptic" and "membrane") are not contradictory and both of them may take part in learning processes.

Another independent indirect support of the "synaptic" hypothesis is provided by studies of long-lasting hippocampal PTP, discovered by Bliss and Lomo (10). It was suggested (82, 89, 92) that elementary cellular mechanisms of hippocampal PTP are similar to or even identical with those of CRs. This suggestion was based on properties of the hippocampal PTP including longevity (19), specificity for stimulated pathway (1, 48, 63), "extinction" and self-restoration (82, 92). "Cooperative" properties (50) with involvement of "reinforcing" systems during conditioning tetanization (82, 92) are other features of the hippocampal PTP similar to those of behavioral conditioning. "Specificity" of the PTP (1, 48, 63) as well as data received with recording of "minimal" evoked potentials (93) and PSPs (1, 65, 89, 91) and morphological measurements (24) suggest the synaptic nature of the hippocampal PTP.

In conclusion, our and literature electrophysiological data seem to support "synaptic" hypothesis of learning (see Introduction) and Konorski's (41, 42) theoretical models. Combination of simple, quickly elaborated motor CRs (of LCSR type) with recording of unitary cortical PSPs seems to be the most prominent and direct method in further study of conditioning mechanisms at cellular level.

REFERENCES

1. ANDERSEN, P., SUNBERG, S. H., SVEEN, O. and WIGSTROM, H. 1977. Specific long-lasting potentiation of synaptic transmission in hippocampal slices. Nature 266: 736–737.
2. ASANUMA, H. and ROSEN, I. 1973. Spread of mono- and polysynaptic connections within cat's motor cortex. Exp. Brain Res. 16: 507–520.
3. ASRATYAN, E. A. 1970. Ocherki po fiziologii ulovnykh refleksov. Izdat. Nauka, Moscow.
4. BALASHOVA, A. N. 1972. Unit activity of sensorimotor cortex at an early stage of conditioning (in Russian). Zh. Vyssh. Nervn. Deyat. im. I. P. Pavlova 22: 822–827.
5. BARANYI, A. and FEHER, O. 1978. Conditioned changes of synaptic transmission in the motor cortex of the cat. Exp. Brain Res. 33: 283–298.
6. BAŠKIS, A. V., VORONIN, L. L. and DEREVYAGIN, V. I. 1978. Analysis of mechanisms of frequency potentiation of synaptic responses in hippocampal neurones (in Russian). Dokl. Akad. Nauk SSSR 230: 234–237.
7. BAUMGARTEN, R. Von. 1970. Plasticity in the nervous system at the unitary level. In F. O. Schmitt (ed.), The Neurosciences: second study program. Rockefeller Univ. Press, New York, p. 260–271.
8. BAUMGARTEN, R. Von and JAHAN-PARVAR, B. 1967. Beitrag zum Problem der heterosynaptischen Facilitation in Aplysia Californica. Pflügers Arch. 295: 328–346.
9. BERGER, T. W. and THOMPSON, R. F. 1978. Identification of pyramidal cells as the critical elements in hippocampal neuronal plasticity during learning. Proc. Natl. Acad. Sci. USA 75: 1572–1576.
10. BLISS, T. V. P. and LØMO, T. 1973. Long-lasting potentiation of synaptic transmission in the dentate area of the anaesthetized rabbit following stimulation of the perforant path. J. Physiol. (Lond) 232: 331–356.
11. BRINDLEY, G. S. 1967. The classification of modifiable synapses and their use in models for conditioning. Proc. R. Soc. 168B: 361–376.
12. BUREŠ, J. and BUREŠOVÁ, O. 1970. Plasticity in single neurones and neural populations. In G. Horn and R. A. Hinde (ed.), Short-term changes in neural activity and behaviour. Univ. Press, Cambridge, p. 363–403.
13. BURKE, W. 1966. Neuronal models for conditioned reflexes. Nature 210: 269–271.
14. BURNS, B. D. 1958. The mammalian cerebral cortex. Edward Arnold Co., London.
15. BUSER, P., ASCHER, P. BRUNER, J., JASSIK-GERSCHENFELD, D. and

SINBERG, R. 1963. Aspects of sensorimotor reverberation to acoustic and visual stimuli. The role of primary specific cortical areas. Prog. Brain Res. 1: 294–322.
16. CAJAL, S. R. 1955. Histologie du Systeme Nerveaux de l'Homme et des Vertebres. Vol. 2, Instituto Ramon y Cajal, Madrid.
17. CHOCHOLLE, R. 1963. Temps de reaction. *In* P. Fraisse and J. Piaget (ed.), Traite de Psychologie Experimentale. 11. Sensation et Motricite. Presses Universitaires de France, Paris.
18. DOTY, R. W. and GIURGEA, C. 1961. Conditioned reflexes established by coupling electrical excitation of two cortical areas. *In* J. F. Delafresnaye (ed.), Brain mechanisms and learning. Blackwell, London, p. 133–152.
19. DOUGLAS, M. R. and GODDARD, G. V. 1975. Long-term potentiation of the perforant path-granule cell synapse in the rat hippocampus. Brain Res. 86: 205–215.
20. DUNIN-BARKOWSKI, V. L. 1978. Informatsionnye protsessy v neĭronnykh strukturakh. Izdat. Nauka, Moscow.
21. ECCLES, J. C. 1953. The neurophysiological basis of mind. Clarendom Press, Oxford, 326 p.
22. ECCLES, J. C. 1972. Possible synaptic mechanisms subserving learning. *In* A. G. Karczmar and J. C. Eccles (ed.), Brain and human behavior. Springer-Verlag, Berlin, p. 39–61.
23. EVARTS, E. V. 1973. Motor cortex reflex associated with learned movement. Science 179: 501–503.
24. FIFKOVA, E. and VAN HARREVELD, A. 1977. Long-lasting morphological changes in dendritic spines of dentate granular cells following stimulation of the entorhinal area. J. Neurocytol. 6: 211–230.
25. GASSANOV, U. G. 1972. Vnutrennee tormozhenie. Izdat. Nauka, Moscow.
26. GARDNER-MEDWIN, A. R. 1969. Modifiable synapses necessary for learning. Nature 223: 916–919.
27. GIURGEA, K. M. 1959. Neurophysiological analysis of conditioned responses to direct electrical stimulation of the cerebral cortex. *In* D. A. Biriukov (ed.), Nekotorye voprosy sovremennoĭ fiziologii. Medgiz, Leningrad, p. 53–58.
28. GLOBUS, A. and SCHEIBEL, A. B. 1967. Synaptic loci on parietal cortical neurones: terminations of corpus callosum fibers. Science 156: 1127–1129.
29. GRASTYAN, E. 1961. The significance of the earliest manifestations of conditioning in mechanisms of learning. *In* J. F. Delafresnaye (ed.), Brain mechanisms and learning. Blackwell, London, p. 243–263.
30. GRASTYAN, E. BAUER, M., PORCZI, J. and SZABO, I. 1962. Über die Rolle der "Zuckreaktion" (Startle-Reaktion) im Mechanismus des bedingten Reflexes. Acta Physiol. Acad. Sci. Hung. Suppl. 20: 71.
31. GRIFFITH, J. S. 1966. A theory of the nature of memory. Nature 211: 1160–1163.
32. GRIFFITH, J. S. 1971. Mathematical neurobiology (An introduction to the mathematics of the nervous system). Acad. Press, London.
33. HEBB, D. O. 1949. The organization of behavior. John Wiley and Sons, New York.
34. HENNEMAN, E. 1974. Principles governing distribution of sensory input to motor neurones. *In* F. O. Shmitt and F. G. Worden (ed.), The neuroscience. Third study program. The MIT Press, Cambridge, p. 281–291.

35. JANKOWSKA, E. and ROBERTS, W. J. 1972. Synaptic actions of single neurones mediating reciprocal Ia inhibition of motoneurones. J. Physiol. (Lond) 222: 633–642.
36. KANDEL, E. R. 1976. Cellular basis of behavior. W. H. Freeman and Co., San Francisco.
37. KANDEL, E. R. and SPENCER, W. A. 1968. Cellular neurophysiological approaches in the study of learning. Physiol. Rev. 48: 65–134.
38. KANDEL, E. R. and TAUC, L. 1965. Mechanism of heterosynaptic facilitation in the giant cell of the abdominal ganglion of Aplysia depilans. J. Physiol. (Lond) 181: 28–47.
39. KHANANASHVILI, M. M., ZARKESHEV, A. G. and SILAKOV, V. L. 1971. Conditioning of single unit activity by intracortical electric stimulation of neuronally isolated cortex. Fiziol. Zh. SSSR im. I. M. Sechenova 57: 490–496.
40. KHOLODOV, Y. A. 1972. Creation of a model of the conditional reflex in neuronally isolated slab of the cerebral cortex. *In* O. S. Adrianov (ed.), XXIII Soveshchanie po problemam vyssheĭ nervnoĭ deiatel'nosti. Part II, Gorki.
41. KONORSKI, J. 1948. Conditioned reflex and neuron organization. Cambridge Univ. Press, London, 267 p.
42. KONORSKI, J. 1967. Integrative activity of the brain. An interdisciplinary approach. Univ. Chicago Press, Chicago, 531 p.
43. KOTLYAR, B. I. 1977. Mekhanizmy formirovaniya vremennoĭ svyazi. MGU, Moscow.
44. KUNO, M. 1971. Quantum aspects of central and ganglionic synaptic transmission in vertebrates. Physiol. Rev. 51: 647–678.
45. LANDIS, C. and HUNT, W. A. 1939. The startle pattern. Farrar and Rinehart, New York.
46. LARSSON, L. E. 1956. The relation between the startle reaction and the nonspecific EEG response to sudden stimuli with a discussion on the mechanism of arousal. Electroencephalogr. Clin. Neurophysiol. 8: 631–644.
47. LIVANOV, M. N. and KOROLKOVA, T. A. 1951. Influence of inadequate stimulation of the cortex by induction current on cortical bioelectrical rhythms and conditioned activity. Zh. Vyssh. Nervn. Deyat. im. I. P. Pavlova 1: 332–346.
48. LYNCH, G. S., DUNWIDDIE, T. and GRIBKOFF, V. 1977. Heterosynaptic depression: a postsynaptic correlate to long-term potentiation. Nature 266: 737–739.
49. MCILWAIN, J. T. and CREUTZFELDT, O. D. 1967. Microelectrode study of synaptic excitation and inhibition in the lateral geniculate nucleus of the cat. J. Neurophysiol. 30: 1–21.
50. MCNAUGHTON, B. L., DOUGLAS, R. M. and GODDARD, G. V. 1978. Synaptic enhancement in fascia dentata: cooperativity among coactive afferents. Brain Res. 157: 277–293.
51. OLDS, J. 1969. The central nervous system and the reinforcement of behavior. Am. Psychol. 24: 114–132.
52. OLDS, J., DISTERHOFT, J. F., SEGAL, M., KORNBLITH, C. L. and HIRSH, R. 1972. Learning centers of rat brain mapped by measuring latencies of conditioned unit responses. J. Neurophysiol. 35: 202–219.

53. PAVLOV, I. P. 1951. Polnoe sobranie sochinenii. Vol. 3–5, Acad. Nauk SSSR, Moscow.
54. PICKENHAIN, L. and KLINGBERG, F. 1969. Hirnmechanismen und Verhalten: Elektrophysiologische und Verhaltensuntersuchungen an der Ratte. Fischer, Jena.
55. PROSSER, C. L. and HUNTER, W. S. 1936. The extinction of startle responses and spinal reflexes in the white rat. Am. J. Physiol. 117: 609–618.
56. RABINOVICH, M. Ia. 1975. Zamykatelnaya funktziya mozga. Meditzina, Moscow.
57. ROSANOV, S. I. 1973. The neuron and memory (in Russian). Usp. Sovrem. Biol. 76: 447–456.
58. ROSENBLATT, F. 1962. Principles of neurodinamics. D. C., Spartan Books, Washington.
59. RUSINOV, V. S. 1969. Dominanta. Meditzina, Moscow.
60. RUTLEDGE, L. T. 1965. Facilitation: electrical response enhanced by conditional excitation of cerebral cortex. Science 148: 1246–1248.
61. SAVCHENKO, E. I., MAIOROV, V. I. and KOTLYAR, B. I. 1976. Neuronal activity in the cat motor cortex during conditioned reaction of the forepaw on a support (in Russian). Zh. Vyssh. Nervn. Deyat. im. I. P. Pavlova 26: 65–72.
62. SHAPOVALOV, A. I. and KOZHANOV, V. M. 1978. Disynaptic brainstem — propriospinal projections to mammalian motoneurones. Neuroscience 3: 105–108.
63. SHARONOVA, I. N., VORONIN, L. L. and SKREBITSKI, V. G. 1976. Long-lasting posttetanic potentiation of hippocampal evoked potentials to the stimulation of Schaffer collaterals (in Russian). Zh. Vyssh. Nervn. Deyat. im. I. P. Pavlova 26: 214–217.
63a. SHVYRKOV, V. B. 1968. Form of participation of cortical somatosensory projection neurons in the development of the conditioned defence reflex. Vestn. USSR Akad. Med. Sci., 23 (7) 125–139 (Translated from Vestn. Akad. Med. Nauk SSSR, 23 (7): 80–89).
64. SKREBITSKI, V. G. and CHEPKOVA, A. N. 1973. The role of time intervals between "conditioned" and "unconditioned" stimuli in the elaboration of neuronal analogs of conditioned reflexes (in Russian). Zh. Vyssh. Nervn. Deyat. im. I. P. Pavlova 23: 576–584.
65. SKREBITSKI, V. G. and VOROBYEV, V. S. 1979. A study of synaptic plasticity in hippocampal slices. Acta Neurobiol. Exp. 39: 632–642.
66. SKREBITSKI, V. G. and VORONIN, L. L. 1966. Intracellular records of single unit activity of the visual cortex in nonanesthetized rabbits (in Russian). Zh. Vyssh. Nervn. Deyat. im. I. P. Pavlova 16: 864–873.
67. SOKOLOV, E. N. 1977. Brain functions: neuronal mechanisms of learning and memory. Ann. Rev. Psychol. 28: 85–112.
68. STENHOUSE, D. and ECCLES, R. M. 1971. Activity in descending tract fibres in cats anaesthetized with chloralose. Brain Res. 35: 127–135.
69. SZABO, I. 1965. Analysis of the muscular action potentials accompanying the acoustic startle reaction. Acta Physiol. Acad. Sci. Hung. 27: 167–178.
70. THOMPSON, R. F., PATTERSON, M. M. and TEYLER, T. J. 1972. The neurophysiology of learning. Ann. Rev. Psychol. 23: 73–104.
71. VORONIN, L. L. 1967. Postsynaptic potentials in motor cortex of waking

rabbit. Neurosci. Transl. 2: 177–184. (Translated from Fiziol. Zh. SSSR im. I. M. Sechenova 53: 623–631).
72. VORONIN, L. L. 1969. Changes in the excitability of individual neurones after electrical stimulation of the cortical surface. Dokl. Biol. Sci. 186: 474–477 (Translated from Dokl. Acad. Nauk SSSR, 186 (5): 1213–1216.
73. VORONIN, L. L. 1970. Neurophysiological mechanisms of trace phenomena (in Russian). Usp. Fiziol. Nauk, 1 (1): 111–136.
74. VORONIN, L. L. 1970. Synaptic responses of sensorimotor cortical neurones to direct cortical stimulation. Bull. Exp. Biol. Med. 70: 1238–1244 (Translated from Biull. Eksp. Biol. Med. 70 (11): 15–19).
75. VORONIN, L. L. 1970. Effects of local polarization on afterdischarges of cortical neurones. Neurophysiology 2: 346–352. (Translated from Neĭrofiziologiya 2: 460–468).
76. VORONIN, L. L. 1970. Post-tetanic changes in intracellular response to direct cortical stimulation. Neurophysiology 2: 454–461. (Translated from Neirofiziologiia 2: 601–610).
77. VORONIN, L. L. 1970. Microelectrode investigations of cellular analogs of learning. Soviet Neurol. Psychiatr. 4: 99–125 (Translated from Neĭronnye mekhanizmy obucheniia, E. N. Sokolov (ed.), MGU, Moscow, p. 5–25).
78. VORONIN, L. L. 1971. Facilitation of reactions of cortical neurones to cortical and peripheral stimulation after tetanization of the brain surface. Neurophysiology 3: 1–8 (Translated from Neirofiziologiya 3: 3–12).
79. VORONIN, L. L. 1971. Microelectrode study of cellular analogs of conditioning. Proc. IUPS, Vol. 8, 25 Int. Congr. Physiol. Sci., Munich, 199–200.
80. VORONIN, L. L. 1971. Microelectrode study of the cellular analog of a conditioned reflex to time (in Russian). Zh. Vyssh. Nervn. Deyat. im. I. P. Pavlova 21: 1283–1246.
81. VORONIN, L. L. 1976. Microelectrode study of neurophysiological mechanisms of conditioning. *In* C. D. Woody (ed.), Soviet research reports, Vol. 2. Brain information service, Brain Research Institute, UCLA, Los Angeles.
82. VORONIN, L. L. 1976. Cellular mechanisms of conditioned activity (in Russian). Zh. Vyssh. Nervn. Deyat. im. I. P. Pavlova 26: 705–719.
83. VORONIN, L. L. 1978. Involvement of cortical neurones in conditioned and unconditioned startle reflex. Neuroscience 3: 133–137.
84. VORONIN, L. L., GERSTEIN, G. Y., IOFFE, S. V. and KUDRYASHOV, I. E. 1973. A rapidly elaborated conditioned reflex with simultaneous recording of neuronal activity (in Russian). Zh. Vyssh. Nervn. Deyat. im. I. P. Pavlova 23: 636–639.
85. VORONIN, L. L., GERSTEIN, G. L., KUDRYASHOV, I. E. and IOFFE, S. V. 1975. Elaboration of a conditioned reflex in single experiment with simultaneous recording of neural activity. Brain Res. 92: 385–403.
86. VORONIN, L. L. and IOFFE, S. V. 1974. Changes in unit postsynaptic responses at sensorimotor cortex with conditioning in rabbits. Acta Neurobiol Exp. 34: 505–513.
87. VORONIN, L. L. and KOZHEDUB, R. G. 1971. Cellular analog of a conditioned reflex to electrical stimulation of the cerebral cortex. Analysis of spike activity (in Russian). Zh. Vyssh. Nervn. Deyat. im. I. P. Pavlova 21: 775–783.
88. VORONIN, L. L. and KOZHEDUB, R. G. 1971. Analysis of postsynaptic po-

tential changes in a cellular analog of conditioned reflex (in Russian). Zh. Vyssh. Nervn. Deyat. im. I. P. Pavlova 21: 997–1005.
89. VORONIN, L. L. and KUDRYASHOV, I. E. 1977. Long-lasting hippocampal posttetanic potentiation with particular reference to mechanisms of conditioned reflex. Proc. of IUPS, Vol. XIII, 27 Inter. Congr. Physiol. Sci., Paris.
90. VORONIN, L. L. and KUDRYASHOV, I. E. 1978. Excitatory postsynaptic potentials of hippocampal neurones and their habituation with repetitive stimulation (in Russian). Neirofiziologiya 10: 3–12.
91. VORONIN, L. L. and KUDRYASHOV, I. E. 1979. Unit responses in hippocampus during long-lasting posttetanic potentiation (in Russian). Zh. Vyssh. Nervn. Deyat. im. I. P. Pavlova 29: 141–150.
92. VORONIN, L. L., KUDRYASHOV, I. E. and IOFFE, S. V. 1974. Posttetanic potentiation in hippocampus and its relation with mechanisms of the conditioned reflex (in Russian). Dokl. Akad. Nauk SSSR 217: 1453–1456.
93. VORONIN, L. L., KUDRYASHOV, I. E., SKREBITSKI, V. G., UVAROV, V. G. and SHARONOVA, I. N. 1977. "Minimal" evoked potentials in the hippocampus and their posttetanic changes (in Russian). Neirofiziologiya 9: 124–134.
94. WATT, D. G. D., STAUFER, E. K., TAYLOR, A., REINKING, R. M. and STUART, D. G. 1976. Analysis of muscle receptor connections by spike-triggered averaging. I. Spindle primary and tendon organ afferents. J. Neurophysiol. 39: 1375–1392.
95. WOODY, C. D. and BLACK-CLEWORTH, P. 1973. Differences in excitability of cortical neurones as a function of motor projection in conditioned cats. J. Neurophysiol. 36: 1104–1116.
96. WOODY, C. D. and ENGEL, J. Jr. 1972. Changes in unit activity and thresholds to electrical microstimulation at coronal-pericruciate cortex of cat with classical conditioning of different facial movements. J. Neurophysiol. 35: 230–241.
97. WOODY, C. D., VASSILEVSKY, N. N. and ENGEL, J. Jr. 1970. Conditioned eye blink: unit activity at coronal-precruciate cortex of the cat. J. Neurophysiol. 33: 851–864.
98. WRIGHT, C. G. and BARNES, C. D. 1972. Audio-spinal reflex responses in decerebrate and chloralose anesthetized cats. Brain Res. 36: 307–331.
99. ZUCKER, R. S. 1972. Crayfish escape behavior and central synapses. II. Physiological mechanisms underlying behavioral habituation. J. Neurophysiol. 35: 621–637.

L. L. VORONIN, Brain Research Institute, Academy of Medical Sciences of the USSR, Per. Obucha 5. Moscow 107 120, USSR.

## LIST OF ABBREVIATIONS

| | |
|---|---|
| CNS | central nervous system |
| CR | conditioned reflex |
| CS | conditional stimulus |
| EMG | electromyogram |
| EPSP | excitatory postsynaptic potential |
| IPSP | inhibitory postsynaptic potential |
| LCSR | local conditioned startle reaction |
| LHR | lateral hypothalamic reinforcement |
| PSP | postsynaptic potential |
| PTP | posttetanic potentiation |
| SR | startle reaction |
| UCS | unconditional stimulus |

Lecture delivered at the Warsaw Colloquium on Instrumental
Conditioning and Brain Research
May 1979

# PARTICIPATION OF CAUDATE NUCLEUS IN DIFFERENT FORMS OF VOLUNTARY ACTIVITY IN CATS

O. S. ADRIANOV, L. N. MOLODKINA, E. I. MUKHIN, N. P. SHUGALEV and N. G. YAMSHIKOVA

Laboratory of Morphophysiology of Conditioned Reflex, Brain Research Institute, USSR Academy of Medical Sciences and Laboratory of Physiology and Genetics of Behavior, Moscow State University
Moscow, USSR

*Abstract.* Different forms of behavior (instrumental conditioned reflexes, generalization tests, extrapolatory reflexes) were investigated before and after lesions of the caudate nucleus or pharmacological influence on its cholinergic systems. Postoperative deficit was small in well-elaborated forms of behavior and large in complex tasks. Chemical stimulation of the caudate nucleus was followed by depression of acquired forms of alimentary behavior.

## INTRODUCTION

In spite of the abundance of data concerning the physiology of the caudate nucleus (CN), its role in controlling different forms of behavior its little known. A number of investigators, including one of us (1, 4), showed that lesions of the brain may have different effects on conditioned-reflex activity, depending on the methods of investigation. This is also true in relation to the caudate nucleus, as it was shown in our laboratory by Andreev (7), and later in Suvorov's laboratory (22).

It is well known that the CN has an inhibitory influence on several brain structures. Such effect on motor reflexes is observed during electrical stimulation of the CN (8). Inhibition of motor behavior is apparently mediated by pallidum (where the majority of the CN efferents terminate, 24), and directly by the motor cortex. We have shown (5) that auditory evoked potentials in the caudate nucleus are decreased during the elaboration of alimentary instrumental conditioned reflex to sound and increased during the elaboration of differentiation. Inhibition of motor conditioned reflexes was also observed following neuro-chemical stimulation of the CN cholinergic (Ch) structures (9).

However, CN participation in integrative activity is not limited to its inhibitory influences. There are some data pointing to the important role of CN in the regulation of stimulus perception and voluntary movements (5, 6, 23).

In this paper we investigate qualitative and quantitative deficits of voluntary activity based on food reinforcement in the cat following lesions or pharmacological stimulation of caudate nucleus.

## METHODS

All experiments were done on cats. Three behavioral methods were used.

1. In the method of alternative choice the cats were first trained to choose the side reinforced by food. First, two circles of different diameters were presented: bigger on the right side and smaller on the left (the relative size was 3 : 1). At the beginning of training the figures were exposed along the edges of the screen, behind which (to the right and left) there were food-trays (one with food), separated by partition. In case of incorrect "umweg" response the cat could correct itself, going round the screen from the other side (Fig. 1A). The cats were trained till more than 50% of correct responses from 10 presentations were obtained in one experimental day. Thereupon new figures were immediately presented to animals. These figures were exposed in the center of the screen for 5 s. Besides, on different experimental days the cats were successively given different series of tasks with initial alternation: to distinguish figures by form (the so-called "geometrical rour"); to distinguish contours of various animals (differing in form and size); to distinguish the number of similar components and also to solve tasks with elementary "abstractness". Thus, the cats had to go to the right food-tray at succesive presentations of (a) larger figures of various forms (5 pairings in all); (b) larger contours of different animals (5 pairings in all); (c) greater number of similar components (5 stars); (d) longer delay time in the starting box (30 s). They were meant to go to the

left food-tray at the presentation of the figures of smaller size, smaller contours of animals, smaller number of similar components (2 stars), shorter delay time in the starting box (10 s). Thus we could evaluate the animal's ability to establish the relation between separate components of the program, beginning with simple qualitative generalization, piking out quantitative parameter, and passing to elementary "abstractness".

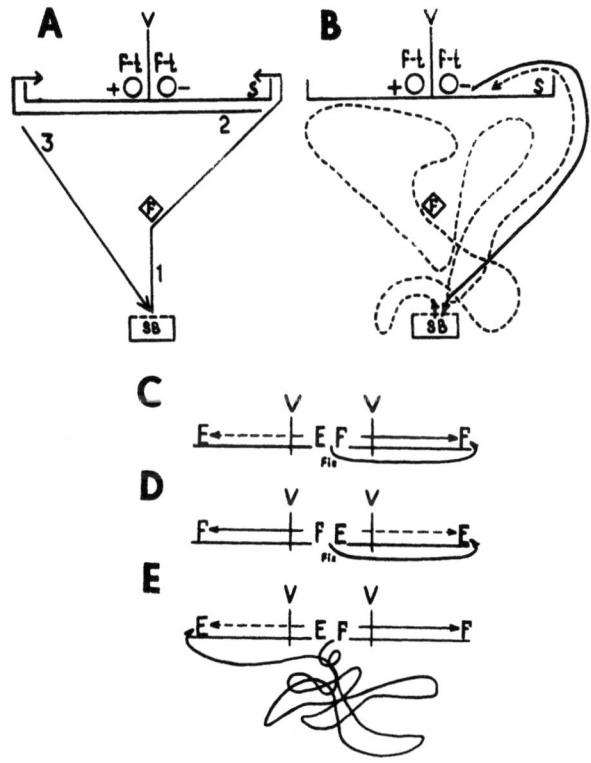

Fig. 1. The scheme of experiments in the alternative choice tasks (A, B) and in extrapolatory tasks (C, D, E). A: trajectory of an intact cat. 1, from starting box (SB) to the figure (F) and then to the empty bowl (F-t) behind the screen (S); 2. "correction" of an error; 3, return to the starting box; V, partition separating the food-tray. B: trajectory of a lesioned cat after leaving the starting box (dotted line) and returning in the case of an error. C and D: the schemes of responses to the side of food movement (F) or empty bowl (E). E: accidental undirected umweg response (in this case to the empty bowl). Horizontal line, the opaque screen with fissure (Fis) in the center; V, valves; F, bowl with food; E, empty bowl.

2. The cat's ability to extrapolate the direction of disappearing food stimulus using Krushinsky's method (13, 14). The animal was required to find food behind one side of an opaque screen after looking of the

moving food-tray (Fig. 1C and D). The results were measured by "Fi" method (18) and Plokhinsky's formulas (2). The animal was getting food and information about the direction of its movement for 2–3 s through the slit in the center of the opaque screen. Both bowls (with food and empty-control) were drawn back behind partitions which were 20 cm from the slit (length of the screen, 2 m, height, 0.75 m).

3. Alimentary instrumental conditioned reflexes were elaborated. The positive CS was 3 clicks presented with frequency 1/s and the negative CS was 3 clicks with frequency 2/s. The instrumental response was pressing a pedal with right forepaw.

Analysis of the material of the given method was done separately for the first 10 pairings of presentations (in each of the series) and for further presentations (10 in each experiment). Nonparametric statistics (Sign Test of Wilcoxon and Fisher) were used to evaluate the significance of results.

Bilateral CN lesions were made electrolytically using stereotaxic atlas (10) by platinum or silver electrodes with the current of 3 mA for 20–120 s in three or four points of each side. Drugs were injected through implanted microcannula into the head of CN.

## RESULTS

### The effects of lesions

General behavior of lesioned cats was normal. The cats found the way from vivarium to the experimental room, adequately reacted to call, to the sight of a dog and so on. High alimentary motivation was observed in all operated animals. The cats preserved natural alimentary reflexes (they could eat meat from the food-tray), and also sexual, defensive and hunting reflexes.

*Alternative choice methods.* Two groups of cats were investigated: the first (3 cats) with preoperative experience and the second (7 cats) without such experience. The first group mastered all tasks preoperatively on the level of more than 50%. This was characteristic for intact (15) and sham operated (3) cats. After lesions the cats of the first group solved the tasks on initial figures (circles of different size) without marked changes (the percentage of correct responses was equal to 63%, Fig. 2A-1). The second group of CN cats in comparison with the first group and intact animals showed a deterioration of ability to make adequate "umweg" responses on initial figures from the side of reinforcement. Correct responses did not exceed 25% (Fig. 2B-1).

No considerable difficulties in task performance were observed in series of different figures, presented separately by one pairing in the experimental day. The percentages of correct choices in both groups were respectively 65 and 58% (Fig. 2A-2 and B-2).

In the series of tasks involving geometric pattern when all the figures were presented successively, but not combined in pairings in one experiment, there was a significant ($P \leqslant 0.01$) predominance of correct umweg responses in comparison with the correct ones in all CN cats. Adequate responses formed on the average only 33% (Fig. 2A-3 and B-3).

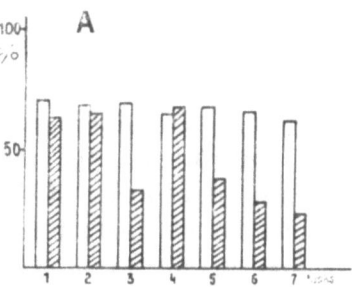

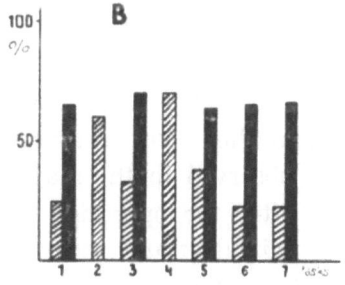

Fig. 2. Results of alternative choice task realization of lesioned animals with preoperative experience (A) and without it (B). Shaded bars, the percentage of correct responses in sequence with presentation of different tasks during first 10 trials after caudatoectomy (1–7); light bars, preoperative testing; black bars, testing after retraining. In tasks 3, 5–7 in A and in tasks 1, 3, 5–7 in B the differences were significant ($P < 0.01$).

The following series, in which cats were given tasks with a new kind of generalization — contours of different animals, did not present serious difficulties for CN cats of both groups. An average percentage of correct reactions to pairs of presented contours in each experiment was respectively 67 and 70% (Fig. 2A-4 and B-4). But presentation of the contours of all animals, not combined in pairs, in one experimental day, changed for the worse the performance of animals in both groups. Adequate responses decreased to 37–40% (Fig. 2A-5 and B-5).

In the following experimental series stimuli with a new kind of cues were presented — "quantity" (two and five stars) and timing parameters (different delay time of animal in the starting box). Considerable difficulties in all CN cats in comparison with intact and "control" ones were marked one more in these series. Correct responses of "quantitative" task decreased to 27–24% (Fig. 2A-6 and B-6), and in "timing" task to 21–24% (Fig. 2A-7 and B-7), i.e. in the majority of trials the animals gave incorrect responses.

To attain adequate performance unequal but systematic number of additional trials were required. For solving relatively simple tasks (differentiation of figures of "geometrical rour" and contours of different animals) the cats required from 10 to 30 trials; for solving "quantitative" tasks — 40–60, and for "timing" tasks — more than 150–200 trials. So, deficit produced by CN lesion was related to task complexity.

*Extrapolatory method.* Experiments on cats without preoperative experience ($n = 9$) showed that these animals did not make "umweg" responses on first presentation of the task more often (56%) than intact cats (27%). The results of the first trial were of special importance for the appreciation of the operated animal's ability of extrapolation — the anticipation of food withdrawal without initial training. In this case CN lesions did not disturb extrapolatory ability itself: correct umweg responses predominated over errors in 56% (confidence level near to $P \leqslant 0.10$). Animals of the operated group, like intact ones, were able to associate the direction of food-tray movement with the side of the screen which hid it.

*Motor performance.* This study was made with the use of the methods of alternative choice and extrapolatory reflex. Changes in behavior, in the form of indeterminate umweg responses (Fig. 1B and E), were observed in the majority of animals in the first 1–2 experimental days after the operation. Latency of reactions of alternative choice in both animal groups increased in comparison with intact cats (from 1–3 to 3–6 s) and in facts remained increased during the whole course of experiments, decreasing to "norm" only following additional training.

The timing of separate components of extrapolatory reaction changed in CN cats without previous experience in comparison with control animals (3). For example, the duration of the first umweg response in CN cats (42 s) was significantly greater ($P < 0.01$) than in intact animals (22 s). Duration of input of information about direction of food movement, when the animal was reaching out for food or looking in the direction of food movement, was somewhat smaller (3 s) in operated cats than in intact ones (5 s). When the animals solved the task "unilaterally" before the operation, the duration of correct umweg responses after the operation increased (from 9 to 12 s). But this was not true for animals which adequately solved the extrapolatory task before the operation (Fig. 3).

Further presentations of extrapolatory task to these cats showed that CN destruction did not prevent correct responses. After the operation "adequate" (3 cats), as well as "unilateral" (6 cats) performance was observed; in the latter case the animal went round one side of the screen in spite of food movement alternation. The results on animals given an extrapolatory task before CN destruction ($n = 11$), showed that after the operation the type of response ("adequate" or "unilateral") was preserved ($P \leqslant 0.05$).

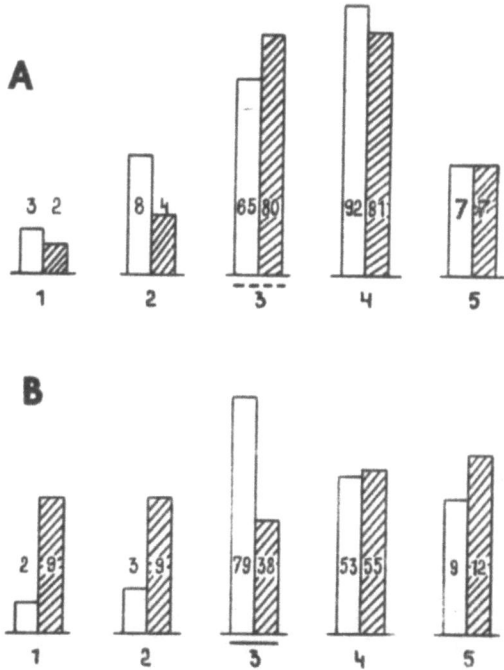

Fig. 3. Caudatoectomical effect on trajectory of responses in "adequate" (A) and "unilateral" (B) cats prior operation (on extrapolatory methods). 1, appearance of the first legible directed umweg response (presentation N); 2, legible directed, not less than 2 times one after another (presentation N); 3, general number of correct responses (%); 4 and 5, number and time of appearance, respectively, of correct directed responses. Shaded collumns, after surgery; unshaded, before surgery; solid line $P \leqslant 0.05$, dotted line $P \leqslant 0.01$.

The experiments with CN lesion during instrumental training showed an increased number of intertrial reactions, i.e. disinhibitory effect of caudatoectomy on motor behavior.

### The effects of cholinergic stimulation

The role of CN cholinergic structures in the above mentioned behavioral reactions was investigated. Proserin, known as acetylcholinesterase inhibitor, and atropine (cholinolytic) were used.

Injections of these drugs into CN, like its damage, did not disturb natural alimentary reflexes. Bilateral proserin injections (2–4 µg) into head of CN inhibited the alimentary instrumental conditioned reflex to sound stimuli. After in 3–5 min the latency increased and after 10–12 min conditioned reactions disappeared (Fig. 4). After 50–60 min conditioned intertrial reactions recovered. After a minimum dose (2 µg) conditioned instrumental reactions were preserved, though their latency increased and pedal pressure considerably diminished.

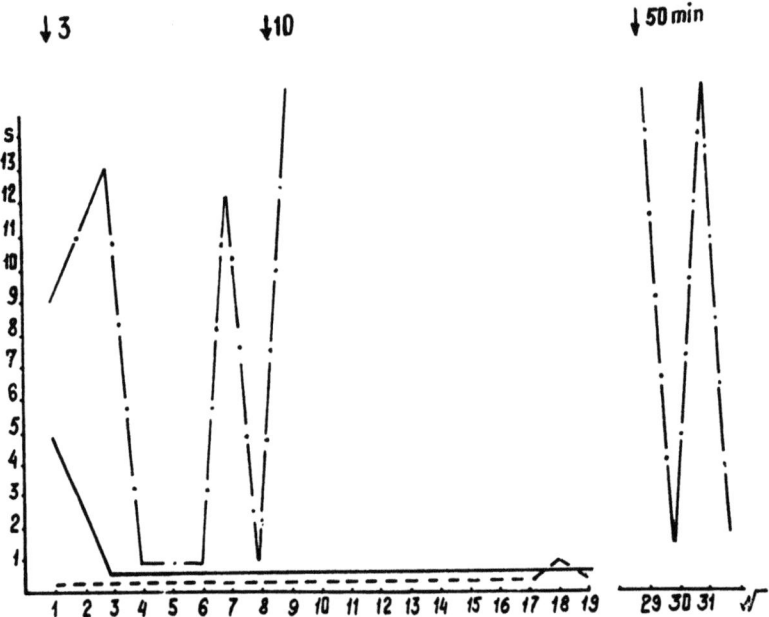

Fig. 4. Changes of latencies of conditioned instrumental reflexes after injections of proserin (dashed line with dots) or atropine (dotted line) into CN head. Continuous line, control. Latency registration in norm and following injections of each drug was made on different experimental days. Ordinate, latency in seconds; abscissa, conditioned signal presentations in one day. Figures above diagram, time after the moment of injection.

Proserin injections in CN resulted also in considerable behavioral disturbances during alternative choice of reinforcing side (Fig. 5). The effect of drug action increased to 10–12 min after its injection. Meanwhile the first manifested changes of motor reaction were observed: running to stimulus followed by coming-back to the starting box, and changes of goal-seeking behavior. Later the animals "refused" to solve presented tasks (Fig. 5 Bb). Thirty minutes after injection a gradual recovery of correct umweg responses was observed (Fig. 5Bb). In the case of this habit preservation after proserin injections (Fig. 5Ab), tasks performance on generalization (geometrical rour) was seriously disturbed.

As preliminary experiments showed, unidirectional disturbances of task performance on extrapolation were not noted after proserin injections in CN. In this case umweg responses were preserved, adequate performance predominating.

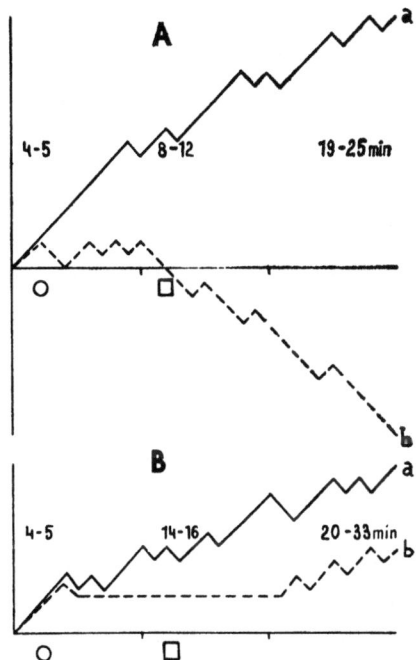

Fig. 5. Results of task realization dynamics during alternative choice after the injections of atropine (a) and proserin (b). Direction of lines upwards, correct performance; downwards, incorrect; horizontal, absence of performance. Circle and square, tasks with different figures. Figures, time in minutes after injection of drugs. A, cat 1; B, cat 2.

Low dose injections of central cholinolytic-atropine (5–7 μg) into CN head did not markedly influence the adequate performance of the above types of cat behavior (Fig. 5Aa and Ba). The general behavior of the animal did not change noticeably. Figure 5Aa and Ba shows a correct performance of alternative choice task on different programs (geometrical rour) following the drug injection. High doses of atropine (10 μg) resulted in a disturbance of conditioned instrumental reflex to the sound stimulus, accompanied by stereotyped movements and general motor excitation.

DISCUSSION

Our complex approach showed various deficits of alimentary voluntary activity in CN animals. Data on pharmacological stimulation of CN cholinoreactive structures supported these results.

Disturbances of alternative choice task depended on complexity of differentiation. The ability to differentiate simple task invariants, such as pairings of different figures and contours of animals, was preserved in CN cats. Task information is analogous to alternation tasks on model patterns, or reflexes on attitude, which are used as tests to study animal ability of elementary generalization (11, 12, 15–17, 19, 20). Higher level of behavior, involving analytical–synthetical activity with wide "generalization effect" (21) in the form of task with "magnitude" (12) and a large number of various elements, presented in one experience (countours and figures), was seriously disturbed in operated animals. Higher integrative activity was more severely impaired following CN lesions in cats, when they acquired definite ability to transfer from analysis of concrete signs (figures, contours, "multitude") to elementary abstractness (tasks with different timing delay). More trials were necessary for training during transition from the performance of "qualitative" to "quantitative" tasks and then to tasks with abstractness from concrete stimuli. Hence it may be assumed, that CN injury deprived the animals of ability to cope with rather complicated relations between the phenomena, requiring from them a high level of generalization and elementary abstractness from concrete signs of real objects. Nevertheless, the ability to acquire conditioned reflexes and simple forms of generalization were preserved.

The animal's ability of simple activity in extrapolatory test (guessing the direction of food stimulus movement) did not change much following CN destruction. Cats preserved the ability to solve this task either adequately, or "unilaterally", similarly, to intact animals. In addition, motor aspects of extrapolatory reflex and clarity of realization of presented task with respect to spatial-timing indices were disturbed. Results of extrapolatory task performance by CN cats, as reported earlier (3), were more comparable with results on intact animals, but differed considerably from prefrontally ablated ones. These facts permit to conclude that in distinction from prefrontal cortex, CN is not a structure closely connected with ability to extrapolate food movement direction, but participates mainly in motor aspects of extrapolatory activity.

Stimulation of CN cholinergic structures by means of proserin microinjections depressed the acquired behavior, such as conditioned instrumental reflex and umweg responses. However, natural food reflexes were preserved.

Summing up, the sensory performance in CN cats was especially disturbed with respect to elementary abstractness. As to the extrapolatory activity, it turned out to be less "vulnerable" to CN destruction, than generalization function. Finally, instrumental reflex, although preserved, was "overgrown" by a large number of intersignal reactions. The differences in sensory and motor aspects of voluntary alimentary behavior of different complexity (elementary abstractness, elementary rational activity, instrumental reflex) in CN animals may be explained by different "sensitivity" of the behavioral methods. Functional changes following CN destruction or pharmacological stimulation, resulted in similar disturbances of inhibitory influence of this structure on different forms of food goal-seeking behavior.

REFERENCES

1. ADRIANOV, O. S. 1976. O printsipakh organizatsii integrativnoĭ deyatelnosti mozga. Izdat. Meditsina, Moscow, 279 p.
2. ADRIANOV, O. S. and MOLODKINA, L. N. 1974. Decision of the extrapolatory task by prefrontally lobectomized cats (in Russian). Zh. Vyssh. Nervn. Deyat. im. I. P. Pavlova 24: 957–967.
3. ADRIANOV, O. S., MOLODKINA, L. N., SHUGALEV, N. P., YAMSHIKOVA, N. G. and BUTENKO, O. B. 1978. Comparative analysis of ability to extrapolation in cats after destruction of caudate nucleus and prefrontal lobectomy (in Russian). Zh. Vyssh. Nervn. Deyat. im. I. P. Pavlova 28: 913–919.
4. ADRIANOV, O. S. and POPOVA, N. S. 1963. On the structural premisses and functional peculiarities of dynamics of nervous processes in visual, auditory and cutaneous analyzers (in Russian). In I. S. Beritashvili (ed.), Gagrskie Besedy. Publ. Georgian Acad. Sci., Tbilisi, 4: p. 13–26.
5. ADRIANOV, O. S. and SHUGALEV, N. P. 1978. Mechanisms of participation of caudate nuclei in the organization of conditioned motor reflexes. Neuroscience 3: 139–141.
6. ALBE-FESSARD, D., OSWALDO-CRUZ, E. and ROCHA-MIRANDA, C. 1960. Activités évoquées dans le noyau caudé du chat en résponse á des types divers d'afférences. I. Étude macrophysiologique. Electroencephalogr. Clin. Neurophysiol. 12: 405–420.
7. ANDREEV, L. N. 1966. Influence of partial destruction of caudate nuclei on the conditioned-reflex activity of the dogs (in Russian). In XXI Sovesh. Probl. Vyssh. Nervn. Deyat., Moscow, p. 15–16.
8. BUCHWALD, N. A., WYERS, E. J., LAUPRECHT, C. W. and HEUSER, G. 1961. The caudate-spindle IV. A behavioral index of caudato-induced inhibition. Electroencephalogr. Clin. Neurophysiol. 13: 531–537.
9. HULL, C. D., BUCHWALD, N. A. and LING, G, 1967. Effects of direct cholinergic stimulation of forebrain structures. Brain Res. 6: 22–35.
10. JASPER, H. H. and AJMONE-MARSAN, C. 1954. A stereotaxic atlas of the diencephalon of the cat. Natl. Res. Council, Ottawa, Canada.
11. KÖHLER, W. 1921. Intellingenzprufungen an Menschenaffen. Berlin, 207 p.
12. KÖHLER, O. 1953. Vom unbenannten Denken. Lebendiges Wissen. 99: 291–299.
13. KRUSHINSKY, L. V. 1958. Extrapolation reflexes as elementary rational activity in animals (in Russian). Dokl. Akad. Nauk SSSR 121: 762–765.

14. KRUSHINSKY, L. V. 1977. Biologicheskie osnovy rassudochnoĭ deyatelnosti. Izdat. Moskow. Gos. Univ., Moscow, 270 p.
15. LADIGINA-KOTS, N. N. 1923. Issledovanie poznavatelnykh sposobnosteĭ shimpanze. GIIS, Moscow, 150 p.
16. LADYGINA-KOTS, N. N. 1959. Konstruktivnaya i orydiĭnaya deyatelnost' vysshikh obezyan (shimpanze). Izdat. Acad. Nauk SSSR, Moscow.
17. PAVLOV, I. P. 1949. Pavlovskie sredy. Protokol i stenogr. Fiziol. Besed. Akad. Nauk SSSR, Moscow, p. 1–3.
18. PLOKHINSKY, N. A. 1970. Biometriya. Izdat. Moskow. Gos. Univ., Moscow, 368 p.
19. PROTOPOPOV, V. P. 1950. Issledovanie vyssheĭ nervnoĭ deyatelnosti v estestvennom experimente. Gosmedizdat Ukrainian SSR, Kiev, 392 p.
20. ROGINSKY, G. Z. 1945. Psikhika chelovekoobraznykh obezyan. Lenizdat, Leningrad.
21. SECHENOV, I. M. 1935. Elementy mysli. Selected papers. Izdat. Inst. Exp. Med., Moscow, SSSR, p. 272–426.
22. SUVOROV, N. F. 1975. Problemy vyssheĭ nervnoĭ deyatelnosti i neĭrofiziologii. Izdat. Nauka, Leningrad, 187 p.
23. TOLKUNOV, B. F. 1978. Striatum i sensornaya spetsializatsiya neĭronalnoĭ seti. Izdat. Nauka, Leningrad, 176 p.
24. VONEIDA, T. 1960. An experimental study of course and destination of fibers arising in the head of the caudate nucleus in the cat and monkey. J. Comp. Neurol. 115: 75–88.

O. S. ADRIANOV, L. N. MOLODKINA, I. E. MUKHIN, N. P. SHUGALEV and N. G. YAMSHIKOVA, Brain Research Institute, Academy of Medical Sciences of the USSR, Per. Obukha 5, Moscow 107120, USSR.

Lecture delivered at the Warsaw Colloquium on Instrumental
Conditioning and Brain Research
May 1979

# FUNCTIONS OF THE NEOSTRIATUM: CORTEX-DEPENDENT OR AUTONOMOUS?

Ivan DIVAC

Institute of Neurophysiology, University of Copenhagen
Copenhagen, Denmark

*Abstract.* Studies of animals with ablations of varying amounts of the neocortex, the neostriatum, or both, are reviewed in an attempt to establish to which extent functions of the neostriatum are dependent on its cortical input. Scarce and inconclusive evidence does not allow firm conclusions. It seems well established that the neostriatum shares some functions with the neocortex. In the rat and infant monkeys these striatal functions appear to be cortex-independent, whereas in cats and adult monkeys they seem to be more cortex-dependent. It is possible, and even likely, that the neostriatum also has functions which are not shared with the cortex and which are cortex-independent. The degree to which the neostriatum is able to contribute to the integration of behavior in the absence of its cortical input is species- and age-specific.

## INTRODUCTION

The functions of the neostriatum (NS), divided in some species by the capsula interna into the caudate nucleus and the putamen, are most commonly described as motor or inhibitory and considered to be essentially unrelated to those of the neocortex (12). On the other hand, much evidence has accumulated showing a close anatomical and functional relation of the neocortex and NS (4, 5, 9, 26, 29). Some evidence even indicated that in the absence of the cortex, NS no longer mediates the functions shared by the two formations (6, 39). The aim of this paper is to see whether the available evidence supports Fulton's (17 p. 491) suggestion that "once the cerebral cortex has been removed the basal ganglia cease to contribute significantly to functional integration", or whether the functions of NS are essentially independent from the neocortex as implicit in the writings of several authors (e.g., 1, 2, 23, 24). The possible role of NS in the organization of the prosencephalon has been discussed elsewhere (8) and will not be considered here.

## DECORTICATE VS. "THALAMIC" PREPARATIONS

As early as 1924 Dresel (14) suggested that the functions of NS can be studied by comparing decorticated and "thalamic" preparations. (The quotation marks should serve as a reminder that after ablation of both the cortex and NS, the thalamus degenerates almost entirely and that consequently the highest grossly preserved part of the brain is the hypothalamus). Presumably the abilities found in decorticate, but not in "thalamic" preparations could be attributed to NS functioning. This idea at first appears attractive, but the results of such experiments are not easy to interpret. Perfect decortication, particularly in gyrencephalic brains, is a formidable task. In attempts to remove the cortex or the neostriatum, or both, it has been particularly difficult to preserve the so-called limbic system. In the past 50 years there have been a number of attempts to follow Dresel's paradigm, but only two studies, in which direct comparisons of the decorticate and "thalamic" preparations were made, provided satisfactory histological reports. In one of these experiments, cats in which NS was ablated together with the cortex, unlike decorticated animals, failed to eat spontaneously, to vocalize, and to clean themselves or other cats. On the other hand, "thalamic" cats were more active than the decorticated ones, and they showed both "obstinate progression" (a tendency to walk forward until an obstacle is encountered and then keep pushing against it as if attempting to go through) and sham rage (38). These results were partly confirmed in two other studies which furthermore demonstrated that "thalamic" cats do not mate (15), but can escape punishment (35). The latter authors found that extensive decortications alone also produced strong "obstinate progression".

A similar experiment performed on rats, provided similar results except that "thalamic" rats showed some elements of grooming, better coordination of locomotion, and an ability to gnaw, although they never ate spontaneously (32). Decorticated rats showed very little motor impairment (32, 34).

Thus, in both rats and cats, the decorticate preparation is more viable than the "thalamic" one, but rat preparations of either kind function better than comparable cat preparations. Unfortunately, the preserved functions in decorticated animals may have been mediated by parts of the limbic system which were damaged much more in "thalamic" animals than in decorticated ones. While the hippocampal contribution to the studied functions seems negligable (34), the amygdaloid nuclei may play some role. Nevertheless, these results may indicate that NS contributes at least something to the superior performance of decorticated animals as compared to the "thalamic" ones. Since these preparations have not been investigated with "cognitive" tasks, we do not know how their abilities for discrimination, learning and memory compare.

## DECORTICATE VS. DENEOSTRIATE PREPARATIONS

Inferences about NS functions can be made from comparisons of the consequences of ablations of either NS or the cortex. This approach faces serious technical difficulties: ablation of the entire NS alone has not yet been achieved. The technique of kainic acid lesions (11) is promising but still imperfect (10). Mettler's group had tried to ablate the caudate nucleus, but extensive additional damage to nonstriatal formations such as the cortex, the centrum semiovale and the fundus striati (33) renders these studies virtually uninterpretable. Similar criticism can be levelled against the recent attempt of Denny-Brown and Yanagisawa (3) to destroy the putament in monkeys. Ablation of the entire caudate nucleus with sparing of the putamen and small damage to nonstriatal tissue has been achieved by Villablenca's group (35–37). In their experiments the caudate ablated cats were behaviorally compared to cats in which a large portion of the frontal cortex had been removed. It is not known, however, whether the ablated cortical area "corresponds" to the ablated neostriatal tissue (Fig. 1) and damage to the cingulum bundle must be considered in the interpretation of the results. Inspite of these difficulties, the experiments of Villablanca's group are valuable. They have shown that cats without the caudate nucleus do not necessarily die nor turn into "real vegetables" as had been claimed by Mettler's group. These cats do not have profound and lasting motor impairments. The only such persistent impairment is an absence of the placing reaction. The most striking symptoms of caudate ablation in cats are: (i) a transient "obstinate progression", less pronounced than after extensive neodecortications (35); the same symptom was found also after thalamectomy by Villablanca and Salinas-Zeballos

(37), (ii) a syndrome which was named the "compulsory approach" (stereotyped approaching and following persons, cats or objects, accompanied by signs of pleasant emotion: purring, kneading and rubbing against the followed object) and interpreted as caused by emotional changes (36), and (iii) impairments in a series of learning tasks such as bar pressing alternation and spatial reversal (27).

Thus, the cats without caudate nuclei did not have any "extrapyramidal" symptoms, but instead were incapable of mastering certain learning tasks, and appeared to have changed emotional reactions. These results agree with the hypothesis of NS as a stage of processing or cortical output (8), but not with the accepted notion of NS as a major component of the "extrapyramidal motor system". It should be noted that cats without caudate nucleus eat spontaneously about a week after surgery (36). The same is true for decorticated preparations, but not for "thalamic" animals (32, 38). These results suggest that permanent aphagia in the latter group may have resulted either from ablations in the posterolateral region of the hemisphere (of putamen or portions of the limbic system; e.g., amygdala (16), or nucleus basalis (7) or from the additive effects of cortical, limbic and NS ablations. The same may be said about sham rage and the absence of vocalization (see above).

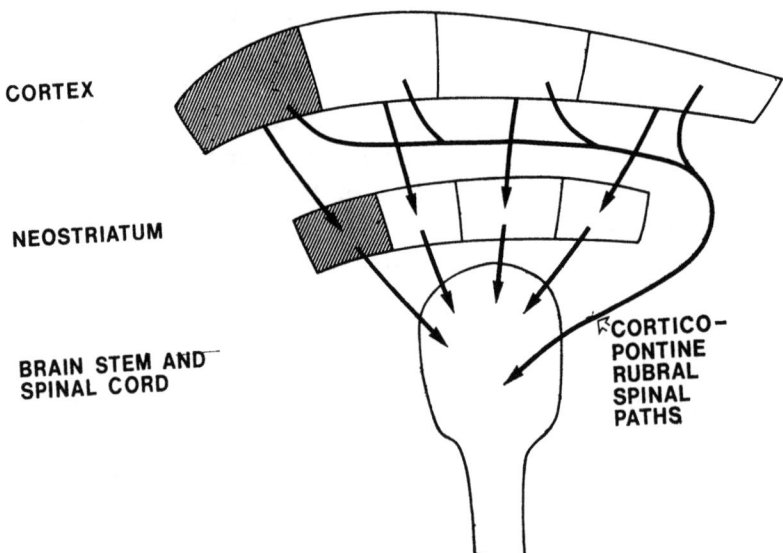

Fig. 1. Schematic illustration of the experimental paradigms reviewed in this paper. Either cortex, or the neostriatum, or both, at once or in sequence, are ablated. The ablation includes the whole formation or one of its parts. In the latter case, meaningful results are obtained if the cortical and neostriatal ablated regions "correspond" to each other (shaded areas).

The "compulsory approach" (different from "obstinate progression", see above), which is seen after large caudate lesions but not after ablations of frontal cortical area or of the entire cortex, may be taken to suggest that the neostriatum mediates functions which are not mediated by the cortex. It is, however, conceivable that the "compulsory approach" may result from a loss of a particular group of functions which could be eliminated also by removal of the cortical area which corresponds to the ablated portion of the neostriatum. This syndrome may alternatively be the expression of unrestrained control of the final common path by limbic and posterior cortical areas via the ventral striatum (21) or the putamen, respectively. Normally these mechanisms may be balanced by the output from the anterior cortex and the head of the caudate nucleus and hence only occasionally reach the final common path (Fig. 1). (Experiments can be easily designed to test the feasibility of these suggestions). In conclusion, the differences of effects of NS, and of neocortical ablations may be interpreted to indicate different functions of these two formations, but other interpretations are possible (Fig. 1).

### ABLATIONS WITHIN A NEOCORTICO-NEOSTRIATAL COUPLE

The third approach in studying the degree of dependence of NS functions on the neocortex consists of manipulations of a restricted neocortical area and the related NS region (Fig. 1). Area-to-region coupling between the neocortex and NS has been extensively discussed elsewhere (4, 5, 9, 25, 29, 30). It suffices here to say that the evidence suggests a close relation between a neocortical area and a NS region. This relation is anatomical, neurobehavioral and metabolic-functional: each cortical area has one principal target region in NS, this target region shares at least some functions with the associated cortical area (11), and artificial activation of a cortical area produces coactivation in its principal target region in NS (9). If a NS region is associated anatomically and functionally with one cortical area (for a detailed discussion see Divac and Diemer 9), one can study cortico-NS relations concentrating only on one couple (Fig. 1). This approach is technically simpler and allows for more detailed analysis, but it is also not without methodological and interpretational difficulties and therefore cannot replace the already discussed paradigms. The most serious criticism is that even partial lesions aimed at NS produce accidental damage to the cortical connections. The evidence relevant for this problem is discussed in Öberg and Divac (26). Another criticism is that it is not exactly known which part of NS corresponds to a particular cortical area. This problem, however, seems possible to solve by the use of labelled 2-deoxyglucose

(9), unless the relation between the cortical and neostriatal regions changes with different functional requirements.

Lesions in the part of NS associated with the prefrontal cortex, if made in animals which mastered a delayed response-type task after prefrontal cortical ablations, produce different outcomes in different species. In Old World monkeys, ablations of either the prefrontal cortex (31) or NS (13) produce impairments in delayed response-type tasks. Since monkeys with total prefrontal cortical ablations acquired after infancy usually do not relearn these tasks, one must conclude that their NS is not capable of mediating their behavior in the absence of cortical input. Monkeys operated in infancy and tested in delayed response tasks at the age of 24 mo were not inferior to unoperated animals of the same age. At this age, however, all monkeys are less efficient than "young adults" (18). Lesions in infant monkeys involving the related neostriatal tissue alone (20) or along with the prefrontal ablation (22) do impair this performance. These and other results suggest that the neostriatum in the monkey is capable of mediating delayed response-type behavior before the related neocortex. The neostriatum never reaches the proficiency of the lateral prefrontal cortex, and moreover appears to loose its ability to independently organize delayed responding at an advanced age even if the prefrontal cortex had been removed in infancy (19).

In cats with prefrontal (gyrus proreus) ablations, the added NS lesion produced neither retention loss (39), nor appeared to play a role in relearning of delayed response type tasks (6). In two other studies the technique of sequential lesions showed some residual participation of the prefrontal neostriatal region deprived of its neocortical input in the prefrontal functions (28, 40). How delayed responding is mediated in cats with ablations of both the prefrontal cortex and the associated NS region is not known.

In rats without the critical part of the prefrontal cortex, the added NS lesion prouced a relapse of the impairment which lasted as long as the animals were tested (40).

These results suggest that the relation between the prefrontal cortex and the associated NS region, as well as the relation of this couple to the rest of the brain, differ among species. Thus, the "within-the-couple" relation is strictly serial in the adult monkey and apparently also in the cat. In both species, in the absence of the prefrontal cortex, the associated part of the NS no longer takes part or has a negligeable role in "prefrontal functions". (However, from this evidence it cannot be concluded that the "prefrontal region" of NS is left without any function when deprived of its principal cortical input). The cited results also suggest a persisting involvement of the "prefrontal portion of the NS" in mediation of delayed alternation in rats with previous prefrontal cortical ablations.

## DISCUSSION

The meager evidence available at present does not allow any firm conclusions about the degree of dependence of functions of the neostriatum on its neocortical input. If the data discussed in the first two sections are interpreted as showing, respectively, that the neostriatum remains partly functioning after decortication and that it subserves some functions which are not mediated by the cortex, then the following picture about neocortico-neostriatal relations, based also on the review in the third section, can be sketched:[1] The neostriatum has certain functions which it shares with the neocortex. Possibly, the neostriatum normally processes cortical input and sends the resulting computation further toward the final common path (8). In addition, the neostriatum may have functions which do not require cortical input. (The roles of other inputs to NS in functions of this formation will not be discussed here). The hypothetical cortex-independent functions of NS may belong to a category quite different from the cognitive functions which are mentioned in Section 3 and were extensively discussed in Öberg and Divac (26). However, in the rat at least, some cognitive functions of NS, normally shared with the neocortex, seem to be cortex-independent. Data reviewed in Sections 1 and 3 do not allow even tentative conclusions of whether cortex-independent functions of NS are qualitatively different from the cortical functions, but the presence of "compulsory approach syndrome" (Section 2), which has not been produced by cortical ablations, may indicate the presence of such NS functions.

Any description of NS functions, both cortex-dependent and cortex-independent, would be premature at present, except in the form of rephrasing the syndrome (see also 26).

One conclusion of this review seems unquestionable: the rat and the cat clearly differ both in the viability of "thalamic" preparations and in the behavioral repertoire following decortications; in each case the rat is superior to the cat. In other words, both the neostriatum and

---

[1] Functions of NS may be dependent or independent on the cortical input, and on the other hand, they can be cortex related or unrelated. These classifications are independent of each other. In the first instance the criterion is preservation of NS functions in the absence of its cortical input; in the second instance it is qualitative similarity of cortical nad NS functions.

the brain stem seem to be more independent from the more rostral regions of the brain in the rat than in the cat and adult monkey. One remains wondering whether the lower brain mechanisms, capable of mediating rather complex behavior in simpler brains, have to be suppressed or undone in order to permit control from even more sophisticated higher mechanisms in more complex brains. Anyhow, it appears that the neostriatum has more cortex-independent functions in the rat than in the cat or monkey.

It seems necessary to emphasize again that although both the cortex and neostriatum participate in the mediation of some common functions, they need not be — and most likely are not — equipotential. Judging from their respective architectures and connections, they almost certainly contribute different operations towards the same goal. In the rat, however, the neostriatal mechanism can be sufficient to maintain correct behavior in the situations which normally require participation of the prefrontal cortex.

R. Gunilla E. Öberg gave valuable comments on an earlier draft of this paper.

## REFERENCES

1. BARBEAU, A. 1973. Biology of the striatum. In G. E. Gaul (ed.), Biology of brain disfunction. Vol. 2. Plenum Press, New York, p. 333–350.
2. DENNY-BROWN, D. 1963. The basal ganglia. Oxford Univ. Press, Oxford.
3. DENNY-BROWN, D. and YANAGISAWA, N. 1976. The role of the basal ganglia in the initiation of movement. In M. D. Yahr (ed.), The basal ganglia. Raven Press, New York, p. 115-159.
4. DIVAC, I. 1968. Functions of the caudate nucleus. Acta Biol. Exp. 28: 107–120.
5. DIVAC, I. 1972. Neostriatum and functions of prefrontal cortex. Acta Neurobiol. Exp. 32: 461–477.
6. DIVAC, I. 1974. Caudate nucleus and relearning of delayed alternation in cats. Physiol. Psychol. 2: 104–106.
7. DIVAC, I. 1975. Magnocellular nuclei of the basal forebrain project to neocortex, brain stem, and olfactory bulb. Review of some functional correlates. Brain Res. 93: 385–398.
8. DIVAC, I, 1977. Does the neostriatum operate as a functional entity? In A. R. Cools, A. H. M. Lohman and J. H. L. van den Bercken (ed.), Psychobiology of the striatum. Elsevier North-Holland, Amsterdam, p. 21–30.
9. DIVAC, I. and DIEMER, N. H. The prefrontal system in the rat visualized by means of labelled deoxyglucose. Further evidence for functional heterogeneity of the neostriatum. J. Comp. Neurol. (in press).
10. DIVAC, I., DIEMER, N. H., MØLLER, M., SØRENSEN, K. E. and JOHNSTON, G. A. R. 1978. Effects of injections of acidic aminoacids into the rat neostriatum. Neurosci. Lett. Suppl. 1. p. 256.

11. DIVAC, I., MARKOWITSCH, H. J. and PRITZEL, M. 1978. Behavioral and anatomical consequences of small interastriatal injections of kainic acid in the rat. Brain Res. 151: 523–532.
12. DIVAC, I. and ÖBERG, R. G. E. 1979. Current concepts of neostiriatal functions. History and an evaluation. In I. Divac and R. G. E. Öberg (ed.), The neostriatum. Pergamon Press, Oxford, p. 215–230.
13. DIVAC, I., ROSVOLD, H. E. and SZWARCBART, M. 1967. Behavioral effects of selective ablation of the caudate nucleus. J. Comp. Physiol. Psychol. 63: 186–190.
14. DRESEL, K., 1924. Die Funktionen eines grosshirn- und striatum- losen Hundes. Klin. Wschr. 3: 2231–2233.
15. EMMERS, R., CHUN, R. W. M. and WANG, G. H. 1965. Behavior and reflexes of chronic thalamic cats. Arch. Ital. Biol. 103: 178–193.
16. FONBERG, E. 1974. Amygdala functions within the alimentary system. Acta Neurobiol. Exp. 34: 435–466.
17. FULTON, J. F. 1951. Physiology of the nervous system. Oxford Univ. Press, New York.
18. GOLDMAN, P. S. 1971. Functional development of the prefrontal cortex in early life and the problem of neuronal plasticity. Exp. Neurol. 32: 366–387.
19. GOLDMAN, P. S. 1974. An alternative to developmental plasticity. Heterology of CNS structures in infants and adults. In D. G. Stein, J. J. Rosen and N. Butters (ed.), Plasticity and recovery of function in the central nervous system. Acad. Press, New York, p. 149–174.
20. GOLDMAN, P. S. and ROSVOLD, H. E. 1972. The effects of selective caudate lesions in infant and juvenile rhesus monkeys. Brain Res. 43: 53–66.
21. HEIMER, L. and VAN HOESEN, G. 1979. Ventral striatum. In I. Divac and R. G. E. Öberg (ed.), The neostriatum. Pergamon Press, Oxford, p. 147–158.
22. KLING, A. and TUCKER, T. J. 1968. Sparing of function following localized brain lesions in neonatal monkeys. In R. Isaacson (ed.), The neuropsychology of development. John Wiley and Sons, New York, p. 121–145.
23. KORNHUBER, H. H. 1974. Cerebral cortex, cerebellum, and basal ganglia: An introduction to their motor functions. In F. O. Schmitt and F. G. Worden (ed.), The Neurosciences, Third Study Program. M. I. T. Press, Cambridge (Mass.), p. 267–280.
24. MARTIN, J. P. 1967. The basal ganglia and posture. Pitman Medical Publ. Co., London.
25. ÖBERG, R. G. E. and DIVAC, I. 1975. Dissociative effects of selective lessions in the caudate nucleus of the cats and rats. Acta Neurobiol. Exp. 35: 675–689.
26. ÖBERG, R. G. E. and DIVAC, I. 1979. "Cognitive" functions of the neostriatum. In I. Divac and R. G. E. Öberg (ed.), The neostriatum. Pergamon Press, Oxford, p. 291–313.
27. OLMSTEAD, C. E., VILLABLANCA, J. R., MARCUS, R. J. and AVERY, D. L. 1976. Effects of caudate nuclei or frontal cortex ablations in cats. IV. Bar pressing, maze learning, and performance. Exp. Neurol. 53: 670–693.
28. ROSENKILDE, C. E. and DIVAC, I. 1976. Time discrimination performance in cats with lesions in the prefrontal cortex and the caudate nucleus. J. Comp. Physiol. Psychol. 90: 343–352.
29. ROSVOLD, H. E. 1968. The prefrontal cortex and caudate nucleus. A system

for effecting correction in response mechanisms. *In* C. Rupp (ed.), Mind as a tissue. Harper and Row, New York, p. 21–38.
30. ROSVOLD, H. E. 1972. The frontal lobe system: cortical-subcortical interrelationships. Acta Neurobiol. Exp. 32: 439–460.
31. ROSVOLD, H. E. and SZWARCBART, M. K. 1964. Neural structures involved in delayed response performance. *In* J. M. Warren and K. Akert (ed.), The frontal granular cortex and behavior. McGraw-Hill, New York, p. 1–15.
32. SORENSON, C. A. and ELLISON, G. D. 1970. Striatal organization of feeding behavior in the decorticate rat. Exp. Neurol. 29: 162–174.
33. THOMPSON, R. L. and METTLER, F. A. 1963. Permanent learning deficit associated with lesions in the caudate nuclei. Am. J. Ment. Defic. 67: 126–134.
34. VANDERWOLF, C. H., KOLB, B. and COOLEY, R. K. 1978. Behavior of the rat after removal of the neocortex and hippocampal formation. J. Comp. Physiol. Psychol. 92: 156–175.
35. VILLABLANCA, J. and MARCUS, R. 1972. Sleep-wakefulness, EEG and behavioral studies on chronic cats without neocortex and striatum: the "diencephalic" cat. Arch. Ital. Biol. 110: 348–382.
36. VILLABLANCA, J., MARCUS, R. J. and OLMSTEAD, C. E. 1976. Effects of caudate nuclei or frontal cortical ablations in cats. I. Neurology and gross behavior. Exp. Neurol. 52: 389–420.
37. VILLABLANCA, J. and SALINAS-ZEBALLOS, M. E. 1972. Sleep-wakefulness, EEG and behavioral studies of chronic cats without the thalamus: the "thalamic" cat. Arch. Ital. Biol. 110: 383–411.
38. WANG, G. H. and AKERT, K. 1962. Behavior and reflexes of chronic striatal cats. Arch. Ital. Biol. 100: 48–85.
39. WIKMARK, R. G. E. and DIVAC, I. 1973. Absence of effect of caudate lesions on delayed responses acquired after large frontal ablations in cats. Isr. J. Med. Sci. Suppl. 9. 92–97.
40. WIKMARK, R. G. E. and DIVAC, I. 1973. Comparative studies of functional relations in the frontal lobes. Paper presented at the 81st Annual Convention of the American Psychological Association, Montreal.
41. WIKMARK, R. G. E., DIVAC, I. and WEISS, R. 1973. Retention of spatial delayed alternation in rats with lesions in the frontal lobes. Implications for a comparative neuropsychology of the prefrontal system. Brain Behav. Evol. 8: 329–339.

Ivan DIVAC, Institute of Neurophysiology, Panum Institute, Blegdamsuej 3, 2200 Copenhagen, Denmark.

Lecture delivered at the Warsaw Colloquium on Instrumental
Conditioning and Brain Research
May 1979

# THE EMOTIOGENIC BRAIN STRUCTURES IN CONDITIONING MECHANISMS: CONDITIONED EVOKED POTENTIALS AND MOTOR RESPONSES

R. Yu. ILYUTCHENOK

Institute of Physiology, Siberian Branch of the Academy of Medical
Sciences of the USSR, Novosibirsk, USSR

*Abstract*. The emotiogenic morphofunctional control system consists of the amygdaloid complex (AM), the zona incerta, the peri- and paraventricular nuclei of the hypothalamus and the midbrain central gray matter (CG). The neuronal relationships between the structures of this system were established. Lesions of these structures prevented one-trial learning, whereas electrical stimulation of the AM or the CG permitted retrieval of a trace which was lower than threshold. AM stimulation accelerated learning by 5–10 times. The possible mechanisms of the emotiogenic control system of memory are discussed. The contribution of the identified structures of the emotiogenic control system of memory was quantitatively estimated, by employing a matrix of the interaction of these structures during the performance of conditioned neurographic responses of the radial nerve. The approach established the role of the emotiogenic control system in the spatial-temporal organization of brain structures needed for the retrieval of conditioned motor responses. Weakly trained cats, in which the AM-CG had been stimulated, did not differ from well trained ones in the patterns and correlation matrices of the conditioned evoked potentials. AM-CG activation may accelerate learning by reproducing such spatial-temporal relationships that are characteristic of well trained animals.

Ample evidence for the important role of emotions in memory has come from recent studies. However, it is not clear how emotions contribute to this process. Specifically what makes the mechanisms of the emotions-memory interaction difficult to identify? Gaps exist in our knowledge of:

1. The spatial-temporal patterns of the system of emotiogenic structures involved in memory control.

2. The relative contribution of each emotiogenic structure and the interaction of these structures during conditioning.

Further, there remain open questions of:

1. What changes are elicited by the excitation of emotiogenic structures, and in which brain regions?

2. What determines the influence of emotiogenic structures on memory: the activation of the emotiogenic structures during the presentation of the unconditioned stimulus, or the brief residual process in these structures, or even long-term retention (perhaps, for life) of memory in the emotiogenic structures?

It is difficult to provide an accuate assessment of the control system of memory because each brain structure contributes specifically to conditioning. To evaluate the structures composing this system, a more rigorous criterion is needed than the degree of changes in conditioned responses acquired as the result of repeated pairings. The limit of memory trace control is information fixation at its first presentation, that is one-trial learning. Our analysis of one-trial learning was based on a model of the passive avoidance conditioned response. As a result, we identified those lesions of brain structures which disturb the elaboration of this response.

It has been reported that the amygdaloid complex (AM) (11, 14) and the ventrolateral region of the midbrain central gray matter (CG) (7) are substrates for one-trial learning. In our laboratory L. Loskutova and I. Vinnitsky (unpublished) have shown that rats with lesions of CG become immuned to one-trial learning when shocked with moderately (0.75 mA) or even very painful (1.5 mA) current. This has also been observed in amygdalectomized rats. It has been suggested that these structures may be integrated into a structural-functional emotiogenic system that controls memory processes.

What structures and pathways compose the amygdaloid complex-central gray (AM-CG) system? Fibers, which connect the AM with the subcortical brain structures (Fig. 1), are the ventral amygdalofugal pathway (VAF) and the stria terminalis (ST). The VAF ends in the lateral preoptic and anterior-hypothalamic regions. The ST ends in the medial preoptic and anterior-hypothalamic regions. L. Loskutova and

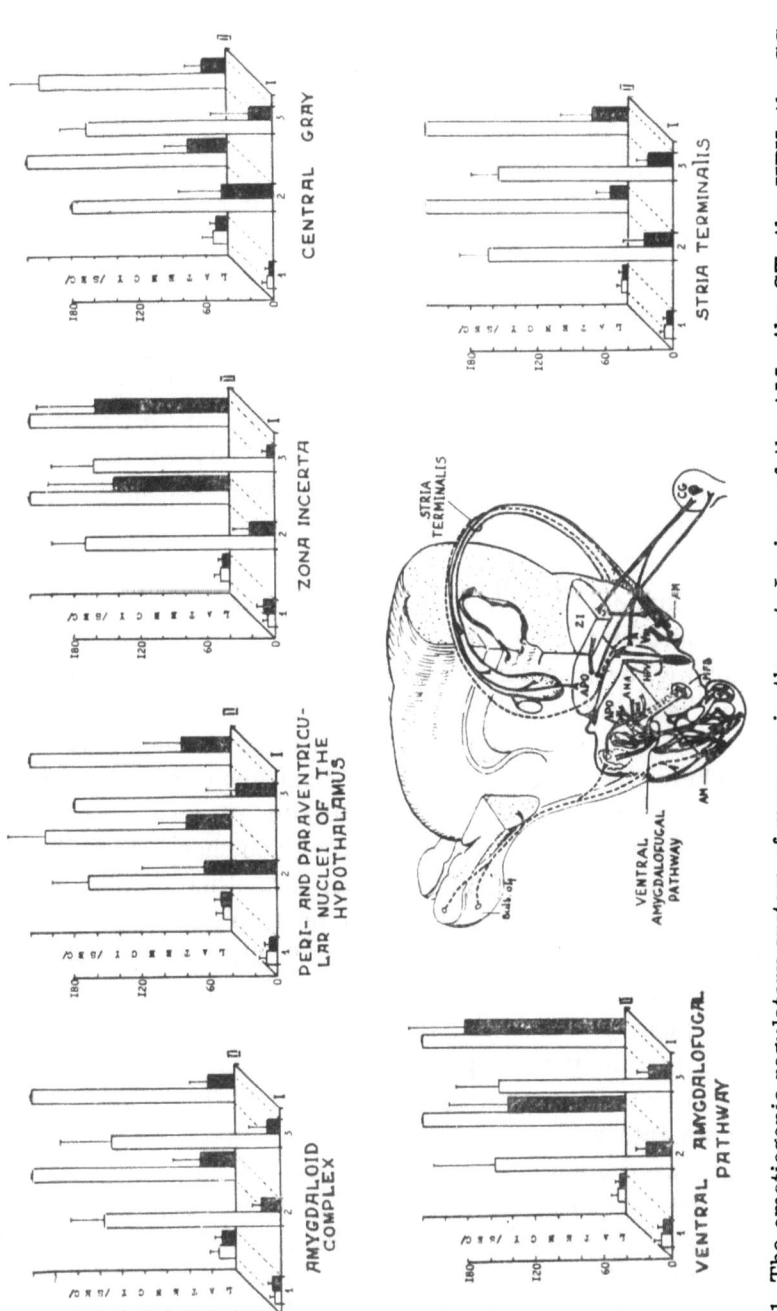

Fig. 1. The emotiogenic regulatory system of memory in the rat. Lesions of the AM, the ST, the HPV, the CG completely prevent one-trial learning. Lesions of the VAF and the ZI prevent it only under moderately aversive stimulation. Current intensity: I, 0.75 mA; II. 1.5 mA. 1, third day of familiarization; 2. testing 24 h after learning; 3, testing 48 h after learning. White columns, for control rats; black column, for lesioned rats.

I. Vinnitsky demonstrated that sections of either of these two pathways make rats incapable of one-trial learning.

Thus, it suffices to transect the VAF at the level of the preoptic area and the anterior hypothalamus to prevent the emotional response and the elaboration of the passive avoidance conditioned response in rats that received moderate footshock (0.75 mA). The integrity of the VAF is most probably needed for the appearance of the emotional response and its autonomic component. There are other data supporting this suggestion. Thus, stimulation of the sites of passage of the VAF fibers elicits a defensive response with components of alertness and fear. Sections of the VAF in the preoptic and the anterior hypothalamic regions prevent the development of the response to AM stimulation (4).

However, in our experiments, section of the VAF at the level of the preoptic area did not prevent one-trial learning when a strong footshock (1.5 mA) was delivered. Rats also did not lose the capacity for multi-trial learning after section of the VAF. Section of the VAF at the level of the caudal hypothalamus had no effect on one-trial learning. Clearly, the VAF is not the only pathway through which the AM influences memory.

Bilateral sections of the ST prevented one-trial learning as well (Fig. 1). It is important to note that such a section made rats incapable of one-trial learning, even under the strong footshock condition (1.5 mA) that resulted in learning in all animals of the control group. Indeed, it is noteworthy that the peri- and paraventricular nuclei, that is the primary sites where the ST ends, are the critical nuclei of the preoptic-hypothalamic region in one-trial learning. Lesion of these nuclei prevented one-trial learning too, in spite of strong footshock (Fig. 1).

The integrity of zona incerta (subthalamus) is another significant condition ensuring one-trial learning. It should be emphasized that lesions of zona incerta prevented one-trial learning only when a moderate footshock was delivered (Fig. 1). Under stronger stimulation, (1.5 mA current), one-trial learning was still possible; however the elaboration of some autonomic and motor components of the conditioned defensive response was poorer, as observed by N. Volf and S. Tsvetovsky (unpublished) in our laboratory.

Thus, the system of emotiogenic nuclei essential for one-trial learning (Fig. 1), and comprising the AM, preoptic area; z. incerta, peri- and paraventricular hypothalamic nuclei, the CG (the AM-CG system). In the case of the fixation of emotionally colored information, its biological meaning is perceived at the first presentation. The role of the AM-CG system is manifest in the course of linkage between memory trace and the retrieval program.

In the case of repetitive trials, the biological information is perceived as meaningful only after a series of presentations, and here the role of the system is not so crucial. Even amygdalectomized rats are capable of multi-trial learning. However, AM-CG control is not excluded entirely.

What is decisive in the regulatory function of the AM-CG system? Not its participation in the registration of reinforcement and not the evaluation of the biological meaning of the information, although they are important initial steps. The biologically meaningful information has to be retained in the brain after its first input. Consequently, the role of the AM-CG system is a determinant in that it produces conditions for the rapid fixation of a memory trace. Possibly, the activation of the AM-CG system gives rise to two parallel processes, that is the appearance of emotions and the reorganization of the functional properties of the neurons of other brain structures providing the rapid fixation of a memory trace. However, it cannot be excluded that such a reorganization of the functional properties of neurons may be a consequence of emotions.

To understand the role of emotiogenic brain structures in memory control, we attempted to evaluate the relative contribution of each nuclear structure and to determine how the structures interact functionally. We proceeded from the assumption that one-trial learning is one extreme of the continuum of engram control. It so, the AM-CG system can presumably regulate the fixation and retrieval of a memory trace in a wide range from null to one-trial learning.

Therefore, repetitive pairings of conditioned and unconditioned stimuli (CS–US) can also be used to evaluate the contribution of the structures identified. For this purpose, M. Gilinsky and I. Pukhov (unpublished) used the conditioned evoked potential (CEP) as suggested by Adam (1). The CEP method has been described in detail earlier (5).

The CEP, which is an electrographic manifestation of a memory trace in brain structures, was compared with the conditioned neurographic response of the motor nerve, an effector component of the conditioned response. It was impossible to establish any unambiguous functional relationship between the manifestation of the conditioned evoked potential and the conditioned neurographic response. For this reason, these relationships were expressed as binary signs reduced to a common table. This approach, developed in cooperation with the Computer Centre of the Siberian Branch of the USSR Academy of Sciences (V. Drobishev, T. Freidin, unpublished), permitted us to establish statistically the relationships between the manifestations of the conditioned neurographic response and the respective set of the CEP in various brain structures.

Quantitative estimates of the relative contribution of each brain structure demonstrated that the appearance of the CEP in the z. incerta, the preoptic area and the auditory cortex is most frequently accompanied by a conditioned neurographic response. This made necessary the building of a matrix of the interaction of the various brain structures

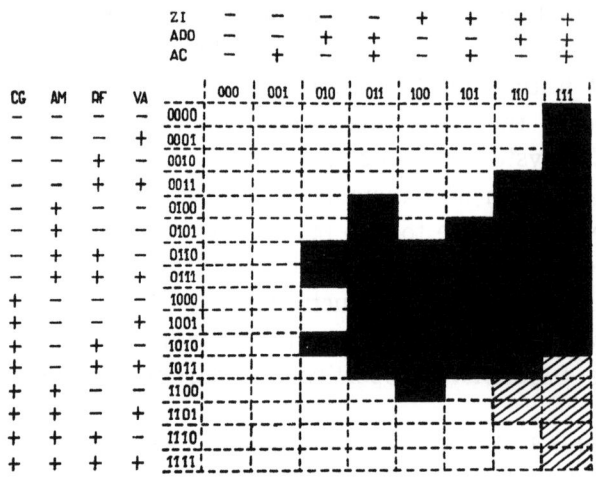

Fig. 2. A matrix of the relationship between the distribution of the conditioned evoked potential (CEP) and the neurographic response. The conditioned neurographic response is constantly elicited when the CEP appears simultaneously in the ZI, the APO and the AC. Black areas are those in which the CEP is distributed during the registration of the neurographic response. Dashed areas indicate the expected area of the neurographic response. +, evoked conditioned response; —, no response.

during the manifestation of the conditioned neurographic response (Fig. 2). Analysis of this interaction indicated that the conditioned neurographic response was a constant concomitant of the CEP in the zona incerta, the preoptic area and the auditory cortex. The response also appeared, but with much lower probability, when the CEP was elicited simultaneously in various other combinations with the involvement of some of these structures.

Taken collectively the analysis of data on the brain structures providing one-trial learning, the CEP probability of occurrence in these structures and their correlation matrices, confirm that the AM-CG system (the amygdaloid complex, the preoptic area, the zona incerta, the peri- and paraventricular nuclei of the hypothalamus, the midbrain central gray) indeed controls memory. In the control of versatile memory processes, this system undoubtedly interacts with the thalamo-cortical and other brain structures.

The AM-CG control system of the formation and retrieval of memory trace was also made apparent in studies of the stimulation of the AM.

In experiments with immobilized cats, M. Gilinsky and I. Pukhov (unpublished) showed that AM stimulation (60 pulses/s, pulse duration of 0.2 ms 3 V, stimulation for 3–5 min) before learning greatly accelerates it. Stimulated cats learned after 20–60 pairings of the CS–US (Fig. 3); whereas nonlesioned cats required 120–300 pairings for acquisi-

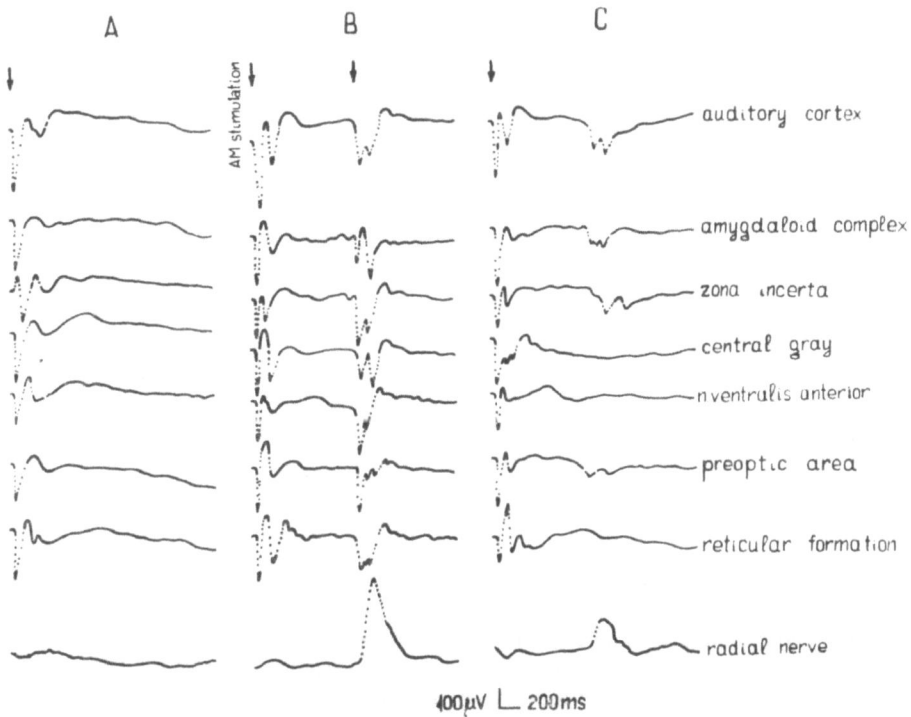

Fig. 3. Accelerated learning after amygdala stimulation in cats. The conditioned evoked potentials after amygdala stimulation as the result of 20–60 pairings of CS and US, as compared with 120–300 pairings in the control. A, habituation; B, 20 CS–US pairings after AM stimulation; C, testing after 20 min. Arrows indicate the time when the CS-US were presented.

tion. Moreover, by artificially modifying the rhythm of the AM neurons, one-trial learning was achieved in a situation requiring repetitive trials for learning (3, 5). A similar effect was observed only when current in the frequency range of 40–60 pulses/s was applied, that is when a rhythm of burst electrical AM activity developed during emotional tension (9, 10). It follows that the activation of one of the influential structures of the system, the AM, greatly accelerates and facilitates the formation of a memory trace.

An obvious question might now be asked: does the AM-CG system participate in the control of memory trace retieval? M. Gilinsky, G. Abuladze, V. Masycheva and I. Pukhov (unpublished) studied the spatial distribution of the CEP in brain structures and the retrieval of the conditioned neurographic response after the stimulation of the AM and the CG. The AM was stimulated (3–5 min train of pulses, 0.2 ms duration at 60 Hz, 3 V) or the CG (3–5 min train of pulses, 0.1 ms duration at

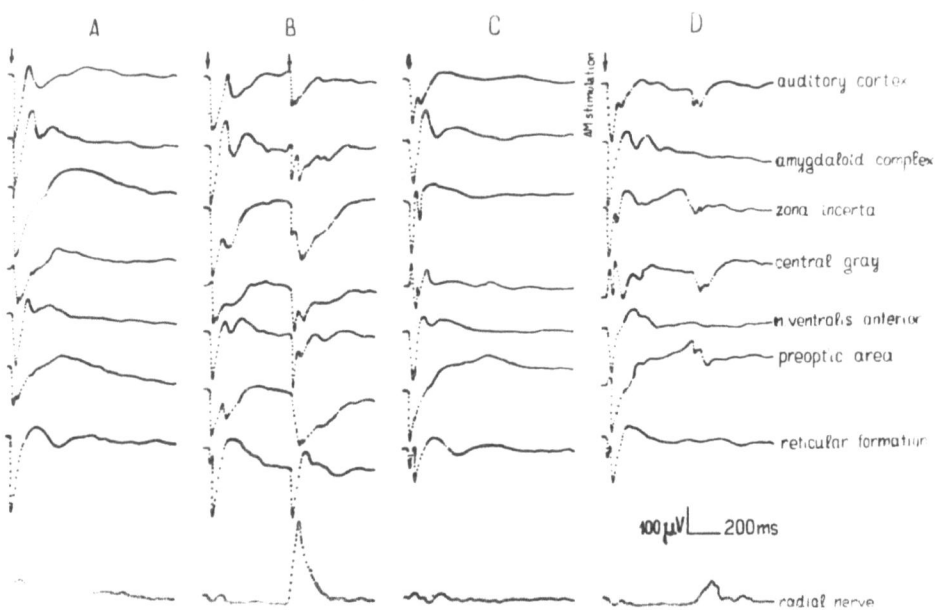

Fig. 4. Facilitation of the retrieval of the conditioned responses by amygdala stimulation. When trained cats do not show conditioned response, AM stimulation facilitates the retrieval of the conditioned neurographic responses and CEP in the AC, the ZI, the CG, the APO. A, habituation; B, 181 to 200 paired presentations; C, testing promptly after 200 CS–US pairings; D, 30 min after AM stimulation. Arrows indicate the time when the CS–US were presented.

100 Hz, 3 V) only in the case when the first test after learning (100–200 CS–US pairings) did not elicit the neurographic response. In the first test for the CEP, the potential either did not appear at all or was restricted to one, two or three brain structures. The conditioned neurographic response appeared in 75% of cats repetitively tested 20–30 min after the AM stimulation. It is remarkable that the conditioned neurographic response was registered simultaneously with the CEP in the auditory cortex, the zona incerta and the preoptic area (Fig. 4). As already mentioned, this simultaneity is of decisive importance in the performance of the behavioral response. Stimulation of the CG resulted in retrieval of the conditioned neurographic response in 62.5% of cats; the CEP was registered in the zona incerta, the preoptic area and, as a rule, in the

AM. Hence, the activation of the structures of the emotiogenic control system of memory creates a spatial-temporal organization of brain structures that promotes the appearance of the effector conditioned response.

Through what mechanisms does the emotiogenic control system act on memory? There are three lines of experimental evidence of relevance to the understanding of the influence of emotiogenic structures on memory trace formation.

1. There is an increase in the number of cortical neurons with polysensory properties after AM stimulation (8). The enhancement of polysensory properties is characteristic of the acquisition of a conditioned response under the effect of a biologically meaningful reinforcement (6). However, the switching of converging mechanisms for the provision of the interaction of stimuli is insufficient for their stable linkage (12).

2. The shortening of the autonomic residual responses to an US after amygdalectomy, observed in the experiments of Volf and Tsvetovsky (unpublished), may be evidence for the important role of the AM in residual processes, especially in the retention of emotional responses. The response of amygdalectomized animals to the direct action of stimuli is retained. The involvement of the AM in the activation of convergence and residual processes does testify to its significance in the control over trace formation. However, the capacity for one-trial learning cannot be explained by such an influence of the AM.

3. The most weighty line of evidence concerns inhibitory amygdalofugal influences. The capacity of inducing a rhythm of neuronal discharges in the AM for prolonging inhibition of impulse flow in the CG is indicative of a modulating type of AM influence on the CG (2). Conceivably, the major influence of the emotiogenic control system on memory trace formation is to facilitate the emergence of a dominat focus and to act on direct (lateral) and recurrent inhibition underlying interference. Activating afferent flows exciting the structures of the emotiogenic control system arise during the stimulation of CG (where the affective-motivational component of an aversive stimulus are first perceived). According to the data of Dubrovina, 86% of reactive neurons of the zona incerta are excited during CG stimulation (Fig. 5).

It may be thought that the delivery of emotionally colored information activates the CG, the initial component of the system which modulates affective behavior. This activation may spread over to other nuclei of the emotiogenic control system (those of the AM, the zona incerta, the peri- and paraventricular nuclei of the hypothalamus). This activation of the entire emotiogenic control system can provide conditions for the dominant determining of the activity of the neuronal centers at a particular point in time. A dominant focus itself is one of the initial steps of the elaboration of conditioned responses as demonstrated by Rusinov (13). There are also data (6) indicating that the dominant state produced by a biologically meaningful reinforcement is of importance in the rate of neuronal activity reorganization of a signal type and in the maintenance of established relationships.

Subsequent amygdalofugal inhibition may maintain the dominant state. Stimulation of the AM inhibited 72% of reactive neurons in the

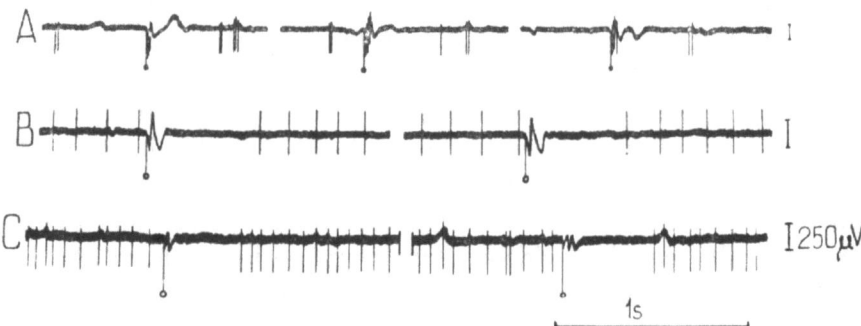

Fig. 5. Neuronal activity in the amygdaloid complex — midbrain central gray matter system. The CG has predominantly an excitatory ascending effect on the ZI neurons (A), whereas the AM has an inhibitory effect on the neurons of the ZI (B) and the CG (C).

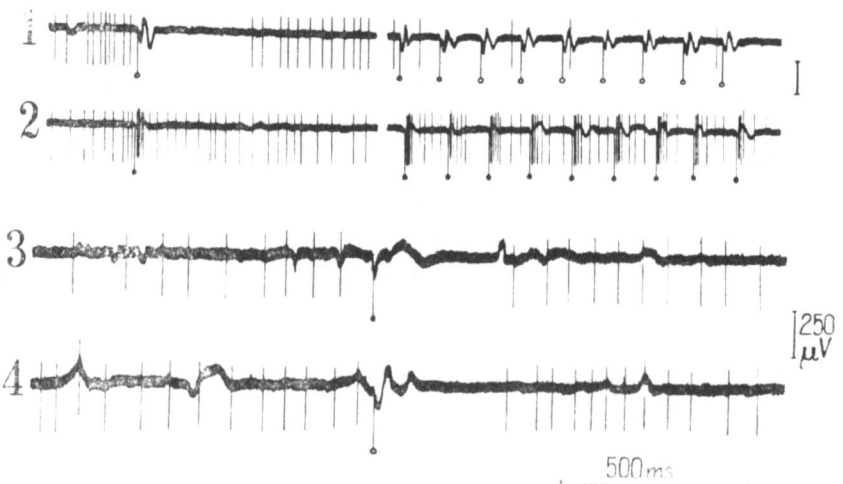

Fig. 6. Functional relationships in the amygdaloid complex — the midbrain central gray matter system. Convergence of inhibitory influences descending from the AM (1) and of excitatory influences ascending from the CG (2) at the ZI neurons. Convergence of predominantly inhibitory influences descending from the AM (3) and the ZI (4) at the CG neurons.

CG and 64% of those in the z. incerta (Fig. 5). It is noteworthy that impulses from the AM and the zona incerta have predominantly an inhibitory effect (71%) on the same CG neurons (Fig. 6); convergence of excitative impulses was observed only in 20% of neurons (2). As to zona incerta, 57% of the neurons show an inhibitory response to the influence of the AM and an excitatory one to that of the CG; 30% of its neurons respond only by excitation to the influences of either the AM or the CG.

Possibly, after the registration of biologically meaningful information, which was evaluated earlier at the level of the CG and the AM, amygdalofugal influences inhibit the structures of the emotiogenic control system. The emerging pattern of retroactive and proactive inhibition may facilitate the formation of a dominant, thereby promoting the fixation of a single information input. Rapid and secure fixation of a trace and its linkage with the established retrieval program are provided by the weakening of the effects of preceding traces by retroactive inhibition, the enhancement of the dominant and the weakening of the interfering effects of subsequent emotionally colored sensory input by proactive inhibition.

However, the feasibility of subsequent trace formation after different time intervals depends on whether the trace is readable, but not on how it is fixed (short- or long-term memory). Interference may be one of the causes of the rapid passage of a fixed trace to a subthreshold state which impedes its retrieval; this is outwardly seeming short-term memory. This memory is short for the only reason that its trace has become subthreshold and difficult to retrieve. One is confronted here not with the disappearance of a short-term memory trace, but rather with its rapid passage to a subthreshold level. Many factors may be causative: retroactive inhibition due to interference of subsequent signals, insufficient reinforcement, attention concentration, general setting and emotional concomitants, among others. The presentation of a single stimulus, without emotions is promptly followed by a sort of information loss, forgetting.

These time intervals of trace retention can be subdivided into shorter ones. We can thus obtain transient, ultrashort and other "versions" of memory, which explains why the so-called short term memory lasts from 10 s to several hours according to data in the literature.

In multi-trial learning, the biological meaning of a neutral stimulus becomes apparent only after its repeated pairings with reinforcement. The production of a dominant is facilitated by the summation of incoming stimuli. Each subsequent stimulus can reinforce the preestablished trace making it more stable. This may promote the maintenance of a trace at a threshold level in which it remains readable for a long time, that is one is confronted here with long-term memory.

Long retention of a memory trace is also possible in one-trial learning, provided that information input was emotionally associated. The activation of the emotiogenic control system of memory at the time of one-trial learning may result in stronger retroactive inhibition attenuating the effects of the preceding traces. This may promote the rise of a new dominant. In turn, the promoted dominant helps the trace to remain superthreshold longer in the structures of this emotiogenic system. Thus, conditions may arise favorable for subsequent retrieval. One of these major conditions is the constant availability of a memory trace in the structures of the emotiogenic control system.

The trace of emotional memory is not erased, nor is it subjected to amnesia (5). The subsequent emotions activate these traces making them easier to read in the emotiogenic structures, and, as a result, the whole adaptive program, preestablished and trace-linked in these structures, is efficiently retrieved.

## REFERENCES

1. ADAM, G. 1967. Interoception and behaviour. Acad., Kiado, Budapest.
2. DUBROVINA, N. I. and ILYUTCHENOK, R. Yu. 1978. Neuronal activity of midbrain central gray substance under amygdaloid and subthalamical stimulation (in Russian). Neurophysiologiya 3: 245–251.
3. GOLD, P. E., HANKINS, L., EDWARDS, R. M., CHESTER, J. and McGAUGH, J. L. 1975. Memory interference and facilitation with posttrial amygdala stimulation effect on memory varies with footshock level. Brain Res. 86: 509–513.
4. HILTON, S. M. and ZBROŻYNA, A. W. 1963. Amygdaloid region for defence reactions and its efferent pathway to the brain stem. J. Physiol. (Lond.) 165: 160–173.
5. ILYUTCHENOK, R. Yu., VINNITSKY, I. M. and LOSKUTOVA, L. W. 1977. Neirokhimicheskie mekhanizmy mozga i pamyat'. Izdat. Nauka, Novosibirsk.
6. KOTLYAR, B. I. 1977. Mekhanizmy formirovanija vremennoi' svyazi: Izdat. Gosud. Moscow Univ., Moscow.
7. LIEBMAN, J. M., MAYER, F. J. and LIEBESKIND, J. C. 1970. Mesencephalic central gray lesions and fear-motivated behavior in rats. Brain Res. 23: 353–370.
8. MAKAROV, V. A. 1970. The role played by the amygdala in the mechanism of convergence of stimulations of different sensory modality on the neurons of the large hemisphere cortex (in Russian). Rep. Acad. Sci. Ukr. SSR. 194, 6: 1454–1457.
9. McLENNAN, H. and CRAYSTONE, P. 1965. The electrical activity of the amygdala and its relationship to that of the olfactory bulb. Can. J. Physiol. Pharmacol. 43: 1009–1017.

10. ONIANI, T. N. and ORJONIKIDZE, Ts. A. 1968. Changes in electrical activity of some brain structures of the cat during general behavioural reactions. *In* Contemporary problems of activity and structure of the central nervous system. Mezniereba, Tbilisi, p. 5–13.
11. PELLEGRINO, L. 1968. Amygdaloid lesions and behavioral inhibition in the rat. J. Comp. Physiol. Psychol. 65: 483–491.
12. RABINOVICH, M. Ya. 1975. Zamykatelnaya funktsiya mozga. Izdat. Meditsina, Moscow.
13. RUSINOV, V. S. 1969. Dominanta. Izdat. Meditsina, Moscow.
14. VINNITSKY, I. M. and ILYUTCHENOK, R. Yu. 1973. Elaboration of defensive conditioned reflexes in amygdalectomized rats. Zh. Vyssh. Nervn. Deyat. Im. I. P. Pavlova 23, 4: 766–770.

R. Yu. ILYUTCHENOK, Institute of Physiology, Siberian Branch of the Academy of Medical Sciences of the USSR, Zolotodolinskaya 101, 630090 Novosibirsk, USSR.

## LIST OF ABBREVIATIONS

| | |
|---|---|
| AC | auditory cortex |
| AHA | anterior hypothalamic area |
| AM | amygdaloid complex |
| APO | preoptic area |
| bed n. Str. t. | bed nucleus of the stria terminalis |
| Br | diagonal band of Broca |
| Bulb. olf. | bulbus olfactorius |
| CG | midbrain central gray matter |
| HPV | peri- and paraventricular nuclei of the hypothalamus |
| MFB | medial forebrain bundle |
| RT | midbrain reticular formation |
| ST | stria terminalis |
| VA | nucleus ventralis anterior |
| VAF | ventral amygdalofugal pathway |
| Vm | ventromedial hypothalamus |
| ZI | zona incerta |

Lecture delivered at the Warsaw Colloquium on Instrumental
Conditioning and Brain Research
May 1979

# RELATIONSHIPS BETWEEN PHENOMENA OF PARADOXICAL SLEEP AND THEIR COUNTERPARTS IN WAKEFULNESS

Adrian R. MORRISON

Laboratories of Anatomy, School of Veterinary Medicine and Institute
of Neurological Sciences, University of Pennsylvania
Philadelphia, Pennsylvania, USA

*Abstract.* Our studies suggest that paradoxical sleep is a state in which the brainstem is in a functional mode normally associated with presentations of novel stimuli. Furthermore, the combination of atonia and an internal state of activation indicates sustained activity of a brainstem mechanism designed to dampen responses to sudden, novel stimuli in wakefulness lest the animal over react and run blindly into danger prior to stimulus analysis. This conclusion stems from two sets of data. Studies of large-amplitude waves, which characteristically occur spontaneously just prior to and during paradoxical sleep, have demonstrated that the waves are signs of alerting. These waves, termed ponto-geniculo-occipital (PGO) spikes, can be induced during both slow wave and paradoxical sleep by external stimuli at a threshold below actual arousal. Whenever cats are confronted with novel stimuli during wakefulness, eye movement potentials, which are recorded from the same sites as PGO spikes but differ from them in several characteristics, assume all the characteristics of PGO spikes. These observations indicate that the central nervous system during paradoxical sleep is in a "peculiar" state of activation which is not behaviorally expressed. Other experiments have focused on the muscle atonia of paradoxical sleep. Small, bilateral dorsolateral pontine tegmental lesions create the dramatic phenomenon of paradoxical sleep without atonia which is characterized as follows: After slow wave sleep, when paradoxical sleep with muscle atonia would normally appear, cats raise their heads, make body righting movements, exhibit alternating movements of the limbs, and even attempt to stand. Throughout an episode, which shows all other aspects of paradoxical sleep, including unresponsiveness to visual stimuli, cats act as if they are being startled, searching and sometimes attacking an object. In wakefulness they show minor cerebellar signs. Presumably the lesions disrupt both the pontine excitation of the medullary inhibitory area and the inhibition of a brainstem system for mobilizing activity normally in force during paradoxical sleep. Studies of the same cats during wakefulness have shown that there in an increase in exploratory locomotor activity of 23 to 127% as measured in an open-field test. The existence of parallel effects on motor control in wakefulness and paradoxical sleep produced by pontine lesions suggests that the atonia of paradoxical sleep is a reflection of excessive activity of a brainstem response dampening mechanism which operates more subtly during wakefulness to produce appropriately modulated responses to various unexpected stimuli.

Until 1949 the view that sleep was a passive process resulting from the withdrawal of sensory input via the long sensory pathways necessary for wakefulness dominated thinking about the relationships of sleep to

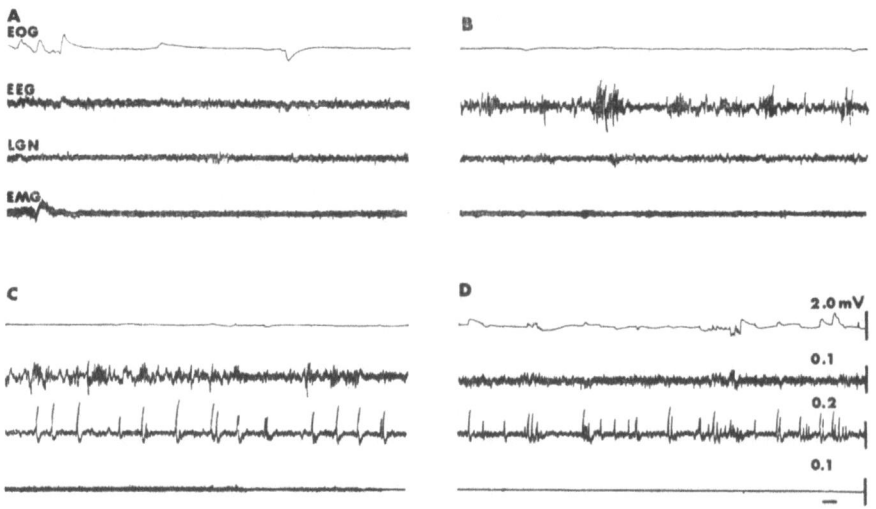

Fig. 1. Characteristics of the behavioral states of (A) quiet wakefulness; (B) slow wave sleep; (C) transition from slow wave sleep to paradoxical sleep; (D) paradoxical sleep. EOG, eye movements recorded on the electro-oculogram; EEG, electroencephalogram; LGN, recording of PGO spikes in the lateral geniculate body in C and D, but eye movement potentials practically absent in A; EMG, electromyographic activity of the dorsal cervical muscles which disappears at the end of C and in D. Time calibration = 1 s. From Morrison (33).

wakefulness (42). The landmark discovery of Moruzzi and Magoun (43), however, showed that the brainstem reticular formation was involved in EEG activation. This work presaged discoveries of greater complexities in the organization of sleep and wakefulness, which demonstrated that the process of falling asleep and remaining asleep involved active mechanisms (41). The primary stimulus for the modern era of sleep research (since 1953) was, of course, the "unusual" nature of paradoxical sleep, first detected by Aserinsky and Kleitman (1). That is, instead of the high-amplitude, slow wave activity recorded on the electroencephalogram (EEG), which had until then been regarded as characteristic of sleep, low-amplitude, high-frequency activity quite similar to that of wakefulness characterized the EEG of paradoxical sleep (Fig. 1). Paradoxical sleep also differs from wakefulness in a number of ways of course. In cats, which will be the subject of this paper and which exhibit a representative mammalian sleep pattern, responsiveness is reduced; the pupils are miotic; the nictitating membranes are relaxed; neuronal firing is more burst-like; and the skeletal musculature is atonic although phasic twitches of the distal limb musculature, facial musculature and extraocular muscles occur (22). The latter are responsible for the well-known rapid eye movements of this sleep stage. Such observations are made on cats implanted with chronic electrodes and sleeping in a lighted chamber (33).

In spite of these dramatic differences, some similarities exist between paradoxical sleep and alert wakefulness in particular in addition to the EEG pattern. In both states brain temperature is higher than in slow wave sleep (45), and theta rhythm is present in the hippocampus (22). Before proceeding, a significant point must be made regarding comparisons between these states. In comparing sleep and wakefulness sleep researchers, who have done so much to emphasize the importance of studying behavioral state changes as they relate to sleep, have, ironically, compared the various phenomena measured during slow wave and paradoxical sleep to those of wakefulness as if the latter were a homogeneous state. Indeed, the qualifiers "alert" and "quiet" have been indiscriminately applied to the state of wakefulness in reporting on differences in neuronal firing, reflexes and other aspects of the sleep states and wakefulness. For obvious technical reasons workers have tended to study quiescent animals, which naturally limits the range of the wakeful state subdivisions available for analysis.

Yet, it has become increasingly clear to us that much of paradoxical sleep, at least, becomes understandable if one examines its various aspects from the point of view that the central nervous system is in a state which one would call a state of "orientation" were the organism interacting with its environment. More specifically, our experiments have led us to conclude that paradoxical sleep resembles a state in which the brainstem is in a functional mode normally associated with presentations of novel stimuli. Furthermore, the combination of atonia and an internal

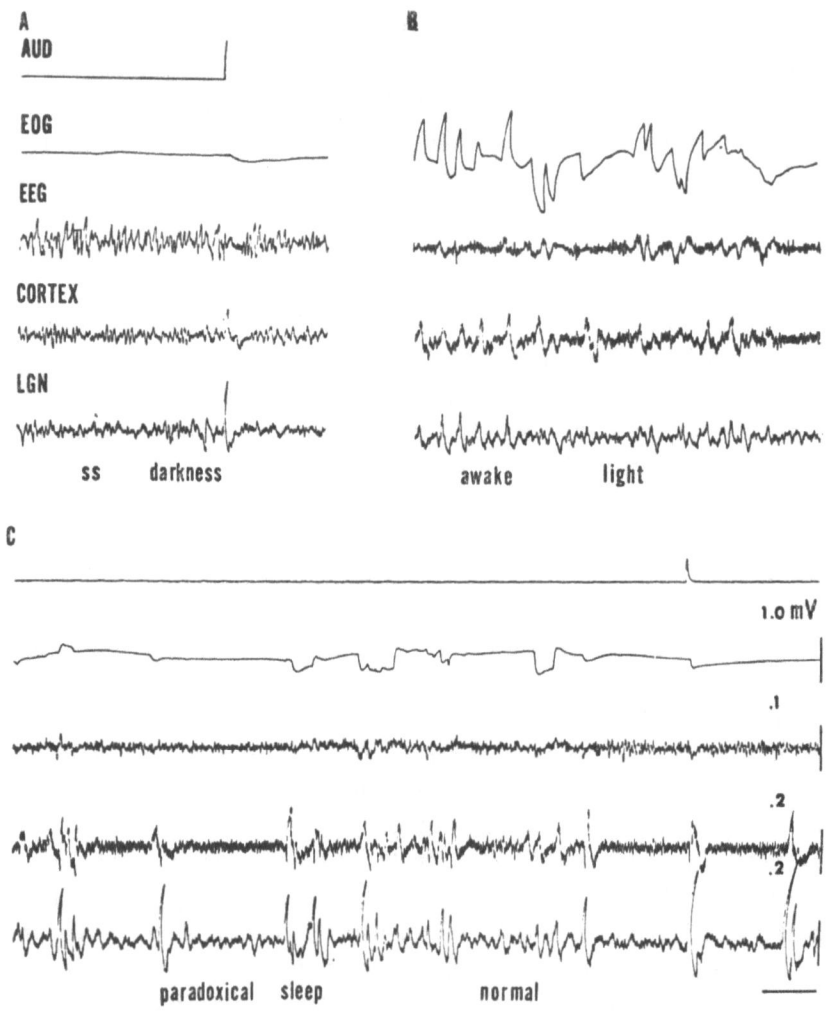

Fig. 2. (A) Normal cat in slow wave sleep (ss) exhibiting a sound-induced PGO spike in lateral geniculate body and visual cortex while in a darkened recording chamber. Compare amplitude of the induced spike with those of spikes in B and C. A transient EEG change also occurred, but the cat then continued in slow wave sleep. AUD, auditory stimulus marker; CORTEX, visual cortex. (B) Same animal during wakefulness in the light illustrating spontaneous eye movements accompanied by LGN and cortical eye movement potentials. (C) Normal cat in paradoxical sleep illustrates spontaneous PGO spikes and a sound-induced PGO spike (AUD marker) recorded from the LGN and visual cortex. Time calibration = 1 s. From Bowker and Morrison (5).

state of activation can be viewed as sustained activity of a brainstem mechanism designed to dampen responses to sudden, novel stimuli in wakefulness lest the animal over react and run blindly into danger prior to stimulus analysis (33, 36).

This concept began to evolve with our observation (5, 35) that normally spontaneous events of paradoxical sleep, ponto-geniculo-occipital (PGO) spikes, could also be elicited by environmental stimuli, not only during paradoxical sleep, but also during slow wave sleep (Fig. 2). PGO spikes are large-amplitude waves recorded by means of macroelectrodes in the pontine tegmentum, lateral geniculate body and occipital cortices 10–30 s prior to and during paradoxical sleep (Fig. 1, C and D) (23). They were first recorded in the pons by Jouvet and Michel (24) and then in the lateral geniculate body by Mikiten et al. (31). Continued work by others demonstrated (1) that the pathways responsible for the waves course through the reticular formation to both the lateral geniculate body and visual cortices in parallel (3, 8, 26, 27) but not to the sensory motor cortex (7) and (2) that activity of the serotonergic raphe nuclei seemed responsible for confining the waves to paradoxical sleep and the immediately preceding transition period (56). Morrison and Pompeiano (39) showed that there are two types of PGO spikes — one which occurs singly in the transition period and throughout an episode (Type I) and the other (Type II) which appears in clusters associated with rapid eye movement bursts. Bilateral vestibular nuclear lesions eliminated both rapid eye movement bursts and Type II clusters but did not affect the Type I spikes. It therefore appeares that the Type II spikes represent activity in the vestibulo-oculomotor system activated internally during paradoxical sleep since rapid eye movements continue after bilateral vestibular nerve section (48). Type I spikes seem a more fundamental part of the paradoxical sleep state, and it is these individual spikes which are evoked by auditory or tactile stimuli during both slow wave and paradoxical sleep (5, 35).

Both natural laboratory sounds and light hair touch were found to evoke waves in the lateral geniculate body during both paradoxical and slow wave sleep. The waves appeared identical in shape to PGO spikes. In order to study this phenomenon in detail pure tone bursts were applied, and recordings were made in both the lateral geniculate body and marginal gyrus within the primary visual area. Several revealing observations were made which led us to conclude that PGO spikes are signs of activation of the reticular alerting network: (1) The waves could be evoked during slow wave sleep at a threshold lower than that which would elicit generalized EEG activation or behavioral evidence of arousal. (2) The latency of the waves was usually 20 to 40 ms; whereas that of the evoked response in the auditory cortex was 10 ms (3) If stimuli were presented at frequent intervals in slow wave sleep, e.g. every 5 ms, the waves diminished in amplitude and then disappeared after a few repetitions. During paradoxical sleep, on the other hand, waves could be evoked at the same latency as those in slow wave sleep at least 60% of the time. (4) Cats with cerebellar cortical lesions exhibit spontaneous flexor or extensor jerks of the forelimbs depending upon the location of the lesion during the transition period between slow wave and paradoxical sleep, and these jerks occur synchronously with PGO spikes until muscle atonia ensues (34). Auditory stimuli elicited the same jerks in addition to PGO spikes (35).

The evidence indicated that the evoked waves were the result of nonspecific activation. The fact that the evoked waves and spontaneous PGO spikes have at least three properties in common argues for the view that they are both merely a sign of reticular activation: (i) the waveforms have an identical appearance, (ii) they both can occur in paradoxical sleep, and (iii) they are both associated with limb jerks in cerebellar-lesioned animals. The last is easily explained. PGO-associated waves occur spontaneously in the cerebellar cortex during the transition to and during paradoxical sleep (46). Because the deep cerebellar nuclei are excited by collaterals of fibers extending to the cerebellar cortex, they would be affected by this brainstem-generated input (19). An auditory stimulus directly excites both deep cerebellar nuclear neurons and Purkinje cells (40). If the overlying inhibitory Purkinje neurons (19) are removed, any input to the deep cerebellar nucleus normally modulated by the damaged cortex would have an exaggerated effect on cells to which the deep cerebellar nuclei project. Thus, damage to the midline vermal region results in extensor jerks because such neurons as those of the lateral vestibular nucleus, which primarily excite extensor motor neurons (28), are unchecked; and a paravermal lesion leads to flexor jerks because the red nucleus, which facilitates flexor motor neuronal activity, is powerfully driven by the released interpositus neurons (63). During wakefulness a sudden external stimulus will incite a flexion movement or an exaggerated extensor posture depending upon whether the lesion is in the paravermal or vermal cortex (35, 63).

These relationships led to the experiments which were the next step in the evolution of the hypothesis that paradoxical sleep can be viewed as a period during which the central nervous system acts as if there were a continual influx of unexpected, or novel, stimuli. During wakefulness potentials occur in the same locations as PGO spikes (Fig. 2B); and because of their association with eye movements, they have been named eye movement potentials (14). Brooks and Gershon (10), however, have recognized their essential similarity to PGO spikes and have proposed the terms $PGO_W$ and $PGO_{REM}$ in recognition of this. The fact that both are eliminated during reversible cooling of the dorsolateral pontine tegmentum supports this idea (27). At the other extreme, Sakai and Cespuglio (52) have stated that there are two classes of waves in paradoxical sleep determined on the basis of latency of appearance in the lateral geniculate body following a wave in the abducens nucleus. Those with a longer latency were termed eye movement potentials of paradoxical sleep because of similarities to those of wakefulness. All workers, however, consider that there are certain differences between the two waves, the most significant probably being the effects of levels of illumination on the amplitudes of the waves. Cortical PGO spikes are unaffected by light, but amplitudes of eye movement potentials are greatly diminished in darkness (10, 20).

Primed by our observations that alerting stimuli could elicit waves which seemed in every respect to be PGO spikes, we postulated that the critical factor responsible for the amplitudes of eye movement potentials was the animal's state of alertness, not level of illumination (6). Even in total darkness a tone burst increased the amplitude of eye movement potentials to the same levels of those of PGO spikes for a few seconds

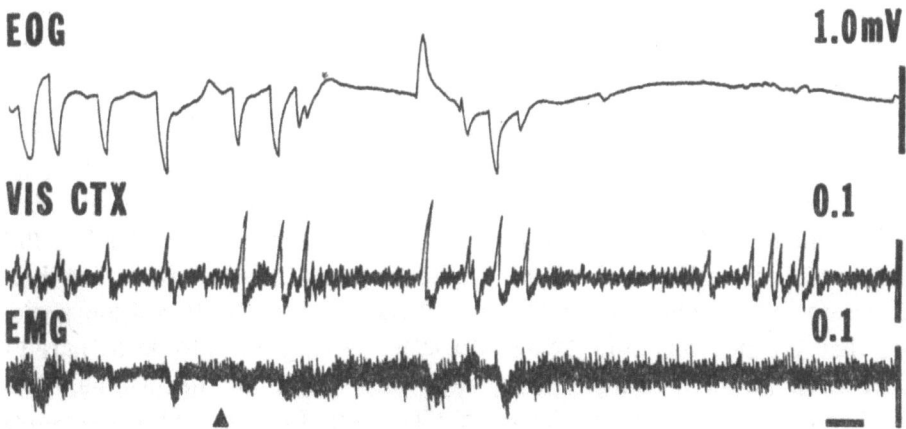

Fig. 3. Recording of a cat during wakefulness while in darkness. When the cat was alerted by a noise (triangle), the EMP increased in amplitude, then became attenuated as quiet wakefulness ensued. VIS CTX, visual cortex; time calibration = 1 s. From Bowker and Morrison (5). Vigilance: An important determinant of cortical eye movement potentials. From Sleep Res. 5: 37.

(Fig. 3). The determining factor seems to be the novelty of the stimulus, because the odor of fish would also increase for only a brief period the eye movement potential amplitudes of a food-deprived cat housed in darkness. Furthermore, Bowker (4) has demonstrated that once cats have learned the significance of a visual stimulus, eye movement potentials greatly diminish in amplitude even though eye movements remain. The amplitudes return to their original level if a novel stimulus is interjected.

The effect of novel stimuli on the visual system at the cellular level is probably to enhance transmission at a time when this would be quite useful, in the presence of novel stimuli. Indeed, there is considerable evidence that reticular stimulation acts to disinhibit geniculate and cortical cells, and such stimulation produces a PGO-like geniculate wave (57).

Although similarities among eye movement potentials, PGO spikes evoked in slow wave sleep, and spontaneous and evoked PGO spikes of paradoxical sleep have been stressed thus far in this paper, there is a difference, previously mentioned, between the first two waves and those occurring in paradoxical sleep, which is important to recognize if one is to explain the nature of the paradoxical sleep state. During paradoxical sleep the waves exhibit little tendency toward habituation. Bizzi and Brooks (3) showed that stimulation within the pontine region from which PGO spikes were recorded would elicit PGO spikes within the lateral geniculate during paradoxical sleep with every stimulation, but that stimulation during slow wave sleep would not evoke them at all. Several studies have strongly implicated the serotonin-containing raphé nuclei as modulators of PGO spike appearance. Interference with serotonin metabolism by administration of reserpine or para-chlorphenylalanine leads to continual occurrence of PGO spikes (23). Indeed, it was the remark that cats under the influence of para-chlorphenylalanine seemed to orient or be startled with appearance of each PGO spike (12) that led us to formulate the hypothesis that PGO spikes were a sign of activation of the startle reflex network (5, 35) [later generalized to alerting (6)]. Single-unit recordings of dorsal raphé neurons have shown that here is a cessation of firing just prior to a PGO spike in the transition period and that the neurons stop firing altogether during an episode of paradoxical sleep (30, 61). Finally, in an ingenious sequel to the original Bizzi and Brooks studies (3, 9), Simon et al. (56) studied the effects of parasagittal brainstem sections on the "gating" effects of the raphé nuclei on PGO spike appearance. They demonstrated that pontine stimulation lateral to the cut would elicit PGO spikes in the lateral geniculate body during both slow wave sleep and paradoxical sleep but that stimulations on the opposite, intact side still only produced the waves in paradoxical sleep. This study indicates that raphé neuronal activity does probably gate PGO spike occurrence and that the circuitry involved in PGO spike regulation is reasonably precise.

The foregoing leads one to the conclusion that paradoxical sleep is a period during which the reticular formation, at the very least, operates as if it were incapable of habituating to novel stimuli. But, of course, the organism itself gives no evidence of interacting with the environment; for except for twitches of various muscles, especially those of the distal limbs and face, paradoxical sleep is characterized by atonia, i.e. muscle paralysis. Descending pathways, stemming presumably from the medullary inhibitory area (29), inhibit the spinal motor neurons postsynaptically to produce the atonia (15, 38). It is this muscular paralysis, coupled with a state of apparent EEG activation, which has generated the genuine puzzlement about the nature of this sleep state. Our experiments have led us to the conclusion that the extreme motor inhibition is a natural accompaniment to the activation expressed by an EEG arousal pattern and the continual signs of alerting in the form of PGO spikes. Both the activation and motor inhibition can be viewed as an exaggeration of the mechanism which must operate almost continually, but with varying intensity on the efferent side, to identify new and/or unexpected stimuli in daily life and then to regulate the response.

The alerting response we have studied in the form of PGO spikes during slow wave sleep and high-amplitude eye movement potentials seems to have the characteristics which would make it a component of the orienting reflex (58). A useful accompaniment to exposure to a novel stimulus would be a brief period of hesitation to allow time for analysis of significance of the stimulus provided that the animal is not forced into immediate action as when too close to a predator (50). The inhibitory area of the medullary reticular formation (29) may act in normal life at the most basic level to provide the element of supraspinal control of spinal motor neurons necessary to allow a brief time for stimulus analysis. Rather than the overwhelming inhibition that electrical stimulation induces, the degree of inhibition should be just enough to reduce the reactivity of motor neurons to other influences to prevent movement, not produce collapse of antigravity muscles.

If paradoxical sleep is an exaggeration of this mechanisms, it is probably not the only one. In response to a sudden strong stimulus such as being grasped by a predator, a variety of animals exhibit a seemingly paradoxical response; they "freeze" (59). Variously called animal hypnosis, tonic immobility, or the still reaction, the behavior is characterized by depression of spinal reflexes and hypotonia, as is paradoxical sleep; but the EEG may be either synchronous or asynchronous (11). Significantly, as the novelty of the stimulus wears off in test experiments by human handlers, the probability of inducing tonic immobility diminishes (50). As Ratner has suggested (50) this reaction, occurring when an animal is held by a predator, is really an extreme variant of the behavior (freezing) which accompanies encounters with novel or unexpected stimuli throughout the day. Depending upon the nature of the stimulus, an animal will fight or escape at intervening distances.

Furthermore, it is not unlikely that narcolepsy represents a pathological variation on this theme. Narcolepsy is a sleep disorder characterized by sleep-onset paradoxical sleep episodes, rather than the usual sequences of slow wave sleep followed by paradoxical sleep, and episodes of partial or total muscular paralysis known as cataplexy, which may or may not be accompanied by some alteration of consciousness. Germane to this discussion is the fact that intense stimuli incite the attacks, e.g. laughter, anger, excitement, suprise (16), rather than the low-intensity, monotonous stimuli which will induce slow wave sleep (49).

The lower brainstem from which PGO spikes, signs of a highly activated reticular formation, originate (3) is probably the seat of the rather primitive behaviors we have been discussing. In cats transected in the

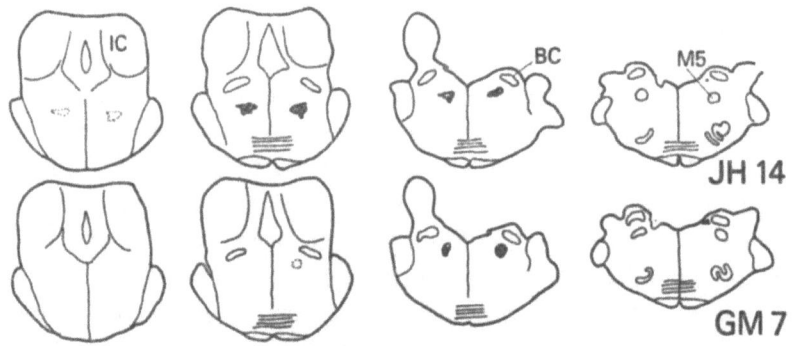

Fig. 4. Lesions which produced paradoxical sleep without atonia. Lesions are denoted by black areas sometimes surrounding areas of cavitation. Dotted circles indicate gliosis. Cross-sections are, from left to right, from stereotaxic levels P1 to P4. Both exhibited raising of the head, body righting, searching, and locomotor movements. JH14 also displayed attack. BC, brachium conjunctivum; IC, inferior colliculus; M5, motor nucleus of the trigeminal nerve.

lower mesencephalon peripheral elements of paradoxical sleep — muscle atonia, twitches and rapid eye movements — will appear (21, 62). Local application of the acetylcholine agonist, carbachol, into the pontine tegmentum will induce prolonged bouts of atonia with other elements of paradoxical sleep (60). Raphé lesions will induce abrupt transitions from wakefulness to paradoxical sleep characteristic of narcolepsy (23, 56). Small, bilateral pontine lesions (Fig. 4), however, will eliminate the atonia of paradoxical sleep so that cats exhibit hallucinatory-like behavior with PGO spikes (18, 24) and possess the other aspects of paradoxical

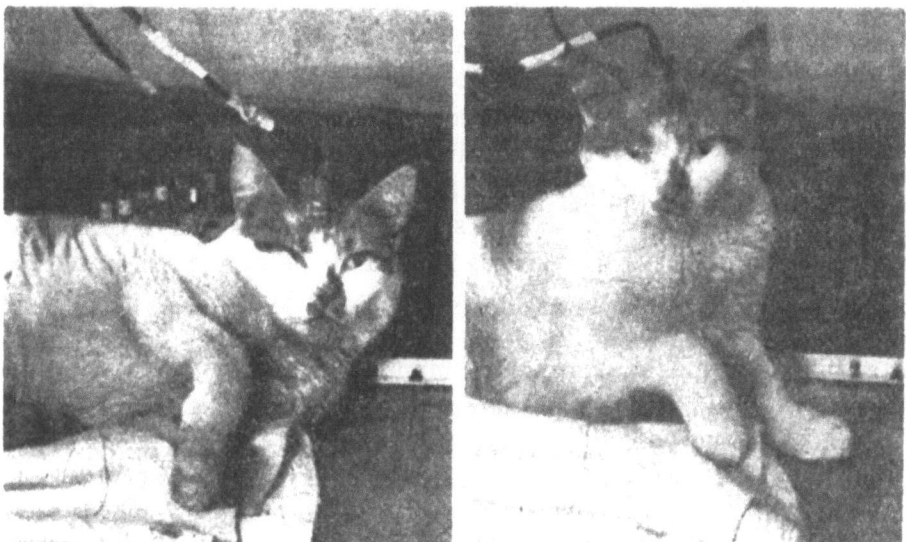

Fig. 5. A cat with bilateral pontine lesions at the initiation (A) of an episode of paradoxical sleep without atonia and fully erect (B) during the same episode. Note the relaxed nicitiating membranes. Frames from a movie taken with floodlamps 1 mo after lesioning. From Henley (18).

sleep which have been measured — activated EEG, rapid eye movements, muscle twitches, miotic pupils and relaxed nictitating membranes, hippocampal theta rhythm, and increased brain temperature (Fig. 5) (17). During wakefulness such animals show no evidence of generalized hypertonia as in decerebrate regidity so that the lesions do not merely produce an increase in antigravity muscle tone across behavioral states. The behavior exhibited during episodes of paradoxical sleep without atonia consists mainly of orienting and searching with the head and forequarters, attempts at locomotion and, in some cases, actual rage and attack behavior.

Because of the behavior released during paradoxical sleep and the essentially normal muscle tone of the cats during wakefulness, we have postulated that the lesions do more than simply eliminate the normal inhibition of spinal motor neurons (33, 36, 37). In our view, the lesions interfere, at the very least, with mechanisms normally limiting locomotor behavior during paradoxical sleep. Such a limitation would seem to be an integral part of the motor regulation accompanying novel stimulus presentation.

In pursuing the argument that paradoxical sleep presents us with an exaggeration of this mechanism, we proposed that the release of orienting and exploratory behavior during paradoxical sleep by the pontine lesions should be paralleled by some increase in exploratory behavior during wakefulness (33, 37). In order to test this hypothesis we studied exploratory behavior before and after bilateral pontine lesions in 9 cats (Fig. 4). The animals were observed during sleep, examined neurologically, observed in their free-ranging behavior and also submitted to open-field tests. For the open-field test each cat was placed into a room marked into squares with 23 cm sides for 5 one-hour sessions before and after the lesions. Each cat was studied at the same hour of the day, usually 11 a.m. Activity was measured by marking each entrance into a new square on a map of the room. Postoperatively the cats were tested after they were able to walk with minimal ataxia, normally within 2 wk. All cats exhibiting paradoxical sleep without atonia also increased their exploratory activity when observed moving about the laboratory as well as in the open-field test. Increases of 23 to 127% were noted in the cats. The cat with the lowest increase was an old cat which walked around very little preoperatively. Two cats, which never demonstrated paradoxical sleep without atonia postoperatively, had slight decreases in waking activity. Their lesions lay caudally and dorsally to those of the others. Cats which increased their activity did not just aimlessly pace; rather, they often ran to investigate sounds or objects they saw. The cats also were more restless when in a given square although this was not recorded quantitatively.

We conclude that the lesions disrupt a mechanism which modulates the cat's motor output, particularly when the nervous system is in a highly activated state as it is in paradoxical sleep or in wakefulness when confronted with a continuously changing environment.

We are frankly unsure at this time precisely which systems are interrupted to produce the changes in paradoxical sleep and wakefulness. Results with both waking and sleeping animals suggest that the phenomenon of paradoxical sleep without atonia depends upon destruction of a minimum of two systems and undoubtably more. Figure 6 depicts the minimum. The lesion must interrupt a pathway which either facilitates the activity of the medullary inhibitory area of the reticular formation (29) and/or directly inhibits the spinal motor neurons. The former alternative is depicted because pontine stimulation produced minimal direct effects on spinal motor neurons; whereas monosynaptic inhibition of antigravity motor neurons results from caudal medullary stimulation (47). A discrete pathway from the region of the locus coeruleus and ventromedial to it to the medulla has been described with the aid of axonal transport techniques (53). It appears to be related to the lateral tegmento-reticular tract described earlier by Russel (51). Interruption of this tract can eliminate atonia in paradoxical sleep (54). The diagram, however, depicts the suggestion that an additional system(s) must be involved in the release of the complex behaviors observed when the paralysis of paradoxical sleep is lifted because the increased exploratory behavior in wakefulness dictates this. Locomotor activity is basic to most of an animal's behaviors, and it is seen with the smallest lesions during

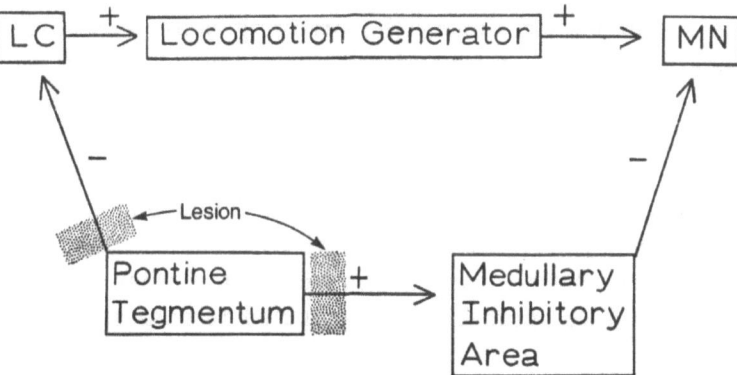

Fig. 6. Diagram of the explanation proposed for the phenomenon of paradoxical sleep without atonia. Shaded areas represent the pontine lesion which is thought to interrupt an inhibitory (—) influence on a brainstem locomotor center (LC) and an excitatory (+) influence on the medullary inhibitory area, which normally inhibits spinal motor neurons (MN). The brainstem locomotor center is believed to exert a tonic facilitatory influence on a locomotion generator, probably located in the spinal cord (55). From Morrison (33).

paradoxical sleep (Fig. 4, GM7). A large number of studies (32, 55, 59) have demonstrated that primitive locomotion in acutely decerebrated cats can be generated by electrical stimulation of a region extending from the nucleus cuneiformis, which underlies the colliculi, through the noradrenergic neurons of the locus coeruleus into the lateral pontine tegmentum. This system may be released in paradoxical sleep and wakefulness as a result of the lesions and would be part of the primitive control mechanism set into action in exaggerated fashion by the activation of paradoxical sleep. Since decerebrate cats periodically express all of the peripheral elements of paradoxical sleep (21, 62), it will be important to determine whether pontine lesions in decerebrate preparations can release elementary locomotor behavior during such episodes.

More elaborate behavior, including attack, also occurs in the case of more ventrally extending lesions (Fig. 4, JH14). When intact cats are stimulated laterally to the lesion site in the region of the locomotor system they exhibit species-specific prey-killing behavior if a rat is present (2). (Interestingly, they merely locomote in the absence of a rat.) During wakefulness our lesioned cats exhibit a tendency to be aggressive toward other cats, and we are beginning to study these interactions. It is probable that disruption of fibers from rostral limbic areas contributes to this behavior (44).

In conclusion, the idea that paradoxical sleep is a highly activated state of the brain is not new; it was suggested by early workers (13). Our work has led us to propose further, however, that seemingly disconnected events, PGO spikes and atonia, are part of a unified mechanism, which is set in motion during the orienting state in wakefulness but which operates in exaggerated form during the special state of brain activation that is paradoxical sleep. The existence of parallel effects on motor control in wakefulness and paradoxical sleep produced by pontine lesions suggests that the atonia of paradoxical sleep is a reflection of excessive activity of a brainstem response dampening mechanism triggered by extreme reticular activation, which is signified by constant PGO spike occurrence. During wakefulness the same mechanism operates more subtly to produce appropriately modulated responses to various unexpected stimuli.

It is with great pleasure that I acknowledge the very important contributions of my colleagues, Robert Bowker, Joan Hendricks and Graziella Mann, to the work reported and to many of the ideas. Financial support was provided by National Institutes of Health grants NS-13110, NS-08377, GM-02051, MH 13767, and RR-05464, and a grant from the American Philosophical Society's Penrose Fund.

## REFERENCES

1. ASERINSKY, E. and KLEITMAN, N. 1953. Regularly occurring periods of eye motility, and concomitant phenomena during sleep. Science 118: 273–274.
2. BERNTSON, G. G. 1973. Attack, grooming and threat elicited by stimulation of the pontine tegmentum. Physiol. Behav. 11: 81–87.
3. BIZZI, E. and BROOKS, D. C. 1963. Functional connections between pontine reticular formation and lateral geniculate nucleus during deep sleep. Arch. Ital. Biol. 101: 666–680.
4. BOWKER, R. M. 1979. The biological significance of eye movement potentials of wakefulness and the PGO spikes of sleep. Unpublished Ph. D. thesis, University of Pennsylvania.
5. BOWKER, R. M. and MORRISON, A. R. 1976. The startle reflex and PGO spikes. Brain Res. 102: 185–190.
6. BOWKER, R. M. and MORRISON, A. R. 1977. The PGO spike: An indicator of hyperalertness. In W. P. Koella and P. Levin (ed.), Sleep research-1976. Karger, Basle, p. 23–27.
7. BROOKS, D. C. 1978. Localization and characteristics of the cortical waves associated with eye movement in the cat. Exp. Neurol. 22: 603–613.
8. BROOKS, D. C. 1977. The PGO phenomenon: Inputs to the lateral geniculate nucleus from the cortex and brain stem. In W. P. Koella and P. Levin (ed.), Sleep research-1976. Karger, Basle, p. 15–18.
9. BROOKS, D. C. and BIZZI, E. 1963. Brain-stem electrical activity during deep sleep. Arch. Ital. Biol. 101: 648–665.

10. BROOKS, D. C. and GERSHON, M. D. 1971. Eye movement potentials in the oculomotor and visual systems of the cat: A comparison of reserpine-induced waves with those present during wakefulness and rapid eye movement sleep. Brain Res. 27: 223–239.
11. CARLI, G. 1969. Dissociation of electrocortical activity and somatic reflexes during rabbit hypnosis. Arch. Ital. Biol. 107: 219–234.
12. DEMENT, W. C. 1969. The biological role of REM sleep (circa 1968). In A. Kales (ed.), Sleep: physiology and pathology. J. B. Lippincott Co., Philadelphia, p. 245–265.
13. DEMENT, W. C. 1973. Commentary on, the biological role of REM sleep (circa 1968). In W. B. Webb (ed.), Sleep: an active process. Scott, Foresman, Glenview, Ill., p. 48–58.
14. FELDMAN, M. and COHEN, B. 1968. Electrical activity in the lateral geniculate body of the alert monkey associated with eye movements, J. Neurophysiol. 31: 455–466.
15. GASSEL, M. M., MARCHIAFAVA, P. L. and POMPEIANO, O. 1965. An analysis of supraspinal influences acting on motoneurons during sleep in the unrestrained cat. Modification of the recurrent discharge of the alpha motoneurons during sleep. Arch. Ital. Biol. 103: 25–44.
16. GUILLEMINAULT, C. 1976. Cataplexy In C. Guilleminault, W. C. Dement, and P. Passouant, (ed.), Narcolepsy. Spectrum, New York, p. 125–143.
17. HENDRICKS, J. C., BOWKER, R. M. and MORRISON, A. R. 1977. Functional characteristics of cats with pontine lesions during sleep and wakefulness and their usefulness for sleep research. In W. P. Koella and P. Levin, (ed.), Sleep research-1976. Karger, Basel, p. 207–210.
18. HENLEY, K. and MORRISON, A. R. 1974. A reevaluation of the effects of lesions of the pontine tegmentum and locus coeruleus on phenomena of paradoxical sleep in the cat. Acta. Neurobiol. Exp. 34: 215–232.
19. ITO, M., YOSHIDA, M., OBATA, K., KAWI, N. and UDO, M. 1970. Inhibitory control of intracerebellar nuclei by the Purkinje cell axons. Exp. Brain Res. 10: 64–80.
20. JEANNEROD, M. and SAKAI, K. 1970. Occipital and geniculate potentials related to eye movements in the unanaesthetized cat. Brain Res. 19: 361–377.
21. JOUVET, M. 1962. Recherches sur les structures nerveuses et les mécanismes responsables des différentes phases du sommeil physiologique. Arch. Ital. Biol. 100: 125–206.
22. JOUVET, M. 1967. Neurophysiology of the states of sleep. Physiol Rev. 47: 117–177.
23. JOUVET, M. 1972. The role of monoamines and acetylcholine-containing neurons in the regulation of the sleep-waking cycle. In Ergebnisse der Physiologie: neurophysiology and neurochemistry of sleep and wakefulness. Springer-Verlag, New York, 64: 96–117.
24. JOUVET, M. and DELORME, G. 1965. Locus coeruleus et sommeil paradoxal. C. R. Soc. Biol. 159: 895–899.
25. JOUVET, M. and MICHEL, F. 1959. Corrélations électromyographique du sommeil chez le chat décortiqué et mésencephalique chronique. C. R. Soc. Biol. (Paris) 153: 422–425.
26. LAURENT, J. CESPUGLIO, R. and JOUVET, M. 1974. Délimitation des voies ascendantes de l'activité ponto-géniculo-occipitale chez le chat. Brain Res. 65: 29–52.

27. LAURENT, J. and GUERRERO, F. A. 1975. Reversible suppression of pontogeniculo-occipital waves by localized cooling during paradoxical sleep in cats. Exp. Neurol. 49: 356–369.
28. LUND, S. and POMPEIANO, O. 1968. Monosynaptic excitation of alpha motoneurones from supraspinal structures in the cat. Acta. Physiol. Scand. 73: 1–21.
29. MAGOUN, H. W. and RHINES, R. 1946. An inhibitory mechanism in the bulbar reticular formation. J. Neurophysiol. 9: 165–171.
30. McGINTY, D. and HARPER, R. M. 1976. Dorsal raphé neurons: depression of firing during sleep in cats. Brain Res. 101: 569–575.
31. MIKITEN, T. H., NIEBYL, P. H. and HENDLEY, C. D. 1961. EEG desynchronization during behavioral sleep associated with spike discharges from the thalamus of the cat. Fed. Proc. 20: 327.
32. MORI, S., SHIK, M. L., and YAGODNITSYN, A. S. 1977. Role of pontine tegmentum for locomotor control in mesencephalic cat. J. Neurophysiol. 40: 284–295.
33. MORRISON, A. R. 1979. Brainstem regulation of behavior during sleep and wakefulness. In J. M. Sprague and A. N. Epstein (ed.), Progress in psychobiology and physiological psychology, 8. Academic Press, New York, p. 91–131.
34. MORRISON, A. R. and BOWKER, R. M. 1973. Cerebellar and spinal contributions to the regulation of muscle tone and movement during sleep. In U. J. Jovanović (ed.), The nature of sleep. Gustav Fischer, Stuttgart, p. 270–277.
35. MORRISON, A. R. and BOWKER, R. M. 1975. The biological significance of PGO spikes in the sleeping cat. Acta. Neurobiol. Exp. 35: 821–840.
36. MORRISON, A. R., HENDRICKS, J. C. and BOWKER, R. M. 1977. A new role for the locus coeruleus. Soc. Neurosci. Absts. 3: 256.
37. MORRISON, A. R., MANN, G., HENDRICKS, J. C. and STARKWEATHER, C. 1979. Release of exploratory behavior in wakefulness by pontine lesions which produce paradoxical sleep without atonia. Anat. Rec. 193: 628.
38. MORRISON, A. R. and POMPEIANO, O. 1965. An analysis of supraspinal influences acting on motoneurons during sleep in the unrestrained cat. Responses of the alpha motoneurons to direct electrical stimulation during sleep. Arch. Ital. Biol. 103: 497–516.
39. MORRISON, A. R. and POMPEIANO, O. 1966. Vestibular influences during sleep IV. Functional relations between vestibular nuclei and lateral geniculate nucleus during desynchronized sleep. Arch. Ital. Biol. 104: 425–458.
40. MORTIMER, J. 1973. Temporal sequence of cerebellar Purkinje and nuclear activity in relation to the acoustic startle response. Brain Res. 50: 457–462.
41. MORUZZI, G. 1963. Active processes in the brainstem during sleep. Harvey Lect. 58: 233–297.
42. MORUZZI, G. 1964. The historical development of the deafferentation hypothesis of sleep. Proc. Am. Philos. Soc. 108: 19–28.
43. MORUZZI, G. and MAGOUN, H. W. 1949. Brainstem reticular formation and activation of the EEG. Electroenceph. Clin. Neurophysiol. 1: 455–473.
44. NAUTA, W. J. H. 1960. Some neural pathways related to the limbic system. In E. R. Ramey and D. S. O'Doherty (ed.), Electrical studies on the unanesthetized brain. Hoeber, New York, p. 1–16.
45. PARMEGGIANI, P. L. 1977. Interaction between sleep and thermoregulation. Waking and Sleeping 1: 123–132.

46. PELLET, J., TARDY, F., DUBROCARD, S. and HARLAY, F. 1974. Étude de l'activité électrique phasique du cortex du cervelet au cours des états de veille et de sommeil. Arch. Ital. Biol. 112: 163-195.
47. PETERSON, B. W. 1979. Reticulospinal projections to spinal motor nuclei. Ann. Rev. Physiol. 41: 127-140.
48. POMPEIANO, O. and MORRISON, A. R. 1965. Vestibular influences during sleep. I. Abolition of the rapid eye movements of desynchronized sleep following vestibular lesions. Arch. Ital. Biol. 103: 569-595.
49. POMPEIANO, O. and SWETT, J. E. 1962. EEG and behavioral manifestations of sleep induced by cutaneous nerve stimulation in normal cats. Arch. Ital. Biol. 100: 311-342.
50. RATNER, S. C. 1967. Comparative aspects of hypnosis. In Handbook of clinical and experimental hypnosis, J. E. Gorden (ed.), MacMillan, New York, p. 550-587.
51. RUSSEL, G. V. 1955. The nucleus locus coeruleus (dorsolateralis tegmenti). Tex. Rep. Biol. Med. 13: 939-988.
52. SAKAI, K. and CESPUGLIO, R. 1976. Evidence for the presence of eye movement potentials during paradoxical sleep in cats. Electroencephalogr. Clin. Neurophysiol. 41: 37-48.
53. SAKAI, K., SASTRE, J. P., SALVERT, D., TOURET, M., TOHYAMA, M. and JOUVET, M. 1978. Tegmento-reticular projections responsible for muscle atonia during paradoxical sleep in the cat. Sleep Res. 7: 42.
54. SASTRE, J. P., SAKAI, K. and JOUVET, M. 1978. Bilateral lesions of the dorso-lateral pontine tegmentum. II — Effect upon muscle atonia. Sleep Res. 7: 44.
55. SHIK, M. L. and ORLOVSKY, G. N. 1976. Neurophysiology of locomotor automatism. Physiol. Rev. 56: 465-501.
56. SIMON, R. P., GERSHON, M. D. and BROOKS, D. C. 1973. The role of raphé nuclei in the regulation of ponto-geniculo-occipital wave activity. Brain Res. 58: 313-330.
57. SINGER, W. 1977. Control of thalamic transmission by corticofugal and ascending reticular pathways in the visual system. Physiol. Rev. 57: 386-420.
58. SOKOLOV, E. N. 1963. Higher nervous functions: The orienting reflex. Ann. Rev. Physiol. 25: 545-580.
59. STEEVES, J. D., JORDAN, L. M. and LAKE, N. 1975. The close proximity of catecholamine-containing cells to the mesencephalic locomotor region (MLR). Brain Res. 100: 663-670.
60. STERIADE, M. and HOBSON, J. A. 1976. Neuronal activity during the sleep-waking cycle. Prog. in Neurobiol. 6: 155-376.
61. TRULSON, M. E. and JACOBS, B. L. 1978. Raphé unit activity in freely moving cats: correlation with level of behavioral arousal. Brain Res. 163: 135-150.
62. VILLABLANCA, J. 1966. Behavioral and polygraphic study of "sleep" and "wakefulness" in chronic decerebrate cats. Electroencephalogr. Clin. Neurophysiol. 21: 562-577.
63. YU, J., TARNECKI, R., CHAMBERS, W. W., LIU, C. N. and KONORSKI, J. 1973. Mechanisms mediating ipsilateral limb hyperflexion after cerebellar paravermal cortical ablation or cooling. Exp. Neurol. 38: 144-156.

Adrian R. MORRISON, Laboratories of Anatomy, School of Veterinary Medicine, University of Pennsylvania, Philadelphia, Pa. 19104, USA.

Lecture delivered at the Warsaw Colloquium on Instrumental
Conditioning and Brain Research
May 1979

# THE ROLE OF STRIATUM IN THE ACQUISITION OF INSTRUMENTAL DEFENSIVE REACTIONS IN DOGS

K. B. SHAPOVALOVA and S. I. BAZHENOVA

Pavlov Institute of Physiology, Academy of Sciences of the USSR
Leningrad, USSR

*Abstract.* The model of instrumental defensive reflexes, connected with the maintenance of a certain posture, was used in a chronic experiment on dogs. The motor components of "coordinating" and "cognitive" programs of defensive reflexes were analysed before and after stimulation of the head of the caudate nucleus (CNH). Predominantly inhibitory influences of the striatum on the reflexes were demonstrated. The degree of inhibitory effects (inhibition of "coordinating" program or elimination of the defensive reflexes, "caudate stopping") depended on the parameters and time of application of caudate stimuli. Preliminary subthreshold high frequency stimulation of intralaminar thalamic nuclei significantly increased the inhibitory effects elicited by CNH stimulation. It is suggested that the CNH controls the acquisition of motor behavioral tasks. The degree of elimination of caudate inhibitory influences may be determined by the level of activation of nonspecific thalamic nuclei.

## INTRODUCTION

The caudate nucleus was traditionally considered as part of the extrapyramidal system primarily because caudate pathology leads to complex motor disturbances. The clinical pathological studies of humans and the results of experiments on animals show the importance of the basal ganglia in the regulation of movement and posture (13, 14, 22, 32, 40). Other data (4, 10, 11, 15, 19, 25, 40, 43, 48, 50, 53) allow the conclusion that the striatum, including the caudate nucleus, is involved in the integrative actions of the brain.

The stimulation of the caudate nucleus by electrical stimuli of varied frequency and intensity is one of the most frequently used experimental approaches for the investigation of the possible participation of the caudate nucleus in the initiation and aquisition of different forms of motor behavior (24, 29, 42, 43, 47, 52). The inhibitory effect of caudate nucleus stimulation and so-called "caudate stopping" during the aquisition of conditioned motor reactions has been shown in many studies (24, 42, 43, 47).

It was interesting to suggest an experimental approach where it is possible to separate the effects of stimulation of the caudate nucleus in the decision of behavioral task from purely motor effects. We used the model of instrumental defensive reactions (IDR) related to the maintenance of certain posture in an experiment on dogs. This model of voluntary motor reactions (36) has advantages for resolving issues raised by us. The IDR has a characteristic configuration, latency, duration and EMG pattern. Using this model we were able to compare the influence of the head of caudate nucleus (CNH) stimulation on the motor components of the IDR with the effect on the acquisition of the instrumental task itself, involving the avoidance of electric shock. The investigation of peculiarities of CNH stimulation on voluntary reactions of a tonic type, also was of interest because the striatum is considered important in the acquisition of the slow movements of the tonic type (5, 26).

This experiment addressed the following questions: (1) What is the manifestation of the inhibitory effect of preliminary stimulation of CNH if the voluntary movement is not arrested? (2) What is the difference between the inhibitory effect of CNH on motor components of IDR and on solution of the instrumental task itself? (3) In what way does CNH stimulation act on the initiation of learned movement? (4) What characterizes the mechanographic (MCG) and electromyographic (EMG) components of "caudate stopping", compared with the MCG and EMG components of stopping movement under normal conditions at the end of the conditioned signal? (5) What determines the degree of the inhibitory effects of CNH?

## METHODS

The experiment used 9 adult dogs. The animals were trained to avoid shock, applied to the hind left limb by raising its leg to a certain height necessary for breaking the contacts and maintaining this posture during the entire duration of the conditioned signal, 10 s. The shock was activated at the 5th s of conditioned signal action and lasted together with the signal for the remaining 5 s.

Bipolar stimulating electrodes were implanted in the CNH of the trained animals under sterile conditions under hexenal anesthesia (nichrome, glass insulation 30–40 k$\Omega$). The coordinates of the points of implantation of the electrodes and their length were calculated stereotaxically (28). The electrodes were implanted bilaterally in the dorsolateral and ventromedial parts of the CNH. The localization of the electrodes was histologically verfied.

The procedure of the main experiment consisted of 3 to 5 successive presentations of the conditioned signal at 1 min. intervals. Stimulation of the CNH was done 5 or 10 s. prior to the conditioned signal (preliminary stimulation) or during IDR usually at the 5th s. of signal duration. The stimulation of CNH was repeated every 5–6 trials. In a single experiment a session 30 to 40 trials and 3 to 4 stimulation were presented. Experimental sessions with stimulation of CNH were not repeated more than 2 or 3 times a week.

The duration of the conditioned signal, a metronome of 2 Hz, the parameters of the stimulating current, the EMG of m. semitendinosus and m. rectus femoris of both hind limbs (i.e. the "working" left and supporting right limbs) and mechanograms (MCG) of the left hind limb were simultaneously registered on photopaper of a K-115 oscilloscope. We were able to perform the quantative mesurement of the IDR motor components during stimulation of CNH to compare them with control test data. The results were processed at the Wang computer.

RESULTS

*The effects of CNH stimulation prior to the IDR initiation*

Figure 1 shows typical example of instrumental defensive reaction. The dogs executed this task with a large store of time (i.e. the latency of the IDR, the duration of the leg raising) and space (the height of raising). In spite of several individual pecularities in MCG and EMG components of the IDR determined by the general excitability of animals, the EMG activity of supporting and working muscles and the criteria of execution of IDR latencies followed a regular, stable pattern (45).

These experimental series were carried out with low frequency stimulation (2 Hz) of the rostral section of CNH by the stimuli of different intensities (from 50 $\mu$A, to 2 mA). The stimulation was applied

10 s prior to conditioned stimulus. The right side and left side stimulation of the rostral section of CNH was done on different days. The effects of the preliminary low frequency stimulation of the CNH of the left or right hemisperes on the EMG and MCG components of the IDR were positively correlated with the intensity of the stimulating current.

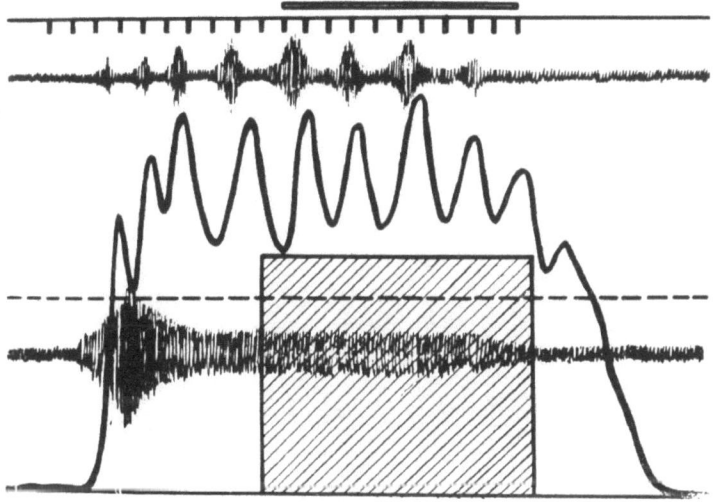

Fig. 1. A typical example of instrumental defensive reaction. Up to down: unconditioned stimulus, conditioned stimulus, EMG of m. semitendinosus. EMG of m. rectus femoris, mechanogram of instrumental movement (MCG). The hatched area shows the required movements value, the "cognitive" part of the defensive instrumental reactions. The level of relay switching on is denoted by the interrupted line.

The stimulation of the CNH of 50 µA was not accompanied by any changes of amplitude, configuration, duration and latency of the IDR, and the EMG pattern of the working, left limb did not change. The only change was a selective inhibition of the earliest component of the EMG, that is, a burst of activity of the m. semitendinosus of the supporting limb usually registered 500 ms before the beginning of the IDR, which is the EMG component connected with the posture change prior to the movement (Fig. 2). The increase of intensity of the stimulating current led to the change in the EMG pattern of the working limb muscles.

With the increase of CNH stimulus intensity, the EMG of m. semitendinosus of the working limb significantly increased its latency (r.c. 0.49, $P < 0.005$). The latency of the MCG, however, increased drastically only with large values of the stimulating current (above 700 µA). This finding may indicate that under the influence of preliminary low frequency stimulation, the CNH changes the coordinating pattern of the voluntary movement, performed with the activation of other muscles group. The change in intensity of preliminary, low frequency CNH stimulation affected some components of the MCG of the IDR (Figs. 2 and 3). Stimuli of average intensity (250–700 µA) "smoothed out" the

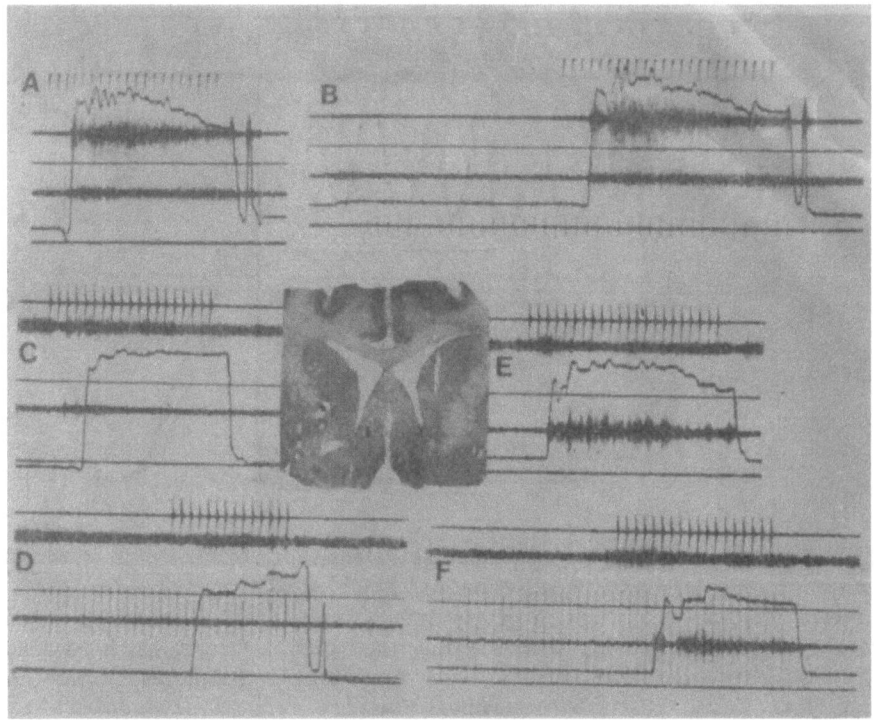

Fig. 2. The influence of preceding low-frequency CNH stimulation on the instrumental defensive reflexes. Dog 1. Control (without stimulation) — A, C, E. The stimulating current: 60 µA (B), 120 µA (C), 300 µA (F), 2 Hz, 0.5 ms. Up to down: conditioned stimuli; EMG m. semitendinosus of "working" limb. (A, B), EMG m. rectus femoris supporting limb (C–F); EMG m. semitendinosus supporting limb (A–D); EMG m. rectus femoris of "working" limb (E–F); MCG of instrumental reaction; the mark of CNH stimulation. Stimulating electrode localization in right CNH.

usually present separate phasic jerk movements superimposed on the tonic component of the IDR. This phenomenon was accompanied with the inhibition of the EMG activity of the m. semitendinosus of the working limb. This muscle discharges by bursts simultaneously with every separate leg jerk. Further increase of current in turn decreased the general IDR amplitude, and increased both the latency of movement and the time of reaching the "safe" zone. In order to inhibit the instrumental response — the "cognitive" program of movement (Fig. 1, hatched area), stimulation of the CNH of considerable amplitude (above 1 mA) was necessary (Fig. 3).

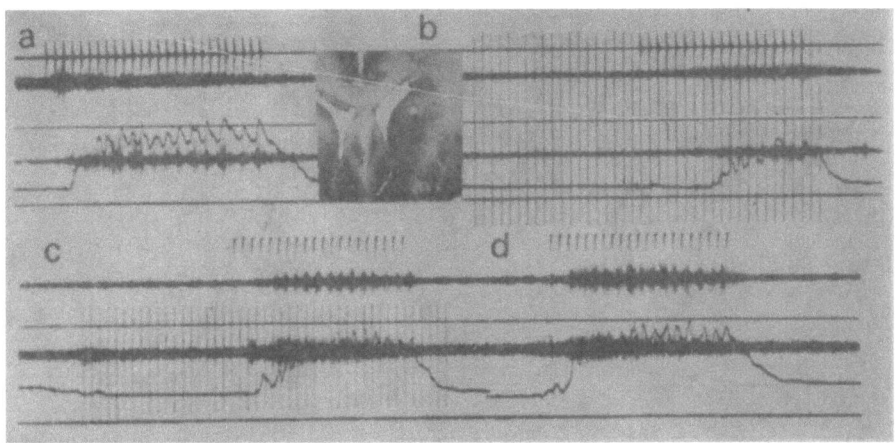

Fig. 3. The influence of preceding low-frequency CNH stimulation on the instrumental defensive reflexes. Dog 2. Control (without stimulation) — a, d. The stimulating current: 450 μA (c), 850 μA (b) 2 Hz, 0.5 ms. Up to down: conditioned stimuli; EMG m. rectus femoris (a, b) of supporting limb; EMG m. semitendinosus of "working" limb (c, d); EMG m. semitendinosus (a, b); EMG m. rectus femoris (c, d) of "working" limb; MCG of instrumental reflexes; the mark of stimulation. Stimulating electrode is located in the right CNH.

*The CNH stimulation effects during IDR performance.*

The low-frequency stimulation during IDR produced an effect on the EMG and MCG components which depended on the amplitude and duration of the stimulating current (44, 47). Current above 1 mA applied at the end of conditioned stimuli could cause the cessation of the IDR, i.e., so-called "caudate stopping" was observed. It is usually difficult to obtain "caudate stopping" in defensive situations (44). Nevertheless, stimulation at 50 Hz at certain amplitude values, the "caudate stopping" threshold, always caused IDR cessation. Therefore the experiment used the high frequency stimulation of the CNH. The data, characterizing the motor components of the cessation of leg movements at the end of the conditioned signal, were compared with the characteristics of movement cessation produced by CNH stimulation ("caudate stopping"). The comparison of the components (duration and speed, measured by angle of cessation) of the natural cessation of movement with the end of the conditioned signal and EMG components of this movement showed that not only lifting of the leg during IDR is programmed, but also cessation of this movement (Fig. 4).

The data for 56 „caudate stopping" trials caused by high frequency stimulation of the CNH of the contralateral hemispere, with respect to the "working" limb, were processed. It was found that all parameters of this movement, speed (angle) and duration of the cessation of mo-

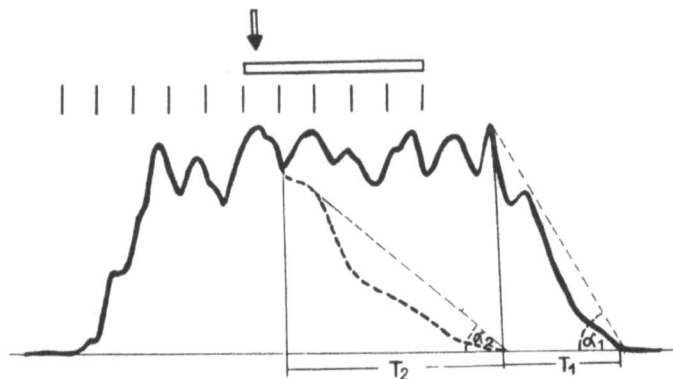

Fig. 4. The scheme of the dog's limb placing under control conditions (normal cessation of movement at the end of conditioning stimuli) and during high frequency stimulation of CNH (the arrow). The normal cessation of movement: $T_1$, $\alpha_1$. Caudate stopping: $T_2$, $\alpha_2$. Up to down: the time course of CNH stimulation; the unconditioned stimuli; the conditioned stimuli; the thick solid line shows MCG of movement. The thick interrupted line indicates "caudate stopping".

vements, were significantly different from those of the cessation of movements after termination of the conditioned signal. Specifically following differences were observed: (1) In "caudate stopping" the duration of leg cessation increased and the speed of the movement decreased (Table I, Figs. 4 and 5). (2) The mean error and standard deviation

TABLE I

The speed (angle) of mechanograms (MCG) of voluntary movement cessation in control ($x_1$) and during CNH stimulation ($x_2$). $n$ = trials; $S_x$ = standard error; SD = standard deviation (S).

| Dogs (no.) | $n_1 = n_2$ | Control $x_1 \pm S_{x_2}$ | With CNH stimulus $x_2 \pm S_{x_2}$ | P | SD |
|---|---|---|---|---|---|
| 2 | 14 | 56.0±1.75 | 48.14±2.63 | < 0.018 | $S_1 = 3.79$ $S_2 = 5.68$ |
| 3 | 15 | 74.21±1.33 | 50.50±4.08 | < 0.001 | $S_1 = 2.89$ $S_2 = 8.83$ |
| 8 | 10 | 46.0±2.73 | 21.75±3.32 | < 0.001 | $S_1 = 8.71$ $S_2 = 10.58$ |
| 6 | 17 | 58.70±2.69 | 54.75±3.05 | < 0.65 | $S_1 = 5.72$ $S_2 = 6.52$ |

values increased sharply (Fig. 5, Table I), which may indicate the absence of stable leg cessation movements program during CNH stimulation. (3) The cessation of movements caused by the high frequency CNH stimulation of the high intensity (above 600 µA) was followed

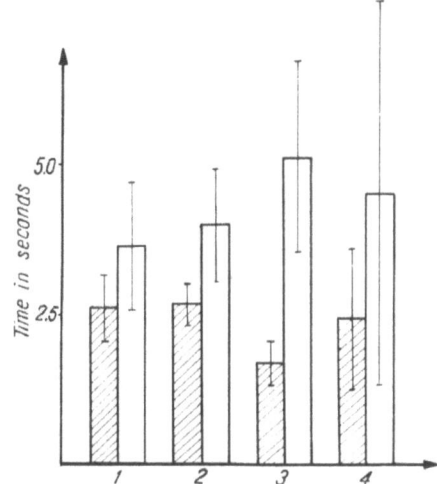

Fig. 5. The time of movement cessation under control conditions (hatched columns) and with CNH stimulation, "caudate stopping" (white columns). Vertical lines show standart deviations. 1–4, dogs.

by an increase of the m. rectus femoris EMG activity, i.e., by hypertonus of the knee joint bending muscle (Fig. 6).

The most obvious feature of movement during the CNH stimulation was the absence of phasic components usually superimposed on the instrumental reaction. The cessation of the IDR, caused by CNH high frequency stimulation in the majority of cases was manifested in the sharp decrease of movement amplitude and in the "stiffening" the leg in a slight raised position (Fig. 6). The selective inhibition of phasic movements was always combined with supression of the EMG activity of the m. semitendinosus, the bursts of activity which correlated with phasic leg jerking.

With increases of intensity of high frequency stimulation of the CNH (Fig. 6), the cessation of movements was shown by different patterns of the EMG activity of m. rectus femoris (Fig. 6 A–D). The results obtained by the high intensity (above 1 mA) of high frequency stimulation of CNH (Fig. 6 C, D) showed in these cases the hypokinesia (the inhibition of phasic movements, the decrease of the IDR amplitude and the "stiffening" with raised limb) combined with hypertonus, i.e., with sharp increase of activity of m. rectus femoris. The motor reactions of this type are similar in their appearance to some forms of stria pathology. This observation allows us to conclude that the degree of triggering of the caudate inhibitory mechanisms is determined by the level of

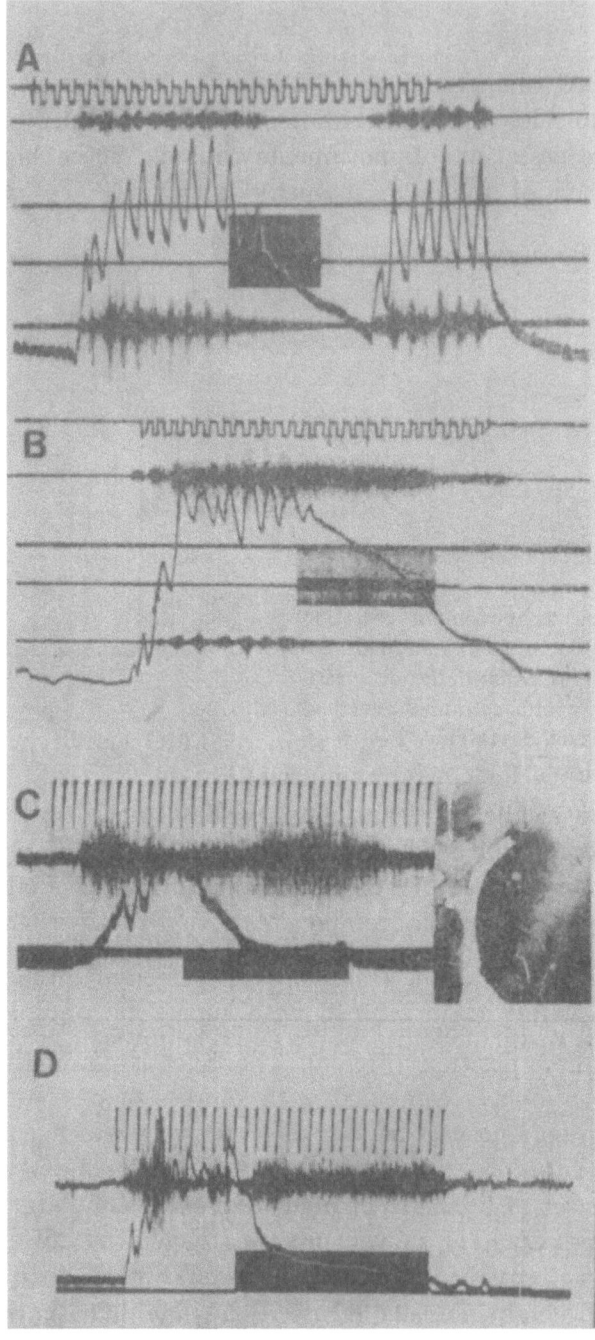

Fig. 6. Some examples of "caudate stopping" in several dogs. High frequency stimulation of CNH: 400 μA (A), 500 μA (B), 1 mA (C, D), 50 Hz, 0.5 ms. Up to down: the time course of conditioned signal; EMG of m. rectus femoris of "working" limb (A–D); EMG of m. semitendinosus of "working" limb (A, B); MCG of instrumental reflexes. In C localization of stimulating electrodes in right CNH.

striatum activation. The latter, in turn, may be dependent on the degree of activation of intralaminar thalamic nuclei giving projections into the entire caudate nucleus (18, 41). The nonspecific afferentation enters the caudate nucleus through this canal from all sensory systems (1, 9, 31). The stimulation of intralaminar thalamic nuclei including CM, according to microelectrode investigations, activates the CNH neurons, causing mostly synaptic responses of the type EPSP (7, 39).

*The influence of preliminary activations of N.CM thalamus on the realization of "caudate stopping" caused by stimulation of different parts of the CNH.*

The electric stimulation of the N.CM thalamus (15–30 µA) with a frequency of 200 Hz brings about reactions of activation in cortical and subcortical structures (27). With the preliminary activation of intralaminar nuclei including N.CM in our experiment, the activation of the motor components of the IDR was observed, which manifested itself in different ways in different dogs. In most cases the decrease of latency of EMG activity in the "working" limb, m. rectus femoris, was observed. The amplitude of the IDR significantly increased (Table II); the number

TABLE II

Influence of preliminary subthreshold stimulation of intralaminar nuclei of thalamus with frequency 200 Hz on amplitude of instrumental defesive reflexes

| Dogs (no.) | $n_1 = n_2$ | Before stimulation | After stimulation | SD | P |
|---|---|---|---|---|---|
| 5 | 30 | 100 | 107.00 | 4.9 | < 0.005 |
| 8 | 13 | 100 | 109.58 | 6.9 | < 0.005 |

of phasic raising movements, superimposed on the IDR increased, and interstimulus movements were observed in some animals during preliminary stimulation of intralaminar nuclei of the thalamus (15–30 µA) with a frequency of 200 Hz. The duration, amplitude and EMG components of these movements differed slightly from the instrumental responce (Fig. 7).

Four dogs were subjected to 466 stimulations of ventromedial and dorsolateral segments of CNH by stimuli that were subthreshold for "caudate stopping", and in 100 tests the stimulation applied to the CNH was preceded by subthreshold activation of intralaminar thalamic nuclei with a frequency of 200 Hz. The effect of the interaction of thalamic and striatal influences depended on the localization of the stimulating electrode in the CNH (Table III). As may be seen from Table III, in the

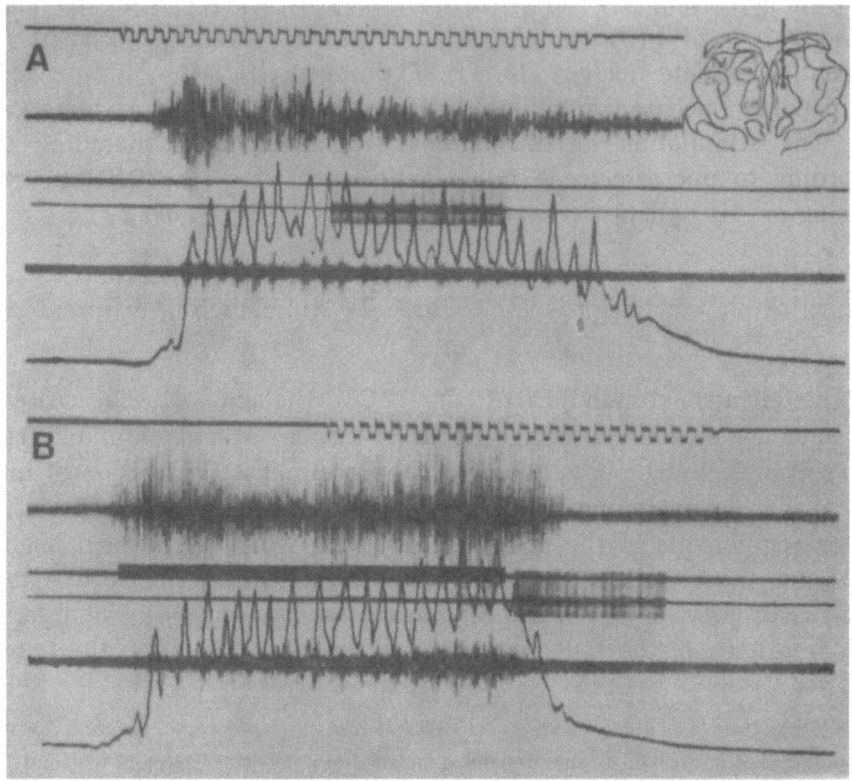

Fig. 7. The influence of preliminary nucleus CM of the thalamus stimulation on inhibition with CNH stimulation. Dog 6. A, The influence of CNH subthreshold stimulation alone on the instrumental defensive reactions performance (without preceding nucleus CM of the thalamus stimulation). B, The same with preliminary nucleus CM of the thalamus stimulation. Up to down: the time course of conditioned stimuli. EMG m. rectus femoris, recording of nucleus CM stimulation (B), recording of CNH stimulation, EMG m. semitendinosus, MCG of instrumental reactions.

case of stimulation of the dorsolateral segment of the CNH, the preliminary stimulation of the intralaminar nuclei of the thalamus of the same hemispere intensified the caudate inhibitory effect by 60.3%. In most cases it was found that CNH stimulation that was previously subthreshold for "caudate stopping", now reached the threshold value and was accompanied by characteristic behavioral effects, typical for "caudate stopping" of the MCG and EMG changes (Figs. 7 and 8). Only in 40.4% of the cases, did high frequency stimulation of the intralaminar thalamic nuclei of the contralateral hemisphere intensify caudate inhibitory effects caused by the stimulation of the ventral segment of the CNH of the same hemispere, leaving them unchanged (40.4%) or re-

TABLE III

The influence of preliminary intralaminar thalamic nuclei high frequency stimulation on the effect of stimulation of several parts of the CNH

| Dogs (no.) | Stimulation of dorsal part of CNH | | | | Stimulation of ventral part of CNH | | | |
|---|---|---|---|---|---|---|---|---|
| | The general number of coupled stimul. (CM+CNH) | The influence on "caudate stopping" | | | The general number of coupled stimulus (CM+CNH) | The influence on "caudate stopping" | | |
| | | positive | zero | negative | | positive | zero | negative |
| 5 | 27 | 16 | 9 | 2 | 17 | 7 | 6 | 4 |
| 6 | 23 | 14 | 4 | 4 | — | — | — | — |
| 4 | 8 | 5 | 3 | 0 | 15 | 6 | 8 | 1 |
| 7 | — | — | — | — | 10 | 4 | 3 | 3 |
| Total | 58 | 35 | 16 | 6 | 42 | 17 | 17 | 8 |
| in % | 100 | 60.3 | 27.2 | 10.3 | 100 | 40.4 | 40.4 | 19.2 |

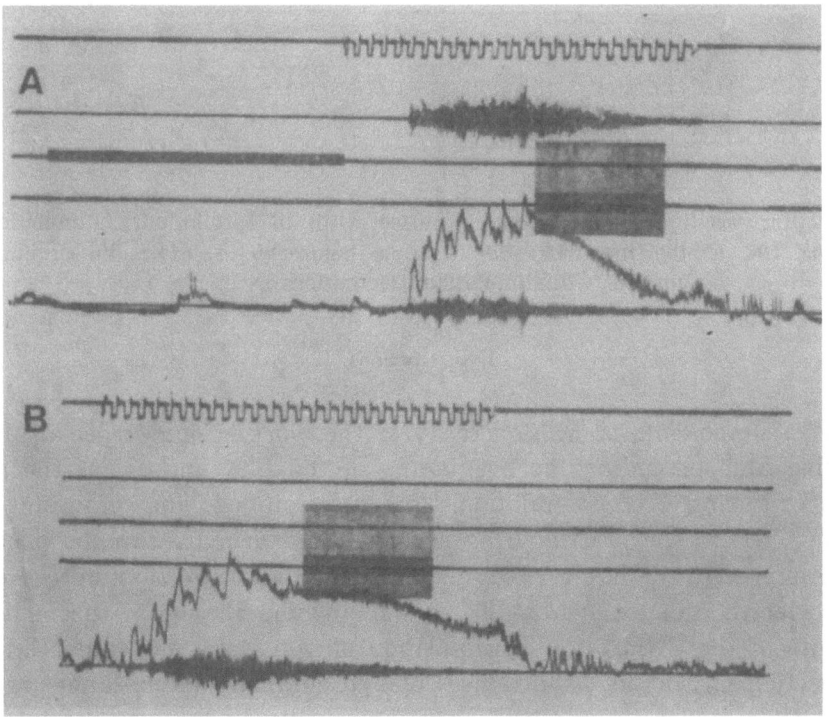

Fig. 8. Increase in caudate inhibitory effect with the preliminary stimulation of nucleus CM of the thalamus (A). Dog 5. Denotations as in Fig. 7.

ducing them (19.2%). Intensification of caudate inhibitory effects was expressed not only in that stimuli formerly subthreshold for "caudate stopping" reached the threshold after preliminary activation of the thalamus nuclei, but also in the prolongation of "caudate stopping" duration and intensification of its characteristic changes at EMG and MCG, in those cases when preliminary tests of CNH threshold stimulation for causing "caudate stopping" were applied (Fig. 9).

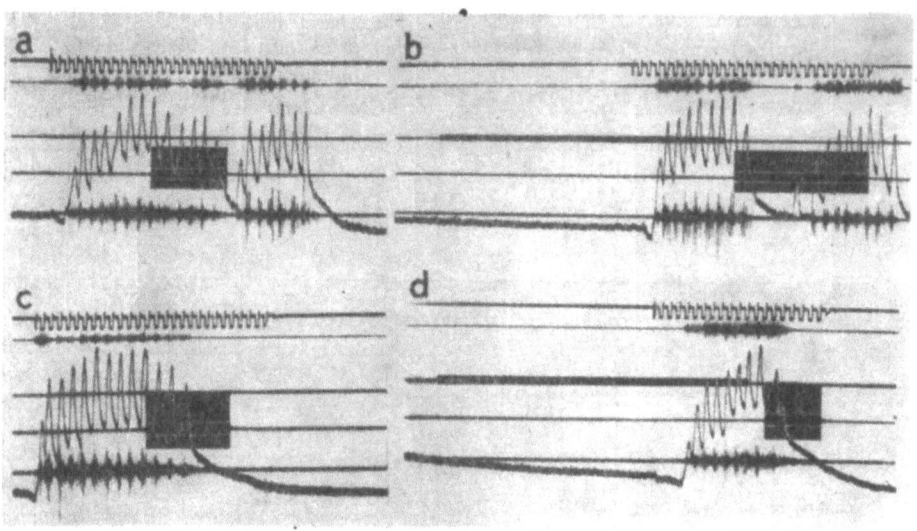

Fig. 9. Increase in "caudate stopping" effect with the preliminary stimulation of nucleus CM of the thalamus. Dog 6. *a, c,* before; *b, d*, after the preliminary nucleus CM stimulation. Denotations as in Fig. 7.

## DISCUSSION

The components of EMG activity of the working and especially supporting limb preceding the begginning of the IDR underwent the most changes during CNH stimulation. This fact points to the important role of the CNH in the pretriggering processes and primarily in the postural change preceding the learned movement. This suggestion agrees with microelectrode experiments (16, 35) showing the change in the neuronal activity of the CNH prior to acquisition of voluntary movement. In this connection it should be stressed that preliminary stimulation of the CNH by a moderate amplitude current selectively inhibits the interstimulus spontaneous raising of limbs in defensive conditioning of dogs (44) and cats (2), leaving the conditioned reflex itself without change.

According to the hypothesis of Kornhüber (26), basal ganglia (strionigro-pallidum) are the generators of slow ("ramp" type) movements. As an argument in favor of this view Kornhüber refers to the data showing that in the paresis of cortical origin, movements of this type are not affected. Nor are they affected by cerebellar lesions. On the contrary, the fast ("ballistic") movements, disturbed mainly with cerebellum pathology, do not change with parkinsonism, while the slow smooth movements are affected. It was Konorski (25) who proposed the possibility of the striatum participation in programming of voluntary reactions in animals. However, there is no direct evidence for this. Brooks (5) and Phillips and Porter (37) supposed that the programs of two types of movements, one slow, closely related to the mechanisms of the posture control and another fast of a "ballistic" type are specified through a common way that includes the associative cortex, basal ganglia (or cerebellum), nucleus VL of the thalamus and sensorimotor cortex (Fig. 10). There are some data (13, 17, 33, 51), that indicate the

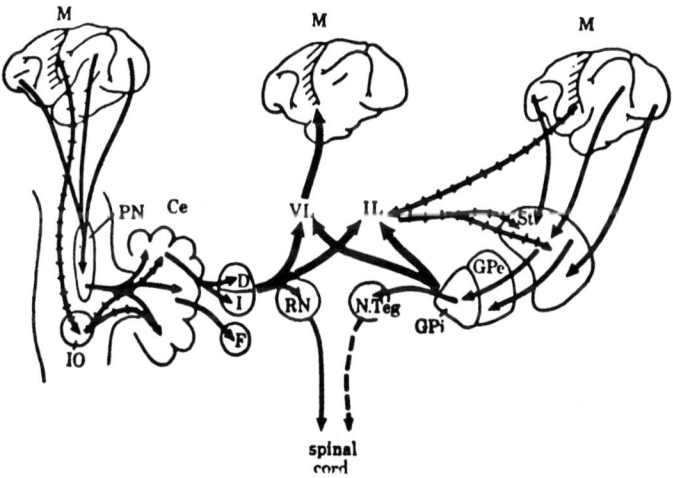

Fig. 10. Diagram of anatomical connections stressing similarities of cerebral links with the cerebellum and basal ganglia. Interrupted arrow indicates possible connection. Ce, cerebellar cortex; D, dentate nucleus; GPe, Globus pallidus, external segment; GPi, Globus pallidus, internal segment; I, interpositus nucleus; IL, intralaminar nucleus of thalamus; M, motor cortex; ST, striatum; VL, ventrolateral nucleus of thalamus. From Kemp and Powell, quoted by Brooks (5).

possibility of this hypotheses having a neuronal basis. But in reality it is difficult to study any of these movement in "pure form". Every voluntary movement, whether slow or fast, includes both types of motor program, manifesting themselves in various degree and in various combinations. An example of this may be seen in presently described IDR, where the slow movements ("ramp type") are superimposed on the fast phasic movements. The stimulation of the CNH caused primarily the selective inhibition of phasic movements. The slow tonic component of the motor response, the organization of which the striatum plays an important role, on the contrary increased. This took place in CNH stimulation by low frequency stimuli as well as by high frequency stimulation of the CNH applied during IDR acquisition.

Is the CNH really taking part in the initiation of "slow" voluntary movement as suggested by Kornhüber (26)? At present we have no convincing data to confirm this conclusion. Microelectrode investigations (35) have shown the discharge of CNH neurons 110-150 ms preceding EMG activity of the limb during the simple choice and bar-press reactions of monkeys, while PT neurons discharge 50-100 ms prior to the EMG activity. CNH is included in the parallel system to modulate the activity of the motor system and primarily the corticospinal system, on the basis of following findings: (1) Before the onset of voluntary movements, the activity of prefrontal cortical neurons is prior to the activity of CNH neurons (35). (2) The activity of striatum neurons during performance of more complex tasks increases (12). (3) The activity of CNH neurons also increases when any signal like the conditioned stimulus is activated during interstimulus intervals, i.e. when the conditions of the experiment are changed (8). (4) The present motor component analysis confirmes that CNH is not structurally included in program of acquisition of IDR, but rather belongs to a parallel system controlling all stages of instrumental reactions. The range of caudate inhibitory influences on motor behavior is very wide, starting from the inhibition of early EMG components connected with the posture change, to the correction of the IDR itself and ending with the complete abolition of voluntary movement.

The significant differences in speed (angle) and duration of the leg cessation and EMG patterns of these movement in control and during high frequency CNH stimulation allow us to assume that "caudate stopping" is not the model of voluntary cessation of the movement. However, this does not mean that under normal conditions, the CNH does not participate in the sudden cessation of the movements.

Buchwald et al. (6) formulated the concept of the so-called "caudate loop" (Fig. 11). According to this hypotheses, nonspecific afferentation from the nucleus CM thalamus entering the caudate nucleus acquires an inhibitory character and in that form moves through nucleus VA and nucleus VL of the thalamus into the cortex. The synaptic mechanisms of transformation of activating influences of nonspecific nuclei of the thalamus on the inputs of caudate nuclei cells into inhibitory effects were investigated in a series of microelectrode studies (7, 23, 39). In this connection it should be noted that against other afferent inputs whose stimulation caused within the caudate nucleus initial synaptic reactions of EPSP-IPSP succession, or IPSP only, in response to stimulation of CM thalamus, not only the succession of EPSP-IPSP, but also "pure" EPSP are registered (7, 38, 39). According to Buchwald et al. (7) this finding means that thalamic intralaminar nuclei are projected on the striatum in some other way and play a special role in caudate effects. The latter is also corroborated by the dependence of the degree of spindle intensity in the cortex, caused by the stimulation of the caudate nucleus on the level of "arousal" shown by the activation of nonspecific nuclei of the thalamus (6). The depression of single discharges of muscular flexor and extensor caused by caudate stimulation triggered this spindle activity in the sensorimotor cortex (21).

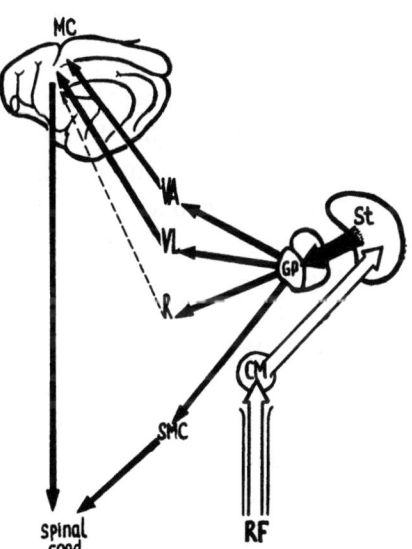

Fig. 11. The possible paths of realization of caudate inhibitory influences. The detailes are in text. MC, motor cortex; ST, striatum; SMC, subcortical motor centers; CM, centromedian nucleus of thalamus.

These data, as well as observations obtained by as, show the intensification of inhibitory effects from the caudate nucleus during activation of nonspecific thalamic nuclei. In behavioral aspects this phenomenon may be compared with the stopping of movements by a sudden superfluous flow of sensory information (e.g., a sudden noise). The limitation of motor activity by caudate nuclei up to complete stopping seems to be going on constantly and plays an especially important role in adapting at the right decision in a difficult situation (47).

The mechanism of "caudate stopping" is related to the hyperfunction of the caudate nucleus caused by disturbances of nigrostrial dopaminergic interrelations (4). Our data suggest that "caudate stopping" may be conditioned also by the intensification of activating thalamo-strial nonspecific influences. The analogy to activation of nonspecific thalamic nuclei may be its high frequency stimulation causing the desynchronization in cortical and subcortical structures (27, 30, 49), and as our investigations showed, in general motor activation including the activation of motor components of the IDR. But an increase in amplitude of high frequency stimulation applied to nonspecific thalamic structures (several times higher than threshold values) also causes movement stopping (20, 30). Based on this evidence, it is possible that the stopping is also connected with the involvement of the caudate inhibitory mechanism.

The present data show that the effect of the interaction of thalamic and strial influences depends on the localization of the stimulating electrodes in the CNH. This finding agrees with the results of microelectrode investigations which demonstrated that the type of neuronal response in the CNH with the interaction of thalamic and caudate impulses, depends on the properties of certain points in striatum (16). In addition to the dorsolateral part of the CNH, according to our data and of other investigators (3), "caudate stopping" is caused at lower values of the stimulating current amplitude. The ways of realization "caudate stopping" are being discussed (4, 34). It may be a consequence of the inhibition of effector neurons of the sensorimotor cortex going on through the canal: of the caudate nucleus, the globus pallidus and the nucleus VL thalamus (Fig. 11). The "extrapyramidal" way should also be considered, the inhibition of superior bulbar and mesencephalic motor centers (34), because "caudate stopping" may be observed in animals with bilateral lesions of the sensorimotor cortex (3). In this regard the activation of caudate nucleus neurons by the afferent impulses coming from nonspecific thalamic structures and, above all, the N. CM thalamus, is of much more importance than that discussed by Buchwald et al. (6). The caudate inhibitory influences in the motor spinal centers amplified by such activation, were realized not only through the cortico-spinal, but also through subcortico-spinal ways (Fig. 11).

## REFERENCES

1. ALBE-FESSARD D., OSWALDO-CRUZ C. and ROCHA-MIRANDA C. 1958. Convergences vers le noyan caude du signauxs d'origines corticale et heterosensorielle. J. Physiol. (Paris) 50: 105–108.
2. ARUSHANYAN E. B. and BELOSERTSEV Yu. A. 1974. On the mechanisms of the caudate nucleus participation in the control of behavior (in Russian). Zh. Vyssh. Nervn. Deyat. im. I. P. Pavlova 24: 55–63.
3. ARUSHANYAN E. B., BELOSERTSEV Yu. A. and TOLPYSHEV K. A. 1972. The analysis of the arrest reaction during the n. caudatus stimulation (in Russian). Zh. Vyssh. Nervn. Deyat. im. I. P. Pavlova 22: 361–367.
4. ARUSHANYAN E. B. and OTELLIN W. A. 1976. Khvostatoe yadro. Izdat. Nauka, Leningrad, 223 p.

5. BROOKS V. 1975. Role of Cerebellum and Basal Ganglia in initiation and control of movement. J. Canad. Sci. Neurol., (August) 265–277.
6. BUCHWALD N. A., WYERS E. J., LAUPRECHT C. W. and HEUSER G. 1961. The caudate spindle. IY. A behavioral index of caudate induced inhibition. EEG Clin. Neurophysiol. 13: 531–537.
7. BUCHWALD N. A., PRICE D. P., VERNON L. and HULL C. D. 1973. Caudate intracellular response to thalamic and cortical inputs. Exp. Neurol. 38: 311–323.
8. BUSER P., POUNDEROUX G. and MEREAUX J. 1974. Single unit recording in the caudate nucleus during session with eleborate movements in the awake monkey. Brain Res. 71: 337–345.
9. BUTHUSI S. M. 1971. Elektrofiziologichzeskoe issledovanie funktsii khvostatogo yadra. Metsniereba, Tbilisi.
10. CHERKES V. A. 1969. Basal ganglia (in Russian). In Rukovodstvo po fiziologii. Obschaya i chastnaya fiziologiya nervnoi sistemy. Izdat. Nauka, Leningrad, p. 387–422.
11. CHERKES V. A. 1978. Perednii mozg i elementy povedeniya. Naukova Dumka, Kiev, 174 p.
12. DE LONG M. R. 1972. Activity of basal ganglia neurons during movements. Brain Res. 40: 127–135.
13. DE LONG M. R. and STRICK P. L. 1974. Relation of basal ganglia, cerebellum and motor cortex units to ramp and ballistic limb movements. Brain Res. 71: 327–335.
14. DENNY-BROWN D. 1962. The basal ganglia and their relation to disordes of movement. Oxford Univ. Press, London, 144 p.
15. DIVAC I., MARKOWITSCH H. and PRITZEL M. 1978. Behavioral and anatomical consequences of small intrastriatal injections of kainic acid in the rat. Brain Res. 151: 523–532.
16. DRYAGIN Yu. M. 1977. Neuronal activity of the caudate nucleus head during formation positive and inhibitory conditioning reflexes (in Russian). Zh. Vyssh. Nervn. Deyat. im. I. P. Pavlova 27: 764–769.
17. ECCLES J. 1973. Review lecture: The cerebellum as computer: patterns in space and time. J. Physiol. (London) 229: 1–32.
18. GORBACHEVSKAJA A. I. 1973. The mutual projections of different parts of the caudate nucleus and some nuclei of thalamus (in Russian). In N. F. Suvorov (ed.), Striopallidarnaya sistema. Izdat. Nauka, Leningrad, p. 111–118.
19. GYBELS J. M., MEULDERS M., CALLENS M. and COLLE J. 1967. Disturbances of visio-motor integration in cats with small lesion of caudate nucleus. Arch. Inter. Physiol. Biochem. 75: 283–302.
20. HUNTER J. and JASPER H. H. 1949. Effects of thalamic stimulation in unanesthetized animals. EEG Clin. Neurophysiol. 1: 305–324.
21. HONGO T., KUBOTA K. and SHIMAZU H. 1963. EEG spindle and depression of gamma motor activity. J. Neurophysiol. 26: 568–580.
22. JUNG R. and HASSLER R. 1960. The extrapyramidal system. In Handbook of physiology. 2. Neurophysiology 863–927.
23. KAJI S., NAITO H. and SATO S. 1971. Responses of single unit in the caudate nucleus to thalamic stimulation. Exp. Neurol. 39: 447–459.
24. KITSIKIS A. and ROUGEL A. 1968. Effect of caudate stimulation on conditioned motor behavior in monkeys. Physiol. Behav. 6: 609–611.

25. KONORSKI J. 1970. Integrativnaya deyatelnost' mozga. Izdat. Mir, Leningrad, 412 p.
26. KORNHÜBER H. H. 1971. Motor function of cerebellum and basal ganglia. Kybernetik 8: 157–162.
27. KRATIN Yu. G. ANDREEVA V. N. and IRGASHEV M. S. 1975. The extintion of EEG reactions to direct electrical stimulation of the brain in normal awaking cats (in Russian), Fiziol. Zh. SSSR im. I. M. Sechenova, 61: 321–330.
28. LIM R., LIU Ch. and MOFFIT R. 1960. A stereotaxic atlas of the dogs brain. Charles C. Thomas, Springfield.
29. LINEBERRY C. D. and SIEGEL J. 1971. EEG synchronization, behavioral inhibition and mesencephalic unit effects produced by stimulation of orbital cortex, basal forebrain and caudate nucleus. Brain Res. 34: 143–161.
30. LISHAK K. and ANDYAN L. 1974. The dependence between the electrical brain activity and cessation of the movement, caused by thalamic stimulation. In M. N. Livanov (ed.), Osnovnye problemy elektrofiziologii golovnogo mozga. Izdat. Nauka, Leningrad p. 189–196.
31. LOBANOVA L. W. and SMIRNOV S. I. 1973. The convergention visceral, somatic and auditory impulses in the nucleus caudatus in cats. In N. F. Suvorov (ed.), Striopallidarnaya sistema. Izdat. Nauka, Leningrad, p. 60–68.
32. MARTIN J. P. 1967. The basal ganglia and posture. Pitmen Med. Publ. London.
33. MASSION J. and SMITH A. 1974. Ventrolateral thalamic neurons related to posture during a modified placing reaction. Brain Res. 71: 358–369.
34. NEWTON R. A. and PRICE D. 1975. Modulation of cortical and pyramidal tract induced motor responses by electrical stimulation of the basal ganglia. Brain Res. 85: 413–422.
35. NIKI H., SAKAI M. and KUBOTA K. 1972. Delayed alternations performance and unit acivity of the caudate head and medial orbitofrontal gyrus in the monkey. Brain Res. 38: 343–353.
36. PETROPAVLOVSKI V. P. 1934. The method of conditional movement reflexes (in Russian). Fiziol. Zh. SSSR im. I. M. Sechenova 17: 217.
37. PHILLIPS C. G. and PORTER R. 1978. Corticospinal neurons. Their role in movement. Acad. Press, London, 450 p.
38. PURPURA D. P. 1975. Physiological organization of the basal ganglia. In M. D. Jhar (ed.), The basal ganglia, Raven Press, New York 91–114.
39. PURPURA D. and MALLIANI A. 1967. Intracellular studies of the corpus striatum. Synaptic potentials and discharge characteristics of caudate neurons activated by thalamic stimulation. Brain Res. 6: 325–340.
40. ROSVOLD H. E., MISHKIN M. and SZWARCBART M. 1958. Effects of subcortical lesions in visual discrimination and single-alternations performance. J. Comp. Physiol. Psychol. 51: 437–443.
41. ROYCE G. J. 1978. Autoradiographic evidence for a discontinuous projection on the caudate nucleus from the centromedian nuclei in the cat. Brain Res. 146: 145–150.
42. SHAPOVALOVA K. B. 1978. Instrumental defensive reaction during the stimulation of the head of the caudate nucleus in dogs. Neuroscience 3: 143–146.
43. SHAPOVALOVA K. B. 1978. Rol korkovykh i podkorkovykh struktur v sensomotornoi' integratsii. Izdat. Nauka, Leningrad, 181 p.
44. SHAPOVALOVA K. B. and BAZHENOVA S. I. 1975. The comparison of the

effect of electrical stimulation of the nucleus caudatus head with that one of somatosensory cortex on the realization of the instrumental defensive reaction in dogs (in Russian). Zh. Vyssh. Nervn. Deyat. im. I. P. Pavlova, 25: 717–725.
45. SHAPOVALOVA K. B. and BAZHENOVA S. I. 1979. The effect of low frequency stimulation of the nucleus caudatus head on the realization of instrumental defensive reaction in dogs (in Russian). Zh. Vyssh. Nervn. Deyat. im. I. P. Pavlova. 29: 549–556.
46. SOŁTYSIK S. 1960. The influence of nucleus caudatus lesion on conditional movement reflex (in Russian). In E. A. Asratyan (ed.), Tsentral'nye i perifericheskie mekhanizmy dvigatel'noĭ deyatel'nosti zhivotnykh. Izdat. Akad. Nauk SSSR, Moscow, p. 300–309.
47. SUVOROV N. Ph., SHAPOVALOVA K. B. and SHUSTOV V. N. 1977. The role of the head of nucleus caudatus in sensorimotor control of different behavior patterns (in Russian). Zh. Vyssh. Nervn. Deyat. im. I. P. Pavlova 27: 747–754.
48. SUVOROV N. Ph. and SUVOROV V. V. 1975. Kholinoreaktivnaya sistema bazalnykh ganglisv i uslovnoreflektornaya deyatelnost. Izdat. Nauka, Leningrad, 95 p.
49. TISSOT R. and MONNIER M. 1959. Dualite du systeme thalamique de projection diffuse. EEG Clin. Neurophysiol. 11: 675–686.
50. TOLKUNOV B. Ph. 1978. Striatum i sensornaya spetsializatsiya neironalnoĭ seti. Izdat. Nauka, Leningrad, 175 p.
51. UNO M., OZAWA N. and YOSHIDA M. 1978. The mode of pallidothalamic transmission investigated with intracellular recording from cat thalamus. Exp. Brain Res. 33: 493–507.
52. WILLBURN M. W. and KESNER R. P. 1974. Effects of caudate nucleus stimulation upon initiation and performance of a complex motor task. Exp. Neurol. 45: 61–71.
53. WINOKUR G. and MILLS J. A. 1969. Effects of caudate lesions on avoidance in rats. J. Comp. Physiol. Psychol. 68: 552–557.

K. B. SHAPOVALOVA and S. I. BAZHENOVA, Pavlov Institute of Physiology, Academy of Sciences of the USSR, Makarova 6, Leningrad V-164, USSR.

Lecture delivered at the Warsaw Colloquium on Instrumental
Conditioning and Brain Research
May 1979

# RELATIONSHIPS OF PRECENTRAL, PREMOTOR AND PREFRONTAL CORTEX TO THE MEDIODORSAL AND INTRALAMINAR NUCLEI OF THE MONKEY THALAMUS

Konrad AKERT and Kurt HARTMANN-VON MONAKOW

Brain Research Institute, University of Zürich
Zürich, Switzerland

*Abstract.* The projections of precentral, premotor and prefrontal cortical areas to the mediodorsal and intralaminar thalamic nuclei have been reexamined by means of the anterograde labelling technique in 21 macaque monkeys. Three principles which have been already mentioned in previous reports could be confirmed within limits: (i) The dichotomy of agranular and granular frontal cortices related to lateral and medial thalamic nuclear division respectively, (ii) the reciprocity of fronto-thalamic connections and (iii) the thalamic matrix consisting of longitudinally arranged cell columns (Kievit and Kuypers 1977) from which the cortical connections originate and where they end. One main exception to these principles involves the intralaminar and paralaminar nuclear continuum which seems to disrupt the nearly parallel system of thalamically relayed information channels by forming complex patterns of converging and diverging connectivities with cortical, subcortical and possibly intrathalamic regions. Another exception concerns the partial bilaterality of fronto-thalamic projections which is not reciprocated as far as we know by the thalamo-cortical counterparts. Major emphasis is laid upon the fact that agranular frontal cortex, i.e., precentral and premotor (including supplementary motor) areas not only project to the ventrolateral nuclear complex and to intralaminar nuclei but in addition to MD (paralaminar zone) whose efferent connections are known to be directed mainly towards the frontal eyefield and the caudate nucleus and to a lesser degree to the agranular frontal cortex.

INTRODUCTION

The history of experimental investigations on thalamo-frontal relationships opens with the degeneration studies of Monakow (44) and his collaborators Rutishauser (66), Fukuda (17) and Minkowski (43). These have been the basis of subsequent pioneering studies which have appeared in various original articles and reviews (15, 25, 33, 57, 65, 74, 77, 78).

*Thalamo-cortical connections.* Le Gros Clark and Boggon (9), Walker (77), Mettler (40), Rose and Woolsey (64) Pribram et al. (59), Akert (2), Batuev et al. (6) and Adrianov (1) emphasized the topological order and correspondence between lateral and medial nuclear groups of the thalamus with agranular and granular areas of the frontal cortex respectively. Narkiewicz and Brutkowski (46), Tanaka (72), Tobias (74) as well as Leichnetz and Astruc (35) presented evidence that the dorsomedial prefrontal cortex — considered to be athalamic in the monkey by Walker (76) and Akert (2) — maintains reciprocal connections with the mediodorsal nucleus (MD). Roberts and Akert (63) underlined the sparing of the center median (CM) from retrograde degeneration after systematic segmental ablations of the frontal lobes including the opercular and insular areas, while Murray (45) presented convincing evidence that intralaminar nuclei (not including CM) project upon extended areas of the fronto-parietal cortex.

*Cortico-thalamic projections.* It is generally accepted that cortico-thalamic relationships are organized in parallel to thalamo-cortical projections (see 26, 27, 62, 71, 72). Important exceptions to this rule were observed by Künzle (28, 30) in the course of a systematic investigation of the efferent connections of precentral, premotor and prefrontal cortex. Not only could he confirm and extend earlier observations by Kuypers and Pandya (32), Mehler (38), DeVito (13) and Petras (55) that area 4 is connected bilaterally to CM, a nucleus which seemed to lack corticopetal relations, but he also established corticofugal projections from area 4 and 6 to MD which is focussing its afferents upon prefrontal and not on premotor and precentral areas.

Significant impulses for the continuing research on the organization of relationships between thalamus and frontal cortex arose from the advent of modern marker techniques (see 11) and it gives the present authors the special pleasure to mention in this context the basic contributions on anterograde and retrograde axoplasmic transport by

Lubińska (36) who gave us the privilege of her presence at this memorable symposium. Of special interest is the work of Kievit and Kuypers (24) carried out by means of retrograde labelling with the enzyme horseradish peroxidase; they confirmed and extended the principle of laminar organization already discussed by one of us (K.A.). at the Pennstate Symposium (79) some 15 years ago.

It is the main purpose of this communication (1) to re-examine the projections of the frontal lobe to the mediodorsal thalamic nucleus (MD) and the adjacent intralaminar thalamic nuclei, and (2) to give a short account on their reciprocity. Special attention will be given to the connectivity of the paralaminar region of MD.

## MATERIAL AND METHODS

Twenty one adult monkeys (*Macaca fascicularis*) were given local injections of the following amino acids (Radiochemical Centre, Amersham, England) into precentral, premotor and prefrontal cortex: L-[5 $^3$H] proline (spec. activity 18 Ci/mmol) and L-[3,4 $^3$H] leucine (spec.

TABLE I

| Case number | Injection site | Radioactive label | | | Survival time (days) |
|---|---|---|---|---|---|
| | | Label | Quantity $\mu$l | Concentration $\mu$Ci/$\mu$l | |
| 72–448 | Area 4 | P | 3.0 | 13 | 4 |
| 78–335 R | | L, P (1:1) | 0.6 | 80 | 5 |
| 78–526 | | L, P (1:1) | 0.6 | 40 | 2 |
| 72–450 | | P | 5.0 | 10 | 4 |
| 72–449 | | P | 5.2 | 13 | 4 |
| 72–451 | | P | 5.2 | 10 | 8 |
| 74–163 | Area 6 | A, L, P (1:2:2) | 1.5 | 50 | 5 |
| 74–678 | | L, P (1:4) | 1.6 | 50 | 4 |
| 78–882 | | L, P (1:1) | 0.6 | 50 | 2 |
| 74–162 | | A, L, P (1:2:2) | 1.5 | 50 | 5 |
| 74–626 | | L, P (1:4) | 2.0 | 50 | 3 |
| 72–770 | | P | 2.0 | 35 | 2 |
| 75–212 | | L, P (1:1) | 1.1 | 45 | 5 |
| 73–228 | Area 8, 9, 10, 13 | P | 2.2 | 20 | 3 |
| 73–234 | | P | 2.0 | 20 | 3 |
| 78–354 B | | L, P (1:1) | 0.8 | 50 | 2 |
| 74–885 | | A, L, P (1:1:1, 3) | 1.0 | 40 | 4 |
| 75–496 | | A, L, P (1:1, 5:1, 5) | 1.1 | 50 | 4 |
| 75–497 | | A, L, P (1:1, 5:1, 5) | 1.1 | 50 | 4 |
| 78–354 A | | L, P (1:1) | 0.6 | 50 | 2 |
| 78–331 | | L, P (1:1) | 0.6 | 50 | 2 |

activity 54 Ci/mmol) or L-[4,5 ³H] leucine (spec. activity 51 Ci/mmol). Table I gives the localization and quantitative characteristics of the injected marker and survival times. In one animal (no. 78-354B) the injection of radioactive amino acids was combined with a 25% solution of horseradish peroxidase (HRP).

Frozen or paraffin embedded sections (8-30 µm thickness) were prepared from the brain at regular intervals ranging between 100-400 µm. These sections were dipped in Kodak NTB 2 nuclear emulsion diluted 1:1 with water. The exposure time was 8 wk at 4°C. in total darkness. After development and fixation the sections were counterstained with cresylviolet and examined with light and darkfield microscopy. Camera lucida drawings and microphotographs were prepared from selected sections at appropriate levels. The thalamic nomenclature of Olszewski (52) was adopted. Brodmann's (7) cytoarchitectural definitions of frontal lobe cortex have been used. A comparison of different parcellations of frontal cortex was given by Akert (2).

RESULTS

*Precentral motor cortex (area 4 of Brodmann).* Figure 1A shows the reconstruction of cortical injection sites arranged according to the classical map of precentral gyrus localization patterns by Woolsey et al. (80). The thalamic projection sites of these loci have been recently dealt with by Künzle (28) and shall not be reported hare in detail. Suffice it to mention briefly the terminations as seen in Fig. 3 where the arm area was injected with radioactive compounds. Five major and several minor thalamic projection sites can be observed. The major ones are (1) the VL, (2) R, (3) CL-PCn, (4) CM and (5) MD. Whilst the VL projection is somatotopically organized with relatively sharply distinct foci (3, 71, 76), the projections into the intralaminar and MD nuclei lack such discrimination, in fact these projections seem to fan out from CM and radiate into the various nuclear districts without clear evidence of somatotopic organization. It should be mentioned that the MD and CM projections from the facial area are situated ventrally and are by far the strongest inspite the fact that less label had been injected in this case (72-448) than in the cases representing areas of the body (Figs. 3 and 4). The typical projection site of the precentral gyrus in MD is located in and around Olszewski's (52) pars multiformis, a segment of MD which is often designated "paralaminaris". Note that both the intralaminar and paralaminar neurons receive substantial projections from area 4 (Fig. 4) and these projections extend longitudinally into the caudal paralaminar region called by Olszewski (52): MDde (Fig. 7).

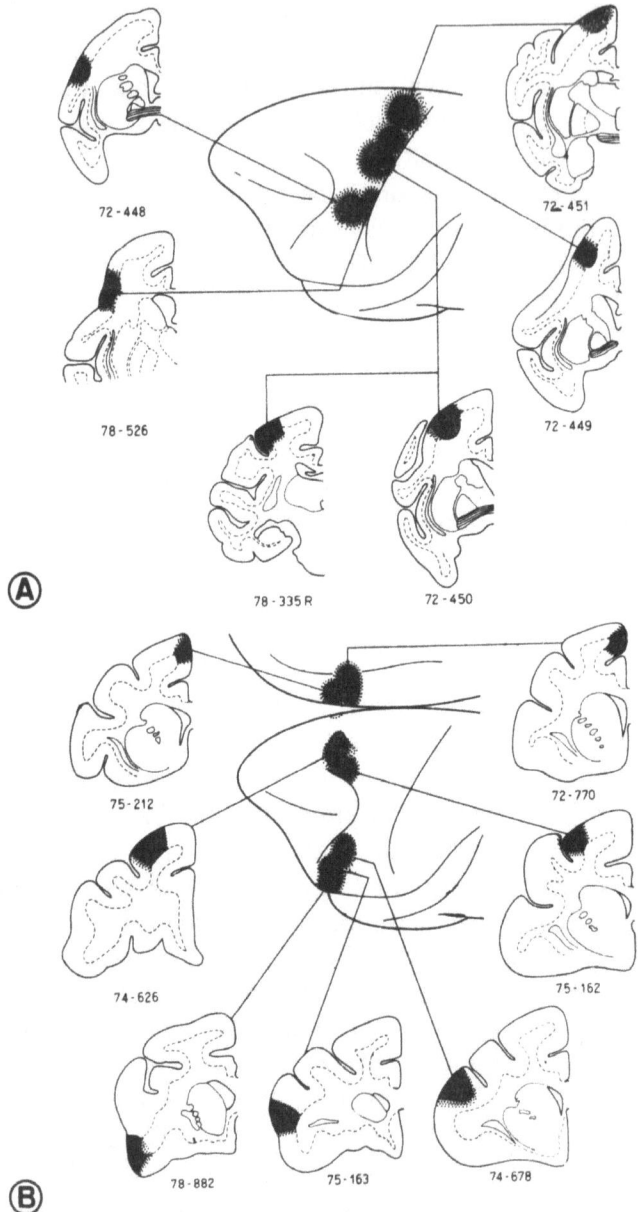

Fig. 1. Site and extent of injection fields, reconstructed according to the criteria applied previously (28). Cells lying within the filled circle (blackened area) have taken up, incorporated and transported radioactive amino acids into their terminal ramification, while cells within the stippled area may have contributed to a faint labelling of the pathways. A: Injection sites in the precentral gyrus corresponding to precentral motor cortex representation patterns of Woolsey et al. (80). B: Site and extent of the injection fields in the medial, dorsal and ventral parts of premotor cortex (Brodmann's area 6) including supplementary motor cortex.

Contralateral mirror foci were found in CM in all cases (28), in MDmf only after injections into the facial representation area. The labelling was clearly less intensive on the contralateral side.

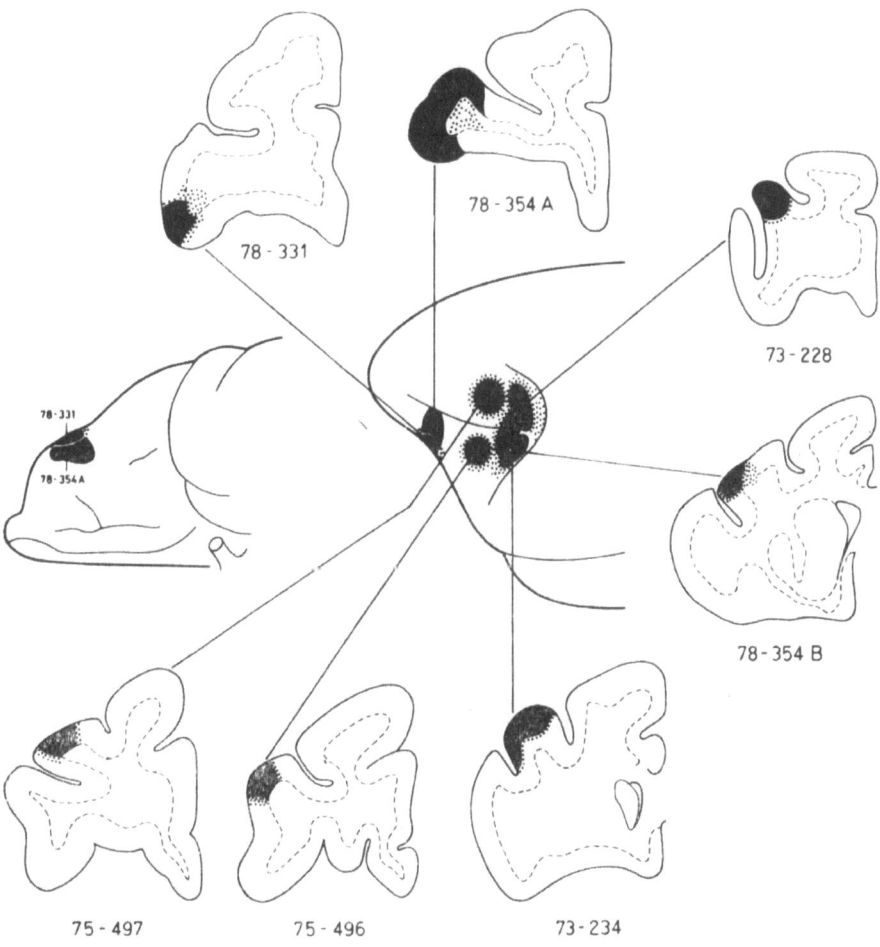

Fig. 2. Site and extent of injection fields in the prefrontal cortex (granular frontal cortex, Brodmann's areas 8, 9, 10 and 11).

*Premotor cortex (area 6 of Brodmann)*. Figure 1B shows the injection sites in the premotor areas: medial, dorsal (lateral) and ventral (lateral). As a general rule the premotor projections occupy similar thalamic nuclei as the precentral projections. However, there is a small but noticeable shifting overlap between precentral and premotor projections, particularly between areas 4 and 6m (= supplementary motor cortex) within the intralaminar nuclei as well as in the paralaminar region of MD with area 6 moving medially. The premotor projections to MD seem to be more marked than the precentral. They occupy not only the

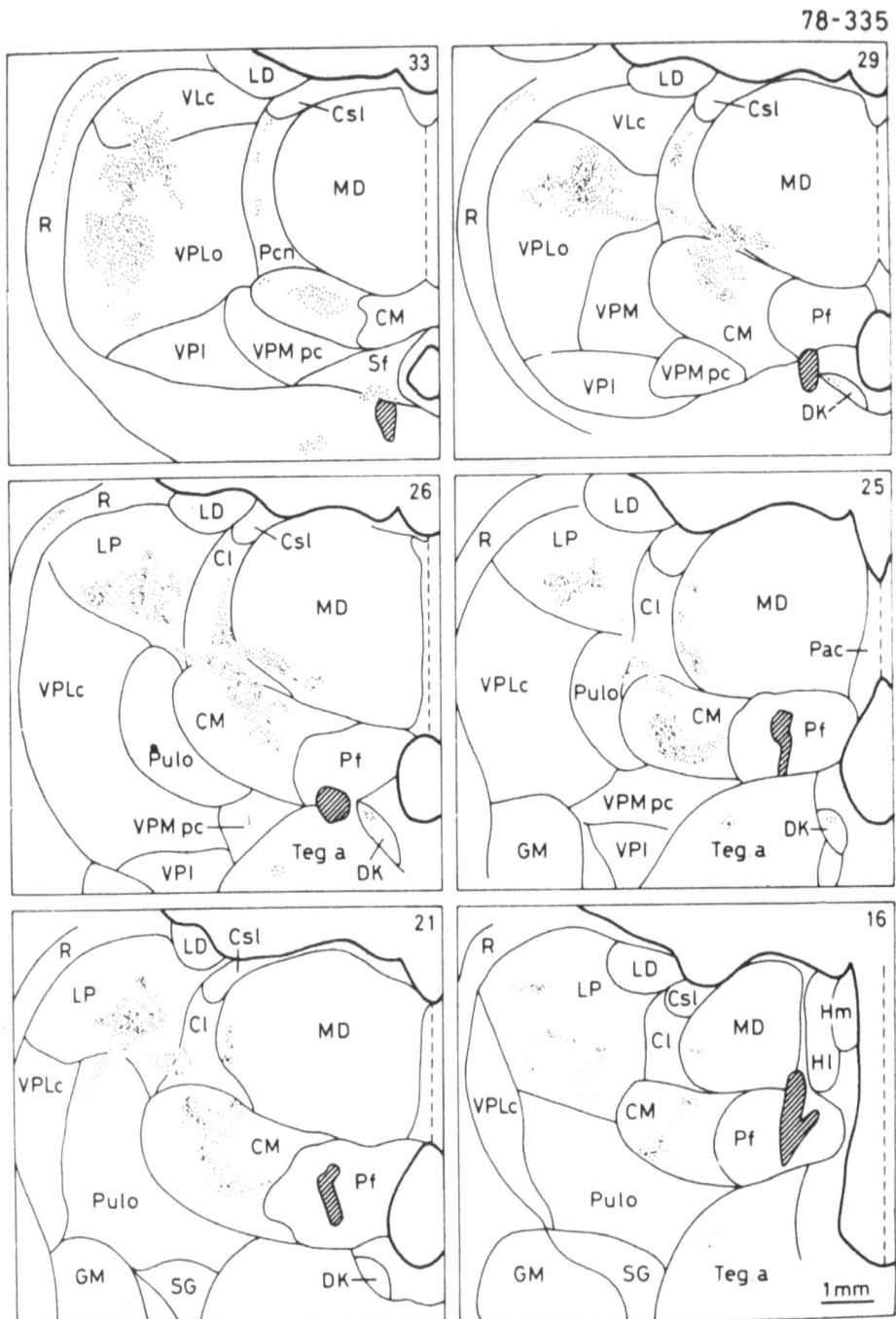

Fig. 3. Anterograde labelling of thalamic projection sites from precentral motor cortex, arm area (case 78–335R). Note that the projections into MD and intralaminar nuclei (Pcn, Cl and CM) are rather patchy. Label in paralaminar MD is immediately adjacent to CM projection, and partly continuous with Pcn and Cl.

lateral but even more intensively the ventral segments of MD (Fig. 5). Area 6v seems to have the strongest affinity to ventral MD and extends its projections contralaterally and in addition terminates in Cif bilaterally. As a general rule, area 6d projects more dorsally and 6v more ventrally into a lamina of MD which lies medial and more caudal to that of area 4 (Fig. 7). One should also note that CM in contrast to MD receives considerably less input from premotor than from precentral motor cortex. The only substantial projection to CM originates from area 6v. On the other hand, as already mentioned by Künzle (30) the superior district of CL (Csl) and Pf receive afferents from area 6 but not from area 4 (Fig. 7).

*Prefrontal cortex.* The injection sites in the prefrontal areas are shown in Fig. 2. Only the posterior lateral and lateral orbital regions have been covered, i.e., cortex corresponding to areas 8, 9, 10 and partly 11 of Brodmann. Yet, the thalamic projection sites as shown in Fig. 6 seem to cover a fairly large portion of MD. The ventrolateral nuclei are spared and the intralaminar nuclei including CM only minimally labelled. Area 8 projections lie laterally (including pars multiformis) and area 11 projections medially and ventrally in MD with areas 9 and 10 occupying intermediate regions. One gains the impression that there is relatively little overlap between the prefrontal thalamic projections within MD, but rather distinct topological arrangements (Fig. 6).

There is, however, some overlap between the thalamic projections originating in the precentral, the supplementary motor (area 6m) and the frontal eyefield (area 8) within the paralaminar region. An additional convergence is noted in the ventromedial MD district between cortical projections from the ventral area 6 and those from area 10/11.

Areas 10 and 11 differ from the other prefrontal regions by the bilaterality of MD projections which are particularly prominent in area 11. The contralateral terminations are limited to MD mc; area 11 projections spread into Cif bilaterally. In contrast, these two areas (10 and 11) fail to project into Pf, while areas 8 and 9 have substantial connections with this nucleus (Fig. 8).

DISCUSSION

*Comparison between cortico-thalamic and thalamo-cortical connections*

It has been generally assumed that the connections between frontal cortex and thalamus are reciprocal as pointed out by Nauta (48) who summarized his preliminary data on prefronto-thalamic degenerations as follows: "In each case the degeneration in the dorsomedial nucleus appeared to correspond in distribution to that subdivision of the nucleus which is known to project to the general area of the cortical lesion." He continued to say that "tentatively, these findings could be interpreted as evidence of fronto-thalamic projections with an organizational mosaic duplicating in detail that of the reciprocating thalamo-prefrontal connection." He cautioned however, against pitfalls of interpretations by saying "it remains to be investigated, however, to what extent the apparent prefronto-dorsomedial projection could be merely mirrored by retrograde axon changes accompanying the rapid cell degeneration in the dorsomedial nucleus following prefrontal lesion."

The modern methods using axoplasmic transport as vehicle for tracing nerve fibers in the central nervous system seem to provide sufficient safety in distinguishing retrograde from anterograde labelling, although recent experiments by Condé and Condé (10) pointed to the fact that misinterpretations of perikaryal vs. axonal labelling are not completely ruled out at the light microscopic level, particularly in instances where the HRP granules are situated in axo-somatic and absent in axo-dendritic terminals.

Kievit and Kuypers (24) made a systematic investigation on the thalamo-frontal connections and prepared maps from frontal and horizontal serial sections. They divided the frontal cortex into seven segments which seem to correspond to some degree to the classical cytoarchitectural divisions which we have adopted. We shall now consider these data in detail and compare them with our own observations on the cortico-thalamic projections. In order to perform this comparison it was necessary to transpose our findings with obvious risks of misinterpretations into the maps used by Kievit and Kuypers (24).

*Precentral motor cortex* (segments 7 and 6 according to Kievit and Kuypers) receives afferents from the VL complex and lateral VPM. Few afferents were recorded from the intralaminar nuclei (including CM) and none from MD.

From our material (see also 28) it is quite clear that the precentral motor cortex sends efferents to the VL complex and that these are most likely reciprocal to the thalamo-cortical connections reported by Kievit and Kuypers (24). In addition, it projects rather substantially to intralaminar nuclei (especially to CM) and MDmf, and these connections seem to have only minor relations in the opposite direction.

*Premotor cortex* (segment 5 according to Kievit and Kuypers 24) receives afferents mainly from VA and medial VL (area "X" of Olszewski), a fact which has been established by Akert (2). Further afferents are derived from paralaminar MD. These three areas of termination form a narrow longitudinal strip which is bisected by PCn and CL; the latter are nearly free of retrogradely labelled neurons.

Our own findings on cortico-thalamic connections of this segment are consistent with the reciprocity principle with the notable exception of ipsilateral intralaminar terminations of area 6 efferents in CL, CLs and Pf and contralateral projections to MD (Fig. 5).

It should be emphasized that area 6 projections to MD are considerably more intense than those which emanate from area 4.

*The prefrontal cortex* is composed of several segments.

1. Segment 4 according to Kievit and Kuypers (24) corresponds approximately to area 8 and area 9 and receives afferents from VA, from a narrow paralaminar border zone of area "X" as well as from lateral and intermediate MD. Again, it is quite clear that this cortical area receives thalamic afferents from a nuclear region disrupted by intralaminar nuclei which according to these authors hardly participate in thalamo-prefrontal projections.

Our own findings on cortico-thalamic connections of area 8 to VA and MD parellel the observations of Kievit and Kuypers (24) in the opposite direction. The reciprocity of connections was tested and confirmed in a special case where both retrograde and anterograde labels have been used (Fig. 9). It turned out that the thalamic cells containing HRP granules in the perikarya were located within the area labelled with maximal intensity by silver grains. It seems from our data in Figs. 6 and 8 that while CL, Pcn and CM are free of cortico-thalamic terminals, Pf receives a projection from area 8 (see also 31).

From area 9 we record efferents into MD and VA, thus sparing the intralaminar nuclei except for Pf (30). Thus, the reciprocity with respect to Kievit and Kuypers' (24) data is positive with the exception of the intralaminar nuclei and the fact that cortico-thalamic connections seem to be more prominent than the thalamo-cortical ones.

2. Segment 3 according to Kievit and Kuypers (24) represents parts of areas 10 and 11 and receives afferents from VA and MD. Again no involvement of the intralaminar nuclei was observed.

Our own though limited data on the projections of area 10 and 11 are more or less consistent with Kievit and Kuypers' MD projections since the pattern of cortico-thalamic fibers follow the same topological order. Nevertheless, cortico-thalamic projections emanating from area 10 and especially from area 11 are situated more medially and extend partially across the midline (18).

3. Unfortunately, we fail to have data for comparison with Kievit and Kuypers' (24) segment 1 and 2.

In conclusion, the comparison between thalamo-cortical and cortico-thalamic connections reveals the following deviations from the reciprocity principle:

1. Projections from cortex to intra- and paralaminar nuclei are substantial, particularly as concerns areas 4, 6 and 8 of Brodmann. Projections are relaively minor or absent in the opposite direction.

2. Projections from convexity and orbital regions of the frontal lobe to specific nuclei of the intralaminar and midline group tend to be bilateral. The contralateral mirror foci concern the medioventral MD and midline nuclei including CM/Pf complex. No bilateral thalamo-frontal connections are known up until now.

3. Within limits a "matrix of longitudinally arranged cell columns" as suggested by Kievit and Kuypers (24) for the thalamo-cortical system may be present also in the cortico-thalamic projections.

*The convergence of cortical and subcortical afferents to the intra- and paralaminar thalamic region*

The reciprocity of connections between ventral lateral tier of thalamic nuclei to agranular frontal cortex (areas 4 and 6), and between dorsal medial tier to granular frontal cortex (areas 8–12) has been generally accepted. Between these two one can identify an intermediate region, which is characterized by exceptionally complex convergence/divergence patterns of connectivity. This intermediate region concerns the intralaminar (CL, Pcn, CM and Pf) and paralaminar (MDmf and MDdc) nuclei whose exact cytoarchitectural boundaries are still controversial. A major characteristic of this nuclear "continuum" is the fact that it receives not only converging cortical inputs but also a wealth of ascending and possibly intrathalamic afferents (60). We shall discuss these problems briefly in the following sections.

*Cortical afferents* reported in this series of exeriments converge onto intra- and paralaminar nuclei mainly from areas 4, 6 and 8. CL receives afferents from area 4 (13, 28, 55, 67), from area 6 (14, 30) and from area 8 (5, 31); these observations were confirmed and extended by the present data. In the monkey, CM is the main target of precentral cortical connections (in agreement with earlier findings by Mehler (38), Kuypers and Pandya (32), DeVito (13) and Künzle (28). As one proceeds with injection sites from precentral to prefrontal areas the terminal labelling diminishes in CM progressively and increases in Pf. Paralaminar MD receives substantial inputs from areas 8 (5, 31) and area 6 (30, 67), and less intensive ones from area 4.

The precentral connections to MD are now established by Marchi (27), and Nauta methods (13) as well as by the axoplasmatic flow technique (this paper, 4, 28). Area 4 projection to MD have also been reported in the chimpanzee (56) and in the cat (50, 61). The anterograde marker technique seems to exclude the possible confusion of precentral with prefrontal fibers which could have been damaged unintentionally in the course of degeneration experiments. The only issue that may be raised in relation to this somewhat unexpected connection is that of precentral cortical terminations on CM dendrites which may extend dorsally and medially into paralaminar MD. However, the results of Golgi studies by Hazlett et al. (19) in the monkey seem to preclude this possibility.

Additionally, convergent afferents from lateral orbital cortex to intra- and paralaminar nuclei have been reported in studies based on anterograde degeneration techniques (16, 35).

Rinvik (61) using anterograde degeneration with Nauta-Gygax techniques in cats stated with respect to MD connection: "that the cruciate sulcus marks differences in the cortico-thalamic projection is also evident from the fact that a lesion anterior to this sulcus results in degeneration in MD, while this nucleus is completely free of degeneration following lesions posterior to the sulcus". This situation is rather consistent with that found in the present investigation on fronto-thalamic connections in the monkey.

*Subcortical afferents.* Intralaminar afferents are known to originate in the spinal cord. Déjérine (12) stated: "Le centre médian de Luys représente donc une sorte de relais ganglionnaire entre le neurone bulbo-thalamique du Ruban de Reil qui s'y termine et les neurones thalamo-cortical et cortico-thalamique qui y prennent leur origine ou y arrivent". This pathway has been experimentally investigated by several authors and led to a modification of Déjérine's conclusion. The termination of spinal afferents in the intralaminar region was confirmed, but it seems to involve primarily CL (and not CM) and paralaminar MD (29, 39, 49). Itoh and Mizuno (21) have identified CL neurons forming a link between a spino-thalamic and a thalamo-cortical pathway terminating in area 4 of the cat.

Other important sources of afferents to the intra- and paralaminar thalamus are the medial reticular formation of the rostral midbrain tegmentum (49, cat), the cerebellum (54), the superior colliculus (51, 73, cat) and the pretectal area (20, cat). Carpenter and coworkers (8) described a nigro-thalamic pathway in the monkey, part of which seems to end in the intralaminar but not in the paralaminar region. This point may have to be re-examined, however, since in all the other instances there is a clearcut convergence to both intra- and paralaminar thalamic neurons.

*Diverging outputs from intra- and paralaminar thalamic cell groups*

The most significant efferent connection of the intralaminar nuclei is directed towards the striatum (see 58) while that of the paralaminar MD focuses primarily upon the frontal cortical area 8 (see 69). These data based on retrograde degeneration studies have been repeatedly confirmed and they received added support by recent observations made with the aid of the retrograde tracer technique. Earlier data on the cortical projections of the intralaminar nuclei (45, 53, 76) have also been substantiated in the same way. For instance, Strick (70) observed labelling of CL cells following HRP injections into the precentral motor arm area in the monkey, and Itoh and Mizuno (21) as well as Macchi et al. (37) reported analogous findings in the cat.

The fact that paralaminar MD not only projects to the frontal eye field, but to other cortical areas and to the striatum is less well known. In the older literature one can find scattered information that MD, especially its lateral portion, may undergo retrograde degeneration after ablation of the precentral gyrus (76). Along the same line, Khalifeh et al. (23) reported anterograde degeneration into the presigmoid gyrus after circumscribed MD lesions in the cat. However, in our own collection of cases with precentral lesions in the monkey (63, 80) we were unable to ascertain cell changes in MD, and Kievit and Kuypers (24) who used the retrograde labelling technique mentioned that after HRP injections into the precentral gyrus of the monkey "a few sporadic labelled neurons also occurred caudally in the ventral paralaminar MD". Thus, it seems that this issue of MD projections upon the precentral motor cortex (area 4) needs further examination.

The projections of MD to premotor area and striatum seem to be on safer grounds. Itoh and Mizuno (21) and Vedovato (75) have obtained labelled cells in paralaminar MD after injecting HRP into area 6 of the cat. Sato et al. (68) found numerous labelled cells not only in CL and CM, but also in the lateral segment of MD after HRP injections into the caudate nucleus of the cat. Similar observations were made by Nauta jr. et al. (47) and by Jones and Leavitt (22) in the rat.

It seems reasonable to conclude that the diverging fiber connections of paralaminar neurons may consist both of primary axons and collateral branches of thalamo-cortical and thalamo-striatal connections respectively. In this context, the MD connection with the caudate nucleus would be of considerable interest if confirmed in the monkey, because the cortical projection site of MD in this species, the frontal eye field, is also known to send information to the caudate (31).

In summary then, the intra-paralaminar continuum not only shares complex patterns of converging afferents but also represents the origin of a spectrum of diverging cortical and subcortical connectivities of which the dichotomy towards specific areas of the frontal lobe and the corpus striatum seems most prominent.

This work has been supported by grants from the Swiss National Science Foundation No. 3.636.75 and 3.611.75, the Dr. Eric Slack-Gyr Stiftung in Zürich. Dr. Heinz Künzle generously put his experimental neuroanatomical collection of monkey brains with cortical injections of radioactive amino acids at our disposal. The competent assistance by R. Emch, A. Fäh, H. Hauser, W. Lang, D. Savini and E. Schneider is gratefully acknowledged.

## REFERENCES

1. ADRIANOV, O. 1977. The problem of organization of thalamo-cortical connections. J. Hirnforsch. 18: 191–221.
2. AKERT, K. 1964. Comparative anatomy of frontal cortex and thalamofrontal connections. In J. M. Warren and K. Akert (ed.), The frontal granular cortex and behavior. McGraw-Hill, New York, p. 372–396.
3. AKERT, K. and WOOLSEY, C. N. 1954. Ventrolateral nuclear group of the thalamus and its projection upon the precentral cortex of the monkey. Fed. Proc. 13: 1–2.
4. AKERT, K., HARTMANN-VON MONAKOW, K. and KÜNZLE, H. 1979. Projection of precentral motor cortex upon nucleus medialis dorsalis thalami in the monkey. Neurosci. Lett. 11 : 103-106.
5. ASTRUC, J. 1971. Corticofugal connections of area 8 (frontal eye field) in Macaca mulatta. Brain Res. 33 : 241–256.
6. BATUEV, A. S., MALYUKOVA, I. V. and HOHRYAKOVA, I. M. 1974. Structural and functional bases for frontal lobe participation in the organization of complex behavior in cats. Brain Behav. Evol. 10 : 290–306.
7. BRODMANN, K. 1905. Beiträge zur histologischen Lokalisation der Grosshirnrinde. J. Psychol. Neurol. 4 : 177–226.
8. CARPENTER, M. B., NAKANO, K. and KIM, R. 1976. Nigrothalamic projections in the monkey demonstrated by autoradiographic technics. J. Comp. Neurol. 165 : 401-416.
9. CLARK, W. E. LE GROS and BOGGON, R. H. 1935. The thalamic connections of the parietal and frontal lobes of the brain in the monkey. Philos. Trans. R. Soc. Lond. B. Biol. Sci. 224 : 313–359.
10. CONDE, F. and CONDE, H. 1979. Observations on the orthograde and retrograde transport of horseradish peroxidase in the cat. J. Hirnforsch. 20 : 35–46.
11. COWAN, E. M. and CUENOD, M. (ed.). 1975. The use of axonal transport for studies of neuronal connectivity. Elsevier, Amsterdam, 365 p.
12. DEJERINE, J. 1901. Structure et connexions de la couche optique. In Anatomie des Centres Nerveux. Vol. 2. J. Rueff, Paris, p. 344–391.
13. DeVITO, J. L. 1969. Projections from the cerebral cortex to intralaminar nuclei in monkey. J. Comp. Neurol. 136 : 193–202.
14. DeVITO, J. L. and SMITH, O. A. 1959. Projections from the mesial frontal

cortex (supplementary motor area) to the cerebral hemispheres and brain stem of the *Macaca mulatta*. J. Comp. Neurol. 111 : 261–277.
15. DOMESICK, V. B. 1972. Thalamic relationships of the medial cortex in the rat. Brain Behav. Evol. 6 : 457–483.
16. FALLON, J. H. and BENEVENTO, L. A. 1978. Projections of lateral orbital cortex to sensory relay nuclei in the rhesus monkey. Brain Res. 144 : 149–154.
17. FUKUDA, T. 1919. Ueber die faseranatomischen Beziehungen zwischen den Kernen des Thalamus opticus und den frontalen Windungen (Frontalregion des Menschen). Schweiz. Arch. Neurol. Psychiatr. 5 : 325–377.
18. GOLDMAN, P. S. 1979. Contralateral projections to the dorsal thalamus from frontal association cortex in the rhesus monkey. Brain Res. 166 : 166–171.
19. HAZLETT, J. C., DUTTA, Ch. R. and FOX, C. A.. 1976. The neurons in the centromedian-parafascicular complex of the monkey (*Macaca mulatta*): a Golgi study. J. Comp. Neurol. 168 : 41–74.
20. ITOH, K. 1977. Efferent projections of the pretectum in the cat. Exp. Brain Res. 30 : 89–105.
21. ITOH, K. and MIZUNO, N. 1977. Topographical arrangement of thalamocortical neurons in the cenrolateral nucleus (CL) of the cat, with special reference to a spino-thalamo-motor cortical path through the CL. Exp. Brain Res. 30 : 471–480.
22. JONES, E. G. and LEAVITT, R. Y. 1974. Retrograde axonal transport and the demonstration of non-specific projections to the cerebral cortex and striatum from thalamic intralaminar nuclei in the rat, cat and monkey. J. Comp. Neurol. 154 : 349–378.
23. KHALIFEH, R. R., KAELBER, W. W. and INGRAM, W. R. 1965. Some efferent connections of the nucleus medialis dorsalis. An experimental study in the cat. Am. J. Anat. 116 : 341–354.
24. KIEVIT, J. and KUYPERS, H. G. J. M. 1977. Organization of the thalamo-cortical connexions to the frontal lobe in the Rhesus monkey. Exp. Brain Res. 29 : 299–322.
25. KNOOK, H. L. 1965. The fibre-connections of the forebrain. Royal VanGorcum Ltd., Assen, 477 p.
26. KRETTEK, J. E. and PRICE, J. L. 1977. The cortical projections of the mediodorsal nucleus and adjacent thalamic nuclei in the rat. J. Comp. Neurol. 171 : 157–192.
27. KRIEG, W. J. S. 1954. Connections of the frontal cortex of the monkey. C. C. Thomas, Springfield, Ill., 320 p.
28. KÜNZLE, H. 1976. Thalamic projections from the precentral motor cortex in *Macaca fascicularis*. Brain Res. 105 : 253–267.
29. KÜNZLE, H. 1977. Evidence for selective axon-terminal uptake and retrograde transport of label in cortico- and rubrospinal system after injection of 3H-proline. Exp. Brain Res. 28 : 125–132.
30. KÜNZLE, H. 1978. An autoradiographic analysis of the efferent connections from premotor and adjacent prefrontal regions (areas 6 and 9) in *Macaca fascicularis*. Brain Behav. Evol. 15 : 185–234.
31. KÜNZLE, H. and AKERT, K. 1977. Efferent connections of cortical area 8 (frontal eye field) in *Macaca fascicularis*. A reinvestigation using the autoradiographic technique. J. Comp. Neurol. 173 : 147–163.

32. KUYPERS, H. G. J. M. and PANDYA, D. 1966. Comments of the cortical projections to the center median in chimpanzee. In D. P. Purpura and M. D. Yahr (ed.), The thalamus. Columbia University Press, New York, p. 122–126.
33. LEONARD, C. M. 1972. The connections of the dorsomedial nuclei. Brain Behav. Evol. 6 : 524–541.
34. LEICHNETZ, G. R. and ASTRUC, J. 1975. Efferent connections of the orbitofrontal cortex in the marmoset (Saguinus oedipus). Brain Res. 84 : 169–180.
35. LEICHNETZ, G. R. and ASTRUC, J. 1976. The efferent projections of the medial prefrontal cortex in the squirrel monkey (Saimiri scireus). Brain Res. 109 : 455–472.
36. LUBIŃSKA, L. 1964. Axoplasmic streaming in regenerating and in normal nerve fibres. In M. Singer and J. P. Schadé (ed.), Mechanisms of neural regeneration. Prog. Brain Res. 13 : 1–71.
37. MACCHI, G., BENTIVOGLIO, M., D'ATENA, C., ROSSINI, P. and TEMPESTA, E. 1977. The cortical projections of the thalamic intralaminar nuclei restudied by means of the HRP retrograde axonal transport. Neurosci. Lett. 4 : 121–126.
38. MEHLER, W. R. 1966. Further notes on the center median nucleus of Luys. In D. P. Purpura and M. D. Yahr (ed.) The thalamus. Columbia University Press, New York, p. 109–127.
39. MEHLER, W. R., FEFERMAN, M. E. and NAUTA, W. J. H. 1960. Ascending axon degeneration following anterolateral cordotomy. An experimental study in the monkey. Brain 83 : 718–750.
40. METTLER, F. A. 1947. Extracortical connections of thy primate frontal cerebral cortex. I. Thalamo-cortical connections. J. Comp. Neurol. 86 : 95–117.
41. METTLER, F. A. 1947. Extracortical connections of the primate frontal cerebral cortex. II. Corticofugal connections. J. Comp. Neurol. 86 : 119–166.
42. METTLER, F. A. 1972. The corticothalamic projection: the structural substrate for the control of the thalamus by the cerebral cortex. In T. Frigyesi, E. Rinvik and M. D. Yahr (ed.), Corticothalamic projections and sensorimotor activities. Raven Press, New York, p. 1–19.
43. MINKOWSKI, M. 1923. Etude sur les connexions anatomiques des circonvolutions rolandiques, pariétales et frontales. Schweiz. Arch. Neurol. Neurochir. Psychiatr. 12 : 71–104.
44. MONAKOW, v., C. 1881. Ueber einige durch Extirpation circumscripter Hirnrindenregionen bedingte Entwicklungshemmungen des Kaninchengehirns. Arch. Psychiatr. Nerrenkr. 12 : 141–156.
45. MURRAY, M. 1966. Degeneration of some intralaminar thalamic nuclei after cortical removals in the cat. J. Comp. Neurol. 127 : 341–368.
46. NARKIEWICZ, O. and BRUTKOWSKI, St. 1967. The organisation of projections from thalamic MD nucleus to the prefrontal cortex of the dog. J. Comp. Neurol. 129 : 361–374.
47. NAUTA, H. J. W., Jr., PRITZ, M. B. and LASEK, R. J. 1974. Afferents to the rat caudoputamen studied with horseradish peroxidase. An evaluation of a retrograde neurochemical research method. Brain Res. 67 : 219–238.
48. NAUTA, W. J. H. 1964. Some efferent connections of the prefrontal cortex in

the monkey. *In.* J. M. Warren and K. Akert (ed.), The frontal granular cortex and behavior. McGraw-Hill, New York, p. 397–409.
49. NAUTA, W. J. H. and KUYPERS, H. G. J. M. 1958. Some ascending pathways in the brain stem reticular formation. *In* Jasper et al. (ed.), The reticular formation. Little Brown, Boston, p. 3–30.
50. NIIMI, K., KISHI, S., MIKI, M. and FUJITA, S. 1963. An experimental study of the course and termination of the projection fibers from cortical areas 4 and 6 in the cat. Folia Psychiatr. Neurol. Jap. 17 : 167–216.
51. NIIMI, K., MICHIHORO, M. and KAWAMURA, S. 1970. Ascending projections of the superior colliculus in the cat. Okajimas Folia Anat. Jap. 47 : 269–287.
52. OLSZEWSKI, J. 1952. The Thalamus of the *Macaca mulatta*. An atlas for use with the stereotaxic instrument. Karger, Basel.
53. PEACOCK, J. H. and COMBS, C. M. 1965. Retrograde cell degeneration in diencephalic and other structures after hemidecortication of rhesus monkey. Exp. Neurol. 11 : 367–399.
54. PERCHERON, G. 1977. The thalamic territory of cerebellar afferents and the lateral region of the thalamus of the macaque in streotaxic ventricular coordinates. J. Hirnforsch. 18 : 375–400.
55. PETRAS, J. M. 1969. Some efferent connections of the motor and somatosensory cortex of simian primates and felid, canid and procyonid carnivores. Ann. N. Y. Acad. Sci. 167 : 469–505.
56. PETRAS, J. M. 1972. Corticostriate and corticothalamic connections in the chimpanzee. *In* T. L. Frigyesi, E. Rinvik and M. D. Yahr (ed.), Corticothalamic projections and sensorimotor activities. Raven Press, New York, p. 201–216.
57. POLYAK, S. 1932. The main afferent fiber systems of the cerebral cortex in primates. Univ. of California Press, Berkeley, 370 p.
58. POWELL, T. P. S. and COWAN, W. M. 1956. A study of the thalamo-striate relations in the monkey. Brain 79 : 364–390.
59. PRIBRAM, K. H., CHOW, K. L. and SEMMES, J. 1953. Limit and organization of the cortical projection from the medial thalamic nucleus in monkey J. Comp. Neurol. 98 : 433–448.
60. PURPURA, D. P. 1972. Synaptic mechanisms in coordination of activity in thalamic internuncial common paths. *In* T. Frigyesi, E. Rinvik and M. D. Yahr (ed.), Corticothalamic Projections and sensorimotor activities. Raven Press, New York, p. 21–56.
61. RINVIK, E. 1968. The corticothalamic projection from the precruciate and coronal gyri in the cat. An experimental study with silver-impregnation methods. Brain Res. 10 : 79–119.
62. RINVIK, E. 1972. Organization of thalamic connections from motor and somatosensory cortical areas in the cat. *In* T. Frigyesi, E. Rinvik and M. D. Yahr (ed.), Corticothalamic projections and sensorimotor activities. Raven Press, New York, p. 57–90.
63. ROBERTS, T. S. and AKERT, K. 1963. Insular nad opercular cortex and its thalamic projection in Macaca mulatta. Schweiz. Arch. Neurol. Neurochir. Psychiatr. 92 : 1–43.
64. ROSE, J. E. and WOOLSEY, C. N. 1948. The orbitofrontal cortex and its connections with the mediodorsal nucleus in rabbit, sheep and cat. Res. Publ. Assoc. Res. Nerv. Ment. Dis. 27 : 210–232.

65. ROSE, J. E. and WOOLSEY, C. N. 1949. Organization of the mammalian thalamus and its relationships to the cerebral cortex. EEG and Clin. Neurophysiol. 1 : 391–404.
66. RUTISHAUSER, F. 1899. Experimenteller Beitrag zur Stabkranzfaserung im Frontalhirn des Affen. Monatsschr. Psychiatr. Neurol. 5 : 161–179.
67. SAKAI, S. 1967. Some observations on the corticothalamic fiber connections in the monkey. Proc. Jap. Acad. 43 : 822–826.
68. SATO, M., ITOH, K. and MIZUNO, N. 1979. Distribution of thalamocaudate neurons in the cat as demonstrated by horseradish peroxidase. Exp. Brain Res. 34 : 143–153.
69. SCOLLO-LAVIZZARI, G. and AKERT, K. 1963. Cortical area 8 and its thalamic projections in Macaca mulatta. J. Comp. Neurol. 121 : 259–269.
70. STRICK, P. L. 1975. Multiple sources of thalamic input to the primate motor cortex. Brain Res. 88 : 372–377.
71. STRICK, P. L. 1976. Anatomical analysis of ventrolateral thalamic input to primate motor cortex. J. Neurophysiol. 39 : 1020–1031.
72. TANAKA, D., Jr. 1976. Thalamic projectons of the dorsomedial prefrontal cortex in the rhesus monkey (Mocaca mulatta). Brain Res. 110 : 21–38.
73. TASIRO, S. 1940. Experimentell-anatomische Untersuchung über die efferenten Bahnen aus den Vierhügeln der Katze. Z. Mikrosk.-Anat. Forsch. Leipz 47 : 1–32.
74. TOBIAS, T. J. 1975. Afferents to prefrontal cortex from the thalamic mediodorsal nucleus in the rhesus monkey. Brain Res. 83 : 191–212 (1975).
75. VEDOVATO, M. 1978. Identification of afferent connections to cortical area 6aβ of the cat by means of retrograde horseradish peroxidase transport. Neurosci. Lett. 9 : 303–310.
76. WALKER, A. E. 1938. The primate thalamus. Univ. of Chicago Press, Chicago, 321 p.
77. WALKER, A. E. 1940. The medial thalamic nucleus. A comparative anatomical, physiological and clinical study of the nucleus medialis dorsalis thalami. J. Comp. Neurol. 73 : 87–115.
78. WALKER, A. E. 1949. Afferent connections. In P. C. Bucy (ed.) The Precentral Motor Cortex. Univ. Illinois Press, Urbana, Ill. p. 111–132.
79. WARREN, J. M. and AKERT, K. (ed.). 1964. The frontal granular cortex and behavior. McGraw-Hill, New York, 492 p.
80. WOOLSEY, C. N., SETTLAGE, P. H., MEYER, D. R., SENCER, W., HAMUY, T. P. and TRAVIS, A. M. 1952. Patterns of localization in precentral and 'supplementary' motor areas and their relation to the concept of a premotor area In Patterns of organization in the central nervous system. Res. Publ. Assoc. Res. Nerv. Ment. Dis. 30 : 238–264.

Konrad AKERT and Kurt HARTMANN-VON MONAKOW, Brain Research Institute. University of Zürich, August Forelstrasse 1, CH-8029 Zürich, Switzerland.

## LIST OF ABBREVIATIONS

| | |
|---|---|
| Cif | Nc. centralis inferior |
| CL | Nc. centralis lateralis |
| CM | Nc. centrum medianum |
| Csl | Nc. centralis superior lateralis |
| DK | Nc. Darkschewitsch |
| GM | Nc. geniculatus medialis |
| Hl | Nc. habenularis lat. |
| Hm | Nc. habenularis med. |
| IC | Nc. interstitialis Cajal |
| LD | Nc. lateralis dorsalis |
| LP | Nc. lateralis posterior |
| MDde | Nc. medialis dorsalis, pars densocellularis |
| MDmc | Nc. medialis dorsalis, pars magnocellularis |
| MDmf | Nc. medialis dorsalis, pars multiformis |
| MDpc | Nc. medialis dorsalis, pars parvocellularis |
| Pcn | Nc. paracentralis |
| Pf | Nc. parafascicularis |
| Pulo | Nc. pulvinaris oralis |
| R | Nc. reticularis |
| Reg. prT. | Regio praetectalis |
| Sf | Nc. subfascicularis |
| SG | Nc. suprageniculatus |
| Teg. a. | Area tegmentalis ant. |
| VLc | Nc. ventralis lateralis, pars caudalis |
| VPI | Nc. ventralis posterior, pars inferior |
| VPLc | Nc. ventralis posterior, lateralis, pars caudalis |
| VPLo | Nc. ventralis posterior lateralis, pars oralis |
| VPM | Nc. ventralis posterior medialis |
| VPMpc | Nc. ventralis posterior medialis, pars parvocellularis |

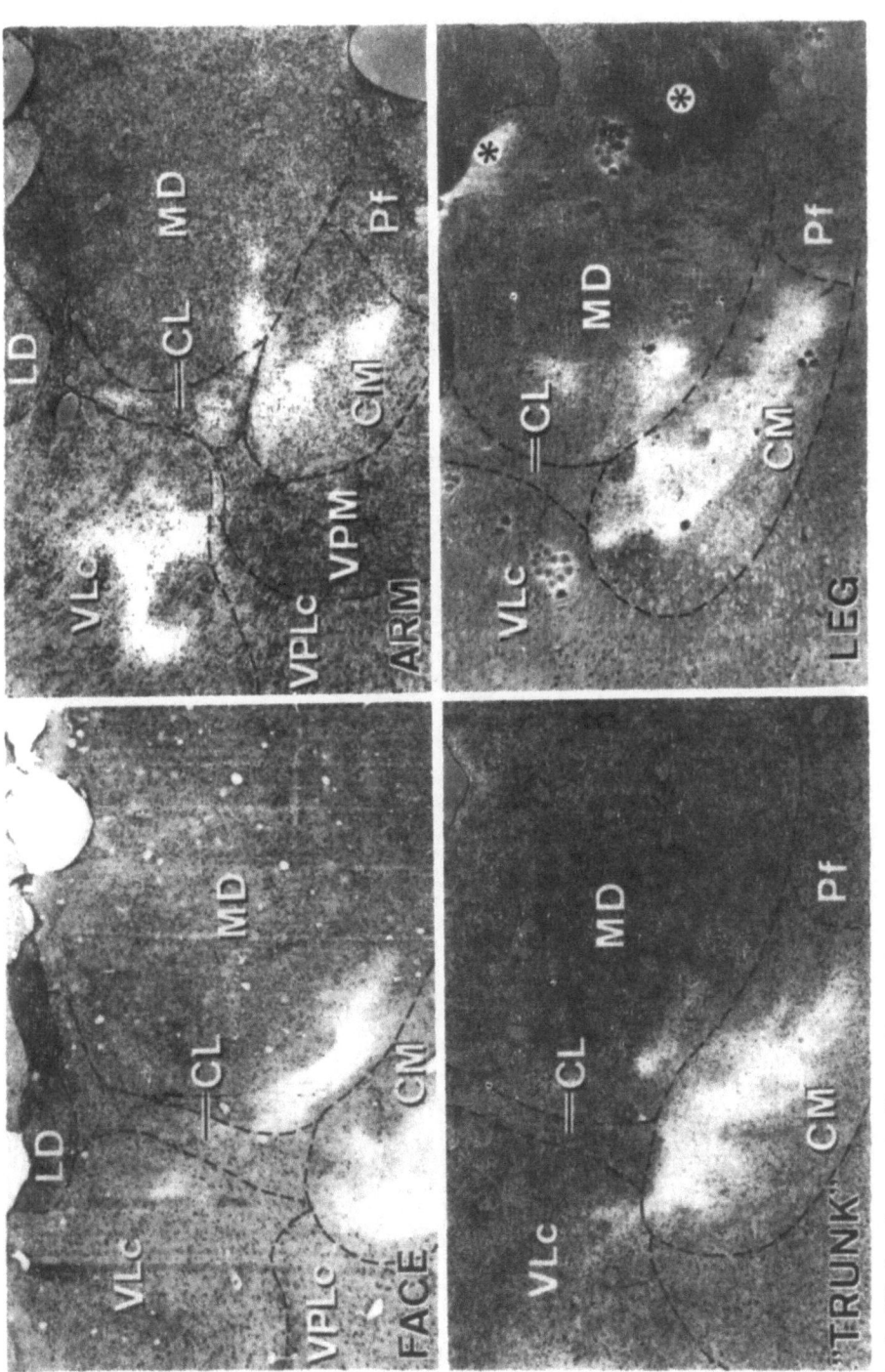

Fig. 4. Thalamic projection sites (anterograde labelling) from various parts of precentral gyrus (see Fig. 1). Main terminations are in CM and paralaminar MD. No clear-cut somatotopic arrangement of projections is distinguishable between leg (72–451), arm (78–335R), and face (72–448) areas. The cortical injection field in case 72–449 designated "trunk" area involves the more ventral portion in the dorsal third of the precentral gyrus and may involve more than axial musculature representation. * = artifact.

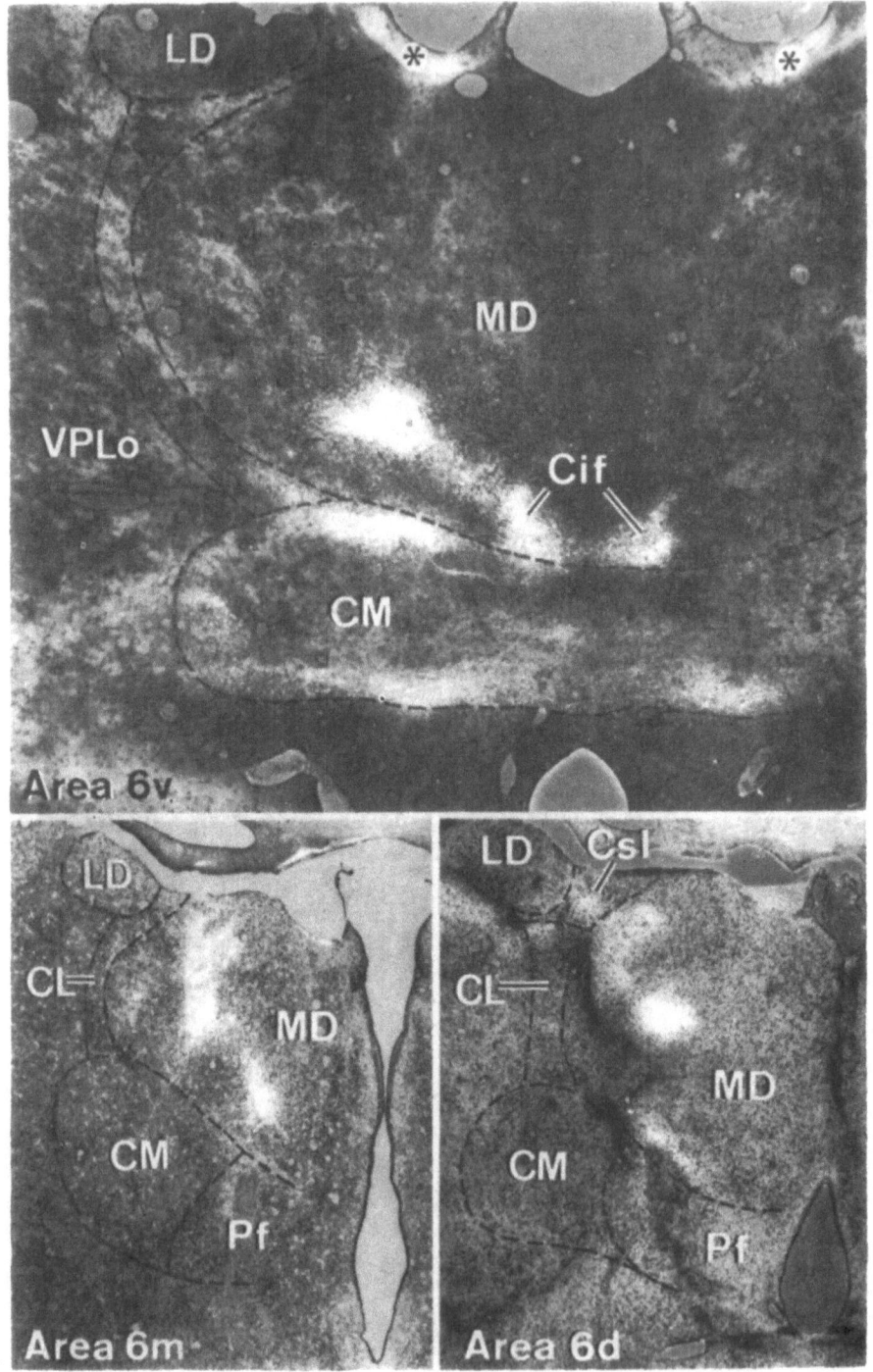

Fig. 5. Thalamic projection sites (anterograde labelling) from various parts of premotor cortex. Note that medial (= supplementary motor cortex) 6m (74–626) and dorsal 6d (75–212) areas project dorsally, ventral areas 6 V (78–882) more ventrally and bilaterally in the thalamus. Only ventral area 6 (6 V) has a clear-cut projection to CM. The projection into MD is more intensive than in area 4 cases (Fig. 4) and shifted slightly more medially. * = artifact.

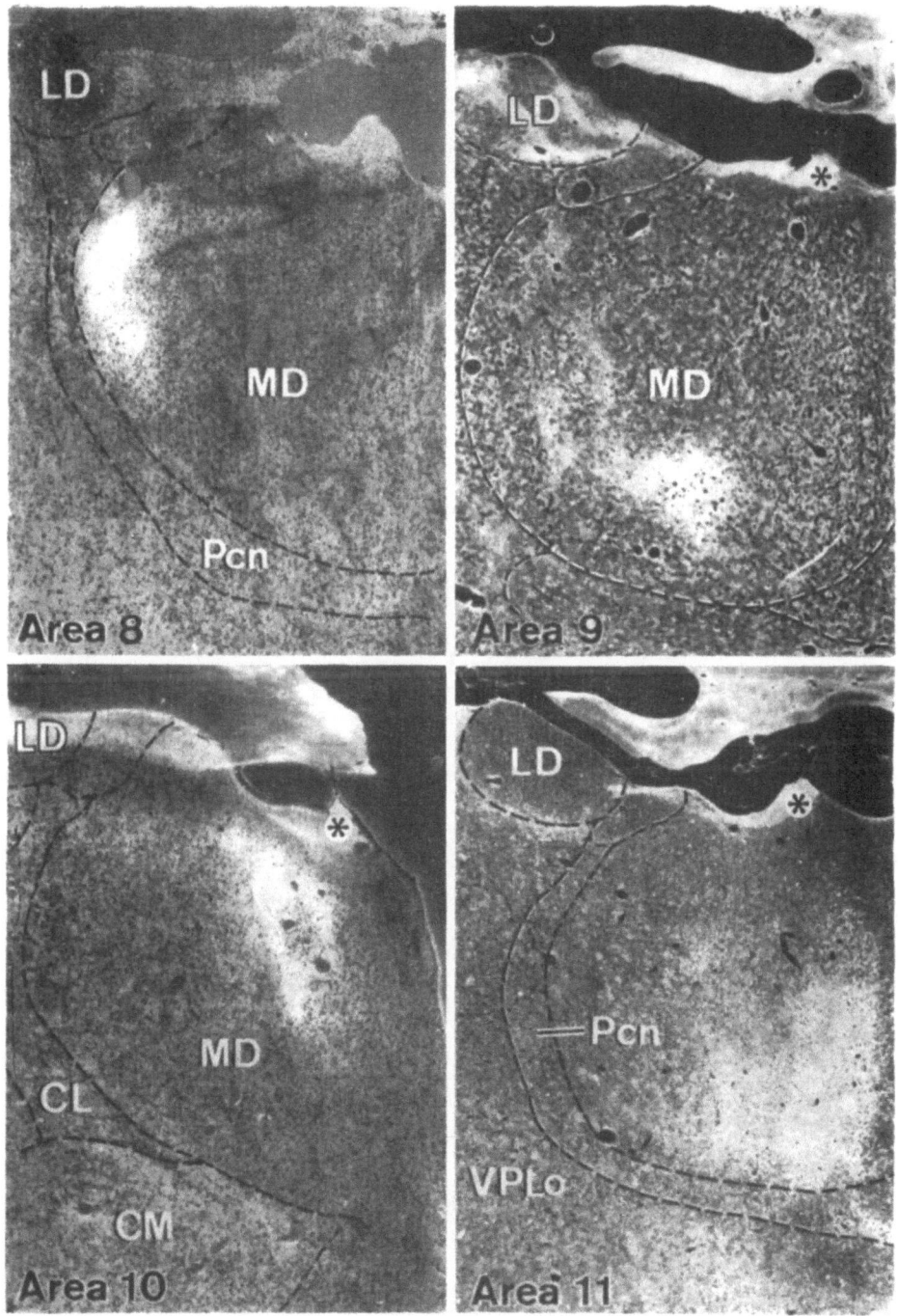

Fig. 6. Thalamic projection sites (anterograde labelling) from various regions of prefrontal cortex. Note that the 4 areas project differentially within MD, area 8 most laterally and area 11 most medially and ventrally. Area 8 = 73–228, area 9 = 75–496, area 10 = 78–331, area 10/11 = 78–354A, * = artifact.

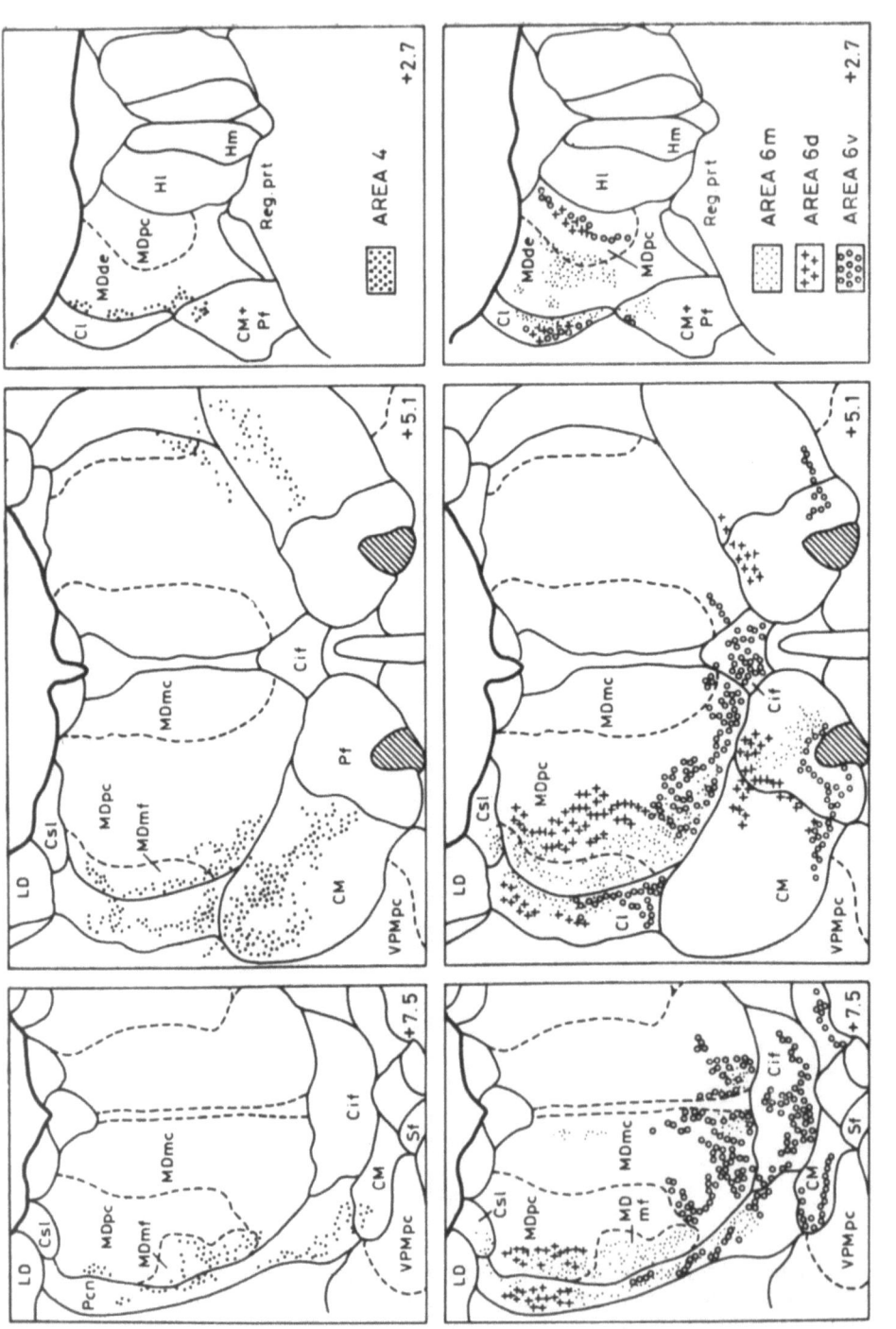

**Fig. 7.** Cumulative plot of precentral (above) and premotor (below) cortical projections to MD and intralaminar thalamic nuclei. Note considerable convergence in the paralaminar and intralaminar region.

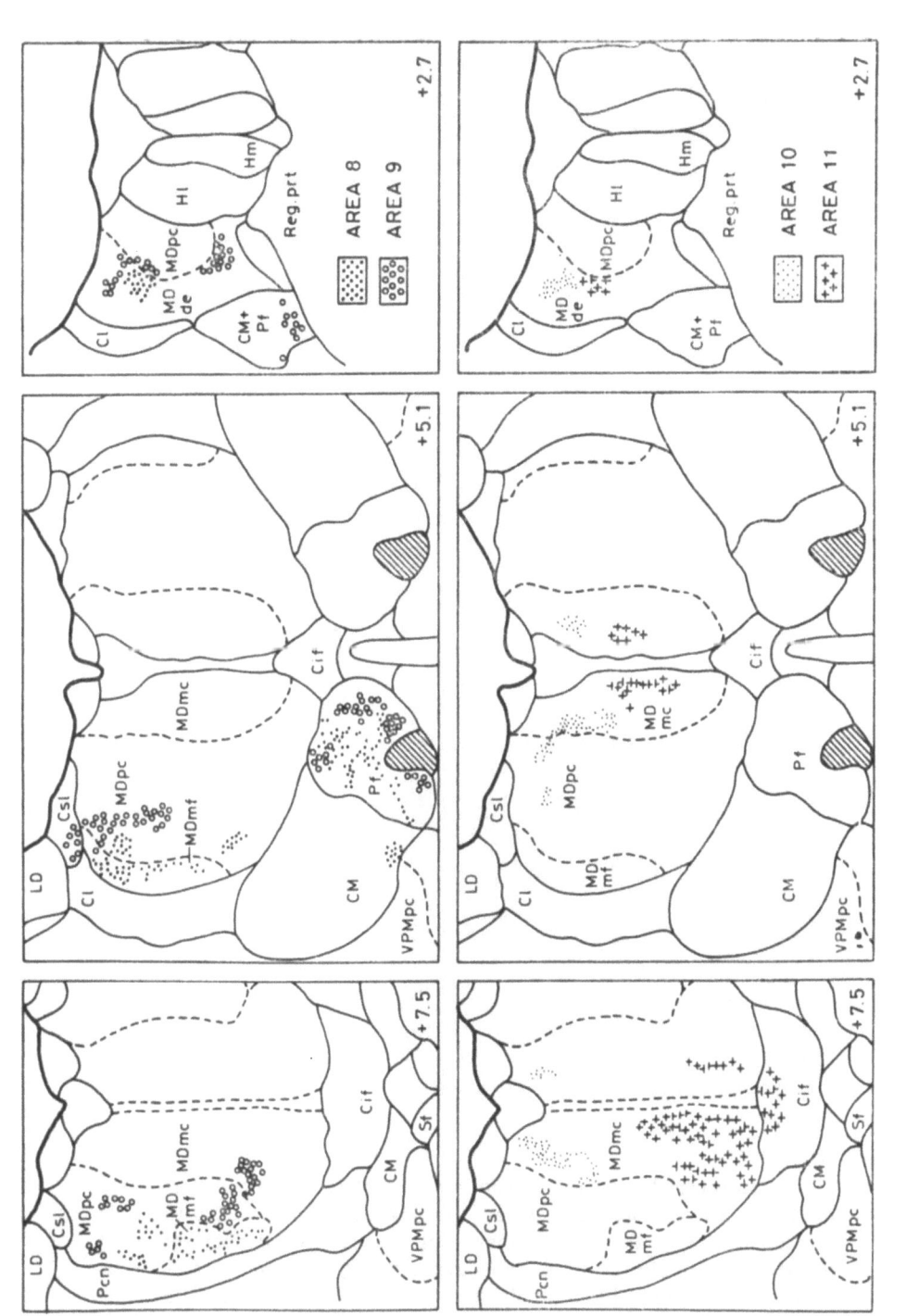

Fig. 8. Cumulative plot of prefrontal cortical projections to MD and intralaminar thalamic nuclei.

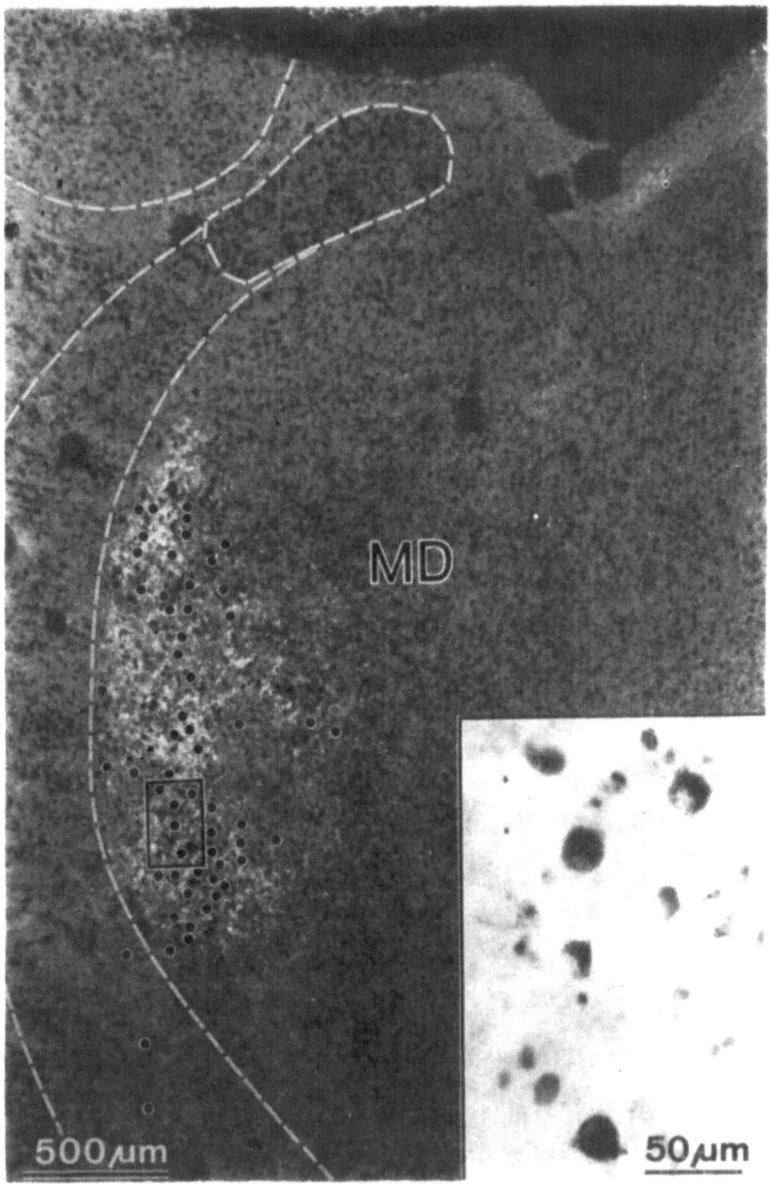

Fig. 9. Anterograde and retrograde labelling of area 8 connections (case 78/354B). Autoradiographic label demonstrates the cortico-thalamic projection site in the paralaminar zone of MD. Black dots represent retrogradely (HRP) labelled cells representing thalamo-cortical connections. Note that both labels are found within exactly the same boundary within paralaminar MD. Framed rectangle corresponds to inset which exposes the HRP labelled cells at higher magnification.

Lecture delivered at the Warsaw Colloquium on Instrumental
Conditioning and Brain Research
May 1979

# THALAMOAMYGDALOID CONNECTIONS STUDIED BY THE METHOD OF RETROGRADE TRANSPORT

Liliana NITECKA, Leszek AMERSKI, Jadwiga PANEK-MIKUŁA
and Olgierd NARKIEWICZ

Department of Anatomy, Institute of Medical Biology, School of Medicine
Gdańsk, Poland

*Abstract.* On the basis of retrograde horseradish peroxidase transport from nuclei ot the amygdaloid body of the rat to the thalamus, it was found that several groups of thalamic nuclei send fibers to the amygdala. These are: (i) nuclei of posterior region of thalamus and neighbouring area of the tegmentum — peripeduncular nucleus, suprageniculate-limitans nucleus, (ii) midline nuclei — paraventricular nucleus, parataenial nucleus, nucleus reuniens, (ii) intralaminar nuclei — central medial nucleus, parafascicular nucleus, (iv) mediodorsal nucleus. There are two main systems of thalamoamygdaloid connections. One of them arising in the posterior region of the thalamus terminates in the lateral nucleus of the amygdala and the lateral part of its central nucleus. The other system begins in the intralaminar and midline nuclei and in the mediodorsal nucleus of the thalamus. It reaches the remaining nuclei of the amygdala. Amygdalopetal connections of the interlaminar and middline nuclei of the thalamus, especially those arising in the paraventricular and parataenial nucleus, are mostly bilateral.

## INTRODUCTION

There are only few data about thalamoamygdaloid connections. Those running in the opposite direction are much better known, although it was not untill aurodiagraphic tracing of connections was employed that it was possible to determine that they are amygdalothalamic and not corticothalamic fibers passing through the amygdaloid complex (3, 9, 10, 11-13, 18).

It is generally supposed that amygdalopetal connections of the thalamus originate mainly in the mediodorsal nucleus (10, 12). Only recently another system of thalamoamygdaloid connections has been shown which arises in neurons of the posterior thalamic region (6, 7, 15). Moreover, some investigations (13) have suggested that perhaps other thalamic nuclei are also connected with nuclei of the amygdala.

Since the problem of the connections of the thalamus with amygdaloid nuclei is still controversial, we tried to study them by means of retrograde axonal transport of horseradish peroxidase (HRP) following injection into the amygdaloid body. There is general agreement as to the cytoarchitectonic division of the thalamus of the rat (1, 8) and certain divergences concern only the posterior part. For this reason we briefly give the topography of the nuclei of this region (Figs. 1 and 2).

In the posterior thalamic region of the rat we have distinguished the posterior nucleus, pulvinar, and the suprageniculate-limitans nucleus. Peripeduncular nucleus may also be included in this region although there are some data suggesting that it is a part of the tegmentum.

*Posterior nucleus.* In frontal sections through the posterior thalamic region, this nucleus is seen as a large complex of generally small cells, densely separated by bundles of fibres of the corticotectal tract. Large oval cells brightly stained with cresyl violet, loosely scattered are visible among small cells. In frontal sections through the posterior commissure, the posterior nucleus is triangular in shape with the base resting on the zona incerta and medial lemniscus. Laterally and anteriorly it borders with the pulvinar, posteriorly with the pulvinar and the medial geniculate complex, and medially with the pretectal area.

*Pulvinar.* It is composed mainly of rather large oval cells of medium intensity staining with cresyl violet. The anterior part of the pulvinar is seen in frontal sections passing through the posterior pole of the habenular complex. The anterior parts is a homogenous cellular mass and perhaps correspond to the anterior pulvinar of higher mammals (2). In the posterior part of the pulvinar in frontal sections through the posterior commissure, one may distinguish two areas — the ventrolateral and dorsomedial. The ventrolateral area is divided into several cell groups by bundles of fibers of the corticotectal tract passing through here. In the dorsomedial area there is often a thick bundle of fibers probably running mainly from the lateral geniculate complex to the pretectal area.

*Suprageniculate-limitans nucleus.* These two nuclei form a joint complex of sparsely scattered large cells darkly stained with cresyl violet. In the rat as in other mammals there is no clear-cut boundary between the two nuclei, which are limited medially by the pretectal area, laterally by the medial geniculate complex, and dorsally by the posterior part of the pulvinar. Ventrally this nuclear complex passes into the peripeduncular nucleus without any clear-cut boundary.

*Peripeduncular nucleus.* Some authors consider this nucleus as a part of the mesencephalon (2), others as a nucleus belonging to the posterior thalamic region (4, 17). According to our observations its medial part, characterized by large cells of medium intensity staining with cresyl violet, is a posterior extension of the ventromedial area of the pulvinar. In the lateral part of the peripeduncular nucleus there is a small area occupied by smaller funsiform cells, stained darkly with cresyl violet. In its anterior part the peripeduncular nucleus adheres medially to the posterior nucleus and dorsally to the medial geniculate complex. In its posterior part it borders dorsally with the suprageniculate-limitans nucleus, medially with mesencephalic reticular formation and ventrally with substantia nigra.

The subdivision of the amygdaloid body into nuclei is shown in Fig. 2 (upper row). A detailed description of it was presented in earlier publications (14, 16).

## MATERIAL AND METHODOS

Our studies were carried out on the brains of 50 Wistar rats, of both sexes, weighing 200-250 g. Injections of 30% solution (Sigma VI) horseradish peroxidase were made in the ı uclei of the amygdala through a glass micropipette connected to a Han ilton syringe 0.05–0.5 mol/µl). The postoperative survival time was 1–2 days. Next the brains were perfused at room temperature in a 0.4% solution of formaldehyde and 1.25% glutaraldehyde in 0.1 M phosphate buffer of pH–7.2. After removal from the skull the brains were fixed in the same solution for 24 h and then placed for another 24 h in a 7.2 phosphate buffer containing 5% sucrose. The brains were cut in a frontal plane after which sections 50 µm in thickness were incubated in a 0.05% solution of 3.3 — diaminobenzidine tetrachloride in 0.05 M in Tris HCl buffer (ph 7.6), at room temperature for 30 min. Additional incubation (at room temp. for 30 min) was carried out in the same solution plus $H_2O_2$ (0.3 ml 3% $H_2O$/100 ml solution). Sections were stained with cresyl violet.

We used two different stereotaxic approaches to the amygdaloid nuclei:

1. The micropipette was inserted from above passing through the

cortex covering the upper surface of the cerebral hemisphere and through the striatum. In this type of procedure, there is very often a slight leakage of HRP into the striatum in which diffusion of HRP is generally large.

2. The micropipette was inserted through the piriform or periamygdaloid cortex. In this case the ventrolateral surface of the brain was reached either from the infratemporal fossa or from the orbit.

The micropipette was inserted into the amygdala at the level of the middle cerebral artery or below it, between its frontal and temporal branches. In some cases (injections of cortical and basal ventral nuclei) the micropipette was inserted from the ventral surface of the piriform lobe after lifting it slightly upwards.

## RESULTS

Our investigations showed that the neurons of various thalamic nuclei are the source of amygdalopetal fibers. We present here a few typical cases for each group of animals with injections encompassing either most of the amygdaloid body or its particular nuclei.

After a large injection of HRP which involved most of the amygdaloid nuclei (R63; Figs. 2 and 3) labeled cells were seen: (i) in the nuclei of the posterior thalamic region and neighboring area of the tegmentum-pulvinar, peripeduncular nucleus the suprageniculate-limitans nucleus, the medial geniculate complex, (ii) in the nuclei of the midlineparaventricular, the parataenial and nucleus reuniens, (iii) in the intralaminar nuclei — central medial and parafascicular, (iv) in the mediodorsal nucleus.

Figure 2 shows the localization of HRP labeled neurons in the thalamic nuclei. Many of them were seen in the dorsomedial area of the posterior part of the pulvinar, the peripeduncular nucleus, the paraventricular nucleus, the parataenial nucleus and in the anterior part of nucleus reuniens. Quite a large number of labeled cells was found in the suprageniculate-limitans nuclei, the parafascicular nucleus, and the posterior part of the nucleus reuniens. The smallest number of labeled cells was observed in the mediodorsal nucleus, the medial geniculate complex, and the posterior nucleus of the thalamus. In the paraventricular nucleus, the parataenial nucleus, and the nucleus reuniens, the neurons labeled by retrograde axonal transport occurred contralaterally also. In the remaining nuclei the labeled neurons were found only ipsilaterally. Because the micropipette filled with HRP passed through the cortex surrounding the amygdala or through the striatum, we also traced the localization of thalamic neurons which send axon to these regions.

After an HRP injection to a large area of the piriform cortex (R98; Fig. 4) a much smaller number of labeled neurons was observed in thalamic nuclei than following injections to amygdala. They were localized in the parataenial nucleus, the mediodorsal nucleus and the central medial nucleus. Cells with HRP granules appeared sporadically in the parafascicular nucleus and the posterior part of nucleus reuniens.

After an HRP injection in the cortex in the vicinity of the rhinal sulcus (R19; Fig. 4) labeled cell somata appeared in large numbers in the posterior nucleus of the thalamus, sporadically in the pulvinar, peripeduncular nucleus and in the suprageniculate-limitans nucleus.

In rat R59 (Fig. 4) the injection involved a small ventral part of the striatum and globus pallidus as well as, to a slight degree, the cortex around the rhinal sulcus. Labeled cells were found in the posterior nucleus of the thalamus, in the pulvinar, in the peripeduncular nucleus and in the suprageniculate-limitans nucleus.

*Amygdalopetal connections of nuclei of the posterior thalamic region (R33, Fig. 5).* The injection included the lateral nucleus and, to a large

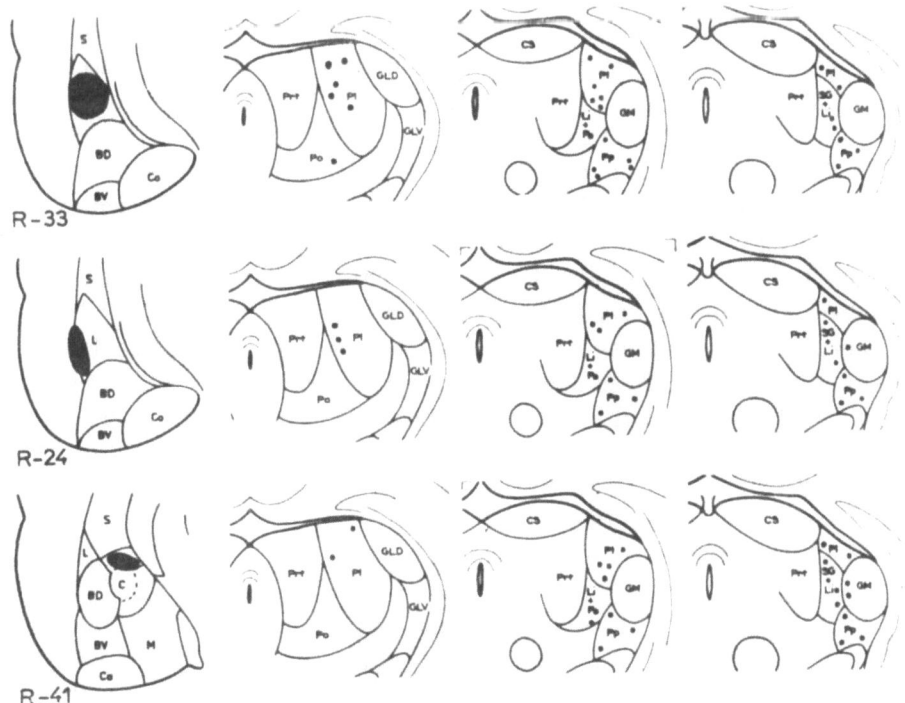

Fig. 5. Distribution of labeled cells in the nuclei of posterior thalamic region following injections into the lateral nucleus of the amygdala (R33 and R24) and lateral part of its central nucleus (R41). Denotations as in Fig. 2.

degree, the lateral part of the central nucleus of the amygdala. A slight leakage was observed in the cortex around the rhinal sulcus. HRP labeled cells were seen in the various nuclei of the posterior region of the thalamus: in the dorsomedial area of the posterior part of the pulvinar, the peripeduncular nucleus, the suprageniculate-limitans nucleus, and sporadically in the posterior nucleus of the thalamus as well as in the posterior part of medial geniculate complex.

R24 (Fig. 5) — The injection site was limited to the posterolateral part of the lateral nucleus of the amygdala. Cells containing HRP labeled granules were seen in the pulvinar, the peripeduncular and suprageniculate-limitans nucleus. In the pulvinar, cells labeled by retrograde transport were located slightly more posteriorly than in the previous animal R33. This finding is probably the result of a different localization of the injection in the area of the lateral nucleus of the amygdala. There were no HRP labeled cells in the posterior nucleus, only single ones appeared in the posteromedial part of medial geniculate nucleus.

R41: (Fig. 5) — The HRP injection area was very small, almost entirely limited to the lateral part of the central nucleus of the amygdala. Cells with HRP granules were observed in the dorsomedial area of the posterior part of the pulvinar, the peripeduncular nucleus and in the suprageniculate-limitans nucleus. It seems that in these last two nuclei, more labeled cells were visible than in rat R24, in which the HRP injection involved only the lateral nucleus of the amygdala.

In both rats R41 and R24 there were single labeled cells in the posteromedial part of the medial geniculate complex. In rats in which HRP injections involved all the nuclei of the amygdala, with the exception of the lateral nucleus, and the lateral part of the central nucleus, no labeled cells were found in the posterior thalamic region (Fig. 6).

On the basis of these cases and all of our experimental material, we found that only in these samples when the HRP injection to the amygdaloid body involved the lateral nucleus of the amygdala or lateral part

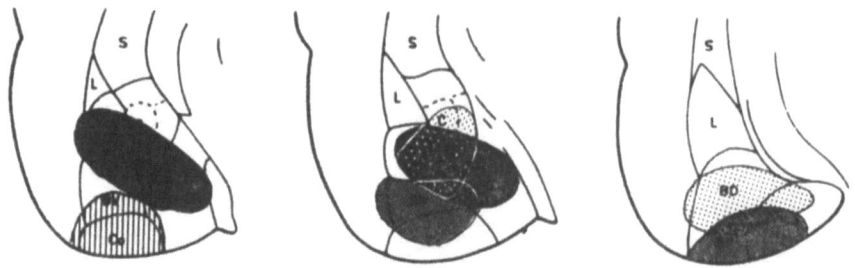

Fig. 6. Injections into the amygdala without evident HRP retrograde axonal transport to neurons of the posterior thalamic region.

of its central nucleus, were there labeled cells in the respective nuclei of the posterior thalamic region. A localization of labeled neurons, similar to that following HRP injections limited to the lateral nucleus and to the lateral part of the central nucleus of the amygdala, was observed after injection involving the anteroventral segment of striatum and the globus pallidus (R59). Both the results presented here as well as those obtained in previous work (15) would appear to show that the posterior nucleus of the thalamus is not the source of amygdalopetal projections.

*Amygdalopetal connections of midline thalamic nuclei (R95, Fig. 7).* After only the basal dorsal nucleus was infiltrated by injections of HRP, labeled cells were found in large numbers in the paraventricular nucleus, mainly in its anterior part. Similarly, in the parataenial nucleus and the anterior segment of nucleus reuniens a significant number of cells containing HRP labeled granules were observed.

R79 (Fig. 7) — The HRP injection was made in the posterior part of the basal ventral nucleus and the cortical nucleus. HRP labeled cells were numerous in the paraventricular nucleus, the parataenial nucleus and in the anterior part of nucleus reuniens.

R74 (Fig. 7) — The HRP injection was limited to the basal ventral nuclcus. Cells containing granules were found in the paraventricular

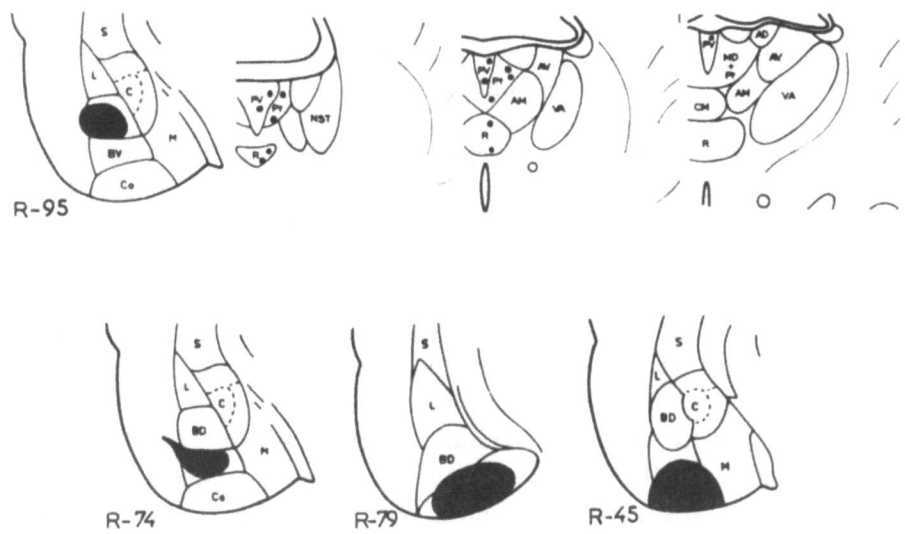

Fig. 7. Distribution of labeled cells in the midline nuclei of the thalamus following injection into the basal dorsal nucleus (upper row — R95). In the lower row injections in which similar localization of labeled cells in the midline thalamic nuclei was found. Denotations as in Fig. 2.

nucleus, the parataenial nucleus and the anterior part of nucleus reuniens. The number of labeled cells was definitely smaller than found in rat R79.

R45 (Fig. 7) — The HRP injection involved the anterior segment of the cortical nucleus and the basal ventral nucleus. The localization of labeled cells was the same as in R74 and R79.

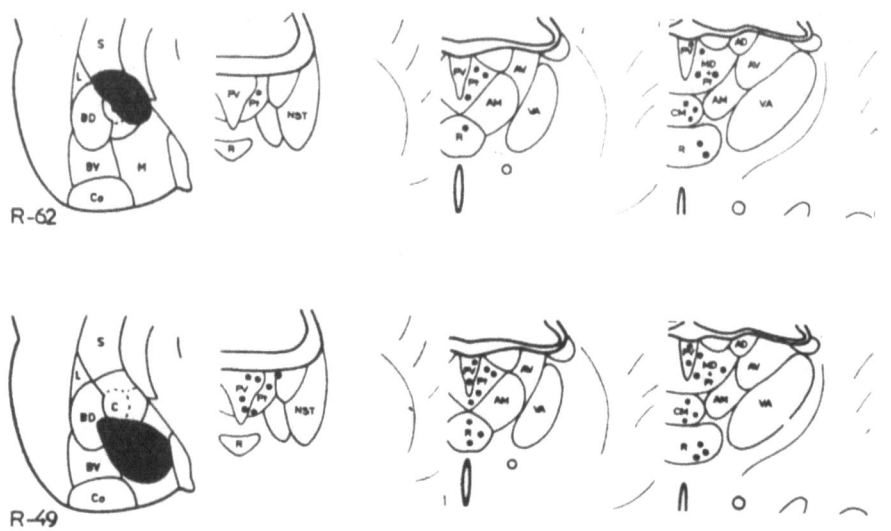

Fig. 8. Distribution of labeled cells in the thalamic midline nuclei following injection involving all parts of the central nucleus of the amygdala — R62 and medial nucleus of amygdala — R49. Legend as in Fig. 2.

R62 (Fig. 8) — After HRP injection involving all of the parts (lateral, intermediate and medial) of the central nucleus, labeled neurons were found in the parataenial nucleus and the posterior segment of the nucleus reuniens. No labeled neurons were found in the paraventricular nucleus except for a few cells in its posterior part.

R49 (Fig. 8) — The main injection was located in the medial nucleus of the amygdala, but it also encroached upon the basal dorsal nucleus and a small area of the central nucleus (pars medialis). Many labeled cells were found in the paraventricular nucleus and the parataenial nucleus, as well as in the nucleus reuniens of the thalamus. In the latter cells with HRP granules were found both in its anterior as welle as posterior segments. With respect to the paraventricular nucleus, it should be noted that only in cases where the HRP injection to the amygdaloid body involved the medial nucleus, labeled neurons appeared in both anterior and posterior parts of this nucleus; in all other cases they were found only in the anterior part. In no case did we succeed in limiting

HRP injection to only medial nucleus of the amygdala. But taking into consideration the intense labelling of HRP cells in rat R49, we may assume that midline thalamic nuclei project to the medial nucleus also. In rats (Fig. 9) in which injection of HRP was limited to the area of the lateral nucleus and the lateral part of the central nucleus, no labeled cells were found in the midline thalamic nuclei.

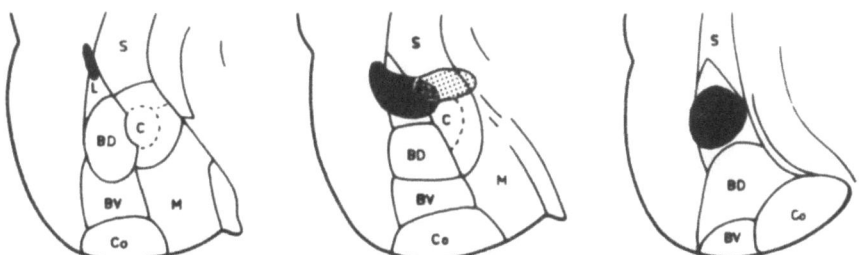

Fig. 9. Injections into the amygdala without evident HRP retrograde transport to neurons of the thalamic midline nuclei.

These data suggest that almost all nuclei of the amygdaloid body receive afferent fibers from midline thalamic nuclei. The central nucleus of the amygdala probably does not receive fibers from the anterior part of the paraventricular nucleus and not many from its posterior part. After injections limited to the area of the central nucleus, no neurons with HRP labeled granules were found in the paraventricular nucleus, except for a few cells in its posterior part. Moreover, it would seem that neurons of the anterior part of nucleus reuniens project mainly to the basal dorsal, basal ventral and cortical nuclei, whereas the neurons of its posterior part project mainly to the central and medial nuclei. Following large injections of HRP to most of the nuclei of the amygdaloid body, labeled granules were found in cells of the midline thalamic nuclei bilaterally, with very definite preponderance on the side of the injection. After injections limited to small areas of particular nuclei, labeled cells appeared contralaterally only rarely. For this reason we do not suppose the bilateral presence of labeled neurons in the midline thalamic nuclei to be connected in a particular way with the HRP injection to any of the amygdaloid body nuclei.

*Amygdalopetal connections of intralaminar thalamic nuclei.* R60 (Fig. 10) — The HRP injection involved the entire central nucleus of the amygdala. Labeled neurons appeared in large numbers in the central medial nucleus: relatively fewer labeled cells were found in the parafascicular nucleus.

R47 (Fig. 10) — The injection involved mainly the medial nucleus of the amygdala, with a slight leakage in the medial part of the central nucleus and the basal ventral nucleus. Numerous HRP labeled cells appeared in the central medial nucleus and the parafascicular nucleus. The large number of labeled cells in these thalamic nuclei of R47, suggests that besides the central amygdala nucleus, the medial nucleus also receives amygdalopetal fibers, from neurons of central medial nucleus and the parafascicular nucleus.

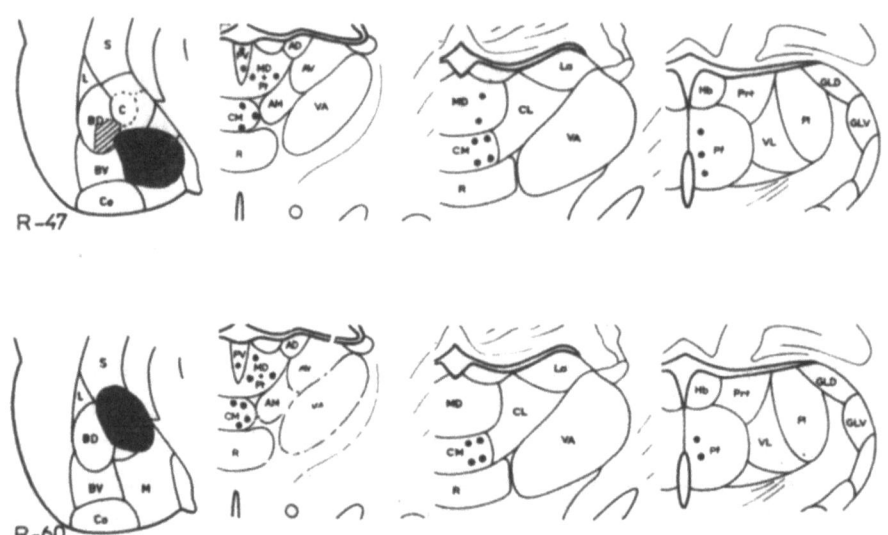

Fig. 10. Distribution of labeled cells in the intralaminar nuclei of the thalamus following injections mainly involving the central nucleus — R60 and medial nucleus of the amygdala — R47. Legend as in Fig. 2.

It the HRP injections involved any other nuclei of the amygdaloid body (Fig. 11), labeled cells were not found in the intralaminar nuclei or there were very few of them. A great number of HRP labeled cells were found in the intralaminar nuclei mentioned above, after injections of HRP to the piriform cortex. Neurons of the central medial nucleus

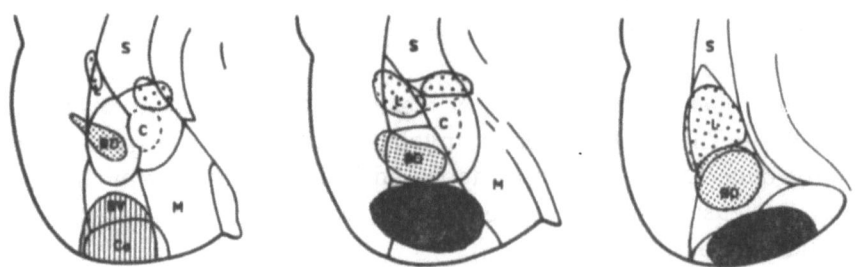

Fig. 11. Injections into the amygdala without evident HRP retrograde transport to neurons of the intralaminar nuclei.

and parafascicular nucleus of the thalamus seem to project mainly to the central and medial nuclei of the amygdala. Perhaps they send fibers to other amygdaloid nuclei also, but probably in fewer numbers. Moreover, it is not inconceivable that the occurrence of a small number of labeled cells in the above mentioned intralaminar nuclei following HRP injections to nuclei of the amygdaloid body (besides central nucleus

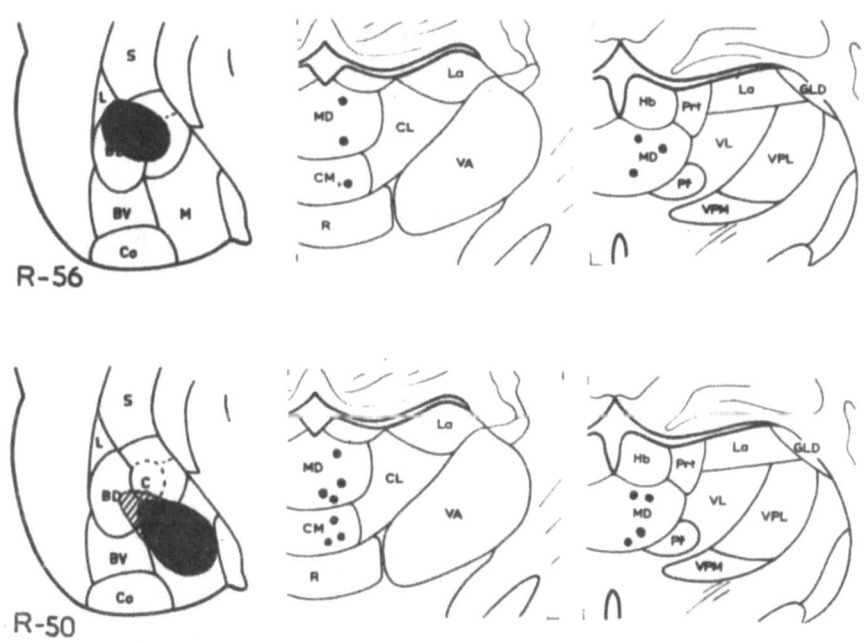

Fig. 12. Distribution of labeled neurons in the mediodorsal nucleus of the thalamus following injections involving the anterior part of the basal dorsal nucleus and additionally central — R56 or medial nucleus of the amygdala — R50. Denotations as in Fig. 2.

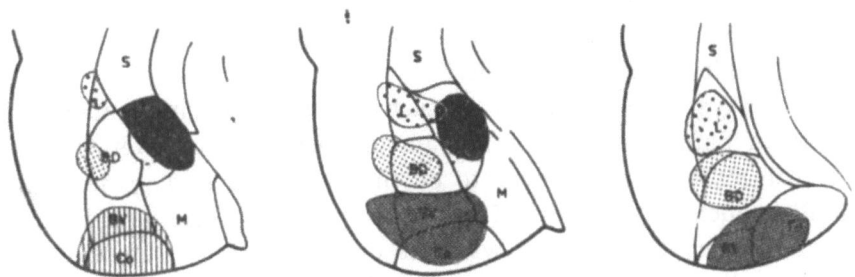

Fig. 13. Various injections into the amygdala without evident HRP retrograde transport to neurons of the mediodorsal nucleus of the thalamus.

and medial nucleus) may be caused by diffusion of HRP to the piriform cortex.

*Amygdalopetal connections of mediodorsal nucleus of the thalamus.* After injections of HRP to the amygdaloid body, labeled neurons appeared in relatively small numbers in the medial part of the mediodorsal nucleus. The labeled neurons were found only when the injection involved the anteromedial area of basal dorsal nucleus. Unfortunately we failed in our attempts to limit the HRP injection only to this area. Labeled neurons in thalamic mediodorsal nucleus definitely appeared in rat R-56 (Fig. 12), in which the injection passed through the medioanterior area of the basal dorsal nucleus and the central nucleus of the amygdala. But there were no labeled neurons in the thalamic mediodorsal nucleus when the HRP injection involved even larger areas of the basal dorsal nucleus or central nucleus (Fig. 13). Exceptionally numerous neurons were observed in the thalamic mediodorsal nucleus in rat R50 (Fig. 12), in which the HRP injection involved the anterior pole of the basal dorsal nucleus and the entire medial nucleus. For this reason we may assume that the mediodorsal nucleus also projects to the medial nucleus of the amygdala.

## CONCLUSIONS AND DISCUSSION

From our results obtained by retrograde transport of HRP from nuclei of the amygdaloid body to the thalamus we may assume that:

1. In the thalamus of the rat there are several groups of nuclei sending fibers to the amygdaloid body. These are nuclei of: (i) the posterior thalamic region-pulvinar, the suprageniculate-limitans nucleus, as well as the peripeduncular nucleus, (ii) the midline thalamic nuclei-paraventricular nucleus, the parataenial nucleus, and nucleus reuniens, (iii) the intralaminar nuclei — central medial nucleus and the parafascicular nucleus, (iv) the mediodorsal nucleus. With but a few exceptions, the thalamic nuclei which project to the amygdaloid body are the nuclei formerly considered as non-specific.

2. The amygdalopetal fibers of each group of thalamic nuclei may terminate in definite nuclei of the amygdaloid body or over its entire extent. The most diffuse appear to be the amygdalopetal projections of the midline nuclei. The fibers of the latter probably terminate in all the nuclei of the amygdaloid body with the exception of the lateral nucleus and the lateral part of the central nucleus. The amygdalopetal fibers which arise in the posterior thalamic and the intralaminar nuclei seem to end in a more limited area of the amygdaloid body, the former in the lateral

nucleus and lateral part of central nucleus, the latter mainly in the medial nucleus and in the intermediate and medial parts of the central nucleus. The amygdalopetal fibers of the thalamic mediodorsal nucleus also appear to have a limited field of projection.

*Projections from the mediodorsal nucleus.* In our material we found, following injection of the amygdala fewer HRP labeled cels in the mediodorsal nucleus than in the adjacent part of the paraventricular and parataenial nuclei. Due to the fact that there is no clear-cut boundary between the posterior part of the parataenial nucleus and the anterior pole of the mediodorsal nucleus, there are sometimes difficulties in deciding to which of the nuclei the labeled cells belong. It seems that the neurons of the mediodorsal nucleus project to the anteromedial area of the basal dorsal and medial nucleus of the amygdala. But we cannot exclude the possibility that the mediodorsal nucleus of the thalamus also projects to other nuclei of the amygdala. The cells of this nucleus project to the piriform cortex also. Krettek and Price (10) using autoradiography showed that the thalamic mediodorsal nucleus in the rat sends quite a large amygdalopetal projection. Previously Nauta (13) using the silver degeneration method in monkeys showed that amygdalopetal connections arise in the neurons of the mediodorsal nucleus of the thalamus, but he mentioned that lesions also partly involved the paraventricular nucleus. Most data indicate the existence of a reciprocal projections between the amygdala and the thalamic mediodorsal nucleus (9, 13, 18). All authors (9, 12, 13) studying the efferent connections of the amygdala by silver degeneration techniques mentioned that terminal degenerations in the mediodorsal nucleus of the thalamus following lesions in the amygdala might have been the result of injury to fibers which originate in the piriform cortex. Lately Krettek and Price (10, 11) and Siegel et al. (18) showed these amygdalofugal connections by means of anterograde and retrograde axonal transport.

*Projections from the posterior thalamic region.* There are some data indicating that neurons of the posterior thalamic region are another source of amygdalopetal connections (5-7, 15). By means of autoradiography and silver degeneration methods it was shown that the pulvinar and peripeduncular nucleus send amygdalopetal projections to the lateral nucleus and the lateral part of the central nucleus. The present results allowed us to ascertain that in the rat the neurons sending axons to the amygdaloid body are located in the dorsomedial area of the posterior part of pulvinar, in the peripeduncular nucleus and in the suprageniculate-limitans nucleus. Nuclei of the posterior thalamic region also send fibers to the ventral part of the striatum adhering to the amygdala. This was to be expected on the basis of previous experiments

(15), carried out by silver degeneration methods, which showed that nuclei of the posterior thalamic region send a large projection terminating in the lateral nucleus and the lateral part of the central nucleus of amygdala, and, in addition, in the anteroventral part of the striatum and in the neocortex above the rhinal sulcus.

It would seem that labeled neurons found sporadically in the thalamic posterior nucleus of some rats appeared not as result of HRP injection to the amygdaloid body, but were caused by the diffusion of this enzyme to the cortex around the rhinal sulcus, just as, after an HRP injections to the cortex of this region a large number of labeled neurons were observed also in the thalamic posterior nucleus.

*Projections from the midline and intralaminar nuclei.* In the literature there are no data concerning any thalamoamygdaloid or amygdalothalamic connections — besides those arising from the mediodorsal nucleus and nuclei of the posterior region. Only Nauta (13), admittedly, suggested that there are amygdalopetal connections from the midline and intralaminar nuclei of the monkey.

As our experiments show, the midline nuclei of the thalamus give off amygdalopetal projections, the fibers of which probably end in the majority of amygdaloid nuclei. Some of the fibers terminate in the piriform and enthorhinal cortex (1a). The projection of intralaminar nuclei to the amygdala is rather diffuse, similar to their cortical projection. There is a certain topographic correlation between the localization of thalamic midline neurons sending amygdalopetal fibers and the localization of their terminals in the nuclei of the amygdaloid body. It seems that the neurons of midline nuclei situated more anteriorly project mainly to the basal nuclei and to the cortical nucleus of the amygdala. Those localized more posteriorly send their axons mostly to the medial nucleus and central nucleus.

The next source of thalamic amygdalopetal connections are probably intralaminar nuclei: central medial and parafascicular. These nuclei seem to project mainly to the central and medial nucleus of the amygdala. Judging by the number of labeled cells, although this may not be conclusive, the central medial nucleus of the thalamus probably sends fibers mainly to the central nucleus of the amygdala, while the parafascicular nucleus does so to the medial. We cannot entirely exclude the possibility that intralaminar nuclei, to a certain extent, also project to other nuclei of the amygdaloid body. The labeled neurons found there sporadically after injections to other nuclei of the amygdala might have been the result of some diffusion of the HRP to the piriform cortex, which receives rather numerous afferents from intralaminar nuclei.

Thalamic midline and intralaminar nuclei emanate thalamostriatal and non-specific thalamocortical connections (8). It seems that the amygdala is an additional third region of projection of these nuclei.

Taking into consideration the thalamoamygdaloid connections as a whole and comparing them with the thalamocortical, we may presume that there are two systems of these connections. One system reaches the frontal cortex and all nuclei of the amygdala with the exception of the lateral nucleus and the lateral part of the central nucleus. The frontal cortex of the rat and the majority of amygdaloid nuclei (basal dorsal, basal ventral, cortical, medial and central) receive afferent fibers arising in the same midline and intralaminar nuclei, as well as in the mediodorsal nucleus of thalamus. The only midline nucleus which deviates from this scheme is the paraventricular nucleus, which does not send fibers to the frontal cortex, but gives an evident projection to the amygdala.

Another system terminates in the temporal cortex (2), the lateral nucleus and lateral part of central nucleus of amygdala. Fibers of this system arise in the nuclei of the posterior thalamic region — pulvinar, peripeduncular and suprageniculate-limitans nuclei (6, 7, 15).

On analyzing our data, we may distinguish two groups of nuclei in the amygdala. One comprises most of the amygdaloid nuclei excluding the lateral nucleus and the lateral part of the central nucleus. In general they receive similar afferent connections from thalamic nuclei, as the frontal cortex and also the entorhinal cortex and the piriform cortex. The second group consists of the lateral nucleus and the lateral part of the central nucleus, both related to nuclei of the posterior thalamic region. These nuclei also seem to differ from other amygdaloid nuclei in regard to their function; it is not excluded that they play a role in conveying sensory information to the limbic system.

## REFERENCES

1. ALBE-FESSARD, D., STUTINSKY, F., LIBOUBANS, S. 1966. Atlas stereotaxique du diencephale du rat blanc. Centre National de la Recherche Scientifique. Paris.
1a. BECKSTEAD, R. M. 1978. Afferent connections of the entorhinal area in the rat as demonstrated by retrograde cell-labeling with horseradish peroxidase. Brain Res. 152: 249–264.
2. BURTON, H. and JONES, E. G. 1976. The posterior thalamic region and its cortical projection in New World and Old World monkeys. J. Comp. Neurol. 168: 249–301.
3. De OLMOS, J. S. 1972. The amygdaloid projection field in the rat as studied with the cupric silver method. In B. E. Eleftheriou (ed.), The neurobiology of the amygdala. Plenum Press, New York, 145–204.
4. EMMERS, R. and AKERT, K. 1963. A stereotaxic atlas of the brain of the squirrel monkey (*Saimiri sciureus*). Univ. Wisconsin Press, Madison.

5. GRAYBIEL, A. M. 1970. Some thalamocortical projections of the pulvinar-posterior system of the thalamus in the cat. Brain Res. 22: 131–136.
6. JONES, E. G. and BURTON, H. 1976. A projection from the medial pulvinar to the amygdala in primates. Brain Res. 104: 142–147.
7. JONES, E. G., BURTON, H., SAPER, C. and SWANSON, L. W. 1976. Midbrain, diencephalic and cortical relationships of the basal nucleus of Meynert and associated structures in primates. J. Comp. Neurol. 167: 385–419.
8. JONES, E. G. and LEAVITT, R. Y. 1974. Retrograde axonal transport and demonstration of non-specific projections to the cerebral cortex and striatum from thalamic intralaminar nuclei in the rat, cat and monkey. J. Comp. Neurol. 154: 349–377.
9. KOSMAL, A. 1973. Efferent projection of the amygdaloid complex in the dog (in Polish). Ph. D. Thesis. Nencki Inst. Exp. Biol., Warsaw.
10. KRETTEK, J. E. and PRICE, J. L. 1977. Projections from the amygdaloid complex to the cerebral cortex and thalamus in the rat and cat. J. Comp. Neurol. 172: 687–721.
11. KRETTEK, J. E. and PRICE, J. L. 1974. A direct input from the amygdala to the thalamus and the cerebral cortex. Brain Res. 67: 169–174.
12. NAUTA, W. J. H. 1962. Neural associations of the amygdaloid complex in the monkey. Brain Res. 85: 505–522.
13. NAUTA, W. J. H. 1961. Fibre degeneration following lesions of the amygdaloid complex in the monkeys. J. Anat. 95: 515–531.
14. NITECKA, L. 1975. Comparative anatomic aspects of localization of acetylcholinesterase activity in the amygdaloig body. Folia Morphol. (Warsz.) 34: 167–185.
15. NITECKA, L. 1979. Connections of the posterior thalamus with the amygdaloid body of the rat. Acta Neurobiol. Exp. 39: 49–55.
16. NITECKA, L., NARKIEWICZ, O. and ZAWISTOWSKA, H. 1971. Acetylocholinesterase activity in the nuclei of the amygdaloid complex in the rat. Acta Neurobiol. Exp. 31: 383–389.
17. ROCKEL, A. J., HEATH, C. J. and JONES, E. G. 1972. Afferent connections to the diencephalon in the marsupial phalanger and the question of sensory convergence in the "posterior group" of the thalamus. J. Comp. Neurol. 145: 105–130.
18. SIEGEL, A., FUKUSHIMA, T., MEIBACH, R., BURKE, L., EDINGER, H. and WEINER, S. 1977. The origin of the afferent supply to the mediodorsal thalamic nucleus: enhancement of HRP transport by selective lesions. Brain Res. 135: 11–23.

Liliana NITECKA, Leszek AMERSKI, Jadwiga PANEK-MIKUŁA and Olgierd NARKIEWICZ, Institute of Medical Biology, School of Medicine, Dębinki 1, 80-211 Gdańsk, Poland.

Note added in proof: Since the manuscript was finished some new data about thalamoamygdaloid connections have been published: Veening, J. G. 1978. Subcortical afferents of the amygdaloid complex in the rat: an HRP study. Neuroscience Letters 8: 192-202.

## LIST OF ABBREVIATIONS

| | |
|---|---|
| AD | nucleus anterior dorsalis thalami |
| AM | nucleus anterior medialis thalami |
| AV | nucleus anterior ventralis thalami |
| BD | nucleus basalis dorsalis corporis amygdaloidei |
| BV | nucleus basalis ventralis corporis amygdaloidei |
| C | nucleus centralis corporis amygdaloidei |
| CL | nucleus centralis lateralis thalami |
| CM | nucleus centralis medialis thalami |
| Co | nucleus corticalis corporis amygdaloidei |
| CS | colliculus superior |
| GLD | corpus geniculatum laterale pars dorsalis |
| GLV | corpus geniculatum laterale pars ventralis |
| GM | corpus geniculatum mediale |
| Hb | nuclei habenulae |
| L | nucleus lateralis corporis amygdaloidei |
| La | nucleus lateralis thalami |
| Li | nucleus limitans |
| M | nucleus medialis corporis amygdaloidei |
| MD | nucleus medialis dorsalis thalami |
| Nst | nucleus striae terminalis |
| Pf | nucleus parafascicularis thalami |
| Pl | pulvinar |
| Pld | pulvinar — area dorsomedialis |
| Plv | pulvinar — area ventrolateralis |
| Po | nucleus posterior thalami |
| Pp | nucleus peripeduncularis thalami |
| Prt | area pretectalis |
| Pt | nucleus parataenialis thalami |
| Pv | nucleus paraventricularis thalami |
| R | nucleus reuniens thalami |
| S | striatum |
| SG | nucleus suprageniculatus thalami |
| VA | nucleus ventralis anterior thalami |
| VL | nucleus ventralis lateralis thalami |
| VPL | nucleus ventralis posterior lateralis thalami |
| VPM | nucleus ventralis posterior medialis thalami |

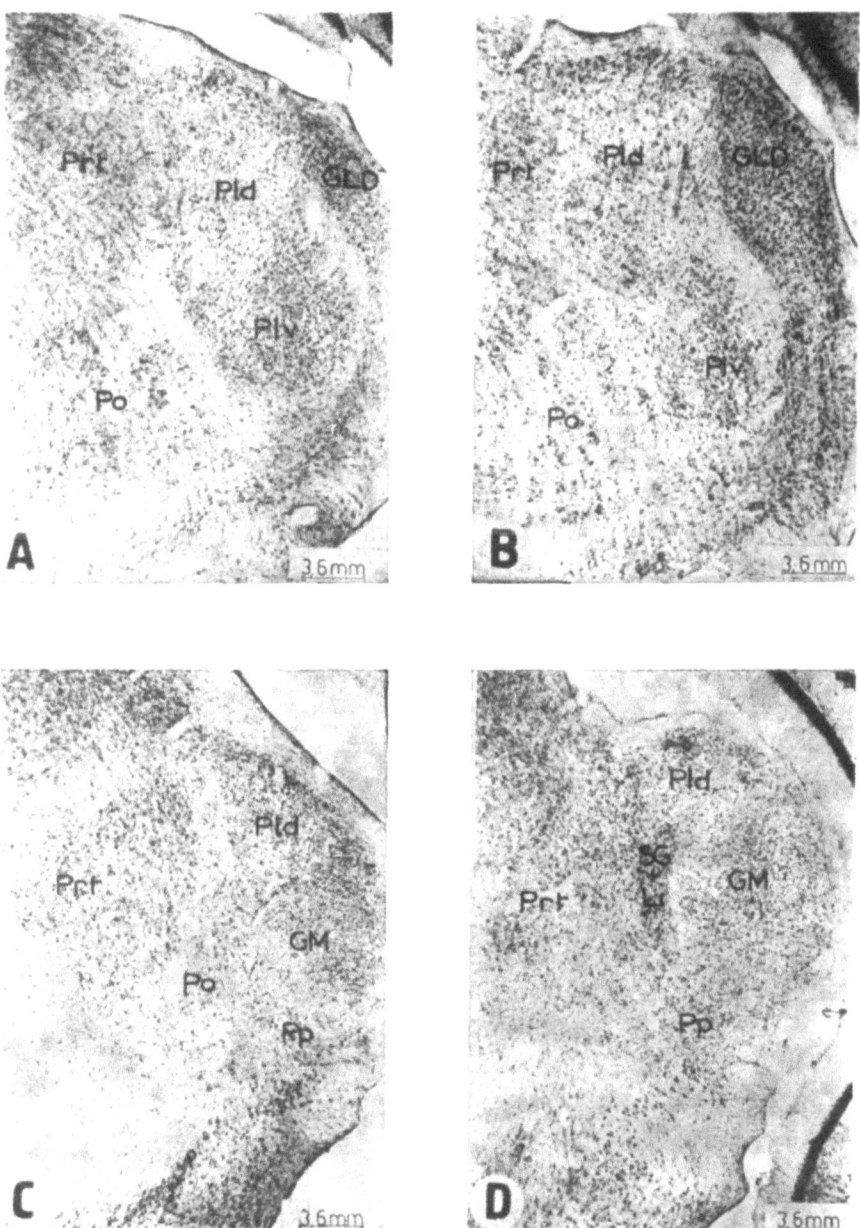

Fig. 1. Low power microphotographs of the posterior thalamic region. Frontal sections *A-D* are set out in rostrocaudal order. Cresyl violet.

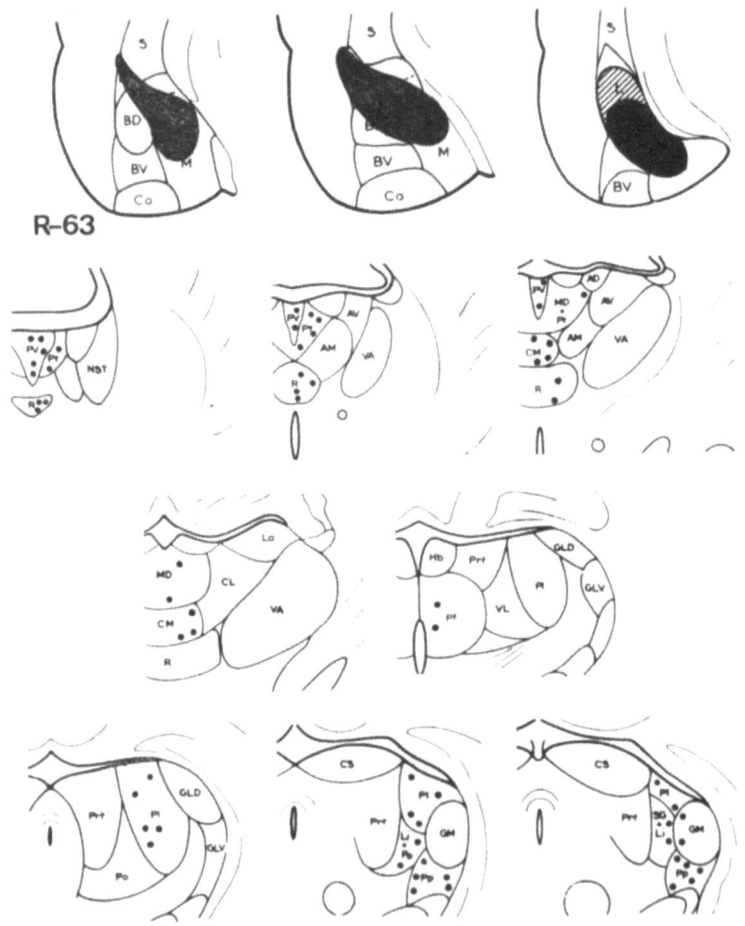

Fig. 2. The distribution of HRP labeled cells in the thalamic nuclei following injection involving most of the amygdala R63. The injection site is marked in black. Slight diffusion of the enzyme is indicated by hatches. Dots represent labeled cells. Frontal sections. Description in the text.

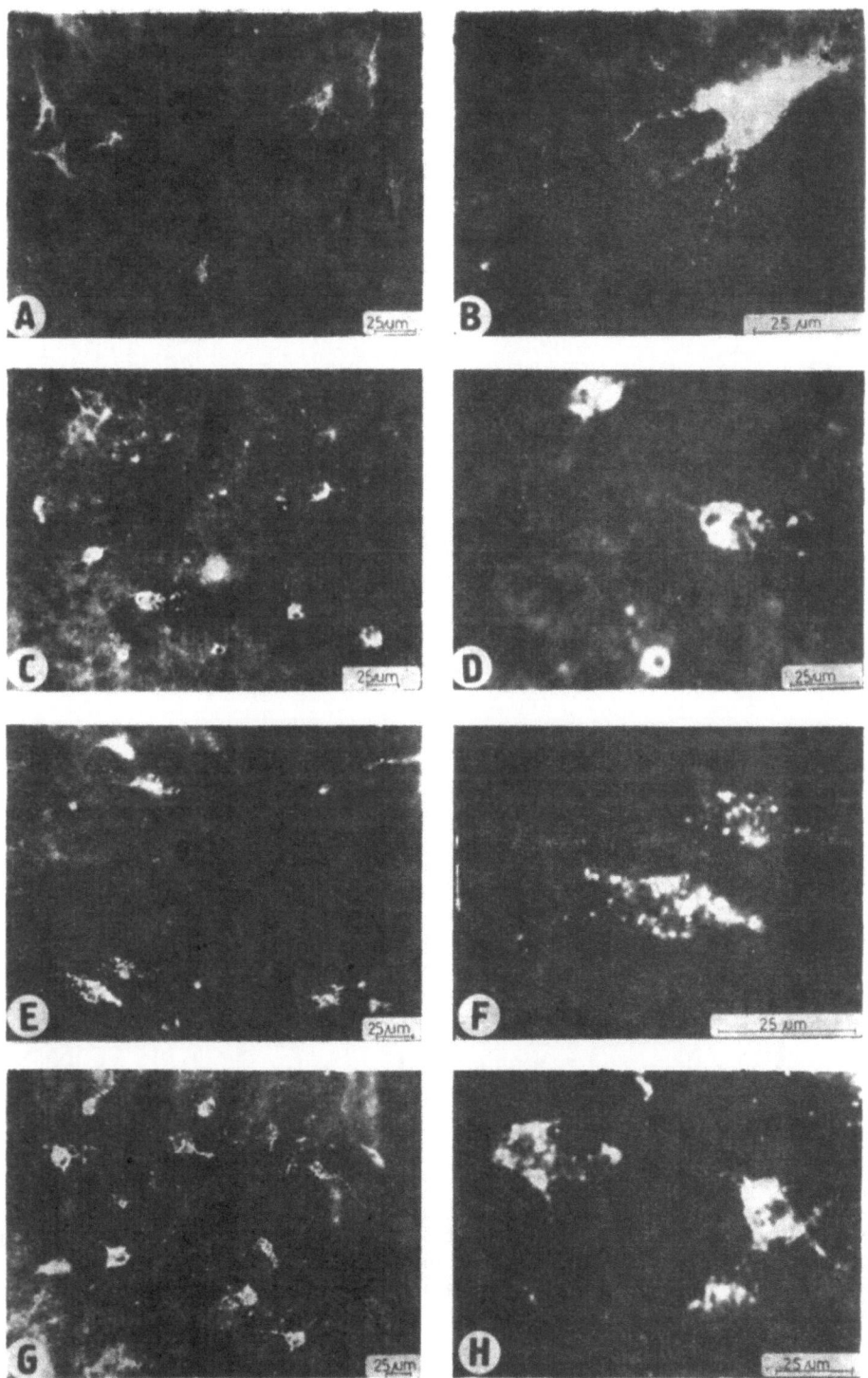

Fig. 3. Dark field microphotographs of labeled cells in the thalamic nuclei following HRP injections into amygdala. A, paraventricular nucleus (left side) and parataenial nucleus (right side); B, paraventricular nucleus; C, central medial nucleus; D, central medial nucleus; E and F, peripeduncular nucleus; G and H, posterior part of the pulvinar.

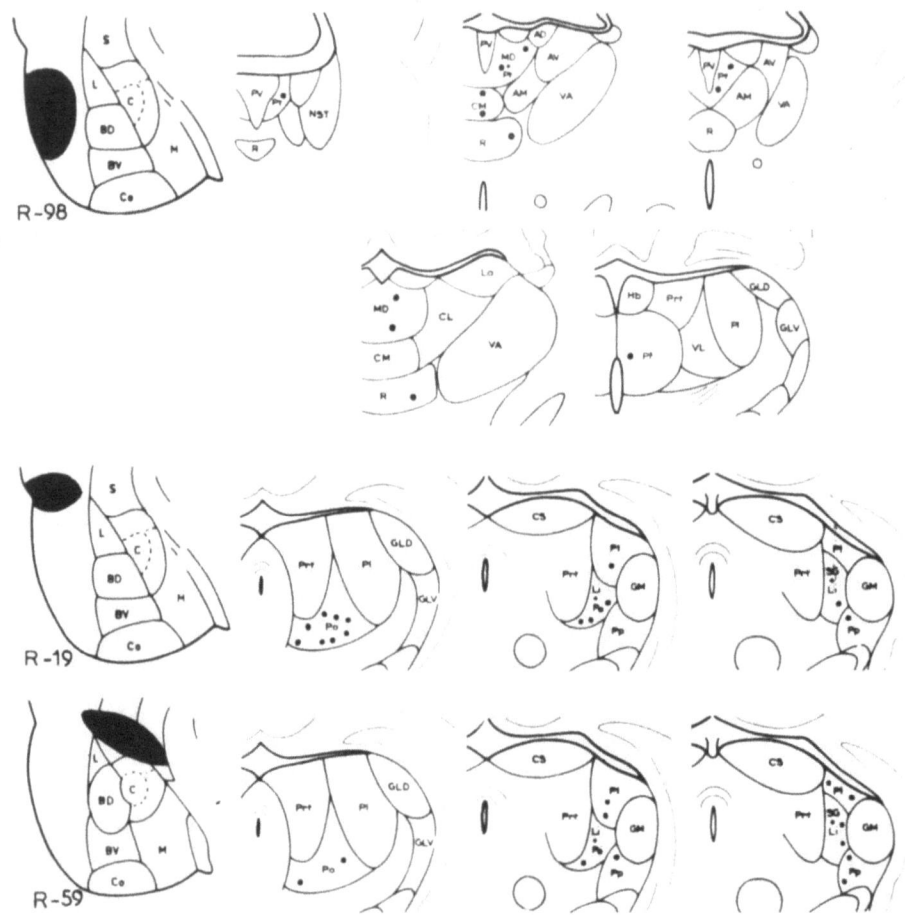

Fig. 4. Distribution of labeled cells in thalamic nuclei following injection involving piriform cortex (R98); cortex adjacent to rhinal sulcus (R19), and anteroventral portion of the striatum (R59). Denotations as in Fig. 2.

# SUBJECT INDEX

The page numbers refer to the first page of the article in which the index term appears

Alternation tests, 299, 339
Amygdaloid complex
    electrical stimulation, 181
    role in instrumental conditioning, 150, 219, 629
    thalamic afferents, 699
Attentive state, 299
Auditory cortex
    neuronal activity, 181, 482
Autoshaping, 78
Aversive classical (Pavlovian) conditioning, see classical defensive conditioning
Avoidance conditioning, 5, 150, 219, 284, 321, 339, 369, 384, 402, 419, 428, 449, 482, 660, see also Differentiation go, no-go
    effects of Locus coeruleus lesions, 419
    effects of pituitary-adrenal hormones 369
    passive, see Punishment

Backward conditioned connections, 525
Backward conditioning, 643

Classical conditioning
    associations in, 5, 207, 571
    alimentary, 5, 525
    defensive, 5, 428, 449, 561, 643
Classical-instrumental interactions, 5, see also Conditioned suppression
Compound stimuli, 42, 321
Conditioned emotional response (CER), see Conditioned suppression
Conditioned inhibitor, 428, 643
Conditioned response decrement, 384
Conditioned response latency, 78, 89, 150, 170, 262, 419, 493, 561, 571, 607, 629, 660
    mathematical modelling of, 402
Conditioned stimulus intensity effect, 339
Conditioned suppression, 207, 321, 384
Corpus callosum
    effects of transection, 32, 262

Decortication, 49, 619
Delayed matching-to-sample test, 63, 299
Delayed response tests, 63, 299, 339, 459, 607, 619
Differentiation
    go, no-go, 207, 321, 561
    go, no-go with symmetrical and asymmetrical reinforcement, 321
    go-left, go-right, 32, 181, 493, 607
    of complex spatial stimuli, 105
    of symmetrical tactile stimuli, 32
Disinhibition, 321, 607
Drive, 207, 219, 321, 629, see also Emotions, Motivation
Drugs effect on behavior, 5, 49, 321, 607, 643

Early experience, 339
Electroconvulsive shock, 49
Emotions 150, 219, 419, see also Drive
Extinction, 150, 207, 331, 339, 419, 428, 449, 561
    as a new positive learning, 181, 561
    with reinforcement, 384, 561

Fimbria-fornix lesions, 78
Frontal cortex, see also Motor cortex, Prefrontal cortex, Premotor cortex

connections with visual cortex, 262
lesions effects, 321, 339, see also Response inhibitory deficit
role in instrumental conditioning, 219

Habituation, 145, 181, 339, 384, 571, 643
Hedonistic learning theories, 150, 219, 231, 250, 449
Hippocampus
    electrical stimulation, 181, 339, 369
    lessions effects, 181, 339, 369
    role in instrumental conditioning, 219, 369
    theta activity, 78, 181, 643
Hormonal influences on avoidance conditioning, 369
Hypothalamus, 339
    and secretion of hormones, 369
    electrical stimulation, 145, 181, 250, 571
    role in instrumental conditioning, 150, 219, 629

Imprinting, 231
Inferotemporal cortex
    electrical stimulation, 299
Inhibition of delay, 321, 384, 459
Instrumental conditioning, see also Avoidance conditioning, Omission training. Punishment, Reward training
    a cellular analog of, 571
    varieties of, 5

Maze learning, 63, 170
Memory, see Short-term memory
Midbrain central gray matter
    role in instrumental conditioning, 629
Motivation, 150, 181, 207, 219, 231, 250, 449, 539, see also Drive, Self-stimulation
Motor act inhibition
    and drug effect, 321
Motor cortex, 262, 660
    electrical stimulation, 284
    neuronal activity, 181, 459, 505, 525, 561
    thalamic efferents, 680

Need, 219, see also Drive
Neuropeptides, 250, 369
Neurotransmitters, 150, 207, 321, 339, 369, 419, 643
Nystagmus, 49, 145

Omission training, 5
Open field activity, 339
Opponent-process theory, 231, 384
Orienting reaction
    conditioned (targeting) 32, 78, 145, 181, 207
    unconditioned, 78, 145

Paradoxical sleep, 643
Partial reinforcement, 5, 207
Partial reinforcement extinction effect, 321
Pericruciate motor cortex, see Motor cortex
Pericruciate frontal cortex, see Premotor cortex
Placing reaction, 284, 505, 619
Plasticity of interneuronal connections, 482, 493, 525, 561, 571
Polarization
    cathodal, 219
    anodal, 299
Postsynaptic potentials, 551, 561, 571
Post-tetanic potentiation, 551, 571
Prefrontal cortex
    lessions effects, 181, 619
    neuronal activity, 459, 660
    role in motor programming, 299, 459
    thalamic efferents, 680
Premotor cortex, 78, 262, 299
    thalamic efferents, 680
Pretrigeminal preparation, 145
Proprioceptive and kinesthetic stimuli, 5, 459
Pseudoconditioning, 5, 339, 525, 561, 571
Punishment, 5, 17, 321, 339, 629

Reaction time, see Conditioned response latency
Response blocking procedure, 181, 339, 419, 449

Response inhibitory deficit, 339
Reward training, 5, 17, 89, 150, 181, 207, 250, 284, 428, 449, 607

Satiation
 effects on conditioned responses, 150, 181, 262, 459
Self-stimulation, 207, 219, 250, 539
Septum
 lesions effects, 181, 339
Short-term memory, 63, 170, 459, 561
Sleep, *see* Paradoxical sleep
Somatosensory cortex, 493, 660
 neuronal activity, 561, 571
Spatial (dis)contiguity
 of CS and reinforcement, 78
 of CS and response, 89, 459
Spreading depression, 49
Stimulus generalization, 105, 339, 607
Striatum, 339, 619, 680
 electrical stimulation, 607, 660
 lesions effects, 607
Synaptic plasticity in hippocampal slices, 551

Striate-peristriate cortex, *see* Visual cortex
Steady potential shifts, 299

Targeting reflex, *see* Orienting reaction
Thalamic nuclei
 and prefrontal cortex, 339, 680
 connections with amygdala, 699
 electrical stimulation, 660
 lesions effects, 262
 neuronal activity, 505
"Thalamic" preparation, 619
Theta rythm, *see also* Hippocampus
 and orienting response, 78
 and paradoxical sleep, 643

Vestibulo-oculomotor system, 49, 643
Visual cortex
 connections with frontal cortex, 262
 lesions effects, 42, 105, 145, 262
 microstimulation, 17
 neuronal activity, 17, 525
Visuomotor guidance, 262

# AUTHORS' INDEX

The page numbers refer to the references

Abe, K., 378
Abzianidze, E. V., 205
Abzug, C., 523
Acuna, C., 205, 281
Adam, G., 640
Adrianov, O. S., 617, 693
Aghajanian, G. K., 168
Ahlskog, J. E., 164
Aivazashvili, I. M., 203
Ajmone-Marsan, C., 281, 617
Akert, K., 283, 320, 363, 383, 479, 628, 693, 694, 696, 697, 713
Albe-Fessard, D., 279, 281, 479, 523, 617, 676, 713
Albert, M., 336
Albert, M. L., 319
Albus, K., 141
Alexeiev, M. A., 297
Alley, K. E., 61
Allikmets, L. H., 164
Allman, J. M., 141
Alloway, T., 61
Altman, J., 363
Amarel, D. G., 426
Amassian, V. E., 279, 523
Amato, G., 279, 283
Amorico, L., 336, 337
Amzelm, A., 203
Anand, B. K., 164
Andersen, P., 559, 600
Anderson, R. N., 380
Anderssen, B., 203
Andre, C., 280
Andreev, L. N., 617
Andreeva, V, N., 678
Andreev, A. E., 298
Andriessen, J. J., 141
Andyan, L., 678
Angaut, P., 279, 280, 281
Anlezark, G. M., 426

Annau, Z., 399
Anokhin, P. K., 203, 399, 570
Anschel, C., 217
Anschel, S., 217
Antonioni, A., 260
Arbuthnott, G. W., 426
Arushanyan, E. B., 676
Asanuma, H., 29, 31, 523, 524, 600
Ascher, P., 600
Aserinsky, E., 656
Ashby, W. R., 141
Asratyan, E. A., 203, 447, 491, 537, 570, 600
Astorga-O., L., 40
Astruc, J., 693, 695
Asuma, M., 481
Atencio, F. W., 141
Atkin, A., 503
Atwood, H. L., 503
Avery, D. L., 627
Axelrod, L. R., 378

Backer, R., 167
Baker, M. A., 491
Balashova, A. N., 600
Ball, G. G., 380
Baranyi, A., 600
Barbeau, A., 626
Bard, P., 523
Barker, L. M., 59
Barlow, H. B., 29
Barnes, C. D., 605
Barnes, G. W., 164, 166
Barnes, J., 164
Barone, R. J., 364
Barrett, J. E., 249
Bartlett, F., 457
Bartlett, J. R., 29, 31
Baškis, A. V., 600
Bastaracho, E., 379

Batini, C., 148
Batuev, A. S., 479, 480, 693
Bauer, M., 601
Bauer, R. H., 480
Baum, M., 363, 456
Baumann, T. P., 279
Baumgarten, R., von, 600
Baure, R. H., 479
Bazhenova, S. I., 678, 679
Beaubaton, D., 279, 283
Beale, I. L., 40
Beardsley, J. V., 457
Bechterew, W., 59
Beck, J., 164
Beckstead, R. H., 363
Beckstead, R. M., 713
Beer, B., 427
Bekhterev, V., 203
Belenkiy, V. E., 297
Bellhorn, R. W., 141
Belluzzi, J. D., 427
Belosertsev, Yu. A., 676
Ben-Ari, Y., 164
Benevento, L. A., 141, 694
Beninger, R. J., 217
Benita, M., 279, 282
Bentivoglio, M., 695
Berger, B. D., 335, 338, 427
Berger, T. W., 600
Bergland, R. M., 169, 378
Bergmann, F., 60
Beritov, I. S., 76, 204
Berkley, M., 142
Berkley, M. A., 141, 143, 144
Berlucchi, G., 141, 143, 144, 260, 282
Berlyne, D. E., 164, 180
Berntson, G. G., 656
Berry, R. N., 141
Besley, S., 365
Best, M. R., 59
Best, P. J., 503
Bettinger, L. A., 282
Białowąs, J., 382
Bidzinski, A., 426
Bignall, K., 280
Bignall, K. E., 279
Bignami, G., 335, 336, 337, 338
Billings-Gagliardi, S. M., 30
Bindra, D., 217

Birch, H., 282
Birch, J., 31
Bird, E. D., 164
Bishop, P. O., 141, 142, 559
Bizzi, E., 656
Björklund, A., 365, 427
Black, A. H., 378
Black-Cleworth, P., 605
Blake, R., 141
Blakemore, C., 143
Bliss, D., 282
Bliss, T. V. P., 559, 600
Bloom, F. E., 383
Bloom, M., 141
Blumfield, T. M., 204
Boakes, R. A., 87
Bogen, J. E., 40
Bogolepov, N. N., 560
Bohdanecky, Z., 549
Bohus, B., 261, 378, 379, 380, 383
Bolles, R. C., 363
Boltz, R. L., 29
Boman, K. K. A., 30
Bouma, H., 141
Bowker, R. M., 656, 657, 658
Bowsher, D., 279
Brady, J. V., 217
Bragin, A. G., 559
Brandt, Th., 60
Braun, J. J., 143, 144, 283
Bregadze, A. N., 456
Brennan, J. F., 363, 364, 367, 368
Briggs, R. M., 104
Brindley, G. S., 30, 600
Broadbent, D. E., 549
Brodmann, K., 693
Brodwick, M., 367
Brookhart, J. M., 297
Brooks, D. C., 656, 657, 659
Brooks, F. H., 261
Brooks, V., 677
Brooks, V. B., 279, 280, 503
Brown, G. W., 204
Brown, M. L., 204
Brown, P. K., 30
Brown, T. S., 382
Brownstein, M. T., 381
Bruner, J., 600
Brush, F. R., 381

Brust-Carmona, H., 31
Brutkowski, S., 204, 364, 695
Bryant, L. H., 491
Bryant, R. W., 260
Bubis, E., 261
Büchele, W., 60
Buchwald, N. A., 381, 456, 503, 617, 677
Buckingham, J. L., 379
Budohoska, W., 48
Bull, J. A., 399, 400
Bunt, A. H., 31
Burden, J., 381
Burden, J. L., 381
Bureš, J., 60, 61, 76, 600
Burešová, O., 60, 61, 600
Burke, L., 714
Burke, W., 559, 600
Burlachkova, N. I., 297
Burns, B. D., 600
Burns, R. A., 399
Burrows, G. R., 142
Burton, H., 713, 714
Buser, P., 280, 281, 282, 600, 677
Butenko, O. B., 617
Buthusi, S. M., 677
Buzsáki, G., 87
Bykov, K. M., 40

Cabanac, M., 164, 260
Cajal, S. R., 601
Callens, M., 143, 677
Campbell, B. A., 364, 366
Campbell, E., 364
Campbell, F. W., 142
Campbell, J. F., 217
Cannizzaro, G., 336
Capaldi, E. J., 447
Carli, G., 657
Carlton, P. L., 204
Carpenter, D. O., 492
Carpentier, M. B., 693
Carro-Ciampi, G., 336
Casady, R. L., 379
Cavonius, C. R., 30
Celiński, M., 48
Cespuglio, R., 657, 659
Chambers, R. A., 142, 280
Chambers, W. W., 504, 659

Chan-Palay, V., 30
Chaymovitz, M., 60
Chepkova, A. N., 603
Cherkes, V. A., 677
Chesher G., 336
Chester, J., 640
Chocholle, R., 601
Chodkiewicz, P., 283
Chorążyna, H., 447
Chow, K. L., 696
Chujo, T., 560
Chun, R. W. M., 627
Ciofalo, V. B., 338
Clark, C., 364
Clark, W. E. LeGros, 31, 693
Clayton, K. N., 218
Clemente, C. D., 456
Cohen, A. I., 30
Cohen, B., 60, 61, 62, 657
Cohen, B. D., 204
Cohen, L. B., 504
Cohen, S. M., 319
Colle, J., 677
Collewijn, H., 60
Collison, C., 76
Combs, C. M., 696
Conde, F., 693
Conde, H., 279, 280, 282, 693
Conforti, N., 379, 380
Conner, R., 337
Conomy, J., 144
Conover, W. I., 180
Conrad, B., 280
Contrucci, J., 166
Cook, L., 336, 338
Cool, S. J., 141
Cooley, R. K., 628
Coover, G. D., 379
Corballis, M. C., 40
Corbit, J. D., 260, 400
Cornsweet, T. N., 142
Cotman, C. W., 559
Cottee, L. J., 142
Coulmance, M., 297
Cowan, E. M., 693
Cowan, W. M., 382, 383, 696
Cowey, A., 30, 144
Cox, D. R., 549
Cragg, B., 503

Crawford, F. T., 141
Crawford, M., 364
Crawford, M. L. J., 141
Craystone, P., 640
Creutzfeldt, O. D., 602
Critchley, M., 280
Critchlow, V., 379
Crow, T. J., 164, 426
Cuenod, M., 693
Cynader, M., 144
Cytawa, J., 164, 165, 230
Czarkowska, J., 368

Dąbrowska, J., 204, 336, 364, 382
Dafny, N., 380
Dahlström, A., 426
Dantzer, R., 336, 337
Darling, D. A., 417
Das, G. D., 363
D'Atena, C., 695
Davidova, E. K., 537
Davidson, J. M., 379, 380
Davila, H. V., 504
Davis, J. L., 282
Davis, R., 559
Davis, W. J., 504
De Acetis, L., 337
Deadwyler, S. A., 559
Dec, K., 148, 149
DeGroot, J., 379
Dejerine, J., 693
Delacour, J., 379
Delgado, J. M. R., 167, 204
Della-Fera, M. A., 261
De Long, M. R., 677
Delorme, G., 657
Dember, W., 367
Dember, W. N., 180
Dembińska, M., 218
Dement, W. C., 657
Demianenko, G. P., 480
Denenberg, V. H., 368
Denny-Brown, D., 142, 280, 626, 677
De Olmos, J. S., 713
De Recondo, J., 282
Derevyagin, V. I., 600
De Ryck, M., 167
Desiraju, T., 480
Desmedt, L. K. C., 168

Deutsch, J. A., 217
De Valois, R. L., 30
DeVito, J. L., 693
Dewar, A. J., 383
Dewar, W., 281
deWied, D., 261, 379, 380, 383
Diamond, I. T., 40, 141, 142
Di Berardino, A., 144, 282
Dichgans, J., 60
Dicks, D., 166
Dickson, J. N., 491
Dickson, J. W., 503
Diemer, N. H., 626
Disterhoft, J. F., 602
Divac, I., 364, 365, 366, 368, 626, 627, 628, 677
Dobelle, W. H., 30, 31
Dobrzecka, C., 40
Domesick, V., 365
Domesick, V. B., 694
Domjan, M., 59
Donhofer, H., 87
Dormont, J. F., 279, 280, 282
Doty, R. W., 29, 30, 31, 142, 149, 206, 601
Douglas, R. J., 204
Douglas, R. M., 559, 601, 602
Dow, B. M., 280
Dreher, B., 142, 144, 148, 149
Dresel, K., 627
Dryagin, Yu. M., 677
Dubner, R., 280
Dubrocard, S., 659
Dubrovina, N. I., 640
Dudek, F. E., 559
Duensing, F., 60
Dufossé, M., 60
Duncan, C. P., 141
Dunin-Barkowski, V. L., 601
Dunn, J., 379
Dunwiddie, T., 559, 602
Dupont, A., 379
Dutta, Ch. R., 694

Eccles, J. C., 601, 677
Eccles, R. M., 603
Edinger, H., 714
Edwards, R. M., 640
Edwards, S. B., 143
Egger, G. J., 365

Eiserer, L. A., 249
Ellison, G. D., 628
Elner, A. M., 298
Elul, R., 148, 549
Emmers, R., 627, 713
Endroczi, E., 379, 380, 382
Engel, J., 481, 524, 570, 605
Engwall, D. B., 365
Enloe, L. J., 166
Epstein, A. N., 260, 261
Ernits, T., 260
Ervin, G. N., 61
Espinoza-V., B., 40
Estes, W. K., 399
Evans, E. F., 104
Evans, J. R., 30
Evarts, E. K., 503
Evarts, E. V., 204, 280, 480, 504, 524, 570, 601
Eydt, K. M., 31

Fabre, M., 280
Falk, J. L., 249
Fallon, J.H., 694
Farthing, G. W., 365
Faugier-Grimaud, S., 280
Favare, B., 337
Feferman, M. E., 695
Feher, O., 600
Feigley, D. A., 365
Fekete, T., 380
Feldman, M., 657
Feldman, S., 379, 380, 381
Feenny, D., 537
Feller, W., 76
Fergusson, D. A. N., 381
Fessard, A., 479
Fetz, E. E., 491, 524
Fibiger, H. C., 427
Fifkova, E., 559, 601
Finan, J. L., 319
Finley, C., 60
Finocchio, D. V., 524
Flexner, J. B., 261
Flexner, L. B., 261
Fluharty, S. J., 260
Flynn, J., 538
Folga, J., 148
Fonberg, E., 165, 167, 456, 457, 627
Foote, W., 538

Forman, R., 503
Forner, S. D., 504
Fortier, C., 379, 380
Foss, J. A., 426
Fouriezos, G., 165
Fowler, H., 180
Fox, C. A., 694
Fox, S., 365
Fox, S. S., 524, 570
Fraley, S. M., 380
Freeman, W. J., 503
Frenois, C., 280
Frey, P. W., 447
Frieman, J. P., 365
Frommer, G., 180
Frommer, G. P., 218
Frontali, M., 336, 337, 338
Frumkes, T. E., 31
Frydrychowski, A., 164
Fuchs, A. F., 31, 320
Fuentes, C., 280
Fujita, S., 696
Fukuda, T., 694
Fukushima, T., 714
Fuller, J. H., 61
Fuller, J. L., 165
Fuller, R. W., 165
Fulton, J. F., 627
Fuster, J. M., 319
Fuster, M., 479
Fuster, Y. M., 480
Fuxe, K., 426

Gadotti, A., 320
Gahéry, Y., 297
Galashina, A. G., 503, 537
Gallistel, C. R., 261
Gandelman, R., 218
Gantt, W. H., 457
Gardner-Medwin, A. R., 601
Garey, L. J., 142
Garrigan, H. A., 399
Gassanov, U. G., 503, 537, 601
Gassanova, R. L., 338, 537
Gassel, M. M., 657
Gatti, G. L., 336
Gazzaniga, M. S., 40
Georgopulos, A., 205, 281
Gerlach, J. L., 382
Gershon, M. D., 657, 659

Gerstein, G., 537
Gerstein, G. L., 491, 503, 504, 537, 604
Gerstein, G. Y., 604
Ghelarducci, B., 60
Gibbs, J., 261
Gibbs, M. E., 60
Gibson, E. J., 16
Gikhman, I. I., 418
Gilbert, C., 142
Gilbert, C. D., 143
Gilbert, P. F. C., 280
Gimeno Alava, A., 280
Grivin, J. P., 30
Gispen, W. H., 380
Giurgea, C., 601
Giurgea, K. M., 601
Glakin, T., 319
Glickstein, M., 142
Globus, A., 601
Głowacka, R., 16
Goddard, G. V., 559, 601, 602
Gold, P. E., 205, 640
Goldman, L., 379
Goldman, P. S., 319, 365, 627, 694
Goldstein, M. L., 399
Gonshor, A., 60
Gorbachevskaja, A. I., 677
Goto, A., 480
Grabowska, A., 48
Grangetto, A., 279, 282, 283
Grastyán, E., 87, 88, 104, 204, 399, 601
Gray, J., 297
Gray, J. A., 337, 380, 426
Graybiel, A. M., 280, 714
Grądkowska, M., 382
Green, D. G., 142
Greene, D., 261
Greenway, A. P., 426
Greven, H. M., 380, 383
Gribkoff, V., 559, 602
Grice, R. G., 418
Griffith, J. S., 601
Grillner, S., 297
Grinvald, A., 504
Gromova, E. A., 165
Grossman, S. P., 204
Guerrero, F. A., 658
Guillemin, R., 380, 381
Guilleminault, C., 657
Guillery, R. W., 142

Gurfinkel, V. S., 297, 298
Gurowitz, E. M., 166
Gustafson, B., 30
Gutman, J., 60
Gybels, J. M., 677

Haaxma, R., 280
Hackett, J. T., 457
Hager, J. L., 61
Halaris, A. E., 382
Halas, E. S., 457
Halgren, C. R., 337
Hall, J. F., 399
Hamilton, K. L., 365
Hamilton, L. W., 365
Hammond, L. J., 337
Hankins, L., 640
Hansson, P., 165
Harlay, F., 659
Harlow, H. F., 165
Harper, R. M., 658
Harris, K., 447
Harrison, J. M., 40, 104
Hartmann-Von Monakow, K., 693
Harvey, J. A., 166
Harwerth, R. S., 29
Hassler, R., 143, 280, 677
Haubenreiser, J., 87
Hauptmann, M., 426
Hawkins, R. C., 261
Hayhow, W. R., 142
Hays, M., 381
Hazlett, J. C., 694
Hearst, E., 88, 365
Heath, C. J., 280, 714
Hebb, D. O., 447, 503, 601
Hécaen, H., 280, 319
Heimer, L., 627
Heise, G. A., 337
Helstrom, C. W., 418
Hendersen, R. W., 447
Hendley, C. D., 658
Hendricks, J. C., 657, 658
Hendrickson, A. E., 31
Hendry, D. P., 400
Henley, K., 657
Henn, V., 60, 62
Henneman, E., 492, 601
Henry, G. H., 141, 142
Hensel, H., 30

Herman, C., 279
Heth, D. C., 399
Heuser, G., 381, 617, 677
Hickey, T., 142
Hilgard, E. R., 399
Hillhouse, E., 381
Hillman, D. E., 281
Hilton, S. M., 640
Hiroshige, T., 378
Hirsh, R., 602
Hobson, J. A., 659
Hodges, J. R., 379
Hodges, L. A., 367
Hoebel, B. C., 164
Hoebel, B. G., 165
Hoffman, H. S., 249
Hoffman, P. L., 261
Hoffmann, J., 143
Hoffmann, K. P., 142
Hohryakova, I. M., 480, 693
Holgate, V., 427
Hollander, A., 142
Holliday, M. S., 204
Hongo, T., 677
Horridge, G. A., 503
Horth-Edel, S., 366
Horvatt, F. E., 456, 503
Hovland, C. I., 399
Howarth, C. I., 217
Hoyle, G., 503
Hrabrich B., 337
Hubel, D. H., 30, 142, 503
Hughes, A., 149
Hull, C. D., 503, 617, 677
Hull, C. L., 165, 204, 399, 417, 447
Hull, E., 365
Hull, E. M., 30
Humphrey, N. R., 142
Hunt, W. A., 602
Hunter, J., 677
Hunter, W. S., 76, 603
Hynek, K., 218
Hyvarinen, J., 280

Ikegami, S., 149
Ilyutchenok, R. Yu., 165, 640, 641
Indra, M., 549
Ingram, W. R., 694
Innes, J. R. M., 368
Ioffé, M. E., 297, 298
Ioffe, S. V., 604, 605

Iorio, L. C., 338
Irgashev, M. S., 678
Isaacson, R., 364, 367, 368
Isaacson, R. L., 204, 381
Ito, M., 60, 281, 657
Itoh, K., 694, 697
Iversen, L. L., 164
Iversen, S. D., 104, 319, 427
Iwahara, S., 337
Iwamoto, I., 319

Jabłonowska, K., 48
Jacobs, B. L., 659
Jacobs, G. H., 30
Jacobsen, S., 365
Jacobson, C. F., 319
Jahan-Pravar, B., 600
Jakubowska, E., 365
James, D. F., 426
Jankowska, E., 30, 602
Janssen, P. A. J., 218
Jarrard, L., 365
Jasper, H. H., 281, 617, 677
Jassik-Gerschenfeld, D., 600
Jastreboff, P. J., 60, 149
Jaworska, K., 457
Jaynes, J., 364
Jeannerod, M., 60, 657
Jenkins, H. M., 88
Jennings, S., 180
Jerlicz, M., 426
John, E. R., 457
Johnston, E. C., 164
Johnston, G. A. R., 626
Johnston, J. C., 448
Jonason, K. R., 166
Jones, E. G., 280, 281, 694, 713, 714
Jones, M. T., 381
Jordan, L. M., 659
Jouvet, M., 657, 659
Juhasz, G., 503
Jung, R., 677

Kaas, J. H., 141
Kączkowska, E., 149
Kaelber, W. W., 694
Kajaia, D. V., 282
Kaji, S., 677
Kamin, L. J., 399, 447, 457, 458
Kandel, E. R., 559, 602
Kant, G. J., 427

Kawaguchi, S., 282
Kawamura, H., 149
Kawamura, S., 142, 696
Kawarasaki, A., 282
Kawi, N., 657
Kayama, Y., 31
Kearly, R. C., 381
Keating, E. G., 281
Keesey, A. E., 218
Kekesi, F., 204
Kelleher, R. T., 337
Kelley, A. E., 427
Kelley, J. P., 142
Kelley, P. H., 426
Keough, T. E., 400
Keshelava, M. V., 205
Kesner, R. P., 205, 679
Kety, S. S., 261, 427
Khalifeh, R. R., 694
Khananashvili, M. M., 602
Kholodov, Y. A., 602
Kievit, J., 281, 694
Kim, R., 693
Kimble, D., 365
Kimble, G. A., 204, 365, 457
Kimmel, H. D., 399
King, F. A., 149
King, R. A., 142
Kinzley, H. Jr., 381
Kirby, R. H., 365
Kish, G. B., 164, 166
Kishi, S., 696
Kislyakov, V. A., 60
Kissileff, H. R., 261
Kitai, S. T., 41
Kitskis, A., 677
Klein, S. B., 365
Kleitman, N., 656
Kling, A., 166, 627
Kling, J. W., 218
Klingberg, F., 603
Klippel, R. A., 427
Kluger, A., 31
Knige, K. M., 381
Knook, H. L., 366, 694
Knott, P. D., 218
Knudsen, E. I., 104
Koob, G. F., 427
Koenig, I. D. V., 164
Kogan, A. B., 503
Kogan, J., 166

Kohler, I., 60
Köhler, O., 617
Köhler, W., 617
Kojima, S., 480
Kolb, B., 167, 366, 367, 628
König, J. T., 427
Konishi, M., 104
Konorski, J., 16, 40, 60, 76, 88, 104,
    149, 166, 204, 218, 319, 337, 366,
    399, 401, 447, 457, 480, 492, 549,
    602, 659, 678
Kobb, G. F., 427
Koranyi, L., 380
Koriakin, M. F., 298
Koridze, M. G., 205
Kornblith, C. L., 602
Kornhuber, H. H., 627, 678
Korolkova, T. A., 602
Koryukin, V. E., 61
Kosmal, A., 166, 365, 714
Kosowski, S., 218
Kossut, M., 149
Kostandov, E. A., 537
Kostopoulos, G. K., 381
Kostowski, W., 426, 427
Koszul, M. F., 281, 282
Kotlyar, B. I., 524, 602, 603, 640
Kotlyar, B. M., 570
Kovacs, S., 380
Kovner, R., 319
Kowalska, D., 364
Kowalska, M., 457, 458
Kozhanov, V. M., 603
Kozhedub, R. G., 604
Kozlovskaya, I. B., 503
Kozlovskaya, M., 230
Kozma, F., Jr., 249
Krames, L., 61
Kratin, Yu. G., 678
Kregule, J., 76
Kreiner, J., 205
Krettek, J. E., 694, 714
Krieg, W. J. S., 694
Krieger, D. T., 381
Kristan, W. B., 503
Krivoy, W. A., 381
Królicki, L., 149
Krushinsky, L. V., 617, 618
Kubota, K., 319, 480, 677, 678
Kudryashov, I. E., 604, 605
Kuleshova, T. F., 61

Kuno, M., 602
Künzle, H., 693, 694
Kuwahara, E., 143
Kuypers, H. G. J. M., 280, 281, 282, 694, 695, 696
Kwan, H. C., 524

Ladigina-Kots, N. N., 618
Lagerspetz, K. M. J., 166
Lagutina, N. I., 205
Lake, N., 659
Lamoreaux, R. R., 458
La Motte, R. M., 281
Landfield, P. W., 205
Landis, C., 602
Lanier, L. P., 381
Lanoir, J., 281
Lánsky, P., 549
Lapin, I. P., 164
Larsen, J. K., 366
Larson, L. E., 602
Lasek, R. J., 695
Lashley, R. L., 249
Laties, A., 142
Latreille, J., 282
Lauprecht, C. W., 617, 677
Laurent, J., 657, 658
Ławicka, W., 88, 104, 205, 319
Leavitt, R. Y., 694, 714
Lee, B. B., 29
LeGrand, Y., 143
Leichnetz, G. R., 695
Lenard, L. G., 427
Lennie, P., 143
Lenzer, I. I., 218
Leonard, C., 366
Leonard, C. M., 695
Leslie, J., 337
Leukel, F., 166
LeVay, S., 143
Levey, A. B., 400
Levick, W. R., 29
Levine, S., 180, 379
Levy, J., 144
Lewin, W. S., 30
Lewis, M., 524
Lewis, P. A. W., 549
Lewis, P. R., 164
Libby, A., 400
Liboubans, S., 713
Lickey, M., 365

Liebeskind, J. C., 640
Liebman, J. M., 640
Lilie, N. L., 337
Lim, R., 678
Linch, G. S., 559
Lindsey, B. G., 503
Lindsley, D. B., 549
Lindvall, O., 365, 427
Lineberry, C. D., 678
Ling, G., 617
Ling, G. M., 381
Liotto, A., 381
Lipscomb, H. S., 380
Lishak, K., 678
Lissak, K., 87, 204, 379, 380
Liu, Ch., 678
Liu, C. N., 659
Liu, C. W., 504
Livanov, M. N., 537, 602
Livesey, P. J., 365
Livingston, R. B., 457
Ljungberg, T., 168
Llinas, R., 60, 281
Lobanova, L. W., 678
LoLordo, V. M., 400, 448
Lömo, T., 559, 600
Londgen, A., 164
Long, C. N. H., 380
Lorente De No, R., 61, 480
Lorenz, K., 249
Loskutova, L. W., 640
Lubińska, L., 695
Lubow, R. E., 366
Łukaszewska, I., 180, 366
Luna, R. G., 143
Lund, J. S., 31
Lund, R. S., 31
Lund, S., 658
Luria, A. C., 48
Luria, A. R., 319, 480
Luszawski, D., 164
Lykken, D. T., 400
Lynch, G. S., 366, 602
Lynch, J. C., 205, 281


Mabry, P. D., 364, 366
Macchi, G., 695
Machne, X., 166
Mackay, A. V. P., 164
Mackensen, G., 61

Mackintosh, N. J., 427, 448
Mac Nicholl, E. F. Jr., 31
Madarasz, J., 87
Magni, S., 76
Magoun, H. W., 658
Maier, S. F., 400
Maiorov, V. I., 524, 603
Makarov, V. A., 640
Malliani, A., 678
Malyukova, I. V., 693
Malyukova, J. V., 480
Malmo, R. B., 319
Mandelbrod, I., 381
Mann, G., 658
Marchiafava, P. L., 148, 149, 657
Marcos, R. A., 491
Marcus, A., 31
Marcus, R. J., 627, 628
Marczyński, T. J., 457
Margules, D. L., 337
Mark, R. F., 60
Markowitsch, H., 677
Markowitsch, H. J., 366, 627
Markowska, A., 180
Marks, N., 381
Marks, W. B., 31
Marquis, D. G., 399
Marr, D., 281
Marrazo, M. S., 367
Marschall, J. F., 166
Martin, I., 400
Martin, J. P., 627, 678
Martin, R. A., 337
Marty, R., 280
Maser, J. D., 337
Mason, J. W., 382
Mason, S. T., 427
Massion, J., 280, 281, 282, 297, 298, 678
Masterton, R. B., 40
Matsuda, Y., 282
Matsuo, V., 60
Matthies, H., 206
Maulsby, R. L., 457
Mayer, F. J., 640
Mayers, K. S., 282
McClearn, G. E., 165
McCleary, R. A., 337
McDonough, J. H., 205
McEwen, B. S., 382
McGaugh, J. L., 205, 640
McGinty, D., 658

McIlwain, J. T., 602
McIntire, R. W., 218
McKearney, J. W., 337
McLeary, R. A., 382
McLennan, H., 640
McNaughton, B. L., 602
McNaughton, N., 426
Mead, W. R., 30
Meadows, J. C., 31
Mehler, W. R., 695
Meibach, R., 714
Meibach, R. C., 382
Meikle, T. H., 282
Meikle, T. H. Jr., 144
Mekhedova, A. Y., 230
Melis, W., 168, 218
Melvill Jones, G., 60
Mendell, P., 366
Mendell, R., 366
Menzel, E. W., 76
Mereaux, J., 677
Merzhanova, G. Ch., 537
Mettler, F. A., 628, 695
Meulders, M., 677
Meyer, D. R., 697
Meyer, P. M., 166
Meyers, R. D., 166
Mgaloblishvili, M. M., 205
Micco, D. J., 382
Michałek, H., 336
Michalski, A., 148, 149
Michel, F., 657
Michihoro, M., 696
Miki, M., 696
Mikiten, T. H., 658
Milar, K. S., 337
Miles, F. A., 61
Milgram, N. W., 61
Millbrook, B. A., 180
Millenson, J. R., 337, 400
Miller, J., 142
Miller, N. E., 204, 457
Miller, S., 16, 166, 218, 457, 549
Mills, J. A., 679
Mills, L. E., 261
Milner, A. D., 281
Milner, B., 205
Milner, P., 261, 368, 549
Milner, P. M., 205, 217, 549
Mink, W. D., 503
Minkowski, M., 695

Mirsky, A. F., 167
Mis, F. W., 400
Miskhin, M., 319, 365, 678
Miyashita, Y., 60
Mizuno, N., 282, 694, 697
Mladejovsky, M. G., 30
Mod, L., 87
Moffit, R., 678
Mogenson, G. J., 218
Moldov, R., 382
Moll, L., 281
Møller, M., 626
Molnar, P., 87
Molodkina, L. N., 617
Monakow, C., von, 695
Monnier, M., 679
Montgomery, K. C., 166
Moore, G. P., 549
Moore, J. W., 366, 400
Moore, R. Y., 382
Mora, F., 166
Morgan, H. C., 30
Mori, S., 658
Mormede, P., 337
Moroz, B., 149
Morrell, F., 457
Morrison, A. R., 656, 657, 658, 659
Morrow, M. C., 400
Morse, W. H., 337
Mortimer, J., 658
Moruzzi, G., 148, 658
Mountcastle, V. B., 205, 281, 503
Mowrer, O. H., 458
Murphy, H. M., 382
Murphy, J. T., 524
Murray, M., 695
Mush-Clement, K., 280
Myers, M. L., 149
Myers, R. E., 40

Nachman, M., 166
Nadel, L., 76, 378
Naidel, A. V., 297
Naito, H., 677
Nakano, K., 693
Narikashvili, S. P., 282
Narkiewicz, O., 382, 695, 714
Naumann, G. L., 40
Nauta, H. J. W. Jr., 695
Nauta, W. J. H., 367, 382, 480, 481, 658, 695, 696, 714

Neafsey, E. J., 503
Neame, R. L. B., 381
Negishi, K., 504
Negrao, N., 29, 30
Nelson, P. G., 104
Neverov, V. P., 60, 61
Newby, V., 249
Newton, R. A., 678
Niebyl, P. H., 658
Niemegeers, C. J. E., 218
Niemer, W. T., 524
Nieoullon, A., 297
Nigito, S., 336
Niimi, K., 142, 143, 696
Niki, H., 319, 320, 480, 481, 678
Ninteman, F., 167
Nishioka, S., 149
Nissen, H. W., 320
Nitecka, L., 166, 714
Nonneman, A., 366, 367
Nonneman, A. J., 366
Nonneman, J., 366
Norgren, R., 166
Nowles, V., 320
Nyakas, C., 382
Nyakas, Cs., 380
Nyquist, J. K., 524

Obata, K., 657
Öberg, R. G. E., 627
O'Brien, J. H., 570
Ockleford, E. M., 281
O'Connel, R., 180
Oderfeld-Nowak, B., 382
Ohye, C., 280
Oka, H., 282
O'Keefe, J., 76, 378
Olds, J., 166, 218, 261, 503, 549, 602
Oliff, G., 141
Olley, J. E., 338
Olmstead, C. E., 627, 628
Olszewski, J., 696
Oltmans, G. A., 166
Olton, D. S., 76, 77
Oniani, T. N., 205, 230, 641
Orban, G. A., 143
Orem, J., 537
Orjonikidze, Ts. A., 641
Orlov, A. A., 480, 481
Orlovsky, G. N., 298, 659
Osborne, S. R., 249

Oswaldo-Cruz, E., 617, 676
Otellin, W. A., 676
Otsuka, R., 143
Ott, T., 206
Overman, W. H. Jr., 29
Overmier, J. B., 399, 400
Owen, F., 164
Ozawa, N., 679
Ozonas, G., 283

Pack, K., 400
Pacut, A., 418
Page, R. B., 378
Palay, S. L., 30
Palestini, M., 148
Palfai, T., 217
Palkovitz, M., 382
Palmer, L. A., 143, 144
Paltsev, E. I., 297
Pandya, D. N., 282, 695
Panksepp, J., 218
Papez, J. W., 382
Parmeggiani, P. L., 658
Parmeggiani, W. A., 297
Parsons, P. J., 365
Pasik, P., 143
Pasik, T., 143
Passingham, R. E., 365
Patel, J. B., 338
Patterson, M. M., 603
Pavlov, I. P., 16, 205, 400, 448, 458, 492, 538, 603, 618
Pavlova, O., 218
Peacook, J. H., 696
Pearlman, A. L., 31
Pellegrino, L., 641
Pellet, J., 659
Percheron, G., 696
Perez-Cruet, J., 218
Perkel, D., 537
Perkel, D. H., 503, 549
Perkins, C. C. Jr., 400
Petersen, W. A., 297
Peterson, B. W., 659
Petras, J. M., 696
Petropavlovski, V. P., 678
Pfaff, G., 382
Pfafmann, C., 166
Phillips, C. B., 31
Phillips, C. G., 678
Phillips, D. S., 282

Phillips, M. J., 261
Phillis, J. W., 381
Phinney, R. L., 368
Pickenhein, L., 603
Pickering, V. L., 249
Pigareva, M. L., 230
Pirogov, A. A., 480, 481
Pi-Sunyer, F. X., 261
Pliskoff, S. S., 218
Płaźnik, A., 427
Plokhinsky, N. A., 618
Polit, A., 298
Polson, M. C., 30
Polyak, S., 31, 104, 696
Pompeiano, O., 657, 658, 659
Popova, N. S., 617
Poranen, A., 280
Porczi, J., 601
Porter, R., 31, 524, 678
Porter, R. W., 380
Poschel, B. P. H., 167
Posluns, D., 205
Pounderoux, G., 677
Powell, D. A., 31
Powell, E., 367
Powell, E. A., 364
Powell, T. P. S., 142, 281, 382, 696
Prado-Alcala, R., 458
Pratt, S. R., 104
Premack, D., 458
Preobrazhenskaya, L. A., 230
Pribram, K. H., 165, 320, 481, 696
Price, D., 678
Price, D. P., 677
Price, J. L., 694, 714
Pritz, M. B., 695
Pritzel, M., 366, 627, 677
Prokasy, W. F., 400
Prosser, C. L., 603
Protopopov, V. P., 618
Provenzano, P. M., 336
Purpura, D., 678
Purpura, D. P., 696

Quinton, E., 166

Rabinovich, M. Ya., 570, 603, 641
Radil-Weiss, T., 40, 149, 549, 550
Raisman, G., 367, 382
Ranck, J. B., 383
Randall, P. K., 164

Randich, A., 400
Raphan, T., 60, 61
Raskin, L. A., 364
Ratner, A. M., 249
Ratner, S. C., 659
Ravizza, R., 40
Reading, H. W., 383
Reid, L., 458
Rein, H., 366
Reinking, R. M., 605
Rescorla, P. A., 205
Rescorla, R. A., 16, 367, 399, 400, 448
Rezak, M., 141
Rhines, R., 658
Ribadeau-Dumas, J. L., 282
Ricciardi, K. M., 418
Riccio, D. C., 364, 365, 367
Richard, D., 281, 282
Riesen, A. H., 320
Riggs, L. A., 141
Rinaldi, P., 503
Rinvik, E., 696
Riso, R. R., 31
Rispal-Padel, L., 281, 282
Ritter, S., 167, 427
Robbins, D. O., 30
Roberts, D. C. S., 427
Roberts, T. D. M., 298
Roberts, T. S., 30, 696
Roberts, W., 367
Roberts, W. A., 77
Roberts, W. J., 602
Roberts, W. W., 204
Robertson, R. T., 282
Robins, E. D., 31
Robinson, A. D., 320
Robinson, D. A., 62
Rocha-Miranda, C., 617, 676
Rockel, A. J., 714
Rodgers, R. J., 167
Rodriguez, E. M., 383
Roginsky, G. Z., 618
Rohrbaugh, M., 365, 367
Rolls, E. T., 261
Rondot, P., 282
Ropartz, P., 61
Rosanov, S. I., 603
Rose, D., 143
Rose, J. E., 696, 697
Rose, W. N., 504
Roselli, L., 365

Rosellini, R. A., 249
Rosen, A. J., 457
Rosen, I., 29, 600
Rosen, S. C., 320
Rosenblatt, F., 603
Rosenblum, M., 523
Rosenkilde, C. E., 627
Rosenquist, A. C., 143, 144
Rosić, N., 336, 338
Rossi, G. F., 148
Rossini, P., 695
Rosvold, H. E., 165, 167, 319, 365, 367, 627, 628, 678
Rougeul, A., 279, 677
Royce, G. J., 678
Rozhanski, N. A., 205
Rożkowska, E., 167
Rudell, A. P., 524
Ruf, K., 383
Rush, H., 458
Rusinov, V. S., 320, 603, 641
Russel, G. V., 659
Rutledge, L. T., 603
Rutishauser, F., 697

Sadowski, B., 168, 218
Sakai, K., 657, 659
Sakai, M., 481, 678
Sakai, S., 697
Sakata, H., 29, 205, 281, 282
Salinas-Zeballos, M. E., 628
Salvert, D., 659
Salzberg, B. M., 504
Samuelson, R. J., 77
Sanderson, K. J., 143
Sandrew, B. D., 320
Sanides, F., 143
Santibañez-H., G., 40, 104, 148, 149
Saper, C., 714
Sar, M., 383
Sarmiento, R. F., 31
Sarna, M., 149
Sasaki, K., 282
Sastre, J. P., 659
Sato, M., 697
Sato, S., 677
Savchenko, E. I., 524, 603
Sawa, M., 167
Schaefer, K. P., 60
Schallert, T., 167
Schally, A. V., 380

Scheibel, A. B., 601
Schlag, J., 281
Schmid, R., 60
Schmied, A., 279, 282
Schneider, G. E., 104
Schoenfeld, W. N., 458
Schotman, P., 380
Schuckman, H., 31
Schulenburg, J., 367
Schultz, W., 168
Schwartz, L. S., 382
Schwartz, M., 167
Schwartzkroin, P. A., 559
Schwing, R. C., 549
Scollo-Lavizzari, G., 697
Scott, J. P., 249
Scott, T. R., 31
Searle, J. L., 249
Sears, R. J., 447
Sechenov, I. M., 618
Sechzer, J. A., 61
Segal, M., 383, 602
Segundo, J. P., 166, 549
Seligman, M. E. P., 61, 448
Semenyuk, E. F., 570
Semmes, J., 696
Sencer, W., 697
Sepinwall, J., 336, 338
Sequndo, J. P., 491
Serdjuchenko, V. M., 537
Sessions, G. R., 427
Settlage, P. H., 697
Seward, J. P., 218
Shapovalov, A. I., 603
Shapovalova, K. B., 678, 679
Sharonova, I. N., 603, 605
Shatz, C. J., 143
Sheffield, F. D., 167, 458
Sheperd, G., 367
Sherman, J. E., 400
Sherman, S. M., 141, 143
Sherrington, Ch., 298
Shibutani, H., 282
Shik, M. L., 298, 658, 659
Shimazu, H., 677
Shimokochi, M., 457
Shimono, T., 282
Shipley, J. E., 167
Shkolnik-Yarros, E. G., 31
Shalaer, R., 149
Shugalev, N. P., 617

Shulgina, G. I., 570
Shumilina, N. I., 298
Shustov, V. N., 679
Shute, C. C., 164
Shvyrkov, V. B., 570, 603
Sidman, M., 400, 458
Siegel, A., 382, 714
Siegel, J., 678
Siegel, S., 249
Siegert, A. J. F., 417, 418
Siegmund, H., 40
Silakov, V. L., 602
Silva, M. T. A., 382
Silvestri, R., 367
Simon, P., 338
Simon, R. P., 659
Simoni, A., 260
Simonov, P. V., 167, 205, 230
Simpson, J. I., 61
Sinberg, R., 601
Singer, P., 279
Singer, W., 144, 659
Singh, R., 366
Skinner, B. F., 16, 167, 205, 399
Skinner, J. E., 367
Skolasińska, K., 149
Skorokhod, A. V., 418
Skrebitsky, V. G., 560, 603, 605
Ślósarska, M., 149
Śmiałowski, A, 168
Smirnov, S. I., 678
Smith, A., 678
Smith, E., 31
Smith, E. L., 29
Smith, G. P., 61, 261
Smith, H. E., 282
Smith, O. A., 693
Snider, R. S., 524
Snyder, D., 368
Sokolov, E. N., 603, 659
Solomon, R., 167
Solomon, R. L., 249, 400, 458
Sołtysik, S., 338, 448, 457, 458, 679
Somjen, G., 492
Sørensen, K. E., 626
Sorenson, C. A., 628
Soubrieg, P., 338
Spatz, W. B., 31
Spear, N. E., 364, 365
Spear, P. D., 143, 144, 279, 283
Spence, K. W., 367, 418

Spencer, W. A., 559, 602
Speransky, A. D., 40
Sperry, R. W., 40
Spokes, E. G., 164
Sprague, J., 141
Sprague, J. M., 142, 143, 144, 260, 282
Staats, S. R., 180
Stafstrom, C. E., 504
Stamm, J. S., 319, 320
Starkweather, C., 658
Starr, A., 40
Starr, M. D., 249
Staufer, E. K., 605
Steele, D., 458
Steeves, J. D., 659
Stein, D. G., 280
Stein, L., 167, 168, 218, 261, 335, 337, 338
Stein, L. C., 427
Steiner, F., 383
Steiner, F. A., 383
Stellar, E., 167, 261
Stellar, J. R., 261
Stenhouse, D., 603
Stępień, I., 88, 320, 400
Stępień, L., 40, 88
Steriade, M., 659
Sterman, M. B., 456
Stevens, J., 504
Stevens, S. S., 31
Stevenson, J. A. F., 168
Stiglick, A., 168
Stikes, E. R., 367
Stone, J., 142, 144
Stone, S. A., 297
Stoney, S. D., 524
Stoney, S. D. Jr., 31
Stoney, S. D. A. Jr., 523
Stratton, G. M., 61
Stratton, J. W., 249
Straus, E., 261
Strick, P. L., 677, 697
Stricker, E. M., 168
Storozhuk, V. M., 570
Strozzi-V., L., 40
Stuart, D. G., 605
Stumpf, W. E., 383
Stutinsky, F., 713
Subramianian, K. N., 503
Sudakov, K. V., 205

Sundberg, S. H., 559, 600
Sutherland, N. S., 180
Suvorov, N. F., 618, 679
Suvorov, V. V., 679
Suzuki, H., 319, 480, 481
Sveen, O., 559, 600
Swanson, L. W., 383, 714
Swett, J. E., 297, 659
Sychowa, B., 40, 205
Syka, J., 40
Szabo, I., 601, 603
Szczechura, J., 88, 104
Szentagothai, J., 61
Szwarcbart, M. K., 367, 627, 628, 678
Szwejkowska, G., 205, 457

Takaoka, Y., 282
Takemori, S., 62
Talbot, W. H., 281
Tanaka, D. Jr., 697
Tanaka, T. Yu. H., 41
Tanji, J., 503, 504, 524
Tapp, J. T., 458
Tardy, F., 659
Tarnecki, R., 148, 659
Tasiro, S., 697
Tatton, W. G., 504
Tauc, L., 602
Taylor, A., 605
Taylor, A. N., 379
Taylor, C. J., 168
Taylor, M. J., 504
Teghtsoonian, R., 364
Teitelbaum, H., 368
Teitelbaum, P., 165, 166, 167
Telegdy, C., 168
Temperley, H., 503
Tempesta, E., 695
Terman, M., 218
Teyler, T. J., 282, 603
Thach, W. T., 280, 524
Thompson, G. I., 168
Thompson, R., 368
Thompson, R. F., 282, 600, 603
Thompson, R. L., 628
Thompson, R. W., 368
Thompson, W. D., 31, 524
Thornton, J. C., 261
Tigges, J., 31
Tigges, M., 31

Tigter, H., 383
Timchenko, A. S., 282
Tiptaft, E. M., 381
Tissot, R., 679
Tobias, T. J., 697
Todt, J. C., 380
Toffey, S., 249
Tohyama, M., 659
Tolkunov, B. F., 618, 679
Tolman, E. C., 77
Tolpyshev, K. A., 676
Tomie, A., 400
Touret, M., 659
Towe, A. L., 524
Towfighi, J. T., 168
Trapold, M. A., 400
Travis, A. M., 697
Travis, R. P., 549
Tretter, F., 144
Trevarthen, C., 104
Trojniar, W., 164, 165, 230
Trouche, E., 279, 283
Trowill, L. A., 210
Trulson, M. E., 659
Tucker, T. J., 627
Tunkl, J., 141, 143
Turlejski, K., 148
Turner, B. H., 168
Turner, H. B., 166
Turner, L. H., 400
Tusa, R., 144
Tusa, R. J., 143, 144
Tveritskaya, I. N., 87
Twanoto, P., 480
Tyner, C. F., 524
Tzavaras, A., 283

Udo, M., 657
Ueda, M., 61
Uemura, T., 62
Ungersted, U., 168
Uno, M., 503, 679
Updyke, B. V., 144
Urban, I., 379
Uvarov, V. G., 605
Uyeda, A., 218

Vahing, V. A., 164
Valasek, J., 550

Vanderwold, C. H., 205
Vanderwolf, C. H., 628
Vanegas, T., 142
Vanetsian, G. L., 537
Van Harrenveld, A., 504, 559, 601
VanHartesveldt, C., 381
Van Hoesen, G., 627
VanRee, J. M., 261, 379
VanRiezen, H., 383
Van Sommers, P., 169
VanWimersma Greydanus, T. B., 383
Vasilevskii, N. N., 481, 524, 570, 605
Vedovato, M., 697
Venegas, H., 538
Verderevskaya, N., 149
Vereczkei, L., 88, 104, 399
Vermes, I., 168
Vernon, L., 677
Versteeg, D. H. G., 261
Verzeano, M., 503, 504
Vest, B., 319
Vicedomini, J. P., 364
Villablanca, J. R., 627, 628, 659
Vinnitsky, I. M., 640, 641
Vinogradova, O. S., 206, 559
Vitikova, T., 336
Voigt, J., 367
Voneida, T., 618
Vorobyev, V. S., 603
Voronin, L. L., 320, 524, 570, 600, 603, 604, 605

Waddington, J. L., 338
Waespe, W., 62
Wagner, A. R., 448
Wakefield, C., 456
Walasek, G., 401
Wald, G., 30
Waldman, A., 230
Walker, A. E., 697
Wallach, G., 382
Walter, R., 261
Walton, K., 60, 281
Wang, G. H., 627, 628
Wang, R. Y., 168
Warburton, D. M., 338
Ward, J. P., 141
Warmath, D. S., 141
Warren, H. B., 283

Warren, J. M., 283, 320, 697
Watkins, D. W., 141, 144
Watson, J. B., 368
Watt, D. G. D., 605
Wauquier, A., 168, 218
Wegener, J. G., 320
Weiner, H. A., 523
Weiner, S., 714
Weingarten, H., 168
Weiskrantz, L., 144
Weiss, B., 168
Weiss, B. M., 381
Weiss, J. M., 382
Weiss, R., 368, 628
Welzel, W., 206
Wendlandt, S., 426
Wenzel, B. M., 168
Wenzel, J., 560
Werz, M. A., 76
Wester, K., 559
Whimbey, A. E., 368
Whishaw, I. O., 167
White, N., 168, 230
Whitfield, I. C., 41
Wicke, J. D., 549
Wideman, C. H., 382
Wiegmann, O., 61
Wiens, T. J., 504
Wieraszko, A., 382
Wiersma, C. A. G., 504
Wiesel, T. N., 30, 142, 503
Wigstrom, H., 600
Wikmark, R., 368
Wikmark, R. G. E., 628
Willburn, M. W., 679
Williams, D. R., 401
Williams, H., 401
Wilson, D. M., 504
Wilson, L. M., 368
Winans, S. S., 144
Wingström, H., 559
Winiczai, Z., 87
Winokur, G., 679
Wise, C., 427
Wise, C. D., 167, 427
Wise, C. M., 168
Wise, R. A., 165, 427
Wong, V. C., 524
Wong-Riley, M. T. T., 144
Wood, C. C., 144, 283

Woodbury, D. M., 383
Woodruff, M. L., 381
Woods, J. W., 168
Woods, P., 180
Woods, P. S., 363
Woody, C. D., 481, 605
Woody, C. G., 570
Woolsey, C. N., 283, 693, 696, 697
Wright, C. G., 605
Wulff, J. J., 167
Wyers, E. J., 456, 617, 677
Wynne, L. C., 458
Wyrwicka, W., 168, 203, 206, 230, 456, 458

Yagi, N., 60
Yagodnitsyn, A. S., 658
Yalow, R. S., 382
Yalow, R. W., 261
Yamaga, K., 30
Yamamoto, C., 560
Yamshikova, N. G., 617
Yanagisawa, N., 626
Yee, R. D., 62
Yin, T. C. T., 281
Yong, L. R., 60
Yoon, M., 29
Yoshida, M., 657, 679
Young, P. T., 169
Yu, J., 659
Yung, R., 538

Zanchetti, A., 148
Zarkeshev, A. G., 602
Zarzecki, P., 31
Zawistowska, H., 714
Zbrożyna, A. W., 640
Zee, D. S., 62
Zelig, S., 60
Zeman, W., 368

Żernicki, B., 148, 149, 401, 550
Zieliński, K., 16, 104, 206, 338, 364, 365, 368, 401, 418, 448, 550
Zigmond, M. J., 168
Zigmond, R. E., 164
Zoladek, L., 77
Zornetzer, S. E., 205
Zucker, R. S., 605
Zvartau, E., 230

MIX
Papier aus verantwortungsvollen Quellen
Paper from responsible sources
**FSC® C105338**

If you have any concerns about our products,
you can contact us on
**ProductSafety@springernature.com**

In case Publisher is established outside the EU,
the EU authorized representative is:
**Springer Nature Customer Service Center GmbH
Europaplatz 3, 69115 Heidelberg, Germany**

Printed by Libri Plureos GmbH
in Hamburg, Germany